An Invitation to Operator Theory

Y. A. Abramovich

Indiana University-Purdue University Indianapolis

C. D. Aliprantis

Purdue University

Graduate Studies in Mathematics

Volume 50

American Mathematical Society

Providence, Rhode Island

2000 *Mathematics Subject Classification.* Primary 46Axx, 46Bxx, 46Gxx, 47Axx, 47Bxx, 47Cxx, 47Dxx, 47Lxx, 28Axx, 28Exx, 15A48, 15A18.

Library of Congress Cataloging-in-Publication Data

Abramovich, Y. A. (Yuri A.)
An invitation to operator theory / Y. A. Abramovich, C. D. Aliprantis.
p. cm. — (Graduate studies in mathematics, ISSN 1065-7339 ; v. 50)
Includes bibliographical references and index.
ISBN 0-8218-2146-6 (alk. paper)
1. Operator theory. I. Aliprantis, Charalambos D. II. Title. III. Series.

QA329.A25 2002
515′.724—dc21 2002074420

∞ The paper used in this book is acid-free and falls within the guidelines
established to ensure permanence and durability.
Visit the AMS home page at http://www.ams.org/

10 9 8 7 6 5 4 3 2 1 07 06 05 04 03 02

To Alla, Julia, and Jane

YAA

To Bernadette, Claire, and Dionissi

CDA

Contents

Foreword ix

Chapter 1. Odds and Ends 1

§1.1. Banach Spaces, Operators, and Linear Functionals 1

§1.2. Banach Lattices and Positive Operators 13

§1.3. Bases in Banach Spaces 27

§1.4. Ultrapowers of Banach Spaces 37

§1.5. Vector-valued Functions 44

§1.6. Fundamentals of Measure Theory 52

Chapter 2. Basic Operator Theory 69

§2.1. Bounded Below Operators 69

§2.2. The Ascent and Descent of an Operator 79

§2.3. Banach Lattices with Order Continuous Norms 84

§2.4. Compact and Weakly Compact Positive Operators 88

Chapter 3. Operators on AL- and AM-spaces 93

§3.1. AL- and AM-spaces 94

§3.2. Complex Banach Lattices 103

§3.3. The Center of a Banach Lattice 112

§3.4. The Predual of a Principal Ideal 119

Chapter 4. Special Classes of Operators 123

§4.1. Finite-rank Operators 123

§4.2. Multiplication Operators 135

§4.3. Lattice and Algebraic Homomorphisms 142
§4.4. Fredholm Operators 155
§4.5. Strictly Singular Operators 170

Chapter 5. Integral Operators 179
§5.1. The Basics of Integral Operators 180
§5.2. Abstract Integral Operators 193
§5.3. Conditional Expectations and Positive Projections 211
§5.4. Positive Projections and Lattice-subspaces 228

Chapter 6. Spectral Properties 237
§6.1. The Spectrum of an Operator 237
§6.2. Special Points of the Spectrum 248
§6.3. The Resolvent of a Positive Operator 253
§6.4. Functional Calculus 256

Chapter 7. Some Special Spectra 271
§7.1. The Spectrum of a Compact Operator 272
§7.2. Turning Approximate Eigenvalues into Eigenvalues 281
§7.3. The Spectrum of a Lattice Homomorphism 287
§7.4. The Order Spectrum of an Order Bounded Operator 291
§7.5. The Essential Spectrum of a Bounded Operator 298

Chapter 8. Positive Matrices 315
§8.1. The Banach Lattices $M_n(\mathbb{R})$ and $M_n(\mathbb{C})$ 316
§8.2. Operators on Finite Dimensional Spaces 320
§8.3. Matrices with Non-negative Entries 329
§8.4. Irreducible Matrices 332
§8.5. The Perron–Frobenius Theorem 339

Chapter 9. Irreducible Operators 347
§9.1. Irreducible and Expanding Operators 347
§9.2. Ideal Irreducibility and the Spectral Radius 357
§9.3. Band Irreducibility and the Spectral Radius 365
§9.4. Krein Operators on $C(\Omega)$-spaces 370

Chapter 10. Invariant Subspaces 381
§10.1. A Smorgasbord of Invariant Subspaces 383
§10.2. The Lomonosov Invariant Subspace Theorem 393

§10.3. Invariant Ideals for Positive Operators 398
§10.4. Invariant Subspaces of Families of Positive Operators 409
§10.5. Compact-friendly Operators 425
§10.6. Positive Operators on Banach Spaces with Bases 436
§10.7. A Characterization of Non-transitive Algebras 440
§10.8. Comments on the Invariant Subspace Problem 449

Chapter 11. The Daugavet Equation 455
§11.1. The Daugavet Equation and Uniform Convexity 456
§11.2. The Daugavet Property in AL- and AM-spaces 467
§11.3. The Daugavet Property in Banach Spaces 471
§11.4. The Daugavet Property in $C(\Omega)$-spaces 477
§11.5. Slices and the Daugavet Property 487
§11.6. Narrow Operators 493
§11.7. Some Applications of the Daugavet Equation 500

Bibliography 505

Index 521

Foreword

Since the appearance of the seminal monograph of Stefan Banach [**51**] on (linear) operators between normed spaces, the study of operators has been at the heart of research in analysis. For the most part, this research was confined to the topological structures of the spaces and to the topological properties of the operators. At the same time there was one more aspect of operator theory that was developed in parallel with the topological and algebraic study of operators. This natural aspect involved the order structures of the spaces between which the operators acted.

Almost every classical Banach space is equipped with a natural order that is compatible with the algebraic and topological structures of the space. Operators that preserve the order structures are known as *positive operators*. They were first introduced and studied by F. Riesz [**278, 279**], H. Freudenthal [**129**], and L. V. Kantorovich [**172, 173**] in the mid 1930's. For some peculiar reason, the investigations of the topological and order properties of operators were not connected, and for quite a while it seemed that they had little in common. The monographs written by L. V. Kantorovich, B. Z. Vulikh and A. Pinsker [**175**] in the Soviet Union and by H. Nakano [**251, 252**] in Japan contained most of the theoretical background on operators between ordered spaces accumulated by 1950. A partial list of books and monographs that deal with Banach spaces and operators between them without emphasizing in depth the order structures includes: [**56, 85, 102, 105, 106, 113, 118, 149, 152, 199, 205, 235, 250, 266, 300, 305, 306, 317**]. For a glimpse of the modern state of the art in Banach space theory and operator theory, see the special volume [**164**] edited by W. B. Johnson and J. Lindenstrauss.

From the early 1960's it was realized that the topological and order structures of operators were related and that they should be studied together. Since then much research has aimed at producing a theory that would encompass these structures in a unified manner. The "unification" of the order and topological properties of operators is due to the efforts of many mathematicians in various countries. The period of this unification can be roughly divided into two parts; the pre-1980 period and the post-1980 period. Let us describe here the basic achievements done in each period.

The pre-1980 period can be described as the period where most of the foundations of the order structure of vectors spaces and the lattice properties of operators were laid down. In this era the theory of partially ordered vector spaces was developed through the efforts of mathematicians from many countries. There were several monographs written on the subject. Those written by A. L. Peressini [**264**] and G. Jameson [**161**] and the last part of H. H. Schaefer's book [**289**] were devoted entirely to ordered vector spaces. The books by W. A. J. Luxemburg and A. C. Zaanen [**225**], B. Z. Vulikh [**328**], and S. Kaplan [**176**] studied the order structure of Riesz spaces, while D. H. Fremlin [**128**] and C. D. Aliprantis and O. Burkinshaw [**29**] studied topological properties of Riesz spaces. Monographs on Banach lattices and positive operators were written by M. A. Krasnoselsky et al. [**191**], H. J. Krieger [**194**], H. E. Lacey [**197**], J. Lindenstrauss and L. Tzafriri [**205, 206**], H. H. Schaefer [**290**], and H.-U. Schwarz [**298**].

The post-1980 period is characterized by the study of positive operators from all points of view. Special attention was paid to the properties of operators dominated by positive operators having some compactness property. We also find in this era a complete study of integral operators in terms of their lattice structure. Monographs regarding positive operators were written in this period by C. D. Aliprantis and O. Burkinshaw [**30**], W. Arendt et al. [**41**], A. V. Bukhvalov et al. [**78**], P. Meyer-Nieberg [**240**], and A. C. Zaanen [**343, 344**]. The list of very recent monographs on the subject of positive operators includes the books by Y. A. Abramovich, E. L. Arenson, and A. K. Kitover [**21**], Y. A. Abramovich and A. K. Kitover [**22**], A. G. Kusraev [**195**], and W. Wnuk [**338**].

Besides its important internal role in mathematics, the theory of Riesz spaces and positive operators has important applications to several disciplines. The areas where the theory of Riesz spaces and positive operators has been found to be useful include: game theory [**182**]; finance [**151**, Chapter 31]; economics [**26**]; nuclear reactor theory [**64, 65**]; statistical decision theory [**201**]; and structured population dynamics [**80, 248**].

The objective of this book is to present at the graduate level in a self-contained manner the theory of linear operators between Banach spaces and Banach lattices by exploiting their topological and order structures, as well as the topological and order properties of the spaces of operators. The subject matter of the book is divided into eleven chapters, which to some extent can be read independently. Each chapter is subdivided into sections. At the end of each section there is a list of exercises of varying degrees of difficulty designed to help the reader comprehend the material in the section. There are more than six hundred exercises in the book. Hints to selected exercises are also given. Our companion book *Problems in Operator Theory* contains complete solutions to all exercises.

Exercises play a considerable role in our presentation. They serve several purposes. First of all, we often relegate some technical details of proofs to the exercises. These details may not be critical for understanding the proofs but they are certainly essential for their validity, and we want to provide the student with an accurate and complete account of how such details should be presented. Second, these exercises not only allow the student to learn how the results presented in the book work out but they also offer a considerable amount of additional material and further developments. There is one more feature to these exercises. There are many useful, almost classical, results whose proofs are not readily available in the literature (for instance, the duality between the uniformly convex and uniformly smooth spaces, the equivalence between the differentiability and other properties of the norm, the explicit closed form of the resolvent of the Volterra operator, the description of the spectrum of the averaging operator $f \mapsto \frac{1}{t}\int_0^t f\,d\mu$, etc.). The reader will find detailed solutions and discussions of many of these results among our problems.

Some of the exercises are new, and some are known. Whenever possible, we indicate the sources of the known exercises. On the other hand, for the standard results or for some variations of classical results, we often omit the sources of their origins.

The prerequisites of the book are the usual graduate introductory courses in real analysis, general topology, measure theory, and functional analysis. Trying to make the book as self-contained as possible, we have provided detailed proofs to most theorems in the text. This, coupled with the inclusion of the exercises, makes the book an ideal text for a graduate or advance course in operator theory and functional analysis. Moreover, since many chapters contain the latest results (some of which have not been published before), we hope that the book will be useful not only to graduate students but also to research mathematicians in operator theory and functional analysis, as well as to scientists in other disciplines. Each of Chapters 5, 7, 9, 10, and 11 can be used for an advanced seminar and each of them will also present the students with possibilities and directions to start on independent research. A brief description of the material in each chapter should give a good idea of the topics covered in the book.

- Chapter 1 presents an introduction to Banach spaces and operator theory. Also, it introduces Banach lattices, positive operators, and vector-valued functions. The chapter culminates with a very brief but, nevertheless, rather comprehensive presentation of the fundamentals of measure theory.

- Chapter 2 reviews the basic structural properties of operators with special emphasis on positive operators. It also studies operators bounded from below and discusses the ascent and descent of an operator.

- Chapter 3 deals with AM- and AL-spaces and demonstrates their importance in operator theory. It also studies the center of a Banach lattice and presents a careful exposition of the complexification of a Banach lattice.

- Chapter 4 presents detailed accounts regarding the following classes of operators: finite-rank operators, multiplication operators, lattice and algebraic homomorphisms, Fredholm operators, and strictly singular operators.

- Chapter 5 deals exclusively with integral operators. It presents their basic properties from the measure-theoretic and abstract points of view. It also studies conditional expectation operators and demonstrates their connection with positive projections.

- Chapter 6 introduces the spectrum of a bounded operator and discusses the basic geometrical properties of the special parts of the spectrum. The study specializes in the spectral properties of positive operators. A comprehensive study of the functional calculus is also here.

- Chapter 7 describes the spectra of compact operators and lattice homomorphisms. It also contains the method of turning approximate eigenvalues to eigenvalues. In addition, this chapter introduces and studies the order spectrum of an order bounded operator as well as the essential spectrum of a bounded operator.

- Chapter 8 specializes in the study of operators to finite dimensional spaces with special emphasis on matrices with non-negative entries. It studies irreducible matrices and presents a thorough discussion on the classical Perron–Frobenius theorem.

- Chapter 9 investigates irreducible operators on Banach lattices and presents several conditions for the strict positivity of the spectral radius. This chapter also studies Krein operators between $C(\Omega)$-spaces.
- Chapter 10 deals with the famous invariant subspace problem. It starts with several classical invariant subspace theorems, but its central topic is the invariant subspace problem for positive and related operators. The results obtained for individual operators are then generalized to collections of positive operators.
- A bounded operator T on a Banach space satisfies the Daugavet equation if $\|I+T\| = 1 + \|T\|$. Chapter 11 presents a comprehensive study of operators on Banach spaces that satisfy the Daugavet equation.

We regret that the size of the book has precluded us from including a number of relevant topics, in particular, disjointness preserving operators and their spectral properties, interpolation of positive operators, and the domination problem. There is one more omission that we would like to mention. The book deals with the general theory of operators and, as a result, Hilbert space operators occur in our discussions only sporadically. There are many excellent books devoted exclusively to Hilbert spaces and operators on them, and so anyone interested in this particular subject will be able to find it easily in the literature; see for instance the monographs [**94, 148**]. At present, the realm of operator theory is so vast that any book can cover only a portion of it.

In writing the book we have received help and benefited from discussions with many colleagues and friends. We are grateful and express our sincerest thanks to all of them: O. Burkinshaw (for many fruitful conversations), W. A. J. Luxemburg (for his support and encouragement), H. Radjavi (for insightful comments on the invariant subspace problem), B. Randrianantoanina (for reading the book, correcting a number of mistakes, and providing us with many valuable suggestions), G. Sirotkin (for a number of useful remarks and problems), T. Oikhberg (for contributing several problems), K. Podgorski (for discussions concerning conditional expectations), S. Klimek and E. Mukhin (for helping us to incorporate graphics in LaTeX), S. I. Gelfand (for constant support at all stages during the production of this book), the AMS Copy Editor Arlene O'Sean (for her professional reading of the manuscript and her many recommendations for its improvement), and two anonymous referees (for their constructive criticism and comments).

Our special thanks go to A. K. Kitover and V. G. Troitsky. They read the entire book, corrected mistakes, modified the old and provided some new proofs, and suggested many interesting improvements. We are also deeply indebted to W. Geller who read most of the book and, apart from fixing some mathematical irregularities, corrected a great number of irregularities in our English. Without his help our writing would have been considerably less idiomatic.

Finally, we would like to thank our wives, Alla and Bernadette, for coping with our idiosyncrasies during the many years it took us to assemble the material, write, and typeset this text. To accomplish this project without their patience and understanding would not have been possible.

Y. A. Abramovich and C. D. Aliprantis

Indianapolis and West Lafayette, May 2002

Chapter 1

Odds and Ends

The material included in this chapter is reflected correctly in its title. We introduce here some basic concepts related to Banach spaces and state a few principal theorems of functional analysis that will be used throughout the book. These concepts include linear operators, linear functionals, and bases in Banach spaces.

Special attention is given to the complexification of real Banach spaces. Section 1.5 is devoted to vector-valued functions and it will serve as the background needed to study the spectral properties of operators on Banach spaces and Banach lattices. Section 1.6 presents a brief but nevertheless complete introduction to measure theory. Some of this material will be necessary for a comprehensive investigation of integral operators undertaken in Chapter 5. Measure theory is studied in conjunction with the measure-theoretic properties of the Riesz space of all measurable functions defined on a measure space.

We present complete proofs to many results and omit mainly the proofs of several well-known ones. As usual, the exercises at the end of the sections contain a wealth of supplemental material that is of independent interest in its own right.

1.1. Banach Spaces, Operators, and Linear Functionals

Here we briefly state a few basic definitions and facts regarding normed spaces and operators that will be used repeatedly throughout this monograph. Although we shall include many proofs, for more details and extensive treatments of the material in this chapter we refer the reader to the books [**30, 31, 118, 205, 317**].

A function $\|\cdot\|\colon X \to \mathbb{R}$ on a (real or complex[1]) vector space is said to be a **norm** whenever it satisfies the following properties:

(1) (**Positivity**) $\|x\| \geq 0$ for all $x \in X$ and $\|x\| = 0$ if and only if $x = 0$.

(2) (**Homogeneity**) $\|\alpha x\| = |\alpha| \cdot \|x\|$ for all $x \in X$ and all scalars α.

(3) (**The Triangle Inequality**) $\|x + y\| \leq \|x\| + \|y\|$ for all $x, y \in X$.

A **normed space** is a vector space equipped with a norm. To avoid trivialities, we shall assume throughout this monograph that all normed spaces are non-trivial, i.e., different from the trivial vector space $\{0\}$.

Every norm $\|\cdot\|$ on a vector space X induces a metric on X via the formula $d(x, y) = \|x - y\|$. A **Banach space** is a normed space which is a complete metric space under the metric induced by its norm.

A mapping $T\colon X \to Y$ between two vector spaces (over the same field) is said to be a **linear operator**, or simply an **operator**, whenever

$$T(\alpha x + \beta y) = \alpha T(x) + \beta T(y)$$

holds for all $x, y \in X$ and all scalars α, β. For simplicity, the value $T(x)$ will often be denoted by Tx. If X and Y are normed spaces and $T\colon X \to Y$ is an operator, then its (**operator**) **norm** is defined by

$$\|T\| = \sup_{\|x\| \leq 1} \|Tx\| = \sup_{\|x\| = 1} \|Tx\| .$$

If $\|T\| < \infty$, then T is called a **bounded operator**, and when $\|T\| = \infty$, the operator T is called an **unbounded operator**. It easily follows that if T is bounded, then

$$\|T\| = \min\bigl\{M \geq 0\colon\ \|Tx\| \leq M\|x\| \text{ for each } x \in X\bigr\} .$$

Also, it is easy to check that an operator $T\colon X \to Y$ between normed spaces is bounded if and only if T is continuous.

An operator $T\colon X \to Y$ between two normed spaces is called an **isometry** (or, more precisely, a **linear isometry**) if $\|Tx\| = \|x\|$ for each $x \in X$. Two normed spaces are said to be **isometric** if there exists a surjective linear isometry from one space onto the other.

The collection of all bounded operators from a normed space X into another normed space Y will be denoted by $\mathcal{L}(X, Y)$. We shall write $\mathcal{L}(X)$ instead of $\mathcal{L}(X, X)$. It turns out that $\mathcal{L}(X, Y)$ with the usual algebraic operations and with the operator norm is a normed space. As a matter of fact, we have the following.

[1] The symbols $\mathbb{R}$ and $\mathbb{C}$ will be used exclusively to denote the set of real and complex numbers, respectively. Also, as usual, $\mathbb{N}$ will denote the set of natural numbers, $\mathbb{Q}$ the set of rational numbers, and $\mathbb{Z}$ the set of integers.

Theorem 1.1. *If X is a normed space and Y is a Banach space, then the vector space $\mathcal{L}(X,Y)$ equipped with the operator norm is a Banach space.*

Recall that an **algebra** is a vector space $\mathcal{A}$ that is equipped with an extra binary operation $(x,y) \mapsto xy$ (called **multiplication**) that satisfies for all $x,y,z \in \mathcal{A}$ and all scalars α the following properties:

(1) $(xy)z = x(yz)$ and $(\alpha x)y = x(\alpha y) = \alpha(xy)$.

(2) $x(y+z) = xy + xz$ and $(x+y)z = xz + yz$.

An algebra $\mathcal{A}$ is called **unital** if there exists a vector e (which is uniquely determined and is called the **unit** of $\mathcal{A}$) such that $xe = ex = x$ holds for all $x \in \mathcal{A}$.

A **Banach algebra** is an algebra $\mathcal{A}$ which is also a Banach space under a norm $\|\cdot\|$ that satisfies $\|xy\| \leq \|x\| \cdot \|y\|$ for all $x, y \in \mathcal{A}$.

Theorem 1.2. *If X is a Banach space, then $\mathcal{L}(X)$ with the composition operation as multiplication is a unital Banach algebra with the identity operator as its unit.*

For a real (resp. complex) normed space X, the Banach space $\mathcal{L}(X,\mathbb{R})$ (resp. $\mathcal{L}(X,\mathbb{C})$) will be denoted by X^* and will be referred to as the **norm dual** (or simply the **dual**) of X. The Banach space $X^{**} = (X^*)^*$ is known as the **second** (or **double**) **dual** of X. Considering every $x \in X$ as a linear functional on X^* via the formula $x(x^*) = x^*(x)$, we get a natural isometric embedding of X into X^{**}, i.e., X can be viewed as a subspace of X^{**}.

As usual, we say that a sequence $\{x_n\}$ in a Banach space X **converges weakly** to some vector $x \in X$ (in symbols, $x_n \xrightarrow{w} x$ or $x_n \xrightarrow{\sigma(X,X^*)} x$) if $x^*(x_n) \to x^*(x)$ holds for each $x^* \in X^*$. Similarly, a sequence $\{x_n^*\}$ in the dual X^* of a Banach space X **weak* converges** to some $x^* \in X^*$ (in symbols, $x_n^* \xrightarrow{w^*} x^*$ or $x_n^* \xrightarrow{\sigma(X^*,X)} x$) if $x_n^*(x) \to x^*(x)$ for each $x \in X$.

If X is a normed space, then the closed unit balls of X, X^*, and X^{**} will be denoted by U, U^*, and U^{**}, respectively. That is,

$$U = \{x \in X \colon \ \|x\| \leq 1\}, \quad U^* = \{x^* \in X^* \colon \ \|x^*\| \leq 1\}, \quad \text{and}$$
$$U^{**} = \{x^{**} \in X^{**} \colon \ \|x^{**}\| \leq 1\}.$$

If clarity requires the normed space X to be indicated, then the above closed unit balls will be denoted by U_X, U_{X^*}, and $U_{X^{**}}$, respectively. It follows that for each $x^* \in X^*$ we have

$$\|x^*\| = \sup_{x \in U} |x^*(x)| = \sup_{x \in U} \langle x^*, x \rangle .$$

Also, for each $x \in X$ we have

$$\|x\| = \sup_{x^* \in U^*} |x^*(x)| = \sup_{x^* \in U^*} \langle x^*, x \rangle .$$

The **adjoint** of a bounded operator $T\colon X \to Y$ between two arbitrary normed spaces is the bounded operator $T^*\colon Y^* \to X^*$ defined via the formula $(T^*y^*)(x) = y^*(Tx)$ for $y^* \in Y^*$ and $x \in X$. It is customary to write the formula defining T^* by means of the duality expression

$$\langle T^*y^*, x\rangle = \langle y^*, Tx\rangle .$$

The adjoint of T^*, i.e., the operator $T^{**}\colon X^{**} \to Y^{**}$ defined by

$$\langle T^{**}x^{**}, y^*\rangle = \langle x^{**}, T^*y^*\rangle$$

for all $x^{**} \in X^{**}$ and all $y^* \in Y^*$, is called the **double adjoint** of the operator T. It is not difficult to see that $\|T\| = \|T^*\| = \|T^{**}\|$. Indeed,

$$\begin{aligned} \|T^*\| &= \sup_{\|y^*\|\le 1} \|T^*y^*\| = \sup_{\|y^*\|\le 1} \Big[\sup_{\|x\|\le 1} \langle T^*y^*, x\rangle \Big] \\ &= \sup_{\|x\|\le 1} \Big[\sup_{\|y^*\|\le 1} \langle y^*, Tx\rangle \Big] \\ &= \sup_{\|x\|\le 1} \|Tx\| = \|T\| . \end{aligned}$$

From now on, unless otherwise stated, X and Y will denote two arbitrary normed spaces. A subset $\mathcal{A}$ of $\mathcal{L}(X,Y)$ is said to be **pointwise bounded** if for each $x \in X$ there exists some $M_x > 0$ (depending upon x) satisfying $\|Tx\| \le M_x$ for each $T \in \mathcal{A}$.

Theorem 1.3 (The Uniform Boundedness Principle)**.** *Assume that X is a Banach space and Y is a normed space. Then a subset of $\mathcal{L}(X,Y)$ is norm bounded if and only if it is pointwise bounded.*

The Uniform Boundedness Principle is also known in the literature as the **Banach–Steinhaus theorem**. For bounded operators between Banach spaces there are two more important results that will be stated next. They too were discovered by S. Banach and are referred to as the "open mapping" and "closed graph" theorems.

Theorem 1.4 (The Open Mapping Theorem)**.** *Any surjective operator between two Banach spaces is an open mapping.*

Recall that the **graph** of a function $f\colon A \to B$ is the subset G_f of the Cartesian product $A \times B$ defined by $G_f = \{(a, f(a))\colon\ a \in A\}$. If $T\colon X \to Y$ is an operator between vector spaces, then its graph G_T is a vector subspace of $X \times Y$. If $X_1, \ldots, X_n$ are normed spaces, then their Cartesian product $X_1 \times \cdots \times X_n$ is also a normed space under the norm $\|(x_1, \ldots, x_n)\| = \big(\sum_{i=1}^n \|x_i\|^2\big)^{\frac{1}{2}}$. This normed space is also called the **direct**

sum normed space and is denoted $X_1 \oplus \cdots \oplus X_n$. If $X_1, \ldots, X_n$ are all Banach spaces, then $X_1 \oplus \cdots \oplus X_n$ is likewise a Banach space.[2]

Theorem 1.5 (The Closed Graph Theorem). *An operator between two Banach spaces is bounded if and only if it has a closed graph.*

Corollary 1.6. *Let (X, p) and (Y, q) be two normed spaces and let there exist two complete norms p_X and q_Y on X and Y, respectively, such that $p_X \geq p$ and $q_Y \geq q$. Then each (p,q)-continuous operator from X to Y is also (p_X, q_Y)-continuous.*

Proof. Assume that $T\colon X \to Y$ is a (p,q)-continuous operator. To show that this operator is (p_X, q_Y)-continuous it suffices to show that its graph is closed. To this end, take any sequence $\{x_n\}$ in X such that $p_X(x_n) \to 0$ and $q_Y(Tx_n - y) \to 0$ for some $y \in Y$. Since p_X dominates p, it follows that $p(x_n) \to 0$, and since T is (p,q)-continuous, we have $q(Tx_n) \to 0$. Similarly, $q_Y(Tx_n - y) \to 0$ implies $q(Tx_n - y) \to 0$, and consequently we see that $y = 0$. That is, the operator $T\colon (X, p_X) \to (Y, q_Y)$ has a closed graph, and hence it is continuous. ■

We close the section with a discussion of the complexification of a real vector space. If X is a real vector space, then the **complexification** of X is the complex vector space

$$X_{\mathbf{C}} = X \oplus \imath X = \{x + \imath y\colon \ x, y \in X\},$$

whose vector space operations are defined by

$$\begin{aligned} (x_1 + \imath y_1) + (x_2 + \imath y_2) &= x_1 + x_2 + \imath(y_1 + y_2), \text{ and} \\ (\alpha + \imath\beta)(x + \imath y) &= \alpha x - \beta y + \imath(\beta x + \alpha y). \end{aligned}$$

The real vector space X itself is usually identified with the (real) subspace $X + \imath\{0\}$ of $X_{\mathbf{C}}$. That is, by identifying every vector $x \in X$ with $x + \imath 0 \in X_{\mathbf{C}}$, we can consider the vectors of X as vectors of $X_{\mathbf{C}}$.

If X is also a normed space with norm $\|\cdot\|$, then we can extend the norm $\|\cdot\|$ to a norm on $X_{\mathbf{C}}$ via the formula

$$\|z\| = \sup_{\theta \in [0, 2\pi]} \|x \cos\theta + y \sin\theta\|$$

for each $z = x + \imath y \in X_{\mathbf{C}}$.[3] Clearly, $\|x\| = \|x + \imath 0\|$ for each $x \in X$. Notice that

$$\tfrac{1}{2}\big(\|x\| + \|y\|\big) \leq \|z\| \leq \|x\| + \|y\|$$

---

[2] Two other frequently used equivalent norms on the direct sum vector space $X_1 \oplus \cdots \oplus X_n$ are $\|(x_1, \ldots, x_n)\| = \sum_{i=1}^n \|x_i\|$ and $\|(x_1, \ldots, x_n)\| = \max_{1 \leq i \leq n} \|x_i\|$.

[3] For simplicity we write $x \cos\theta + y \sin\theta$ instead of the standard expression $(\cos\theta)x + (\sin\theta)y$.

for all $z = x + \imath y \in X_{\mathbf{C}}$. Consequently, $z_n = x_n + \imath y_n \to x + \imath y$ in $X_{\mathbf{C}}$ if and only if $x_n \to x$ and $y_n \to y$ in X. This implies that if X is a Banach space, then $X_{\mathbf{C}}$ is also a Banach space.

Every operator $T\colon X \to Y$ between two real vector spaces gives rise naturally to a complex linear operator $T_{\mathbf{C}}\colon X_{\mathbf{C}} \to Y_{\mathbf{C}}$ defined via the formula

$$T_{\mathbf{C}}(x + \imath y) = Tx + \imath Ty\,.$$

Lemma 1.7. *If X and Y are normed spaces and $T\colon X \to Y$ is a bounded operator, then the operator $T_{\mathbf{C}}\colon X_{\mathbf{C}} \to Y_{\mathbf{C}}$ is also bounded and satisfies $\|T_{\mathbf{C}}\| = \|T\|$.*

Proof. First note that the inequality $\|T_{\mathbf{C}}\| \geq \|T\|$ is obvious. On the other hand, for each $z = x + \imath y$ and any angle θ we have

$$\begin{aligned} \big\|(Tx)\cos\theta + (Ty)\sin\theta\big\| &= \big\|T(x\cos\theta + y\sin\theta)\big\| \\ &\leq \|T\|\big(\|x\cos\theta + y\sin\theta\|\big) \\ &\leq \|T\|\cdot\|z\|\,. \end{aligned}$$

This implies $\|T_{\mathbf{C}}z\| \leq \|T\|\cdot\|z\|$ or $\|T_{\mathbf{C}}\| \leq \|T\|$. So, $\|T_{\mathbf{C}}\| = \|T\|$. ■

In the sequel, quite often, we shall identify T with $T_{\mathbf{C}}$ without any mention. When we do so the identification will be clear to the reader from the context.

It is very common to denote the vectors $x + \imath y$ of $X_{\mathbf{C}}$ as column vectors $\begin{bmatrix} x \\ y \end{bmatrix}$. With this notation in mind, it is not difficult to see that every operator $\mathcal{T}\colon X_{\mathbf{C}} \to X_{\mathbf{C}}$ can be represented by a unique operator matrix of the form $\mathcal{T} = \begin{bmatrix} T & -S \\ S & T \end{bmatrix}$, where $S, T\colon X \to X$ are linear operators, and as usual

$$\mathcal{T}(x + \imath y) = \begin{bmatrix} T & -S \\ S & T \end{bmatrix}\begin{bmatrix} x \\ y \end{bmatrix} = \begin{bmatrix} Tx - Sy \\ Sx + Ty \end{bmatrix} = Tx - Sy + \imath(Sx + Ty)\,.$$

Notice that in this notation, if $T\colon X \to X$ is an operator, then the operator $T_{\mathbf{C}}$ is represented by the matrix

$$T_{\mathbf{C}} = \begin{bmatrix} T & 0 \\ 0 & T \end{bmatrix}.$$

Any operator $\mathcal{T}\colon X_{\mathbf{C}} \to X_{\mathbf{C}}$ for which $S = 0$ is called a **real operator on** $X_{\mathbf{C}}$. Of course, an operator $\mathcal{T}\colon X_{\mathbf{C}} \to X_{\mathbf{C}}$ is real if and only if $\mathcal{T} = T_{\mathbf{C}}$ for some operator $T\colon X \to X$. Also, it should be noticed that for an operator $\mathcal{T}\colon X_{\mathbf{C}} \to X_{\mathbf{C}}$ to be real it is necessary and sufficient that $\mathcal{T}$ leaves X invariant, i.e., $\mathcal{T}(X) \subseteq X$.

Thus, every operator $\mathcal{T} \in \mathcal{L}(X_{\mathbf{C}}, Y_{\mathbf{C}})$ can be identified with a matrix and also with a pair of operators $T, S \in \mathcal{L}(X, Y)$. Equivalently, every operator $\mathcal{T}$

in $\mathcal{L}(X_{\mathbf{C}}, Y_{\mathbf{C}})$ can be identified with the vector $T + \imath S$ in $\mathcal{L}(X,Y) \oplus \imath\mathcal{L}(X,Y)$. Moreover, the action of $T + \imath S$ on $X_{\mathbf{C}}$ can be computed formally (and mnemonically!) by the usual multiplication procedure on complex numbers. That is,

$$(T + \imath S)(x + \imath y) = (Tx - Sy) + \imath(Sx + Ty)\,.$$

Thus, we have established the following result.

Theorem 1.8. *If X and Y are two real normed spaces, then*

$$\mathcal{L}(X_{\mathbf{C}}, Y_{\mathbf{C}}) = \mathcal{L}(X,Y)_{\mathbf{C}} = \mathcal{L}(X,Y) \oplus \imath\mathcal{L}(X,Y)\,,$$

where every $T + \imath S \in \mathcal{L}(X,Y)_{\mathbf{C}}$ acts on $X_{\mathbf{C}}$ via the formula

$$(T + \imath S)(x + \imath y) = (Tx - Sy) + \imath(Sx + Ty)\,.$$

If, in addition, Z is another real normed space, $T_1 + \imath S_1 \in \mathcal{L}(Y_{\mathbf{C}}, Z_{\mathbf{C}})$ and $T + \imath S \in \mathcal{L}(X_{\mathbf{C}}, Y_{\mathbf{C}})$, then the operator $(T_1 + \imath S_1)(T + \imath S) \in \mathcal{L}(X_{\mathbf{C}}, Z_{\mathbf{C}})$ is given by

$$(T_1 + \imath S_1)(T + \imath S) = (T_1 T - S_1 S) + \imath(T_1 S + S_1 T)\,.$$

There is one more important connection between complex vector spaces and real vector spaces. To discuss this connection, let X be a complex vector space. Then X can also be considered as a vector space over the field of real numbers $\mathbb{R}$. We shall denote this real vector space by $X_{\mathbb{R}}$.

Now let $f\colon X \to \mathbb{C}$ be a linear functional. Then for each $x \in X$ we can write

$$f(x) = g(x) + \imath h(x)\,,$$

where $g(x)$ and $h(x)$ are real numbers. An easy argument shows that the functions $g, h\colon X_{\mathbb{R}} \to \mathbb{R}$ are in fact both linear functionals. From the homogeneity of f we have $f(\imath x) = \imath f(x)$ or $g(\imath x) + \imath h(\imath x) = -h(x) + \imath g(x)$, and so $h(x) = -g(\imath x)$. Thus, we have shown that if $f\colon X \to \mathbb{C}$ is a linear functional, then there exists a unique linear functional $g\colon X_{\mathbb{R}} \to \mathbb{R}$ (denoted $\mathrm{Re} f$, i.e., $g = \mathrm{Re} f$, and called the **real part** of f) such that

$$f(x) = g(x) - \imath g(\imath x) \tag{$\star$}$$

holds for each $x \in X$. Conversely, if $g\colon X_{\mathbb{R}} \to \mathbb{R}$ is a linear functional, then by means of $(\star)$ the functional g gives rise to a linear functional $f\colon X \to \mathbb{C}$ such that $\mathrm{Re} f = g$.

Now assume that X is a complex normed space and that $f\colon X \to \mathbb{C}$ is a continuous linear functional. Write $f(x) = g(x) - \imath g(\imath x)$ and note that if $x \in X$ satisfies $\|x\| \le 1$, then from the inequality $|g(x)| \le |f(x)| \le \|f\|$, it follows that $\|g\| = \sup_{\|x\|\le 1}|g(x)| \le \|f\|$. On the other hand, if $x \in X$ satisfies $\|x\| \le 1$, then write $f(x) = |f(x)|e^{\imath\theta}$, and note that

$$|f(x)| = e^{-\imath\theta} f(x) = f\big(e^{-\imath\theta}x\big) = g\big(e^{-\imath\theta}x\big) \le \|g\|\,.$$

Therefore, $\|f\| = \sup_{\|x\| \le 1} |f(x)| \le \|g\|$, and so $\|f\| = \|g\|$. That is, we have shown that each $f \in X^*$ satisfies $\|f\| = \|\mathrm{Re} f\|$.

The preceding discussion shows that the mapping $f \mapsto \mathrm{Re} f$, from $(X^*)_{\mathbb{R}}$ to $(X_{\mathbb{R}})^*$, is a linear isometry. In other words, we have established the following result.

Theorem 1.9. *If X is a complex normed space, then the mapping $f \mapsto \mathrm{Re} f$ is a (real) surjective linear isometry—and so, up to this linear isometry, we can write $(X^*)_{\mathbb{R}} = (X_{\mathbb{R}})^*$.*

Using the previous result, we can also establish the following.

Theorem 1.10 (Bohnenblust–Hahn–Banach). *Every continuous linear functional defined on a subspace of a complex normed vector space X has a continuous linear extension to all of X with preservation of its norm.*

Recall that if V is a subspace of a vector space W, then the **quotient vector space** W/V is the vector space $\{[w]\colon\ w \in W\}$ of all equivalence classes, where as usual for each $w \in W$ its equivalence class is defined by $[w] = w + V = \{w + v\colon\ v \in V\}$. The addition and multiplication in W/V are given by

$$[x] + [y] = [x + y] \quad \text{and} \quad \lambda[x] = [\lambda x]\,.$$

The mapping $w \mapsto [w]$ is called the **quotient map** of W onto W/V.

The following result summarizes some basic properties of quotient normed spaces. Its proof is left to the reader.

Theorem 1.11. *Let Y be a closed subspace of a normed space X, and define the function $|||\cdot|||\colon X/Y \to \mathbb{R}$ by*

$$|||[x]||| = d(x, Y) = \inf_{y \in Y} \|x - y\| = \inf_{z \in [x]} \|z\|\,,$$

where $d(x, Y)$ is the distance from x to Y. Then:

(a) *The function $|||\cdot|||$ is a norm on the quotient vector space X/Y (called the **quotient norm**).*

(b) *The set $U_{X/Y} = \{[x]\colon\ x \in U_X\}$ is the unit ball of X/Y under the quotient norm.*

(c) *The quotient map is an open contraction, i.e., it is an open mapping whose norm is at most one.*

(d) *If X is a Banach space, then $(X/Y, |||\cdot|||)$ is also a Banach space.*

If V is a vector subspace of a normed space X, then its **annihilator** is the norm closed subspace of X^* defined by

$$V^{\perp} = \{x^* \in X^*\colon\ x^*(v) = 0 \text{ for all } v \in V\}\,.$$

In terms of annihilators the norm duals of a subspace and of a quotient space are described as follows.

Theorem 1.12. *For a closed subspace Y of a normed space X we have*

$$(X/Y)^* = Y^\perp \qquad \text{and} \qquad Y^* = X^*/Y^\perp\,,$$

subject to the following surjective linear isometries.

(1) *The mapping $\phi\colon Y^\perp \to (X/Y)^*$, defined via the formula*

$$\phi(x^*)([x]) = x^*(x)\,, \quad x^* \in Y^\perp, \ \ [x] \in X/Y\,,$$

is a well-defined surjective linear isometry.

(2) *The mapping $\psi\colon X^*/Y^\perp \to Y^*$, defined via the formula*

$$\psi([x^*]) = x^*|_Y \ \ (\text{the restriction of } x^* \text{ to } Y)$$

for each $[x^*] \in X^*/Y^\perp$, is a well-defined surjective linear isometry.

Exercises

1. If two vectors x and y in a normed space satisfy $\|x+y\| = \|x\| + \|y\|$, then show that $\|\alpha x + \beta y\| = \alpha\|x\| + \beta\|y\|$ for all scalars $\alpha, \beta \geq 0$. [HINT: If the scalars α and β satisfy $\alpha \geq \beta$, then note that

$$\|\alpha x + \beta y\| = \|\alpha(x+y) + (\beta - \alpha)y\| \geq \alpha\|x+y\| - (\alpha - \beta)\|y\|\,.]$$

2. Let $T\colon V \to W$ be a surjective one-to-one operator between two vector spaces. Show that T^{-1} (the inverse of T) is a linear operator.

3. If $T\colon X \to Y$ is a bounded operator between normed spaces, then show that

$$\|T\| = \min\{M \geq 0\colon \ \|Tx\| \leq M\|x\| \text{ for all } x \in X\}\,.$$

4. Prove Theorem 1.1.

5. Prove the following converse of Theorem 1.1. If X is a non-trivial normed space, Y is another normed space and $\mathcal{L}(X,Y)$ is a Banach space, then Y is likewise a Banach space.

6. Show that the complexification $X_{\mathbf{c}}$ of a real vector space X under the operations

$$\begin{aligned}(x_1 + \imath y_1) + (x_2 + \imath y_2) &= x_1 + x_2 + \imath(y_1 + y_2), \text{ and}\\ (\alpha + \imath\beta)(x + \imath y) &= \alpha x - \beta y + \imath(\beta x + \alpha y)\end{aligned}$$

is a complex vector space.

7. Let X be a real normed space with norm $\|\cdot\|$. Show that the function $\|\cdot\|_{\mathbf{c}}\colon X_{\mathbf{c}} \to \mathbb{R}$ defined via the formula

$$\|z\|_{\mathbf{c}} = \sup_{\theta \in [0,2\pi]} \|x\cos\theta + y\sin\theta\|, \quad z = x + \imath y \in X_{\mathbf{c}}\,,$$

is a norm on $X_{\mathbf{c}}$ that extends $\|\cdot\|$. Also, show that

$$\tfrac{1}{2}(\|x\| + \|y\|) \le \|z\|_{\mathbf{c}} \le \|x\| + \|y\|$$

for each $z = x + \imath y \in X_{\mathbf{c}}$. Use the preceding inequality to verify that:

(a) A sequence $\{z_n\}$, where $z_n = x_n + \imath y_n$, satisfies $\|z_n\|_{\mathbf{c}} \to 0$ if and only if $\|x_n\| \to 0$ and $\|y_n\| \to 0$.

(b) If X is a Banach space, then $X_{\mathbf{c}}$ is also a Banach space.

8. Let X be a real vector space with complexification $X_{\mathbf{c}} = X \oplus \imath X$. Establish the following properties.

(a) If V is a vector subspace of X, then $V_{\mathbf{c}} = V \oplus \imath V = \{x + \imath y\colon\ x, y \in V\}$ is a vector subspace of $X_{\mathbf{c}}$.

(b) If X is also a normed space and V is a vector subspace of X, then the vector subspace $V_{\mathbf{c}} = V \oplus \imath V$ of $X_{\mathbf{c}}$ is closed in $X_{\mathbf{c}}$ if and only if V is a closed subspace of X.

(c) Let $T\colon X \to X$ be an operator. A subspace V of X is said to be **invariant under** T (or simply T-**invariant**) whenever $T(V) \subseteq V$. A vector subspace V of X is T-invariant if and only if the vector subspace $V_{\mathbf{c}} = V \oplus \imath V$ of $X_{\mathbf{c}}$ is $T_{\mathbf{c}}$-invariant.

9. Let X be a real vector space. It is very common to denote the vectors $x + \imath y$ of $X_{\mathbf{c}}$ as column vectors $\begin{bmatrix} x \\ y \end{bmatrix}$. Under this notation the vector space operations of $X_{\mathbf{c}}$ are given by

$$\begin{bmatrix} x_1 \\ y_1 \end{bmatrix} + \begin{bmatrix} x_2 \\ y_2 \end{bmatrix} = \begin{bmatrix} x_1 + x_2 \\ y_1 + y_2 \end{bmatrix} \quad \text{and} \quad (\alpha + \imath\beta) \begin{bmatrix} x \\ y \end{bmatrix} = \begin{bmatrix} \alpha x - \beta y \\ \beta x + \alpha y \end{bmatrix}.$$

Now assume that X and Y are two real vector spaces. Show that a mapping $\mathcal{T}\colon X_{\mathbf{c}} \to Y_{\mathbf{c}}$ is a linear operator if and only if there exist two (uniquely determined) operators $S, T\colon X \to Y$ such that $\mathcal{T}$ has the matrix representation $\mathcal{T} = \begin{bmatrix} T & -S \\ S & T \end{bmatrix}$, where as usual

$$\mathcal{T}(x + \imath y) = \begin{bmatrix} T & -S \\ S & T \end{bmatrix} \begin{bmatrix} x \\ y \end{bmatrix} = \begin{bmatrix} Tx - Sy \\ Sx + Ty \end{bmatrix}.$$

Also, show that if X and Y are normed spaces, then $\mathcal{T}$ is a bounded operator if and only if both S and T are bounded operators.

10. A bounded operator $T\colon X \to Y$ between normed spaces is said to be **invertible** if there exists (a uniquely determined) bounded operator $S\colon Y \to X$ satisfying $ST = I_X$ and $TS = I_Y$; the operator S is called the inverse of T and is denoted T^{-1}.

For an operator $T\colon X \to Y$ between real vector spaces establish the following properties.

(a) The operator T is surjective if and only if $T_{\mathbf{c}}$ is surjective.

(b) The operator T is one-to-one if and only if $T_{\mathbf{c}}$ is one-to-one.

(c) The operator T is invertible if and only if $T_{\mathbf{c}}$ is invertible. Also, show that if T is invertible, then $(T_{\mathbf{c}})^{-1} = (T^{-1})_{\mathbf{c}}$.

11. Let $X = X_1 \oplus \cdots \oplus X_n$ be the direct sum Banach space. Show that a mapping $A\colon X \to X$ is a bounded linear operator if and only if there exist bounded linear operators $A_{ij}\colon X_j \to X_i$ $(i, j = 1, 2, \dots, n)$ (uniquely determined) such that A has the matrix representation $A = [A_{ij}]$, where for each $x = x_1 \oplus \cdots \oplus x_n$ we have

$$\begin{aligned} Ax &= \begin{bmatrix} A_{11} & A_{12} & \cdots & A_{1n} \\ A_{21} & A_{22} & \cdots & A_{2n} \\ \vdots & \vdots & \ddots & \vdots \\ A_{n1} & A_{n2} & \cdots & A_{nn} \end{bmatrix} \begin{bmatrix} x_1 \\ x_2 \\ \vdots \\ x_n \end{bmatrix} = \begin{bmatrix} \sum_{j=1}^n A_{1j}x_j \\ \sum_{j=1}^n A_{2j}x_j \\ \vdots \\ \sum_{j=1}^n A_{nj}x_j \end{bmatrix} \\ &= \Big(\sum_{j=1}^n A_{1j}x_j\Big) \oplus \Big(\sum_{j=1}^n A_{2j}x_j\Big) \oplus \cdots \oplus \Big(\sum_{j=1}^n A_{nj}x_j\Big). \end{aligned}$$

12. Recall that a (not necessarily surjective) operator $T\colon X \to Y$ between normed spaces is said to be an **isometry** (or more precisely a **linear isometry**) if $\|Tx\| = \|x\|$ holds for each $x \in X$.

Show that a bounded operator $T\colon X \to Y$ between two normed spaces is an isometry if and only if its double adjoint $T^{**}\colon X^{**} \to Y^{**}$ is likewise an isometry. [HINT: Assume that $T\colon X \to Y$ is an isometry. Clearly, $\|T\| = \|T^{**}\| = 1$ and so $\|T^{**}x^{**}\| \le \|x^{**}\|$ for all $x^{**} \in X^{**}$. Now fix $x^* \in X^*$ with $\|x^*\| \le 1$. Then the formula $y^*(Tx) = \langle x, x^* \rangle$ defines a continuous linear functional on the range of T such that $\|y^*\| \le 1$. Let y^* also denote a continuous linear extension of y^* to all of Y such that $\|y^*\| \le 1$; clearly $T^*y^* = x^*$. Now note that for each $x^{**} \in X^{**}$ we have

$$\begin{aligned} |\langle x^*, x^{**} \rangle| &= |\langle T^*y^*, x^{**} \rangle| = |\langle y^*, T^{**}x^{**} \rangle| \\ &\le \|y^*\| \cdot \|T^{**}x^{**}\| \le \|T^{**}x^{**}\|, \end{aligned}$$

and so $\|x^{**}\| = \sup_{\|x^*\| \le 1} |\langle x^*, x^{**} \rangle| \le \|T^{**}x^{**}\|$.]

13. Let $T\colon X \to X$ be a bounded operator on a normed vector space. A glance at the inequality $\|Tx - Ty\| \le \|T\| \cdot \|x - y\|$ implies that T is a uniformly continuous mapping. So, T has a unique (linear) continuous extension $\hat{T}\colon \hat{X} \to \hat{X}$, where $\hat{X}$ denotes the norm completion of X. Establish the following properties:

(a) If T is an isometry on X, then so is $\hat{T}$ on $\hat{X}$.

(b) If T is an invertible isometry with inverse S, then S is also a (linear) isometry.

(c) If T is invertible, then so is $\hat{T}$, and we have $(\hat{T})^{-1} = (T^{-1})\hat{}$.

(d) If $\lambda - T$ is invertible on X for some scalar λ, then $\lambda - \hat{T}$ is likewise invertible on $\hat{X}$.[4]

(e) If $\hat{T}$ is an invertible isometry on $\hat{X}$, is T necessarily an invertible isometry on X?

[HINT: For (e) let X denote the vector space of all polynomials defined on $[0, 1]$ equipped with the sup norm. Clearly, $\hat{X} = C[0, 1]$. Now consider the operator $T\colon X \to X$ defined by $Tp(t) = p(t^2)$ for all $p \in X$ and all $t \in [0, 1]$. Note that T is a non-invertible isometry. However, it should be clear that $\hat{T}\colon C[0, 1] \to C[0, 1]$ satisfies $\hat{T}x(t) = x(t^2)$ for all $x \in C[0, 1]$ and each $t \in [0, 1]$, which is an invertible isometry on $\hat{X}$.]

[4] In this context, as usual, the scalar λ stands for the operator λI.

14. Show that for bounded operators $X \xrightarrow{T} Y \xrightarrow{S} Z$ between normed spaces (all over the same field) we have $\|ST\| \le \|S\| \cdot \|T\|$. Use this inequality to show that:

(a) The map $(S,T) \mapsto ST$, from $\mathcal{L}(Y,Z) \times \mathcal{L}(X,Y)$ to $\mathcal{L}(X,Z)$, is jointly norm continuous.

(b) The Banach space $\mathcal{L}(U,V)$ of all bounded operators from a normed space U to a Banach space V is a unital Banach algebra.

15. Let X and Y be two normed spaces over the same field. Show that the mapping $T \mapsto T^*$, from $\mathcal{L}(X,Y)$ into $\mathcal{L}(Y^*,X^*)$, is a linear isometry.

16. Prove Theorem 1.10.

17. Let A be a non-empty subset of a normed space X and let $x_0 \in X$. We say that there is a **nearest** (or a **closest**) **point** to x_0 in A if there exists some $a_0 \in A$ such that $\|x_0 - a_0\| \le \|x_0 - a\|$ for all $a \in A$. Establish the following.

(a) The norm function $x \mapsto \|x\|$ is weakly lower semicontinuous, i.e., $x_\alpha \xrightarrow{w} x$ in X implies $\|x\| \le \liminf \|x_\alpha\|$.

(b) If X is a reflexive Banach space and A is a non-empty weakly closed subset of X, then for every point $x_0 \in X$ there exists a nearest point to x_0 in A.

(c) Assume that the norm of X is **uniformly convex**, that is, for each $0 < \epsilon \le 2$ there exists some $0 < \delta < 1$ such that whenever $x, y \in U_X$ satisfy $\|x - y\| \ge \epsilon$, then $\|\frac{x+y}{2}\| < 1 - \delta$. A normed space is called **uniformly convex** if its norm is uniformly convex. (It is easy to verify that Hilbert spaces are uniformly convex.)
If A is a non-empty, convex, and closed subset of a uniformly convex Banach space X, then for every $x_0 \in X$ there exists exactly one point in A that is nearest to x_0. In other words, there exists a mapping $\pi\colon X \to A$ defined by

$$\big\|x - \pi(x)\big\| = \min_{y \in A} \|x - y\|.$$

The mapping π is called the **nearest point mapping** from X to A.

(d) Let C be a non-empty closed convex subset of a uniformly convex Banach space X. Show that the nearest point mapping π is continuous.

18. For a non-empty closed convex subset C of a Hilbert space X establish the following properties of the nearest point mapping $\pi\colon X \to C$.

(a) If $x \in C$, then $\pi(x) = x$.

(b) If $x \notin C$, then $\pi(x) \in \partial C$.

(c) For each $x \in X$ and each $y \in C$ we have $\big(x - \pi(x), y - \pi(x)\big) \le 0$.

(d) For each $x, y \in X$ we have $\big(\pi(x) - \pi(y), x - y\big) \ge 0$.

(e) For all $x, y \in X$ we have $\|\pi(x) - \pi(y)\| \le \|x - y\|$.

(f) If C is a closed vector subspace of X, then for each $x \in X$ the vector $x - \pi(x)$ is orthogonal to C, i.e., $\big(x - \pi(x), c\big) = 0$ for all $c \in C$.

Remark: Recall that a non-empty subset A of a topological space Ω is called a **retract** of Ω if there exists a continuous mapping $f\colon \Omega \to A$, called a **retraction of Ω onto** A, such that $f(a) = a$ for each $a \in A$. The preceding properties show that π is a retraction of X onto C.

19. (Support Points). In this problem X will denote a real Banach space, H is a real Hilbert space, and C is a non-empty closed convex subset of H. The function $\pi\colon H \to C$ is the nearest point mapping as defined in Exercise 17 above. Recall that a point a_0 in a subset A of X is called a **support point** of A if there exists some non-zero $x^* \in X^*$ such that $x^*(a_0) \le x^*(a)$ for each $a \in A$, i.e., if x^* attains its minimum value on A at a_0. Establish the following.

(a) Only boundary points of a set can be support points.
(b) If a closed convex subset of X has a non-empty interior, then every boundary point of the set is a support point.
(c) A point $c_0 \in \partial C$ is a support point of C if and only if there exists some $x \notin C$ such that $\pi(x) = c_0$.
(d) If $H = \mathbb{R}^k$, then every boundary point of C is a support point.
(e) (Bishop–Phelps [**66**]) The set of support points of C is dense in the boundary ∂C of C.[5]
(f) Give an example of a non-empty closed convex subset of ℓ_2 for which not every boundary point is a support point.

20. (The Finite Dimensional Separation Theorem). Recall that a non-zero linear functional $x^* \in X^*$ **separates** two non-empty subsets A and B of a Banach space X if $x^*(a) \le x^*(b)$ for all $a \in A$ and all $b \in B$.

Show that every pair of disjoint non-empty convex subsets of $\mathbb{R}^k$ can always be separated. [HINT: Assume that A and B are two non-empty disjoint convex subsets of $\mathbb{R}^k$. Now consider the non-empty convex set $C = A - B$ and note that $0 \notin C$. If $0 \notin \overline{C}$, then $x^* = \pi(0)$ separates A and B, where π is the nearest point mapping from $\mathbb{R}^k$ onto $\overline{C}$. If $0 \in \overline{C}$, then verify that $0 \in \partial\overline{C}$.]

21. Prove Theorem 1.11.

22. Prove Theorem 1.12.

1.2. Banach Lattices and Positive Operators

In this section, we review very briefly a few basic properties of ordered vector spaces and positive operators. For omitted proofs and details we refer the reader to the books [**29, 30, 225, 240, 290, 343**]. Recall that a reflexive, antisymmetric, and transitive relation on a set is known as an **order relation**. That is, a binary relation $\ge$ on a set is an **order relation** whenever it satisfies the properties:

(1) (**Reflexivity**) $x \ge x$ for each x.
(2) (**Antisymmetry**) $x \ge y$ and $y \ge x$ imply $x = y$.

[5] This famous result of E. Bishop and R. R. Phelps is true for any real Banach space; for a complete discussion of the Bishop–Phelps theorem see [**25**, Section 8.9]. V. I. Lomonosov [**213**] has proven recently that the Bishop–Phelps theorem fails for complex Banach spaces.

(3) (**Transitivity**) $x \geq y$ and $y \geq z$ imply $x \geq z$.

As usual, the symbol $y \leq x$ is equivalent to $x \geq y$. The notation $x > y$ (or $y < x$) means $x \geq y$ and $x \neq y$. Any set equipped with an order relation is called a **partially ordered set**.

A **vector space order** on a real vector space X is an order relation $\geq$ on X that is compatible with the algebraic structure of X in the sense that it satisfies the following two extra properties.

(a) If $x \geq y$, then $x + z \geq y + z$ for each $z \in X$.

(b) If $x \geq y$, then $\alpha x \geq \alpha y$ for each scalar $\alpha \geq 0$.

A **partially ordered vector space** (or simply an **ordered vector space**) is a real vector space equipped with a vector space order.

If X is an ordered vector space, then the set $X^+ = \{x \in X \colon\ x \geq 0\}$ is referred to as the **positive cone**, or simply the **cone**, of X. The cone X^+ satisfies the following properties:

(i) $X^+ + X^+ \subseteq X^+$.

(ii) $\alpha X^+ \subseteq X^+$ for each scalar $\alpha \geq 0$.

(iii) $X^+ \cap (-X^+) = \{0\}$.

Any subset C of a real vector space X that satisfies properties (i), (ii), and (iii) above is referred to as a (convex) **cone** of X. The following important relationship between cones and vector space orders is true: *If C is a cone in a real vector space X, then the binary relation $\geq$ defined by letting $x \geq y$ whenever $x - y \in C$ is a vector space order on X such that $X^+ = C$.*

An ordered vector space X has a **generating cone** if $X = X^+ - X^+$, i.e., if every vector can be written as a difference of two positive vectors (or equivalently, if for each $x \in X$ there exists some $y \in X^+$ satisfying $y \geq x$).

Definition 1.13. *An operator $T \colon X \to Y$ between two ordered vector spaces is called:*

(1) ***positive*** *(in symbols, $T \geq 0$ or $0 \leq T$) if $x \geq 0$ implies $Tx \geq 0$;*

(2) ***strictly positive*** *if $x > 0$ implies $Tx > 0$; and*

(3) ***regular*** *if T can be written as a difference of two positive operators.*

The collection of all regular operators between two ordered vector spaces X and Y will be denoted $\mathcal{L}_r(X, Y)$. The reader should notice that under the pointwise algebraic operations and under the natural order generated by the cone of positive operators the set $\mathcal{L}_r(X, Y)$ is, in fact, an ordered vector space.

A subset A of a partially ordered vector space X is said to be **order bounded** if there exist $x, y \in X$ such that $x \leq a \leq y$ holds for each $a \in A$.

Definition 1.14. *An operator $T\colon X \to Y$ between two ordered vector spaces is called* **order bounded** *if T carries order bounded subsets of X to order bounded subsets of Y.*

Clearly, every positive operator (and hence every regular operator) is order bounded. However, an order bounded operator need not be regular; see [**30**, Example 1.11, p. 10]. The collection of all order bounded operators from X into Y will be denoted $\mathcal{L}_b(X,Y)$. Under the pointwise algebraic operations and pointwise order, $\mathcal{L}_b(X,Y)$ is an ordered vector space that includes the vector space of all regular operators, i.e, $\mathcal{L}_r(X,Y) \subseteq \mathcal{L}_b(X,Y)$.

The positive operators are characterized by means of their additivity property on the positive cone.

Theorem 1.15 (Kantorovich). *Let X and Y be two ordered vector spaces and let $T\colon X^+ \to Y^+$ be an additive mapping, i.e., $T(x+y) = Tx+Ty$ holds for all x, y in X^+. If the cone of X is generating and Y is Archimedean,*[6] *then T extends uniquely to a positive operator from X to Y. Denoting this unique linear extension by T again, we have*

$$Tx = Tx_1 - Tx_2\,,$$

where $x = x_1 - x_2$ is any representation of x as a difference of two positive vectors.

A lattice ordered vector space E is called a **Riesz space** (or a **vector lattice**). That is, an ordered vector space E is a Riesz space if every pair of vectors has a **least upper bound** (**supremum**) and a **greatest lower bound** (**infimum**). Following the standard lattice notation, the supremum and infimum of a pair of vectors $\{x,y\}$ will be denoted by $x \vee y$ and $x \wedge y$, respectively. That is,

$$x \vee y = \sup\{x,y\} \quad \text{and} \quad x \wedge y = \inf\{x,y\}\,.$$

For an element x in a Riesz space, its **positive part**, its **negative part**, and its **absolute value** are defined by

$$x^+ = x \vee 0, \quad x^- = (-x) \vee 0 \quad \text{and} \quad |x| = x \vee (-x)\,,$$

respectively. We have the following two important identities:

$$x = x^+ - x^- \quad \text{and} \quad |x| = x^+ + x^-\,. \tag{$\star$}$$

The functions $(x,y) \mapsto x \vee y$, $(x,y) \mapsto x \wedge y$, $x \mapsto x^+$, $x \mapsto x^-$, and $x \mapsto |x|$ are referred to collectively as the **lattice operations** of a Riesz space. From $(\star)$, it follows that every Riesz space has a generating cone.

A Riesz space is said to be **Dedekind** (or **order**) **complete** whenever every non-empty subset of the space that is bounded from above has a least

[6] An ordered vector space Y is said to be **Archimedean** if $nx \le y$ for all n implies $x \le 0$.

upper bound. Similarly, a Riesz space is said to be **σ-Dedekind** (or **σ-order**) **complete** whenever every non-empty countable subset of the space that is bounded from above has a least upper bound.

When E and F are Riesz spaces with F Dedekind complete, an operator $T\colon E \to F$ is order bounded if and only if it is a regular operator. The details of this important fact are included in the next result whose proof can be found in [**30**, p. 12].

Theorem 1.16 (F. Riesz–Kantorovich). *If E and F are two Riesz spaces with F Dedekind complete, then the ordered vector space $\mathcal{L}_r(E,F)$ is a Dedekind complete Riesz space satisfying $\mathcal{L}_r(E,F) = \mathcal{L}_b(E,F)$. Moreover, its lattice operations are given by*

(1) $T^+(x) = \sup\{Ty\colon\ 0 \le y \le x\}$,

(2) $T^-(x) = \sup\{-Ty\colon\ 0 \le y \le x\}$, *and*

(3) $|T|(x) = \sup\{Ty\colon\ -x \le y \le x\}$,

for all $T \in \mathcal{L}_r(E,F)$ and all $x \in E^+$.

The operator $|T| = T \vee (-T)$ is known as the **modulus** of the operator T. Formulas (1), (2), and (3) in Theorem 1.16 are called the **Riesz–Kantorovich formulas**. Two operators $S, T \in \mathcal{L}_r(E,F)$ are said to be **disjoint** if $S \wedge T = 0$.

When E is also Dedekind complete, there is a useful modification of the Riesz–Kantorovich formulas in terms of "components." Recall that if x is a positive vector in a Riesz space, then any positive vector y satisfying $y \wedge (x-y) = 0$ is referred to as a **component** of x.

Theorem 1.17 (Abramovich [**1**]). *If $T\colon E \to F$ is an order bounded operator between Dedekind complete Riesz spaces, then*

(1) $T^+(x) = \sup\{Ty\colon\ y \in E \text{ and } y \wedge (x-y) = 0\}$,

(2) $T^-(x) = \sup\{-Ty\colon\ y \in E \text{ and } y \wedge (x-y) = 0\}$, *and*

(3) $|T|x = \sup\{Ty - T(x-y)\colon\ y \in E \text{ and } y \wedge (x-y) = 0\}$

hold true for each $x \in E^+$.

A net $\{x_\alpha\}_{\alpha \in A}$ in a partially ordered set is said to be **decreasing** (in symbols, $x_\alpha \downarrow$) if $\alpha_1 \ge \alpha_2$ implies $x_{\alpha_2} \le x_{\alpha_1}$. The notation $x_\alpha \downarrow x$ means $x_\alpha \downarrow$ and $x = \inf\{x_\alpha\}$. The symbols $x_\alpha \uparrow$ and $x_\alpha \uparrow x$ have analogous meanings.

Definition 1.18. *A net $\{x_\alpha\}_{\alpha \in A}$ in a partially ordered set is said to be* ***order convergent*** *to an element x, in symbols, $x_\alpha \xrightarrow{o} x$, whenever there exist two nets $\{y_\beta\}_{\beta \in B}$ and $\{z_\gamma\}_{\gamma \in \Gamma}$ such that:*

(1) $y_\beta \uparrow x$ *and* $z_\gamma \downarrow x$.

(2) *For each* $\beta \in B$ *and* $\gamma \in \Gamma$ *there exists some* $\alpha_0 \in A$ *(depending on* β *and* γ*) satisfying* $y_\beta \le x_\alpha \le z_\gamma$ *for all* $\alpha \ge \alpha_0$.

The element x *is called the* ***order limit*** *of the net* $\{x_\alpha\}$.

The following observation regarding order limits will be useful. If in the definition of $x_\alpha \xrightarrow{o} x$ we consider the product index set $\Lambda = B \times \Gamma$ and define the two nets $\{u_\lambda\}_{\lambda\in\Lambda}$ and $\{v_\lambda\}_{\lambda\in\Lambda}$ by

$$u_\lambda = y_\beta \quad \text{and} \quad v_\lambda = z_\gamma$$

for each $\lambda = (\beta, \gamma) \in \Lambda$, then we can replace the nets $\{y_\beta\}_{\beta\in B}$ and $\{z_\gamma\}_{\gamma\in\Gamma}$ by $\{u_\lambda\}_{\lambda\in\Lambda}$ and $\{v_\lambda\}_{\lambda\in\Lambda}$, respectively. That is, we can assume in the definition of order convergence that the index sets B and Γ are the same.

It should be clear that if $x_\alpha \uparrow x$ or $x_\alpha \downarrow x$ in a partially ordered set, then $x_\alpha \xrightarrow{o} x$. We list below two basic properties of order limits.

- *A net in a partially ordered set can have at most one order limit.*
- *If a net* $\{x_\alpha\}$ *in a partially ordered set satisfies* $x_\alpha \uparrow$ *(resp.* $x_\alpha \downarrow$*) and* $x_\alpha \xrightarrow{o} x$*, then* $x_\alpha \uparrow x$ *(resp.* $x_\alpha \downarrow x$*).*

(See Exercise 4 at the end of the section.) The next result deals with order convergence in Riesz spaces.

Lemma 1.19. *For a net* $\{x_\alpha\}_{\alpha\in A}$ *in a Riesz space* E *we have:*

(a) *If there exists a net* $\{u_\alpha\}_{\alpha\in A}$ *(with the same index set) such that* $u_\alpha \downarrow 0$ *and* $|x_\alpha - x| \le u_\alpha$ *for each* $\alpha \in A$*, then* $x_\alpha \xrightarrow{o} x$.

(b) *If* E *is Dedekind complete and* $\{x_\alpha\}_{\alpha\in A}$ *is order bounded, then* $x_\alpha \xrightarrow{o} x$ *in* E *if and only if there exists a net* $\{u_\alpha\}_{\alpha\in A}$ *(with the same index set) such that* $u_\alpha \downarrow 0$ *and* $|x_\alpha - x| \le u_\alpha$ *for each* $\alpha \in A$.

Proof. (a) Assume that there exists a net $\{u_\alpha\}_{\alpha\in A}$ satisfying $u_\alpha \downarrow 0$ and $|x_\alpha - x| \le u_\alpha$ for each $\alpha \in A$. Let $y_\alpha = x - u_\alpha$ and $z_\alpha = x + u_\alpha$. Then it should be clear that $y_\alpha \uparrow x$, $z_\alpha \downarrow x$, and $y_\alpha \le x_\alpha \le z_\alpha$ holds for each $\alpha \in A$. Therefore, $x_\alpha \xrightarrow{o} x$.

(b) Suppose that E is Dedekind complete, $\{x_\alpha\}_{a\in A}$ is order bounded, and that $x_\alpha \xrightarrow{o} x$. Pick two nets $\{y_\beta\}_{\beta\in B}$ and $\{z_\gamma\}_{\gamma\in\Gamma}$ such that $y_\beta \uparrow x$, $z_\gamma \downarrow x$, and for each $\beta \in B$ and $\gamma \in \Gamma$ there exists some $\alpha(\beta,\gamma) \in A$ so that $y_\beta \le x_\alpha \le z_\gamma$ for each $\alpha \ge \alpha(\beta,\gamma)$. For each $\alpha \in A$, let

$$v_\alpha = \inf_{\alpha'\ge\alpha} x_{\alpha'} \qquad \text{and} \qquad w_\alpha = \sup_{\alpha'\ge\alpha} x_{\alpha'}\,.$$

The above suprema and infima exist since the net $\{x_\alpha\}$ is order bounded and E is Dedekind complete. Clearly, $v_\alpha \uparrow$ and $w_\alpha \downarrow$. Moreover, since for each $\beta \in B$ and $\gamma \in \Gamma$ we have $y_\beta \le x_{\alpha'} \le z_\gamma$ for all $\alpha' \ge \alpha(\beta,\gamma)$, it follows that $y_\beta \le v_\alpha \le z_\gamma$ for all $\alpha \ge \alpha(\beta,\gamma)$. So, if $v_\alpha \uparrow v$, then $y_\beta \le v \le z_\gamma$ holds for all $\beta \in B$ and $\gamma \in \Gamma$. From $y_\beta \uparrow x$ and $z_\gamma \downarrow x$, we get $v = x$. Similarly, $w_\alpha \downarrow x$. To finish the proof notice that $|x_\alpha - x| \le w_\alpha - v_\alpha$ for each α and that $w_\alpha - v_\alpha \downarrow 0$. ■

Since order convergent sequences in Riesz spaces are automatically order bounded, the sequential analogue of Lemma 1.19 can be stated as follows.

Lemma 1.20. *In a σ-Dedekind complete Riesz space a sequence $\{x_n\}$ is order convergent to a vector x if and only if there exists a sequence $\{u_n\}$ such that $u_n \downarrow 0$ and $|x_n - x| \le u_n$ for each $n \in \mathbb{N}$.*

Historically, the conclusion of Lemma 1.20 is taken as the definition of order convergence of sequences. To be consistent with the literature, we also follow this tradition. That is, for sequences we define order convergence as follows.

Definition 1.21. *A sequence $\{x_n\}$ in a partially ordered set is said to be **order convergent** to some element x, in symbols, $x_n \xrightarrow{o} x$, whenever there exist two sequence $\{y_n\}$ and $\{z_n\}$ such that $y_n \uparrow x$, $z_n \downarrow x$, and $y_n \le x_n \le z_n$ for all n.*

For Riesz spaces we have the following:

- *A sequence $\{x_n\}$ in a Riesz space is order convergent to a vector x if and only if there exists a sequence $\{u_n\}$ such that $u_n \downarrow 0$ and $|x_n - x| \le u_n$ for each $n \in \mathbb{N}$.*

According to Lemma 1.20, for σ-Dedekind complete Riesz spaces this definition is, of course, equivalent to the general definition of order convergence of nets as given in Definition 1.18.

Definition 1.22. *A subset D of a partially ordered set is said to be **order closed** (resp. **σ-order closed**) if $\{x_\alpha\} \subseteq D$ and $x_\alpha \xrightarrow{o} x$ imply $x \in D$ (resp. $\{x_n\} \subseteq D$ and $x_n \xrightarrow{o} x$ imply $x \in D$).*

We now turn our attention to order continuous mappings.

Definition 1.23. *A mapping $f\colon X \to Y$ between partially ordered sets is:*

(1) ***σ-order continuous*** *if $x_n \xrightarrow{o} x$ in X implies $f(x_n) \xrightarrow{o} f(x)$ in Y,*

(2) ***order continuous*** *if $x_\alpha \xrightarrow{o} x$ in X implies $f(x_\alpha) \xrightarrow{o} f(x)$ in Y.*

Simple examples of order continuous functions are provided by the lattice operations of a Riesz space. For instance, in any Riesz space $x_\alpha \xrightarrow{o} x$ implies $x_\alpha^+ \xrightarrow{o} x^+$ and $|x_\alpha| \xrightarrow{o} |x|$. (See Exercise 4 at the end of the section.)

The proof of the following characterization of order continuity of positive operators is left as an exercise.

Lemma 1.24. *A positive operator $T\colon X \to Y$ between two ordered vector spaces is order continuous if and only if $x_\alpha \downarrow 0$ in X implies $Tx_\alpha \downarrow 0$ in Y.*

Similarly, a positive operator $T\colon X \to Y$ is σ-order continuous if and only if $x_n \downarrow 0$ in X implies $Tx_n \downarrow 0$ in Y.

The order continuity of order bounded operators is characterized as follows.

Theorem 1.25. *For an order bounded operator $T\colon E \to F$ between Riesz spaces with F Dedekind complete the following statements are equivalent.*

(1) *T is order continuous.*
(2) *If $x_\alpha \downarrow 0$ in E, then $Tx_\alpha \xrightarrow{o} 0$ in F.*
(3) *If $x_\alpha \downarrow 0$ in E, then $\inf_\alpha |Tx_\alpha| = 0$ in F.*
(4) *T^+ and T^- are both order continuous.*
(5) *$|T|$ is order continuous.*

Proof. (1)$\Longrightarrow$(2) If $x_\alpha \downarrow 0$, then $x_\alpha \xrightarrow{o} 0$, and so $Tx_\alpha \xrightarrow{o} 0$.

(2)$\Longrightarrow$(3) Assume that $x_\alpha \downarrow 0$ and let some $y \in F$ satisfy $y \le |Tx_\alpha|$ for each α. Since $Tx_\alpha \xrightarrow{o} 0$, there exists a net $\{u_\lambda\}_{\lambda\in\Lambda}$ such that $u_\lambda \downarrow 0$ and for each $\lambda \in \Lambda$ there exists some α_0 satisfying $|Tx_\alpha| \le y_\lambda$ for each $\alpha \ge \alpha_0$. This implies $y \le |Tx_{\alpha_0}| \le u_\lambda$, and so $y \le u_\lambda$ for each $\lambda \in \Lambda$. From $u_\lambda \downarrow 0$, we get $y \le 0$. This shows that $\inf_\alpha |Tx_\alpha| = 0$ in F.

(3)$\Longrightarrow$(4) It suffices to establish that T^+ is order continuous. To this end, assume that $x_\alpha \downarrow 0$ in E. Let $0 \le u \le T^+x_\alpha$ for each α. To show that T^+ is order continuous, we must prove that $u = 0$.

Fix some index α_0, and let $x = x_{\alpha_0}$. Clearly, for each $0 \le y \le x$ and all $\alpha \ge \alpha_0$ we have $0 \le y - y \wedge x_{\alpha_0} = y \wedge x - y \wedge x_\alpha \le x - x_\alpha$. This implies

$$Ty - T(y \wedge x_\alpha) = T(y - y \wedge x_\alpha) \le T^+(x - x_\alpha) = T^+x - T^+x_\alpha\,,$$

and consequently

$$0 \le u \le T^+x_\alpha \le T^+x + |T(y \wedge x_\alpha)| - Ty \qquad (\star)$$

for each $\alpha \ge \alpha_0$ and all $0 \le y \le x$. Since for each $0 \le y \le x$ we have $y \wedge x_\alpha \downarrow_{\alpha\ge\alpha_0} 0$, it follows from our hypothesis that $\inf_{\alpha\ge\alpha_0} |T(y \wedge x_\alpha)| = 0$. Now a glance at $(\star)$ yields $0 \le u \le T^+x - Ty$ for each $0 \le y \le x$. Taking into consideration the Riesz–Kantorovich formula $T^+x = \sup_{0\le y\le x} Ty$, the latter shows that $u = 0$, as desired.

(4)$\Longrightarrow$(5) Obvious.

(5)$\Longrightarrow$(1) Let $x_\alpha \xrightarrow{o} 0$ in E. Pick two nets $\{y_\beta\}_{\beta\in B}$ and $\{z_\gamma\}_{\gamma\in\Gamma}$ such that $y_\beta \uparrow x$, $z_\gamma \downarrow x$, and for each $\beta \in B$ and $\gamma \in \Gamma$ there exists some $\alpha(\beta,\gamma) \in A$ satisfying $y_\beta \le x_\alpha \le z_\gamma$ for each $\alpha \ge \alpha_0$. It is easy to see that $|x_\alpha| \le z_\gamma - y_\beta$ holds for all $\alpha \ge \alpha(\beta,\gamma)$.

If we put $\Lambda = B \times \Gamma$ and for each $\lambda = (\beta,\gamma)$, we let $u_\lambda = z_\gamma - y_\beta \ge 0$, then $|Tx_\alpha| \le |T|(u_\lambda)$ holds for all $\alpha \ge \alpha(\beta,\gamma) = \alpha(\lambda)$. The order continuity of $|T|$ implies $|T|(u_\lambda) \downarrow 0$, and so $Tx_\alpha \xrightarrow{o} 0$ in F. The proof is finished. ■

The corresponding result for σ-order continuous operators is as follows.

Theorem 1.26. *For an order bounded operator $T\colon E \to F$ between Riesz spaces with F Dedekind complete the following statements are equivalent.*

(1) *T is σ-order continuous.*

(2) *If $x_n \downarrow 0$ in E, then $Tx_n \xrightarrow{o} 0$ in F.*

(3) *If $x_n \downarrow 0$ in E, then $\inf_n |Tx_n| = 0$ in F.*

(4) *T^+ and T^- are both σ-order continuous.*

(5) *$|T|$ is σ-order continuous.*

Let E be a Riesz space. A subset A of E is said to be **solid** if $|x| \le |y|$ and $y \in A$ imply $x \in A$. A solid vector subspace of E is called an **ideal**. The **ideal generated by a non-empty subset** A of E is the smallest (with respect to inclusion) ideal containing A; it coincides with the intersection of all ideals that contain A and is given by

$$E_A = \Big\{x \in E\colon\ \exists\, x_1,\dots,x_n \in A \text{ and } \lambda_1,\dots,\lambda_n \in \mathbb{R}^+ \text{ with } |x| \le \sum_{i=1}^n \lambda_i|x_i|\Big\}.$$

If $A = \{x\}$, then $E_x = E_{\{x\}}$ is called the **principal ideal** generated by the vector x. Note that

$$E_x = \big\{y \in E\colon\ \text{there exists } \lambda \ge 0 \text{ such that } |y| \le \lambda|x|\,\big\}.$$

The principal ideals will play an important role in this book. A vector $e > 0$ is called an **order unit** (or a **strong unit** or simply a **unit**) if $E_e = E$, i.e., if for each $x \in X$ there exists some $\lambda \ge 0$ such that $|x| \le \lambda e$.

An order closed ideal is called a **band**. Since a solid subset D in a Riesz space E is order closed if and only if $\{x_\alpha\} \subseteq D \cap E^+$ and $x_\alpha \uparrow x$ in E imply $x \in D$ (see Exercise 5 at the end of the section), it follows that an ideal A is a band if and only if $\{x_\alpha\} \subseteq A$ and $0 \le x_\alpha \uparrow x$ imply $x \in A$.

If D is a non-empty subset of a Riesz space E, then the **band generated** by D is the smallest (with respect to inclusion) band B_D that contains D. If $D = \{x\}$, then $B_x = B_{\{x\}}$ is called the **principal band** generated by the vector x. The band generated by an ideal is described as follows.

Theorem 1.27. *If J is an ideal in a Riesz space E, then the band B_J generated by J is given by*

$$B_J = \big\{x \in E\colon \ \exists \ \{x_\alpha\} \subseteq J \ \text{ such that } \ 0 \le x_\alpha \uparrow |x| \big\}\,.$$

Moreover, for each $x \in E$ the principal band B_x generated by x is given by

$$B_x = \big\{y \in E\colon \ \ |y| \wedge n|x| \uparrow |y| \big\}\,.$$

A vector $e > 0$ is said to be a **weak unit** if $B_e = E$, i.e., if $x \wedge ne \uparrow x$ holds for each $x \in E^+$. If E is Archimedean, then a vector $e > 0$ is a weak unit if and only if $|x| \wedge e = 0$ implies $x = 0$. Clearly, every order unit is a weak unit.

The **disjoint complement** of a non-empty subset A of E is defined by

$$A^{\mathrm{d}} = \big\{x \in E\colon \ \ |x| \wedge |a| = 0 \ \text{ for each } \ a \in A \big\}\,.$$

The order continuity of the lattice operations guarantees that every disjoint complement is a band. In Archimedean Riesz spaces every band is a disjoint complement.

Theorem 1.28. *In an Archimedean Riesz space, the band generated by a non-empty set A coincides with $A^{\mathrm{dd}} = (A^{\mathrm{d}})^{\mathrm{d}}$, i.e., $B_A = A^{\mathrm{dd}}$. In particular, every band B in an Archimedean Riesz space satisfies $B = B^{\mathrm{dd}}$.*

A band B in a Riesz space E is said to be a **projection band** whenever $E = B \oplus B^{\mathrm{d}}$ holds. Clearly, each projection band B gives rise to a natural projection[7] with range B which is referred to as a **band projection**. In a Dedekind complete Riesz space every band is a projection band.

The collection of all order continuous regular operators from an ordered vector space X into another ordered vector space Y will be denoted by $\mathcal{L}_{\mathrm{n}}(X,Y)$. Similarly, the collection of all σ-order continuous regular operators will be denoted by $\mathcal{L}_{\mathrm{c}}(X,Y)$. Both $\mathcal{L}_{\mathrm{n}}(X,Y)$ and $\mathcal{L}_{\mathrm{c}}(X,Y)$ are vector subspaces of the vector space of all regular operators $\mathcal{L}_r(X,Y)$, and $\mathcal{L}_{\mathrm{n}}(X,Y) \subseteq \mathcal{L}_{\mathrm{c}}(X,Y)$.

Theorem 1.29 (Ogasawara)**.** *If E and F are Riesz spaces with F Dedekind complete, then $\mathcal{L}_{\mathrm{n}}(E,F)$ and $\mathcal{L}_{\mathrm{c}}(E,F)$ are both bands in $\mathcal{L}_r(E,F)$. In particular, we have the decompositions*

$$\mathcal{L}_r(E,F) = \mathcal{L}_{\mathrm{n}}(E,F) \oplus [\mathcal{L}_{\mathrm{n}}(E,F)]^{\mathrm{d}} = \mathcal{L}_{\mathrm{c}}(E,F) \oplus [\mathcal{L}_{\mathrm{c}}(E,F)]^{\mathrm{d}}\,.$$

If E is a Riesz space, then the Dedekind complete Riesz space $\mathcal{L}_r(E,\mathbb{R})$ is called the **order dual** of E and is denoted $E^\sim$, i.e., $E^\sim = \mathcal{L}_r(E,\mathbb{R})$. The band $E^\sim_{\mathrm{n}} = \mathcal{L}_{\mathrm{n}}(E,\mathbb{R})$ is referred to as the **order continuous dual** of E. If $E^\sim_{\mathrm{n}}$ separates the points of E, then E is called a **normal Riesz space**.

[7] Recall that a linear operator $P\colon V \to V$ on a vector space is called a **projection** if $P^2 = P$.

Recall also that a Riesz space E is said to have the **countable sup property** if for every subset D of E having a supremum there exists an at most countable subset C of D satisfying $\sup C = \sup D$. The countable sup property allows us to work with sequences rather than nets. For example, if E satisfies the countable sup property, then $\mathcal{L}_c(E, F) = \mathcal{L}_n(E, F)$.

A seminorm p on a Riesz space is said to be a **lattice seminorm** whenever $|x| \le |y|$ implies $p(x) \le p(y)$. A **normed Riesz space** is a Riesz space equipped with a lattice norm. A normed Riesz space which is also norm complete (i.e., a Banach space) is called a **Banach lattice**. In a normed Riesz space, the lattice operations are uniformly continuous functions. In particular, this implies that every band is a closed subspace. Every band projection on a normed Riesz space is always a contraction. We also have the following simple but very useful result.

Lemma 1.30. *In a normed Riesz space E:*

(a) *The cone E^+ is norm (and hence weakly) closed.*

(b) *The order intervals are norm (and hence weakly) closed.*

(c) *If a net $\{x_\alpha\} \subseteq E$ satisfies $x_\alpha \uparrow$ and $\|x_n - x\| \to 0$, then $x_\alpha \uparrow x$.*

Our next result informs us that positive operators between Banach lattices are continuous; for a proof see [**30**, p. 175].

Theorem 1.31. *Every order bounded operator from a Banach lattice to a normed Riesz space is continuous (and hence every positive and every regular operator on a Banach lattice is also continuous).*

Now consider two Banach lattices E and F with F Dedekind complete. For a regular operator $T\colon E \to F$ its r**-norm** is defined by $\|T\|_r = \big\| |T| \big\|$. Note that we have

$$\|T\|_r = \inf\{\|S\|\colon\ \pm T \le S\}.$$

This formula allows us to define the r-norm for regular operators with values in a non-Dedekind complete Banach lattice. For the proof of the next result see [**30**, Theorem 15.2, p. 248].

Theorem 1.32. *If E and F are two Banach lattices with F Dedekind complete, then the Dedekind complete Riesz space $\mathcal{L}_r(E, F)$ of all regular operators from E to F equipped with the r-norm is a Banach lattice.*

Since every continuous linear functional on a normed Riesz space E is automatically order bounded, it follows that $E^* \subseteq E^\sim$. In fact, it follows that the operator norm coincides with the r-norm on E^*; see Exercise 20 at the end of the section. In other words, we have the following important result.

Theorem 1.33. *If E is a normed Riesz space, then its norm dual E^* is a Banach lattice and its lattice operations are given by*

$$\begin{aligned} (x^*)^+(x) &= \sup\{x^*(y)\colon\ 0 \le y \le x\}, \\ (x^*)^-(x) &= \sup\{-x^*(y)\colon\ 0 \le y \le x\}, \text{ and} \\ |x^*|(x) &= \sup\{x^*(y)\colon\ |y| \le x\} \end{aligned}$$

for each $x^* \in E^*$ and each $x \in E^+$. Moreover, E^* is an ideal in the order dual $E^\sim$ and if E is a Banach lattice, then $E^* = E^\sim$.

We continue our discussion by introducing some additional terminology. A positive operator $T\colon E \to F$ between two Riesz spaces is said to be a **lattice** (or a **Riesz**) **homomorphism** if $T(x \vee y) = Tx \vee Ty$ for all x and y in E. A one-to-one lattice isomorphism is known as a **lattice** (or a **Riesz**) **isomorphism**. A lattice isomorphism T between two Banach lattices is said to be a **lattice isometry** if $\|Tx\| = \|x\|$ for each $x \in E$. Two Banach lattices are called **lattice isometric** if there exists a lattice isometry from one onto the other.

The most useful characterizations of lattice homomorphisms are included in the next result.

Theorem 1.34. *For a positive operator $T\colon E \to F$ between two Riesz spaces the following statements are equivalent.*

(a) *T is a lattice homomorphism.*

(b) *$T(x^+) = (Tx)^+$ for each $x \in E$.*

(c) *$T(x \wedge y) = Tx \wedge Ty$ for all $x, y \in E$.*

(d) *$|Tx| = T|x|$ for each $x \in E$.*

(e) *$x \wedge y = 0$ in E implies $Tx \wedge Ty = 0$ in F.*

Proof. (a) $\Longrightarrow$ (b) We have

$$T(x^+) = T(x \vee 0) = T(x) \vee T(0) = T(x) \vee 0 = (Tx)^+ .$$

(b) $\Longrightarrow$ (c) Using the identity $x \wedge y = y - (y - x)^+$, we get

$$\begin{aligned} T(x \wedge y) &= T(y) - T\big((y-x)^+\big) = T(y) - [T(y-x)]^+ \\ &= T(y) - \big[T(y) - T(x)\big]^+ = T(x) \wedge T(y) . \end{aligned}$$

(c) $\Longrightarrow$ (d) The identity $|x - y| = x + y - 2(x \wedge y)$ yields:

$$\begin{aligned} T(|x-y|) &= T(x) + T(y) - 2T(x \wedge y) \\ &= T(x) + T(y) - 2[T(x) \wedge T(y)] = |T(x) - T(y)| . \end{aligned}$$

Letting $y = 0$ gives $|T(x)| = T(|x|)$ for all $x \in E$.

(d) $\Longrightarrow$ (e) Let $x \wedge y = 0$. The identity $u \wedge v = \frac{1}{2}(u + v - |u - v|)$ implies

$$\begin{aligned} T(x) \wedge T(y) &= \tfrac{1}{2}\big[T(x) + T(y) - |T(x) - T(y)|\big] \\ &= \tfrac{1}{2}\big[T(x) + T(y) - T(|x - y|)\big] \\ &= T\big(\tfrac{1}{2}(x + y - |x - y|)\big) = T(x \wedge y) = T(0) = 0\,. \end{aligned}$$

(e) $\Longrightarrow$ (a) From $(x - x \wedge y) \wedge (y - x \wedge y) = 0$, we get $T(x \wedge y) = T(x) \wedge T(y)$. Now use the identity $x \vee y = x + y - x \wedge y$ to obtain

$$\begin{aligned} T(x \vee y) &= T(x) + T(y) - T(x \wedge y) \\ &= T(x) + T(y) - [T(x) \wedge T(y)] = T(x) \vee T(y)\,. \end{aligned}$$

This completes the proof. ■

A positive operator $T\colon E \to E$ is said to be **interval preserving** if $T[0, x] = [0, Tx]$ for each $x \in E^+$. There are some nice duality properties between interval preserving operators and lattice homomorphisms. They are included in the next result whose proof can be found in [**30**, p. 92]; see also Exercise 11 in Section 4.3.

Theorem 1.35 (Kim [**183**]; Andô–Lotz [**216**]). *For an arbitrary positive operator $T\colon E \to F$ between two Banach lattices we have the following.*

(1) *If T is interval preserving, then its adjoint $T^*\colon F^* \to E^*$ is a lattice homomorphism.*

(2) *T is a lattice homomorphism if and only if T^* is interval preserving.*

Building on the work and ideas of D. Maharam [**228, 229, 230**] on integral representations of linear operators, W. A. J. Luxemburg and B. de Pagter [**222**] have also established the following remarkable result regarding extensions of positive operators to interval preserving operators.

Theorem 1.36 (Luxemburg–de Pagter). *Assume that E is an Archimedean Riesz space, F is a Dedekind complete Riesz space, and $T\colon E \to F$ is a positive operator. Then there exists a Dedekind complete Riesz space L containing E as a Riesz subspace and an order continuous interval preserving operator $\widehat{T}\colon L \to F$ that extends T.*

Exercises

1. Let C be a cone in a real vector space X. Show that the relation $\geq$ on X defined by $x \geq y$ whenever $x - y \in C$ is a vector space order such that $X^+ = C$.

2. Prove Theorem 1.15.

3. Show that order complete Riesz spaces and normed Riesz spaces are Archimedean.

4. Establish the following properties of order convergence.
 (a) A net in a partially ordered set can have at most one order limit.
 (b) If $x_\alpha \uparrow x$ or $x_\alpha \downarrow x$ in a partially ordered set, then $x_\alpha \xrightarrow{o} x$.
 (c) If a net $\{x_\alpha\}_{\alpha \in A}$ in a partially ordered set satisfies $x_\alpha \uparrow$ (resp. $x_\alpha \downarrow$) and $x_\alpha \xrightarrow{o} x$, then $x_\alpha \uparrow x$ (resp. $x_\alpha \downarrow x$).
 (d) A positive operator $T\colon X \to Y$ between two ordered vector spaces is order continuous if $x_\alpha \downarrow 0$ in X implies $Tx_\alpha \downarrow 0$ in Y.
 (e) A net $\{x_\alpha\}_{\alpha \in A}$ in a Riesz space satisfies $x_\alpha \xrightarrow{o} x$ if and only if there exists another net $\{u_\lambda\}_{\lambda \in \Lambda}$ such that $u_\lambda \downarrow 0$ and for each $\lambda \in \Lambda$ there exists some α_0 (depending on λ) such that $|x_\alpha - x| \leq u_\lambda$ holds for each $\alpha \geq \alpha_0$.
 (f) The lattice operations in any Riesz space are order continuous.

5. Show that a solid subset D of a Riesz space E is order closed if and only if $\{x_\alpha\} \subseteq D \cap E^+$ and $x_\alpha \uparrow x$ in E imply $x \in D$.

6. Prove Theorem 1.27.

7. Prove Theorem 1.28.

8. Show that a vector $e > 0$ in an Archimedean Riesz space is a weak unit if and only if $|x| \wedge e = 0$ implies $x = 0$.

9. Show that a positive operator $T\colon X \to Y$ between two partially ordered vector spaces is order continuous if and only if $x_\alpha \downarrow 0$ in X implies $Tx_\alpha \downarrow 0$ in Y.

10. Show the following:
 (a) If Ω is a compact topological space, then in $C(\Omega)$ each component of the constant function $\mathbf{1}$ is of the form χ_V for some clopen set V.
 (b) If μ is a finite measure, then in $L_1(\mu)$ each component of the constant function $\mathbf{1}$ is of the form χ_A for some measurable set A.

11. If E and F are Riesz spaces with F Dedekind complete, then show that the lattice operations of any two operators $S, T \in \mathcal{L}_b(E, F)$ are given by

$$\begin{aligned}(S \vee T)(x) &= \sup\{Sy + Tz\colon\ y, z \in E^+ \text{ and } y + z = x\}, \text{ and}\\ (S \wedge T)(x) &= \inf\{Sy + Tz\colon\ y, z \in E^+ \text{ and } y + z = x\}\end{aligned}$$

for each $x \in E^+$.

12. Prove Lemma 1.30.

13. Show that if $x_n \to x$ holds in a Banach lattice, then there exist a subsequence $\{y_n\}$ of $\{x_n\}$ and some $u \geq 0$ satisfying $|y_n - x| \leq \frac{1}{n}u$ for each

n. [HINT: Pick a subsequence $\{y_n\}$ of $\{x_n\}$ satisfying $\|y_n - x\| \le \frac{n}{2^n}$ for each n and let $u = \sum_{n=1}^{\infty} n|y_n - x|$.]

14. Prove Theorem 1.31.

15. Show that for a given Riesz space E there is, up to an equivalence, at most one lattice norm which makes E a Banach lattice. [HINT: Assume that a Riesz space E is a Banach lattice under the lattice norms $\|\cdot\|_1$ and $\|\cdot\|_2$. Then the identity operator $I\colon (E, \|\cdot\|_1) \to (E, \|\cdot\|_2)$ is positive, and hence (by Theorem 1.31) it must be a homeomorphism.]

16. Show that a normed Riesz space E is a Banach lattice if and only if every increasing norm Cauchy sequence of E^+ is norm convergent. [HINT: Assume that the condition is satisfied and let $\{x_n\}$ be a norm Cauchy sequence in E. Without loss of generality, we can suppose that $\|x_{n+1} - x_n\| < \frac{1}{2^n}$ for each n. Now consider the sequences $\{y_n\}$ and $\{z_n\}$ of E^+ defined by

$$y_n = \sum_{i=1}^{n} (x_{i+1} - x_i)^+ \quad \text{and} \quad z_n = \sum_{i=1}^{n} (x_{i+1} - x_i)^-,$$

and note that both sequences are increasing norm Cauchy sequences and $x_{n+1} - x_1 = y_n - z_n$.]

17. Prove Theorem 1.32.

18. Give an example of a regular operator $T\colon E \to F$ between two Banach lattices with F Dedekind complete satisfying $\|T\| < \|T\|_r$.

19. Show that a band projection P on a normed Riesz space is a contraction (i.e., $\|P\| \le 1$).

20. If E is a normed Riesz space, then show that $\|x^*\| = \|x^*\|_r = \|\,|x^*|\,\|$ holds for each $x^* \in E^*$. [HINT: From

$$|x^*(x)| \le |x^*|(|x|) \le \|\,|x^*|\,\| \cdot \|x\|,$$

we infer that $\|x^*\| \le \|\,|x^*|\,\|$. On the other hand, if $\|x\| \le 1$ and $\epsilon > 0$ are fixed, then from $|x^*|(|x|) = \sup\{|x^*(y)|\colon\ |y| \le |x|\,\}$, we see that there exists some $y \in E$ with $|y| \le |x|$ such that $|x^*|(|x|) - \epsilon < |x^*(y)| \le \|x^*\|$. Therefore,

$$\|\,|x^*|\,\| = \sup\{|x^*|(|x|)\colon\ \|x\| \le 1\} \le \|x^*\| + \epsilon,$$

for each $\epsilon > 0$. This implies $\|\,|x^*|\,\| \le \|x^*\|$.]

21. A Dedekind complete Riesz space $\hat{E}$ is said to be a **Dedekind completion** of a Riesz space E if there exists a lattice isomorphism $\pi\colon E \to \hat{E}$ such that

$$\begin{aligned} \hat{x} &= \sup\{\pi(v)\colon\ v \in E \text{ and } \pi(v) \le \hat{x}\,\} \\ &= \inf\{\pi(w)\colon\ w \in E \text{ and } \hat{x} \le \pi(w)\,\} \end{aligned}$$

for each $\hat{x} \in \hat{E}$. We can identify E with $\pi(E)$ so that E can be viewed as a Riesz subspace of $\hat{E}$. It should be clear that only Archimedean Riesz spaces can have Dedekind completions. As a matter of fact, every

Archimedean Riesz space E has a unique (up to a lattice isomorphism) Dedekind completion E^δ; for details see [**225**, pp. 185–196].

Now assume that E is a Banach lattice (and so E is an Archimedean Riesz space) and let E^δ be its Dedekind completion. On E^δ define the norm

$$|||\hat{x}||| = \inf\{\|x\|:\ x \in E^+ \text{ and } |\hat{x}| \le x\}.$$

Show that $|||x||| = \|x\|$ for each $x \in E$ and that $(E^\delta, |||\cdot|||)$ is a Banach lattice. [HINT: We shall denote $|||\cdot|||$ by $\|\cdot\|$ again. For the norm completeness of E^δ, we use Exercise 16 of this section. Assume that a sequence $\{\hat{x}_n\}$ in E^δ satisfies $0 \le \hat{x}_n \uparrow$ and $\|\hat{x}_{n+1} - \hat{x}_n\| < \frac{1}{2^n}$ for all n. For each n pick some $y_n \in E^+$ such that $0 \le \hat{x}_{n+1} - \hat{x}_n \le y_n$ and $\|\hat{x}_{n+1} - \hat{x}_n\| < \|y_n\| < \frac{1}{2^n}$. If we let $x_n = \sum_{i=1}^n y_i$, then $\{x_n\}$ is a norm Cauchy sequence in E and so $\{x_n\}$ is norm convergent to $x = \sum_{n=1}^\infty y_n$ in E. Clearly, $x_n \uparrow x$ in E. In particular, if we let $z_n = \sum_{i=n}^\infty y_i \in E$, then $\|z_n\| \to 0$. From

$$\hat{x}_n = \sum_{i=1}^n (\hat{x}_i - \hat{x}_{i-1}) \le \sum_{i=1}^n y_{i-1} = x_{n-1}$$

(here we assume $\hat{x}_0 = 0$ and $x_0 \ge |\hat{x}_1|$), we see that $\hat{x}_n \le x$ for each n. Since E^δ is Dedekind complete, there exists some $\hat{x} \in E^\delta$ such that $\hat{x}_n \uparrow \hat{x}$. Next, note that for each n and p we have

$$0 \le \hat{x}_{n+p} - \hat{x}_n = \sum_{i=n}^{n+p-1} (\hat{x}_{i+1} - \hat{x}_i) \le \sum_{i=n}^{n+p-1} y_i \le \sum_{i=n}^{\infty} y_i = z_n .$$

So, by letting $p \to \infty$, we get $0 \le \hat{x} - \hat{x}_n \le z_n$ for each n. This implies $\|\hat{x} - \hat{x}_n\| \le \|z_n\|$, and from this we get $\|\hat{x}_n - \hat{x}\| \to 0$.]

22. (Nakano [**251**]) Let E be a Riesz space and let $f \in E^\sim$. The **null ideal** of the functional f is the ideal $N_f = \{x \in E:\ |f|(|x|) = 0\}$. The band $C_f = N_f^{\mathrm{d}}$ is called the **carrier** of f.

Assume that E is Archimedean and $f, g \in E_{\mathrm{n}}^\sim$. Show that $|f| \wedge |g| = 0$ if and only if $C_f \cap C_g = \{0\}$. [HINT: We can assume that $0 \le f, g \in E_{\mathrm{n}}^\sim$. Suppose that $f \wedge g = 0$. Pick $0 \le x \in C_f = N_f^{\mathrm{d}}$ and fix some $\epsilon > 0$. Next, pick a sequence $\{x_n\} \subseteq [0, x]$ such that $f(x_n) + g(x - x_n) < 2^{-n}\epsilon$ holds for each n. If $y_n = \bigwedge_{i=1}^n x_i$, then note that $y_n \downarrow 0$. This implies $0 \le g(x - y_n) = g(\bigvee_{i=1}^n (x - x_i)) \le \sum_{i=1}^n g(x - x_i) < \epsilon$. Therefore, $0 \le g(x) \le \epsilon$ holds for each $\epsilon > 0$ or $g(x) = 0$. Thus, $C_f \subseteq N_g$ and so $C_f \cap C_g = \{0\}$.]

1.3. Bases in Banach Spaces

In this section X will denote a Banach space. Our objective is to review a few basic properties of bases in Banach spaces. For extensive treatments of bases in Banach spaces, we refer the reader to [**205**], [**305**], and [**306**]. For an excellent quick review of bases see [**160**].

A sequence $\{x_n\}$ in X is said to be a **Schauder basis** (or simply a **basis**) in X if for every $x \in X$ there exists a unique sequence of scalars $\{\alpha_n\}$ such that x has the series representation $x = \sum_{n=1}^{\infty} \alpha_n x_n$, where the series is assumed to converge in the norm. The uniqueness of the sequence $\{\alpha_n\}$ guarantees that $x_n \neq 0$ for each n and that the vectors $x_1, x_2, \ldots$ are linearly independent. Every Banach space X with a basis is automatically separable. Indeed, if we let

$$D = \Big\{ x \in X \colon \ \exists \ r_1, \ldots, r_n \in \mathbb{Q} \ \text{ such that } \ x = \sum_{i=1}^{n} r_i x_i \Big\},$$

then D is a countable set that is dense in X. This means that only separable Banach spaces can possibly have bases. The **basis problem** was the following long standing question: *Does every separable Banach space have a basis?* In 1972 P. Enflo [**120**] produced a separable Banach space without a basis, and so he provided a negative answer to the basis problem.

Now let $\{x_n\}$ be a basis in a Banach space X, and for each vector $x \in X$ let $x = \sum_{n=1}^{\infty} \alpha_n x_n$ be its series representation. The **n^{th}-coordinate functional** c_n^* of the basis $\{x_n\}$ is the linear functional defined by

$$c_n^*(x) = c_n^*\Big(\sum_{n=1}^{\infty} \alpha_n x_n\Big) = \alpha_n .$$

It should be clear that c_n^* is indeed a linear functional. Also, there exists a sequence $\{P_n\}$ of natural projections on X that are associated with the given basis $\{x_n\}$. The range of the projection $P_n \colon X \to X$ is the vector subspace generated by the set $\{x_1, x_2, \ldots, x_n\}$ and P_n is defined by

$$P_n(x) = \sum_{i=1}^{n} \alpha_i x_i = \sum_{i=1}^{n} c_i^*(x) x_i .$$

The projections P_n and the coordinate functionals c_n^* are all continuous.

Theorem 1.37. *Assume that X is a Banach space with a basis $\{x_n\}$ and for each vector $x \in X$ let $x = \sum_{n=1}^{\infty} \alpha_n x_n = \sum_{n=1}^{\infty} c_n^*(x) x_n$ be its series representation with respect to the basis $\{x_n\}$. Then:*

(a) *The function $|||\cdot||| \colon X \to \mathbb{R}$, defined by*

$$|||x||| = \sup_{n \in \mathbb{N}} \Big\| \sum_{i=1}^{n} \alpha_i x_i \Big\|,$$

is a norm that is equivalent to $\|\cdot\|$ (the norm of X).

(b) *Each projection P_n is continuous and $\kappa = \sup_n \|P_n\| < \infty$.*

(c) *Each coordinate functional c_n^* is continuous and $\|c_n^*\| \leq \frac{2\kappa}{\|x_n\|}$.*

Proof. (a) The verification that $|||\cdot|||$ is a norm satisfying $|||\cdot||| \geq \|\cdot\|$ is trivial. Verifying the completeness of $|||\cdot|||$ is cumbersome and is left as an exercise.

(b) Since $|||\cdot|||$ is equivalent to $\|\cdot\|$ there exists a constants $C > 0$ such that $|||x||| \leq C\|x\|$ for each $x \in X$. This implies

$$\|P_n(x)\| = \Big\|\sum_{i=1}^{n} \alpha_i x_i\Big\| \leq |||x||| \leq C\|x\|,$$

and so P_n is a continuous projection satisfying $\|P_n\| \leq C$ for each n. Therefore, $\kappa = \sup_n \|P_n\| \leq C < \infty$.

(c) Observe that $\|c_1^*\| = \frac{\|P_1\|}{\|x_1\|}$. Furthermore, if $n > 1$, then

$$|c_n^*(x)| = |\alpha_n| = \tfrac{1}{\|x_n\|}\|\alpha_n x_n\| = \tfrac{1}{\|x_n\|}\|P_n(x) - P_{n-1}(x)\| \leq \tfrac{2\kappa}{\|x_n\|}\|x\|$$

holds for each $x \in X$. So, each c_n^* is continuous and $\|c_n^*\| \leq \frac{2\kappa}{\|x_n\|}$. ■

Definition 1.38. *If $\{x_n\}$ is a basis in a Banach space, then the positive number $\kappa = \sup_n \|P_n\| < \infty$ is called the* ***basis constant*** *of $\{x_n\}$. If $\kappa = 1$, then $\{x_n\}$ is called a* ***monotone basis****.*

Corollary 1.39. *If $\{x_n\}$ is a basis in a Banach space, then there exists an equivalent norm in which the basis $\{x_n\}$ is monotone.*

Proof. The norm $|||\cdot|||$ in Theorem 1.37 is the desired norm. ■

A basis $\{x_n\}$ is said to be **normalized** if $\|x_n\| = 1$ for each n. From part (c) of Theorem 1.37 we have the following.

Corollary 1.40. *If $\{x_n\}$ is a normalized basis in a Banach space, then every coordinate functional c_n^* of the basis $\{x_n\}$ satisfies $\|c_n^*\| \leq 2\kappa$.*

The next result presents a characterization of the sequences in a Banach space that are bases. Its proof is left to the reader.

Theorem 1.41. *A sequence $\{x_n\}$ of non-zero vectors in a Banach space X is a basis if and only if it satisfies the following properties:*

1. *The linear span of the set $\{x_1, x_2, \ldots\}$ is dense in X.*
2. *There exists some $K > 0$ such that if $m > n$ and $\alpha_1, \ldots, \alpha_m$ are arbitrary scalars, then*

$$\Big\|\sum_{i=1}^{n} \alpha_i x_i\Big\| \leq K\Big\|\sum_{i=1}^{m} \alpha_i x_i\Big\|.$$

A sequence $\{x_n\}$ of non-zero vectors that satisfies property (2) of Theorem 1.41 is called a **basic sequence**. From Theorem 1.41 it should be immediate that a basic sequence $\{x_n\}$ is a basis in the closed linear span of $\{x_n\}$.

Next, we shall prove that every Banach space has a basic sequence. To establish this important result, we need a lemma due to S. Mazur.

Lemma 1.42 (Mazur). *Let X be an infinite dimensional Banach space, let Y be a finite dimensional vector subspace of X, and let $\epsilon > 0$. Then there is a unit vector $x \in X$ such that $\|y\| \leq (1+\epsilon)\|y+\alpha x\|$ for all $y \in Y$ and all scalars α.*

Proof. Assume $0 < \epsilon < 1$. It suffices to show that there exists some unit vector $x \in X$ such that $1 \leq (1+\epsilon)\|y+\alpha x\|$ for all $y \in Y$ with $\|y\| = 1$.

Since the closed unit ball of Y is norm compact, there exist unit vectors $y_1, \ldots, y_n$ in Y such that for each unit vector $y \in Y$ there exists some i so that $\|y - y_i\| < \frac{\epsilon}{2}$. For each $1 \leq i \leq n$ choose some $x_i^* \in X^*$ satisfying $\|x_i^*\| = 1$ and $x_i^*(y_i) = 1$. Since X is infinite dimensional we can find a unit vector $x \in X$ such that $x_i^*(x) = 0$ for each $1 \leq i \leq n$, that is, $x \in \bigcap_{i=1}^n \operatorname{Ker} x_i^*$. Now if α is any scalar and $y \in Y$ satisfies $\|y\| = 1$, then choose some $1 \leq i \leq n$ with $\|y - y_i\| < \frac{\epsilon}{2}$, and note that

$$\|y+\alpha x\| \geq \|y_i + \alpha x\| - \tfrac{\epsilon}{2} \geq x_i^*(y_i + \alpha x) - \tfrac{\epsilon}{2} = 1 - \tfrac{\epsilon}{2} \geq \tfrac{1}{1+\epsilon}\,.$$

Therefore, $1 \leq (1+\epsilon)\|y+\alpha x\|$ for all unit vectors $y \in Y$, as desired. ■

Theorem 1.43 (Banach). *Every infinite dimensional Banach space has a basic sequence.*

Proof. Let X be an infinite dimensional Banach space. Fix some real number $\epsilon > 0$ and then choose a sequence $\{\epsilon_n\}$ of positive real numbers such that $K = \prod_{n=1}^{\infty}(1+\epsilon_n) \leq 1+\epsilon$. Pick any unit vector $x_1 \in X$ and then use Mazur's Lemma 1.42 to select inductively a sequence of unit vectors $x_2, x_3, \ldots$ such that for each n we have

$$\|z\| \leq (1+\epsilon_n)\|z + \alpha x_{n+1}\|$$

for all scalars α and all z in the vector subspace generated by $\{x_1, \ldots, x_n\}$. Now note that if $m > n$, then for any choice of scalars $\alpha_1, \ldots, \alpha_m$ we have

$$\Big\|\sum_{i=1}^{n} \alpha_i x_i\Big\| \leq K \Big\|\sum_{i=1}^{m} \alpha_i x_i\Big\|\,.$$

This shows that $\{x_n\}$ is a basic sequence in X. ■

Let $\{x_n\}$ be a basis in a Banach space X. Then $c_i^*(x_j) = \delta_{ij}$ holds for all i and j. For this reason the sequence of coordinate functionals $\{c_n^*\}$ is also called the sequence of **biorthogonal functionals** associated with the basis $\{x_n\}$. Notice that if $m > n$ and $\lambda_1, \ldots, \lambda_m$ are arbitrary scalars, then for each $x = \sum_{i=1}^{\infty} \alpha_i x_i \in X$ with $\|x\| = 1$ we have

$$\begin{aligned}\Big\langle x, \sum_{j=1}^{n} \lambda_j c_j^* \Big\rangle &= \sum_{j=1}^{n} \lambda_j \alpha_j = \Big\langle \sum_{i=1}^{n} \alpha_i x_i, \sum_{j=1}^{m} \lambda_j c_j^* \Big\rangle \\ &= \Big\langle P_n x, \sum_{j=1}^{m} \lambda_j c_j^* \Big\rangle = \Big\langle x, P_n^* \Big(\sum_{j=1}^{m} \lambda_j c_j^* \Big) \Big\rangle \\ &\leq \|P_n^*\| \cdot \Big\| \sum_{j=1}^{m} \lambda_j c_j^* \Big\| \cdot \|x\| \leq \kappa \Big\| \sum_{j=1}^{m} \lambda_j c_j^* \Big\| .\end{aligned}$$

This implies $\big\|\sum_{j=1}^{n} \lambda_j c_j^*\big\| \leq \kappa \big\|\sum_{j=1}^{m} \lambda_j c_j^*\big\|$, and therefore $\{c_n^*\}$ is a basic sequence in X^*.

The basic sequence $\{c_n^*\}$ need not be a basis in X^*. For instance, if X^* is not separable, then the biorthogonal sequence $\{c_n^*\}$ cannot be a basis. However, if X is reflexive, then $\{c_n^*\}$ is a basis in X^*.

We shall close the section with a discussion of unconditional bases. To introduce the notion of an unconditional basis we need the concept of an unconditionally convergent series.

Definition 1.44. *A series $\sum_{n=1}^{\infty} x_n$ in a Banach space X is said to* ***converge unconditionally*** *if for every permutation*[8] *π of $\mathbb{N}$ the series $\sum_{n=1}^{\infty} x_{\pi(n)}$ is norm convergent in X.*

The proof of the next result that characterizes the unconditionally convergent series is left to the reader.

Lemma 1.45. *For a series $\sum_{n=1}^{\infty} x_n$ in a Banach space the following statements are equivalent.*

(1) *The series $\sum_{n=1}^{\infty} x_n$ converges unconditionally.*
(2) *For any sequence $\{s_n\}$ of signs (i.e., $s_n = \pm 1$ for each n) the series $\sum_{n=1}^{\infty} s_n x_n$ is norm convergent.*
(3) *For every strictly increasing sequence $\{k_n\}$ of natural numbers the series $\sum_{n=1}^{\infty} x_{k_n}$ is norm convergent.*
(4) *For each $\epsilon > 0$ there exists a natural number k such that for each finite subset A of $\mathbb{N}$ with $\min A \geq k$ we have $\big\|\sum_{n \in A} x_n\big\| < \epsilon$.*

Corollary 1.46. *If a series $\sum_{n=1}^{\infty} x_n$ in a Banach space converges unconditionally, then $\sum_{n=1}^{\infty} x_n = \sum_{n=1}^{\infty} x_{\pi(n)}$ for each permutation π of $\mathbb{N}$.*

[8] Recall that a **permutation** of a set Ω is any one-to-one and surjective function $\pi\colon \Omega \to \Omega$.

It is an immediate consequence of Lemma 1.45 that if a series $\sum_{n=1}^{\infty} x_n$ converges unconditionally, then for each choice of signs $\{s_n\}$ the series $\sum_{n=1}^{\infty} s_n x_n$ likewise converges unconditionally.

We are now ready to introduce the notion of an unconditional basis.

Definition 1.47 (James [**159**]). *A basis $\{u_n\}$ in a Banach space is said to be **unconditional** if for each vector x its series expansion $x = \sum_{n=1}^{\infty} \alpha_n u_n$ converges unconditionally.*

Now assume that $\{u_n\}$ is an unconditional basis in a Banach space X. Let $\mathcal{S} = \{-1, 1\}^{\mathbb{N}}$ be the collection of all signs. We consider $\{-1, 1\}$ equipped with the discrete topology, and so $\mathcal{S}$ is a compact metric space.

Given an arbitrary sign sequence $s = (s_1, s_2, \dots) \in \mathcal{S}$, we define a mapping $M_s \colon X \to X$ by

$$M_s(x) = \sum_{n=1}^{\infty} s_n \alpha_n u_n \,,$$

where $x = \sum_{n=1}^{\infty} \alpha_n u_n$. It should be clear that M_s is a linear operator. If $\{c_n^*\}$ denotes the sequence of biorthogonal functionals with respect to the basis $\{u_n\}$ and $T_n \colon X \to X$ is the (bounded) finite-rank operator defined by $T_n(x) = \sum_{k=1}^{n} s_k c_k^*(x) u_k$, then we have

$$T_n(x) = \sum_{k=1}^{n} s_n c_k^*(x) u_k = \sum_{k=1}^{n} s_n \alpha_k u_k \xrightarrow[n \to \infty]{} M_s(x) \,.$$

By the Uniform Boundedness Principle, the operator M_s is also bounded.

Next, we claim that for each fixed $x = \sum_{n=1}^{\infty} \alpha_n u_n \in X$ the mapping $s \mapsto M_s(x)$, from $\mathcal{S}$ to X, is continuous. To see this, assume that a sequence $\{\mathbf{s}_n\} \subseteq \mathcal{S}$ satisfies $\mathbf{s}_n = (s_1^n, s_2^n, \dots) \to \mathbf{s} = (s_1, s_2, \dots)$ in $\mathcal{S}$ and let $\epsilon > 0$. Since the series $\sum_{n=1}^{\infty} s_n \alpha_n u_n$ converges unconditionally, there exists some k such that for any non-empty finite subset A of $\mathbb{N}$ that satisfies $\min A \geq k$ we have $\left\| \sum_{n \in A} s_n \alpha_n u_n \right\| < \epsilon$. Next, pick some $n_0 > k$ such that $s_i^n = s_i$ holds for all $n \geq n_0$ and all $i = 1, \dots, k$. For each p let

$$A_p = \left\{ m \in \mathbb{N} \colon \; k \leq m \leq k + p \text{ and } s_i^m = -s_i \right\},$$

and note that for each $n \geq n_0$ we have

$$\left\| \sum_{i=k}^{k+p} (s_i \alpha_i x_i - s_i^n \alpha_i x_i) \right\| = 2 \left\| \sum_{i \in A_p} s_i \alpha_i x_i \right\| < 2\epsilon \,.$$

Letting $p \to \infty$, we get $\left\| \sum_{i=k}^{\infty} (s_i \alpha_i x_i - s_i^n \alpha_i x_i) \right\| \leq 2\epsilon$ for all $n \geq n_0$. So, for $n \geq n_0$ we have

$$\|M_{s_n}(x) - M_s(x)\| = \left\| \sum_{i=k}^{\infty} (s_i \alpha_i x_i - s_i^n \alpha_i x_i) \right\| \leq 2\epsilon \,.$$

This shows that the mapping $s \mapsto M_s(x)$ is continuous.

Since $\mathcal{S}$ is a compact metric space, it follows that for each $x \in X$, the set of vectors $\{M_s(x)\colon\ s \in \mathcal{S}\}$ is norm bounded—in fact, a compact set. That is, the family of bounded operators $\{M_s\colon\ s \in \mathcal{S}\}$ is pointwise bounded. Consequently, by the Uniform Boundedness Principle, the family $\{M_s\colon\ s \in \mathcal{S}\}$ is norm bounded.

Definition 1.48. *If $\{u_n\}$ is an unconditional basis in a Banach space, then the positive number*

$$\kappa_u = \sup_{s\in\mathcal{S}} \|M_s\|$$

is called the ***unconditional constant*** *of the basis $\{u_n\}$.*

The following result reveals a fundamental property of an unconditional basis.

Lemma 1.49. *Assume that $\{u_n\}$ is an unconditional basis in a (real) Banach space with unconditional constant κ_u. Let F be a non-empty finite subset of $\mathbb{N}$, let $\{\alpha_i\colon\ i \in F\}$ be an arbitrary collection of scalars, and let $\{\gamma_i\colon\ i \in F\}$ be another collection of scalars such that $|\gamma_i| \le 1$ holds for each $i \in F$. Then we have*

$$\Big\|\sum_{i\in F}\gamma_i\alpha_i u_i\Big\| \le \kappa_u \Big\|\sum_{i\in F}\alpha_i u_i\Big\| .$$

Proof. Assume that F and the scalars α_i and γ_i are as in the statement of the lemma. Choose a continuous linear functional $x^* \in X^*$ of norm one satisfying $x^*\big(\sum_{i\in F}\gamma_i\alpha_i u_i\big) = \big\|\sum_{i\in F}\gamma_i\alpha_i u_i\big\|$. Next, pick any sequence of signs $s = (s_1, s_2, \ldots)$ such that for each $i \in F$ we have $s_i = 1$ if $\alpha_i x^*(u_i) \ge 0$ and $s_i = -1$ if $\alpha_i x^*(u_i) < 0$. This implies

$$\begin{aligned}\Big\|\sum_{i\in F}\gamma_i\alpha_i u_i\Big\| &= x^*\Big(\sum_{i\in F}\gamma_i\alpha_i u_i\Big) \le \sum_{i\in F}|\gamma_i||\alpha_i x^*(u_i)| \le \sum_{i\in F} s_i\alpha_i x^*(u_i)\\ &= x^*\Big(M_s\Big(\sum_{i\in F}\alpha_i u_i\Big)\Big) \le \kappa_u\Big\|\sum_{i\in F}\alpha_i u_i\Big\|,\end{aligned}$$

as desired. ■

If we replace κ_u by $2\kappa_u$, then the conclusion of the previous lemma remains true for complex Banach spaces as well. Indeed, letting $\gamma_i = a_i + \imath b_i$ we have $|a_i| \vee |b_i| \le 1$, and hence

$$\Big\|\sum_{i\in F}\gamma_i\alpha_i u_i\Big\| \le \Big\|\sum_{i\in F}a_i\alpha_i u_i\Big\| + \Big\|\sum_{i\in F}b_i\alpha_i u_i\Big\| \le 2\kappa_u\Big\|\sum_{i\in F}\alpha_i u_i\Big\| .$$

Corollary 1.50. *If $\{u_n\}$ is an unconditional basis in a Banach space X, then there exists an equivalent norm in which the unconditional constant of the basis $\{u_n\}$ is one.*

Proof. Define $\|\cdot\|_1 \colon X \to [0,\infty)$ by $\|x\|_1 = \sup_{s\in\mathcal{S}} \|M_s x\|$. It is easy to verify that this formula defines a norm on X. Moreover, note that

$$\|x\| \le \|x\|_1 \le \kappa_u \|x\|$$

for each $x \in X$. This shows that $\|\cdot\|$ and $\|\cdot\|_1$ are equivalent norms. To finish the proof, notice that the unconditional constant of $\{u_n\}$ with respect to the norm $\|\cdot\|_1$ is one. ■

The unconditional bases are characterized as follows.

Theorem 1.51. *For a basis $\{u_n\}$ in a Banach space X the following statements are equivalent.*

(1) *The basis $\{u_n\}$ is unconditional.*
(2) *For any permutation π of $\mathbb{N}$ the sequence $\{u_{\pi(n)}\}$ is a basis.*
(3) *If for some sequence $\{\alpha_n\}$ of scalars the series $\sum_{n=1}^\infty \alpha_n u_n$ converges in norm, then for any strictly increasing sequence $\{k_n\}$ of natural numbers the series $\sum_{n=1}^\infty \alpha_{k_n} x_{k_n}$ is norm convergent.*
(4) *If for some sequence $\{\alpha_n\}$ of scalars the series $\sum_{n=1}^\infty \alpha_n u_n$ converges in norm and another sequence $\{\beta_n\}$ of scalars satisfies $|\beta_n| \le |\alpha_n|$ for each n, then $\sum_{n=1}^\infty \beta_n u_n$ is norm convergent.*

Proof. (1) $\Longleftrightarrow$ (2) If (1) is true, then for any permutation π of $\mathbb{N}$ the sequence $\{u_{\pi(n)}\}$ is clearly an unconditional basis.

For the converse, assume that (2) is true and fix a permutation π of $\mathbb{N}$. Also, let $\{c_n^*\}$ and $\{f_n\}$ be the sequences of biorthogonal functionals with respect to the bases $\{u_n\}$ and $\{u_{\pi(n)}\}$, respectively. Since $c^*_{\pi(n)}$ agrees with f_n on the vector subspace V generated by $\{u_1, u_2, \ldots\}$ and V is norm dense in X, it follows that $f_n = c^*_{\pi(n)}$ holds for each n.

Now fix some $x \in X$ and let $x = \sum_{n=1}^\infty \alpha_n u_n$ be its (unique) series expansion with respect to the basis $\{u_n\}$. Also, let $x = \sum_{n=1}^\infty \lambda_n u_{\pi(n)}$ be the series expansion of x with respect to the basis $\{u_{\pi(n)}\}$. Therefore,

$$x = \sum_{n=1}^\infty \lambda_n u_{\pi(n)} = \sum_{n=1}^\infty f_n(x) u_{\pi(n)} = \sum_{n=1}^\infty c^*_{\pi(n)}(x) u_{\pi(n)} = \sum_{n=1}^\infty \alpha_{\pi(n)} u_{\pi(n)}\,.$$

This shows that $\{u_n\}$ is an unconditional basis.

(1) $\Longleftrightarrow$ (3) This equivalence follows immediately from the equivalence of (1) and (3) of Lemma 1.45.

(4) $\Longleftrightarrow$ (1) Assume that (4) is true and let $s = (s_1, s_2, \ldots)$ be a sequence of signs. If $x = \sum_{n=1}^\infty \alpha_n u_n$, then by letting $\beta_n = s_n \alpha_n$ we have $|\beta_n| = |\alpha_n|$ for each n, and so $\sum_{n=1}^\infty s_n \alpha_n u_n = \sum_{n=1}^\infty \beta_n u_n$ is norm convergent. This shows that $\{u_n\}$ is an unconditional basis.

To see that (1) implies (4) consider a norm convergent series $\sum_{n=1}^{\infty} \alpha_n u_n$ and let a sequence $\{\beta_n\}$ of scalars satisfy $|\beta_n| \leq |\alpha_n|$ for each n. Put $\gamma_n = \frac{\beta_n}{\alpha_n}$ if $\alpha_n \neq 0$ and $\gamma_n = 0$ if $\alpha_n = 0$. An easy application of Lemma 1.49 shows that the series $\sum_{n=1}^{\infty} \gamma_n \alpha_n u_n$ is norm convergent. To finish the proof note that $\sum_{n=1}^{\infty} \beta_n u_n = \sum_{n=1}^{\infty} \gamma_n \alpha_n u_n$. ∎

If a sequence $\{v_n\}$ of vectors in a Banach space is an unconditional basis in the closed linear span of $\{v_n\}$, then this sequence is called an **unconditional basic sequence**. It is easy to see that an analogue of Theorem 1.51 is true for unconditional basic sequences as well. It is important to note that the analogue of Theorem 1.43 for unconditional basic sequences is not true. That is, the existence of an unconditional basic sequence in a given infinite dimensional Banach space was a long standing open problem. Only recently W. T. Gowers and B. Maurey [**137**] have answered this problem in the negative by constructing an infinite dimensional Banach space without an unconditional basic sequence. For Banach lattices the situation is more pleasant.

Lemma 1.52. *Every infinite dimensional Banach lattice has an unconditional basic sequence.*

Proof. Let E be an infinite dimensional Banach lattice. Take any sequence $\{u_n\}$ of pairwise disjoint positive unit vectors in E. If α_n and β_n are any scalars such that $|\beta_n| \leq |\alpha_n|$ for each n, then for each m we have

$$\Big|\sum_{n=1}^{m} \beta_n u_n\Big| = \sum_{n=1}^{m} |\beta_n| u_n \leq \sum_{n=1}^{m} |\alpha_n| u_n = \Big|\sum_{n=1}^{m} \alpha_n u_n\Big| .$$

Therefore, if the series $\sum_{n=1}^{\infty} \alpha_n u_n$ converges, then the series $\sum_{n=1}^{\infty} \beta_n u_n$ converges too. Hence, in view of Theorem 1.51, $\{u_n\}$ is an unconditional basic sequence. ∎

For a special connection between unconditional bases and Banach lattices see Exercise 2 in Section 10.6.

Theorem 1.51 has a useful consequence regarding projections. If $\{u_n\}$ is an unconditional basis in a Banach space X, then every non-empty subset A of $\mathbb{N}$ gives rise to a bounded projection $P_A \colon X \to X$ defined by

$$P_A x = \sum_{n \in A} \alpha_n u_n$$

for each $x = \sum_{n=1}^{\infty} \alpha_n u_n \in X$. If $A = \{1, 2, \ldots, n\}$, then P_A is simply the projection P_n that was introduced at the beginning of the section. Now if A is a non-empty subset of $\mathbb{N}$ and $s \in \{-1, 1\}^{\mathbb{N}}$ is the sequence of signs defined by $s_n = 1$ if $n \in A$ and $s_n = -1$ if $n \in A^c$, then it is easy to see that $P_A = \frac{1}{2}(I + M_s)$, where I is the identity operator on X. This implies $\|P_A\| \leq \frac{1}{2}(\|M_s\| + I) \leq \kappa_u$. On the other hand, if $s \in \{-1,1\}^{\mathbb{N}}$ is any sequence of signs and we put $A = \{n \in \mathbb{N}\colon \ s_n = 1\}$, then we have $M_s = P_A - P_{A^c}$. The above imply the following inequalities.

Lemma 1.53. *If $\{u_n\}$ is an unconditional basis in a Banach space, then*

$$\kappa \leq \sup_A \|P_A\| \leq \kappa_u \leq 2 \sup_A \|P_A\|,$$

where the suprema are taken over all non-empty subsets A of $\mathbb{N}$.

Exercises

1. Prove part (a) of Theorem 1.37
2. Prove Theorem 1.41.
3. This exercise describes a basis for the Banach spaces $L_p[0,1]$ for $1 \leq p < \infty$. Consider the sequence of functions $\{h_n\}$ defined by $h_1(t) = 1$ for each $t \in [0,1]$ and

$$h_{2^k+\ell}(t) = \begin{cases} 1 & \text{if } \frac{2\ell-2}{2^{k+1}} \leq t < \frac{2\ell-1}{2^{k+1}} \\ -1 & \text{if } \frac{2\ell-1}{2^{k+1}} \leq t \leq \frac{2\ell}{2^{k+1}} \\ 0 & \text{otherwise}, \end{cases}$$

for $k = 0,1,2,\ldots$ and $1 \leq \ell \leq 2^k$. The sequence $\{h_n\}$ is known as the **Haar system**. Establish the following.
   (a) $h_n \neq 0$ for each n.
   (b) The linear span of the set $\mathcal{H} = \{h_1, h_2, \ldots\}$ is norm dense in $L_p[0,1]$.
   (c) For any collection of scalars $\lambda_1, \ldots, \lambda_n, \lambda_{n+1}$ we have

$$\Big\|\sum_{i=1}^{n} \lambda_i h_i\Big\|_p \leq \Big\|\sum_{i=1}^{n+1} \lambda_i h_i\Big\|_p.$$

   (d) The Haar system $\{h_n\}$ is a monotone basis in each $L_p[0,1]$.
4. This exercise describes a basis in the Banach space $C[0,1]$. Let $\{h_n\}$ be the Haar system that was introduced in the preceding exercise. Consider the sequence of functions $\{s_n\}$ in $C[0,1]$ defined by

$$s_1 = \mathbf{1} \quad \text{and} \quad s_n(t) = \int_0^t h_{n-1}(\tau)\,d\tau, \text{ for } n = 2,3,4,\ldots.$$

The sequence $\{s_n\}$ is called the **Schauder system**. Show that $\{s_n\}$ is a monotone basis for $C[0,1]$.[9]
5. (Bessaga–Pelczynski [**61**]) Let $\{x_n\}$ be a sequence of vectors in an infinite dimensional Banach space satisfying $x_n \xrightarrow{w} 0$ and $\liminf \|x_n\| > 0$. Show that $\{x_n\}$ has a subsequence that is a basic sequence.

[9] It is well known that neither $C[0,1]$ nor $L_1[0,1]$ has an unconditional basis; see Corollary 11.61.

6. Assume that X is a reflexive Banach space with a basis $\{x_n\}$. Show that the biorthogonal sequence $\{c_n^*\}$ is a basis in X^*.

7. Let $\sum_{n=1}^{\infty} x_n$ be a series in a Banach space X.
  (a) If X is finite dimensional, show that $\sum_{n=1}^{\infty} x_n$ converges unconditionally if and only if $\sum_{n=1}^{\infty} \|x_n\| < \infty$.
  (b) Give an example of an unconditionally convergent series $\sum_{n=1}^{\infty} x_n$ such that $\sum_{n=1}^{\infty} \|x_n\| = \infty$.

8. Prove Lemma 1.45.

9. Prove Corollary 1.46.

10. Let $\sum_{n=1}^{\infty} x_n$ be an unconditionally convergent series in a Banach space X. If $(\alpha_1, \alpha_2, \ldots) \in \ell_\infty$, then show that the series $\sum_{n=1}^{\infty} \alpha_n x_n$ is norm convergent. Moreover, show that the mapping $T\colon \ell_\infty \to X$, defined by

$$T(\alpha_1, \alpha_2, \ldots) = \sum_{n=1}^{\infty} \alpha_n x_n ,$$

is a bounded linear operator.

1.4. Ultrapowers of Banach Spaces

In this section we shall discuss a technique that allows one to construct a special enlargement of a Banach space that is useful in many contexts. To do this, we need the concept of a filter.

Definition 1.54. *Let Δ be an infinite set. A non-empty collection $\mathcal{F}$ of subsets of Δ is said to be a **filter** if:*

(a) $\emptyset \notin \mathcal{F}$,

(b) *$\mathcal{F}$ is closed under finite intersections, that is, $A, B \in \mathcal{F}$ implies $A \cap B \in \mathcal{F}$,*

(c) *$\mathcal{F}$ is closed under supersets, i.e., $A \in \mathcal{F}$ and $A \subseteq B$ imply $B \in \mathcal{F}$.*

Since a filter $\mathcal{F}$ is a non-empty collection, it follows from (c) that $\Delta \in \mathcal{F}$ and that every filter is closed under arbitrary unions.

A filter $\mathcal{F}$ is said to be an **ultrafilter** if it is maximal with respect to inclusion. That is, a filter $\mathcal{F}$ is an ultrafilter if it is not properly included in any other filter, i.e., if $\mathcal{F} \subseteq \mathcal{G}$ and $\mathcal{G}$ is a filter, then $\mathcal{G} = \mathcal{F}$. It follows from Zorn's lemma that every filter is included in an ultrafilter.

The proof of the next result that characterizes the ultrafilters among filters is straightforward.

Lemma 1.55. *For a filter $\mathcal{F}$ the following statements are equivalent.*

(1) *$\mathcal{F}$ is an ultrafilter.*

(2) *If $A \cup B \in \mathcal{F}$, then either $A \in \mathcal{F}$ or $B \in \mathcal{F}$.*

(3) *If* $A \cap B \neq \emptyset$ *for all* $B \in \mathcal{F}$, *then* $A \in \mathcal{F}$.

An immediate consequence of the identity $A \cup A^c = \Delta \in \mathcal{F}$ and the above result is the following.

Corollary 1.56. *If* $\mathcal{F}$ *is an ultrafilter, then for any subset* A *of* Δ *either* $A \in \mathcal{F}$ *or (exclusively)* $A^c \in \mathcal{F}$.

It is not difficult to see that if $\mathcal{U}$ is an ultrafilter and $\bigcap_{A \in \mathcal{U}} A \neq \emptyset$, then there exists a unique point $\delta \in \Delta$ such that

$$\mathcal{U} = \{A \subseteq \Delta \colon\ \delta \in A\}.$$

Any ultrafilter of the above form is called a **fixed ultrafilter**. An ultrafilter $\mathcal{U}$ that is not fixed, i.e., $\bigcap_{A \in \mathcal{U}} A = \emptyset$, is referred to as a **free ultrafilter**.

There is a special filter that plays a crucial role in the study of free ultrafilters. This is the filter Φ of all co-finite subsets of Δ, i.e., the filter

$$\Phi = \{A \subseteq \Delta \colon\ A^c \text{ is a finite set}\}.$$

A straightforward verification shows that Φ is a filter—called the **Fréchet filter** on Δ. It turns out that every free ultrafilter includes the Fréchet filter Φ. To see this, let $\mathcal{U}$ be a free ultrafilter and let F be a finite subset of Δ. If $F = \bigcup_{x \in F} \{x\} \in \mathcal{U}$, then it follows from Lemma 1.55 that for some $x \in F$ we have $\{x\} \in \mathcal{U}$. This implies $\mathcal{U} = \mathcal{U}_x$ and so $\mathcal{U}$ is fixed, a contradiction. Consequently, $\Phi \subseteq \mathcal{U}$ for each free ultrafilter $\mathcal{U}$.

The above conclusion yields the following properties for an ultrafilter $\mathcal{U}$.

- *No finite subset of* Δ *is a member of* $\mathcal{U}$.
- *If* $A \in \mathcal{U}$, *then* $A \setminus F = A \cap F^c \in \mathcal{U}$ *for each finite subset* F *of* Δ.

By considering the collection of all filters that include the Fréchet filter Φ and using Zorn's lemma, we can easily conclude the following.

Lemma 1.57. *Every infinite set admits a free ultrafilter.*

For the rest of the discussion in this section $\mathcal{U}$ will denote a free ultrafilter on $\mathbb{N}$. By means of the ultrafilter $\mathcal{U}$ we now introduce a new notion of sequential convergence in a topological space.

Definition 1.58. *A sequence* $\{x_n\}$ *in a topological space* Ω ***converges to some*** $x \in \Omega$ ***along the ultrafilter*** $\mathcal{U}$, *in symbols,* $x_n \xrightarrow{\mathcal{U}} x$ *or* $x = \lim_{\mathcal{U}} x_n$, *if for each neighborhood* V *of* x *the set* $\{n \in \mathbb{N} \colon\ x_n \in V\}$ *belongs to* $\mathcal{U}$. *The point* x *is called a* $\mathcal{U}$***-limit*** *of the sequence* $\{x_n\}$.

Here are a few properties of the convergence along an ultrafilter.

Lemma 1.59. *For a sequence $\{x_n\}$ in a topological space (Ω, τ) we have:*

(1) *If Ω is Hausdorff, then $\{x_n\}$ can have at most one $\mathcal{U}$-limit. That is, in Hausdorff spaces the $\mathcal{U}$-limits are uniquely determined.*

(2) *If $x_n \xrightarrow{\tau} x$, then $x_n \xrightarrow{\mathcal{U}} x$.*

(3) *If Ω is metrizable and $x_n \xrightarrow{\mathcal{U}} x$, then there exists a subsequence $\{x_{k_n}\}$ of $\{x_n\}$ such that $x_{k_n} \xrightarrow{\tau} x$.*

(4) *If $\{x_n\}$ lies in a compact set K, then $x_n \xrightarrow{\mathcal{U}} x$ for some $x \in K$.*

Proof. (1) Assume Ω is Hausdorff, $x_n \xrightarrow{\mathcal{U}} x$ and $x_n \xrightarrow{\mathcal{U}} y$ with $x \neq y$. Pick two neighborhoods V and W of x and y, respectively, such that $V \cap W = \emptyset$. Now consider the sets

$$A = \{n \in \mathbb{N}\colon\ x_n \in V\} \quad \text{and} \quad B = \{n \in \mathbb{N}\colon\ x_n \in W\}$$

and note that $A, B \in \mathcal{U}$. This implies that $\emptyset = A \cap B \in \mathcal{U}$, which is impossible. Hence, a sequence in a Hausdorff space can have at most one limit along $\mathcal{U}$.

(2) If V is a neighborhood of x, then there exists some n_0 such that $x_n \in V$ for all $n > n_0$. Therefore, $\{1, 2, \ldots, n_0\}^c \subseteq \{n \in \mathbb{N}\colon\ x_n \in V\}$. This implies that the set $\{n \in \mathbb{N}\colon\ x_n \in V\}$ belongs to $\mathcal{U}$, and so $\lim_{\mathcal{U}} x_n = x$.

(3) Let d be a distance generating the topology τ and assume that $x_n \xrightarrow{\mathcal{U}} x$. For each n the set $A_n = \{k \in \mathbb{N}\colon\ d(x_k, x) < \frac{1}{n}\}$ belongs to the ultrafilter $\mathcal{U}$. Clearly, $A_{n+1} \subseteq A_n$ for each n. Next, define the sequence $\{m_n\}$ of natural numbers recursively by

$$m_1 = \min A_1 \quad \text{and} \quad m_{n+1} = \min\left[A_{n+1} \setminus \{m_1, \ldots, m_n\}\right] \ \text{ for } n \geq 1\,.$$

Note that $m_n < m_{n+1}$ for each n and that the subsequence $\{x_{m_n}\}$ of $\{x_n\}$ satisfies $d(x_{m_n}, x) < \frac{1}{n}$ for each n. In particular, $x_{m_n} \xrightarrow{\tau} x$.

(4) Let $\{x_n\}$ be a sequence in a compact set K. If $\{x_n\}$ does not converge along $\mathcal{U}$ to any point in K, then for each $y \in K$ there exists a neighborhood V_y of y such that the set $A_y = \{n \in \mathbb{N}\colon\ x_n \in V_y\}$ does not belong to $\mathcal{U}$. Since K is compact, there exist points $y_1, y_2, \ldots, y_k$ in K such that $K \subseteq \bigcup_{i=1}^k V_{y_i}$. This implies $\bigcup_{i=1}^k A_{y_i} = \mathbb{N} \in \mathcal{U}$. Consequently, $A_{y_i} \in \mathcal{U}$ must be true for at least one i, which is impossible. This contradiction shows that the sequence $\{x_n\}$ converges along $\mathcal{U}$ to some point in K. ■

It should be noted that a $\mathcal{U}$-limit need not be a topological limit. As a matter of fact, according to statement (4) of the preceding result, every bounded sequence in $\mathbb{R}^n$ has a unique $\mathcal{U}$-limit. For instance, consider the non-convergent sequence $\{x_n\}$ of $\mathbb{R}$ defined by $x_n = (-1)^n$. From $\{1, 3, 5, \ldots\} \cup \{2, 4, 6, \ldots\} = \mathbb{N}$, it follows that either $\{1, 3, 5, \ldots\} \in \mathcal{U}$ or

$\{2,4,6,\ldots\} \in \mathcal{U}$. If $\{1,3,5,\ldots\} \in \mathcal{U}$, then $x_n \xrightarrow{\mathcal{U}} -1$ and if $\{2,4,6,\ldots\} \in \mathcal{U}$, then $x_n \xrightarrow{\mathcal{U}} 1$.

In topological vector spaces the $\mathcal{U}$-limits also satisfy the following properties.

Lemma 1.60. *If in a topological vector space $x_n \xrightarrow{\mathcal{U}} x$ and $y_n \xrightarrow{\mathcal{U}} y$, then $x_n + y_n \xrightarrow{\mathcal{U}} x + y$ and $\lambda x_n \xrightarrow{\mathcal{U}} \lambda x$ for each scalar λ.*

Now we turn our attention to Banach spaces. Let X be a (real or complex) Banach space. As usual, we denote by $\ell_\infty(X)$ the vector space of all bounded sequences in X. That is, $x = (x_1, x_2, \ldots)$ belongs to $\ell_\infty(X)$ if and only if $\{x_n\}$ is a bounded sequence. Under the sup norm

$$\|x\|_\infty = \sup_{n\in\mathbb{N}} \|x_n\|,$$

the vector space $\ell_\infty(X)$ is a Banach space. If X is a Banach lattice, then under the pointwise lattice operations $\ell_\infty(X)$ is likewise a Banach lattice.

We shall denote by $\mathcal{N}_\mathcal{U}$ the vector space of all sequences in $\ell_\infty(X)$ that converge to zero along the free ultrafilter $\mathcal{U}$. That is,

$$\mathcal{N}_\mathcal{U} = \left\{ x = (x_1, x_2, \ldots) \in \ell_\infty(X)\colon\ x_n \xrightarrow{\mathcal{U}} 0 \right\}.$$

A glance at Lemma 1.60 guarantees that $\mathcal{N}_\mathcal{U}$ is a vector subspace of $\ell_\infty(X)$. Moreover, we claim that $\mathcal{N}_\mathcal{U}$ is also a closed subspace. To see this, assume that some $x = (x_1, x_2, \ldots)$ belongs to the closure of $\mathcal{N}_\mathcal{U}$ and let $\epsilon > 0$. Choose some $y = (y_1, y_2, \ldots) \in \mathcal{N}_\mathcal{U}$ with $\|x - y\|_\infty < \epsilon$. Since $y \in \mathcal{N}_\mathcal{U}$, the set $A = \{n \in \mathbb{N}\colon\ \|y_n\| < \epsilon\}$ belongs to $\mathcal{U}$. Now use the inequalities

$$\|x_n\| \le \|x_n - y_n\| + \|y_n\| \le \|x - y\|_\infty + \|y_n\| < \epsilon + \|y_n\|$$

to conclude that $A \subseteq \{n \in \mathbb{N}\colon\ \|x_n\| < 2\epsilon\}$. So, $\{n \in \mathbb{N}\colon\ \|x_n\| < 2\epsilon\} \in \mathcal{U}$. This shows that $x_n \xrightarrow{\mathcal{U}} 0$, i.e., $x \in \mathcal{N}_\mathcal{U}$. Thus, $\mathcal{N}_\mathcal{U}$ is a closed subspace.

It should be clear that when X is a Banach lattice, the closed subspace $\mathcal{N}_\mathcal{U}$ is also an ideal. In particular, if X is a Banach lattice, then the quotient Banach space $\ell_\infty(X)/\mathcal{N}_\mathcal{U}$ is likewise a Banach lattice.

Definition 1.61. *The quotient Banach space $\ell_\infty(X)/\mathcal{N}_\mathcal{U}$ is called the* ***ultrapower*** *of X with respect to the free ultrafilter $\mathcal{U}$ and is denoted $X_\mathcal{U}$, i.e., $X_\mathcal{U} = \ell_\infty(X)/\mathcal{N}_\mathcal{U}$.*

Here is an important property of the quotient norm of an ultrapower. As usual, for each $x = (x_1, x_2, \ldots) \in \ell_\infty(X)$ we shall denote its equivalence class in $X_\mathcal{U}$ by $[x]$.

Lemma 1.62. *If $x = (x_1, x_2, \ldots) \in \ell_\infty(X)$, then $\big\|[x]\big\| = \lim_\mathcal{U} \|x_n\|$.*

Proof. Since for each $u = (u_1, u_2, \ldots) \in \ell_\infty(X)$ the sequence of real numbers $\{\|u_n\|\}$ is bounded (and so it lies in a compact subset of $\mathbb{R}$), it follows from Lemma 1.59(4) that $\lim_{\mathcal{U}} \|u_n\|$ exists in $\mathbb{R}$.

Now fix $x = (x_1, x_2, \ldots) \in \ell_\infty(X)$ and put $a = \lim_{\mathcal{U}} \|x_n\|$. Let $y \in [x]$. From $\|y_n\| - \|x_n\| \leq \|y_n - x_n\|$, the fact that $\mathcal{U}$-limits preserve inequalities (see Exercise 5 at the end of the section), and $y - x \in \mathcal{N}_{\mathcal{U}}$, it follows that

$$\lim_{\mathcal{U}} \|y_n\| - \lim_{\mathcal{U}} \|x_n\| = \lim_{\mathcal{U}} \big[\, \|y_n\| - \|x_n\| \,\big] \leq \lim_{\mathcal{U}} \|y_n - z_n\| = 0\,.$$

This implies $\lim_{\mathcal{U}} \|y_n\| \leq \lim_{\mathcal{U}} \|x_n\|$. By the symmetry of the situation, $\lim_{\mathcal{U}} \|x_n\| \leq \lim_{\mathcal{U}} \|y_n\|$ is also true, and so $\lim_{\mathcal{U}} \|y_n\| = \lim_{\mathcal{U}} \|x_n\| = a$ for each $y \in [x]$. From $\|y_n\| \leq \|y\|_\infty$, we also get $a = \lim_{\mathcal{U}} \|y_n\| \leq \|y\|_\infty$. Since $y \in [x]$ is arbitrary, we conclude that $a \leq \inf_{y \in [x]} \|y\|_\infty = \big\|[x]\big\|$.

For the reverse inequality fix $\epsilon > 0$. It follows that the (infinite) set $B = \big\{n \in \mathbb{N}\colon\ a - \epsilon < \|x_n\| < a + \epsilon\big\}$ belongs to $\mathcal{U}$. Now consider the sequence $y = (y_1, y_2, \ldots) \in \ell_\infty(X)$ defined by $y_n = x_n$ if $n \in B$ and $y_n = 0$ if $n \notin B$. From the inclusion

$$B \subseteq \big\{n \in \mathbb{N}\colon\ \|x_n - y_n\| = 0\big\}\,,$$

we see that $\big\{n \in \mathbb{N}\colon\ \|x_n - y_n\| = 0\big\} \in \mathcal{U}$. This implies $x - y \in \mathcal{N}_{\mathcal{U}}$, and so $y \in [x]$. Now notice that

$$\big\|[x]\big\| \leq \|y\|_\infty = \sup_{n \in \mathbb{N}} \|y_n\| = \sup_{n \in B} \|y_n\| = \sup_{n \in B} \|x_n\| \leq a + \epsilon\,.$$

Since $\epsilon > 0$ is arbitrary, we get $\big\|[x]\big\| \leq a$. Thus, $\big\|[x]\big\| = a$, and the proof is finished. ■

There is a natural embedding of X into its ultrapowers. This is the mapping $x \mapsto [(x, x, \ldots)]$. From Lemma 1.62, it easily follows that the mapping $x \mapsto [(x, x, \ldots)]$, from X to $X_{\mathcal{U}}$, is a linear isometry—which is also a lattice isometry if X is a Banach lattice. This means that if X is identified with its image in $X_{\mathcal{U}}$, then X becomes a closed vector subspace of $X_{\mathcal{U}}$. In other words, we can view the ultrapowers of X as enlargements of X. We summarize the above observations in the following result.

Theorem 1.63. *Any ultrapower of a Banach space X contains X as a closed vector subspace. If X is also a Banach lattice, then any ultrapower of X contains X as a closed vector sublattice.*

The importance of ultrapowers to operator theory stems from the following result.

Theorem 1.64. *If $T\colon X \to Y$ is a bounded operator between two Banach spaces, then for each free ultrafilter $\mathcal{U}$ on $\mathbb{N}$ the mapping $T_{\mathcal{U}}\colon X_{\mathcal{U}} \to Y_{\mathcal{U}}$, defined for each $(x_1, x_2, \ldots) \in \ell_\infty(X)$ by*

$$T_{\mathcal{U}}\big([(x_1, x_2, \ldots)]\big) = [(Tx_1, Tx_2, \ldots)],$$

is a well-defined bounded operator that extends T and satisfies $\|T_{\mathcal{U}}\| = \|T\|$.

In other words, the mapping $T \mapsto T_{\mathcal{U}}$, from $\mathcal{L}(X,Y)$ to $\mathcal{L}(X_{\mathcal{U}}, Y_{\mathcal{U}})$, is a linear isometry.

Proof. Assume $T\colon X \to Y$ is a non-zero bounded operator. Let us verify first that $T_{\mathcal{U}}$ is well defined. To this end, fix $x = (x_1, x_2, \ldots) \in \ell_\infty$ and let $y \in [x]$. If $\epsilon > 0$, then the set $A = \big\{n \in \mathbb{N}\colon\ \|y_n - x_n\| < \frac{\epsilon}{\|T\|}\big\}$ belongs to $\mathcal{U}$. From the inclusion $A \subseteq \big\{n \in \mathbb{N}\colon\ \|Tx_n - Ty_n\| < \epsilon\big\}$, it follows that $\big\{n \in \mathbb{N}\colon\ \|Tx_n - Ty_n\| < \epsilon\big\} \in \mathcal{U}$. So, $[(Tx_1, Tx_2, \ldots)] = [(Ty_1, Ty_2, \ldots)]$ in $Y_{\mathcal{U}}$, and consequently the mapping $T_{\mathcal{U}}$ is well defined.

It should also be obvious that the mapping $T_{\mathcal{U}}\colon X_{\mathcal{U}} \to Y_{\mathcal{U}}$ is a linear operator. Moreover, if $x \in X$, then from $T_{\mathcal{U}}\big([(x, x, \ldots)]\big) = [(Tx, Tx, \ldots)]$, we see that the operator $T_{\mathcal{U}}\colon X_{\mathcal{U}} \to Y_{\mathcal{U}}$ is an extension of $T\colon X \to Y$.

To check that $T_{\mathcal{U}}$ is bounded, notice that according to Lemma 1.62 for each $x = (x_1, x_2, \ldots) \in \ell_\infty(X)$ we have

$$\begin{aligned} \big\|T_{\mathcal{U}}[x]\big\| &= \big\|[(Tx_1, Tx_2, \ldots)]\big\| = \lim_{\mathcal{U}} \|Tx_n\| \le \lim_{\mathcal{U}} \|T\| \cdot \|x_n\| \\ &= \|T\| \lim_{\mathcal{U}} \|x_n\| = \|T\| \cdot \big\|[x]\big\|. \end{aligned}$$

This shows that $T_{\mathcal{U}}$ is a bounded operator and that $\|T_{\mathcal{U}}\| \le \|T\|$. Now if $x \in X$ satisfies $\|x\| \le 1$, then

$$\|Tx\| = \big\|[(Tx, Tx, \ldots)]\big\| = \big\|T_{\mathcal{U}}[(x, x, \ldots)]\big\| \le \|T_{\mathcal{U}}\| \cdot \big\|[(x, x, \ldots)]\big\| \le \|T_{\mathcal{U}}\|.$$

Hence, $\|T\| = \sup_{\|x\| \le 1} \|Tx\| \le \|T_{\mathcal{U}}\|$ is also true, and so $\|T_{\mathcal{U}}\| = \|T\|$. ■

Regarding bounded operators between Banach lattices we have the following result whose straightforward proof is left for the reader.

Theorem 1.65. *For a bounded operator $T\colon X \to Y$ between two Banach lattices the following statements are true.*

(1) *T is a positive operator if and only if $T_{\mathcal{U}}$ is a positive operator.*

(2) *T is a lattice homomorphism if and only if $T_{\mathcal{U}}$ is a lattice homomorphism.*

(3) *If T is interval preserving, then so is $T_{\mathcal{U}}$.*

Corollary 1.66. *If X is a Banach space, then the mapping $T \mapsto T_{\mathcal{U}}$, from $\mathcal{L}(X)$ to $\mathcal{L}(X_{\mathcal{U}})$, is an algebraic homomorphism and a linear isometry which is a positive operator if X is a Banach lattice.*

Proof. This conclusion follows immediately from Theorems 1.64 and 1.65 and the (easily checked) fact that if $X \xrightarrow{T} Y \xrightarrow{S} Z$ is a scheme of bounded operators between Banach spaces, then $(ST)_{\mathcal{U}} = S_{\mathcal{U}}T_{\mathcal{U}}$.

Notice also that $(I_X)_{\mathcal{U}} = I_{X_{\mathcal{U}}}$. This implies that if $T \in \mathcal{L}(X)$ is an invertible operator, then $T_{\mathcal{U}} \in \mathcal{L}(X_{\mathcal{U}})$ is likewise an invertible operator and $(T^{-1})_{\mathcal{U}} = (T_{\mathcal{U}})^{-1}$. ■

Exercises

1. Prove Lemma 1.55.
2. Show that if an ultrafilter $\mathcal{U}$ on a set Δ satisfies $\bigcap_{A\in\mathcal{U}} A \neq \emptyset$, then there exists a unique $\delta \in \Delta$ such that $\mathcal{U} = \{A \subseteq \Delta\colon\ \delta \in A\}$.
3. Show that every filter coincides with the intersection of all ultrafilters that include it.
4. Prove Lemma 1.60.
5. Let Ω be a topological space, let A be a subset of Ω, and let $\{x_n\}$ be a sequence in Ω such that for some $N \in \mathcal{U}$ we have $x_n \in A$ for each $n \in N$. Show that if $x_n \xrightarrow{\mathcal{U}} x$, then $x \in \overline{A}$.

   Use this conclusion to show that the $\mathcal{U}$-limits preserve inequalities in the following sense: *If in a Banach lattice $x_n \xrightarrow{\mathcal{U}} x$, $y_n \xrightarrow{\mathcal{U}} y$, and for some $N \in \mathcal{U}$ we have $x_n \geq y_n$ for each $n \in N$, then $x \geq y$.*
6. For a non-zero bounded operator $T\colon X \to Y$ between two Banach spaces and a sequence $\{x_n\}$ in X establish the following.
   (a) If $x_n \xrightarrow{\mathcal{U}} 0$ in X, then $Tx_n \xrightarrow{\mathcal{U}} 0$ in Y.
   (b) If $Tx_n \xrightarrow{\mathcal{U}} 0$ in Y and $\lim_{\mathcal{U}} x_n \neq 0$ in X, then there exist some $\eta > 0$ and a subsequence $\{x_{k_n}\}$ of $\{x_n\}$ satisfying $\|x_{k_n}\| \geq \eta$ for each n and $\|Tx_{k_n}\| \to 0$.
7. Let Z be a normed space and let $\mathcal{U}$ be a free ultrafilter on $\mathbb{N}$. If, as usual, we let $\mathcal{N}_{\mathcal{U}} = \{(z_1, z_2, \ldots) \in \ell_\infty(Z)\colon\ z_n \xrightarrow{\mathcal{U}} 0\}$, then $\mathcal{N}_{\mathcal{U}}$ is a closed vector subspace of Z, and so $Z_{\mathcal{U}} = \ell_\infty(Z)/\mathcal{N}_{\mathcal{U}}$ is a normed space. That is, in the definition of the ultrapowers we do not need Z to be complete.

   Suppose that X is a normed space, Y is a subspace of X, and $\mathcal{U}$ is an ultrafilter on $\mathbb{N}$. Also, let $\mathcal{N}_{\mathcal{U}}^Y = \{(y_1, y_2, \ldots) \in \ell_\infty(Y)\colon\ y_n \xrightarrow{\mathcal{U}} 0\}$ and $\mathcal{N}_{\mathcal{U}}^X = \{(x_1, x_2, \ldots) \in \ell_\infty(X)\colon\ x_n \xrightarrow{\mathcal{U}} 0\}$, and define the mapping $J\colon Y_{\mathcal{U}} \to X_{\mathcal{U}}$ by
   $$J(y + \mathcal{N}_{\mathcal{U}}^Y) = y + \mathcal{N}_{\mathcal{U}}^X$$
   for each $y = (y_1, y_2, \ldots) \in \ell_\infty(Y)$. Show that J is a linear isometry—and so, under this isometry, $Y_{\mathcal{U}}$ can be considered as a vector subspace (which is norm closed if Y is also norm closed) of $X_{\mathcal{U}}$.
8. Prove Theorem 1.65.

9. For a bounded operator $T\colon X \to Y$ between two Banach spaces establish the following properties.
(a) If T is surjective, then $T_{\mathcal{U}}$ is surjective.
(b) If T has a closed range and $T_{\mathcal{U}}$ is surjective, then T is surjective.
(c) The operator T is invertible if and only if $T_{\mathcal{U}}$ is invertible—in which case we have $(T_{\mathcal{U}})^{-1} = (T^{-1})_{\mathcal{U}}$.

1.5. Vector-valued Functions

In this section, X and Y will denote arbitrary complex Banach spaces. We shall review here a few basic properties of the X-valued functions that are defined on subsets of the complex plane $\mathbb{C}$. We shall pay special attention to the results that will be employed in this book. They are the analogues of the results for complex-valued functions and their proofs can often be deduced from the scalar cases—for more details regarding Banach space-valued functions see [**108**]. We shall refer to the basic results about complex-valued functions of one complex variable as "standard." These results can be found in any book on complex analysis; see for instance [**67**].

We start our discussion with the fundamental properties of formal series in complex Banach spaces.

Theorem 1.67. *Let $\{a_0, a_1, \dots\}$ be a sequence of vectors in X, let $\lambda_0 \in \mathbb{C}$, and consider the formal series*

$$\sum_{n=0}^{\infty} (\lambda - \lambda_0)^n a_n\,.$$

Also, let $R = \frac{1}{\limsup \sqrt[n]{\|a_n\|}}$, where we adhere to the conventions $\frac{1}{0} = \infty$ and $\frac{1}{\infty} = 0$.[10] *Then:*

(1) *The series converges uniformly in norm on every compact subset of the* **open disk** *$D_R(\lambda_0)$.*[11]

(2) *The series does not converge in norm whenever $|\lambda - \lambda_0| > R$.*

(3) *When $|\lambda - \lambda_0| = R$, the series may or may not converge in norm.*

Proof. (1) We consider only the case $0 < R < \infty$ and leave the other two cases for the reader. Let K be an arbitrary compact subset of $D_R(\lambda_0)$. If $\delta = \max_{\lambda \in K} |\lambda - \lambda_0| > 0$, then we have $|\lambda - \lambda_0| \le \delta < R$ for every $\lambda \in K$, and so $\delta \cdot \limsup_{n\to\infty} \sqrt[n]{\|a_n\|} < 1$. Therefore, if we fix some α such that $\delta \cdot \limsup_{n\to\infty} \sqrt[n]{\|a_n\|} < \alpha < 1$, then there exists some n_0 such that

---

[10] As in the standard case, the number $0 \le R \le \infty$ is referred to as the **radius of convergence** of the series.

[11] As usual, if $r > 0$, we define $D_r(\lambda_0) = \{\lambda \in \mathbb{C}\colon\ |\lambda - \lambda_0| < r\}$. Occasionally, the open disk $D_r(\lambda_0)$ will also be denoted $D(\lambda_0, r)$.

$\delta\sqrt[n]{\|a_n\|} < \alpha$ for all $n \geq n_0$. In particular, we have $|\lambda - \lambda_0|\sqrt[n]{\|a_n\|} < \alpha$ for all $n \geq n_0$. So, $|\lambda - \lambda_0|^n \|a_n\| < \alpha^n$ for all $n \geq n_0$ and all $\lambda \in K$, and the desired conclusion follows.

(2) In this case, there exists some n_0 such that $|\lambda - \lambda_0|^n \|a_n\| \geq 1$ for all $n \geq n_0$. So, $|\lambda - \lambda_0|^n \|a_n\| \not\to 0$ and this shows that for $|\lambda - \lambda_0| > R$ the series cannot converge in norm.

(3) The usual examples can be used to establish these claims. ∎

Definition 1.68. *A function $f\colon \mathcal{O} \to X$, where $\mathcal{O}$ is an open subset of $\mathbb{C}$, is said to be **differentiable** at a point $\lambda_0 \in \mathcal{O}$ provided there exists some (uniquely determined) $x_0 \in X$ such that*

$$\lim_{\lambda \to \lambda_0} \left\| \frac{f(\lambda) - f(\lambda_0)}{\lambda - \lambda_0} - x_0 \right\| = 0. \qquad (\star)$$

*As usual, the vector x_0 is called the **derivative** of f at λ_0 and is denoted $f'(\lambda_0)$ or $\frac{df(\lambda_0)}{d\lambda}$, i.e., $f'(\lambda_0) = \frac{df(\lambda_0)}{d\lambda} = x_0$.*

*A function $f\colon \mathcal{O} \to X$ is said to be **analytic** (or **holomorphic**) at a point $\lambda_0 \in \mathcal{O}$ whenever f is differentiable at every point of an open disk about λ_0. If f is differentiable at every point of $\mathcal{O}$, then f is called an **analytic function**.*

*If S is a subset of $\mathbb{C}$, then a function $g\colon S \to \mathbb{C}$ is said to be **analytic** on S if g has an analytic extension to some open set $\mathcal{O}$ containing S.*

The definition of the (Riemann) integral $\int_C f(\lambda)\,d\lambda$ of a piecewise continuous function $f\colon C \to X$ over a continuous rectifiable curve C lying in $\mathbb{C}$ is defined as in the standard case. (The reader should notice that the only property required to guarantee the existence of the integral is the completeness of the range space.) We shall state below the most important results for Banach space-valued functions.

The first theorem relates the Banach space-valued integral to continuous operators and also gives a method of reducing Banach space-valued integrals to standard ones.

Theorem 1.69. *Let $T \in \mathcal{L}(X,Y)$, let C be a continuous rectifiable curve lying in $\mathbb{C}$, and let $f\colon C \to X$ be a continuous function. Then*

$$T\Big(\int_C f(\lambda)\,d\lambda\Big) = \int_C T(f(\lambda))\,d\lambda.$$

In particular, for each $x^* \in X^*$ we have

$$x^*\Big(\int_C f(\lambda)\,d\lambda\Big) = \int_C x^*(f(\lambda))\,d\lambda.$$

Using the definition of the integral, one can prove Cauchy's classical theorem in a similar manner. We state it below for future reference.

Theorem 1.70 (Cauchy's Integral Theorem). *Let $\mathcal{O}$ be a bounded open subset of $\mathbb{C}$ whose boundary $\partial\mathcal{O}$ consists of a finite number of rectifiable simple curves, say $\partial\mathcal{O} = \bigcup_{i=1}^{n} C_i$. Assume also that each curve C_i is oriented in the positive direction—in the usual sense of the theory of complex variables.*

If U is an open subset of $\mathbb{C}$ such that $\overline{\mathcal{O}} \subseteq U$ and $f\colon U \to X$ is an analytic function, then

$$\int_{\partial\mathcal{O}} f(\lambda)\, d\lambda = 0\,.$$

Following the standard terminology, we say that a function $f\colon S \to X$, where S is a subset of $\mathbb{C}$, has a **power series expansion** at some interior point $\lambda_0 \in S$ whenever there exist vectors $a_0, a_1, \ldots$ in X and some positive number $r > 0$ satisfying $D_r(\lambda_0) \subseteq S$ and

$$f(\lambda) = \sum_{n=0}^{\infty} (\lambda - \lambda_0)^n a_n$$

for each $\lambda \in D_r(\lambda_0)$, where (as always) the convergence of the series is assumed to be in the norm of X.

Lemma 1.71. *A function can have at most one power series expansion at any given point.*

Proof. Let $\lambda_0 \in S$, where S is a subset of $\mathbb{C}$, and assume that

$$\sum_{n=0}^{\infty} (\lambda - \lambda_0)^n a_n = \sum_{n=0}^{\infty} (\lambda - \lambda_0)^n b_n$$

holds for each $\lambda \in D_r(\lambda_0) \subseteq S$. If $x^* \in X^*$, then

$$\sum_{n=0}^{\infty} (\lambda - \lambda_0)^n x^*(a_n) = \sum_{n=0}^{\infty} (\lambda - \lambda_0)^n x^*(b_n)$$

for each λ satisfying $|\lambda - \lambda_0| < r$. From the standard uniqueness theorem for complex series, we get $x^*(a_n) = x^*(b_n)$ for each n and each $x^* \in X^*$. This implies $a_n = b_n$ for each n. ■

The following results can also be proven exactly as in the standard case.

Theorem 1.72. *A function $f\colon \mathcal{O} \to X$, where $\mathcal{O}$ is an open subset of $\mathbb{C}$, is analytic if and only if f has a power series expansion at every point in $\mathcal{O}$.*

Theorem 1.73. *Let $f\colon \mathcal{O} \to X$ be an analytic function on a open set $\mathcal{O}$. Then:*

(1) *The function f has derivatives of all orders.*

(2) *If a non-empty open subset V of $\mathcal{O}$ satisfies $\overline{V} \subseteq \mathcal{O}$ and its (positively oriented) boundary consists of a finite number of rectifiable simple curves, then the derivatives of f are given by Cauchy's classical formulas*

$$f^{(n)}(\lambda_0) = \tfrac{n!}{2\pi\imath} \int_{\partial V} \tfrac{f(\lambda)}{(\lambda-\lambda_0)^{n+1}}\, d\lambda\,,$$

for any $n = 0, 1, 2, \ldots$ and any $\lambda_0 \in V$.

(3) *At any $\lambda_0 \in V$ the series expansion of the function f is given by $f(\lambda) = \sum_{n=0}^{\infty}(\lambda - \lambda_0)^n a_n$, where*

$$a_n = \tfrac{f^{(n)}(\lambda_0)}{n!} = \tfrac{1}{2\pi\imath} \int_{\partial V} \tfrac{f(\lambda)}{(\lambda-\lambda_0)^{n+1}}\, d\lambda\,.$$

The above power series converges at least for each $\lambda \in D(\lambda_0, r)$, where $r = d\big(\lambda_0, \mathcal{O}^{\mathrm{c}}\big)$, the distance from λ_0 to $\mathcal{O}^{\mathrm{c}} = \mathbb{C} \setminus \mathcal{O}$.

As in the standard case, a function which is defined and is analytic in $\mathbb{C}$ is called an **entire function**.

Theorem 1.74 (Liouville). *A bounded entire function is constant.*

Proof. Let $f\colon \mathbb{C} \to X$ be a bounded entire function. Then for each $x^* \in X^*$ the function $x^* \circ f\colon \mathbb{C} \to \mathbb{C}$ is a bounded entire function. So (by Liouville's classical theorem), $x^* \circ f$ is a constant for each $x^* \in X^*$. Consequently, $x^*\big(f(\lambda) - f(0)\big) = 0$ for each $x^* \in X^*$ and each $\lambda \in \mathbb{C}$. Hence, $f(\lambda) = f(0)$ for each $\lambda \in \mathbb{C}$, i.e., f is a constant function. ■

In the same manner, we can establish the following uniqueness result for analytic functions.

Theorem 1.75. *If two analytic functions $f, g\colon \mathcal{O} \to X$ coincide on a subset of $\mathcal{O}$ which has an accumulation point in $\mathcal{O}$, then $f(\lambda) = g(\lambda)$ for each $\lambda \in \mathcal{O}$.*

Let us say that a function $f\colon \mathcal{O} \to X$ is **weakly analytic** whenever the (standard) function $x^* \circ f\colon \mathcal{O} \to \mathbb{C}$ is analytic for each $x^* \in X^*$. Every analytic function is obviously weakly analytic. Remarkably, as the next theorem shows, the converse is also true.

Theorem 1.76. *A function $f\colon \mathcal{O} \to X$ defined on an open subset $\mathcal{O}$ of the complex plane $\mathbb{C}$ is analytic if and only if it is weakly analytic.*

Proof. Assume that $f\colon \mathcal{O} \to X$ is weakly analytic. Fix $\lambda_0 \in \mathcal{O}$ and then choose $r > 0$ such that $C = \big\{\lambda \in \mathbb{C}\colon\ |\lambda - \lambda_0| = 2r\big\} \subseteq \mathcal{O}$. Since for each $x^* \in X^*$ the (standard) function $x^* \circ f$ is continuous, it follows that the set $\big\{x^*(f(\lambda))\colon\ \lambda \in C\big\}$ is bounded. Hence, by the Uniform Boundedness Principle, there exists some $M > 0$ satisfying $\big\|f(\lambda)\big\| \le M$ for each $\lambda \in C$.

Now let $\lambda_n \to \lambda_0$ with $\lambda_n \neq \lambda_0$ for each n. By deleting a finite number of terms if necessary, we can suppose that $|\lambda_n - \lambda_0| < r$ for each n. Also, let

$$x_n = \frac{f(\lambda_n) - f(\lambda_0)}{\lambda_n - \lambda_0}, \quad n = 1, 2, \ldots\ .$$

Next, fix some $x^* \in X^*$ with $\|x^*\| \le 1$. From Cauchy's classical formula, we have

$$\begin{aligned} x^*(x_n) &= \tfrac{x^*(f(\lambda_n)) - x^*(f(\lambda_0))}{\lambda_n - \lambda_0} \\ &= \tfrac{1}{2\pi\imath(\lambda_n - \lambda_0)} \int_C x^*(f(\xi)) \big[\tfrac{1}{\xi - \lambda_n} - \tfrac{1}{\xi - \lambda_0}\big]\, d\xi \\ &= \tfrac{1}{2\pi\imath} \int_C \tfrac{x^*(f(\xi))}{(\xi - \lambda_n)(\xi - \lambda_0)}\, d\xi, \end{aligned}$$

where the circle C is considered positively oriented. This implies

$$\begin{aligned} \big|x^*(x_n - x_m)\big| &= \tfrac{1}{2\pi}\Big|\int_C \tfrac{x^*(f(\xi))}{\xi - \lambda_0}\big[\tfrac{1}{\xi - \lambda_n} - \tfrac{1}{\xi - \lambda_m}\big]\, d\xi\Big| \\ &= \tfrac{1}{2\pi}\Big|\int_C \tfrac{x^*(f(\xi))}{\xi - \lambda_0} \cdot \tfrac{\lambda_n - \lambda_m}{(\xi - \lambda_n)(\xi - \lambda_m)}\, d\xi\Big| \\ &\le \tfrac{1}{2\pi} \cdot \tfrac{\|x^*\| M}{2r} \cdot \tfrac{|\lambda_n - \lambda_m|}{r \cdot r} \cdot 4\pi r \\ &\le \tfrac{M}{r^2}\, |\lambda_n - \lambda_m|\,. \end{aligned}$$

Therefore,

$$\big\|x_n - x_m\big\| = \sup_{\|x^*\| \le 1} \big|x^*(x_n - x_m)\big| \le \tfrac{M}{r^2}\, |\lambda_n - \lambda_m|\,,$$

and so $\{x_n\}$ is a norm Cauchy sequence. Let $x_n \to x$ in X, and note that (since the function f is weakly analytic) for each $x^* \in X^*$ we have $x^*(x) = \lim_{n\to\infty} x^*(x_n) = (x^* \circ f)'(\lambda_0)$.

Now if another sequence $\{\mu_n\} \subseteq \mathcal{O}$ satisfies $\mu_n \to \lambda_0$, then as above

$$\lim_{n\to\infty} \tfrac{f(\mu_n) - f(\lambda_0)}{\mu_n - \lambda_0} = y \in X\,.$$

Moreover, $x^*(x) = x^*(y) = (x^* \circ f)'(\lambda_0)$ for each $x^* \in X^*$. This implies $x = y$ and that

$$\lim_{\lambda\to\lambda_0} \tfrac{f(\lambda) - f(\lambda_0)}{\lambda - \lambda_0} = x\,,$$

i.e., f is differentiable at λ_0. Since λ_0 is arbitrary, f is analytic. ∎

An **open annulus** centered at a point $\lambda_0 \in \mathbb{C}$ is any open set of the form

$$A_{R_1,R_2}(\lambda_0) = \left\{\lambda \in \mathbb{C}\colon\ R_1 < |\lambda - \lambda_0| < R_2\right\},$$

where $R_1 = 0$ and $R_2 = \infty$ are allowed. A formal **Laurent series** in X about $\lambda_0 \in \mathbb{C}$ is any series of the form

$$\sum_{n=-\infty}^{\infty} (\lambda - \lambda_0)^n a_n\,,$$

where $a_n \in X$ for each $n = 0, \pm 1, \pm 2, \ldots$. We say that the Laurent series **converges** at some $\lambda \in \mathbb{C}$ whenever the two series $\sum_{n=1}^{\infty}(\lambda - \lambda_0)^{-n} a_{-n}$ and $\sum_{n=0}^{\infty}(\lambda - \lambda_0)^n a_n$ both converge in norm, in which case we write

$$\sum_{n=-\infty}^{\infty} (\lambda - \lambda_0)^n a_n = \sum_{n=1}^{\infty}(\lambda - \lambda_0)^{-n} a_{-n} + \sum_{n=0}^{\infty}(\lambda - \lambda_0)^n a_n\,.$$

The proof of the following fundamental theorem concerning convergence of Laurent series is similar to that of Theorem 1.67.

Theorem 1.77. *Let $\sum_{n=-\infty}^{\infty}(\lambda - \lambda_0)^n a_n$ be a formal Laurent series in a Banach space X and put*

$$R_1 = \limsup \sqrt[n]{\|a_{-n}\|} \qquad \text{and} \qquad R_2 = \frac{1}{\limsup \sqrt[n]{\|a_n\|}}\,,$$

where we adhere to the conventions $\frac{1}{0} = \infty$ and $\frac{1}{\infty} = 0$.

If $R_1 < R_2$, then the Laurent series converges uniformly in norm on every compact subset of the open annulus $A_{R_1,R_2}(\lambda_0)$.

A function $f\colon A \to X$ on an open annulus centered at a point λ_0 is said to have a **Laurent series expansion** if there exists a two-sided sequence of vectors $\ldots, c_{-2}, c_{-1}, c_0, c_1, c_2, \ldots$ in X such that

$$f(\lambda) = \sum_{n=-\infty}^{\infty} (\lambda - \lambda_0)^n c_n$$

holds for each $\lambda \in A$, where the convergence is again assumed to be in the norm.

Theorem 1.78 (Laurent). *Let $f\colon \mathcal{O} \to X$ be an analytic function defined on an open set and let $A_{R_1,R_2}(\lambda_0) \subseteq \mathcal{O}$ be an open annulus (where λ_0 need not necessarily lie in $\mathcal{O}$). Then f has a unique Laurent series expansion on $A_{R_1,R_2}(\lambda_0)$. That is, there exist (uniquely determined) vectors $\ldots, c_{-2}, c_{-1}, c_0, c_1, c_2, \ldots$ in X such that*

$$f(\lambda) = \sum_{n=-\infty}^{\infty} (\lambda - \lambda_0)^n c_n \tag{$\star$}$$

holds for each $\lambda \in A_{R_1,R_2}(\lambda_0)$. *Moreover, if* C *is any positively oriented circle with center at* λ_0 *and radius* R *satisfying* $R_1 < R < R_2$, *then*

$$c_n = \frac{1}{2\pi\imath}\int_C \frac{f(\lambda)}{(\lambda-\lambda_0)^{n+1}}\,d\lambda$$

for each $n = 0, \pm 1, \pm 2, \ldots$.

In addition, the Laurent series expansion given by $(\star)$ *is true for any* λ *in the largest open annulus in* $\mathcal{O}$ *centered at* λ_0 *that contains* $A_{R_1,R_2}(\lambda_0)$.

As in the standard case, if an analytic function $f\colon \mathcal{O} \to X$ has a Laurent series expansion $f(\lambda) = \sum_{n=-\infty}^{\infty}(\lambda - \lambda_0)^n c_n$ in an annulus, then for every simple closed rectifiable positively oriented curve C lying in the annulus we have

$$\int_C f(\lambda)\,d\lambda = 2\pi\imath c_{-1}\,.$$

Exercises

1. Establish the following "product rule" for derivatives: Consider two functions $g\colon \mathcal{O} \to \mathbb{C}$ and $f\colon \mathcal{O} \to X$ (where $\mathcal{O}$ is an open subset of $\mathbb{C}$ and X is a complex Banach space). If both functions f and g are differentiable at some $\lambda_0 \in \mathcal{O}$, then show that the X-valued function $\lambda \mapsto [gf](\lambda) = g(\lambda)f(\lambda)$ is also differentiable at λ_0 and
$$(gf)'(\lambda_0) = g'(\lambda_0)f(\lambda_0) + g(\lambda_0)f'(\lambda_0)\,.$$

2. Morera's theorem is a converse of Cauchy's theorem and it states that: *If a continuous function* $f\colon \mathcal{O} \to \mathbb{C}$ *(where* $\mathcal{O}$ *is an open subset of* $\mathbb{C}$*) satisfies* $\int_C f(\lambda)\,d\lambda = 0$ *for each simple closed rectifiable curve* C *which lies with its interior in* $\mathcal{O}$, *then* f *is an analytic function.*

   Show that Morera's theorem is valid for Banach space-valued functions.

3. Let C be a rectifiable curve lying in $\mathbb{C}$, and let X and Y be two complex Banach spaces. Also, let $A\colon C \to \mathcal{L}(X,Y)$ be a continuous operator-valued function. Show that for each $x \in X$ we have
$$\Big[\int_C A(\lambda)\,d\lambda\Big](x) = \int_C A(\lambda)x\,d\lambda\,.$$
[HINT: Since the vectors in both sides belong to Y, it suffices to verify that for each $y^* \in Y^*$ we have
$$\Big\langle y^*, \Big[\int_C A(\lambda)\,d\lambda\Big](x)\Big\rangle = \Big\langle y^*, \int_C A(\lambda)x\,d\lambda\Big\rangle\,.$$
In view of Theorem 1.69, the right-hand side equals $\int_C y^*(A(\lambda)x)\,d\lambda$. Now consider the functional $y^* \otimes x$ on $\mathcal{L}(X,Y)$ and recall that for each T in

$\mathcal{L}(X,Y)$ we have $\langle y^* \otimes x, T\rangle = \langle y^*, Tx\rangle$. So, on one hand

$$\Big\langle y^* \otimes x, \int_C A(\lambda)\, d\lambda \Big\rangle = \Big\langle y^*, \Big[\int_C A(\lambda)\, d\lambda\Big](x)\Big\rangle,$$

and on the other hand, applying $y^* \otimes x$ to the operator $\int_C A(\lambda)\, d\lambda$ and using again Theorem 1.69 we get

$$\Big\langle y^* \otimes x, \int_C A(\lambda)\, d\lambda \Big\rangle = \int_C \langle y^* \otimes x, A(\lambda)\rangle\, d\lambda = \int_C y^*(A(\lambda)x)\, d\lambda\,.$$

Therefore,

$$\Big\langle y^* \otimes x, \int_C A(\lambda)\, d\lambda \Big\rangle = \Big[\int_C A(\lambda)\, d\lambda\Big](x) = \int_C A(\lambda)x\, d\lambda\,.]$$

4. This is a generalization of Theorem 1.76. Let X be a complex Banach space and let G be a (not necessarily closed) **norming subspace** of X^*, that is, there exists a constant $\gamma > 0$ such that

$$\sup\{|x^*(x)|\colon\ x^* \in G \text{ and } \|x^*\| \le 1\} \ge \gamma\|x\|$$

for each $x \in X$.

Show that a continuous function $f\colon \mathcal{O} \to X$ defined on an open subset $\mathcal{O}$ of the complex plane $\mathbb{C}$ is analytic if and only if the scalar-valued function $x^* \circ f$ is analytic for each $x^* \in G$.

5. For a continuous function $A\colon \mathcal{O} \to \mathcal{L}(X,Y)$, where $\mathcal{O}$ is an open subset of $\mathbb{C}$, show that the following statements are equivalent.

(a) The function $\lambda \mapsto A(\lambda)$ is analytic on $\mathcal{O}$.

(b) For each $x \in X$ the Y-valued function $\phi_x(\lambda) = A(\lambda)x$ is analytic on $\mathcal{O}$.

(c) For every $x \in X$ and $y^* \in Y^*$ the standard complex-valued function

$$\phi_{x,y^*}(\lambda) = \langle A(\lambda)x, y^*\rangle = y^*\big(A(\lambda)x\big)$$

is analytic on $\mathcal{O}$.

6. For each $r > 0$ let $A_r = \{\lambda \in \mathbb{C}\colon\ |\lambda| > r\}$. Assume that X is a complex Banach space and for some $r > 0$ a function $f\colon A_r \to X$ is analytic having the Laurent series expansion $f(\lambda) = \sum_{n=-\infty}^{\infty} \lambda^n c_n$ on the open annulus A_r. Show that $\lim_{\lambda\to\infty} f(\lambda) = 0$ if and only if $c_n = 0$ for each $n = 0, 1, 2, \dots$.

7. Assume that two points a and b in a topological space X are joined by a curve C. If C is covered by a collection $\mathcal{V}$ of open sets, then show that there exist (not necessarily distinct) sets $V_1, V_2, \dots, V_n$ in $\mathcal{V}$ such that:

(a) $a \in V_1$ and $b \in V_n$,

(b) $C \subseteq V_1 \cup V_2 \cup \cdots \cup V_n$, and

(c) $V_i \cap V_{i+1} \ne \emptyset$ for each $i = 1, 2, \dots, n-1$.

(Such a collection $V_1, V_2, \dots, V_n$ of open sets is called a **chain** of open sets covering the curve C.)

1.6. Fundamentals of Measure Theory

The purpose of this section is to present some measure theory that will be needed throughout this book. This material will be employed especially heavily in Sections 5.1 and 5.2 for the development of a self-contained theory of integral operators. The material selected is quite standard and can be found in many texts on measure theory. The book [**31**] is our primary reference here.

We begin by establishing some notation and terminology. The generic measure space in our discussion will be denoted by (Ω, Σ, π), where Ω is a set, Σ is a σ-algebra of subsets of Ω, and π is a σ-additive measure on Σ. A function $f: \Omega \to \mathbb{R}$ is said to be **Σ-measurable** (or simply **measurable**) if $f^{-1}(B) \in \Sigma$ for each Borel subset B of $\mathbb{R}$. The collection of all measurable functions is a vector space. Two functions $f, g: \Omega \to \mathbb{R}$ are said to be **π-almost everywhere equal** (in brief π-a.e. or simply a.e.) whenever $f(\omega) = g(\omega)$ holds for π-almost all ω. That is, f and g are equal almost everywhere if there exists some $A \in \Sigma$ with $\pi(A) = 0$ such that $f(\omega) = g(\omega)$ for all $\omega \notin A$. Almost everywhere equality defines an equivalence relation on the vector space of all measurable function. It is easy to see that under the algebraic and lattice operations defined π-a.e., the set of all equivalence classes is a Riesz space. This is the very important Riesz space $L_0(\pi)$.

Definition 1.79. *The Riesz space of all equivalence classes of measurable real functions on Ω is denoted by $L_0(\Omega, \Sigma, \pi)$ or simply $L_0(\pi)$.*

As usual, we identify the equivalence classes with functions and agree that two arbitrary measurable functions $f, g: \Omega \to \mathbb{R}$ are equal if and only if $f(\omega) = g(\omega)$ for π-almost all ω. With this agreement in mind, a measurable function f can be changed (or can even be left undefined) on a set of π-measure zero and still represent the same class as f.

For a proof of the next result, see Exercise 5 at the end of the section.

Theorem 1.80. *If π is σ-finite, then $L_0(\pi)$ is a Dedekind complete Riesz space with the countable sup property.*

Theorem 1.80 shows that when we deal with σ-finite measure spaces, the various order related notions in $L_0(\pi)$ can be described by their sequential analogues. For instance, the concept of σ-order continuity (for functionals and operators) coincides with the concept of order continuity. When dealing with Riesz spaces satisfying the countable sup property, it is a common practice to use the general standard terms (like *order continuous operators*, *order continuous functionals*, *order closed sets*, etc.) while only sequences are needed to verify their definitions. Let us recall three basic types of convergent sequences.

Definition 1.81. *A sequence $\{x_n\}$ in $L_0(\pi)$:*

(1) ***converges pointwise*** *π-a.e. to some $x \in L_0(\pi)$, in symbols, $x_n \xrightarrow{p} x$ or $x_n \to x$ π-a.e., if $x_n(\omega) \to x(\omega)$ for π-almost all $\omega \in \Omega$,*

(2) ***converges relatively uniformly*** *to some $x \in L_0(\pi)$ if there exists a positive function $u \in L_0(\pi)$ (called the* ***regulator of the convergence****) such that for each $\epsilon > 0$ there exists some k so that $|x_n - x| \le \epsilon u$ holds for all $n \ge k$, and*

(3) ***converges in measure*** *to some $x \in L_0(\pi)$, in symbols, $x_n \xrightarrow{\pi} x$, if for each $\epsilon > 0$ and for each $A \in \Sigma$ with $\pi(A) < \infty$ we have*

$$\lim_{n\to\infty} \pi\big(\{\omega \in A\colon\ |x_n(\omega) - x(\omega)| \ge \epsilon\}\big) = 0\,.^{12}$$

Convergence in measure is characterized as follows.

Theorem 1.82. *Let π be a σ-finite measure space. For a sequence $\{x_n\}$ in $L_0(\pi)$ the following are equivalent.*

(1) *$x_n \xrightarrow{\pi} 0$.*

(2) *Every subsequence of $\{x_n\}$ has a subsequence that converges pointwise π-a.e. to zero.*

(3) *For each $A \in \Sigma$ with $\pi(A) < \infty$ we have $\lim\limits_{n\to\infty} \int_A \frac{x_n}{1+x_n}\, d\pi = 0$.*

Proof. We can assume that $\{x_n\}$ is a positive sequence.

(1)$\Longrightarrow$(2) Fix some $A \in \Sigma$ with $\pi(A) < \infty$, and let $\{y_n\}$ be a subsequence of $\{x_n\}$. Also, for each $\epsilon > 0$ let $A_n(\epsilon) = \{\omega \in A\colon\ y_n(\omega) \ge \epsilon\}$. Since $y_n \xrightarrow{\pi} 0$, it follows that for each $\epsilon > 0$ we have $\pi\big(A_n(\epsilon)\big) \to 0$. In particular, an inductive argument shows that there exists a strictly increasing sequence $\{k_n\}$ of integers such that $\pi\big(A_m(\frac{1}{n})\big) < \frac{1}{2^n}$ holds for all $m \ge k_n$. If we let $B_n = \bigcup_{i=1}^{\infty} A_{k_i}(\frac{1}{i})$, then it is easy to see that $B_n \downarrow \emptyset$ π-a.e. This implies $y_{k_n}(\omega) \to 0$ for π-almost all $\omega \in A$. That is, we have shown that for each $A \in \Sigma$ with $\pi(A) < \infty$ and for each subsequence $\{y_n\}$ of $\{x_n\}$ there exists a subsequence of $\{y_n\}$ that converges π-a.e. to zero on A.

Next, pick a sequence $\{\Omega_n\} \subseteq \Sigma$ with $\pi(\Omega_n) < \infty$ for each n and $\Omega_n \uparrow \Omega$, and let $\{y_n\}$ be a subsequence of $\{x_n\}$. Then, using the preceding conclusion, an easy inductive argument shows that for each n there exists a sequence $\{x_{n,k}\colon\ k = 1, 2, \ldots\}$ such that:

- $\{x_{1,k}\}$ is a subsequence of $\{y_n\}$.

[12] The reader should be warned that this definition of convergence in measure is less restrictive than the standard definition, where a sequence $\{x_n\}$ in $L_0(\pi)$ converges to $x \in L_0(\pi)$ in measure if for each $\epsilon > 0$ we have $\lim_{n\to\infty} \pi\big(\{\omega \in \Omega\colon\ |x_n(\omega) - x(\omega)| \ge \epsilon\}\big) = 0$. In practice, only sets of finite measure are important and this is why we restrict our definition to sets of finite measure.

- $\{x_{n+1,k}\colon\ k=1,2,\ldots\}$ is a subsequence of $\{x_{n,k}\colon\ k=1,2,\ldots\}$.
- $x_{n,k} \xrightarrow[k\to\infty]{p} 0$ on Ω_n.

If we let $z_n = x_{n,n}$ (the diagonal sequence), then $\{z_n\}$ is a subsequence of $\{y_n\}$ that converges π-a.e. to zero on Ω.

(2)$\Longrightarrow$(3) Fix some $A\in\Sigma$ with $\pi(A)<\infty$, and assume contrary to our claim that $\int_A \frac{x_n}{1+x_n}\,d\pi \not\to 0$. Then there exist a subsequence $\{y_n\}$ of $\{x_n\}$ and some $\epsilon>0$ such that $\int_A \frac{y_n}{1+y_n}\,d\pi \ge \epsilon$ for each n. The validity of (2) guarantees that there exists a subsequence $\{z_n\}$ of $\{y_n\}$ such that $z_n \xrightarrow{p} 0$. Since $\frac{z_n}{1+z_n} \xrightarrow{p} 0$ and $0 \le \frac{z_n}{1+z_n} \le \mathbf{1}$ on A, it follows from the Lebesgue Dominated Convergence Theorem that $\int_A \frac{z_n}{1+z_n}\,d\pi \to 0$. However, this contradicts $\int_A \frac{z_n}{1+z_n}\,d\pi \ge \epsilon$ for each n and the validity of (3) has been established.

(3)$\Longrightarrow$(1) Assume that some $A\in\Sigma$ satisfies $\pi(A)<\infty$ and let $\epsilon>0$. Also, let $A_n=\{\omega\in A\colon\ x_n(\omega)\ge\epsilon\}$. From the fact that $t\ge\epsilon$ if and only if $\frac{t}{1+t}\ge\frac{\epsilon}{1+\epsilon}$, it follows that $A_n=\big\{\omega\in A\colon\ \frac{x_n(\omega)}{1+x_n(\omega)}\ge\frac{\epsilon}{1+\epsilon}\big\}$. This implies $\frac{x_n}{1+x_n}\chi_A \ge \frac{\epsilon}{1+\epsilon}\chi_{A_n}$, and consequently

$$\tfrac{\epsilon}{1+\epsilon}\pi(A_n) = \int_{A_n} \tfrac{\epsilon}{1+\epsilon}\chi_{A_n}\,d\pi \le \int_A \tfrac{x_n}{1+x_n}\,d\pi \longrightarrow 0\,.$$

Thus, $\pi(A_n)\to 0$, and this proves that $x_n \xrightarrow{\pi} 0$. ■

Corollary 1.83. *Pointwise convergence in $L_0(\pi)$ implies convergence in measure, i.e., $x_n \xrightarrow{p} x$ in $L_0(\pi)$ implies $x_n \xrightarrow{\pi} x$.*

The next few results contain some basic properties of sequences in $L_0(\pi)$. These results as well as some additional properties can be found in Section 71 of the book by W. A. J. Luxemburg and A. C. Zaanen [**225**], or in Chapter VI of the book by B. Z. Vulikh [**328**].

Lemma 1.84. *If $\{x_n\}\subseteq L_0(\pi)$ and π is σ-finite, then there exists a sequence of positive scalars $\{\lambda_n\}$ such that the sequence $\{\lambda_n x_n\}$ is order bounded in $L_0(\pi)$.*

Proof. Let $\{\Omega_n\}$ be a sequence in Σ such that $\Omega_n\uparrow\Omega$ and $\pi(\Omega_n)<\infty$ for each n, and fix some sequence $\{x_n\}$ in $L_0(\pi)$. Replacing $\{x_n\}$ by $\{|x_n|\}$, we can assume that each x_n is a positive function. Since for each fixed n we have $x_n\wedge k\uparrow_k x_n$ π-a.e., it follows from Egorov's theorem that for each n there exists some $A_n\in\Sigma$ with $A_n\subseteq\Omega_n$ such that $\pi(\Omega_n\setminus A_n)<\frac{1}{2^n}$ and the sequence $\{x_n\wedge k\colon\ k\ge 1\}$ converges uniformly to x_n on A_n. In particular, there exists a strictly increasing sequence $\{k_n\}$ of natural numbers such that $|x_n(\omega)\wedge k_n - x_n(\omega)|<1$ for each $\omega\in A_n$. This implies $x_n(\omega)\le 1+k_n$ for each $\omega\in A_n$. By letting $\lambda_n=\frac{1}{n(1+k_n)}$, we get $\lambda_n x_n(\omega)\le\frac{1}{n}$ for each $\omega\in A_n$.

Now, if we let $A = \liminf A_n = \bigcup_{m=1}^{\infty} \bigcap_{k=m}^{\infty} A_k$, then $A \in \Sigma$ and for each $\omega \in A$ we have $\lambda_n x_n(\omega) \to 0$. Moreover, for each $n, r \in \mathbb{N}$ we have

$$\begin{aligned}
\pi(\Omega_n \setminus A) &= \pi\Big(\Omega_n \setminus \bigcup_{m=1}^{\infty} \bigcap_{k=m}^{\infty} A_k\Big) \leq \pi\Big(\Omega_n \setminus \bigcap_{k=n+r}^{\infty} A_k\Big) \\
&= \pi\Big(\bigcup_{k=n+r}^{\infty} (\Omega_n \setminus A_k)\Big) \leq \sum_{k=n+r}^{\infty} \pi(\Omega_n \setminus A_k) \\
&\leq \sum_{k=n+r}^{\infty} \pi(\Omega_k \setminus A_k) \leq \tfrac{1}{2^{n+r+1}} \,.
\end{aligned}$$

As $r \geq 1$ is arbitrary, $\pi(\Omega_n \setminus A) = 0$, and this guarantees that $\pi(\Omega \setminus A) = 0$. Therefore, if we let $x(\omega) = \sup\{\lambda_n x_n(\omega)\colon\ n \geq 1\}$, then $x \in L_0(\pi)$ and $0 \leq \lambda_n x_n \leq x$ for each n. ■

We continue with one more property of pointwise convergent sequences.

Theorem 1.85. *If $x_n \xrightarrow{p} 0$ holds in $L_0(\pi)$ and π is σ-finite, then there exists a sequence $\{\lambda_n\}$ of positive real numbers such that $\lambda_n \uparrow \infty$ and $\lambda_n x_n \xrightarrow{p} 0$.*

Proof. Pick a sequence $\{\Omega_n\} \subseteq \Sigma$ such that $\Omega_n \uparrow \Omega$ and $\pi(\Omega_n) < \infty$ for each n, and let a sequence $\{x_n\}$ in $L_0(\pi)$ satisfy $x_n \xrightarrow{p} 0$. By Egorov's theorem, for each k there exists some $A_k \in \Sigma$ such that $\pi(\Omega_k \setminus A_k) < \frac{1}{2^k}$ and the sequence $\{x_n\}$ converges uniformly to zero on A_k. An inductive argument shows that there exists a strictly increasing sequence $\{n_k\}$ of natural numbers such that $|x_n(\omega)| \leq \frac{1}{k^2}$ for all $n \geq n_k$ and all $\omega \in A_k$. Let $\lambda_n = 1$ if $n < n_1$ and $\lambda_n = k$ if $n_k \leq n < n_{k+1}$. Then $\lambda_n \uparrow \infty$ and clearly if $\omega \in \bigcap_{k=m}^{\infty} A_k$ and $k \geq m$, then $\lambda_n |x_n(\omega)| \leq \frac{1}{k}$ for all $n \geq n_k$. This implies $\lambda_n x_n(\omega) \to 0$ for each $\omega \in \bigcap_{k=m}^{\infty} A_k$. In particular, we have $\lambda_n x_n(\omega) \to 0$ for each $\omega \in A = \liminf A_n = \bigcup_{m=1}^{\infty} \bigcap_{k=m}^{\infty} A_k$. Since (as in the proof of Lemma 1.84) we have $\pi(\Omega \setminus A) = 0$, the latter shows that $\lambda_n x_n \xrightarrow{p} 0$. ■

In $L_0(\pi)$ the notions of pointwise convergence, order convergence, and relative uniform convergence coincide.

Corollary 1.86. *If π is σ-finite, then for a sequence $\{x_n\} \subseteq L_0(\pi)$ and some $x \in L_0(\pi)$ the following statements are equivalent.*

1. *The sequence $\{x_n\}$ is pointwise convergent to x, i.e., $x_n \xrightarrow{p} x$.*
2. *The sequence $\{x_n\}$ is relatively uniformly convergent to x.*
3. *The sequence $\{x_n\}$ is order convergent to x, i.e., $x_n \xrightarrow{o} x$.*

Proof. (1)$\Longrightarrow$(2) Let $x_n \xrightarrow{p} x$. By Theorem 1.85, there exists a sequence $\{\lambda_n\}$ of positive scalars satisfying $\lambda_n \uparrow \infty$ and $\lambda_n(x_n - x) \xrightarrow{p} 0$. Now consider the function $y = \sup\{\lambda_n|x_n - x|: n \in \mathbb{N}\}$. Then $y \in L_0(\pi)$ and $|x_n - x| \leq \frac{1}{\lambda_n} y$ for each n. This shows that $\{x_n\}$ converges relatively uniformly (with regulator y) to zero.

(2)$\Longrightarrow$(3) Pick a function $u \in L_0(\pi)$ and a sequence $\{k_n\}$ of strictly increasing natural numbers with $k_1 = 1$ such that $|x_m - x| \leq \frac{1}{n} u$ for each $m \geq k_n$. Now for each n let m_n be the unique natural number such that $k_{m_n} \leq n < k_{m_n+1}$ and put $\epsilon_n = \frac{1}{m_n}$. Clearly $\epsilon_n \downarrow 0$ and $|x_n - x| \leq \epsilon_n u$ holds for each n. This shows that $x_n \xrightarrow{o} x$.

(3)$\Longrightarrow$(1) Obvious. ■

In particular, note that $f_n \uparrow f$ in $L_0(\pi)$ is equivalent to saying that $f_n(\omega) \uparrow f(\omega)$ for π-almost all $\omega \in \Omega$. The $L_0(\pi)$-spaces also satisfy the diagonal property.

Theorem 1.87 (The Diagonal Property). *Suppose that π is σ-finite and $\{x_{n,k}\}$ is a double sequence in $L_0(\pi)$ such that for each n we have $x_{n,k} \xrightarrow[k\to\infty]{p} 0$. Then there exists a sequence $\{y_n\} \subseteq L_0(\pi)$ and a strictly increasing sequence $\{k_n\}$ of natural numbers such that $y_n \downarrow 0$ and $|x_{n,k}| \leq y_n$ for all $k \geq k_n$ and each n.*[13]

Moreover, if $|x_{n,k}| \leq x$ holds for all n and k, then we can take the sequence $\{y_n\}$ to satisfy $y_n \leq x$ for each n.

Proof. By Corollary 1.86, for each k there exists some $u_n \in L_0^+(\pi)$ such that the sequence $\{x_{n,k}: k \geq 1\}$ is relatively uniformly convergent to zero with respect to the regulator u_n. Now, by Lemma 1.84, there exists a sequence $\{\lambda_n\}$ of positive scalars and some $u \in L_0^+(\pi)$ such that $\lambda_n u_n \leq u$ for each n. Put $y_n = \frac{1}{n} u$ and note that for each fixed natural number n there exists an integer k_n such that $|x_{n,k}| \leq \frac{\lambda_n}{n} u_n \leq \frac{1}{n} u$ for all $k \geq k_n$. An easy inductive argument shows that $\{k_n\}$ can be taken to be strictly increasing. If $|x_{n,k}| \leq x$ holds for all n and k, then let $y_n = x \wedge \left(\frac{1}{n} u\right)$. ■

The order bounded subsets of $L_0(\pi)$ are characterized as follows.

Theorem 1.88. *When π is σ-finite, then a subset D of $L_0(\pi)$ is order bounded if and only if for each sequence $\{x_n\} \subseteq D$ and for each sequence $\{\lambda_n\}$ of scalars with $\lambda_n \downarrow 0$ we have $\lambda_n x_n \xrightarrow{o} 0$.*

---

[13] A Riesz space satisfying this property is referred to as a Riesz space with the **diagonal property**.

Proof. Let D be a non-empty subset of $L_0(\pi)$. If D is order bounded, $\lambda_n \downarrow 0$ in $\mathbb{R}$ and $\{x_n\} \subseteq D$, then clearly $\lambda_n x_n \xrightarrow{o} 0$. For the converse recall that (by Corollary 1.86) order convergence and pointwise convergence of sequences in $L_0(\pi)$ coincide. Now assume that the set D satisfies the property:

(P) *If* $\{x_n\} \subseteq D$ *and* $\lambda_n \downarrow 0$ *in* $\mathbb{R}$, *then* $\lambda_n x_n(\omega) \to 0$ *holds for* π*-almost all* $\omega \in \Omega$.

We must show that D is order bounded in $L_0(\pi)$. Since D is order bounded if and only if the set $|D| = \{|x|\colon\ x \in D\}$ is order bounded and the set $|D|$ clearly has property (P), replacing D by $|D|$, we can assume that $D \subseteq L_0^+(\pi)$. Also, since D is order bounded if and only if the set $\mathbf{1}+D = \{\mathbf{1}+x\colon\ x \in D\}$ is order bounded and the set $\mathbf{1}+D$ has property (P), replacing D by $\mathbf{1}+D$, we can assume that each $x \in D$ satisfies $x(\omega) \geq 1$ for π-almost all $\omega \in \Omega$. Furthermore, the set $D' = \{x \wedge k\colon\ x \in D \text{ and } k \in \mathbb{N}\}$ satisfies property (P) and has the same upper bounds as the set D. So, replacing D by D', we can also assume that D consists of bounded functions and that each $k \in \mathbb{N}$ dominates some function in D.

Next, let C denote the set consisting of all finite suprema of the functions in D. Clearly, D and C have the same upper bounds, and it can be easily shown that C likewise satisfies property (P); see Exercise 7 at the end of the section. Thus, replacing D by C, we can assume furthermore that D is also closed under finite suprema.

Now for each k let $D_k = \{x \in D\colon\ x \leq k\}$. Since each D_k is order bounded (and non-empty by the preceding discussion), it follows from the countable sup property of $L_0(\pi)$ (see Theorem 1.80) that there exists a sequence $\{x_n^k\colon\ n = 1, 2, \ldots\} \subseteq D_k$ such that $x_n^k \uparrow \sup D_k$. Let $y_n = \bigvee_{i=1}^n x_n^i \in D$. Clearly, $y_n \uparrow$ and we claim that the function y defined by $y(\omega) = \sup_n y_n(\omega)$ belongs to $L_0(\pi)$. If this is not the case, then there exists a $B \in \Sigma$ with $\pi(B) < \infty$ such that $y_n(\omega) \uparrow \infty$ (or, equivalently, $\frac{1}{y_n(\omega)} \downarrow 0$) for each $\omega \in B$. By Egorov's theorem, we can assume that $\frac{1}{y_n}$ converges uniformly to zero on B. This implies that there exists a strictly increasing sequence $\{m_n\}$ of natural numbers such that $\frac{1}{y_{m_n}(\omega)} < \frac{1}{n}$ for all $\omega \in B$ and all n. That is, $\frac{1}{n} y_{m_n}(\omega) > 1$ holds for all $\omega \in B$ and all n. However, the latter conclusion contradicts property (P), according to which $\frac{1}{n} y_{m_n} \xrightarrow{p} 0$. This contradiction shows that $y \in L_0(\pi)$. To finish the proof note that y is an upper bound of D; in fact, $y = \sup D$ in $L_0(\pi)$. ■

Corollary 1.89. *For an operator $T\colon E \to L_0(\pi')$, where π and π' are two σ-finite measures and E is an ideal in $L_0(\pi)$, we have the following.*

(a) *If T is σ-order continuous, then T is regular.*

(b) *If for each order bounded sequence $\{x_n\}$ in E that satisfies $x_n \xrightarrow{p} 0$ we have $Tx_n \xrightarrow{p} 0$, then T is σ-order continuous—and hence it is a regular order continuous operator.*

Proof. (a) Fix $x \in E^+$. For any sequence $\{x_n\} \subseteq [0, x]$ and any sequence $\{\lambda_n\}$ of scalars with $\lambda_n \downarrow 0$, we have $0 \le \lambda_n x_n \le \lambda_n x \downarrow 0$. This shows that $\lambda_n x_n \xrightarrow{o} 0$ in E, and so by the σ-order continuity of T we get $\lambda_n T x_n \xrightarrow{o} 0$. From Corollary 1.86, we obtain $\lambda_n T x_n \xrightarrow{p} 0$, and hence, by Theorem 1.88, the set $T[0, x]$ is order bounded in $L_0(\pi')$. Therefore, in view of Theorem 1.16, T is a regular operator.

(b) Let $x_n \xrightarrow{o} 0$ in E. This implies that $\{x_n\}$ is an order bounded sequence and $x_n \xrightarrow{p} 0$. By our hypothesis, $Tx_n \xrightarrow{p} 0$. Now a glance at Corollary 1.86 shows that $Tx_n \xrightarrow{o} 0$, and so T is σ-order continuous. ■

When a measure space (Ω, Σ, π) is σ-finite, then (as we saw in Theorem 1.80) the Riesz space $L_0(\pi)$ is Dedekind complete, and hence the ideals in $L_0(\pi)$ are Dedekind complete Riesz spaces in their own right. In particular, the classical $L_p(\pi)$-spaces $(0 < p \le \infty)$ are all Dedekind complete. For a proof of the following important density property of order ideals in $L_0(\pi)$ see Exercise 13 at the end of the section.

Lemma 1.90. *If (Ω, Σ, π) is a σ-finite measure space and J is an order dense ideal*[14] *in $L_0(\pi)$, then for each function $x \in L_0(\pi)$ there exists a sequence $\{x_n\} \subseteq J \cap L_1(\pi)$ such that $|x_n| \le |x|$ for each n and $x_n(\omega) \to x(\omega)$ for π-almost all $\omega \in \Omega$.*

If, in addition, $x \in L_0^+(\pi)$, then the above sequence $\{x_n\}$ can be chosen in $J^+ \cap L_1(\pi)$ and increasing, i.e., satisfying $0 \le x_n \uparrow x$ in $L_0(\pi)$.

Recall that if $x\colon \Omega \to \mathbb{R}$ is a function, then the **support** of x is the set $\operatorname{Supp} x = \{\omega \in \Omega\colon\ x(\omega) \ne 0\}$.

Corollary 1.91. *If π is a σ-finite measure and J is an order dense ideal in $L_0(\pi)$, then for each function $x \in L_0(\pi)$ there exists a disjoint sequence $\{V_n\} \subseteq \Sigma$ such that $\chi_{V_n} \in J$, $\pi(V_n) < \infty$ for each n and* $\operatorname{Supp} x = \bigcup_{n=1}^{\infty} V_n$.

In particular, there is a disjoint sequence $\{\Omega_n\}$ in Σ such that $\pi(\Omega_n) < \infty$ for each n, $\Omega = \bigcup_{n=1}^{\infty} \Omega_n$ and each $\chi_{\Omega_n} \in J$.

---

[14] Recall that a Riesz subspace G of a Riesz space H is **order dense** if for each $0 < x \in H$ there exists some $0 < y \in G$ satisfying $0 < y \le x$. If H is Archimedean, G is order dense in H if and only if for each $x \in H^+$ we have $\{y \in G^+ : y \le x\} \uparrow x$ in H; see Exercise 6 at the end of the section.

Proof. Choose a sequence $\{\Omega_n\}$ in Σ satisfying $\pi(\Omega_n) < \infty$ for each n and $\Omega = \bigcup_{n=1}^{\infty} \Omega_n$. Suppose first that $x \in J$. If $A_n^x = \{\omega \in \Omega\colon\ |x(\omega)| \geq \frac{1}{n}\}$, then $\operatorname{Supp} x = \bigcup_{n=1}^{\infty} A_n^x$. From $\chi_{A_n^x} \leq n|x|$, we get $\chi_{A_n^x} \in E$ for each n.

Now let $x \in L_0(\pi)$. By Lemma 1.90 there exists a sequence $\{x_n\} \subseteq E^+$ such that $x_n \uparrow |x|$; we can assume that $x_n(\omega) \uparrow |x(\omega)|$ for each $\omega \in \Omega$. It follows that

$$\operatorname{Supp} x = \bigcup_{n=1}^{\infty} \operatorname{Supp} x_n = \bigcup_{n=1}^{\infty}\bigcup_{k=1}^{\infty} A_k^{x_n} = \bigcup_{n=1}^{\infty}\bigcup_{k=1}^{\infty}\bigcup_{m=1}^{\infty} A_k^{x_n} \cap \Omega_m\,.$$

Since $A_k^{x_n} \cap \Omega_m \subseteq A_k^{x_n}$ and $\chi_{A_k^{x_n}} \in E$, we see that $\chi_{A_k^{x_n} \cap \Omega_m} \in E$ for all n, k, and m. Thus, we can write $\operatorname{Supp} x = \bigcup_{n=1}^{\infty} B_n$ with $\chi_{B_n} \in E$ for each n. Now if $V_1 = B_1$ and $V_{n+1} = B_{n+1} \setminus \bigcup_{i=1}^{n} B_i$ for $n = 1, 2, \ldots$, then $\chi_{V_n} \in J$ and $\pi(V_n) < \infty$ for each n, $V_k \cap V_n = \emptyset$ for $k \neq n$, and $\operatorname{Supp} x = \bigcup_{n=1}^{\infty} V_n$.

The validity of the last statement of the corollary follows from the fact that $\mathbf{1} = \chi_\Omega \in L_0(\pi)$. ■

We continue with a description of the ideals and bands in $L_0(\pi)$.

Theorem 1.92. *For a σ-finite measure space (Ω, Σ, π) and an ideal E in $L_0(\pi)$ we have the following:*

(a) *There exists a smallest (with respect to π-a.e. inclusion) set C_E in Σ such that every $x \in E$ vanishes π-a.e. on $\Omega \setminus C_E$.*

(b) *If E is a band, then*

$$E = \big\{x \in L_0(\pi)\colon\ x = 0 \ \ \pi\text{-a.e. on } \Omega \setminus C_E\big\} = L_0(C_E)\,.$$

(c) *Two bands B_1 and B_2 in $L_0(\pi)$ coincide if and only if $C_{B_1} = C_{B_2}$.*

Proof. (a) We consider the finite and σ-finite cases separately.

CASE I: *π is a finite measure.*

In this case, consider the collection of measurable sets

$$\mathcal{F} = \big\{A \in \Sigma\colon\ x = 0 \ \ \pi\text{-a.e. on } A \text{ for each } x \in E\big\}\,.$$

Clearly, $\mathcal{F}$ is closed under countable unions. Put $m = \sup\{\pi(A)\colon\ A \in \mathcal{F}\}$. Since π is a finite measure on Σ, we have $0 \leq m < \infty$. If for each n we choose some $A_n \in \mathcal{F}$ such that $m - \frac{1}{n} < \pi(A_n)$, then the set $A = \bigcup_{n=1}^{\infty} A_n \in \mathcal{F}$ and satisfies $\pi(A) = m$. We claim that the measurable set $C_E = \Omega \setminus A$ satisfies the desired properties.

To see this, note first that for each $x \in E$ we have $x = 0$ π-a.e. on $C_E = \Omega \setminus A$. Now assume that for some measurable set C we have $x = 0$ π-a.e. on $C^c = \Omega \setminus C$ for each $x \in E$. Then $C^c \in \mathcal{F}$ and from the identity $C^c = (C^c \cap A) \cup (C^c \setminus A)$, we get $C^c \setminus A \in \mathcal{F}$. The definition of A implies

$\pi\big(C^c \setminus A\big) = 0$, and from this we get $C^c \subseteq A$ π-a.e. Hence, $C_E = A^c \subseteq C$, and the proof of this case is finished.

CASE II: *π is a σ-finite measure.*

Fix a disjoint sequence $\{\Omega_n\} \subseteq \Sigma$ with $\bigcup_{n=1}^{\infty} \Omega_n = \Omega$ and $\pi(\Omega_n) < \infty$ for each n. Also, let $\Sigma_n = \{A \cap \Omega_n\colon\ A \in \Sigma\}$ and $E_n = \{f\chi_{\Omega_n}\colon\ f \in E\}$. Then each E_n can be considered as an ideal in $L_0(\Sigma_n, \pi)$. Therefore, from Case I, for each n there exists a unique minimal measurable set $C_{E_n} \subseteq \Omega_n$ such that each function $x \in E_n$ vanishes outside of the set C_{E_n}. Now if we let $C_E = \bigcup_{n=1}^{\infty} C_{E_n}$, then a direct verification shows that this measurable set C_E satisfies the desired properties.

(b) Assume that E is a band, and let

$$B = \big\{x \in L_0(\pi)\colon\ x = 0 \ \ \pi\text{-a.e. on } \Omega \setminus C_E\big\}\,.$$

Clearly, B is a band in $L_0(\pi)$ and $E \subseteq B$. Now let $0 < f \in B$ and consider the set $C = \{\omega \in \Omega\colon\ f(\omega) > 0\}$, and note that $\pi(C) > 0$ and $C \subseteq C_E$. From the definition of the set C_E, it follows that there exists some $g \in J^+$ such that $g \neq 0$ π-a.e. on C. This means that the measurable set $G = \{\omega \in C\colon\ g(\omega) > 0\}$ satisfies $\pi(G) > 0$. Moreover, the sequence $\{f \wedge ng\} \subseteq E$ satisfies $f \wedge ng \uparrow f\chi_G$. Therefore, $f\chi_G \in E$, and thus $0 < f\chi_G \leq f$. This shows that E is order dense in the band B. Since E is a band, $E = B$. (Notice also that E can be identified with $L_0(C_E, \pi)$.)

(c) This follows immediately from part (b). ∎

Definition 1.93. *If E is an ideal in $L_0(\pi)$, then the π-a.e. unique measurable subset C_E determined in part* (a) *of Theorem* 1.92 *is called the* ***carrier*** *(or the* ***support****) of the ideal E.*

It should be noticed that the order dense ideals in $L_0(\pi)$ are precisely the ideals whose carriers coincide (π-a.e.) with Ω.

Now let E be an ideal in $L_0(\pi)$. Since the measurable sets of the measure space (C_E, Σ_{C_E}, μ), where $\Sigma_{C_E} = \{A \cap C_E\colon\ A \in \Sigma\}$, are precisely the restrictions of the measurable subsets of Ω to C_E (see [**32**, Problem 15.7, p. 125]), it follows that the band generated by E in $L_0(\pi)$ can be identified with $L_0(C_E, \pi) = L_0(C_E)$. In other words, we have the following result.

Lemma 1.94. *If E is an ideal in $L_0(\pi)$ and C_E is its carrier, then E is order dense in $L_0(C_E)$.*

We proceed with a brief discussion of Carathéodory's classical extension of measures defined on semirings. Recall that a collection Γ of subsets of a set X is said to be a **semiring** if:

(1) $\emptyset \in \Gamma$.

(2) $A, B \in \Gamma$ imply $A \cap B \in \Gamma$.

(3) If $A, B \in \Gamma$, then $A \setminus B$ can be written as a finite disjoint union of sets from Γ.

A (set) function $m \colon \Gamma \to [0, \infty]$ defined on a semiring Γ is called a **measure** if $m(\emptyset) = 0$ and m is σ-additive. As usual, the latter property means that if $A \in \Gamma$ is a disjoint union of a sequence $\{A_n\}$ in Γ, i.e., if $A = \bigcup_{n=1}^{\infty} A_n$ and $A_n \cap A_k = \emptyset$ for $n \neq k$, then $m(A) = \sum_{n=1}^{\infty} m(A_n)$. The **Carathéodory extension** of m (or the **outer measure generated** by m) is the outer measure $m^* \colon 2^X \to [0, \infty]$ defined by

$$m^*(A) = \inf\Big\{\sum_{n=1}^{\infty} m(A_n)\colon \ \{A_n\} \subseteq \Gamma \ \text{ and } \ A \subseteq \bigcup_{n=1}^{\infty} A_n\Big\},$$

with the convention that $\inf \emptyset = \infty$. Recall that a subset E of X is said to be m-measurable if

$$m^*(A) = m^*(A \cap E) + m^*(A \cap E^c)$$

for each $A \subseteq X$. It is well known that the collection

$$\Sigma_m = \big\{A \in 2^X \colon \ A \text{ is } m\text{-measurable}\big\}$$

is a σ-algebra. The relationships between this σ-algebra and the initial semiring Γ as well as between m^* and m are summarized in the next classical theorem due to C. Carathéodory; for details see [**31**, Chapter 3].

Theorem 1.95 (Carathéodory). *Let $m \colon \Gamma \to [0, \infty]$ be a measure on a semiring. Then the Carathéodory extension m^* of m restricted to Σ_m is a measure that extends m, i.e., $m^*(A) = m(A)$ for each $A \in \Gamma$ and every member of Γ is measurable, i.e., $\Gamma \subseteq \Sigma_m$. Moreover, we have the following three important properties.*

(a) *The measure $m^* \colon \Sigma_m \to [0, \infty]$ is complete, i.e., $m^*(A) = 0$ implies $A \in \Sigma_m$.*

(b) *If another semiring Γ_1 satisfies $\Gamma \subseteq \Gamma_1 \subseteq \Sigma_m$ and we consider m^* restricted to Γ_1, then the outer measure generated by the measure $m^* \colon \Gamma_1 \to [0, \infty]$ is precisely m^*; in particular, $(m^*)^* = m^*$.*

(c) *If m is σ-finite and Γ_1 is another semiring satisfying $\Gamma \subseteq \Gamma_1 \subseteq \Sigma_m$, then the restriction of m^* to Γ_1 is the only extension of m to a measure on Γ_1.*

In order to avoid introducing unnecessary symbols, it is customary to denote (with some abuse of notation) the outer measure m^* on Σ_m by m again. Accordingly, the symbols $L_1(m)$ and $L_1(m^*)$ denote the same Riesz space $L_1(X, \Sigma_m, m^*)$. Similarly the symbols $L_0(m)$ and $L_0(m^*)$ denote the Riesz space $L_0(X, \Sigma_m, m^*)$.

The next result presents a very useful characterization of non-negative functions in terms of their integrals.

Lemma 1.96. *Let $m\colon \Gamma \to [0,\infty]$ be a measure on a semiring. Then for a function $f \in L_1(m)$ we have:*

(1) *$f \geq 0$ m-a.e. if and only if $\int_A f\,dm \geq 0$ holds for all $A \in \Gamma$ with $m(A) < \infty$.*

(2) *$f = 0$ m-a.e. if and only if $\int_A f\,dm = 0$ holds for all $A \in \Gamma$ with $m(A) < \infty$.*

In the study of integral operators, two basic results from the theory of iterated integrals will be needed. To state these results we need to recall the definition of the product measure of two measures. To do so, let $(\mathcal{S}, \Sigma_1, \mu)$ and $(\mathcal{T}, \Sigma_2, \nu)$ be two measure spaces.

Any subset of $\mathcal{S} \times \mathcal{T}$ of the form $A \times B$, where $A \in \Sigma_1$ and $B \in \Sigma_2$, is called a **rectangle**. The collection of all rectangles is denoted by $\Sigma_1 \times \Sigma_2$. This is a semiring and the familiar "length $\times$ width" formula

$$(\mu \times \nu)(A \times B) = \mu(A)\nu(B)$$

defines a measure on $\Sigma_1 \times \Sigma_2$. This measure is called the **product measure** of μ and ν. The Carathéodory extension procedure extends the measure $\mu \times \nu$ from $\Sigma_1 \times \Sigma_2$ to the measure $(\mu \times \nu)^*$ on the σ-algebra $\Sigma_{\mu\times\nu}$ of all $\mu \times \nu$-measurable sets. In accordance with our agreement, we will often write $\mu \times \nu$ instead of $(\mu \times \nu)^*$.

The σ-algebra generated by $\Sigma_1 \times \Sigma_2$ is denoted by $\Sigma_1 \otimes \Sigma_2$ and is referred to as the **product σ-algebra**. We have the following obvious inclusions: $\Sigma_1 \times \Sigma_2 \subseteq \Sigma_1 \otimes \Sigma_2 \subseteq \Sigma_{\mu\times\nu}$. For a detailed discussion of product measures see [**31**, Section 26].

Consider now a function $f\colon \mathcal{S} \times \mathcal{T} \to \mathbb{R}$. We say that the **iterated integral** $\int_{\mathcal{T}} \int_{\mathcal{S}} f(s,t)\,d\mu(s)d\nu(t)$ exists if the following are true:

(1) *For ν-almost all $t \in \mathcal{T}$ the function $f(\cdot, t)$ is μ-integrable.*

(2) *The function $g(t) = \int_{\mathcal{S}} f(s,t)\,d\mu(s)$ is ν-integrable over $\mathcal{T}$.*

The value of the iterated integral $\int_{\mathcal{T}} \int_{\mathcal{S}} f(s,t)\,d\mu(s)d\nu(t)$ is computed by starting with the innermost integration and then continuing with the outside integration as follows:

$$\int_{\mathcal{T}} \int_{\mathcal{S}} f(s,t)\,d\mu(s)d\nu(s) = \int_{\mathcal{T}} \Big[\int_{\mathcal{S}} f(s,t)\,d\mu(s) \Big]\,d\nu(t)\,.$$

The meaning of the iterated integral $\int_{\mathcal{S}} \int_{\mathcal{T}} f(s,t)\,d\nu(t)d\mu(s)$ is analogous. That is,

$$\int_{\mathcal{S}} \int_{\mathcal{T}} f(s,t)\,d\nu(t)d\mu(s) = \int_{\mathcal{S}} \Big[\int_{\mathcal{T}} f(s,t)\,d\nu(t) \Big]\,d\mu(s)\,.$$

Recall that a function $f\colon \mathcal{S}\times\mathcal{T}\to\mathbb{R}$ is **measurable** (or, more precisely, **$\mu\times\nu$-measurable** if reference to the measure is necessary) if f is measurable with respect to the σ-algebra $\Sigma_{\mu\times\nu}$, that is, $f^{-1}(B)\in\Sigma_{\mu\times\nu}$ for each Borel subset B of $\mathbb{R}$. A function $f\colon \mathcal{S}\times\mathcal{T}\to\mathbb{R}$ is said to be **jointly measurable** if f is measurable with respect to the σ-algebra $\Sigma_1\otimes\Sigma_2$. That is, $f^{-1}(B)\in\Sigma_1\otimes\Sigma_2$ for each Borel subset B of $\mathbb{R}$.

The iterated integrals of $\mu\times\nu$-measurable functions are associated with a very important classical result known as Fubini's theorem. For a proof see [**31**, p. 167].

Theorem 1.97 (Fubini). *If $f\colon \mathcal{S}\times\mathcal{T}\to\mathbb{R}$ is a $\mu\times\nu$-integrable function, then both iterated integrals exist and*

$$\int_{\mathcal{S}\times\mathcal{T}} f\,d(\mu\times\nu)=\int_{\mathcal{T}}\int_{\mathcal{S}} f(s,t)\,d\mu(s)d\nu(t)=\int_{\mathcal{S}}\int_{\mathcal{T}} f(s,t)\,d\nu(t)d\mu(s)\,.$$

A very useful converse to Fubini's theorem is known as Tonelli's theorem; see [**31**, p. 168] for a proof.

Theorem 1.98 (Tonelli). *Let $(\mathcal{S},\Sigma_1,\mu)$ and $(\mathcal{T},\Sigma_2,\nu)$ be two σ-finite measure spaces, and let $f\colon \mathcal{S}\times\mathcal{T}\to\mathbb{R}$ be a $\mu\times\nu$-measurable function. If one of the iterated integrals*

$$\int_{\mathcal{S}}\int_{\mathcal{T}} |f(s,t)|\,d\mu(s)d\nu(t)\quad\text{or}\quad\int_{\mathcal{T}}\int_{\mathcal{S}} |f(s,t)|\,d\nu(t)d\mu(s)$$

exists, then the function f is $\mu\times\nu$-integrable—in which case the other iterated integral exists as well and

$$\int_{\mathcal{S}\times\mathcal{T}} f\,d(\mu\times\nu)=\int_{\mathcal{T}}\int_{\mathcal{S}} f(s,t)\,d\mu(s)d\nu(t)=\int_{\mathcal{S}}\int_{\mathcal{T}} f(s,t)\,d\nu(t)d\mu(s)\,.$$

The next result presents some basic measurability properties of $\mu\times\nu$-measurable functions.

Lemma 1.99. *Assume that $(\mathcal{S},\Sigma_1,\mu)$ and $(\mathcal{T},\Sigma_2,\nu)$ are two σ-finite measure spaces and let $f\colon \mathcal{S}\times\mathcal{T}\to\mathbb{R}$ be a $\mu\times\nu$-measurable function. Then:*

(1) *For ν-almost all $t\in\mathcal{T}$ the function $f(\cdot,t)$ is μ-measurable.*

(2) *For μ-almost all $s\in\mathcal{S}$ the function $f(s,\cdot)$ is ν-measurable.*

If $f\colon \mathcal{S}\times\mathcal{T}\to\mathbb{R}$ is jointly measurable, then $f(\cdot,t)$ is Σ_1-measurable for each $t\in\mathcal{T}$ *and $f(s,\cdot)$ is ν-measurable for* each $s\in\mathcal{S}$.

Proof. We establish the validity of (1) only. Since $J=L_1(\mu\times\nu)$ is an order dense ideal in $L_0(\mu\times\nu)$, Lemma 1.90 is applicable to the function f and so there exists a sequence $\{f_n\}\subseteq J$ and a subset A of $\mathcal{S}\times\mathcal{T}$ such that $(\mu\times\nu)(A)=0$ and

$$f_n(s,t)\to f(s,t)\ \text{ for all }\ (s,t)\notin A\,. \tag{$\star$}$$

By Fubini's Theorem 1.97, it follows that for each n the function $f_n(\cdot,t)$ is μ-integrable for ν-almost all t. So, there exists a set $N_1 \in \Sigma_2$ such that $\nu(N_1)=0$ and $f_n(\cdot,t)$ is μ-measurable for all $t \notin N_1$.

Since $(\mu \times \nu)(A) = 0$, it follows (from Fubini's theorem again) that the sets $A_t = \{s \in \mathcal{S}\colon\ (s,t) \in A\}$ are μ-null sets for ν-almost all t. That is, there exists some $N_2 \in \Sigma_2$ such that $\nu(N_2) = 0$ and $\mu(A_t) = 0$ for each $t \notin N_2$. Then $N = N_1 \cup N_2 \in \Sigma_2$ and $\nu(N) = 0$.

Fix any $t \notin N$. For each $s \notin A_t$ we have $(s,t) \notin A$ and so from $(\star)$, we get $f_n(s,t) \to f(s,t)$, i.e, the sequence $\{f_n(\cdot,t)\}$ of μ-measurable functions satisfies $f_n(s,t) \to f(s,t)$ for each $s \notin A_t$. Thus, $f(\cdot,t) \in L_0(\mu)$ for each $t \notin N$, as desired. ■

For a proof of the next result see [**32**, Problem 20.16]. See also Exercise 16 at the end of this section.

Lemma 1.100. *If $(\mathcal{S},\Sigma_1,\mu)$ and $(\mathcal{T},\Sigma_2,\nu)$ are two arbitrary σ-finite measure spaces and $f\colon \mathcal{S}\times\mathcal{T} \to \mathbb{R}$ is a $\mu\times\nu$-measurable function, then there exists a $\Sigma_1\otimes\Sigma_2$-measurable function $g\colon \mathcal{S}\times\mathcal{T} \to \mathbb{R}$ such that $f = g$ $\mu\times\nu$-a.e.*

A real number M is said to be an **essential upper bound** of a function $f\colon \Omega \to [-\infty,\infty]$ if $f(\omega) \le M$ holds for π-almost all $\omega \in \Omega$. The **essential sup** of f (denoted $\operatorname{ess\,sup} f$ or $\operatorname{ess\,sup}_{\omega\in\Omega} f(\omega)$) is defined by

$$\operatorname{ess\,sup} f = \inf\bigl\{M \in \mathbb{R}\colon\ M \text{ is an essential bound of } f\bigr\}.$$

If f does not have any essential upper bounds, then—as usual—we let $\operatorname{ess\,sup} f = \inf\emptyset = \infty$. The function f is **essentially bounded above** (or simply **bounded above**) if it has an essential upper bound (or, equivalently, if $\operatorname{ess\,sup} f < \infty$). The essential infimum of f (denoted $\operatorname{ess\,inf} f$ or $\operatorname{ess\,inf}_{\omega\in\Omega} f(\omega)$) is defined analogously. We have $\operatorname{ess\,inf} f = -\operatorname{ess\,sup}(-f)$.

The **essential sup norm** (or simply the **sup norm**) of a function $f\colon \Omega \to [-\infty,\infty]$ is denoted by $\|f\|_\infty$ and is the essential supremum of the function $|f|$. That is,

$$\|f\|_\infty = \operatorname{ess\,sup}|f| = \inf\bigl\{M \ge 0\colon\ M \text{ is an essential bound of } |f|\bigr\}.$$

A function f is said to be **essentially bounded** (or simply **bounded**) if $|f|$ has an essential upper bound or, equivalently, if $\|f\|_\infty < \infty$. It is easy to verify that the sup norm satisfies the three axioms of a norm. That is, for arbitrary functions $f,g\colon \Omega \to [-\infty,\infty]$, we have:

(1) $\|f\|_\infty \ge 0$ and $\|f\|_\infty = 0$ if and only if $f = 0$ π-a.e.

(2) $\|\alpha f\|_\infty = |\alpha|\cdot\|f\|_\infty$ for each scalar α.

(3) $\|f+g\|_\infty \le \|f\|_\infty + \|g\|_\infty$.

The collection of all essentially bounded functions in $L_0(\pi)$ is denoted $L_\infty(\Omega, \Sigma, \pi)$ or simply $L_\infty(\pi)$. The Riesz space $L_\infty(\pi)$ equipped with the norm $\|\cdot\|_\infty$ is a Dedekind complete Banach lattice.

Lemma 1.101. *The essential sup norm satisfies the following properties.*

(1) *If $0 \le f_n \uparrow f$ in $L_0(\pi)$, then $\|f_n\|_\infty \uparrow \|f\|_\infty$.*

(2) *If $\mathcal{G}$ is a non-empty order bounded from above subset of $L_\infty^+(\pi)$, then we have $\big\|\sup_{g\in\mathcal{G}} g\big\|_\infty = \sup_{g\in\mathcal{G}} \|g\|_\infty$ or, equivalently,*

$$\operatorname*{ess\,sup}_{\omega\in\Omega} \Big[\sup_{g\in\mathcal{G}} g\Big](\omega) = \sup_{g\in\mathcal{G}} \Big[\operatorname*{ess\,sup}_{\omega\in\Omega} g(\omega)\Big] .$$

Proof. (1) Let $\|f_n\|_\infty \uparrow \ell$. Clearly, $\ell \le \|f\|_\infty$. If $f = 0$ π-a.e., then the conclusion is trivial. So, assume that $f > 0$, and let $\delta < \|f\|_\infty$. Obviously, the measurable set $D = \{\omega \in \Omega\colon\ f(\omega) > \delta\}$ has positive measure. If $\|f_n\|_\infty \le \delta$ holds for each n, then it is easy to see that $f_n(\omega) \le \delta$ for π-almost all $\omega \in D$. This shows that $f_n(\omega) \not\uparrow f(\omega)$ for π-almost all $\omega \in D$, which is a contradiction. Hence, $\delta \le \|f_n\|_\infty \le \ell$ must hold for some n. Since $\delta < \|f\|_\infty$ is arbitrary, the latter implies that $\ell \ge \|f\|_\infty$. Consequently, $\ell = \|f\|_\infty$, and hence $\|f_n\|_\infty \uparrow \|f\|_\infty$.

(2) Let $f = \sup_{g\in\mathcal{G}} f$ in $L_\infty(\mu)$. From $0 \le g \le f$, we get $\|g\|_\infty \le \|f\|_\infty$ for each $g \in \mathcal{G}$. Consequently, $\sup_{g\in\mathcal{G}} \|g\|_\infty \le \|f\|_\infty$. If $\|f\|_\infty = 0$, then the conclusion is trivial. So, assume that $\|f\|_\infty > 0$.

Fix $0 < \delta < \|f\|_\infty$. Then the measurable set $D = \{\omega \in \Omega\colon f(\omega) > \delta\}$ must have positive measure. If $\sup_{g\in\mathcal{G}} \|g\|_\infty \le \delta$, then let $h = f\chi_{D^c} + \delta\chi_D$, and note that $g \le h < f$ holds for each $g \in \mathcal{G}$, contradicting the fact that $f = \sup_{g\in\mathcal{G}} g$. Hence, $\sup_{g\in\mathcal{G}} \|g\|_\infty > \delta$ for each $0 < \delta < \|f\|_\infty$. This shows that $\sup_{g\in\mathcal{G}} \|g\|_\infty \ge \|f\|_\infty$, and so $\sup_{g\in\mathcal{G}} \|g\|_\infty = \|f\|_\infty$. ■

A normed Riesz space E is said to satisfy the **Fatou property** whenever $0 \le x_\alpha \uparrow x$ implies $\|x_\alpha\| \uparrow \|x\|$. When E has the countable sup property, the Fatou property is equivalent to the **σ-Fatou property**: if $0 \le x_n \uparrow x$, then $\|x_n\| \uparrow \|x\|$. Lemma 1.101 shows that $L_\infty(\pi)$ has the Fatou property. In general, every AM-space with unit (as defined after Definition 3.1) satisfies the Fatou property.

Exercises

1. Let E be a Riesz space with the countable sup property, and let D be a subset of E such that the supremum $u = \sup D$ exists. Does there exist a countable subset D_1 of D such that for each $x \in D$ there is an element $a \in D_1$ such that $a \geq x$? (Any subset D_1 of D with this property is called **cofinal** in D.)

2. Let E be an Archimedean Riesz space satisfying the countable sup property and having a weak unit. Show that for each non-empty subset D of E there exists a countable subset D_0 of D such that D_0 has the same set of upper bounds as D. That is, whenever $u \in E$ satisfies $u \geq x$ for each $x \in D_0$, then $u \geq x$ for each $x \in D$.

3. Let (Ω, Σ, π) be a measure space, and let $\widehat{\Sigma}$ be the collection of all subsets of Ω that can be written in the form $A \Delta M$, where $A \in \Sigma$ and $M \subseteq N$ for some $N \in \Sigma$ with $\pi(N) = 0$. (As always, $A \Delta M = (A \setminus M) \cup (M \setminus A)$ is the symmetric difference of the sets A and M.) Verify that:
   (a) $\widehat{\Sigma}$ is a σ-algebra and $\widehat{\Sigma} \subseteq \Sigma_\pi$.
   (b) The function $\widehat{\pi}\colon \widehat{\Sigma} \to [0, \infty]$, defined by $\widehat{\pi}(A \Delta M) = \pi(A)$, is well defined and extends π from Σ to a measure on $\widehat{\Sigma}$.
   (c) If some $A \in \Sigma_\pi$ satisfies $\pi^*(A) < \infty$, then $A \in \widehat{\Sigma}$. In particular, $\widehat{\pi}$ is a **complete measure**, that is, if some subset A of Ω satisfies $\pi^*(A) = 0$, then $A \in \widehat{\Sigma}$.
   (d) If π is σ-finite, then $\widehat{\Sigma} = \Sigma_\pi$.

4. Let J be an ideal in some $L_0(\pi)$-space and let $\{x_n\}$ be an order bounded sequence in J. Show that $x_n \xrightarrow{o} x$ in J if and only if $x_n(\omega) \to u(\omega)$ for π-almost all $\omega \in \Omega$. [HINT: Assume $|x_n| \leq u$ for each n and some $u \in J$ and $x_n \xrightarrow{o} x$. Let $y_n = \bigvee_{i=n}^\infty x_i$, $z_n = \bigwedge_{i=1}^\infty x_i$, and $v_n = y_n - z_n$. Then $v_n \downarrow 0$ holds in J and $|x_n - x| \leq v_n$ for each n.]

5. Prove Theorem 1.80. [HINT: Choose a sequence $\{\Omega_n\} \subseteq \Sigma$ such that $0 < \pi(\Omega_n) < \infty$ for each n and $\Omega = \bigcup_{n=1}^\infty \Omega_n$, and define the function $\rho\colon L_0^+(\pi) \to [0, \infty)$ by
$$\rho(x) = \sum_{n=1}^\infty \tfrac{1}{2^n} \cdot \tfrac{1}{\pi(\Omega_n)} \int_{\Omega_n} \tfrac{x}{1+x}\, d\pi\,.$$
Now let D be a bounded from above non-empty subset of $L_0^+(\pi)$. If $y \in L_0^+(\pi)$ satisfies $\rho(y) = \sup_{x \in D} \rho(x)$, then show that $y = \sup D$.]

6. Recall that a Riesz subspace G of a Riesz space L is said to be **order dense** if for each $0 < x \in L$ there exists some $0 < y \in G$ satisfying $0 < y \leq x$.

   Show that a Riesz subspace G of an Archimedean Riesz space L is order dense if and only if for each $x \in L^+$ we have $\{y \in G^+ : y \leq x\} \uparrow x$ in L. [HINT: Let G be an order dense Riesz subspace of an Archimedean Riesz space L, and let $x \in L^+$. Now assume by way of contradiction that there exists some $z < x$ such that $y \in G$ and $y \leq x$ implies $y \leq z$. Put $u = x - z > 0$, and then select some $v \in G$ with $0 < v \leq u < x$. This implies $v \le z$ and so $2v = v + v \le u + z = x$. By induction we get $nv \le x$ for each n. Since L is Archimedean, the latter yields $v = 0$, which is a contradiction.]

7. Consider the following property (P) for a subset D of a Riesz space:
(P) *If* $\{x_n\} \subseteq D$ *and* $\lambda_n \downarrow 0$ *in* $\mathbb{R}$, *then* $\lambda_n x_n \xrightarrow{o} 0$.

Assume that a subset D of a Riesz space E satisfies the above property (P). Prove that the set consisting of all finite suprema of the collection $|D| = \{|d| \colon d \in D\}$ also satisfies (P).

8. For a sequence $\{x_n\}$ in some $L_0(\pi)$-space establish the following.
(a) If $x_n \xrightarrow{\pi} x$ and $x_n \xrightarrow{\pi} y$ in $L_0(\pi)$, then $x = y$.
(b) If $x_n \xrightarrow{\pi} x$ in $L_0(\pi)$ and $\{x_n\}$ is a positive sequence, then $x \ge 0$.

9. We say that a sequence $\{x_n\}$ in some $L_0(\pi)$ is ***$\star$-convergent*** *to* x *(in symbols,* $x_n \xrightarrow{\star} x$*) if every subsequence of* $\{x_n\}$ *has a subsequence that converges pointwise* π*-almost everywhere to* x.

Let (Ω, Σ, π) be a σ-finite measure space and let A be a measurable set of finite measure. Also let $\phi \colon A \to \mathbb{R}$ be a π-integrable function such that $\phi(\omega) > 0$ for π-almost all $\omega \in A$. Show that for each sequence $\{B_n\}$ of measurable subsets of A the following statements are equivalent:
(a) $\chi_{B_n} \xrightarrow{\pi} 0$.
(b) $\chi_{B_n} \xrightarrow{\star} 0$.
(c) $\int_{B_n} \phi \, d\pi \longrightarrow 0$.
(d) $\pi(B_n) \longrightarrow 0$.

10. Assume that (Ω, Σ, π) is a σ-finite measure space and that a sequence $\{\Omega_n\} \subseteq \Sigma$ satisfies $\pi(\Omega_n) < \infty$ for each n and $\bigcup_{n=1}^{\infty} \Omega_n = \Omega$. Suppose also that $\phi \colon [0, \infty) \to [0, \infty)$ is a continuous strictly increasing bounded function with $\phi(0) = 0$; for instance, let $\phi(t) = \frac{t}{1+t}$ or $\phi(t) = 1 - e^{-t}$.

Show that a sequence $\{x_n\} \subseteq L_0(\pi)$ satisfies $x_n \xrightarrow{\pi} 0$ if and only if $\lim_{n\to\infty} \int_{\Omega_k} \phi(|x_n|) \, d\pi = 0$ for each k.

11. Let (Ω, Σ, π) be a σ-finite measure. Fix a sequence $\{\Omega_k\}$ in Ω such that $0 < \pi(\Omega_k) < \infty$ for each k and $\Omega = \bigcup_{k=1}^{\infty} \Omega_k$. Also, define the function $\rho \colon L_0(\pi) \to [0, \infty)$ by

$$\rho(x) = \sum_{k=1}^{\infty} \tfrac{1}{2^k} \cdot \tfrac{1}{\pi(\Omega_k)} \int_{\Omega_k} \tfrac{|x|}{1+|x|} \, d\pi \, .$$

Show that for a sequence $\{x_n\} \subseteq L_0(\pi)$ and a function $x \in L_0(\pi)$ the following statements are equivalent.
(a) $x_n \xrightarrow{\pi} x$.
(b) For each k we have $\lim_{n\to\infty} \int_{\Omega_k} \frac{|x_n - x|}{1+|x_n - x|} \, d\pi = 0$.
(c) $\lim_{n\to\infty} \rho(x_n - x) = 0$.

12. Let $(\mathcal{S}, \Sigma_1, \mu)$ and $(\mathcal{T}, \Sigma_2, \nu)$ be two measure spaces. If $f \colon \mathcal{S} \times \mathcal{T} \to \mathbb{R}$ is jointly measurable, then $f(\cdot, t)$ is Σ_1-measurable for *each* $t \in \mathcal{T}$ and $f(s, \cdot)$ is ν-measurable for *each* $s \in \mathcal{S}$.

13. Prove Lemma 1.90. [HINT: Note that $L_1(\pi)$ is an order dense ideal in $L_0(\pi)$ and that the intersection of two order dense ideals is an order dense ideal.]

14. Prove Lemma 1.96.

15. Let Γ be a semiring of subsets of set Ω and let $\theta, \xi \colon \Gamma \to [0, \infty)$ be two set functions such that $\theta(A) \leq \xi(A)$ holds for each $A \in \Gamma$. If θ is finitely additive and ξ is a measure (i.e., countably additive), then show that θ is also a measure.

16. Establish the following property of the product measure: If the measures μ and ν are σ-finite, then for each set $C \in \Sigma_{\mu\times\nu}$ and for each $\epsilon > 0$ there exists a set of the form $D = \bigcup_{n=1}^{k} A_n \times B_n$, where $A_n \in \Sigma_1$ and $B_n \in \Sigma_2$ for each n, such that $(\mu \times \nu)(C \Delta D) < \epsilon$.

17. If $f \in L_0(\pi)$, then show that

$$\max\{\operatorname{ess\,sup} f, \operatorname{ess\,sup}(-f)\} = \operatorname{ess\,sup}|f| = \|f\|_\infty \,.$$

18. Assume that (Ω, Σ, π) is a measure space and let $\{x_n\} \subseteq L_1^+(\pi)$. Also let $x = \sup_n x_n$ be the π-a.e. pointwise supremum of the sequence $\{x_n\}$. Show that $x \notin L_1(\pi)$ if and only if there exists a pairwise disjoint sequence $\{A_n\} \subseteq \Sigma$ such that $\sum_{n=1}^{\infty} \int_{A_n} x_n \, d\pi = \infty$.

Chapter 2

Basic Operator Theory

This chapter presents the general background needed to study (linear) operators. It gives a concise presentation of the basic structural properties of Banach spaces and Banach lattices and discusses the fundamental properties of operators between these spaces. We pay special attention to the structure of Banach lattices and operators acting on them. Banach lattices are introduced as special cases of ordered vector spaces. We take this opportunity and review the fundamental properties of compact and weakly compact operators between Banach spaces. We also state the basic compactness results on operators dominated by compact or weakly compact operators.

We assume the reader is familiar with a basic course in functional analysis that includes an elementary introduction to operator theory.

2.1. Bounded Below Operators

Recall that a mapping $f\colon A \to B$ is said to be **surjective** (or **onto**) whenever its **range** $R(f)$ coincides with B, i.e., whenever for each $b \in B$ there exists some $a \in A$ with $f(a) = b$. In this section, we shall discuss some important properties of surjective operators between normed spaces.

Definition 2.1. *An operator $T\colon X \to Y$ between normed spaces is said to be* ***bounded below*** *if there exists a constant $\gamma > 0$ such that*

$$\|Tx\| \geq \gamma \|x\|$$

for each $x \in X$. Equivalently, T is bounded below if there exists some $\gamma > 0$ satisfying $\|Tx\| \geq \gamma$ for all $x \in X$ with $\|x\| = 1$.

The negation of the preceding definition can be formulated in terms of sequences as follows.

Lemma 2.2. *An operator $T\colon X \to Y$ between two normed spaces fails to be bounded below if and only if there exists a sequence of unit vectors $\{x_n\}$ in X such that $\|Tx_n\| \to 0$.*

The proof of the next result is left for the reader.

Lemma 2.3. *An operator $T\colon X \to Y$ between two real normed spaces is bounded below if and only if $T_{\mathbf{c}}\colon X_{\mathbf{c}} \to Y_{\mathbf{c}}$ is likewise bounded below.*

In the next simple lemma we collect some properties of bounded below operators which will be useful in our work.

Lemma 2.4. *Let X and Y be normed spaces and let $T \in \mathcal{L}(X,Y)$ be a bounded below operator. Then:*

(1) *T is one-to-one.*

(2) *If $S \in \mathcal{L}(X,Y)$ is close enough to T, then S is also bounded below, i.e., the set of all bounded below operators from X to Y is an open subset of $\mathcal{L}(X,Y)$.*

Proof. (1) Clearly $Tx = 0$ implies $x = 0$.

(2) Let $\|Tx\| \geq \gamma\|x\|$ for each $x \in X$ and suppose that another operator S in $\mathcal{L}(X,Y)$ satisfies $\|S - T\| < \frac{\gamma}{2}$. Now if x is a unit vector in X, then

$$\begin{aligned}\|Sx\| &= \big\|(S-T)x + Tx\big\| \geq \|Tx\| - \big\|(S-T)x\big\| \\ &\geq \gamma - \|S-T\| > \gamma - \tfrac{\gamma}{2} = \tfrac{\gamma}{2} > 0,\end{aligned}$$

which shows that S is also bounded below. ■

The bounded below operators between Banach spaces are characterized as follows.

Theorem 2.5. *A continuous operator $T\colon X \to Y$ between Banach spaces is bounded below if and only if T is one-to-one and has a closed range.*

Proof. Let $T\colon X \to Y$ be a continuous operator between Banach spaces. Assume first that $\|Tx\| \geq \gamma\|x\|$ holds for all $x \in X$ and some $\gamma > 0$. Clearly, T is one-to-one. Next, if $Tx_n \to y$ in X, then from

$$\|Tx_n - Tx_m\| = \|T(x_n - x_m)\| \geq \gamma\|x_n - x_m\|,$$

we see that $\{x_n\}$ is a norm Cauchy sequence. If $x_n \to x$, then $Tx_n \to Tx$, and so $y = Tx$, proving that T has a closed range.

For the converse, assume that T has a closed range and is one-to-one. By the Open Mapping Theorem, the operator $T\colon X \to R(T)$ has a continuous inverse. So, for some $c > 0$ we have $\|T^{-1}y\| \le c\|y\|$ for all $y \in R(T)$. Letting $y = Tx$, we get $\|Tx\| \ge \frac{1}{c}\|x\|$ for all $x \in X$. ■

Corollary 2.6. *If a bounded operator $T\colon X \to Y$ between Banach spaces has a closed range and a finite dimensional null space, then every operator $S \in \mathcal{L}(X,Y)$ close to T has also closed range and a finite dimensional null space such that $\dim N(S) \le \dim N(T)$.*

In particular, the collection of all bounded operators in $\mathcal{L}(X,Y)$ having a finite dimensional null space and a closed range is an open subset of $\mathcal{L}(X,Y)$.

Proof. Assume that a bounded operator $T\colon X \to Y$ between Banach spaces has a closed range and a finite dimensional null space. Since $N(T)$ is finite dimensional, it is complemented in X; see Exercise 6 of Section 4.1. So, there exists a closed subspace V of X such that $N(T) \oplus V = X$. In particular, $T\colon V \to R(T)$ is an isomorphism. Fix some $\epsilon > 0$ such that $2\epsilon\|x\| \le \|Tx\|$ for each $x \in V$. Now let $S \in \mathcal{L}(X,Y)$ be an operator satisfying $\|T-S\| < \epsilon$. Then for each $x \in V$ we have

$$\|Sx\| \ge \|Tx\| - \|Tx - Sx\| \ge 2\epsilon\|x\| - \epsilon\|x\| = \epsilon\|x\| .$$

This shows that the operator S is bounded below on V, and so its range $S(V)$ is a closed subspace of Y and S is one-to-one. Now note that $S(X) = S\big(N(T)\big) + S(V)$ and that $S\big(N(T)\big)$ is a finite dimensional vector subspace. Since $S(V)$ is closed, it follows that S has a closed range; see Exercise 1 in Section 3.4. Finally, a look at Exercise 6 at the end of the section shows that $\dim N(S) \le \dim N(T)$. ■

The notion of an isomorphism between Banach spaces is as follows.

Definition 2.7. *A bounded surjective one-to-one operator between two Banach spaces is called an* ***isomorphism****. Two Banach spaces are said to be* ***isomorphic*** *if there is an isomorphism from one space onto the other.*

It is interesting to know that two complex Banach spaces can be isomorphic as real Banach spaces but they may fail to be isomorphic as complex Banach spaces; see [**70, 169**].

Two characterizations of isomorphisms are given in the next result.

Lemma 2.8. *For a bounded operator $T\colon X \to Y$ between two Banach spaces the following conditions are equivalent.*

(1) *T is an isomorphism, i.e., T is surjective and one-to-one.*

(2) *T is bounded below and $R(T) = Y$.*

(3) *T is bounded below and its range $R(T)$ is dense in Y.*

Proof. The implications (1) $\iff$ (2) $\implies$ (3) are obvious, and (3) $\implies$ (2) follows from Theorem 2.5. ■

Now let $\{T_n\} \subseteq \mathcal{L}(X,Y)$ be a sequence of surjective operators between two Banach spaces, and assume that $\|T_n - T\| \to 0$ for some operator T in $\mathcal{L}(X,Y)$. It should be clear that T may easily fail to be onto. For example, if S is a surjective operator, then the sequence of surjective operators $\{\frac{1}{n}S\}$ converges to zero. However, as we shall show next, the situation is different if we know that T is bounded below. The next result is very important and it will be used quite often in this monograph—it can be found in [**170**] and [**19**].

Theorem 2.9. *Let $T\colon X \to Y$ be a bounded below operator between Banach spaces. If a sequence $\{T_n\} \subseteq \mathcal{L}(X,Y)$ of surjective operators converges in norm to T, then T is necessarily surjective (and hence an isomorphism).*

Proof. Fix some constant $\gamma > 0$ such that $\|Tx\| \geq 2\gamma\|x\|$ for each $x \in X$. Without loss of generality we can assume that $\|T - T_n\| \leq \gamma$ for each n. This easily implies $\|T_n x\| \geq \gamma\|x\|$ for each $x \in X$; see the proof of Lemma 2.4. Next, fix an arbitrary $y \in Y$. To establish that T is surjective, we need to show that there exists some $x \in X$ satisfying $Tx = y$.

Start by choosing a sequence $\{x_n\}$ in X satisfying $T_n x_n = y$ for each n. From $\gamma\|x_n\| \leq \|T_n x_n\| = \|y\|$, we see that $\|x_n\| \leq \frac{1}{\gamma}\|y\|$. We claim that $\{x_n\}$ is a Cauchy sequence. To see this, note that

$$\begin{aligned} 2\gamma\|x_n - x_m\| &\leq \|Tx_n - Tx_m\| \leq \|Tx_n - y\| + \|y - Tx_m\| \\ &= \|Tx_n - T_n x_n\| + \|T_m x_m - Tx_m\| \\ &\leq \tfrac{1}{\gamma}\|y\| \cdot \big[\, \|T - T_n\| + \|T - T_m\| \,\big] \xrightarrow[n,m\to\infty]{} 0 \,. \end{aligned}$$

Let $x_n \to x$ in X and so $Tx_n \to Tx$. On the other hand, from

$$\|Tx_n - y\| = \|Tx_n - T_n x_n\| \leq \tfrac{1}{\gamma}\|T - T_n\| \cdot \|y\| \to 0 \,,$$

it follows that $Tx_n \to y$. Thus, $Tx = \lim Tx_n = y$, as desired. ■

The next result (which can also be found in [**170**] and [**19**]) asserts that the set of all surjective operators is an open set. It should be noticed that its proof resembles the proof of the Open Mapping Theorem.

Theorem 2.10. *If X and Y are Banach spaces, then the set of all surjective operators from X to Y is an open subset of $\mathcal{L}(X,Y)$.*

Proof. Let $T \in \mathcal{L}(X,Y)$ be a surjective operator. By the Open Mapping Theorem we know that T is an open mapping. Therefore, there exists some $0 < c < 1$ such that $2cU_Y \subseteq T(U_X)$ or

$$cU_Y \subseteq T\big(\tfrac{1}{2}U_X\big) \subseteq T(U_X) \,. \qquad (\star)$$

To complete the proof, we shall show that every operator $S \in \mathcal{L}(X,Y)$ satisfying $\|T - S\| < c$ is surjective. To this end, let $S \in \mathcal{L}(X,Y)$ satisfy $\|T - S\| < c$.

Fix $y_0 \in cU_Y$. By $(\star)$, there exists some $x_0 \in U_X$ such that $Tx_0 = y_0$. Let $y_1 = Sx_0$. Then $\|y_0 - y_1\| = \|Tx_0 - Sx_0\| \leq \|T - S\| < c$. So, by $(\star)$ again, there exists some $x_1 \in \frac{1}{2}U_X$ such that $Tx_1 = y_0 - y_1$. Put $y_2 = Sx_1$ and note that $\|y_0 - y_1 - y_2\| = \|Tx_1 - Sx_1\| \leq \frac{c}{2}$.

Now continuing this way, by using $(\star)$ and an inductive argument, we can construct vectors $x_0, x_1, \ldots \in X$ and $y_0, y_1, \ldots \in Y$ such that

$$y_n = Sx_{n-1}, \quad Tx_n = y_0 - \sum_{i=1}^{n} y_i, \quad \|Tx_n\| \leq \tfrac{c}{2^{n-1}}, \quad \text{and} \quad \|x_n\| \leq \tfrac{1}{2^n}$$

for $n = 1, 2, \ldots$.

Since X is a Banach space, the series $\sum_{i=0}^{\infty} x_i$ is norm convergent in X, say $x = \sum_{i=0}^{\infty} x_i$. Moreover, from $\big\|y_0 - \sum_{i=1}^{n} y_i\big\| = \|Tx_n\| \to 0$, we see that

$$y_0 = \sum_{i=1}^{\infty} y_i = \sum_{i=1}^{\infty} S(x_{i-1}) = S\Big(\sum_{i=0}^{\infty} x_i\Big) = Sx\,.$$

This proves that S is a surjective operator. ■

Corollary 2.11. *The collection of all isomorphisms between two Banach spaces X and Y is an open (possibly empty) subset of $\mathcal{L}(X,Y)$.*

Proof. Let $T\colon X \to Y$ be an isomorphism. If an operator $S \in \mathcal{L}(X,Y)$ is close enough to T, then S is surjective by Theorem 2.10 and bounded below by Lemma 2.4(2). Consequently, S is an isomorphism by Lemma 2.8. ■

Corollary 2.12. *If X and Y are Banach spaces, then the set of all surjective and non-invertible operators from X to Y is an open subset of $\mathcal{L}(X,Y)$.*

Proof. Consider the collection of bounded operators

$$\mathcal{C} = \big\{T \in \mathcal{L}(X,Y)\colon\ T \text{ is surjective and not invertible}\big\}\,,$$

and fix some $T \in \mathcal{C}$, i.e., the operator $T \in \mathcal{L}(X,Y)$ is surjective and non-invertible. By Theorem 2.10, T is an interior point in the set of all surjective operators from X to Y. Now suppose by way of contradiction that T is not an interior point of $\mathcal{C}$. This means that there exists a sequence $\{T_n\}$ of invertible operators from X to Y satisfying $T_n \to T$ in $\mathcal{L}(X,Y)$.

Since T is not invertible there exists a non-zero vector $x \in X$ such that $Tx = 0$. Let V be the (one-dimensional) vector subspace generated by x, and let W be a closed subspace of X such that $V \oplus W = X$; see Exercise 6 of Section 4.1. Clearly, the operator $T|_W \in \mathcal{L}(W,Y)$ is surjective and none of the operators $T_n|_W \in \mathcal{L}(W,Y)$ is surjective since each $T_n\colon X \to Y$ is injective. However, from $T_n|_W \to T|_W$ in $\mathcal{L}(W,Y)$ and Theorem 2.10, it follows that eventually the operators $T_n|_W$ must be surjective, which is a contradiction. This contradiction establishes that $\mathcal{C}$ is an open subset of $\mathcal{L}(X,Y)$. ■

If X is finite dimensional and $\dim Y \geq \dim X$, then $\mathcal{C} = \emptyset$. If X is finite dimensional and $\dim Y < \dim X$, then the collection $\mathcal{C}$ is non-empty and coincides with the set of all surjective operators from X onto Y.

We shall finish the section with a few results concerning the closedness of the range of a bounded operator.

Theorem 2.13. *If $T\colon X \to Y$ is a bounded operator between normed spaces, then $[R(T)]^\perp = N(T^*)$. In particular, if the operator T has closed range, then*

$$\dim\,[Y/R(T)]^* = \dim\,[R(T)]^\perp = \dim N(T^*)\,.$$

Proof. For the first part note that:

$$\begin{aligned} y^* \in [R(T)]^\perp &\iff y^*(Tx) = T^*y^*(x) = 0 \text{ for all } x \in X \\ &\iff T^*y^* = 0 \iff y^* \in N(T^*)\,. \end{aligned}$$

The second part follows from Theorem 1.12. ■

Although the range of a bounded operator between Banach spaces need not be closed, there is a way to define a "stronger" norm on the range so that it becomes a Banach space. The details are as follows.

Theorem 2.14. *Let $T\colon X \to Y$ be a bounded operator between Banach spaces, and let $|||\cdot|||$ denote the quotient norm of the quotient Banach space $X/N(T)$. If for each $y = Tx \in R(T)$, we let*

$$\|y\|_T = \|y\| + |||[x]|||\,,$$

then we have the following:

(a) *The function $\|\cdot\|_T$ defines a norm on the range $R(T)$ of T.*

(b) *$(R(T), \|\cdot\|_T)$ is a Banach space.*

(c) *The operator $T\colon (X, \|\cdot\|) \to (R(T), \|\cdot\|_T)$ is continuous.*

(d) *The operator T has closed range (i.e., $R(T)$ is $\|\cdot\|$-closed) if and only if the norms $\|\cdot\|$ and $\|\cdot\|_T$ are equivalent on $R(T)$.*

(e) *Every operator $S\colon R(T) \to R(T)$ which is $\|\cdot\|$-continuous is also $\|\cdot\|_T$-continuous. In particular, every operator $S \in \mathcal{L}(Y)$ that leaves $R(T)$ invariant is $\|\cdot\|_T$-continuous on $R(T)$.*

Proof. (a) For each $x \in X$ we have

$$|||[x]||| = \inf\{\|z\|: \ z \in X \text{ and } Tz = Tx\}.$$

So, if $Tu = Tv$, then $|||[u]||| = |||[v]|||$. This easily implies that $\|\cdot\|_T$ is a well-defined norm on $R(T)$.

(b) Let $\{y_n\}$ be a $\|\cdot\|_T$-Cauchy sequence in $R(T)$. Choose a sequence $\{x_n\}$ in X such that $y_n = Tx_n$. From

$$\|y_n - y_m\|_T = \|y_n - y_m\| + |||[x_n] - [x_m]|||,$$

it follows that $\{y_n\}$ is a $\|\cdot\|$-Cauchy sequence in Y and $\{[x_n]\}$ is a $|||\cdot|||$-Cauchy sequence in $X/N(T)$. Pick $y \in Y$ and $[x] \in X/N(T)$ such that

$$\|y_n - y\| \to 0 \quad \text{and} \quad |||[x_n] - [x]||| \to 0.$$

From the definition of the quotient norm, it follows that there exists a sequence $\{z_n\}$ in X such that $\|z_n\| \to 0$ and $Tz_n = Tx_n - Tx = y_n - Tx$. Since $\|y_n - y\| \to 0$ and $\|Tz_n\| \to 0$, we see that $y = Tx$. So, $y \in R(T)$ and

$$\|y_n - y\|_T = \|y_n - y\| + |||[x_n] - [x]||| \to 0.$$

This implies that $(R(T), \|\cdot\|_T)$ is a Banach space.

(c) This follows immediately from Corollary 1.6.

(d) The identity operator $I\colon (R(T), \|\cdot\|_T) \to (R(T), \|\cdot\|)$ is clearly continuous. If $R(T)$ is closed in X, then $(R(T), \|\cdot\|)$ is a Banach space. But then, according to the Open Mapping Theorem, I is a homeomorphism and this guarantees that the norms $\|\cdot\|_T$ and $\|\cdot\|$ are equivalent.

Conversely, if the norms $\|\cdot\|$ and $\|\cdot\|_T$ are equivalent, then, in view of (b), the range $R(T)$ is $\|\cdot\|$-complete and, hence, closed.

(e) Again, this follows immediately from Corollary 1.6. ■

The next consequence of the preceding theorem includes several elementary characterizations of the closedness of the range of an operator.

Corollary 2.15. *For a bounded operator $T\colon X \to Y$ between two Banach spaces the following statements are equivalent.*

1. *The range of T is closed.*
2. *There exists a constant $c > 0$ such that for each $y \in R(T)$ there exists some $x \in X$ satisfying $y = Tx$ and $\|x\| \le c\|y\|$.*
3. *For any sequence $\{y_n\}$ in $R(T)$ satisfying $\|y_n\| \to 0$, there exists a sequence $\{x_n\}$ in X such that $\|x_n\| \to 0$ and $y_n = Tx_n$ for each n.*

Proof. (1) $\Longrightarrow$ (2) Since $R(T)$ is closed, the norms $\|\cdot\|_T$ and $\|\cdot\|$ are equivalent on $R(T)$. That is, there exists some $\delta > 0$ such that $\|y\|_T \le \delta\|y\|$

holds for all $y \in R(T)$. In particular, if $z \in X$ satisfies $Tz = y$, then we have $\|[z]\| \leq \delta\|y\|$, where $[z] = z + N(T)$. Now notice that there exists some $x \in X$ with $y = Tx$ and $\|x\| \leq (1+\delta)\|y\|$.

(2) $\Longrightarrow$ (3) Obvious.

(3) $\Longrightarrow$ (1) Assume that a sequence $\{y_n\} \subseteq R(T)$ satisfies $\|y_n\| \to 0$. Pick a sequence $\{x_n\}$ in X satisfying $y_n = Tx_n$ for all n and $\|x_n\| \to 0$. From the inequality

$$\|y_n\|_T = \|y_n\| + \big\|[x_n]\big\| \leq \|y_n\| + \|x_n\|,$$

we get $\|y_n\|_T \to 0$. So, the identity operator $I\colon \big(R(T), \|\cdot\|\big) \to \big(R(T), \|\cdot\|_T\big)$ is continuous. By part (d) of Theorem 2.14, $R(T)$ is closed. ■

Another simple—but very useful—characterization of the closedness of the range of a bounded operator is the following.

Theorem 2.16. *A bounded operator $T\colon X \to Y$ between two Banach spaces has a closed range if and only if there exists a closed subspace Z of Y such that $R(T) \cap Z = \{0\}$ and the vector subspace $R(T) \oplus Z$ of Y is closed.*

Proof. If $R(T)$ is closed, then the closed subspace $Z = \{0\}$ of Y satisfies the stated properties. For the converse, assume that there exists a closed subspace Z of Y such that $R(T) \cap Z = \{0\}$ and the vector subspace $V = R(T) \oplus Z$ of Y is closed. Clearly, the vector space V/Z equipped with its quotient norm is a Banach space. Moreover, the mapping $J\colon R(T) \to V/Z$, defined by $J(y) = y + Z$, is a surjective linear isomorphism.

Since $\big\|J(y)\big\| \leq \|y\| \leq \|y\|_T$ holds for each $y \in R(T)$, it follows that the operator $J\colon \big(R(T), \|\cdot\|_T\big) \to \big(V/Z, \|\cdot\|\big)$ is continuous, and hence it is an isomorphism. In particular, there exists a constant $c > 0$ satisfying $c\|y\|_T \leq \big\|J(y)\big\|$ for all $y \in R(T)$. This implies

$$c\|y\|_T \leq \|y\| \leq \|y\|_T$$

for all $y \in Y$. That is, the norms $\|\cdot\|$ and $\|\cdot\|_T$ are equivalent on $R(T)$. Therefore $R(T)$ is a closed subspace of Y. ■

Corollary 2.17. *Let $T\colon X \to Y$ be a bounded operator between two Banach spaces. If the quotient vector space $Y/R(T)$ is finite dimensional, then the range of T is closed.*

Our final characterization of the closedness of the range of a bounded operator is due to S. Banach [**51**, pp. 149–150].

Theorem 2.18 (Banach)**.** *A bounded operator $T\colon X \to Y$ between Banach spaces has closed range if and only if its adjoint operator $T^*\colon Y^* \to X^*$ likewise has closed range.*

Proof. Assume first that $R(T)$ is closed. Since the operator $T\colon X \to R(T)$ is a surjective operator between Banach spaces, there exists (by virtue of the Open Mapping Theorem) some constant $c > 0$ such that $cU_R \subseteq T(U_X)$, where U_R denotes the open unit ball of $R(T)$.

Let $\{x_n^*\} \subseteq R(T)$ satisfy $x_n^* \to 0$. Pick a sequence $\{y_n^*\} \subseteq Y^*$ satisfying $x_n^* = T^*y_n^*$ for each n, and let $y_n^*|_R$ be the restriction of y_n^* to $R(T)$. From

$$\begin{aligned}\|x_n^*\| &= \sup_{x\in U_X} \big|x_n^*(x)\big| = \sup_{x\in U_X} \big|(T^*y_n^*)(x)\big| = \sup_{x\in U_X} \big|y_n^*(Tx)\big| \\ &\geq c \sup_{y\in U_R} \big|y_n^*(y)\big| = c\big\|y_n^*|_R\big\|\end{aligned}$$

and $\|x_n^*\| \to 0$, it follows that $\big\|y_n^*|_R\big\| \to 0$. Let z_n^* be a linear extension of $y_n^*|_R$ to all of Y such that $\|z_n^*\| = \big\|y_n^*|_R\big\|$. Then $T^*z_n^* = T^*y_n^*$ holds for all n and $\|z_n^*\| \to 0$. By Corollary 2.15(3), the range of the operator T^* is closed.

For the converse, assume that $R(T^*)$ is closed. Put $Z = \overline{R(T)}$ and consider the operator $S\colon X \to Z$ defined by $Sx = Tx$ for each $x \in X$. Let $y^* \in Z^*$ be an arbitrary linear extension of z^* to all of Y. Then we have $S^*z^*(x) = z^*(Tx) = y^*(Tx) = T^*y^*(x)$ for all $x \in X$, i.e., $S^*z^* = T^*y^*$. This implies that $R(S^*) = R(T^*)$. In other words, replacing T by S if necessary, we can suppose without loss of generality that the range of T is also dense. Under this extra assumption, we must show that $R(T) = Y$.

Since $\overline{R(T)} = Y$, it easily follows (see Exercise 10 at the end of the section) that T^* is one-to-one. Since $R(T^*)$ is assumed to be closed, by Theorem 2.5, the operator T^* is bounded below. That is, there exists some constant $c > 0$ satisfying $c\|y^*\| \leq \|T^*y^*\|$ for all $y^* \in Y^*$. Next, we claim that we have $\{y \in Y\colon\ \|y\| < c\} \subseteq \overline{T(U_X)}$. If this is done, then (as in the proof of the Open Mapping Theorem) we get $\{y \in Y\colon\ \|y\| < \frac{c}{2}\} \subseteq T(U_X)$, from which it follows that $R(T) = Y$.

To establish the inclusion $\{y \in Y\colon\ \|y\| < c\} \subseteq \overline{T(U_X)}$, pick any $y \in Y$ satisfying $\|y\| < c$ and assume by way of contradiction that $y \notin \overline{T(U_X)}$. Then, by the Separation Theorem, there exists some $y^* \in Y^*$ such that

$$y^*(y) > \sup\big\{|y^*(Tx)|\colon\ x \in U_X\big\} = \|T^*y^*\| \geq c\|y^*\|\,.$$

Since $y^*(y) \leq \|y^*\| \cdot \|y\|$, the preceding inequality implies $\|y\| > c$, a contradiction. So, $\{y \in Y\colon\ \|y\| < c\} \subseteq \overline{T(U_X)}$ is true, as desired. ∎

Exercises

1. For a bounded below continuous operator $T\colon X \to Y$ between normed spaces establish the following:

(a) If X is not a Banach space, then T need not have a closed range.

(b) The unique continuous linear extension of T to the norm completion of X is also bounded below.

[HINT: Let Y be an arbitrary Banach space, X a non-closed subspace of Y and define the operator $T\colon X \to Y$ by $Tx = x$ for each $x \in X$. Then $R(T) = X$ is not closed in Y.]

2. Prove Lemma 2.3.

3. Let $T\colon X \to Y$ be a bounded, one-to-one, and surjective operator between two Banach spaces. Show that T is an isometry if and only if T and T^{-1} are both contractions. (Recall that a bounded operator between two normed spaces is called a **contraction** if its norm is no more than one.)

4. Let $T \in \mathcal{L}(X)$ be bounded below but not invertible. Show that there exists some $\epsilon > 0$ such that the operator $\lambda - T$ is not surjective whenever $|\lambda| < \epsilon$. [HINT: Use Theorem 2.9.]

5. This is a generalization of the previous exercise. Assume that X and Y are Banach spaces and let $T \in \mathcal{L}(X,Y)$ be bounded below but not invertible. Then show that there exists some $\epsilon > 0$ such that the operator $T - S$ is not surjective whenever $S \in \mathcal{L}(X,Y)$ satisfies $\|S\| < \epsilon$. That is, show that the set of all bounded below and non-surjective operators from X to Y is an open subset of $\mathcal{L}(X,Y)$.

6. Establish the following property that was used in the proof of Corollary 2.6. *Assume that $X = V \oplus W$, where V is finite dimensional. If Y is an arbitrary vector space, then any operator $T\colon X \to Y$ that is one-to-one on W satisfies* $\dim N(T) \le \dim V$.

7. Show that the assumption $\dim N(T) < \infty$ in the last conclusion of Corollary 2.6 cannot be replaced by the assumption that $N(T)$ is complemented.

8. Prove Corollary 2.17. [HINT: Let $\{[y_1], [y_2], \ldots, [y_n]\}$ be a basis of $Y/R(T)$. If V is the vector space generated in Y by $\{y_1, y_2, \ldots, y_n\}$, then V is closed, $R(T) \cap V = \{0\}$, and $R(T) \oplus V = Y$. Now invoke Theorem 2.16.]

9. Let $T\colon X \to Y$ be a bounded operator between Banach spaces such that its range $R(T)$ is not closed. If $X = V \oplus W$, where V is closed and W is finite dimensional, then show that $T\colon V \to Y$ is not bounded below.

10. For a bounded operator $T\colon X \to Y$ between Banach spaces establish the following properties regarding the ranges of T and its adjoint T^*.

(a) The range $R(T)$ of the operator T is dense if and only if the adjoint operator $T^*\colon Y^* \to X^*$ (defined as usual via the duality identity $\langle Tx, y^* \rangle = \langle x, T^*y^* \rangle$) is one-to-one. [HINT: Use the Hahn–Banach theorem.]

(b) If the adjoint operator T^* is one-to-one and has closed range, then show that T is surjective. [HINT: Use the preceding part in connection with Theorem 2.18.]

(c) If T carries norm bounded closed sets onto closed sets, then T has a closed range. Moreover, all powers of T also have closed ranges. [HINT: We can assume that T is one-to-one. (Otherwise, we can replace X by $\hat{X} = X/\mathrm{Ker}\,T$ and T by $\hat{T}\colon \hat{X} \to Y$ defined by $\hat{T}[x] = Tx$.) Now assume $Tx_n \to y$. If $\{x_n\}$ has a norm bounded subsequence, say $\{z_n\}$, then y belongs to the closed set $T(\overline{\{z_1, z_2, \ldots\}})$. The case $\|x_n\| \to \infty$ cannot occur. For if $\|x_n\| \to \infty$, then $T(\frac{x_n}{\|x_n\|}) \to 0$, and consequently zero belongs to $T(\{x \in X\colon \|x\| = 1\})$, contrary to the fact that T is one-to-one.]

11. Let X and Y be two Banach spaces, and let $\mathcal{I}(X,Y)$ denote the open subset of $\mathcal{L}(X,Y)$ consisting of all isomorphisms from X to Y. Show that the mapping $T \mapsto T^{-1}$, from $\mathcal{I}(X,Y)$ to $\mathcal{I}(Y,X)$, is (norm) continuous. [HINT: If $S, T \in \mathcal{I}(X,Y)$, then note that

$$\begin{aligned} S^{-1} - T^{-1} &= T^{-1}(T-S)S^{-1} = \big[(T^{-1} - S^{-1}) + S^{-1}\big](T-S)S^{-1} \\ &= (T^{-1} - S^{-1})(T-S)S^{-1} + S^{-1}(T-S)S^{-1}\,.] \end{aligned}$$

12. For a bounded operator $T\colon X \to Y$ between Banach spaces establish the following.

(a) The operator T has a closed range if and only if $R(T^*) = N(T)^\perp$.

(b) If T has a closed range, then $X^*/R(T^*) = [N(T)]^*$.

2.2. The Ascent and Descent of an Operator

Recall that if $T\colon X \to Y$ is an operator between two vector spaces, then $N(T)$ and $R(T)$ denote the **null space** and the **range** of T, respectively. That is,

$$N(T) = \{x \in X\colon\ Tx = 0\} \quad \text{and} \quad R(T) = \{Tx\colon\ x \in X\}\,.$$

The null space of T is, of course, another name for the kernel of T.

In this section, we shall discuss the notions of *ascent* and *descent* of an operator on a vector space. So, fix an operator $T\colon X \to X$ on an arbitrary vector space. The null space of the operator T^k is a T-invariant subspace of X, that is, $T\big(N(T^k)\big) \subseteq N(T^k)$.[1] Indeed, if $x \in N(T^k)$, then $T^k x = 0$, and therefore $T^k(Tx) = T(T^k x) = 0$, i.e., $Tx \in N(T^k)$. We also have the following subspace inclusions:

$$N(T) \subseteq N(T^2) \subseteq N(T^3) \subseteq \cdots\ .$$

The range $R(T^k) = T^k(X)$ of each operator T^k is clearly another T-invariant subspace of X. Moreover:

$$R(T) \supseteq R(T^2) \supseteq R(T^3) \supseteq \cdots\ .$$

[1] If X is a normed space and T is continuous, then $N(T^k)$ is also a T-invariant closed subspace.

Lemma 2.19. *For an operator $T\colon X \to X$ on a vector space we have the following.*

(1) *If $N(T^k) = N(T^{k+1})$ holds for some k, then $N(T^n) = N(T^k)$ for all $n \geq k$.*

(2) *If $R(T^k) = R(T^{k+1})$ holds for some k, then $R(T^n) = R(T^k)$ for all $n \geq k$.*

Proof. We establish the validity of (1) only. To this end, assume that equality $N(T^k) = N(T^{k+1})$ holds for some k. The proof is by induction on n. For the induction step, suppose that $N(T^k) = N(T^{k+n})$ for some $n \geq 1$. If $x \in N(T^{k+n+1})$, then we have $T^{k+1}(T^n x) = T^{k+n+1}x = 0$ or, equivalently, $T^n x \in N(T^{k+1}) = N(T^k)$. This implies $T^{n+k}x = T^k(T^n x) = 0$ or $x \in N(T^{n+k})$. Consequently, $N(T^{n+k+1}) = N(T^{n+k}) = N(T^k)$. ∎

We now introduce the ascent and descent of an operator.

Definition 2.20. *Let $T\colon X \to X$ be an operator on a vector space.*

(1) *The **ascent** $\alpha(T)$ of T is the smallest natural number k such that $N(T^k) = N(T^{k+1})$. If there is no $k \in \mathbb{N}$ with $N(T^k) = N(T^{k+1})$, then we let $\alpha(T) = \infty$.*

(2) *The **descent** $\delta(T)$ of T is the smallest natural number k such that $R(T^k) = R(T^{k+1})$. If there is no $k \in \mathbb{N}$ with $R(T^k) = R(T^{k+1})$, then we let $\delta(T) = \infty$.*

It turns out that if the ascent and descent of an operator are finite, then they are equal. This useful result is stated next.

Lemma 2.21. *If an operator $T\colon X \to X$ on a vector space has finite ascent and descent, then they must coincide, i.e., $\alpha(T) = \delta(T) = p < \infty$, and*

$$X = N(T^p) \oplus R(T^p)\,.$$

Moreover, if X is a Banach space and $T \in \mathcal{L}(X)$, then $R(T^p)$ is a closed subspace.

Proof. Let $k = \alpha(T) < \infty$ and $m = \delta(T) < \infty$. We claim that

$$N(T^p) \cap R(T^k) = \{0\} \quad \text{for} \quad p = 1, 2, 3, \ldots \tag{$\star$}$$

and

$$X = N(T^m) + R(T^q) \quad \text{for} \quad q = 1, 2, 3, \ldots\ . \tag{$\star\star$}$$

To see ($\star$), fix $p \in \mathbb{N}$ and let $y \in N(T^p) \cap R(T^k)$. Then there exists some $x \in X$ such that $y = T^k x$. This implies $T^{k+p}x = T^p y = 0$ or, equivalently, $x \in N(T^{k+p}) = N(T^k)$. Hence, $y = T^k x = 0$, and the validity of ($\star$) has been established.

For $(\star\star)$ fix $q \in \mathbb{N}$ and let $x \in X$. Then $T^m x \in R(T^m) = R(T^{m+q})$, and so there exists some $z \in X$ such that $T^m x = T^{m+q} z$. This implies $T^m(x - T^q z) = 0$ or $x - T^q z \in N(T^m)$. Now notice that $x = (x - T^q z) + T^q z$ belongs to $N(T^m) + R(T^q)$.

Letting $p = m$ in $(\star)$ and $q = k$ in $(\star\star)$, we get

$$X = N(T^m) \oplus R(T^k)\,. \qquad (\star\star\star)$$

Fix $u \in N(T^{m+1})$. From $(\star\star\star)$ we can write $u = u_1 + u_2$ with the vectors $u_1 \in N(T^m) \subseteq N(T^{m+1})$ and $u_2 \in R(T^k)$. But then from $(\star)$, we get $u_2 = u - u_1 \in N(T^{m+1}) \cap R(T^k) = \{0\}$ or $u_2 = 0$. Hence, $u = u_1 \in N(T^m)$ and so $N(T^{m+1}) = N(T^m)$. Thus, $k \le m$.

Next, let $v \in R(T^k)$. Using $(\star\star)$ we can write $v = v_1 + v_2$ with v_1 in $N(T^m)$ and $v_2 \in R(T^{k+1}) \subseteq R(T^k)$. But then it follows from $(\star)$ that the vector $v_1 = v - v_2$ satisfies $v_1 \in R(T^k) \cap N(T^m) = \{0\}$ or $v_1 = 0$. Consequently, $v = v_2 \in R(T^{k+1})$ and so $R(T^k) = R(T^{k+1})$. This implies $m \le k$. Therefore, $m = k = p < \infty$. The closedness of $R(T^p)$ follows from Theorem 2.16. ■

Let $T\colon X \to X$ be a bounded operator on a Banach space. A pair (V, W) of closed subspaces of X is said to be a **reducing pair** for the operator T (or it **reduces** the operator T) if $X = V \oplus W$ and both V and W are T-invariant. We also say that a closed subspace V of X **reduces** T if there exists another closed subspace W of X such that the pair (V, W) reduces T.

Theorem 2.22. *A closed subspace V of a Banach space X reduces a bounded operator $T\colon X \to X$ if and only if there exists a projection $P \in \mathcal{L}(X)$ such that its range is V and P commutes with T, i.e., $TP = PT$.*

Proof. Let $T\colon X \to X$ be a bounded operator on a Banach space X. Assume first that (V, W) is a reducing pair for T, and let $P \in \mathcal{L}(X)$ be the continuous projection with range V along W. If $x = v + w \in V \oplus W = X$, then we have

$$PTx = P(Tv + Tw) = Tv = TPx\,,$$

and so $PT = TP$.

For the converse, assume that V is the range of a continuous projection P on X such that $PT = TP$. Let $W = (I - P)(X)$ and note that W is a closed subspace of X satisfying $V \oplus W = X$. Now if $v \in V$, then $Tv = TPv = P(Tv) \in V$, and so V is T-invariant. Finally, for each $x \in X$ we have

$$T(I - P)x = T(x - Px) = Tx - TPx = Tx - PTx = (I - P)Tx \in W\,.$$

This shows that W is T-invariant too. Therefore, V reduces T. ■

Recall that an operator $T\colon U \to U$ on a vector space is said to be **nilpotent** if $T^k = 0$ for some k. The smallest k satisfying $T^k = 0$ is called the **index of nilpotence** of the operator T.

Theorem 2.23. *A bounded operator $T\colon X \to X$ on a Banach space has finite ascent and descent if and only if T has a reducing pair of closed subspaces (V, W) such that the operator $T\colon V \to V$ is nilpotent and $T\colon W \to W$ is invertible.*

Moreover, if $p = \alpha(T) = \delta(T) < \infty$, then the pair (V, W), where $V = N(T^p)$ and $W = R(T^p)$, is the only reducing pair for the operator T such that T is nilpotent on V and invertible on W.

Proof. Let $T\colon X \to X$ be a bounded operator on a Banach space. If the ascent and descent of T are both finite, then (by Lemma 2.21) they must coincide. Consequently, if we let $\alpha(T) = \delta(T) = p < \infty$, then we have $X = N(T^p) \oplus R(T^p)$. Clearly, $\big(N(T^p), R(T^p)\big)$ is a reducing pair for T. Moreover, T is nilpotent on $N(T^p)$ and invertible on $R(T^p)$.

For the converse, assume that (V, W) is a reducing pair for T such that T is nilpotent on V and invertible on W. Let p be the smallest natural number such that $T^p = 0$ on V, the index of nilpotence of T. We claim that $p = \alpha(T)$.

To see this, let $x \in N(T^p)$. Since $X = V \oplus W$, there exist $v \in V$ and $w \in W$ such that $x = v + w$. From $0 = T^p x = T^p v + T^p w = T^p w$ and the fact that the invertibility of T on W implies the invertibility of T^p on W, we infer that $w = 0$. Thus, $x = v \in V$, and so $N(T^p) \subseteq V$. Since $T^p = 0$ on V, the inclusion $V \subseteq N(T^p)$ is trivially true, and hence $N(T^p) = V$. A similar argument shows that $N(T^{p+n}) = V$ for all $n \geq 0$. This implies $\alpha(T) \leq p < \infty$. Now assume that some $k \in \mathbb{N}$ satisfies $N(T^{k+1}) = N(T^k)$. Then $N(T^k) = N(T^n)$ for all $n \geq k$, and so $N(T^k) = V$. In particular, $T^k = 0$ on V, and therefore $k \geq p$. This establishes that $p = \alpha(T)$.

To finish the proof, we must show that $\delta(T) = p$ and that $W = R(T^p)$. Start by observing that if $x = v + w \in V \oplus W = X$, then the vector

$$T^p x = T^p v + T^p w = T^p w$$

belongs to W, and so $R(T^p) \subseteq W = T^p(W) \subseteq R(T^p)$. That is, $R(T^p) = W$. Similarly, we have $R(T^m) = W$ for all $m \geq p$. In particular, we have $\delta(T) \leq p < \infty$. Since $\delta(T)$ is finite, it follows that $\delta(T) = \alpha(T) = p$, and the proof is finished. ■

Exercises

1. Prove part (2) of Lemma 2.19.

2. Show that every operator on a finite dimensional vector space has finite ascent and descent.

3. Consider the operator $T\colon \mathbb{R}^3 \to \mathbb{R}^3$ defined by the matrix

$$T = \begin{bmatrix} 0 & 0 & 1 \\ 0 & 1 & 0 \\ 0 & 0 & 0 \end{bmatrix}.$$

Find the ascent and descent of T.

4. Give examples of bounded operators $T\colon X \to X$ on a Banach space such that:
(i) $\alpha(T) < \infty$ and $\delta(T) = \infty$.
(ii) $\alpha(T) = \infty$ and $\delta(T) < \infty$.
(iii) $\alpha(T) = \infty$ and $\delta(T) = \infty$.

5. Let $T\colon X \to X$ be an arbitrary bounded operator on a Banach space such that $p = \alpha(T) = \delta(T) < \infty$. Show that the adjoint operator $T^*\colon X^* \to X^*$ also has finite ascent and descent and that $\alpha(T^*) = \delta(T^*) = p$. [HINT: Let $V = N(T^p)$ and $W = R(T^p)$, and note $X = V \oplus W$. Then $X^* = V^* \oplus W^*$, and T^* is nilpotent on V^* and invertible on W^*.]

6. Show that a bounded operator $T\colon X \to X$ on a Banach space is nilpotent if and only if for each $x \in X$ there exists some $n \in \mathbb{N}$ such that $T^n x = 0$. [HINT: Note that $X = \bigcup_{n=1}^{\infty} N(T^n)$.]

7. If a nilpotent operator $T\colon V \to V$ on a vector space has index p, then show that $\alpha(T) = \delta(T) = p$.

8. For an operator $T\colon V \to V$ on a vector space establish the following algebraic properties.
(a) For each $n, m \in \mathbb{N}$ we have $T^n(N(T^{n+m})) \subseteq R(T^n) \cap N(T^m)$.
(b) For each $n, m \in \mathbb{N}$ the operator $T^n\colon N(T^{n+m}) \to R(T^n) \cap N(T^m)$ is surjective—and consequently the vector spaces $N(T^{n+m})/N(T^m)$ and $R(T^n) \cap N(T^m)$ are linearly isomorphic.
(c) If the ascent of T is finite, then $R(T^p) \cap N(T^n) = \{0\}$ for each $n \in \mathbb{N}$ and all $p \geq \alpha(T)$.
(d) If for some $p, n \in \mathbb{N}$ we have $R(T^p) \cap N(T^n) = \{0\}$, then T has finite ascent and $\alpha(T) \leq p$.

9. Let $T\colon V \to V$ be an operator on a vector space. Show that for each pair $n, m \in \mathbb{N}$ the mapping $J\colon V/[N(T^n) + R(T^m)] \to R(T^n)/R(T^{n+m})$, defined by

$$J[v] = T^n v + R(T^{n+m}),$$

is a well-defined surjective linear isomorphism.

2.3. Banach Lattices with Order Continuous Norms

An important connection between the order and topological structures of a Riesz space is provided by the notion of order continuity.

Definition 2.24. *A lattice seminorm p on a Riesz space is said to be* **order continuous** *if $x_\alpha \downarrow 0$ implies $p(x_\alpha) \downarrow 0$. A Banach lattice* **has order continuous norm** *if its norm is order continuous.*[2]

Banach lattices with order continuous norms play a very important role in analysis. Before stating their most useful characterizations, we need to recall a few basic concepts pertaining to Riesz spaces.

First of all, in a partially ordered vector space any subset of the form $[x, y] = \{z\colon\ x \leq z \leq y\}$, where $x \leq y$, is referred to as an **order interval**. A vector subspace A of a Riesz space E is said to be a **Riesz subspace** (or a **vector sublattice**) if for every $x, y \in A$ the supremum $x \vee y$ and the infimum $x \wedge y$ (taken in E) belong to A.[3] Every ideal is a Riesz subspace. If E is a normed Riesz space, then E (embedded naturally in its double dual E^{**}) is always a Riesz subspace of E^{**}. In other words, we have the following important result.

Theorem 2.25. *The natural embedding of a normed Riesz space into its double norm dual is a lattice isometry.*

We are now ready to state the most important characterizations of Banach lattices with order continuous norm; for details see [**30**, pp. 179–183]. For an extensive treatment of Banach lattices with order continuous norms we refer the reader to the well-written monograph of W. Wnuk [**338**].

Theorem 2.26. *For a Banach lattice E the following statements are equivalent.*

(1) *E has order continuous norm.*
(2) *Every continuous linear functional on E is order continuous.*
(3) *E is an ideal in E^{**}.*
(4) *The order intervals in E are weakly compact.*
(5) *Every order bounded disjoint sequence in E is norm convergent to zero.*

Since every ideal in a Dedekind complete Riesz space is a Dedekind complete Riesz space in its own right, the following result should be immediate.

---

[2] The "order continuity" concept is very general. A topology τ on a partially ordered set is said to be **order continuous** if $x_\alpha \xrightarrow{o} x$ implies $x_\alpha \xrightarrow{\tau} x$.

[3] It should be noted that if the supremum of every two vectors in an ordered vector space exists, then the infimum of every pair x and y also exists and $x \wedge y = -[(-x) \vee (-y)]$.

Corollary 2.27. *A Banach lattice with order continuous norm is Dedekind complete.*

An immediate consequence of Theorem 2.26 is that reflexive Banach lattices have order continuous norm. The class of reflexive Banach lattices is a proper subclass of the so-called KB-spaces. Recall that a Banach lattice E is said to be a KB-**space** (Kantorovich–Banach space) if every increasing norm bounded sequence is norm convergent. For details see [**30**, Section 14].

It is well known that for Banach spaces X and Y the mapping $T \mapsto T^*$ from $\mathcal{L}(X,Y)$ to $\mathcal{L}(Y^*,X^*)$ is a linear isometry. In the case of Banach lattices this mapping has additional properties related to the order structure. The following result (which is a special case of a more general theorem due to U. Krengel [**193**] and J. Synnatzschke [**313**]) is quite useful and describes these properties; for a proof see [**30**, p. 65].

Theorem 2.28 (Krengel–Synnatzschke). *If E and F are Banach lattices and F has order continuous norm, then the mapping $T \mapsto T^*$, from $\mathcal{L}_r(E,F)$ to $\mathcal{L}_r(F^*,E^*)$, is a lattice isometry, that is,*

$$|T^*| = |T|^* \quad \textit{and} \quad \|T^*\|_r = \|T\|_r$$

for each $T \in \mathcal{L}_r(E,F)$.

Proof. We shall sketch a proof when E is in addition Dedekind complete. So, assume that E is Dedekind complete, F has order continuous norm, and let $T \in \mathcal{L}_r(E,F)$.

From $\pm T \le |T|$, it follows that $\pm T^* \le |T|^*$, and so

$$|T^*| \le |T|^* .$$

To establish the reverse inequality $|T|^* \le |T^*|$ we must show that

$$\langle |T|^*x^*, x\rangle \le \langle |T^*|x^*, x\rangle$$

holds for all $0 \le x^* \in F^*$ and all $x \in E^+$.

To this end, fix $0 \le x^* \in F^*$ and $x \in E^+$. Since $Tx \in F \subseteq F^{**}$ and F is a Riesz subspace of F^{**}, it follows from Theorem 1.16 that for each $z \in E^+$ we have

$$\begin{aligned} \langle x^*, |Tz|\rangle &= \sup_{|y^*|\le x^*} \langle y^*, Tz\rangle = \sup_{|y^*|\le x^*} \langle T^*y^*, z\rangle \\ &\le \sup_{|y^*|\le x^*} \langle |T^*||y^*|, z\rangle = \langle |T^*|x^*, z\rangle . \end{aligned}$$

That is, for each $z \in E^+$ we have

$$\langle x^*, |Tz|\rangle \le \langle |T^*|x^*, z\rangle . \tag{$\star$}$$

Now let $\mathcal{C}$ be the collection of all vectors in F of the form $\sum_{i=1}^n |Tx_i|$, where $x_i \wedge x_j = 0$ for $i \neq j$ and $\sum_{i=1}^n x_i = x$. It is easy to see that the set is directed upward. Moreover, from $\sum_{i=1}^n |Tx_i| \leq \sum_{i=1}^n |T|x_i = |T|x$ and Theorem 1.17, it follows that $\mathcal{C} \uparrow |T|x$. The order continuity of the norm of F guarantees that

$$\{x^*(c)\colon\ c \in \mathcal{C}\} \uparrow \langle x^*, |T|x\rangle = \langle |T|^*, x\rangle\,. \tag{$\star\star$}$$

From ($\star$), it follows that for each $c = \sum_{i=1}^n |Tx_i| \in \mathcal{C}$ we have

$$\begin{aligned} x^*(c) &= \Big\langle x^*, \sum_{i=1}^n |Tx_i|\Big\rangle = \sum_{i=1}^n \langle x^*, |Tx_i|\rangle \\ &\leq \sum_{i=1}^n \langle |T^*|x^*, x_i\rangle = \Big\langle |T^*|x^*, \sum_{i=1}^n x_i\Big\rangle \\ &= \big\langle |T^*|x^*, x\big\rangle\,. \end{aligned}$$

From this and ($\star\star$), we get $\langle |T|^*x^*, x\rangle \leq \langle |T^*|x^*, x\rangle$.

For the norm identity note that

$$\|T^*\|_r = \||T^*|\| = \||T|^*\| = \||T|\| = \|T\|_r\,,$$

and the proof is complete. ■

We continue with the definition of an atom.

Definition 2.29. *A vector $u > 0$ in a Riesz space is said to be:*

1. *an **atom** whenever $0 \leq x \leq u$, $0 \leq y \leq u$ and $x \wedge y = 0$ imply that either $x = 0$ or $y = 0$ and*
2. *a **discrete vector** whenever for each $0 \leq x \leq u$ there exists some $\lambda \geq 0$ such that $x = \lambda u$, i.e., whenever $[0, u] = \{\lambda u\colon\ 0 \leq \lambda \leq 1\}$.*

For the proof of the next result see Exercise 2 at the end of this section.

Lemma 2.30. *In an Archimedean Riesz space, a positive vector is an atom if and only if it is a discrete vector. Moreover, in this case, the vector space $\{\lambda u\colon\ \lambda \in \mathbb{R}\}$ generated by an atom u is a projection band.*

A Riesz space is said to be **atomless** (or **non-atomic**) if it does not have any atoms. Our next lemma describes two duality properties of atomless Banach lattices.

Lemma 2.31. *For a Banach lattice E the following statements hold:*

1. *If E has order continuous norm and is atomless, then its norm dual E^* is likewise an atomless Banach lattice.*
2. *If E^* is atomless, then E is also atomless.*

Proof. (1) Let E be an atomless Banach lattice with order continuous norm. Assume, by way of contradiction, that E^* has an atom, say $0 < \phi \in E^*$. So, whenever some $\psi \in E^*$ satisfies $0 \leq \psi \leq \phi$, then there exists a constant $\lambda \geq 0$ such that $\psi = \lambda\phi$.

The order continuity of the norm of E implies that the linear functional ϕ is order continuous, and so its null ideal $N_\phi = \{x \in E\colon\ \phi(|x|) = 0\}$ is a band—and, of course, we have $E = N_\phi \oplus N_\phi^{\rm d} = N_\phi \oplus C_\phi$. The assumption $\phi > 0$ implies $C_\phi \neq \{0\}$ (and, moreover, $0 < x \in C_\phi$ implies $\phi(x) > 0$). Thus, there exists some $0 < u \in C_\phi$. We claim that u is an atom of E.

To see this, let $0 < x \leq u$ and $0 \leq y \leq u$ satisfy $x \wedge y = 0$. We shall show that $y = 0$. Let P_x denote the band projection of E onto the band B_x generated by x in E. Clearly, $\phi \circ P_x \in E^*$ and $0 \leq \phi \circ P_x \leq \phi$. So, there is some $\lambda \geq 0$ such that $\phi \circ P_x = \lambda\phi$. From $0 < \phi(x) = [\phi \circ P_x](x) = \lambda\phi(x)$, we infer that $\lambda = 1$ and so $\phi = \phi \circ P_x$. Therefore, $\phi(y) = [\phi \circ P_x](y) = \phi(0) = 0$, and by the strict positivity of ϕ on its carrier C_ϕ, we see that $y = 0$. Hence, u is an atom of E which contradicts our hypothesis. Consequently, E^* is an atomless Banach lattice.

(2) Now let E^* be atomless, and assume that some $0 < u \in E$ is an atom. Then the vector subspace $B = \{\lambda u\colon\ \lambda \in \mathbb{R}\} = \mathbb{R}u$ is a projection band of E; see Lemma 2.30. So, if $A = B^{\rm d}$, then $E = A \oplus B$ holds. (Both bands A and B are, of course, closed vector subspaces.) It easily follows that E is topologically and lattice isomorphic to the Banach lattice $A \oplus \mathbb{R}$, and so E^* is lattice isomorphic to $A^* \oplus \mathbb{R}$. Since $A^* \oplus \mathbb{R}$ has atoms, we see that E^* likewise has atoms, which is a contradiction. Hence, E must be an atomless Banach lattice. ■

Exercises

1. Let Ω be a compact Hausdorff space. Show that the Banach lattice $C(\Omega)$ has order continuous norm if and only if Ω is a finite set. [HINT: Fix any $\omega \in \Omega$ and consider the set

$$D_\omega = \{x \in C(\Omega)\colon\ 0 \leq x \leq \mathbf{1},\ x(\omega) = 1,\ \text{and } x = 0 \text{ outside of some open neighborhood of } \omega\}.$$

Now note that if ω is an accumulation point, then $D_\omega \downarrow 0$ holds in $C(\Omega)$.]

2. Prove Lemma 2.30. [HINT: Assume that $u > 0$ is an atom in an Archimedean Riesz space E and let $0 < x \leq u$. Put $\alpha = \sup\{\beta \geq 0\colon\ \beta x \leq u\}$. Since E is Archimedean, we see that $1 \leq \alpha < \infty$. Moreover, $\alpha x \leq u$, and we claim that $\alpha x = u$.

To see this, assume by way of contradiction that $v = u - \alpha x > 0$. Since $(v - \frac{1}{n}x)^+ \uparrow v$ (here we use again that E is Archimedean), it follows that $(v - \frac{1}{k}x)^+ > 0$ for some k with $\frac{1}{k} < \alpha$. Clearly,

$$(v - \tfrac{1}{k}x)^+ = [u - (\alpha + \tfrac{1}{k})x]^+ \leq u^+ = u \leq 2\alpha u\,.$$

On the other hand, $[u - (\alpha + \frac{1}{k})x]^- > 0$. (Otherwise, $[u - (\alpha + \frac{1}{k})x]^- = 0$ implies $(\alpha + \frac{1}{k})x \leq u$, contrary to the definition of α.) Next, observe that

$$[u - (\alpha + \tfrac{1}{k})x]^- = [(\alpha + \tfrac{1}{k})x - u]^+ \leq 2\alpha x \leq 2\alpha u\,.$$

So, if $y = \frac{1}{2\alpha}[u - (\alpha + \frac{1}{k})x]^+$ and $z = \frac{1}{2\alpha}[u - (\alpha + \frac{1}{k})x]^-$, then we have $0 < y \leq u$, $0 < z \leq u$ and $y \wedge z = 0$, contradicting the fact that u is an atom. Hence, $\alpha x = u$.]

3. Show that $C[0,1]$ and $L_1[0,1]$ are atomless Banach lattices.

4. This exercise establishes that in general the norm dual of an atomless Banach lattice need not be atomless; compare this with Lemma 2.31.

Show that the norm dual of the atomless Banach lattice $E = C[0,1]$ has atoms. [HINT: Every Dirac measure δ_x is an atom of E^*.]

5. If $g, h\colon E \to \mathbb{R}$ are two disjoint non-zero positive linear functionals on a Riesz space, then show that there exist disjoint vectors $u, v \in E^+$ such that $g(u) > 0$ and $h(v) > 0$.

6. Let E be a Riesz space and let $0 < f \in E^\sim$. Show that f is an atom in $E^\sim$ if and only if f is a lattice homomorphism, that is, if and only if $f(x \vee y) = \max\{\, f(x), f(y)\,\}$ for all $x, y \in E$.

7. Let $S, T\colon E \to F$ be two regular operators between Banach lattices with F Dedekind complete. Show that:

(a) $S \leq T$ in $\mathcal{L}_r(E,F)$ implies $S^* \leq T^*$ in $\mathcal{L}_r(F^*, E^*)$.

(b) $|T^*| \leq |T|^*$.

Also, give an example where $|T^*| < |T|^*$ holds true. (Compare this with Theorem 2.28.)

8. Show that a Banach lattice E is a KB-space if and only if E is a band in its double dual E^{**}.

9. (Dodds–Fremlin [**109**]) Let E and F be two Banach lattices with F having order continuous norm. If $x \in E^+$, then show that the set

$$\mathcal{B} = \bigl\{ T \in \mathcal{L}_r(E,F)\colon\ T[0,x] \text{ is norm totally bounded} \bigr\}$$

is a band in $\mathcal{L}_r(E,F)$.

2.4. Compact and Weakly Compact Positive Operators

We shall review here some basic properties of compact and weakly compact operators on Banach spaces and especially on Banach lattices. We start with the definition of compact and weakly compact operators.

Definition 2.32. *An operator $T\colon X \to Y$ between Banach spaces is called:*

(1) ***compact*** *if T carries norm bounded subsets of X to norm totally bounded subsets of Y and*

(2) ***weakly compact*** *if T carries norm bounded subsets of X to relatively weakly compact subsets of Y.*

It should be clear that every compact operator is weakly compact and that weakly compact operators are automatically continuous. It is well known that a bounded operator $T: X \to Y$ between Banach spaces is compact (resp. weakly compact) if and only if for every norm bounded sequence $\{x_n\}$ of X, the sequence $\{Tx_n\}$ has a norm convergent (resp. a weakly convergent) subsequence in Y.

We summarize below the basic properties of compact and weakly compact operators between Banach spaces. Proofs can be found in any book on functional analysis; see, for instance [**30**, Section 16].

- *Sums and scalar multiples of compact (resp. weakly compact) operators are compact (resp. weakly compact) operators.*
- *The composition of operators at least one of which is compact (resp. weakly compact) is a compact (resp. weakly compact) operator.*
- *The norm limit of a sequence of compact (resp. weakly compact) operators is a compact (resp. weakly compact) operator.*
- (Gantmacher [**132**]) *An operator is compact (resp. weakly compact) if and only if its norm adjoint is compact (resp. weakly compact).*

The collection of all compact (resp. weakly compact) operators from a Banach space X into another Banach space Y will be denoted by $\mathcal{K}(X,Y)$ (resp. $\mathcal{W}(X,Y)$). Clearly, $\mathcal{K}(X,Y)$ and $\mathcal{W}(X,Y)$ are both closed subspaces of $\mathcal{L}(X,Y)$.

Now we introduce the notion of domination between operators.

Definition 2.33. *Let $T: E \to F$ be a positive operator between Riesz spaces. We shall say that an operator $S: E \to F$ is* ***dominated*** *by T (or that T* ***dominates*** *S) whenever $|Sx| \le T(|x|)$ holds for each $x \in E$.*

The following simple facts are immediate:

(1) A positive operator T dominates S if and only if $-T \le S \le T$.

(2) If S is positive, then T dominates S if and only if $0 \le S \le T$.

(3) If F is Dedekind complete, then S is dominated by T if and only if $|S| \le T$.

There is an important collection of results regarding compactness properties of operators between Banach lattices that are dominated by compact or weakly compact operators. We list them below and for proofs and details we refer the reader to [**30**]. There is no parallel for these results on general Banach spaces.

Theorem 2.34 (Aliprantis–Burkinshaw [**27**]). *If in the scheme of continuous operators between Banach lattices* $E \xrightarrow{S_1} F \xrightarrow{S_2} G \xrightarrow{S_3} H$ *each operator* S_i *is dominated by a compact positive operator, then* $S_3S_2S_1$ *is a compact operator.*

Corollary 2.35 (Aliprantis–Burkinshaw [**27**]). *If an operator* $S\colon E \to E$ *on a Banach lattice is dominated by a compact positive operator, then* S^3 *is a compact operator—although* S^2 *need not be compact.*

Theorem 2.36 (Aliprantis–Burkinshaw [**27**]). *If in the scheme of continuous operators between Banach lattices* $E \xrightarrow{S_1} F \xrightarrow{S_2} G$ *each operator* S_i *is dominated by a compact positive operator, and either* E^* *or* G *has order continuous norm, then* S_2S_1 *is a compact operator.*

Corollary 2.37 (Aliprantis–Burkinshaw [**27**]). *Assume that* E *is a Banach lattice such that either* E *or* E^* *has order continuous norm. If an operator* $S\colon E \to E$ *is dominated by a compact positive operator, then its square* S^2 *is compact—although* S *need not be compact.*

Theorem 2.38 (Dodds–Fremlin [**109**]). *If* E *and* F *are Banach lattices such that* E^* *and* F *both have order continuous norms, then every operator* $S\colon E \to F$ *that is dominated by a compact positive operator is compact.*

Theorem 2.39 (Aliprantis–Burkinshaw [**28**]). *If in the scheme of continuous operators between Banach lattices* $E \xrightarrow{S_1} F \xrightarrow{S_2} G$ *each operator* S_i *is dominated by a weakly compact positive operator, then* S_2S_1 *is a weakly compact operator—although neither* S_1 *nor* S_2 *need be weakly compact.*

Theorem 2.40 (Wickstead [**335**]). *If* E *and* F *are Banach lattices such that either* E^* *or* F *has order continuous norm, then every positive operator* $S\colon E \to F$ *that is dominated by a weakly compact positive operator is weakly compact.*

Exercises

1. Show that a positive operator $T\colon E \to F$ between two Riesz spaces dominates an operator $S\colon E \to F$ if and only if $-T \leq S \leq T$.
2. Let $T\colon E \to F$ be a positive operator between two Banach lattices. Show that for an operator $S\colon E \to F$ the following statements are equivalent.
   (a) T dominates S.
   (b) T^* dominates S^*
   (c) T^{**} dominates S^{**}.
   [HINT: Use Exercise 1 above.]

3. Prove the following basic properties of compact operators.

(a) Sums and scalar multiples of compact operators are compact operators.

(b) The composition of two (or more) bounded operators at least one of which is compact is likewise a compact operator.

(c) Bounded finite-rank operators are compact. (Recall that an operator between two vector spaces is said to be **finite-rank** if its range is finite dimensional.)

(d) A bounded operator is compact if and only if its adjoint is likewise a compact operator.

(e) The norm limit of a sequence of compact operators is a compact operator.

[HINT for (d): Assume that $T\colon X \to Y$ is compact and consider Y as a closed vector subspace of $C(U_Y^*)$, where U_Y^* is the closed unit ball of Y^* equipped with its w^*-topology. Now fix $y^* \in U_Y^*$ and let $\epsilon > 0$. Since $T(U_X)$ is a norm totally bounded subset of Y, it follows from the classical Ascoli–Arzelà theorem that $T(U_X)$ is an equicontinuous bounded subset of $C(U_Y^*)$. Consequently, there exists a w^*-neighborhood V_{y^*} of y^* satisfying $|Tx(y^*) - Tx(y^*)| = |x(T^*y^* - T^*z^*)| < \frac{\epsilon}{2}$ for all $z^* \in V_{y^*}$ and all $x \in U_X$. Thus, $\|T^*y^* - T^*z^*\| = \sup_{x\in U_X} |x(T^*y^* - T^*z^*)| \le \frac{\epsilon}{2} < \epsilon$ for all z^* in V_{y^*}. If $y_1^*, \ldots, y_n^* \in U_Y^*$ are chosen so that $U_Y^* \subseteq \bigcup_{i=1}^n V_{y_i^*}$, then $T^*(U_Y^*) \subseteq \bigcup_{i=1}^n B(T^*y_i^*, \epsilon)$, proving that $T^*(U_Y^*)$ is a norm totally bounded subset of X^*.]

4. (Gantmacher [**132**]) For a bounded operator $T\colon X \to Y$ between Banach spaces show that the following statements are equivalent.

(a) T is weakly compact.

(b) $T^{**}(X^{**}) \subseteq Y$.

(c) $T^*\colon (Y^*, \sigma(Y^*, Y)) \to (X^*, \sigma(X^*, X^{**}))$ is continuous.

(d) $T^*\colon Y^* \to X^*$ is weakly compact.

5. Prove properties similar to the ones listed in Exercise 3 above for weakly compact operators.

6. Show that the range of an arbitrary compact operator is separable. [HINT: Let $T\colon X \to Y$ be a compact operator between Banach spaces. Then $T(U_X)$ is a norm totally bounded subset of Y and hence a separable metric space. If A is a norm dense countable subset of $T(U_X)$, then $B = \bigcup_{n=1}^\infty nA$ is a norm dense countable subset in $R(T)$.]

7. Show that a subset of a normed space is norm totally bounded if and only if it is contained in the closed convex hull of a sequence that converges to zero.

8. (Terzioglu [**318**]) Show that an operator $T\colon X \to Y$ between two Banach spaces is compact if and only if there exists a sequence $\{x_n^*\}$ in X^* satisfying $\|x_n^*\| \to 0$ and $\|Tx\| \le \sup_n |x_n^*(x)|$ for each $x \in X$. [HINT: Use the preceding exercise.]

9. Show that an operator $T\colon X \to Y$ between Banach spaces is compact if and only if T factors with compact factors through a closed subspace of c_0, i.e., if and only if there exists a closed vector subspace Z of c_0 and compact operators $X \xrightarrow{S} Z \xrightarrow{R} Y$ such that $T = RS$.

10. Let $\tau\colon \Omega \to \Omega$ be a continuous mapping on a compact Hausdorff space, and let $T\colon C(\Omega) \to C(\Omega)$ be the composition operator defined as usual by $Tf = f \circ \tau$. When is T a compact operator? [HINT: Use the Ascoli–Arzelà theorem.]

11. Let $T\colon \ell_1 \to Y$ be a bounded operator, where Y is a Banach space. Show that T is weakly compact if and only if the set $\{T(e_n)\colon\ n = 1, 2, \ldots\}$ is a relatively weakly compact subset of Y (where e_n denotes the sequence whose n^{th} entry is one and every other is zero).

12. For a Banach space X prove the following.

(a) An operator $T \in \mathcal{L}(X, \ell_1)$ is weakly compact if and only if T is compact.

(b) An operator $T \in \mathcal{L}(c_0, X)$ is weakly compact if and only if T is compact.

13. Let $\{y_1, y_2, \ldots\}$ be a countable relatively weakly compact subset of a Banach space Y; in particular, this is so if $\{y_n\}$ is a weakly convergent sequence in Y. Show that the operator $T\colon \ell_1 \to Y$, defined by

$$T(\lambda_1, \lambda_2, \ldots) = \sum_{n=1}^{\infty} \lambda_n y_n,$$

is weakly compact.

Chapter 3

Operators on AL- and AM-spaces

Banach lattices possess a property that has no parallel in the general theory of Banach spaces. It can be described by saying that: *Every Banach lattice is locally "classical."* Specifically, if E is a Banach lattice and x is any vector in E, then the principal ideal E_x generated by x, i.e., the vector subspace

$$E_x = \{y \in E\colon\ |y| \le \lambda|x| \text{ for some } \lambda \ge 0\},$$

can be identified with a $C(\Omega)$-space. This important and fundamental result—which is known as the Kakutani–Bohnenblust–Krein representation theorem—allows one to bring into the study of operators the structure of $C(\Omega)$-spaces. In particular, one can view the vectors of a Banach lattice as functions and introduce a "local" multiplication operation between the vectors of a Banach lattice. These multiplication operations define various classes of "local" multiplication operators that in many instances can be extended to operators on the whole space. This special feature of having many "local" multiplication operators on a Banach lattice is discussed in detail in this chapter.

Abstract Banach lattices that behave like the classical $L_1(\mu)$- and $C(\Omega)$-spaces are called AL- and AM-spaces, respectively. The purpose of this chapter is to introduce AL-spaces ("Abstract Lebesgue" spaces) and AM-spaces ("Abstract Maximum" spaces) and to study their structures as well as the operators acting between these spaces. During our investigation, we shall demonstrate that many important operators between Banach lattices behave "locally" as operators between $L_1(\mu)$- and $C(\Omega)$-spaces. In particular, we shall consider abstract multiplication operators and lattice homomorphisms.

As we shall see, these operators are in essence the "building blocks" for the general theory of positive operators.

3.1. AL- and AM-spaces

Two important classes of Banach lattices are provided by the AL- and AM-spaces. As mentioned before, these Banach lattices play a fundamental role not only in the study of positive operators but in functional analysis in general. They are defined as follows.

Definition 3.1. *A Banach lattice E is said to be:*

(1) *an **AL-space** if $\|x+y\| = \|x\|+\|y\|$ for all $x, y \in E^+$ with $x \wedge y = 0$ and*

(2) *an **AM-space** if $\|x \vee y\| = \max\{\|x\|, \|y\|\}$ for all $x, y \in E^+$ with $x \wedge y = 0$.*

Recall that in an ordered vector space X a vector $e \in X^+$ is said to be an **order unit** (or a **strong unit**, or simply a **unit**) if for each $x \in X$ there exists some $\lambda \geq 0$ such that $x \leq \lambda e$. Notice that if X is a Riesz space, then a vector $e \in X^+$ is an order unit if and only if for each $x \in X$ there exists some $\lambda \geq 0$ such that $|x| \leq \lambda e$. (See also Exercise 1 at the end of this section.)

A Banach lattice E is said to be an ***AM*-space with unit** if in addition to being an AM-space, E also has an order unit. If E is an AM-space with a unit e, then the formula

$$\|x\|_\infty = \inf\{\lambda > 0\colon\ |x| \leq \lambda e\} = \min\{\lambda \geq 0\colon\ |x| \leq \lambda e\} \qquad (\star)$$

defines an equivalent lattice norm on E. Under this new norm, E remains an AM-space and its closed unit ball coincides with the order interval $[-e, e]$; see Exercise 2 at the end of this section. Unless otherwise stated, we will always assume that an AM-space with unit is equipped with the norm given by $(\star)$ for some order unit e.

The norm of a positive operator acting on an AM-space with unit can be computed easily.

Lemma 3.2. *If E is an AM-space with unit e, F is an arbitrary Banach lattice and $T\colon E \to F$ is a positive operator, then $\|T\| = \|Te\|$.*

The AL-spaces and the AM-spaces are dual to each other. For the proof of the first part of the next theorem see [**30**, p. 188].

Theorem 3.3. *A Banach lattice E is an AL-space (resp. an AM-space) if and only if its norm dual E^* is an AM-space (resp. an AL-space). Moreover:*

(a) *If E is an AL-space, then E^* is a Dedekind complete AM-space with unit e^*, where $e^*(x) = \|x^+\| - \|x^-\|$ for each $x \in E$.*

(b) *If E is an AM-space with unit e, then E^{**} is also an AM-space with unit e.*

Proof. We prove parts (a) and (b) only.

(a) Let E be an AL-space. By the first part of the theorem, we know that E^* is an AM-space. Now consider the function $e^*: E^+ \to \mathbb{R}$ defined by $e^*(x) = \|x\|$. Since E is an AL-space, it follows that e^* is additive, and so (by Theorem 1.15) it extends uniquely (via the formula $e^*(x) = \|x^+\| - \|x^-\|$) to a positive linear functional on E, i.e., $e^* \in E^*_+$. We leave the details to the reader to verify that e^* is an order unit of E^*.

(b) Assume that E is an AM-space with unit e. We claim that e is also a unit in the AM-space E^{**}. To see this, note first that $\|x^*\| = x^*(e) = e(x^*)$ for each $x^* \in E^*_+$. Now if $x^{**} \in E^{**}_+$, then for each $x^* \in E^*_+$ we have

$$x^{**}(x^*) \le \|x^{**}\| \cdot \|x^*\| = \|x^{**}\| e(x^*) \, ,$$

and so $x^{**} \le \|x^{**}\| e$, proving that e is also a unit in E^{**}. ∎

Recall that the **principal ideal** generated by a vector u in a Riesz space E is the vector subspace

$$E_u = \{x \in E\colon \text{ there exists some } \lambda \ge 0 \text{ such that } |x| \le \lambda |u|\} \, .$$

It turns out that when E is a Banach lattice, then each principal ideal has the structure of an AM-space. For a proof of the next result see [**30**, p. 187].

Theorem 3.4. *Let E be a Banach lattice and let E_u be the principal ideal generated by a vector $u \in E$. Then E_u under the norm*

$$\|x\|_\infty = \inf\{\lambda \ge 0\colon \ |x| \le \lambda |u|\}$$

is an AM-space with unit $|u|$.

The next two classical results allow us to consider the vectors of AL- and AM-spaces as functions in some familiar function spaces.

Theorem 3.5 (Kakutani–Bohnenblust–Nakano). *A Banach lattice is an AL-space if and only if it is lattice isometric to an $L_1(\mu)$-space.*

Theorem 3.6 (Kakutani–Bohnenblust–Krein). *A Banach lattice E is an AM-space with unit e if and only if E is lattice isometric to some $C(\Omega)$-space for a unique (up to homeomorphism) compact Hausdorff space Ω in such a way that the unit e is identified with the constant function $\mathbf{1}$ on Ω.*

Furthermore, a Banach lattice E is an AM-space if and only if E is lattice isometric to a closed vector sublattice of a $C(\Omega)$-space.

The reader should keep in mind that for a compact Hausdorff space Ω the Banach lattice $C(\Omega)$ is Dedekind complete if and only if Ω is **extremally disconnected** or **Stonean** (i.e., if and only if the closure of every open subset of Ω is also open). For details and more on these classical theorems we refer the reader to [**30**, pp. 192–195].

Corollary 3.7. *Every AL-space has order continuous norm.*

Lemma 3.8. *If E is either a KB-space or a Dedekind complete AM-space with unit, then there is a positive projection $P\colon E^{**} \to E^{**}$ with range E.*

Proof. If E is a KB-space, then E is a band in E^{**} (see Exercise 8 in Section 2.3) and the band projection from E^{**} onto E is the desired positive projection.

Now assume that E is a Dedekind complete AM-space with unit. By Theorem 3.3 the unit of E is a unit of E^{**} too, and so E majorizes[1] E^{**}. In particular, the identity operator $I\colon E \to E$ has a positive extension to all of E^{**}; see [**30**, Theorem 2.8, p. 26]. Any such extension is a positive projection on E^{**} with range E. ■

Theorem 3.9. *Assume that either F is a Dedekind complete AM-space with unit or E is an AL-space and F is a KB-space. Then every continuous operator from E to F is regular, i.e., $\mathcal{L}(E,F) = \mathcal{L}_r(E,F)$. In particular, each $T \in \mathcal{L}(E,F)$ has a modulus $|T|$.*

Proof. Assume first that F is a Dedekind complete AM-space with unit, and take any continuous operator $T\colon E \to F$. Since T maps the unit ball U_E of E to a norm bounded subset of F, the set $T(U_E)$ is necessarily order bounded, and thus T is an order bounded operator. As F is Dedekind complete, T is a regular operator and its modulus $|T|$ exists.

Assume next that E is an AL-space and F is a KB-space. Consider an arbitrary $T \in \mathcal{L}(E,F)$. Then $T^*\colon F^* \to E^*$ and E^* is a Dedekind complete AM-space with unit. According to the preceding case, the operator T^* is regular. This implies that $T^{**}\colon E^{**} \to F^{**}$ is also a regular operator. Pick two positive operators $A, B\colon E^{**} \to F^{**}$ such that $T^{**} = A - B$. By Lemma 3.8 there exists a positive projection P from F^{**} onto F. It remains to notice that PA, PB are positive operators from E^{**} to F and the equality $T = PA - PB$ holds on E. This shows that T is a regular operator. ■

Corollary 3.10. *Assume that $T\colon E \to F$ is a continuous operator, where either F is a Dedekind complete AM-space with unit or E is an AL-space and F is a KB-space. Then $\|T\| = \|T\|_r$, i.e., the operator norm and the r-norm of T coincide.*

---

[1] Recall that a vector subspace Y of an ordered vector space Z **majorizes** Z if for every $z \in Z$ there exists some $y \in Y$ such that $z \le y$.

In particular, in these cases, $\mathcal{L}(E,F)$ equipped with the operator norm is a Dedekind complete Banach lattice.

Proof. Assume first that F is a Dedekind complete AM-space with unit and so, by Theorem 3.6, $F = C(\Omega)$ for some compact Hausdorff space Ω. As established in Theorem 3.9, the operator $|T|\colon E \to F$ exists and so

$$\|T\|_r = \big\|\,|T|\,\big\| = \sup_{\|x\|\le 1} \big\|\,|T|x\,\big\|_\infty .$$

Fix $\epsilon > 0$ and pick some unit vector $x \in E^+$ such that $\big\|\,|T|x\,\big\|_\infty > \|T\|_r - \epsilon$. Since

$$\big\|\,|T|x\,\big\|_\infty = \sup_{\omega\in\Omega}(|T|x)(\omega)\,,$$

we can find a non-empty open subset V of Ω such that

$$(|T|x)(\omega) > \|T\|_r - \epsilon \tag{$\star$}$$

for each $\omega \in V$. The Riesz–Kantorovich formula

$$|T|x = \sup\{Tu\colon\ \ -x \le u \le x\}$$

implies the existence of some $u \in [-x,x]$ and some point $\omega_0 \in V$ such that

$$(|T|u)(\omega_0) > \|T\|_r - \epsilon\,.$$

(Otherwise the values of the function $|T|x$ on the set V would be bounded from above by $\|T\|_r - \epsilon$, contradicting ($\star$).) This implies $\|Tu\|_\infty > \|T\|_r - \epsilon$, and hence $\|T\| > \|T\|_r - \epsilon$. As $\epsilon > 0$ is arbitrary, it follows that $\|T\| \ge \|T\|_r$. The converse inequality $\|T\| \ge \|T\|_r$ is always true, and consequently we have the desired equality $\|T\| = \|T\|_r$.

Now assume that E is an AL-space and F is a KB-space. The modulus $|T|\colon E \to F$ exists by Theorem 3.9. Furthermore, Theorem 2.28 guarantees that $|T|^* = |T^*|$. Since E^* is a Dedekind complete AM-space with unit, the first case yields $\big\|\,|T^*|\,\big\| = \|T^*\|$ and so

$$\|T\|_r = \big\|\,|T|\,\big\| = \big\|\,|T|^*\,\big\| = \big\|\,|T^*|\,\big\| = \|T^*\| = \|T\|\,.$$

The proof is finished. ■

The preceding result combined with Corollary 3.7 and Theorem 2.28 yields the following.

Corollary 3.11. *If E is an AL-space and F is a KB-space, then the mapping $T \mapsto T^*$, from $\mathcal{L}(E,F)$ to $\mathcal{L}(F^*,E^*)$, is a lattice isometry.*

We continue with some basic properties of weakly compact operators on AL- and AM-spaces.

Theorem 3.12. *The following two statements are true.*

(1) *Every weakly compact operator between AL-spaces has a weakly compact modulus.*

(2) *Every weakly compact operator from an AM-space to a Dedekind complete AM-space with unit has a weakly compact modulus.*

Proof. Statement (1) is due to C. D. Aliprantis and O. Burkinshaw [**28**]. For a proof see [**30**, Theorem 17.14, p. 292].

We prove (2) (which is due to K. D. Schmidt [**296**]) by using the factorization technique outlined below. Let $T\colon E \to F$ be a weakly compact operator from an AM-space E to a Dedekind complete AM-space F with unit. Note that $|T^{**}|\colon E^{**} \to F^{**}$ is weakly compact. Indeed, since E^* and F^* are AL-spaces and $T^*\colon F^* \to E^*$ is weakly compact, part (1) guarantees that its modulus $|T^*|\colon F^* \to E^*$ is also weakly compact. Now by Theorem 2.28, we have $|T^{**}| = |T^*|^*$, and so $|T^{**}|$ is weakly compact. To complete the proof, we shall show that the modulus $|T|$ of T admits a factorization of the form

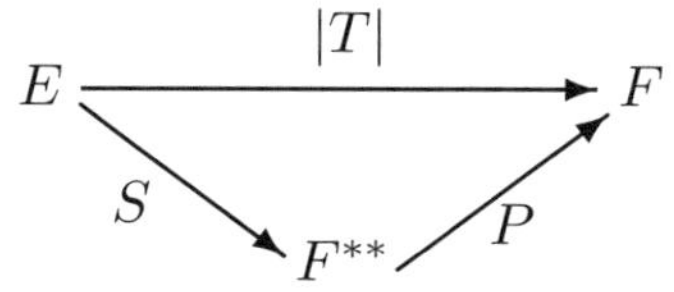

with S a positive weakly compact operator and P a positive projection. The projection $P\colon F^{**} \to F^{**}$ is the one guaranteed by Lemma 3.8. The operator $S\colon E \to F^{**}$ is the restriction of the weakly compact operator $|T^{**}|\colon E^{**} \to F^{**}$ to E.

To see that the equality $|T| = P|T^{**}|$ holds, let $x \in E^+$ be fixed. If $y \in E$ satisfies $|y| \le x$, then from the inequality $Ty = T^{**}y \le |T^{**}|x$, we obtain $Ty = PTy \le P|T^{**}|x \in F$, and so

$$|T|x = \sup\{Ty\colon\ y \in E \text{ and } |y| \le x\} \le P|T^{**}|x\,,$$

where the supremum is taken in F. On the other hand, if $x^{**} \in E^{**}$ satisfies $|x^{**}| \le x$, then there exists (see Exercise 10 at the end of the section) a net $\{x_\alpha\}$ in the interval $[-x, x]$ such that $x_\alpha \xrightarrow{\sigma(E^{**},E^*)} x^{**}$. From the inequality $T^{**}x_\alpha = Tx_\alpha \le |T|x$, we get $T^{**}x^{**} \le |T|x$ and so

$$|T^{**}|x = \sup\{T^{**}x^{**}\colon\ x^{**} \in E^{**} \text{ and } |x^{**}| \le x\} \le |T|x\,,$$

where the supremum is now taken in F^{**}. Thus, $P|T^{**}|x \le P|T|x = |T|x$, and the desired equality $|T|x = P|T^{**}|x$ follows. ∎

In order to obtain the corresponding results for compact operators, we need a lemma.

Lemma 3.13 (Krengel [**193**]). *If A is a norm totally bounded subset of an AM-space E, then:*

(1) *The subset D of E consisting of the suprema of the finite subsets of A is likewise norm totally bounded.*

(2) $\sup A$ *exists in E.*

(3) $\sup A \in \overline{D}$, *the norm closure of D in E.*

Proof. By Theorem 3.6, we can assume that E is a closed Riesz subspace of some $C(\Omega)$-space with Ω compact and Hausdorff. Thus, A can be viewed as a norm totally bounded subset of $C(\Omega)$, and hence, by the Ascoli–Arzelà theorem, A is norm bounded and equicontinuous; see for instance [**31**, p. 66].

Now let $g = \bigvee_{i=1}^{n} f_i \in D$, where $f_i \in A$ for each i. Since A is equicontinuous on Ω, given $\epsilon > 0$ and $\omega \in \Omega$, there exists a neighborhood V of ω such that $|f(t) - f(\omega)| < \epsilon$ for all $t \in V$ and all $f \in A$. This implies

$$\big|g(t) - g(\omega)\big| \leq \max_{1 \leq i \leq n} \big|f_i(t) - f_i(\omega)\big| < \epsilon$$

for all $t \in V$ and all $g \in D$. But then, by the Ascoli–Arzelà theorem again, D is also a norm totally bounded set.

Clearly, D is an upward directed net, and hence D (being norm totally bounded) has a norm convergent subnet to some $g \in E$. Since D is upward directed, it follows that D is norm convergent to g. Hence, $g \in \overline{D}$ and $g = \sup D = \sup A$; see Lemma 1.30(c). ■

We are now ready to state and prove the analogue of Theorem 3.12 for compact operators.

Theorem 3.14. *Let $T\colon E \to F$ be a compact operator between two Banach lattices. If either*

(a) *F is an AM-space, or else*

(b) *E an AL-space and F is a KB-space,*

then T has a compact modulus that is given by the Riesz–Kantorovich formula

$$|T|x = \sup\{Ty\colon\ y \in E \ \ and \ \ |y| \leq x\}.$$

Proof. (a) Let $T\colon E \to F$ be a compact operator. Assume first that F is an AM-space. Notice that Lemma 3.13 guarantees that in the formula

$$|T|x = \sup\big\{Ty\colon\ y \in E \ \text{ and } \ -x \leq y \leq x\,\big\}$$

the supremum exists in F for each $x \in E^+$. This establishes that the operator T has a modulus—even when F is not Dedekind complete!

On the other hand, if U is the closed unit ball of E and D denotes the subset of F consisting of the suprema of the finite subsets of the norm totally bounded set $T(U)$, then (by Lemma 3.13) D is a norm totally bounded subset of F. Moreover, the preceding formula implies that for each $x \in U^+$ we have $|T|x \in \overline{D}$, i.e., $|T|(U^+) \subseteq \overline{D}$. This shows that $|T|$ is a compact operator.

(b) Now assume that E is an AL-space and F is a KB-space. By Theorem 3.9, the modulus $|T|$ of T exists. Next, notice that $T^*\colon F^* \to E^*$ is a compact operator and E^* is a Dedekind complete AM-space with unit. By the preceding case, T^* has a compact modulus $|T^*|$. Finally, Theorem 2.28 implies $|T|^* = |T^*|$, and so $|T|^*$ is a compact operator. The latter guarantees that $|T|$ itself is a compact operator. ■

We close the section by introducing two special classes of positive operators acting on AM- or AL-spaces. These are the classes of Markov and stochastic operators.

Definition 3.15. *A positive operator $T\colon E \to F$ between two AM-spaces with units u and v, respectively, is said to be a* ***Markov operator*** *if $Tu = v$.*

A positive operator $T\colon E \to F$ between two AL-spaces is called a ***stochastic operator*** *if $\|Tx\| = \|x\|$ for each $x \in E^+$.*

For some simple examples of Markov and stochastic operators see Exercise 15 at the end of this section. The Markov and stochastic operators are dual to each other in the following sense.

Theorem 3.16. *A positive operator between two AM-spaces with units is a Markov operator if and only if its norm adjoint is a stochastic operator.*

Similarly, a positive operator between two AL-spaces is a stochastic operator if and only if its norm adjoint is a Markov operator.

Proof. Assume first that $T\colon E \to F$ is a Markov operator between two AM-spaces with units u and v, respectively. So, T is positive and $Tu = v$. Now if $0 \le y^* \in F^*$, then according to Lemma 3.2 we have

$$\|T^*y^*\| = \big\|T^*y^*(u)\big\| = \big\|y^*(Tu)\big\| = \big\|y^*(v)\big\| = \|y^*\|\,.$$

This shows that T^* is a stochastic operator.

Next, suppose that $T\colon E \to F$ is a stochastic operator between two AL-spaces. That is, T is positive and $\|Tx\| = \|x\|$ for each $x \in E^+$. By Theorem 3.3, the norm duals E^* and F^* are both AM-spaces with units u^*

and v^* given by $u^*(x) = \|x^+\| - \|x^-\|$ for $x \in E$ and $v^*(y) = \|y^+\| - \|y^-\|$ for $y \in F$. Thus, if $x \in E^+$, then

$$T^*v^*(x) = v^*(Tx) = \|Tx\| = \|x\| = u^*(x).$$

Therefore, $T^*v^* = u^*$, and so T^* is a Markov operator.

Now let $T\colon E \to F$ be a positive operator between two AM-spaces with units u and v, respectively, and assume that $T^*\colon F^* \to E^*$ is stochastic. Then $T^{**}\colon E^{**} \to F^{**}$ is a Markov operator. Since E^{**} is an AM-space with unit u and F^{**} is an AM-space with unit v (see Theorem 3.3), it follows that $Tu = T^{**}u = v$. Hence, T is a Markov operator.

Finally, assume that the norm dual T^* of a positive operator $T\colon E \to F$ between AL-spaces is a Markov operator. Then $T^{**}\colon E^{**} \to F^{**}$ is a stochastic operator. Consequently, if $x \in E^+$, then $\|x\| = \|T^{**}x\| = \|Tx\|$ proving that T is a stochastic operator. The proof of the theorem is now complete. ■

Exercises

1. A vector u in a subset A of a vector space X is called an **internal point** of A if for each $x \in X$ there exists some $\lambda_0 > 0$ such that $u + \lambda x \in A$ for all $|\lambda| \le \lambda_0$.

   Show that in an ordered vector space X the strong units are precisely the internal points of the positive cone X^+.

2. If E is an AM-space with a unit e, then show that the formula

   $$\|x\|_\infty = \inf\{\lambda > 0\colon\ |x| \le \lambda e\} = \min\{\lambda \ge 0\colon\ |x| \le \lambda e\} \qquad (\star)$$

   defines an equivalent lattice norm on E. Under this norm E remains an AM-space and its closed unit ball coincides with the order interval $[-e, e]$.

3. Prove Lemma 3.2.
4. Prove Theorem 3.3.
5. Prove Theorem 3.4.
6. Prove Corollary 3.7.
7. Show that a Banach lattice E is:
   (a) an AL-space if and only if $\|x+y\| = \|x\| + \|y\|$ for all $x, y \in E^+$ and
   (b) an AM-space if and only if $\|x \vee y\| = \max\{\|x\|, \|y\|\}$ for all $x, y \in E^+$.
8. Show that the lattice operations in any AM-space E are weakly sequentially continuous. [HINT: We can assume (by Theorem 3.6) that $E = C(\Omega)$ for some compact Hausdorff space Ω. By the Riesz Representation Theorem, we know that a sequence $\{f_n\}$ in $C(\Omega)$ satisfies $f_n \xrightarrow{w} f$ if and only if $\{f_n\}$ is norm bounded and $f_n(\omega) \to f(\omega)$ for each $\omega \in \Omega$. Therefore, if $f_n \xrightarrow{w} f$ in $C(\Omega)$, then also $|f_n| \xrightarrow{w} |f|$.]

9. Show that the lattice operations in an AL-space need not be weakly sequentially continuous.

10. Let E be a Banach lattice. For $x \in E^+$ put

$$\begin{aligned} [-x, x] &= \{y \in E\colon\ -x \le y \le x\}\,, \text{ and} \\ [\![-x, x]\!] &= \{y^{**} \in E^{**}\colon\ -x \le y^{**} \le x\}\,. \end{aligned}$$

Establish the following:

(a) $[-x, x]$ and $[\![-x, x]\!]$ are both convex sets.

(b) $[-x, x]$ is $\sigma(E, E^*)$-closed and $[\![-x, x]\!]$ is $\sigma(E^{**}, E^*)$-compact.

(c) $[-x, x]$ is $\sigma(E^{**}, E^*)$-dense in $[\![-x, x]\!]$.

[HINT: To see (c), assume that $[-x, x]$ is not $\sigma(E^{**}, E^*)$-dense in $[\![-x, x]\!]$. Then, by the Separation Theorem, there exists some $x^* \in E^*$ satisfying $x^*(y) = 0$ for each $y \in [-x, x]$ and $x^*(y^{**}) \neq 0$ for some $y^{**} \in [\![-x, x]\!]$. It follows that

$$|x^*|(x) = \sup\{x^*(y)\colon\ y \in E \text{ and } -x \le y \le x\} = 0\,.$$

But then $0 \le |x^*(y^{**})| \le |x^*|(y^{**}) \le |x^*|(x) = 0$, and so $x^*(y^{**}) = 0$, a contradiction.]

11. Show that every regular operator from an AM-space with unit to any ℓ_p-space, where $1 \le p < \infty$, is compact. [HINT: Use the fact that the order intervals of ℓ_p-spaces with $1 \le p < \infty$ are norm compact sets.]

12. Let E be an AM-space. Extend the norm of E to its Dedekind completion E^δ as in Exercise 21 of Section 1.2. Show that the Banach lattice E^δ is also an AM-space. If E is an AM-space with unit e, then E^δ is also an AM-space with unit e.

13. Assume that S and Ω are two compact metric spaces, μ is an arbitrary (finite) Borel measure on S, and $K\colon S \times \Omega \to \mathbb{R}$ is a continuous function. Also let $T\colon C(S) \to C(\Omega)$ denote the **integral operator** with kernel K, i.e., T is defined by

$$Tx(\omega) = \int_S K(s, \omega)x(s)\, d\mu(s)$$

for each $x \in C(S)$ and each $\omega \in \Omega$. Establish the following.

(a) T is a compact operator.

(b) The modulus $|T|\colon C(S) \to C(\Omega)$ of the operator T (which exists according to Theorem 3.14) is also an integral operator satisfying $|T|x(\omega) = \int_S |K(s, \omega)| x(s)\, d\mu(s)$ for all $x \in C(S)$ and all $\omega \in \Omega$.

(c) The norm of T is given by

$$\|T\|_\infty = \sup\{\|Tx\|_\infty\colon\ \|x\|_\infty \le 1\} = \sup_{\omega \in \Omega} \int_S |K(s, \omega)|\, d\mu(s)\,.$$

[HINTS: (a) Let d and ρ be the metrics on S and Ω, respectively, and let $\epsilon > 0$. Fix some $\delta > 0$ such that $d(s_1, s_2) + \rho(\omega_1, \omega_2) < \delta$ implies $|K(s_1, \omega_1) - K(s_2, \omega_2)| < \epsilon$. Now if $x \in C(S)$ satisfies $\|x\|_\infty \le 1$ and ω_1, ω_2 in Ω satisfy $\rho(\omega_1, \omega_2) < \delta$, then

$$|Tx(\omega_1) - Tx(\omega_2)| \le \int_S |K(s, \omega_1) - K(s, \omega_2)| \cdot |x(s)|\, d\mu(s) \le \epsilon\mu(S)\,.$$

This shows that the norm bounded set $\{Tx\colon \ \|x\|_\infty \le 1\}$ is an equicontinuous subset of $C(\Omega)$.

(b) Fix any function $0 \le x \in C(S)$ and let $\omega \in \Omega$. Clearly, we have the inequality $|T|x(\omega) \le \int_S |K(s,\omega)|x(s)\,d\mu(s)$. Denote by σ the sign function of $K(\cdot,\omega)$, i.e., $\sigma(s) = 1$ if $K(s,\omega) \ge 0$ and $\sigma(s) = -1$ if $K(s,\omega) < 0$, and then choose a sequence $\{x_n\} \subseteq C(S)$ such that $x_n(s) \to \sigma(s)$ for μ-almost all s. Replacing x_n by $[x_n \vee (-\mathbf{1})] \wedge \mathbf{1}$, we can assume that $\|x_n\|_\infty \le 1$ for each n. Since $|x_n(s)x(s)| \le x(s)$ for each $s \in S$ and n, it follows from the Lebesgue Dominated Convergence Theorem that

$$\begin{aligned}\int_S |K(s,\omega)|x(s)\,d\mu(s) &= \int_S K(s,\omega)\sigma(s)x(s)\,d\mu(s)\\ &= \lim_{n\to\infty}\int_S K(s,\omega)x_n(s)x(s)\,d\mu(s) \le |T|x(\omega)\,.\end{aligned}$$

(c) Clearly, $\|T\|_\infty \le \sup_{\omega\in\Omega}\int_S |K(s,\omega)|\,d\mu(s)$. For the converse inequality pick some $\omega_0 \in \Omega$ with $\int_S |K(s,\omega_0)|\,d\mu(s) = \sup_{\omega\in\Omega}\int_S |K(s,\omega)|\,d\mu(s)$. Denote by σ the sign function of $K(\cdot,\omega_0)$ and choose a sequence $\{x_n\}$ in $C(S)$ such that $\|x_n\|_\infty \le 1$ for each n and $x_n(s) \to \sigma(s)$ for μ-almost all s. Now use the Lebesgue Dominated Convergence Theorem to see that

$$\begin{aligned}\int_S |K(s,\omega_0)|\,d\mu(s) &= \int_S K(s,\omega_0)\sigma(s)\,d\mu(s)\\ &= \lim_{n\to\infty}\int_S K(s,\omega_0)x_n(s)\,d\mu(s) \le \|T\|_\infty\,.]\end{aligned}$$

14. Show that every Markov operator carries order units to order units.

15. Every Euclidean space $\mathbb{R}^n$ can be considered either as an AM-space with unit $e = (1,1,\dots,1)$ or as an AL-space under the usual norms

$$\|x\|_\infty = \max_{1\le i\le n} |x_i| \quad \text{and} \quad \|x\|_1 = \sum_{i=1}^n |x_i|\,,$$

respectively. An $m \times n$ matrix $A = [a_{ij}]$ with non-negative real entries is called **Markov** (resp. **stochastic**) if $\sum_{j=1}^n a_{ij} = 1$ for each $1 \le i \le m$ (resp. $\sum_{i=1}^m a_{ij} = 1$ for each $1 \le j \le n$).

Now let A be an $m\times n$ matrix with non-negative entries and consider A as an operator from $\mathbb{R}^n$ to $\mathbb{R}^m$. Show the following:

(a) The positive operator $A\colon (\mathbb{R}^n, \|\cdot\|_\infty) \to (\mathbb{R}^m, \|\cdot\|_\infty)$ is a Markov operator if and only if the matrix A is a Markov matrix.

(b) The positive operator $A\colon (\mathbb{R}^n, \|\cdot\|_1) \to (\mathbb{R}^m, \|\cdot\|_1)$ is a stochastic operator if and only if the matrix A is a stochastic matrix.

3.2. Complex Banach Lattices

Apart from the standard norm introduced in Section 1.1, there is another way of defining an equivalent norm on the complexification $E_{\mathbf{C}}$ of an arbitrary real Banach lattice E. This new norm is very important because, as we shall see, it is closely related to the lattice structure of E. To introduce this new norm, assume first that $E = C(\Omega)$, the Banach lattice of all real-valued continuous functions on a compact Hausdorff space Ω.[2] An easy argument shows that for each pair $f, g \in C(\Omega)$ and each $\omega \in \Omega$, we have

$$\begin{aligned}\sup_{\theta\in[0,2\pi]}\big[f(\omega)\cos\theta + g(\omega)\sin\theta\big] &= \sup_{\theta\in\mathbb{R}}\big[f(\omega)\cos\theta + g(\omega)\sin\theta\big] \\ &= \sqrt{[f(\omega)]^2 + [g(\omega)]^2}\,.\end{aligned}$$

Consequently,

$$\begin{aligned}\sup_{\theta\in[0,2\pi]}\big[f\cos\theta + g\sin\theta\big] &= \sup_{\theta\in\mathbb{R}}\big[f\cos\theta + g\sin\theta\big] \\ &= \sqrt{f^2+g^2} = |f + \imath g|\,,\end{aligned}$$

where the supremum is taken in $C(\Omega)$ in the usual lattice sense.

We now turn our attention to the general case of a real Banach lattice E. Let $x, y \in E$ and consider the ideal E_u generated in E by the element $u = |x| + |y|$. By Theorems 3.4 and 3.6, we know that E_u is lattice isometric to a $C(\Omega)$-space. This implies (in view of the preceding discussion) that the supremum

$$\sup_{\theta\in[0,2\pi]}\big[x\cos\theta + y\sin\theta\big] = \sup_{\theta\in\mathbb{R}}\big[x\cos\theta + y\sin\theta\big]$$

exists in E_u (and hence in E). The above supremum is called the **modulus** of the element $z = x + \imath y \in E_{\mathbf{C}}$. That is, the modulus of an arbitrary element $z = x + \imath y \in E_{\mathbf{C}}$ is the positive element $|z|$ of E defined via the formula

$$|z| = \sup_{\theta\in[0,2\pi]}\big[x\cos\theta + y\sin\theta\big]\,.$$

Notice that

$$|z| = \big|\pm x \pm \imath y\big| = \big|\pm|x| \pm \imath|y|\big| = \big|\pm y \pm \imath x\big| = \big|\pm|y| \pm \imath|x|\big| \ \big(= \sqrt{x^2+y^2}\,\big)\,.$$

It can be shown that for arbitrary $z, z_1, z_2 \in E_{\mathbf{C}}$ and $\lambda \in \mathbb{C}$, we have:

(1) $|z| \geq 0$ and $|z| = 0$ if and only if $z = 0$;

(2) $|\lambda z| = |\lambda| \cdot |z|$; and

(3) $|z_1 + z_2| \leq |z_1| + |z_2|$.

So, if we define the function $\|\cdot\|_{\mathbf{C}} \colon E_{\mathbf{C}} \to \mathbb{R}$ via the formula

$$\|z\|_{\mathbf{C}} = \big\|\,|z|\,\big\|\,, \tag{$\star$}$$

[2] If we want to emphasize that we deal with real-valued functions rather than with complex-valued functions, then we shall denote $C(\Omega)$ by $C_{\mathbb{R}}(\Omega)$. The symbol $C_{\mathbb{C}}(\Omega)$ will denote the complex Banach space of all complex-valued continuous functions on Ω equipped with the supremum norm.

then the above properties (1), (2), and (3) guarantee that $\|\cdot\|_{\mathbf{C}}$ is, in fact, a norm on $E_{\mathbf{C}}$. Now notice that if $z = x + \imath y \in E_{\mathbf{C}}$, then the inequalities $|x| \le |z|$, $|y| \le |z|$, and $|z| \le |x| + |y|$ imply that

$$\tfrac{1}{2}\big(\|x\| + \|y\|\big) \le \|z\|_{\mathbf{C}} \le \|x\| + \|y\| .$$

Therefore, not only is $\|\cdot\|_{\mathbf{C}}$ a norm on $E_{\mathbf{C}}$ but it is also equivalent to the standard norm on $E_{\mathbf{C}}$ defined by

$$\|z\| = \sup_{\theta\in[0,2\pi]} \|x\cos\theta + y\sin\theta\| .$$

The reader should also observe that $\|z\|_{\mathbf{C}} = \|z\|$ for all $z \in E$ and $|z_1| \le |z_2|$ implies $\|z_1\|_{\mathbf{C}} \le \|z_2\|_{\mathbf{C}}$.

From now on, we will always assume that on the complexification $E_{\mathbf{C}}$ of a real Banach lattice E the norm is defined by $(\star)$. Specifically, we have the following definition.

Definition 3.17. *Any complex Banach space of the form $E_{\mathbf{C}} = E \oplus \imath E$, where E is a real Banach lattice, will be referred to as a* ***complex Banach lattice****.*

The rest of this section will demonstrate the advantages of this norm. A linear isometry $T\colon E_{\mathbf{C}} \to F_{\mathbf{C}}$ between two complex Banach lattices is said to be a **lattice isometry** if $|Tz| = T|z|$ for each $z \in E_{\mathbf{C}}$. Two complex Banach lattices $E_{\mathbf{C}} = E \oplus \imath E$ and $F_{\mathbf{C}} = F \oplus \imath F$ are said to be **lattice isometric** if there exists a surjective linear isometry $T\colon E_{\mathbf{C}} \to F_{\mathbf{C}}$ satisfying $|Tz| = T|z|$ for all $z \in E_{\mathbf{C}}$. We have the following.

Lemma 3.18. *If $T\colon E \to F$ is a surjective lattice isometry between two real Banach lattices, then $T_{\mathbf{C}}\colon E_{\mathbf{C}} \to F_{\mathbf{C}}$ is also a surjective lattice isometry.*

Proof. Assume that $T\colon E \to F$ is a surjective lattice isometry. To show that $T_{\mathbf{C}}$ is a lattice isometry, pick $z = x + \imath y \in E_{\mathbf{C}}$. If D denotes the subset of all vectors of the form $\bigvee_{i=1}^{n}(x\cos\theta_i + y\sin\theta_i)$, where θ_i is a real number for each i, then $D \uparrow |z|$. Since T is order continuous, it follows that $T(D) \uparrow T|z|$ in F. On the other hand, if $w = \bigvee_{i=1}^{n}(x\cos\theta_i + y\sin\theta_i) \in D$, then $T(w) = \bigvee_{i=1}^{n}\big[(Tx)\cos\theta_i + (Ty)\sin\theta_i\big]$, and so $T(D) \uparrow |T_{\mathbf{C}}z|$ in F. Hence, $|T_{\mathbf{C}}z| = T|z|$ for each $z \in E_{\mathbf{C}}$. Finally, from

$$\|T_{\mathbf{C}}z\| = \big\|\,|T_{\mathbf{C}}z|\,\big\| = \big\|T|z|\big\| = \big\|\,|z|\,\big\| = \|z\| ,$$

we see that $T_{\mathbf{C}}$ is indeed a surjective lattice isometry. ■

Definition 3.19. *A* ***complex AM-space*** *F with unit is the complexification of a (real) AM-space E with unit, i.e., $F = E_{\mathbf{C}} = E \oplus \imath E$. The unit of E will also be referred to as the unit of F.*

Two complex AM-spaces F and G with units u and v, respectively, are said to be ***lattice isometric*** *if there exists a surjective lattice isometry $T\colon F \to G$ satisfying $Tu = v$.*

From Lemma 3.18, it should be clear that the complex version of the Kakutani–Bohnenblust–Krein Representation Theorem 3.6 can be stated as follows.

Theorem 3.20. *If $E_{\mathbf{C}} = E \oplus \imath E$ is a complex AM-space with unit e, then there exists a compact Hausdorff space Ω (which is uniquely determined up to homeomorphism) such that $E_{\mathbf{C}}$ is lattice isometric to the complex AM-space $C_{\mathbb{C}}(\Omega) = C(\Omega) \oplus \imath C(\Omega)$ and e corresponds to the constant function* **1** *on Ω.*

Corollary 3.21. *If u is a positive vector in a real Banach lattice E, then the complexification of the ideal E_u,*

$$E_u \oplus \imath E_u = \{z \in E_{\mathbf{C}}\colon\ |z| \in E_u\},$$

is lattice isometric to a complex Banach lattice of the form $C_{\mathbb{C}}(\Omega)$.

Moreover, for each $z \in E_u \oplus \imath E_u \subseteq E \oplus \imath E$ the modulus of z in the complex Banach lattice $E_u \oplus \imath E_u$ coincides with the modulus of z in the complex Banach lattice $E \oplus \imath E$.

Now let E and F be two real Banach lattices. For an operator T in $\mathcal{L}(E,F)$ its norm $\|T\|_{\mathbf{C}}$ is, by definition, the operator norm of the operator $T_{\mathbf{C}}\colon E_{\mathbf{C}} \to F_{\mathbf{C}}$, where as usual $T_{\mathbf{C}}(x + \imath y) = Tx + \imath Ty$.[3] That is,

$$\|T\|_{\mathbf{C}} = \sup_{\|z\|_{\mathbf{C}} \le 1} \|T_{\mathbf{C}} z\|_{\mathbf{C}}.$$

It should be obvious that for any bounded operator $T\colon E \to E$ we have the estimates $\|T\| \le \|T\|_{\mathbf{C}} \le 2\|T\|$. Moreover, for positive operators the two norms $\|T\|$ and $\|T\|_{\mathbf{C}}$ coincide. This is a consequence of the following result.

Lemma 3.22. *If $T\colon E \to F$ is a positive operator between (real) Banach lattices, then $|T_{\mathbf{C}} z| \le T|z|$ for all $z \in E_{\mathbf{C}}$.*

Proof. Let $z = x + \imath y \in E_{\mathbf{C}}$. Then $T_{\mathbf{C}} z = Tx + \imath Ty$ and from

$$(Tx)\cos\theta + (Ty)\sin\theta = T(x\cos\theta + y\sin\theta) \le T|z|,$$

we get $|T_{\mathbf{C}} z| = \sup_{\theta \in [0,2\pi]} \big[(Tx)\cos\theta + (Ty)\sin\theta\big] \le T|z|$, as claimed. ■

[3] To avoid unnecessary notation, we shall use the symbol T instead of $T_{\mathbf{C}}$ whenever this does not cause any ambiguity.

Corollary 3.23. *If T is a positive operator, then $\|T\|_{\mathbf{c}} = \|T\|$.*

In general, strict inequality $\|T\| < \|T\|_{\mathbf{c}}$ is possible even in the finite dimensional case; see Exercise 9 at the end of this section.

The rest of the section is devoted to the concept of a regular operator on a complex Banach lattice. So, for our discussion here, let E and F be two real Banach lattices. Recall that $\mathcal{L}(E_{\mathbf{c}}, F_{\mathbf{c}}) = \mathcal{L}(E, F) \oplus \imath\mathcal{L}(E, F)$ since every bounded operator from $E_{\mathbf{c}}$ to $F_{\mathbf{c}}$ is of the form $T + \imath S$, where $T, S \in \mathcal{L}(E, F)$, and

$$(T + \imath S)(x + \imath y) = (Tx - Sy) + \imath(Sx + Ty)$$

for all $z = x + \imath y \in E_{\mathbf{c}}$. We shall say that an operator $T + \imath S \in \mathcal{L}(E_{\mathbf{c}}, F_{\mathbf{c}})$ is **regular** if both operators $T, S \in \mathcal{L}(E, F)$ are regular. The collection of all (complex) regular operators will be denoted $\mathcal{L}_r(E_{\mathbf{c}}, F_{\mathbf{c}})$. Clearly, the vector space $\mathcal{L}_r(E_{\mathbf{c}}, F_{\mathbf{c}})$ coincides with the complexification of $\mathcal{L}_r(E, F)$, that is, $\mathcal{L}_r(E_{\mathbf{c}}, F_{\mathbf{c}}) = \mathcal{L}_r(E, F) \oplus \imath\mathcal{L}_r(E, F)$.

From now on, let us assume that F is also Dedekind complete. Then, by the Riesz–Kantorovich Theorem 1.16, $\mathcal{L}_r(E, F)$ is a Dedekind complete Riesz space which is a Banach lattice when equipped with the r-norm; see Theorem 1.32. In particular, $\mathcal{L}_r(E_{\mathbf{c}}, F_{\mathbf{c}}) = \mathcal{L}_r(E, F) \oplus \imath\mathcal{L}_r(E, F)$ is a complex Banach lattice. Moreover, every regular operator $T + \imath S$ has a modulus $|T + \imath S|$ in $\mathcal{L}(E_{\mathbf{c}}, F_c)$. Remarkably, the modulus $|T + \imath S|$ is given (just as in the real case) by the Riesz–Kantorovich formula. The details are included in the next result.

Theorem 3.24. *Let E and F be two real Banach lattices with F Dedekind complete, and let $\mathcal{T} = T + \imath S \in \mathcal{L}_r(E_{\mathbf{c}}, F_{\mathbf{c}}) = \mathcal{L}_r(E, F) \oplus \imath\mathcal{L}_r(E, F)$. Then the modulus $|\mathcal{T}|$ of $\mathcal{T}$ satisfies the Riesz–Kantorovich formula, i.e., for each $u \in E^+$ we have*

$$|\mathcal{T}|u = |T + \imath S|u = \sup_{|z| \leq u} |(T + \imath S)z| = \sup_{|x + \imath y| \leq u} |(T + \imath S)(x + \imath y)|\,.$$

Proof. Let $\mathcal{T} = T + \imath S \in \mathcal{L}_r(E_{\mathbf{c}}, F_{\mathbf{c}})$ and fix $u \in E^+$. It should be clear that $a = \sup_{|z| \leq u} |\mathcal{T}z|$ exists in F. We must show that $a = |\mathcal{T}|u$.

Recall that $|\mathcal{T}| = \sup_{\theta \in \mathbb{R}}[T\cos\theta + S\sin\theta]$, where the supremum is taken in the Dedekind complete Riesz space $\mathcal{L}_r(E, F)$. Consider the subset D of F consisting of all vectors of the form $v = \sum_{i=1}^{n}[(Tu_i)\cos\theta_i + (Su_i)\sin\theta_i]$, where $\theta_1, \ldots, \theta_n \in \mathbb{R}$ and $u_1, \ldots, u_n$ are positive vectors in E satisfying $u_1 + \cdots + u_n = u$. It is well known that $\sup D = |\mathcal{T}|u$; see [**30**, Theorem 1.16, p. 15].

Now let $v = \sum_{i=1}^{n}[(Tu_i)\cos\theta_i + (Su_i)\sin\theta_i] \in D$. Put $x = \sum_{i=1}^{n} u_i \cos\theta_i$ and $y = \sum_{i=1}^{n} u_i \sin\theta_i$ and consider the vector $z = x - \imath y \in E_{\mathbf{C}}$. From

$$\begin{aligned} x\cos\phi - y\sin\phi &= \sum_{i=1}^{n} u_i(\cos\theta_i\cos\phi - \sin\theta_i\sin\phi) \\ &= \sum_{i=1}^{n} u_i\cos(\theta_i + \phi) \le \sum_{i=1}^{n} u_i = u\,, \end{aligned}$$

it follows that $|z| = \sup_{\phi\in\mathbb{R}}[x\cos\phi - y\sin\phi] \le u$. Next, notice that

$$\mathcal{T}z = (T + \imath S)(x - \imath y) = (Tx + Sy) + \imath(Sx - Ty) = v + \imath(Sx - Ty)$$

implies

$$v \le |v| \le \bigl|v + \imath(Sx - Ty)\bigr| = |\mathcal{T}z| \le a$$

for all $v \in D$. Since $\sup D = |\mathcal{T}|u$, we see that

$$|\mathcal{T}|u \le a = \sup_{|z|\le u} |\mathcal{T}z|\,. \qquad (\star)$$

For the reverse inequality, we shall need the following two (easily checked) simple facts:

(1) If $w \in E_{\mathbf{C}}$ satisfies $|w| \le v$ for some $v \in E^+$, then $|\mathcal{T}w| \le (|T| + |S|)v$.

(2) If $v \in E^+$, then $|\mathcal{T}v| \le |\mathcal{T}|v$.

Now let $z \in E_{\mathbf{C}}$ satisfy $|z| \le u$, and fix $\epsilon > 0$. Then there exist positive vectors $u_1, \dots, u_n$ in E and complex scalars $\alpha_1, \dots, \alpha_n$ such that $|\alpha_i| \le 1$ for each $i = 1, \dots, n$, $\sum_{i=1}^{n} u_i = 1$, and $\bigl|z - \sum_{i=1}^{n} \alpha_i u_i\bigr| \le \epsilon u$; see Exercise 12 at the end of the section. Letting $R = |T| + |S|$, we have

$$\begin{aligned} |\mathcal{T}z| &= \Bigl|\mathcal{T}\Bigl(z - \sum_{i=1}^{n}\alpha_i u_i\Bigr) + \mathcal{T}\Bigl(\sum_{i=1}^{n}\alpha_i u_i\Bigr)\Bigr| \\ &\le \Bigl|\mathcal{T}\Bigl(z - \sum_{i=1}^{n}\alpha_i u_i\Bigr)\Bigr| + \Bigl|\mathcal{T}\Bigl(\sum_{i=1}^{n}\alpha_i u_i\Bigr)\Bigr| \\ &\le \Bigl|\mathcal{T}\Bigl(z - \sum_{i=1}^{n}\alpha_i u_i\Bigr)\Bigr| + \sum_{i=1}^{n}\bigl|\alpha_i\mathcal{T}(u_i)\bigr| \le \epsilon R u + \sum_{i=1}^{n}\bigl|\mathcal{T}(u_i)\bigr| \\ &\le \epsilon R u + \sum_{i=1}^{n}|\mathcal{T}|u_i = \epsilon R u + |\mathcal{T}|u\,. \end{aligned}$$

Therefore, $a = \sup_{|z|\le u}|\mathcal{T}z| \le \epsilon R u + |\mathcal{T}|u$ for each $\epsilon > 0$. Hence, $a \le |\mathcal{T}|u$ is also true. This inequality, combined with $(\star)$, establishes the validity of the Riesz–Kantorovich formula. ■

Here are three useful consequences of the preceding results.

Corollary 3.25. *If E and F are two Banach lattices with F Dedekind complete, then every regular operator $\mathcal{T} \in \mathcal{L}(E_{\mathbf{C}}, F_{\mathbf{C}})$ satisfies*

$$|\mathcal{T}z| \leq |\mathcal{T}|\big(|z|\big)$$

for all $z \in E_{\mathbf{C}}$.

Corollary 3.26. *If E is a real Banach lattice, then the norm dual of the complexification $E_{\mathbf{C}}$ of E coincides with the complexification of E^*, i.e., $E^*_{\mathbf{C}} = E^* \oplus \imath E^*$. Moreover, if $f \in E^*_{\mathbf{C}}$, then its modulus satisfies the familiar Riesz–Kantorovich formula, i.e.,*

$$|f|(u) = \sup_{|z| \leq u} |f(z)|$$

for each $u \in E^+$.

Corollary 3.27. *If E is a Dedekind complete Banach lattice, then the complex Banach lattice $\mathcal{L}_r(E_{\mathbf{C}})$ under the r-norm is a complex unital Banach algebra.*

Proof. By Theorem 1.32, we know that $\mathcal{L}_r(E_{\mathbf{C}}) = \mathcal{L}_r(E) \oplus \imath\mathcal{L}_r(E)$ is a complex Banach lattice. Now let $\mathcal{S}, \mathcal{T} \in \mathcal{L}_r(E_{\mathbf{C}})$ and fix some $z \in E_{\mathbf{C}}$ satisfying $|z| \leq u$ for some $u \in E^+$. Then we have

$$\big|(\mathcal{S}\mathcal{T})z\big| = \big|\mathcal{S}(\mathcal{T}z)\big| \leq |\mathcal{S}|\big(|\mathcal{T}z|\big) \leq |\mathcal{S}|\big(|\mathcal{T}|u\big) = (|\mathcal{S}||\mathcal{T}|)u\,,$$

and from this and Theorem 3.24 we get $|\mathcal{S}\mathcal{T}| \leq |\mathcal{S}||\mathcal{T}|$. Consequently,

$$\|\mathcal{S}\mathcal{T}\|_r = \big\||\mathcal{S}\mathcal{T}|\big\| \leq \big\||\mathcal{S}||\mathcal{T}|\big\| \leq \big\||\mathcal{S}|\big\| \cdot \big\||\mathcal{T}|\big\| = \|\mathcal{S}\|_r \cdot \|\mathcal{T}\|_r\,.$$

Hence, $\mathcal{L}_r(E_{\mathbf{C}})$ is a complex unital Banach algebra. ∎

Finally, we can extend the notion of domination to operators between complex Banach lattices. We shall say that a positive operator $T\colon E \to F$ between (real) Banach lattices **dominates** an operator $S\colon E_{\mathbf{C}} \to F_{\mathbf{C}}$ (or that S is **dominated** by T) if $|Sz| \leq T|z|$ for all $z \in E_{\mathbf{C}}$. It should be noted that if an operator between two complex Banach lattices is dominated by a positive operator, then it is automatically continuous.

Exercises

1. If E is a Banach lattice and $z = x + \imath y \in E_{\mathbf{C}}$, then show that
$$|z| = \sup_{\theta \in [0,2\pi]} |x\cos\theta + y\sin\theta| = \sup_{\theta \in \mathbb{R}} |x\cos\theta + y\sin\theta|\,.$$
2. If E is a Banach lattice, then for arbitrary $z, z_1, z_2 \in E_{\mathbf{C}}$ and $\lambda \in \mathbb{C}$ show that:
   (a) $|z| \geq 0$, and $|z| = 0$ if and only if $z = 0$.
   (b) $|\lambda z| = |\lambda| \cdot |z|$.

(c) $|z_1 + z_2| \le |z_1| + |z_2|$.

Use the above properties to verify that the function $\|\cdot\|_{\mathbf{c}}\colon E_{\mathbf{c}} \to \mathbb{R}$, defined by $\|z\|_{\mathbf{c}} = \|\,|z|\,\|$, is a norm.

3. Show that the complexification of $C_{\mathbb{R}}(\Omega)$ coincides with $C_{\mathbb{C}}(\Omega)$. Moreover, show that for each $f, g \in C_{\mathbb{R}}(\Omega)$ we have $|f + \imath g| = \sqrt{f^2 + g^2}$ and hence $\|f + \imath g\|_{\mathbf{c}} = \|\sqrt{f^2 + g^2}\|_\infty$.

4. Show that the complexification of $\mathbb{R}^n$ coincides with $\mathbb{C}^n$. Also, show that all norms on $\mathbb{C}^n$ (or $\mathbb{R}^n$) are equivalent. [HINT: Let $\|\cdot\|$ be a norm on $\mathbb{C}^n$ and let $\|\cdot\|_1$ denote the ℓ_1-norm, i.e., $\|z\|_1 = \sum_{k=1}^n |z_k|$. An easy argument shows that the identity mapping $I\colon (\mathbb{C}^n, \|\cdot\|_1) \to (\mathbb{C}^n, \|\cdot\|)$ is continuous. This implies that the set $K = \{z \in \mathbb{C}^n\colon \|z\|_1 = 1\}$ is a $\|\cdot\|$-compact subset of $\mathbb{C}^n$. The continuity of the mapping $z \mapsto \|z\|$ guarantees that $m = \min_{z\in K} \|z\| > 0$ and $M = \max_{z\in K} \|z\| < \infty$. Now use the identity $\|z\| = \|z\|_1 \cdot \|\frac{z}{\|z\|_1}\|$ to get $m\|z\|_1 \le \|z\| \le M\|z\|_1$ for all $z \in \mathbb{C}^n$.]

5. Show that the complexification of the real $L_p(\mu)$ coincides with the complex $L_p(\mu)$. Also, show that for all real-valued functions f and g we have $|f + \imath g| = \sqrt{f^2 + g^2}$ and that $\|f + \imath g\|_{\mathbf{c}} = \|\sqrt{f^2 + g^2}\|_p$.

6. Recall that $\|\cdot\|_1$- and $\|\cdot\|_\infty$-norms on $\mathbb{C}^n$ are defined via the formulas

$$\|z\|_1 = \sum_{i=1}^n |z_i| \quad \text{and} \quad \|z\|_\infty = \max_{1\le i\le n} |z_i|\,.$$

If $A = [a_{ij}]$ is an $n \times n$ matrix, then the $\|\cdot\|_1$- and the $\|\cdot\|_\infty$-norms of A are defined by

$$\|A\|_1 = \sup_{\|z\|_1\le 1} \|Az\|_1 \quad \text{and} \quad \|A\|_\infty = \sup_{\|z\|_\infty\le 1} \|Az\|_\infty\,.$$

Show that:

$$\|A\|_1 = \max_{1\le j\le n} \sum_{i=1}^n |a_{ij}| \quad \text{and} \quad \|A\|_\infty = \max_{1\le i\le n} \sum_{j=1}^n |a_{ij}|\,.$$

[HINT: We shall indicate how to establish the second formula only. Let $\sigma = \max_{1\le i\le n} \sum_{j=1}^n |a_{ij}|$. If $\|z\|_\infty \le 1$ and $y = Az$, then

$$|y_i| = \Big|\sum_{j=1}^n a_{ij} z_j\Big| \le \sum_{j=1}^n |a_{ij}||z_j| \le \sum_{j=1}^n |a_{ij}| \le \sigma$$

for each i, so that $\|A\|_\infty \le \sigma$. Next, fix some m such that $\sum_{j=1}^n |a_{mj}| = \sigma$ and then for each j pick some θ_j such that $a_{mj} e^{\imath\theta_j} = |a_{mj}|$, where $\imath$ denotes as usual the imaginary unit. Then $z = (e^{\imath\theta_1}, \dots, e^{\imath\theta_n})$ satisfies $\|z\|_\infty = 1$ and if $y = Az$, then $y_m = \sum_{j=1}^n a_{mj} e^{\imath\theta_j} = \sum_{j=1}^n |a_{mj}| = \sigma$, proving that $\|A\|_\infty = \sigma$.]

7. Let E be a real Banach lattice with complexification $E_{\mathbf{c}}$. A vector subspace J of $E_{\mathbf{c}}$ is said to be an **ideal** whenever $|z_1| \le |z_2|$ and $z_2 \in J$ imply $z_1 \in J$.

(a) Show that the ideals in $E_{\mathbf{c}}$ are precisely the vector subspaces of the form $J = J_0 \oplus \imath J_0$, where J_0 is an ideal in E.

(b) A bounded operator $T: F \to F$ on a (real or complex) Banach lattice is said to be **ideal irreducible** if there is no non-trivial closed ideal in F which is T-invariant. That is, T is ideal irreducible if and only if $T(J) \subseteq J$ and J a closed ideal in F imply either $J = \{0\}$ or $J = F$. If $T: E \to E$ is an operator on a real Banach lattice, then show that T is ideal irreducible if and only if $T_{\mathbf{c}}: E_{\mathbf{c}} \to E_{\mathbf{c}}$ is likewise ideal irreducible.

[HINT: Assume that J is an ideal of $E_{\mathbf{c}}$. Let

$$J_0 = \{x \in E\colon \text{ There exists some } y \in E \text{ with } x + \imath y \in J\}.$$

Verify that J_0 is an ideal in E satisfying $J = J_0 \oplus \imath J_0$. For part (b) note that an ideal $J = J_0 \oplus i J_0$ is closed in $E_{\mathbf{c}}$ if and only if J_0 is a closed ideal in E.]

8. Show that all "domination properties" included in the results 2.34–2.40 are also true for complex Banach lattices. [HINT: Let $X_{\mathbf{c}}$ and $Y_{\mathbf{c}}$ be the complexifications of two real vector spaces and let $\mathcal{T}: X_{\mathbf{c}} \to Y_{\mathbf{c}}$ be a linear operator. In view of Exercise 9 in Section 1.1, we know that $\mathcal{T}$ can be represented by a matrix of the form $\mathcal{T} = \begin{bmatrix} T & -S \\ S & T \end{bmatrix}$, where $T, S\colon X \to Y$ are uniquely determined linear operators—which are both bounded if and only if $\mathcal{T}$ is a bounded operator. Now, assume that a bounded operator $\mathcal{S} = \begin{bmatrix} A & -B \\ B & A \end{bmatrix} : E_{\mathbf{c}} \to F_{\mathbf{c}}$ between two complex Banach lattices is dominated by a positive operator $T\colon E \to F$. Show that both A and B are dominated by T.]

9. (Wickstead) Let $E = \mathbb{R}^2$ be equipped with the sup norm $\|\cdot\|_\infty$ (i.e., $\|(x, y)\|_\infty = \max\{|x|, |y|\}$) and let $F = \mathbb{R}^2$ be equipped with the ℓ_1-norm $\|\cdot\|_1$ (i.e., $\|(x, y)\|_1 = |x| + |y|$). Now consider the operator $T\colon E \to F$ defined by

$$T(x, y) = (x + y, x - y).$$

Show that the operator T is regular satisfying $\|T\|_{\mathbf{c}} \geq 2\sqrt{2} > 2 = \|T\|$. (Note that this does not contradict Corollary 3.23 since T is not positive.)

10. Let H be a real Hilbert space with inner product $(\cdot, \cdot)$. Extending the inner product to $H_{\mathbf{c}}$ via the formula

$$(z_1, z_2) = (x_1 + \imath y_1, x_2 + \imath y_2) = [(x_1, x_2) + (y_1, y_2)] + \imath[(y_1, x_2) - (x_1, y_2)],$$

we turn $H_{\mathbf{c}}$ into a (complex) Hilbert space. Show that the standard norm $\|\cdot\|_{\mathbf{c}}$ on $H_{\mathbf{c}}$ defined by

$$\|z\|_{\mathbf{c}} = \|x + \imath y\|_{\mathbf{c}} = \sup_{\theta \in \mathbb{R}} \|x \cos\theta + y \sin\theta\|$$

is not in general an inner product norm—and hence $\|\cdot\|_{\mathbf{c}}$ does not coincide with the norm induced by the above inner product.

11. Let X be a real Banach space and define $N\colon X \times X \to \mathbb{R}$ by

$$N(x, y) = (\|x\|^2 + \|y\|^2)^{\frac{1}{2}}.$$

Establish the following:

(a) The real vector space $X \oplus X$ normed with N is a Banach space.

(b) The same function N now considered defined on the complexification $X_{\mathbf{c}} = X \oplus \imath X$, that is, $N(x + \imath y) = (\|x\|^2 + \|y\|^2)^{\frac{1}{2}}$, is a norm on $X_{\mathbf{c}}$ if and only if X is a real Hilbert space.

12. Let $E_{\mathbf{c}}$ be the complexification of a real Banach lattice E and assume that $|z| \le u$ for some $z \in E_{\mathbf{c}}$ and $u \in E^+$. Show that for each $\epsilon > 0$ there exist positive vectors $u_1, \ldots, u_n$ in E and complex scalars $\alpha_1, \ldots, \alpha_n$ such that $|\alpha_i| \le 1$ for each $i = 1, \ldots, n$, $\sum_{i=1}^n u_i = 1$, and $|z - \sum_{i=1}^n \alpha_i u_i| \le \epsilon u$. [HINT: We can assume that $E_{\mathbf{c}} = C_{\mathbb{C}}(\Omega)$ and $|z| \le \mathbf{1}$. Since $z(\Omega)$ is a compact set, there exist $\omega_1, \ldots, \omega_n \in \Omega$ such that $z(\Omega) \subseteq \bigcup_{i=1}^n D(z(\omega_i), \epsilon)$. So, if $V_i = z^{-1}(D(z(\omega_i), \epsilon))$, then each V_i is a non-empty open subset of Ω and $\Omega = \bigcup_{i=1}^n V_i$. Next, pick a partition of unity $\{u_1, u_2, \ldots, u_n\}$ subordinate to the open cover $\{V_1, \ldots, V_n\}$; see [**31**, Theorem 10.9, p. 85]. Put $\alpha_i = z(\omega_i)$, and note that

$$\begin{aligned} \Big| z(\omega) - \sum_{i=1}^n \alpha_i u_i(\omega) \Big| &= \Big| \sum_{i=1}^n u_i(\omega)[z(\omega) - z(\omega_i)] \Big| \\ &\le \sum_{i=1}^n |u_i(\omega)[z(\omega) - z(\omega_i)]| \le \sum_{i=1}^n u_i(\omega)\epsilon = \epsilon\,.] \end{aligned}$$

3.3. The Center of a Banach Lattice

The purpose of this section is to discuss an important class of operators that generalizes the class of multiplication operators. This is the class of central operators. For details and complete proofs of the results stated here, we refer the reader to [**30**, Sections 8 and 15].

Definition 3.28. *An operator $T\colon E \to E$ on a real or a complex Riesz space is called **central** if it is dominated by a multiple of the identity operator. That is, T is a central operator if and only if there exists some scalar $\lambda > 0$ such that $|Tx| \le \lambda|x|$ holds for all $x \in E$.*

*The collection of all central operators is denoted by $\mathcal{Z}(E)$ and is referred to as the **center** of the Banach lattice E.*

Clearly, each central operator is regular and every positive central operator is a lattice homomorphism. Moreover, it is easy to see that the center $\mathcal{Z}(E)$ is an ideal in $\mathcal{L}_r(E)$, the vector space of all regular operators on E.

The central operators are special examples of operators known as orthomorphisms. An order bounded operator $T\colon E \to E$ on a Riesz space is said to be an **orthomorphism** if $|x| \wedge |y| = 0$ implies $|x| \wedge |Ty| = 0$. Recall also that an operator $T\colon E \to E$ on a Riesz space is **band preserving** if $T(B) \subseteq B$ for every band B of E.

It is worth pointing out that in general a band preserving operator does not need to be order bounded; see [**23, 342**]. However, for Banach lattices the situation is better: Namely, the band preserving operators and the central operators coincide. This is included in the next result; for a proof see [**30**, Theorem 15.4, p. 250].

Theorem 3.29 (Abramovich–Veksler–Koldunov [**23**]). *For a given operator $T\colon E \to E$ on a Banach lattice the following statements are equivalent.*

(1) *T is a central operator.*

(2) *T is an orthomorphism.*

(3) *T is a band preserving operator.*

In spite of the fact that for an arbitrary Riesz space E, the partially ordered vector space $\mathcal{L}_r(E)$ need not be a Riesz space, the center $\mathcal{Z}(E)$ is always a Riesz space. The details are in the next result.

Theorem 3.30 (Bigard–Keimel [**62**]; Conrad–Diem [**93**]). *If E is a Riesz space, then $\mathcal{Z}(E)$ is a Riesz space. Its lattice operations are given by*

$$(S \vee T)(x) = S(x) \vee T(x) \quad \text{and} \quad (S \wedge T)(x) = S(x) \wedge T(x)$$

for all $S, T \in \mathcal{Z}(E)$ and all $x \in E^+$.

Moreover, if $T \in \mathcal{Z}(E)$, then its modulus $|T|$ satisfies

$$|T|(|x|) = |T(|x|)| = |T(x)|$$

for each $x \in E$.

For a proof of Theorem 3.30 see [**30**, pp. 107–109]. In fact more is true. The center of a Banach lattice is always an AM-space with unit.

Theorem 3.31 (Wickstead [**334**]). *For a Banach lattice E the following two statements are true.*

(1) *The center $\mathcal{Z}(E)$ equipped with the operator norm is an AM-space with unit. The unit is the identity operator I, so that for each operator $T \in \mathcal{Z}(E)$ we have*

$$\|T\| = \|\,|T|\,\| = \inf\{\lambda \geq 0\colon\ |T| \leq \lambda I\}\,.$$

Moreover, $\mathcal{Z}(E)$ is a unital Banach algebra.

(2) *If E is Dedekind complete, then the center $\mathcal{Z}(E)$ coincides (as a Riesz space) with the band generated by the identity operator I in $\mathcal{L}_r(E)$; and so*

$$\mathcal{L}_r(E) = \mathcal{Z}(E) \oplus I^{\mathrm{d}}\,.$$

Now let us discuss the center of an AM-space with unit. To start with, recall that (by Theorem 3.6) an AM-space with unit e is lattice isometric to some $C(\Omega)$-space so that the unit e corresponds to the constant function $\mathbf{1}$ on Ω. Every function $x \in C(\Omega)$ gives rise to a central operator (also called a multiplication operator) $M_x\colon C(\Omega) \to C(\Omega)$ defined by $M_x(y) = xy$. A moment's thought reveals that

$$\begin{aligned}\|M_x\| &= \sup_{\|y\|_\infty \le 1} \|M_x(y)\|_\infty = \sup_{\|y\|_\infty \le 1} \|xy\|_\infty \\ &= \|x\|_\infty = \inf\{\lambda \ge 0\colon\ |M_x| \le \lambda I\}\,.\end{aligned}$$

This shows that $x \mapsto M_x$ is a linear isometry from $C(\Omega)$ to $\mathcal{Z}(C(\Omega))$. It turns out that this linear isometry is, in fact, a surjective lattice isometry.

Theorem 3.32 (Zaanen [**342**]). *If Ω is a compact Hausdorff space, then the mapping $x \mapsto M_x$, from $C(\Omega)$ to $\mathcal{Z}(C(\Omega))$, is a surjective lattice isometry; and so, under this identification, we have $\mathcal{Z}(C(\Omega)) = C(\Omega)$.*

Proof. To see that $x \mapsto M_x$ is a lattice isometry, fix $u, v, y \in C(\Omega)$ with $y \ge 0$. Then from Theorem 3.30, we get

$$(M_u \vee M_v)(y) = M_u(y) \vee M_v(y) = (uy) \vee (vy) = (u \vee v)y = M_{u\vee v}(y)\,.$$

This implies $M_u \vee M_v = M_{u\vee v}$, and so $x \mapsto M_x$ is a lattice isometry.

The only thing missing to complete the proof is the fact that $x \mapsto M_x$ is surjective. To establish this we need the following simple property of central operators.

- *A central operator $T\colon C(\Omega) \to C(\Omega)$ is zero if and only if $T\mathbf{1} = 0$.*

To see this, assume that a central operator T satisfies $T\mathbf{1} = 0$. Now if $x \in C(\Omega)$, then choose some $\lambda > 0$ satisfying $|x| \le \lambda\mathbf{1}$, and then use Theorem 3.30 to get

$$\big|T(x)\big| = |T|\big(|x|\big) \le \lambda|T|\mathbf{1} = \lambda\big|T\mathbf{1}\big| = 0\,.$$

Hence $T(x) = 0$ for each $x \in C(\Omega)$, and so $T = 0$.

Finally, let $T \in \mathcal{Z}\big(C(\Omega)\big)$ and put $u = T\mathbf{1} \in C(\Omega)$. Since the multiplication operator M_u is central, the operator $S = T - M_u$ is also central and satisfies $S\mathbf{1} = T\mathbf{1} - u = 0$. By the above, $S = 0$ or $T = M_u$. This shows that $x \mapsto M_x$ is a surjective lattice isometry. ■

Next, consider an AM-space E with unit e. Since E is lattice isometric to a (unique) $C(\Omega)$-space, we can view the vectors in E as continuous functions on Ω. In other words, we can equip E with a unique multiplication having e as a unit. In this setting, we can consider every element $x \in E$ as a central operator M_x on E defined via the formula $M_x(y) = xy$. Moreover, by Theorem 3.32, the mapping $x \mapsto M_x$ is a lattice isometry from E onto

$\mathcal{Z}(E)$. That is, subject to the above identification, we have the following result.

Theorem 3.33. *If E is an AM-space with unit, then $\mathcal{Z}(E) = E$.*

Now consider an arbitrary Banach lattice E and let $0 < u \in E$. By Theorem 3.4, we know that the principal ideal E_u, under the norm

$$\|x\|_\infty = \inf\{\lambda \geq 0 \colon \ |x| \leq \lambda u\},$$

is an AM-space with unit u. So, every element $x \in E_u$ defines a central operator M_x on E_u via the formula $M_x(y) = xy$. As before, every central operator M_x on E_u is continuous with respect to the norm $\|\cdot\|$ of E (in fact $\|M_x\| = \|x\|_\infty$), and hence it extends uniquely to a $\|\cdot\|$-continuous operator on $\overline{E}_u$, the $\|\cdot\|$-closure of E_u in E. We shall denote this extension by M_x again. If E_u happens to be dense in E, then the unique continuous extension of M_x is also a central operator of E. Remarkably, in this case, it turns out that each central operator on E is of this type. This is an important result.

Theorem 3.34. *Let E be a Banach lattice and let $0 < u \in E$ satisfy $\overline{E}_u = E$. Then the mapping $x \mapsto M_x$, from E_u to $\mathcal{Z}(E)$, is a surjective lattice isometry—and so, in this case, the center $\mathcal{Z}(E)$ can be identified with the AM-space E_u.*

Proof. The principal ideal E_u is lattice isometric to some $C(\Omega)$. By virtue of Theorem 3.32, we know $\mathcal{Z}(E_u)$ is lattice isometric to E_u via the mapping $x \mapsto M_x$. As before, we again denote by M_x the unique $\|\cdot\|$-continuous linear extension of M_x to all of E. This extension is also a central operator and preserves the norm of M_x, i.e., $\|M_x\| = \|x\|_\infty$. By the continuity of the lattice operations and the $\|\cdot\|$-denseness of E_u in E, it follows that $x \mapsto M_x$ is also a lattice isometry from E_u to $\mathcal{Z}(E)$. To finish the proof, we shall show that $x \mapsto M_x$ is also surjective.

To this end, let $T \in \mathcal{Z}(E)$. Then $T(E_u) \subseteq E_u$ and so T restricted to E_u is a central operator on E_u. By Theorem 3.32, $T = M_x$ for some $x \in E_u$. By the $\|\cdot\|$-continuity of both T and M_x, it follows that $T = M_x$. This shows that $x \mapsto M_x$ is also surjective, and so we can identify $\mathcal{Z}(E)$ with E_u. ■

There is one more description of the center of a Banach lattice which is very useful in many contexts. To understand it, we need some facts from the representation theory of Riesz spaces. Recall that if Q is a **Stonean space**, i.e., a compact Hausdorff space such that the closure of every open set is open, then the collection $C^\infty(Q)$ of all extended continuous real-valued functions $f \colon Q \to [-\infty, \infty]$ for which the open set $\{q \in Q \colon \ |f(q)| < \infty\}$ is dense in Q, under the pointwise algebraic and lattice operations, is a Dedekind complete Riesz space and an algebra. It turns out that for every Archimedean Riesz space E there is a unique (up to homeomorphism) Stonean space Q and a lattice isomorphism π from E into $C^\infty(Q)$ such that for each $0 < f \in C^\infty(Q)$ there exists some $0 < x \in E$ such that $0 < \pi(x) \leq f$, i.e., $\pi(E)$ is order dense in $C^\infty(Q)$. If E has a weak unit e, we can assume that π carries e to $\mathbf{1}$, i.e., $\pi(e) = \mathbf{1}$, the constant function one on Q. The unique Stonean space Q is referred to as the **Stone space** of E. For all practical purposes, we can identify E with $\pi(E)$ and consider the elements of E as extended real-valued functions on Q. For details about the representation of Riesz spaces we refer the reader to [**225**, Chapter 7]. Summarizing the discussion above, we have the following important result.

Theorem 3.35 (Maeda–Ogasawara–Vulikh). *If E is an Archimedean Riesz space, then there exists a unique—up to homeomorphism—Stonean space Q and an order dense Riesz subspace $\hat{E}$ of the Dedekind complete Riesz space $C^\infty(Q)$ that is lattice isomorphic to E. Moreover:*

(1) *If E has a weak unit e, then the lattice isomorphism can be chosen to carry e to the constant function $\mathbf{1}$ on Q.*

(2) *If E is Dedekind complete, then $\hat{E}$ is an order dense ideal in $C^\infty(Q)$.*

(3) *If E is Dedekind complete with a strong unit e, then $\hat{E} = C(Q)$.*

It is customary to identify $\hat{E}$ with E and we shall follow this custom. The unique Dedekind complete Riesz space $C^\infty(Q)$ is called the **Maeda–Ogasawara–Vulikh completion** of E. It is also known as the **universal completion** of the Riesz space E. For details about the universal completion see [**29**, Chapter 7].

From the general theory of orthomorphisms we also have the following characterization of the center of a Banach lattice. For a proof see [**30**, Theorem 8.28, p. 121].

Theorem 3.36. *If Q is the Stone space of a Banach lattice E, then*

$$\mathcal{Z}(E) = \big\{ f \in C^\infty(Q)\colon\ fg \in E \ \text{ for all } \ g \in E \big\}.$$

Moreover, if E is Dedekind complete, then $\mathcal{Z}(E) = C(Q)$.

We point out here an important difference between a Dedekind complete Banach lattice and a general Banach lattice. From Theorem 3.36, it follows that the center $\mathcal{Z}(E)$ of a Dedekind complete Banach lattice E is always a **non-trivial** AM-space, that is, $\mathcal{Z}(E) \neq \{\lambda I\colon\ \lambda \text{ an arbitrary scalar}\}$. However, a non-Dedekind complete Banach lattice can have a trivial center. An example of such a pathological Banach lattice was provided first by A. Goullet de Rugy [**142**]; see also [**336**].

By Theorem 3.31, we know that if E is a Dedekind complete Banach lattice, then $\mathcal{Z}(E)$ is a band in $\mathcal{L}_b(E)$ and

$$\mathcal{L}_b(E) = \mathcal{L}_r(E) = \mathcal{Z}(E) \oplus I^{\mathrm{d}}.$$

That is, every regular operator $T\colon E \to E$ can be decomposed uniquely in the form $T = T_1 \oplus T_2$, where T_1 is a central operator and T_2 is an operator disjoint from the identity operator. This shows that the formula $P_{\mathcal{Z}}(T) = T_1$ defines a natural projection $P_{\mathcal{Z}}$ defined on $\mathcal{L}_r(E)$ and having the center $\mathcal{Z}(E)$ as its range. This projection is known as the **diagonal projection** of $\mathcal{L}_r(E)$ and $P_{\mathcal{Z}}(T)$ is called the **diagonal part** of T. As a band projection, $P_{\mathcal{Z}}$ is a contraction with respect to the r-norm. Remarkably, J. Voight has shown that $P_{\mathcal{Z}}$ is also a contraction with respect to the operator norm.

Theorem 3.37 (Voight [**327**]). *If E is a Dedekind complete Banach lattice, then the diagonal projection $P_{\mathcal{Z}}$ from $\mathcal{L}_r(E)$ onto the center $\mathcal{Z}(E)$ is a contraction with respect to the r-norm.*

Proof. We must establish that $\|P_{\mathcal{Z}}(T)\| \leq \|T\|$ for each $T \in \mathcal{L}_r(E)$. To this end, fix some $T \in \mathcal{L}_r(E)$. Since $\mathcal{Z}(E)$ is a band in $\mathcal{L}_r(E)$, we can write $T = Z + R$, where $Z = P_{\mathcal{Z}}(T) \in \mathcal{Z}(E)$ and $|R| \wedge I = 0$.

We can assume $Z \neq 0$. Fix $0 < \gamma < 1$ and then select some $0 < u \in E$ with $\|u\| = 1$ and $\|Zu\| > \gamma\|Z\|$. (This is possible since $\big|Z(|z|)\big| = |Zz|$ holds for all $z \in E$; see Theorem 3.30.) Next, we denote by B_u the band generated by u in E and let $C^\infty(Q)$ be the Maeda–Ogasawara–Vulikh completion of B_u, where u corresponds to the constant function $\mathbf{1}$ on Q. Since $Z(B_u) \subseteq B_u$ (see Theorem 3.29), there exists (by Theorem 3.36) some $h \in C(\Omega)$ satisfying $Z(y) = hy$ for each $y \in B_u$. From $\|Zu\| > \gamma\|Z\|$, we get $\|h\|_\infty > \gamma\|Z\|$. So, if V is the closure of the open set $\{q \in Q\colon\ |h(q)| > \gamma\|Z\|\}$, then V is a non-empty clopen subset of Q and $\chi_V \in E$.

Now fix $0 < \delta < \gamma$. Using Exercise 10 at the end of this section, we know that there exists a non-zero component $a = \chi_W$ of χ_V (where W is a clopen subset of V) such that $P_a(|R|a) \leq \delta\|Z\|a$, where P_a denotes the projection onto the band generated by a in E. From

$$\begin{aligned} \big|(Z+R)a\big| &\geq \big|P_a(Z+R)a\big| = \big|P_a(Za) + P_a(Ra)\big| = \big|Za + P_a(Ra)\big| \\ &= \big|ha + P_a(Ra)\big| \geq |h|a - P_a(|Ra|) \geq (\gamma - \delta)\|Z\|a\,, \end{aligned}$$

we get $\|Z+R\| \cdot \|a\| \geq \big\||(Z+R)a|\big\| \geq (\gamma-\delta)\|Z\| \cdot \|a\|$. Consequently, we have $\|Z+R\| \geq (\gamma-\delta)\|Z\|$ for all $\gamma, \delta > 0$ with $0 < \delta < \gamma < 1$. This implies $\|T\| = \|Z+R\| \geq \|Z\|$, as desired. ∎

Exercises

1. Show that a positive operator $T\colon E \to E$ on a Riesz space is an orthomorphism if and only if $I+T$ is a lattice homomorphism.

2. Assume that E is a Dedekind complete Banach lattice. Use Theorem 1.17 to present an alternate proof of Theorem 3.30.

3. Show that every central operator $T\colon E \to E$ on a Banach lattice leaves the principal ideals invariant. That is, show that $T(E_u) \subseteq E_u$ holds for each $u \in E^+$.

4. Prove Theorem 3.31.

5. If μ is a finite measure, show (with appropriate identifications) that $\mathcal{Z}(L_p(\mu)) = L_\infty(\mu)$ for each $1 \le p \le \infty$.

6. Prove the second part of the Maeda–Ogasawara–Vulikh Representation Theorem 3.35 by establishing the following general result for Riesz spaces.

If F is an order dense Riesz subspace of an Archimedean Riesz space E and F is Dedekind complete in its own right, then F is an ideal in E.

7. Show that a central operator on a Banach lattice commutes with all band projections.

8. This exercise justifies the name "diagonal projection" for the band projection $P_{\mathcal{Z}}$ of $\mathcal{L}_r(E)$ onto the center of the Banach lattice E. Let $E = \mathbb{R}^n$, the Euclidean n-dimensional (Riesz) space. Then $\mathcal{L}(E) = \mathcal{L}_r(E)$, the vector space of all $n \times n$ matrices with real entries. If D is a **diagonal matrix** with diagonal entries $d_1, \ldots, d_n$, then we write $D = \operatorname{diag}(d_1, \ldots, d_n)$. Establish the following properties.

(a) If $A = [a_{ij}]$ and $B = [b_{ij}]$ are two $n \times n$ matrices, then $A \ge B$ holds in $\mathcal{L}_r(E)$ if and only if $a_{ij} \ge b_{ij}$ for all i and j.

(b) If $A = [a_{ij}]$ and $B = [b_{ij}]$ are two $n \times n$ matrices, then

$$|A| = [\,|a_{ij}|\,], \quad A \vee B = [a_{ij} \vee b_{ij}] \quad \text{and} \quad A \wedge B = [a_{ij} \wedge b_{ij}]\,.$$

(c) The center of E consists of all diagonal matrices.

(d) If $A = [a_{ij}]$ is an $n \times n$ matrix, then $P_{\mathcal{Z}}(A) = \operatorname{diag}(a_{11}, a_{22}, \ldots, a_{nn})$.

9. Prove the second part of Theorem 3.36. That is, show that if E is a Dedekind complete Banach lattice, then $\mathcal{Z}(E) = C(Q)$, where Q is the Stone space of E.

10. Recall that the **principal band** B_x generated by a vector x in a Riesz space E is the smallest band that contains x. From Theorem 1.27, we know that $B_x = \{y \in E\colon\ |y| \wedge n|x| \uparrow |y|\}$. If E is Dedekind complete, then B_x is always a projection band, i.e., $B_x \oplus B_x^{\mathrm{d}} = E$. The band projection of E onto B_x will be denoted by P_x.

Now let E be a Dedekind complete Banach lattice and let $T, Z\colon E \to E$ be two disjoint regular operators (i.e., $|T| \wedge |Z| = 0$) with $0 \ne Z \in \mathcal{Z}(E)$. Show that for each $0 < x \in E$ and each $0 < \epsilon < \|Z\|$ there exists a non-zero component a of x (i.e., $a \wedge (x-a) = 0$) such that $P_a(|T|a) \le \epsilon a$. [HINT: We can suppose that T and Z are positive operators with Z satisfying $0 < Z \le I$. Fix $0 < \epsilon < \|Z\|$. Replacing E by B_x, we can assume that x is a weak unit. Let Q be the Stone space of E. We identify E with the ideal $C(Q)$ of $C^\infty(Q)$ so that x corresponds to the constant function $\mathbf{1}$ on Q. By Theorem 3.36, there exists some $h \in C(Q)$ satisfying $Z(z) = hz$ for all $z \in E$ and $\|Z\| = \|h\|_\infty$. Now notice that the set $V = \overline{\{q \in Q\colon\ h(q) > \epsilon\}}$ is a clopen subset of Q and so $y = \chi_V \in E$. From Theorem 1.17, we know that

$$0 = (T \wedge Z)y = \inf\{Tz + Z(y-z)\colon\ z \in E^+ \ \text{ and } \ z \wedge (y-z) = 0\}.$$

This implies that there exist a component χ_W of $y = \chi_V$ (where W is a clopen subset of Q) and some $q_0 \in V$ such that

$$T(\chi_W)(q_0) + h(q_0)\chi_{V\setminus W}(q_0) < \epsilon.$$

In view of $h(q) \geq \epsilon$ for each point $q \in V$, we see that $q_0 \in W$. From the inequality $T(\chi_W)(q_0) < \epsilon$, we infer that there exists some clopen set $W_1 \subseteq W$ satisfying $0 \leq T(\chi_{W_1})(q) \leq T(\chi_W)(q) < \epsilon$ for each $q \in W_1$. If we let $a = \chi_{W_1} \neq 0$, then $P_a(Ta) = T(\chi_{W_1})\chi_{W_1} \leq \epsilon\chi_{W_1} = \epsilon a$.]

3.4. The Predual of a Principal Ideal

We shall describe in this section an interesting lattice property of the ideal generated by a positive linear functional in the order dual of a Banach lattice. So, for our discussion, let E be a real Banach lattice and let $0 < \phi \in E^*$ be a fixed positive linear functional.

As usual, E^*_ϕ denotes the principal ideal generated by $\phi \in E^*$, i.e.,

$$E^*_\phi = \{x^* \in E^*\colon\ \exists\, \lambda > 0 \text{ such that } |x^*| \leq \lambda\phi\}.$$

The ideal E^*_ϕ with its $\|\cdot\|_\infty$-norm, where $\|x^*\|_\infty = \inf\{\lambda \geq 0\colon\ |x^*| \leq \lambda\phi\}$, is an AM-space with unit ϕ. Our goal is to describe an AL-space whose dual is E^*_ϕ.

The linear functional ϕ defines a lattice seminorm $\|\cdot\|_\phi$ on E via the formula

$$\|x\|_\phi = \phi(|x|).$$

This is an L-seminorm in the sense that (besides being a lattice seminorm) it also satisfies $\|x+y\|_\phi = \|x\|_\phi + \|y\|_\phi$ for all $x, y \in E^+$. If we consider the **null ideal** of ϕ, i.e., the closed ideal $N_\phi = \{x \in E\colon\ \phi(|x|) = 0\}$, then the quotient seminorm induced on the quotient Riesz space E/N_ϕ by $\|\cdot\|_\phi$, i.e., the lattice seminorm

$$\|[x]\|_\phi = \inf_{y \in [x]} \|y\|_\phi = \phi(|x|),$$

is in fact an L-norm. The norm completion of the normed Riesz space $(E/N_\phi, \|\cdot\|_\phi)$ is an AL-space, which we shall denote by $E(\phi)$. Also, we shall denote the norm dual of $E(\phi)$ by $E(\phi)^*$. Keep in mind that $E(\phi)^*$ is an AM-space with unit e^* defined by the norm of the AL-space $E(\phi)$, i.e., $e^*([x]) = \phi(x^+) - \phi(x^-)$ for each $x \in E$.

There is a natural mapping $\Phi\colon E^*_\phi \to E(\phi)^*$ or $x^* \mapsto \Phi_{x^*}$, defined by

$$\Phi_{x^*}([x]) = x^*(x)\,, \tag{$\star$}$$

for all $x^* \in E^*_\phi$ and all $x \in E$. To see that Φ is well defined we argue as follows. Fix a linear functional $x^* \in E^*_\phi$, and then select a scalar $\lambda > 0$ such that $|x^*| \le \lambda\phi$. Now assume that $y \in [x]$, i.e., $[y] = [x]$, or $\phi(|x-y|) = 0$. From the inequality $\big|x^*(x) - x^*(y)\big| \le |x^*|(|x-y|) \le \lambda\phi(|x-y|) = 0$, we see that $x^*(x) = x^*(y)$, and so ($\star$) is a well-defined linear functional on the quotient Riesz space E/N_ϕ. From

$$\big|\Phi_{x^*}([x])\big| = \big|x^*(x)\big| \le |x^*|(|x|) \le \lambda\phi(|x|) = \lambda\|[x]\|_\phi = \lambda e^*(|x|)\,, \tag{$\dagger$}$$

it follows that Φ_{x^*} is a continuous linear functional on E/N_ϕ. Therefore, Φ_{x^*} has a unique continuous linear extension to the norm completion $E(\phi)^*$ of E/N_ϕ, which we shall denote by Φ_{x^*} again. From ($\dagger$) it should be clear that $\|\Phi_{x^*}\| = \|x^*\|_\infty$ holds for each $x^* \in E^*_\phi$.

We now leave the details of the proof of the following result to the reader.

Lemma 3.38. *The mapping $\Phi\colon E^*_\phi \to E(\phi)^*$ as defined by ($\star$) is a surjective lattice isometry.*

Next, assume that $T\colon E \to E$ is a positive operator such that its norm dual $T^*\colon E^* \to E^*$ leaves the principal ideal E^*_ϕ invariant, i.e., $T^*(E^*_\phi) \subseteq E^*_\phi$. Then there is a (unique) way to define a positive operator $T_\phi\colon E(\phi) \to E(\phi)$ such that the diagram in Figure 1 is commutative.

$$\begin{array}{ccc} E^*_\phi & \xrightarrow{\ \Phi\ } & E(\phi)^* \\ {\scriptstyle T^*}\big\downarrow & & \big\downarrow{\scriptstyle T^*_\phi} \\ E^*_\phi & \xrightarrow{\ \Phi\ } & E(\phi)^* \end{array}$$

Figure 1

The operator T_ϕ is defined as follows. For each $[x] \in E/N_\phi$ we let

$$T_\phi\big([x]\big) = [Tx]\,. \tag{$\star\star$}$$

To see that the value $T_\phi\big([x]\big)$ in ($\star\star$) is well defined, let $y \in [x]$, i.e., let $[y] = [x]$ or $\phi(|x-y|) = 0$. Since T^* leaves E^*_ϕ invariant, there exists some λ_0 satisfying $T^*\phi \le \lambda_0\phi$. This implies

$$0 \le \phi\big(|Tx - Ty|\big) \le \phi\big(T|x-y|\big) = T^*\phi\big(|x-y|\big) \le \lambda_0\phi\big(|x-y|\big) = 0\,,$$

and hence $\phi(|Tx - Ty|) = 0$. This shows $[Tx] = [Ty]$, and so $(\star\star)$ defines a linear operator from E/N_ϕ to E/N_ϕ. From

$$\begin{aligned} \|T_\phi([x])\|_\phi &= \|[Tx]\|_\phi = \phi(|Tx|) \le \phi(T|x|) \\ &= T^*\phi(|x|) \le \lambda_0\phi(|x|) = \lambda_0\|[x]\|_\phi, \end{aligned}$$

we see that $T_\phi\colon E/N_\phi \to E/N_\phi$ is also continuous. Let T_ϕ denote the unique continuous extension of $T_\phi\colon E/N_\phi \to E/N_\phi$ to all of $E(\phi)$. To see that the diagram shown in Figure 1 commutes, note that for all $x^* \in E^*_\phi$ and all $x \in E$, we have:

$$\begin{aligned} \langle [x], (T^*_\phi\Phi)x^*\rangle &= \langle [x], T^*_\phi\Phi_{x^*}\rangle = \langle T_\phi[x], \Phi_{x^*}\rangle = \langle [Tx], \Phi_{x^*}\rangle \\ &= x^*(Tx) = \langle Tx, x^*\rangle = \langle x, T^*x^*\rangle\,, \quad \text{and} \\ \langle [x], (\Phi T^*)x^*\rangle &= \langle [x], \Phi_{T^*x^*}\rangle = T^*x^*(x) = \langle x, T^*x^*\rangle\,. \end{aligned}$$

Thus, we have established the first part of the following result.

Theorem 3.39. *Assume that $T\colon E \to E$ is a positive operator such that for some $0 < \phi \in E^*$ the adjoint operator $T^*\colon E^* \to E^*$ leaves the principal ideal E^*_ϕ invariant. Then there exists a unique positive operator $T_\phi\colon E(\phi) \to E(\phi)$ whose adjoint makes the diagram in Figure 1 commutative. Moreover, if T is a lattice homomorphism, then so is T_ϕ.*

Proof. We need to prove only that T_ϕ is a lattice homomorphism if T is a lattice homomorphism. So, assume that T is a lattice homomorphism. Then, using the fact that the canonical projection $x \mapsto [x]$ of E onto E/N_ϕ is a lattice homomorphism, for each $x, y \in E$ we have

$$T_\phi([x] \vee [y]) = T_\phi([x \vee y]) = [Tx \vee Ty] = [Tx] \vee [Ty] = T_\phi([x]) \vee T_\phi([y])\,.$$

This shows that $T_\phi\colon E/N_\phi \to E/N_\phi$ is a lattice homomorphism, and from this (by continuity) it follows that $T_\phi\colon E(\phi) \to E(\phi)$ is also a lattice homomorphism. ■

Exercises

1. Establish the following useful result: *If Y and Z are closed subspaces of a normed space X, one of which is finite dimensional, then $Y+Z$ is a closed subspace of X.* [HINT: Assume that Y is closed, Z is finite dimensional, and let $\pi\colon X \to X/Y$ denote the quotient map. Then $\pi(Z)$ is a finite dimensional subspace of the normed space X/Y, and so $\pi(Z)$ is a closed subspace of X/Y. Now note that $Y + Z = \pi^{-1}(\pi(Z))$.]

2. Let E be a Riesz space and let J be an ideal in E. Establish the following:
(a) The set $C = \{[x]\colon\ x \in E^+\}$ is a (convex) cone in E/J.
(b) The vector space E/J ordered by the cone C is a Riesz space.
(c) The quotient map $x \mapsto [x]$ is a lattice homomorphism from E onto E/J.

3. Let E be a normed Riesz space and let J be a closed ideal in E. Establish the following.
(a) The quotient norm is a lattice norm—and so E/J is a normed Riesz space.
(b) If E is a Banach lattice, then E/J is also a Banach lattice.

4. Show that the norm completion of a normed Riesz space E is a Banach lattice containing E as a Riesz subspace. [HINT: Let E be a normed Riesz space with norm completion $\hat{E}$. Show that the norm closure of E^+ in $\hat{E}$ is a cone under which $\hat{E}$ is a Banach lattice.]

5. Let E be an arbitrary M-**space**, that is, E is a normed Riesz space such that $\|x \vee y\| = \max\{\|x\|, \|y\|\}$ holds for all $x, y \in E^+$. Show the following.
(a) The norm completion of E is an AM-space.
(b) If E has a unit e and the closed unit ball of E coincides with the order interval $[-e, e]$, then the norm completion of E is an AM-space with unit e.

6. Let E be an arbitrary L-**space**, that is, E is a normed Riesz space such that $\|x+y\| = \|x\| + \|y\|$ holds for all $x, y \in E$ with $x \wedge y = 0$. Show that the norm completion of E is an AL-space.

7. Let X be a normed space and let $\hat{X}$ be its norm completion. For each $f \in X^*$, let $\hat{f}$ denote the unique continuous (linear) extension of f to $\hat{X}$. Show the following.
(a) The mapping $f \mapsto \hat{f}$ is a surjective linear isometry from X^* onto $\hat{X}^*$ (and so, subject to this linear isometry, $X^* = \hat{X}^*$).
(b) If X is a normed Riesz space, then $f \mapsto \hat{f}$ is a surjective lattice isometry from X^* onto $\hat{X}^*$.

8. Let ρ be an arbitrary lattice seminorm on a Riesz space E, and define its **null ideal** by $N_\rho = \{x \in E\colon\ \rho(x) = 0\}$. Establish the following.
(a) N_ρ is indeed an ideal in E.
(b) The function $\hat{\rho}\colon E/N_\rho \to \mathbb{R}$ defined by $\hat{\rho}([x]) = \rho(x)$ is a lattice norm on E/N_ρ and $\hat{\rho}([x]) = \inf_{y \in [x]} \rho(y)$.
(c) If ρ is an M-seminorm (resp. an L-seminorm), then $\hat{\rho}$ is an M-norm (resp. an L-norm).

9. Complete the details of the proof of Lemma 3.38.

Chapter 4

Special Classes of Operators

This chapter studies the properties of several important classes of operators that were discussed in a fragmented way in the previous chapters. These classes are: finite-rank operators, multiplication operators, lattice and algebraic homomorphisms, Fredholm operators, and strictly singular operators. We pay special attention to the order-theoretic properties of these classes of operators.

The class of finite-rank operators is considered in conjunction with the approximation property of Banach spaces. The presentation of multiplication operators is in the framework of $C(\Omega)$-spaces. By means of the classical Kakutani–Bohnenblust–Krein Representation Theorem 3.6, this framework allows us to study multiplication operators on AM-spaces. The section on lattice and algebraic homomorphisms deals with lattice homomorphisms between $C(\Omega)$-spaces and contains a few basic results regarding extension of lattice homomorphisms. The section on Fredholm operators presents their basic properties including proofs of the Index Theorem and of the continuity of the index function.

The last section deals with strictly singular operators. An operator between two Banach spaces is said to be **strictly singular** if its restriction to any infinite dimensional closed subspace is not an isomorphism. After establishing some basic properties of these operators, we investigate their relationships with compact and Fredholm operators.

4.1. Finite-rank Operators

We discuss here some basic properties of finite-rank operators. An operator $T\colon V \to W$ between two vector spaces is said to be a **rank-one operator** if its range is one dimensional. So, if $T\colon V \to W$ is a rank-one operator and w is a non-zero vector in the range of T, then there exists a (uniquely determined) linear functional f on V satisfying $T(v) = f(v)w$ for each $v \in V$. Following the standard tensor-product notation, we shall denote the value $f(v)w$ by $(f \otimes w)(v)$, i.e., we shall write $T = f \otimes w$.

Definition 4.1. *An operator $T\colon V \to W$ between two vector spaces is said to be a **finite-rank operator** if its range is finite dimensional.*

Thus, if $T\colon V \to W$ is a finite-rank operator and $w_1, \ldots, w_k$ is a basis in the range $T(V)$, then there exist (uniquely determined) linear functionals $f_1, \ldots, f_k$ on V such that $T(v) = \sum_{i=1}^k f_i(v)w_i$, i.e., $T = \sum_{i=1}^k f_i \otimes w_i$. It is easy to see that an operator $T\colon V \to W$ between two vector spaces is a finite-rank operator if and only if there exist (not necessarily linearly independent) vectors $u_1, \ldots, u_n$ in W and linear functionals $g_1, \ldots, g_n$ on V such that $T = \sum_{i=1}^n g_i \otimes u_i$. The dimension of the range of a finite-rank operator is called its **rank**. Thus, an **operator of rank** k is a finite-rank operator whose range has dimension k. The zero operator is, of course, a finite-rank operator with zero rank. Here are two immediate properties of finite-rank operators.

- *Sums and scalar products of finite-rank operators are finite-rank.*
- *If in a scheme of operators $V \xrightarrow{\ S\ } W \xrightarrow{\ T\ } U$ between vector spaces either S or T is finite-rank, then TS is likewise finite-rank.*

We now turn our attention to continuous finite-rank operators between Banach spaces. In this case, a bounded operator $T\colon X \to Y$ between two Banach spaces is a finite-rank operator if and only if there exist continuous linear functionals $x_1^*, \ldots, x_n^*$ on X and vectors $y_1, \ldots, y_n$ in Y such that $T = \sum_{i-1}^n x_i^* \otimes y_i$. The latter expression is referred to as a **representation** of T. From now on we shall consider only continuous finite-rank operators between Banach spaces.

The reader should verify the following property of finite-rank operators; for details see Exercise 3 at the end of this section.

Lemma 4.2. *Assume that $T\colon X \to Y$ is an operator of rank k between Banach spaces. If $u_1, \ldots, u_k$ is a basis in the range of T, then T has a unique representation $T = \sum_{i=1}^k x_i^* \otimes u_i$ with the continuous linear functionals $x_1^*, \ldots, x_k^*$ linearly independent.*

Observe that if $T = x^* \otimes y\colon X \to Y$ is a rank-one operator between Banach spaces, then its norm adjoint is $T^* = y \otimes x^*$, where now y is viewed as a continuous linear functional on Y^*. This easily follows from the duality identity

$$\langle (x^* \otimes y)x, y^* \rangle = x^*(x)y^*(y) = \langle x, (y \otimes x^*)y^* \rangle .$$

The above identity yields the following property of finite-rank operators.

Lemma 4.3. *If $T\colon X \to Y$ is an operator of rank k between Banach spaces with a representation $T = \sum_{i=1}^{k} x_i^* \otimes u_i$, then its norm adjoint $T^*\colon Y^* \to X^*$ is also an operator of rank k and $T^* = \sum_{i=1}^{k} u_i \otimes x_i^*$.*

The following result also should be obvious.

Lemma 4.4. *Every finite-rank operator between Banach spaces is compact.*

The collection of all finite-rank operators from a Banach space X into another Banach space Y will be denoted by $\mathcal{F}(X,Y)$. As usual, $\mathcal{F}(X)$ stands for $\mathcal{F}(X,X)$. Clearly, $\mathcal{F}(X,Y)$ is a vector subspace of $\mathcal{L}(X,Y)$ and moreover $\mathcal{F}(X,Y) \subseteq \mathcal{K}(X,Y)$. It was a long standing question whether or not every compact operator between two Banach spaces was the norm limit of a sequence of finite-rank operators. That is, it was not known whether or not $\mathcal{F}(X,Y)$ was norm dense in $\mathcal{K}(X,Y)$. In 1971, P. Enflo [**120**] answered this problem in the negative. A. Szankowski [**315**] showed that even when X and Y are Banach lattices, the vector space $\mathcal{F}(X,Y)$ need not be norm dense in $\mathcal{K}(X,Y)$.

There is another commonly used notation pertaining to finite-rank operators. Let X and Y be two Banach spaces and let Φ be a subspace of X^*. Then the symbol $\Phi \otimes Y$ denotes the vector space of all finite-rank operators $T\colon X \to Y$ which have a representation $T = \sum_{i=1}^{n} x_i^* \otimes y_i$ with $y_1, \ldots, y_n \in Y$ and $x_1^*, \ldots, x_n^* \in \Phi$. In this notation $\mathcal{F}(X,Y) = X^* \otimes Y$.

The vector space of finite-rank operators plays an important role in analysis and so do the spaces of operators that contain it. Next, we shall discuss its role in connection with the approximation property in Banach spaces.

Definition 4.5. *A Banach space X is said to have the* ***approximation property*** *if for every compact subset C of X and every $\epsilon > 0$ there exists a finite-rank operator $T \in \mathcal{L}(X)$ such that $\|Tx - x\| < \epsilon$ holds for each $x \in C$.*

In other words, a Banach space X has the approximation property if and only if the identity operator on X can be approximated uniformly on every compact set by finite-rank operators. The following result, stated in Exercise 7 in Section 2.4, is used frequently in the study of the approximation property.

Lemma 4.6. *A subset of a normed space is totally bounded if and only if it is contained in the closed convex hull of a sequence that is norm convergent to zero.*

Theorem 4.7. *A Banach space with a basis has the approximation property.*

Proof. Let X be a Banach space with a basis $\{e_n\}$. As usual, let $\{P_n\}$ denote the sequence of projections on X defined by $P_n(x) = \sum_{i=1}^{n} \alpha_i e_i$ for each $x = \sum_{i=1}^{\infty} \alpha_i e_i$ in X. Pick a number $M > 0$ such that $\|P_n\| \leq M$ for each n.

Assume that C is a compact subset of X, and let $\epsilon > 0$. According to Lemma 4.6 there exists a sequence $\{x_n\}$ of X satisfying $C \subseteq \overline{\text{co}}\{x_1, x_2, \ldots\}$ and $\|x_n\| \to 0$. Next, pick n_0 such that $\|x_n\| < \frac{\epsilon}{1+M}$ for each $n > n_0$. So, for each $n > n_0$ and for all m we have

$$\|P_m x_n - x_n\| \leq \big[\|P_m\| + \|I\|\big]\|x_n\| = (1+M)\|x_n\| < \epsilon\,.$$

Since $\|P_m x - x\| \to 0$ holds for each $x \in X$, there exists some k such that $m \geq k$ implies $\|P_m x_n - x_n\| < \epsilon$ for each $1 \leq n \leq n_0$. Therefore, for each n and all $m \geq k$ we have $\|P_m x_n - x_n\| < \epsilon$.

Now pick any $y \in \text{co}\{x_1, x_2, \ldots\}$ and write $y = \sum_{j=1}^{m} \lambda_j x_{k_j}$, where $\lambda_j > 0$ for each $1 \leq j \leq m$ and $\sum_{j=1}^{m} \lambda_j = 1$. Then for $n \geq k$ we have

$$\|P_n y - y\| = \Big\|\sum_{j=1}^{m} \lambda_j (P_n x_{k_j} - x_{k_j})\Big\| \leq \sum_{j=1}^{m} \lambda_j \|P_n x_{k_j} - x_{k_j}\| < \sum_{j=1}^{m} \lambda_j \epsilon = \epsilon\,.$$

This implies $\|P_n y - y\| \leq \epsilon$ for each $y \in C$ and all $n \geq k$. Since each P_n is a finite-rank operator, it follows that X has the approximation property. ■

The validity of the converse of Theorem 4.7, i.e., of the statement "every separable Banach space with the approximation property has a basis," was another long standing open problem that was solved in the negative by P. Enflo in [**120**]. The situation does not improve even when X and Y are Banach lattices; see [**315**].

Now let us discuss the connections between the approximation property and the topology of uniform convergence on compact sets. Fix two Banach spaces X and Y, and let $\mathcal{C}_X$ denote the collection of all non-empty norm compact subsets of X. For each $C \in \mathcal{C}_X$ define the seminorm p_C on $\mathcal{L}(X,Y)$ via the formula

$$p_C(T) = \max_{x \in C} \|Tx\|\,.$$

It should be clear that the family of seminorms $\{p_C\}_{C \in \mathcal{C}_X}$ generates a Hausdorff locally convex topology $\tau_c(X,Y)$ on $\mathcal{L}(X,Y)$. Also, it should be obvious that a net $\{T_\alpha\} \subseteq \mathcal{L}(X,Y)$ satisfies $T_\alpha \xrightarrow{\tau_c(X,Y)} T$ in $\mathcal{L}(X,Y)$ if and only if $\{T_\alpha\}$ converges uniformly to T on every compact subset of X. For this reason, the topology $\tau_c(X,Y)$ is called the **topology of uniform convergence on the compact subsets** of X. We write $\tau_c(X)$ for $\tau_c(X,X)$.

Two easy characterizations of the approximation property in terms of the topology of uniform convergence on compact sets are included in the next result.

Lemma 4.8. *For a Banach space X the following statements are equivalent.*

(1) *X has the approximation property.*

(2) *The identity operator $I\colon X \to X$ belongs to the $\tau_c(X)$-closure of the vector subspace $\mathcal{F}(X)$ of all finite-rank operators in $\mathcal{L}(X)$.*

(3) *If an arbitrary $\tau_c(X)$-continuous linear functional ϕ on $\mathcal{L}(X)$ satisfies $\phi(x^* \otimes x) = 0$ for all $x \in X$ and all $x^* \in X^*$, then $\phi(I) = 0$.*

For a more useful characterization of the approximation property, we need to describe the topological dual of $\big(\mathcal{L}(Y,X), \tau_c(Y,X)\big)$.

Theorem 4.9. *A linear functional ϕ on $\mathcal{L}(Y,X)$, where X and Y are Banach spaces, is $\tau_c(Y,X)$-continuous if and only if there exist two sequences $\{y_n\} \subseteq Y$ and $\{x_n^*\} \subseteq X^*$ satisfying $\sum_{n=1}^\infty \|x_n^*\| \cdot \|y_n\| < \infty$ and*

$$\phi(T) = \sum_{n=1}^\infty x_n^*(Ty_n)$$

for each $T \in \mathcal{L}(Y,X)$.

Proof. Let X and Y be two Banach spaces and let ϕ be a linear functional on $\mathcal{L}(Y,X)$. Assume first that there exist two sequences $\{y_n\} \subseteq Y$ and $\{x_n^*\} \subseteq X^*$ satisfying $\sum_{n=1}^\infty \|x_n^*\| \cdot \|y_n\| < \infty$ and $\phi(T) = \sum_{n=1}^\infty x_n^*(Ty_n)$ for each $T \in \mathcal{L}(Y,X)$. Pick an arbitrary sequence of positive scalars $\{\lambda_n\}$ such that $\lambda_n \to \infty$ and $M = \sum_{n=1}^\infty \lambda_n \|x_n^*\| \cdot \|y_n\| < \infty$.[1] Without loss of generality we can assume that $y_n \neq 0$ for each n. Next, let $u_n = \frac{y_n}{\lambda_n \|y_n\|}$ and note that $u_n \to 0$. Therefore, the set $C = \{0, u_1, u_2, u_3, \ldots\}$ is a compact subset of Y. Now, if $T \in \mathcal{L}(Y,X)$, then

$$\begin{aligned} |\phi(T)| &= \Big|\sum_{n=1}^\infty x_n^*(Ty_n)\Big| = \Big|\sum_{n=1}^\infty \lambda_n \|y_n\| x_n^*(Tu_n)\Big| \\ &\leq \Big[\sum_{n=1}^\infty \lambda_n \|x_n^*\| \cdot \|y_n\|\Big] \max_{z \in C} \|Tz\| = Mp_C(T)\,. \end{aligned}$$

This shows that ϕ is dominated by a $\tau_c(Y,X)$-continuous seminorm, and so ϕ is itself $\tau_c(Y,X)$-continuous.

For the converse, suppose that ϕ is $\tau_c(Y,X)$-continuous. This means that there exists some constant $M > 0$ and some non-empty compact subset C of Y such that $|\phi(T)| \leq Mp_C(T)$ holds for each $T \in \mathcal{L}(Y,X)$. In view of Lemma 4.6, we can suppose without loss of generality that

---

[1] If $a_n > 0$ holds for each n and $\sum_{n=1}^\infty a_n < \infty$, then choose a strictly increasing sequence $\{k_n\}$ of natural numbers such that $\sum_{i=k_n}^\infty a_i < \frac{1}{n^3}$. Letting $\lambda_i = 1$ for $1 \leq i < k_1$ and $\lambda_i = n > 0$ for $k_n \leq i < k_{n+1}$, we see that $\lambda_n \to \infty$ and $\sum_{n=1}^\infty \lambda_n a_n < \infty$.

$C = \overline{\text{co}}\{y_1, y_2, \ldots\}$ for some sequence $\{y_n\}$ of Y satisfying $\|y_n\| \to 0$. Now we define a mapping $\mathcal{S}\colon \mathcal{L}(Y,X) \to (X \oplus X \oplus X \oplus \cdots)_0$ via the formula

$$\mathcal{S}(T) = \big(Ty_1, Ty_2, Ty_3, \ldots\big)\,.$$

Clearly, $\mathcal{S}$ is linear and from $\|\mathcal{S}(T)\|_\infty = \sup_{n\in\mathbb{N}} \|Ty_n\| \le \big(\sup_{n\in\mathbb{N}} \|y_n\|\big)\|T\|$, we see that $\mathcal{S}$ is a bounded operator. Next, on the range of $\mathcal{S}$ define the linear functional Φ on $\mathcal{S}\big(\mathcal{L}(Y,X)\big)$ via the formula

$$\Phi\big(\mathcal{S}(T)\big) = \Phi\big((Ty_1, Ty_2, \ldots)\big) = \phi(T)\,.$$

Note that if $T, R \in \mathcal{L}(Y,X)$ satisfy $(Ty_1, Ty_2, \ldots) = (Ry_1, Ry_2, \ldots)$, then $T - R = 0$ on C, and so $|\phi(T) - \phi(R)| = |\phi(T-R)| \le Mp_C(T-R) = 0$ implies $\phi(T) = \phi(R)$. This shows that Φ is a well-defined linear functional. From

$$\big|\Phi(Ty_1, Ty_2, \ldots)\big| = |\phi(T)| \le Mp_C(T) = M\|(Ty_1, Ty_2, \ldots)\|_\infty\,,$$

it follows that Φ is also norm continuous. By the Hahn–Banach theorem, Φ has a norm continuous linear extension (denoted by Φ again) to all of $(X\oplus X\oplus X\oplus\cdots)_0$, i.e., $\Phi \in (X \oplus X \oplus X \oplus \cdots)_0^* = (X^*\oplus X^*\oplus X^*\oplus\cdots)_1$; see Exercise 3 in Section 7.2. This means that there exists a sequence $\{x_n^*\} \subseteq X^*$ satisfying $\sum_{n=1}^\infty \|x_n^*\| < \infty$ and

$$\phi(T) = \Phi\big((Ty_1, Ty_2, \ldots)\big) = \sum_{n=1}^\infty x_n^*(Ty_n)\,.$$

Since $\{y_n\}$ is a bounded sequence, we also have $\sum_{n=1}^\infty \|x_n^*\| \cdot \|y_n\| < \infty$, and the proof is finished. ■

Corollary 4.10 (Grothendieck [**141**]). *A Banach space X has the approximation property if and only if for all sequences $\{x_n^*\} \subseteq X^*$ and $\{x_n\} \subseteq X$ that satisfy $\sum_{n=1}^\infty \|x_n^*\| \cdot \|x_n\| < \infty$ and $\sum_{n=1}^\infty x_n^* \otimes x_n = 0$ in $\mathcal{L}(X)$, we have $\sum_{n=1}^\infty x_n^*(x_n) = 0$.*

Proof. Assume first that X has the approximation property, and let two sequences $\{x_n^*\} \subseteq X^*$ and $\{x_n\} \subseteq X$ satisfy $\sum_{n=1}^\infty \|x_n^*\| \cdot \|x_n\| < \infty$ and $\sum_{n=1}^\infty x_n^* \otimes x_n = 0$. Also, let ϕ denote the $\tau_c(X)$-continuous linear functional on $\mathcal{L}(X)$ defined by $\phi(T) = \sum_{n=1}^\infty x_n^*(Tx_n)$. Then for each rank-one operator $x^* \otimes x$ we have

$$\begin{aligned}
\phi(x^* \otimes x) &= \sum_{n=1}^\infty x_n^*\big((x^* \otimes x)x_n\big) = \sum_{n=1}^\infty x_n^*\big(x^*(x_n)x\big) = \sum_{n=1}^\infty x_n^*(x)x^*(x_n)\\
&= x^*\Big(\sum_{n=1}^\infty x_n^*(x)x_n\Big) = x^*\Big(\Big[\sum_{n=1}^\infty x_n^* \otimes x_n\Big](x)\Big)\\
&= x^*(0) = 0\,.
\end{aligned}$$

Therefore, ϕ vanishes on the vector subspace of all finite-rank operators, and so it vanishes on their $\tau_c(X)$-closure. Since X has the approximation property, this $\tau_c(X)$-closure contains the identity operator I, and hence $\sum_{n=1}^{\infty} x_n^*(x_n) = \phi(I) = 0$.

For the converse, assume that the condition is satisfied, and let ϕ be a $\tau_c(X)$-continuous linear functional on $\mathcal{L}(X)$ that vanishes on the vector subspace of all finite-rank operators. By Theorem 4.9 there exist two sequences $\{x_n^*\} \subseteq X^*$ and $\{x_n\} \subseteq X$ with $\sum_{n=1}^{\infty} \|x_n^*\| \cdot \|x_n\| < \infty$ such that $\phi(T) = \sum_{n=1}^{\infty} x_n^*(Tx_n)$ holds for each $T \in \mathcal{L}(X)$. If $x^* \otimes x$ is an arbitrary rank-one operator, then we have

$$0 = \phi(x^* \otimes x) = \sum_{n=1}^{\infty} x_n^*\big((x^* \otimes x)x_n\big) = \sum_{n=1}^{\infty} x_n^*\big(x^*(x_n)x\big) = x^*\Big(\sum_{n=1}^{\infty}(x_n^* \otimes x_n)x\Big).$$

Since $x^* \in X^*$ and $x \in X$ are arbitrary, it follows that $\sum_{n=1}^{\infty} x_n^* \otimes x_n = 0$. So, from our hypothesis, we have $\phi(I) = \sum_{n=1}^{\infty} x_n^*(x_n) = 0$. This shows that the identity operator I belongs to the $\tau_c(X)$-closure of $X^* \otimes X$, and so by Lemma 4.8 the Banach space X has the approximation property. ■

Corollary 4.11 (Grothendieck [**141**]). *If the norm dual X^* of a Banach space X has the approximation property, then X itself has the approximation property.*

Proof. Assume that X^* has the approximation property, and let two sequences $\{x_n^*\} \subseteq X^*$ and $\{x_n\} \subseteq X$ satisfy $\sum_{n=1}^{\infty} \|x_n^*\| \cdot \|x_n\| < \infty$ and $\sum_{n=1}^{\infty} x_n^* \otimes x_n = 0$. Then, considering each x_n as a vector in X^{**}, it follows from Lemma 4.3 that $\sum_{n=1}^{\infty} x_n \otimes x_n^* = \big(\sum_{n=1}^{\infty} x_n^* \otimes x_n\big)^* = 0$. Since the Banach space X^* has the approximation property, Corollary 4.10 implies $\sum_{n=1}^{\infty} x_n(x_n^*) = \sum_{n=1}^{\infty} x_n^*(x_n) = 0$. But then the same corollary guarantees that X has the approximation property. ■

A. Grothendieck has also shown that a Banach space X has the approximation property if and only if for each Banach space Y the finite-rank operators from Y to X are dense in $\mathcal{K}(Y,X)$, the closed vector subspace of compact operators. This result is stated next.

Theorem 4.12 (Grothendieck [**141**]). *A Banach space X has the approximation property if and only if for each Banach space Y the vector subspace of all finite-rank operators $Y^* \otimes X$ is norm dense in $\mathcal{K}(Y,X)$.*

Proof. Let X be a Banach space with the approximation property and let Y be an arbitrary Banach space. Fix a compact operator $T\colon Y \to X$ and some $\epsilon > 0$. Since the set $C = \overline{T(U_Y)}$ is a compact subset of X and X has the approximation property, there exists a finite-rank operator

$S = \sum_{i=1}^{n} x_i^* \otimes x_i \colon X \to X$ such that $\|Sx - x\| < \epsilon$ holds for each $x \in C$. In particular, for each $y \in U_Y$ we have

$$\Big\| \Big[\sum_{i=1}^{n} (T^* x_i^*) \otimes x_i - T \Big] y \Big\| = \Big\| \sum_{i=1}^{n} x_i^*(Ty) x_i - Ty \Big\| = \|S(Ty) - Ty\| < \epsilon \, . \quad (\star)$$

Now notice that the operator

$$R = ST = \sum_{i=1}^{n} T^* x_i^* \otimes x_i \colon Y \to X$$

is a finite-rank operator, and by $(\star)$ it satisfies $\|R - T\| \le \epsilon$. This shows that the vector subspace of all finite-rank operators $Y^* \otimes X$ is norm dense in $\mathcal{K}(Y, X)$.

For the converse, assume that for each Banach space Y the vector subspace of all finite-rank operators $Y^* \otimes X$ is norm dense in $\mathcal{K}(Y, X)$. Fix a compact subset C of X and let $\epsilon > 0$. By Lemma 4.6, we can assume that there exists some sequence $\{x_n\}$ in X satisfying $\|x_n\| \to 0$ and $C = \overline{\text{co}}\{x_1, x_2, \ldots\}$. Scaling appropriately, we can also suppose without loss of generality that $0 < \|x_n\| \le 1$ holds for each n.

Define the sequence $\{u_n\}$ of X by $u_{2n} = \frac{x_n}{\sqrt{\|x_n\|}}$ and $u_{2n-1} = -\frac{x_n}{\sqrt{\|x_n\|}}$. From $\|u_{2n}\| = \|u_{2n-1}\| = \sqrt{\|x_n\|}$, we see that $\|u_n\| \to 0$, and therefore the set $U = \overline{\text{co}}\{u_1, u_2, \ldots\}$ is a compact subset of X. It is also convex and circled (i.e., $|\lambda| \le 1$ and $x \in U$ imply $\lambda x \in U$). From the identity $x_n = \sqrt{\|x_n\|} u_{2n}$ and $\sqrt{\|x_n\|} \le 1$, we see that $x_n \in U$ for each n, and so $C \subseteq U$.

The preceding properties of U guarantee that $Y = \bigcup_{n=1}^{\infty} nU$ is a vector subspace of X. Moreover, if Y is equipped with the norm $|||\cdot|||$, defined by

$$|||y||| = \inf\{\lambda \ge 0\colon \; y \in \lambda U\} \, ,$$

then Y is a Banach space. Its closed unit ball coincides with U, that is, $\{y \in Y\colon \; |||y||| \le 1\} = U$. Moreover, if we let $J\colon Y \to X$ denote the natural embedding (that is, $Jy = y$ for each $y \in Y$), then from $J(U) = U$, it follows that J is a compact operator. Therefore, according to our hypothesis, there exists a finite-rank operator $\sum_{i=1}^{m} y_i^* \otimes v_i \in \mathcal{L}(Y, X)$ satisfying

$$\Big\| \sum_{i=1}^{m} y_i^*(y) \otimes v_i - y \Big\| < \epsilon \;\text{ for all }\; y \in U \, . \qquad (\star\star)$$

The proof would be finished from $(\star\star)$ if we knew that each $y_i^* \in Y^*$ belongs to X^*. Unfortunately, this might not be the case. However, the proof can be completed if we can verify that for each $y^* \in Y^*$ and each $\delta > 0$ there exists some $x^* \in X^*$ satisfying $|y^*(x) - x^*(x)| < \delta$ for each $x \in C$ or equivalently $|y^*(x_n) - x^*(x_n)| < \delta$ for each n.

To see this, fix a non-zero $y^* \in Y^*$ and let $\delta = 2\eta > 0$. Let Z be the closed vector subspace generated by $\{x_n \colon n \in \mathbb{N}\}$ in X. If $y^* = 0$ on Z, then the zero functional $0 \in X^*$ satisfies $0(x_n) = y^*(x_n)$ for each n. So, we can suppose that for some $z_0 \in Z$ we have $y^*(z_0) = 1$.

From $u_{2n} = \frac{x_n}{\sqrt{\|x_n\|}} \in U$ it follows that $|||x_n||| \leq \sqrt{\|x_n\|}$ for each n. This implies $|||x_n||| \to 0$, and so (since y^* is $|||\cdot|||$-continuous) $y^*(x_n) \to 0$. Pick some k such that $|y^*(x_n)| < \eta$ for each $n > k$.

Now consider the set $K = \frac{1}{\eta}\overline{\text{co}}\{\pm x_n \colon n > k\}$, and note that K is a convex, circled, and $\|\cdot\|$-compact subset of Z. Next, let Z_k denote the (finite dimensional) vector subspace generated by the vectors $\{z_0, x_1, \ldots, x_k\}$, and let $F = \{z \in Z_k \colon y^*(z) = 1\} = z_0 + Z_k \cap \operatorname{Ker} y^*$. Clearly, F is non-empty, and (as y^* is $\|\cdot\|$-continuous on Z_k) F is also a $\|\cdot\|$-closed subset of Z_k and hence of Z. Moreover, it is easy to see that $K \cap F = \emptyset$. Hence, by the classical Separation Theorem (see, for instance [**25**, Theorem 5.58, p. 193]), there exist a non-zero $\|\cdot\|$-continuous linear functional x^* on Z (we can assume that x^* is defined on all of X) and some scalar $\alpha \geq 0$ such that

$$|x^*(c)| \leq \alpha \leq x^*(z) \qquad (\star\star\star)$$

holds for all $c \in K$ and all $z \in F$. (Here we assume that X is a real Banach space; the complex case can be treated similarly.) Since $Z_k \cap \operatorname{Ker} y^*$ is a vector subspace of Z_k, it follows from $(\star\star\star)$ that $Z_k \cap \operatorname{Ker} y^* \subseteq Z_k \cap \operatorname{Ker} x^*$. This shows that there exists some scalar λ such that $x^* = \lambda y^*$ on Z_k. If $\lambda = 0$, then $x^* = 0$ on Z_k, and from $(\star\star\star)$ we see that $x^*(x_n) = 0$ for $n > k$, i.e., $x^* = 0$ on Z, which is a contradiction. Hence, $\lambda \neq 0$, and from $(\star\star\star)$ we get that $\lambda > 0$. Therefore, replacing x^* by $\frac{x^*}{\lambda}$, we can assume that $x^* = y^*$ on F and that $0 \leq \alpha \leq 1$. In particular, $x^*(x_n) = y^*(x_n)$ for all $1 \leq n \leq k$ and for $n > k$ we have $\frac{1}{\eta}|x^*(x_n)| \leq \alpha \leq 1$ or $|x^*(x_n)| \leq \eta$, which implies $|x^*(x_n) - y^*(x_n)| \leq |x^*(x_n)| + |y^*(x_n)| < \eta + \eta = 2\eta = \delta$. The proof of the theorem is now complete. ■

We close the section with two results regarding finite-rank operators between Banach lattices.

Lemma 4.13. *If $x^* \otimes u \colon E \to F$ is a rank-one operator between Banach lattices, then $x^* \otimes u$ has a modulus and*

$$|x^* \otimes u| = |x^*| \otimes |u| .$$

Proof. If $x \in E^+$, then we have

$$\begin{aligned}
\sup_{|y|\le x} |(x^*\otimes u)(y)| &= \sup_{|y|\le x} |x^*(y)u| = \sup_{|y|\le x} |x^*(y)||u| \\
&= \Big[\sup_{|y|\le x} |x^*(y)|\Big]|u| = |x^*|(x)|u| \\
&= \big(|x^*|\otimes|u|\big)(x)\,.
\end{aligned}$$

This implies that $x^*\otimes u$ has a modulus and that $|x^*\otimes u| = |x^*|\otimes|u|$. ∎

Theorem 4.14. *Every finite-rank operator between Banach lattices has a modulus which is compact but not necessarily a finite-rank operator.*

Proof. Let $T = \sum_{i=1}^n x_i^*\otimes u_i \colon E \to F$ be a finite-rank operator between Banach lattices. If $u = \sum_{i=1}^n |u_i|$, then it is easy to see that $T(E) \subseteq F_u$, where F_u is the ideal generated by u in F and F_u is equipped with its u-norm. Clearly, $T\colon E \to F_u$ (as a finite-rank operator) is compact. Now replicate the proof of Theorem 3.14 to conclude that $T\colon E \to F_u$ has a compact modulus $|T|$. Since F_u is an ideal in F, the latter guarantees that $|T|$ is also the modulus of the operator $T\colon E \to F$. Since the topology generated by the norm of F_u is finer than the topology generated on F_u by the norm of F, we infer that $|T|\colon E \to F$ is also a compact operator.

To see that the modulus of a finite-rank operator need not be a finite rank operator, consider the constant function $\mathbf{1}$ and the function $u\colon [0,1] \to \mathbb{R}$ defined by $u(t) = t$. We leave it to the reader to verify that the rank-two operator $T = \mathbf{1}\otimes u - u\otimes\mathbf{1}$ on $L_1[0,1]$ does not have a finite-rank modulus; see Exercise 9 at the end of the section. ∎

Exercises

1. Let $f, f_1, \ldots, f_k$ be linear functionals on a vector space V. Show that f lies in the linear span of $f_1, \ldots, f_k$ if and only if $\bigcap_{i=1}^k \operatorname{Ker} f_i \subseteq \operatorname{Ker} f$.

   In particular, a collection $g_1, \ldots, g_n$ of linear functionals on a vector space V is linearly dependent if and only if $\bigcap_{i\ne j} \operatorname{Ker} g_i \subseteq \operatorname{Ker} g_j$ holds for some j. [HINT: Assume that V is a real vector space (the same arguments show that the result is true for vector spaces over any field) and let $\bigcap_{i=1}^k \operatorname{Ker} f_i \subseteq \operatorname{Ker} f$. We consider the vector subspace of $\mathbb{R}^k$

   $$W = \{(f_1(v), \ldots, f_k(v)) \in \mathbb{R}^k\colon\ v \in V\}$$

   and define the linear functional $\phi\colon W \to \mathbb{R}$ by $\phi(f_1(v), \ldots, f_k(v)) = f(v)$. Now if $(f_1(v), \ldots, f_k(v)) = (f_1(u), \ldots, f_k(u))$, then $v - u \in \bigcap_{i=1}^k \operatorname{Ker} f_i$, and so by our condition $f(v) = f(u)$, which shows that ϕ is well defined. Notice that ϕ extends linearly to all of $\mathbb{R}^k$. So, there exist scalars $\lambda_1, \ldots, \lambda_k$ such that $\phi(t_1, \ldots, t_k) = \sum_{i=1}^k \lambda_i t_i$. This implies $f = \sum_{i=1}^k \lambda_i f_i$.]

2. Show that an operator $T\colon V \to W$ between two vector spaces is a finite-rank operator if and only if there exist vectors $w_1, \ldots, w_n$ in W and linear functionals $g_1, \ldots, g_n$ on V such that $T = \sum_{i=1}^n g_i \otimes w_i$.

3. Let $T\colon V \to W$ be an operator of rank k and let $w_1, \ldots, w_k$ be a basis of the range of T. Then there exist (uniquely determined) linear functionals $f_1, \ldots, f_k$ on V such that $T = \sum_{i=1}^k f_i \otimes w_i$. Show that the linear functionals $f_1, \ldots, f_k$ are linearly independent. [HINT: If $f_1 = \sum_{i=2}^k \lambda_i f_i$, then $T = \sum_{i=2}^k f_i \otimes (w_i + \lambda_i w_1)$ and so the rank of T is at most $k-1$, a contradiction.]

4. Let $T\colon V \to V$ be a finite-rank operator on a vector space and assume that $T = \sum_{i=1}^n f_i \otimes v_i = \sum_{j=1}^m g_j \otimes u_j$ are two representations of T, where the vectors v_i and u_j need not belong to the range of T. Show that

$$\sum_{i=1}^n f_i(v_i) = \sum_{j=1}^m g_j(u_j),$$

i.e., the scalar $\operatorname{tr}(T) = \sum_{i=1}^n f_i(v_i)$ (called the **trace** of T) is independent of the representation of T. [HINT: Fix a basis $w_1, \ldots, w_k$ of the range of T and write $T = \sum_{r=1}^k h_r \otimes w_r$, where the linear functionals h_r are uniquely determined. Now let $T = \sum_{i=1}^n f_i \otimes v_i$ be another representation of T. Assume first that $f_1, \ldots, f_n$ are linearly independent. Then for each i there exists some $x_i \in V$ such that $f_j(x_i) = \delta_{ij}$ holds for each j; see Exercise 1 above. Hence, $v_i \in T(V)$ for each i, and so $v_i = \sum_{r=1}^k \lambda_r^i w_r$ for appropriate scalars λ_r^i. So, $T = \sum_{i=1}^n f_i \otimes v_i = \sum_{r=1}^k (\sum_{i=1}^n \lambda_r^i f_i) \otimes w_r$, and from this we get $h_r = \sum_{i=1}^n \lambda_r^i f_i$ for each $1 \le r \le k$. Thus,

$$\sum_{i=1}^n f_i(v_i) = \sum_{r=1}^k \Big(\sum_{i=1}^n \lambda_r^i f_i\Big)(w_r) = \sum_{r=1}^k h_r(w_r).$$

Now assume the set $\{f_1, \ldots, f_n\}$ is linearly dependent. There exist linearly independent linear functionals $\theta_1, \ldots, \theta_\ell$ such that $f_i = \sum_{\nu=1}^\ell \mu_\nu^i \theta_\nu$ for appropriate scalars μ_ν^i, and so $\sum_{i=1}^n f_i \otimes v_i = \sum_{\nu=1}^\ell \theta_\nu \otimes (\sum_{i=1}^n \mu_\nu^i v_i)$. By the preceding case, we have

$$\sum_{r=1}^k h_r(w_r) = \sum_{\nu=1}^\ell \theta_\nu\Big(\sum_{i=1}^n \mu_\nu^i v_i\Big) = \sum_{i=1}^n \Big(\sum_{\nu=1}^\ell \mu_\nu^i \theta_\nu\Big)(v_i) = \sum_{i=1}^n f_i(v_i).]$$

5. Show that the trace of a finite-rank operator on a Banach space is similarity invariant. That is, show that if $T\colon X \to X$ is a finite-rank operator on a Banach space, then $\operatorname{tr}(T) = \operatorname{tr}(S^{-1}TS)$ for each invertible bounded operator $S\colon X \to X$.

6. Let X be an arbitrary Banach space and assume that $\{x_1^*, x_2^*, \ldots, x_n^*\} \subseteq X^*$ and $\{x_1, x_2, \ldots, x_n\} \subseteq X$ satisfy $x_i^*(x_j) = \delta_{ij}$ for all i and j. Show that the finite-rank operator $P = \sum_{i=1}^n x_i^* \otimes x_i$ has rank n, is a continuous projection, and has range the vector subspace generated by $\{x_1, x_2, \ldots, x_n\}$.

Use this conclusion to show that every finite-dimensional subspace Y of X is **complemented**, i.e., there exists a closed subspace Z of X satisfying $Y \cap Z = \{0\}$ and $Y \oplus Z = X$. [HINT: For the second part let $\{x_1, x_2, \ldots, x_n\}$ be a basis of Y. Pick linear functionals $x_1^*, x_2^*, \ldots, x_n^*$ on Y such that $x_i^*(x_j) = \delta_{ij}$ holds for all i and j. Since the norm topology of X induces the Euclidean topology on Y, each x_i^* is norm continuous. By the Hahn–Banach theorem, we can assume that each x_i^* is defined on all of X, i.e., $x_i^* \in X^*$ for each i. Now consider the continuous projection $P = \sum_{i=1}^n x_i^* \otimes x_i$.]

7. (Grothendieck [**141**]) For a Banach space X establish that the following statements are equivalent.

(a) X has the approximation property.

(b) For every Banach space Y the vector space of finite-rank operators $Y^* \otimes X$ is $\tau_c(Y, X)$-dense in $\mathcal{L}(Y, X)$.

(c) For every Banach space Y the vector space of finite-rank operators $X^* \otimes Y$ is $\tau_c(X, Y)$-dense in $\mathcal{L}(X, Y)$.

8. Let X be a Banach space with the approximation property. If Y is a Banach space and $T\colon Y \to X$ is a compact operator, then show that $T^*\colon X^* \to Y^*$ belongs to the norm closure of the finite-rank operators from X^* to Y^*, i.e., $T^* \in \overline{X^{**} \otimes Y^*}$.

9. Consider the rank-two operator $T = \mathbf{1} \otimes u - u \otimes \mathbf{1}$ on $L_1[0,1]$ introduced in the proof of Theorem 4.14, where $\mathbf{1}(s) = 1$ and $u(s) = s$ for each $s \in [0,1]$. Also, recall that each function $v \in L_\infty[0,1]$ can be considered as a continuous linear functional acting on $L_1[0,1]$ via the duality formula $\langle x, v \rangle = \int_0^1 v(s)x(s)\,ds$. Prove the following.

(a) $Tx(t) = \int_0^1 (t-s)x(s)\,ds$ for all $x \in L_1[0,1]$ and each $t \in [0,1]$.

(b) The modulus of the operator T satisfies $|T|x(t) = \int_0^1 |s-t|x(s)\,ds$ for all $x \in L_1[0,1]$ and $t \in [0,1]$.

(c) $|T|$ is not a finite-rank operator.

[HINT: For (b), see Exercise 13 of Section 3.1. For the validity of (c) note that if $x_k(s) = (k+1)(k+2)s^k$, then

$$|T|x_k(t) = 2t^{k+2} - (k+2)t + k + 1\,.$$

Since the functions $|T|x_1, |T|x_2, \ldots$ are linearly independent, we infer that $|T|$ cannot have a finite dimensional range.]

10. An operator $T\colon X \to Y$ between two Banach spaces is said to be a **nuclear operator** if there exist two sequences $\{x_n^*\} \subseteq X^*$ and $\{y_n\} \subseteq Y$ satisfying $\sum_{n=1}^\infty \|x_n^*\| \cdot \|y_n\| < \infty$ and

$$T(x) = \sum_{n=1}^\infty x_n^*(x) y_n$$

for each $x \in X$, i.e., $T = \sum_{n=1}^\infty x_n^* \otimes y_n$ in $\mathcal{L}(X, Y)$. Show that:

(a) Every nuclear operator is compact.
(b) If X and Y are two arbitrary Banach lattices, $\sum_{n=1}^{\infty} \|x_n^*\| < \infty$ and $\sum_{n=1}^{\infty} \|y_n\| < \infty$, then the nuclear operator $T = \sum_{n=1}^{\infty} x_n^* \otimes y_n$ has a compact modulus.

4.2. Multiplication Operators

In this section, we shall discuss multiplication operators on $C(\Omega)$-spaces. We start with the following result.

Lemma 4.15. *For a positive element $u > 0$ in a Banach lattice E the following statements are equivalent.*

(1) *The principal ideal generated by u is norm dense in E, i.e., $\overline{E}_u = E$.*
(2) *For each $x \in E^+$ we have $\|x \wedge nu - x\| \to 0$.*
(3) *If $0 < x^* \in E^*$, then $x^*(u) > 0$, i.e., the vector u acts as a strictly positive linear functional on E^*.*

Proof. (1) $\Longrightarrow$ (2) Let $x \in E^+$ and fix $\epsilon > 0$. Since E_u is dense in E, there exists some $y \in E_u$ such that $\|y - x\| < \epsilon$. From the lattice inequality $|y^+ \wedge x - x| \leq |y - x|$, we see (by replacing y by $y^+ \wedge x$) that we can assume $0 \leq y \leq x$. Now fix some k such that $0 \leq y \leq ku$ and so $y \leq x \wedge ku$. Hence, if $n \geq k$, then from the inequality $0 \leq x - x \wedge nu \leq x - x \wedge ku \leq x - y$, we see that $\|x - x \wedge nu\| < \epsilon$. Therefore, $\|x - x \wedge nu\| \to 0$.

(2) $\Longrightarrow$ (3) Let $0 < x^* \in E^*$. If $x^*(u) = 0$, then $x^*(x \wedge nu) = 0$ for all n and each $x \geq 0$, and so $x^*(x) = 0$ for all $x \in E^+$. This implies $x^* = 0$, a contradiction. Hence, $x^*(u) > 0$.

(3) $\Longrightarrow$ (1) Assume by way of contradiction that $\overline{E}_u \neq E$. Then, by the Separation Theorem, there exists some non-zero $x^* \in E^*$ that vanishes on E_u. This implies $0 < |x^*|(u) = \sup\{|x^*(y)|: \ |y| \leq u\} = 0$, a contradiction. Hence, $\overline{E}_u = E$. ■

Any positive vector u in a Banach lattice E that satisfies $\overline{E}_u = E$ is called a **quasi-interior point** or a **strictly positive vector**. The name "quasi-interior point" originates from the fact that an order unit of a Banach lattice is simply an interior point of the positive cone. The name "strictly positive vector" is justified by statement (3) of Lemma 4.15. We have the following implications:

Order Unit $\Longrightarrow$ *Quasi-interior Point* $\Longrightarrow$ *Weak Order Unit*

In general no other implication is true. However, in a Banach lattice with order continuous norm a positive vector is a quasi-interior point if and only if it is a weak order unit. We shall say that a complex Banach lattice

$E_{\mathbf{C}} = E \oplus \imath E$ has a quasi-interior point if the Banach lattice E has a quasi-interior point.

How many quasi-interior points can a Banach lattice have? It might have none! For instance, the Banach lattice $\mathcal{M}[0,1]$ of all Borel measures of bounded variation on $[0,1]$ does not have quasi-interior points; see Exercise 5 at the end of the section. On the other hand, if a Banach lattice E has a quasi-interior point, say u, then E has quite many more: λu for each $\lambda > 0$ and any positive vector w satisfying $w \geq \lambda u$ for some $\lambda > 0$ are all quasi-interior points. It follows that the set of all quasi-interior points of a Banach lattice is either empty or else is dense in E^+.

However, in spite of the above remarks, if a Banach lattice has a quasi-interior point, then it is essentially "unique" in the sense that if u and v are two quasi-interior points, then the principal ideals E_u and E_v are lattice isometric. This easily follows from Theorem 3.34. Let us elaborate on this relationship a little bit more.

Let $u > 0$ be a positive vector in a Banach lattice E. Then (by Theorem 3.6) the principal ideal E_u generated by u is an AM-space under the norm

$$\|x\|_\infty = \inf\{\lambda \geq 0\colon\ |x| \leq \lambda u\}\,.$$

Moreover, it is lattice isometric to a concrete $C(\Omega)$-space, where the compact Hausdorff space Ω is uniquely determined up to a homeomorphism. Now assume that u and v are two quasi-interior points in a Banach lattice E such that E_u and E_v have the representations $C(\Omega)$ and $C(\Theta)$, respectively. Then, by Theorem 3.34, $C(\Omega)$ and $C(\Theta)$ are lattice isometric and so Ω and Θ are homeomorphic. The unique (up to a homeomorphism) compact Hausdorff space Ω is called the **structure space** of E and it plays a significant role in the representation theory of Banach lattices. For details, we refer the reader to the book of H. H. Schaefer [**290**, §4, pp. 165–168]. It should be noted that the structure space of a Banach lattice need not coincide with its Stone space. For instance, if $E = C[0,1]$, then the structure space of E is $[0,1]$ while the Stone space Q of E is extremally disconnected—and hence different from $[0,1]$.

As discussed in Section 3.3, as soon as we represent a principal ideal E_u as a $C(\Omega)$-space, we get at our disposal the "operator structure" of $C(\Omega)$. For instance, we have a plethora of multiplication and composition operators on $C(\Omega)$ which can be viewed as operators on E_u—this important feature has no parallel in the general theory of Banach spaces. In other words, we can transfer via the lattice isometry the (pointwise) multiplication of $C(\Omega)$ to E_u. In particular, every element $v \in E_u$ gives rise to a **multiplication operator** $M_v\colon E_u \to E_u$ via the formula

$$M_v(x) = vx\,,\quad x \in E_u\,.$$

Moreover, M_v is continuous with respect to the original norm $\|\cdot\|$ restricted to E_u (in fact $\|M_v\| = \|v\|_\infty$). When u is a quasi-interior point, the multiplication operators extend (by virtue of their uniform continuity) to continuous operators on all of E.

The next two results that will be used extensively to study the invariant subspace problem indicate the importance of quasi-interior points.

Lemma 4.16. *For a (real or complex) Banach lattice E with a quasi-interior point $u > 0$ we have the following.*

(1) *For each non-zero vector $y \in E_u$ there exists an operator M dominated by the identity operator (and hence M is a central operator) satisfying $M(y) > 0$.*

(2) *For every element v satisfying $0 \le v \le u$ there exists an operator $T\colon E \to E$ dominated by the identity operator and such that $Tu = v$.*

Proof. Let $u > 0$ be a quasi-interior point in a (real or complex) Banach lattice E. As before, we can identify E_u with a $C(\Omega)$-space in such a way that the vector u corresponds to the constant function one on Ω.

(1) Fix $y \in E_u$ with $y \neq 0$ and view y as a continuous function on Ω. By scaling, we can suppose that $\|y\|_\infty = \max\{|y(\omega)|\colon\ \omega \in \Omega\} = 1$. Now consider the function $\overline{y} \in C(\Omega)$ (the complex conjugate of y) and denote by M the multiplication operator defined by the function $\overline{y}$ on $C(\Omega)$ (and hence on E_u). That is, $M(x) = \overline{y}x$ for each $x \in C(\Omega)$. Clearly, $M(y) = |y|^2 > 0$ and $|Mx| \le |x|$ holds for each $x \in E_u$. Since $\overline{E_u} = E$, the (unique) continuous extension of M to E satisfies the desired properties.

(2) Again, as above, we view v as a continuous function on Ω and consider the multiplication operator T defined on $C(\Omega)$ (and hence on E_u) by v, i.e, $T(x) = vx$ for each $x \in C(\Omega)$. Clearly, $T(u) = v$ and $|Tx| \le |x|$ for each $x \in E_u$. The (unique) continuous extension of T to all of E also satisfies $|Tx| \le |x|$ for each $x \in E$. ∎

Lemma 4.17 (de Pagter [**260**]). *Let E be a (real or complex) Banach lattice with a quasi-interior point and let $|x| \le |y|$. Then there exists a sequence $\{M_n\}$ of operators on E such that $\|M_n y - x\| \to 0$ and each M_n is dominated by the identity operator.*

Proof. Let $u > 0$ be a quasi-interior point of E. Replacing u by $u + |y|$, we can assume that $|y| \le u$. Now let us view E_u as a $C(\Omega)$-space with u corresponding to the constant function $\mathbf{1}$ on Ω. So, $x, y \in C(\Omega)$.

For each n consider the multiplication operator $M_n\colon E_u \to E_u$ generated by the function $u_n = \frac{x\overline{y}}{(|y|+\frac{1}{n})^2}$, i.e., $M_n z = u_n z$ for each $z \in E_u$. The unique continuous linear extension of M_n to all of E will be denoted by M_n again.

Since $|u_n| \le \mathbf{1}$ in $C(\Omega)$, we see that $|M_n z| \le |z|$ for each $z \in E$, i.e., each M_n is dominated by the identity operator. Moreover, an easy computation shows that $|M_n y - x| \le \frac{2}{n}\mathbf{1}$ holds in $C(\Omega)$. This implies $\|M_n y - x\| \le \frac{2}{n}\|u\|$, and so $\|M_n y - x\| \to 0$. ■

The remainder of the section is devoted to some commutativity properties of multiplication operators on $C(\Omega)$-spaces. Accordingly, for the rest of this section, Ω will denote a fixed Hausdorff compact space. As usual, each $\phi \in C(\Omega)$ generates the multiplication operator $M_\phi \colon C(\Omega) \to C(\Omega)$ defined by $M_\phi(f) = \phi f$. We start our discussion with a useful property of weakly compact multiplication operators.

Lemma 4.18. *If a multiplication operator M_ϕ is weakly compact, then the function ϕ vanishes at every accumulation point of Ω.*

In particular, if Ω does not have isolated points, then the only weakly compact (and hence the only compact) multiplication operator on $C(\Omega)$ is the zero operator.

Proof. Assume that M_ϕ is weakly compact. Suppose by way of contradiction that $\phi(\omega_0) \ne 0$ for some accumulation point $\omega_0 \in \Omega$. Therefore, there is an open neighborhood V of ω_0 and some $\epsilon > 0$ such that $|\phi(v)| \ge \epsilon$ for all $v \in V$. By the Urysohn–Tietze extension theorem there exists a function $\psi \in C(\Omega)$ such that $\psi(\omega) = \frac{1}{\phi(\omega)}$ for each $\omega \in \overline{V}$, the closure of V.

From $M_\psi M_\phi = M_{\psi\phi}$, we see that $M_{\psi\phi}$ is a weakly compact operator on $C(\Omega)$. Since the function $\psi\phi$ equals one at each point of $\overline{V}$, it follows (again by the Urysohn–Tietze extension theorem) that the identity operator I on $C(\overline{V})$ is weakly compact. This implies that $C(\overline{V})$ is reflexive, and hence (by Theorem 2.26) its norm must be order continuous. But then $\overline{V}$ must be a finite set (see Exercise 1 in Section 2.3), and so ω_0 is an isolated point, which is a contradiction. ■

Recall that if $T \colon X \to X$ is an operator on a Banach space, then its **commutant** $\{T\}'$ consists of all bounded operators on X that commute with T, that is,

$$\{T\}' = \{S \in \mathcal{L}(X) \colon \ ST = TS\}.$$

A direct verification shows that the commutant of an operator T is a unital subalgebra of $\mathcal{L}(X)$. In general, it is a very difficult task to describe the commutant of a given operator. Our next goal is to describe some classes of operators that commute with a given multiplication operator. The rest of the material in this section is taken from [**15**].

For a function $\phi \in C(\Omega)$, we shall denote by $\mathcal{A}_\phi$ the closed unital (i.e., containing $\mathbf{1}$) algebra generated by ϕ in $C(\Omega)$, where the product of functions in $C(\Omega)$ is always understood as the pointwise product. This algebra $\mathcal{A}_\phi$ is the uniform closure of the subspace $\{p(\phi)\colon\ p \text{ is a polynomial}\}$. That is,

$$\mathcal{A}_\phi = \{p(\phi)\colon\ p \text{ is a polynomial}\}^{-}.$$

It should be noted that besides being an algebra, $\mathcal{A}_\phi$ is also a Riesz subspace of $C(\Omega)$.

Lemma 4.19. *For a function $\phi \in C(\Omega)$ and an arbitrary bounded operator $T\colon C(\Omega) \to C(\Omega)$ commuting with M_ϕ the following statements are true.*

1. *There exists a function $h \in C(\Omega)$ satisfying $Tx = hx$ for each x in the closed unital algebra $\mathcal{A}_\phi$.*
2. *$TM_f = M_fT$ holds for each $f \in \mathcal{A}_\phi$.*
3. *If ϕ is one-to-one, then T itself is a multiplication operator. In other words, if ϕ is one-to-one, then the commutant of M_ϕ consists of all multiplication operators, i.e, $\{M_\phi\}' = \{M_f\colon\ f \in C(\Omega)\}$.*

Proof. (1) The identity $TM_\phi = M_\phi T$ means that $T(\phi x) = \phi Tx$ for each $x \in C(\Omega)$. By induction, we get $T(\phi^n x) = \phi^n Tx$ for each $n \geq 0$ and all $x \in C(\Omega)$. Thus,

$$T(p(\phi)x) = p(\phi)Tx \tag{$\star$}$$

for each polynomial $p(t)$ of one real variable t and all $x \in C(\Omega)$.

Letting $x = \mathbf{1}$ and $h = T\mathbf{1}$, it follows from ($\star$) that $T(p(\phi)) = hp(\phi)$ for each polynomial p. In other words, $Tx = hx$ for all x in the unital algebra generated by ϕ. Consequently, by continuity, $Tx = hx$ for all x in the closed unital algebra $\mathcal{A}_\phi$ generated by ϕ.

(2) Let $f \in \mathcal{A}_\phi$. If $x \in C(\Omega)$, then from ($\star$) we see that

$$TM_f(x) = T(fx) = fT(x) = M_fT(x),$$

so that $TM_f = M_fT$.

(3) If ϕ is one-to-one, then the Stone–Weierstrass Approximation Theorem guarantees that $\mathcal{A}_\phi = C(\Omega)$. So, by part (1), $T = M_h$. ■

If ϕ is not one-to-one, then the commutant of M_ϕ might include operators other than multiplication operators. For an example take $\Omega = [-1, 1]$ and $\phi(\omega) = \omega^2$. An easy argument shows that $\mathcal{A}_\phi$ consists of all even functions. If we consider the operator $T\colon C[-1,1] \to C[-1,1]$ defined by $Tx(\omega) = x(-\omega)$, then T is not a multiplication operator. Clearly, $M_\phi T = TM_\phi$.

The next surprising result asserts that a continuous function is constant on an open set if and only if its multiplication operator commutes with a non-zero finite-rank operator.

Theorem 4.20. *For a multiplication operator M_ϕ generated by a function $\phi \in C(\Omega)$ the following statements are equivalent.*

(1) *The function ϕ is constant on some non-empty open subset of Ω.*

(2) *M_ϕ commutes with a rank-one positive operator.*

(3) *M_ϕ commutes with a rank-one operator.*

(4) *M_ϕ commutes with a non-zero finite-rank operator.*

Proof. (1) $\Longrightarrow$ (2) Suppose that $\phi(\omega) = c$ for each $\omega \in V$, where V is a non-empty open subset of Ω. Choose a non-zero positive function $\psi \in C(\Omega)$ such that $\psi(\omega) = 0$ for all $\omega \notin V$ and fix some $\omega_0 \in V$. Now define the rank-one operator $K\colon C(\Omega) \to C(\Omega)$ via the formula $Kx = x(\omega_0)\psi$.

We claim that M_ϕ and K commute. Indeed, for $x \in C(\Omega)$ and $\omega \in \Omega$, we have

$$(KM_\phi x)(\omega) = (K\phi x)(\omega) = \phi(\omega_0)x(\omega_0)\psi(\omega) = cx(\omega_0)\psi(\omega).$$

On the other hand, since ψ vanishes off V and $\phi = c$ on V, we also have that

$$(M_\phi Kx)(\omega) = (M_\phi x(\omega_0)\psi)(\omega) = x(\omega_0)\psi(\omega)\phi(\omega) = cx(\omega_0)\psi(\omega).$$

Therefore, $KM_\phi = M_\phi K$. Clearly, K is also a positive operator.

(2) $\Longrightarrow$ (3) $\Longrightarrow$ (4) Obvious.

(4) $\Longrightarrow$ (1) Assume that the multiplication operator M_ϕ commutes with a non-zero finite-rank operator $T = \sum_{i=1}^k x_i^* \otimes \psi_i$, where $\psi_i \in C(\Omega)$ and $x_i^* \in C^*(\Omega)$ for each i. We can suppose that $\psi_1, \dots, \psi_k$ are linearly independent and that $x_i^* \neq 0$ for each i. From Lemma 4.19(2), we know that $M_f T = TM_f$ for each function f in the closed unital algebra $\mathcal{A}_\phi$ generated by ϕ in $C(\Omega)$. That is,

$$\sum_{i=1}^k x_i^*(x)f(\omega)\psi_i(\omega) = \sum_{i=1}^k x_i^*(fx)\psi_i(\omega) = \sum_{i=1}^k M_f^* x_i^*(x)\psi_i(\omega) \qquad (\star)$$

holds for all $f \in \mathcal{A}_\phi$ and for each $\omega \in \Omega$, where as usual $M_f^*\colon C^*(\Omega) \to C^*(\Omega)$ is the adjoint operator of M_f.

We claim that $\bigcap_{n=1}^k \operatorname{Ker} x_n^* \subseteq \operatorname{Ker} M_f^* x_i^*$ for each i and each function $f \in \mathcal{A}_\phi$. To see this, fix $f \in \mathcal{A}_\phi$ and let $x \in \bigcap_{n=1}^k \operatorname{Ker} x_n^*$. Then $(\star)$ yields $\sum_{i=1}^k M_f^* x_i^*(x)\psi_i = 0$, and from the linear independence of $\psi_1, \dots, \psi_k$, it follows that $M_f^* x_i^*(x) = 0$ for each i. Thus, $\bigcap_{n=1}^k \operatorname{Ker} x_n^* \subseteq \operatorname{Ker} M_f^* x_i^*$.[2] Consequently, each linear functional $M_f^* x_i^*$ belongs to the vector subspace

[2] Here we use the conclusion of Exercise 1 in Section 4.1: *A linear functional f belongs to the span of the linear functionals $f_1, \dots, f_n$ if and only if $\bigcap_{i=1}^n \operatorname{Ker} f_i \subseteq \operatorname{Ker} f$.*

generated by $x_1^*, \ldots, x_k^*$. In particular, for each i the linear functionals $M_\phi^* x_i^*, M_{\phi^2}^* x_i^*, \ldots, M_{\phi^{k+1}}^* x_i^*$ must be linearly dependent. That is, for each i there exist constants $\alpha_{i,1}, \ldots, \alpha_{i,k+1}$ not all zero such that

$$\sum_{j=1}^{k+1} \alpha_{i,j} M_{\phi^j}^* x_i^* = M_{p_i(\phi)}^* x_i^* = 0\,,$$

where p_i is the non-zero polynomial defined by $p_i(\lambda) = \sum_{j=1}^{k+1} \alpha_{i,j}\lambda^j$.

Next, consider the non-zero polynomial $p = p_1 p_2 \cdots p_k$ and note that

$$M_{p(\phi)}^* x_i^* = \Big[\prod_{j\neq i} M_{p_j(\phi)}^*\Big] M_{p_i(\phi)}^* x_i^* = 0$$

for each i. Letting $f = p(\phi)$ in $(\star)$, we get $\sum_{i=1}^k x_i^*(x) p(\phi)(\omega)\psi_i(\omega) = 0$ or

$$p(\phi)(\omega) \cdot \Big[\sum_{i=1}^k x_i^*(x)\psi_i(\omega)\Big] = 0 \tag{$\star\star$}$$

for all $x \in C(\Omega)$ and each $\omega \in \Omega$.

Now if we choose $x \in C(\Omega)$ with $x_1^*(x) \neq 0$, then it follows from the linear independence of $\psi_1, \ldots, \psi_k$ that $\sum_{i=1}^k x_i^*(x)\psi_i(\omega_0) \neq 0$ for some $\omega_0 \in \Omega$. Therefore, $\sum_{i=1}^k x_i^*(x)\psi_i(\omega) \neq 0$ for all ω in an open neighborhood V of ω_0. But then $(\star\star)$ yields

$$p\big(\phi(\omega)\big) = p(\phi)(\omega) = 0 \quad \text{for all} \quad \omega \in V\,. \tag{$\star\star\star$}$$

Since $\lambda_0 = \phi(\omega_0)$ is a root of the polynomial p, and since p has a finite number of roots, there exists an open neighborhood N of λ_0 that does not contain any other roots of p. Finally, if we consider the open neighborhood $W = V \cap (p \circ \phi)^{-1}(N)$ of ω_0, then it should be clear from $(\star\star\star)$ that $\phi(\omega) = \lambda_0$ for each $\omega \in W$, and the proof is finished. ■

Exercises

1. Prove the following implications:

   Order Unit $\Longrightarrow$ *Quasi-interior Point* $\Longrightarrow$ *Weak Order Unit*

   Also show that in general no converse implication is true.

2. Show that a Banach lattice E has a weak unit if and only if E has an at most countable maximal subset of pairwise disjoint vectors. [HINT: If $\{x_1, x_2, \ldots\}$ is a maximal countable set of pairwise disjoint positive unit vectors, then $x = \sum_{n=1}^\infty \frac{1}{2^n} x_n$ is a weak unit.]

3. Show that every separable Banach lattice has a quasi-interior point—and hence it also has a weak unit.

4. In a Banach lattice with order continuous norm, show that a positive vector is a quasi-interior point if and only it is a weak unit.

5. Prove that the Banach lattice $\mathcal{M}[0,1]$, of all signed Borel measures on $[0,1]$ of bounded variation, does not have quasi-interior points. [HINT: Notice that since $\mathcal{M}[0,1]$ has order continuous norm, the notions of weak unit and quasi-interior point coincide. Assume that $0 < \mu \in \mathcal{M}[0,1]$ is a weak unit. (This means that $\mu \wedge \nu = 0$ implies $\nu = 0$.) Then $\mu(\{x\}) > 0$ for each $x \in [0,1]$; otherwise $\mu(\{x\}) = 0$ implies $\mu \wedge \delta_x = 0$, where δ_x is the Dirac measure supported at x, which is impossible. However, the latter conclusion shows that $[0,1]$ is countable, a contradiction.]

6. Show that if E is a Banach lattice, then its set of quasi-interior points is either empty or else it is dense in E^+.

7. If $T\colon E \to F$ is a strictly positive operator between two Banach lattices, then show that the adjoint operator $T^*\colon F^* \to E^*$ carries quasi-interior points to quasi-interior points.

8. Fix a function $\phi \in C(\Omega)$ (where Ω is compact and Hausdorff) and let $\mathcal{A}_\phi$ denote the closed unital algebra generated by ϕ in $C(\Omega)$. That is,

$$\mathcal{A}_\phi = \{p(\phi)\colon\ p \text{ is a polynomial}\}^- .$$

Show that $\mathcal{A}_\phi$ is a Riesz subspace of $C(\Omega)$.

9. Let (Θ, Σ, μ) be a σ-finite measure space and let $\phi \in L_\infty(\mu)$. Show that for the multiplication operator $M_\phi\colon L_p(\mu) \to L_p(\mu)$, where $1 \le p \le \infty$, the following statements are equivalent.
   (a) The function ϕ is constant on some measurable set of positive measure.
   (b) M_ϕ commutes with a rank-one positive operator.
   (c) M_ϕ commutes with a rank-one operator.
   (d) M_ϕ commutes with a non-zero finite-rank operator.

10. Let $\tau\colon \Omega \to \Omega$ be a continuous mapping on a compact Hausdorff space Ω. For the **composition operator** $C_\tau\colon C(\Omega) \to C(\Omega)$, defined by the formula $C_\tau(x) = x \circ \tau$, establish the following.
   (a) C_τ is a contractive positive operator satisfying $C_\tau(\mathbf{1}) = \mathbf{1}$.
   (b) C_τ commutes with a rank-one positive operator.

11. Show that for a continuous map $\tau\colon \Omega \to \Omega$, where Ω is compact and Hausdorff, the following statements are equivalent.
   (a) The map τ is surjective.
   (b) The composition operator C_τ is an isometry.
   (c) The composition operator C_τ is one-to-one.

4.3. Lattice and Algebraic Homomorphisms

The material presented so far has already demonstrated the usefulness of $C(\Omega)$-spaces in the study of positive operators. As explained earlier, the importance of $C(\Omega)$-spaces stems, on one hand, from the fact that every principal ideal in a Banach lattice is an AM-space with unit and, on the other hand, from the classical Kakutani–Bohnenblust–Krein Theorem 3.6 on the representation of AM-spaces with unit. In this section, we shall present the basic characterizations of lattice and algebraic homomorphisms between $C(\Omega)$-spaces as well as some extension properties of lattice homomorphisms. We start with the following result.

Theorem 4.21. *Let E be a Banach lattice, let $\mathcal{F}$ be an at most countable subset of E and let $\mathcal{C}$ be an at most countable collection of regular operators on E. Then there exists some $u > 0$ such that:*

(1) *The set $\mathcal{F}$ lies in E_u, i.e., $\mathcal{F} \subseteq E_u$.*
(2) *Every operator in $\mathcal{C}$ leaves the principal ideal E_u invariant, that is, $T(E_u) \subseteq E_u$ holds for each $T \in \mathcal{C}$.*

Proof. The proof will be done in three steps.

Step I: *If $\mathcal{F}$ is at most countable, then $\mathcal{F}$ is included in a principal ideal.*

Assume that $\mathcal{F}$ is a countable set, say $\mathcal{F} = \{x_1, x_2, \ldots\}$. We can assume that $x_i \neq 0$ for each i. If $u = \sum_{n=1}^{\infty} \frac{1}{2^n} \frac{|x_n|}{\|x_n\|}$, then note that $\mathcal{F} \subseteq E_u$.

Step II: *If $\mathcal{F}$ is an at most countable subset of E and $T \in \mathcal{L}(E)$ is a regular operator, then there exists some $u > 0$ such that $\mathcal{F} \subseteq E_u$ and $T(E_u) \subseteq E_u$.*

To see this, assume that T is a non-zero regular operator. So, we can write $T = T_1 - T_2$ with $T_1, T_2 \geq 0$. Put $S = T_1 + T_2 > 0$.

By Step I, there exists some $w > 0$ such that $\mathcal{F} \subseteq E_w$. Next, consider the positive vector $u = \sum_{n=0}^{\infty} \frac{1}{2^n} \frac{S^n w}{\|S\|^n}$. Since $w \leq u$, it follows that $\mathcal{F} \subseteq E_u$. It is now a routine matter to verify that $T(E_u) \subseteq E_u$.

Step III: *The general case.*

Assume that $\mathcal{C}$ is countable, say $\mathcal{C} = \{T_1, T_2, \ldots\}$, where $T_i \neq 0$ for each i. Write $T_i = T_{i1} - T_{i2}$ with $T_{i1}, T_{i2} \geq 0$, and let $S_i = T_{i1} + T_{i2} > 0$.

Next, consider the positive operator $S = \sum_{n=0}^{\infty} \frac{1}{2^n} \frac{S_i}{\|S_i\|}$. By Step II, there exists some $u > 0$ such that $\mathcal{F} \subseteq E_u$ and $S(E_u) \subseteq E_u$. Next, fix any $x \in E_u$ and pick some $\lambda > 0$ such that $|x| \leq \lambda u$. Then for each k we have

$$\begin{aligned} |T_k x| &\leq (T_{k1} + T_{k2})|x| = S_k|x| \leq \lambda S_k u \\ &= \lambda 2^k \|S_k\| \cdot \frac{S_k u}{2^k \|S_k\|} \leq \lambda 2^k \|S_k\| S u \in E_u\,. \end{aligned}$$

So, $T_k x \in E_u$, that is, $T_k(E_u) \subseteq E_u$ holds for each k. ■

The important conclusion from Theorem 4.21 is the following. When one represents E_u as a $C(\Omega)$-space, then all vectors in $\mathcal{F}$ as well as all vectors $T_k x$, $x \in E_u$, can be viewed as continuous functions on the compact Hausdorff space Ω. We can then use the properties of the continuous functions to study the properties of positive operators. In addition, expressions like e^x, $\frac{x}{1+|y|}$, etc. (where x and y are vectors in E_u) as well as $\sqrt[n]{x}$ if $x \geq 0$ make sense as functions in $C(\Omega)$; and hence they are well-defined vectors in E!

It should also be noticed that since E_u is an ideal in E, its complexification $E_u \oplus \imath E_u$ can be identified with $C_{\mathbb{C}}(\Omega)$, the Banach space of all complex-valued continuous functions on Ω.

We now turn our attention to the properties of lattice homomorphisms between $C(\Omega)$-spaces. As we shall see, they are closely related to algebraic homomorphisms that are defined next.

Definition 4.22. *An operator $T\colon C(\Omega) \to C(Q)$ is said to be an* ***algebraic homomorphism*** *(or a* ***multiplicative operator****) if*

$$T(fg) = T(f)T(g)$$

holds for all $f, g \in C(\Omega)$.

We first characterize the lattice homomorphisms from a $C(\Omega)$-space to the real numbers.

Lemma 4.23. *A non-zero linear functional $\phi\colon C(\Omega) \to \mathbb{R}$ is a lattice homomorphism if and only if there exists some $c > 0$ and some $\omega_0 \in \Omega$ (both uniquely determined) such that $\phi = c\delta_{\omega_0}$.*

Proof. If $\phi = c\delta_{\omega_0}$ with $c \geq 0$, then ϕ is clearly a lattice homomorphism, that is, $\phi(f \wedge g) = \min\{\phi(f), \phi(g)\}$ for all $f, g \in C(\Omega)$.

For the converse, assume that $\phi\colon C(\Omega) \to \mathbb{R}$ is a lattice homomorphism. Clearly, ϕ is a positive linear functional. Hence, by the Riesz Representation Theorem, there exists a unique regular Borel measure μ on Ω satisfying $\phi(f) = \int f\,d\mu$ for each $f \in C(\Omega)$. We claim that the support of μ is a singleton. To see this, assume that two distinct points, say s and t, belong to the support of μ. Pick two functions $f, g \in C(\Omega)$ satisfying $f \wedge g = 0$ and $f(s) = g(t) = 1$. This implies

$$0 = \phi(f \wedge g) = \min\{\phi(f), \phi(g)\} = \min\{\textstyle\int f\,d\mu, \int g\,d\mu\} > 0\,,$$

which is a contradiction. Hence, the support of μ consists of a single point, say ω_0. This implies

$$\phi(f) = \int f\,d\mu = \int_{\{\omega_0\}} f\,d\mu = f(\omega_0)\mu(\{\omega_0\}) = \big[\mu(\{\omega_0\})\delta_{\omega_0}\big](f)$$

for each $f \in C(\Omega)$. ■

The algebraic homomorphisms from a $C_{\mathbb{C}}(\Omega)$-space to the complex numbers are characterized in the next result.

Lemma 4.24 (I. M. Gelfand). *A non-zero linear functional $\phi\colon C_{\mathbb{C}}(\Omega) \to \mathbb{C}$ is multiplicative if and only if there exists a (uniquely determined) point $\omega_0 \in \Omega$ such that $\phi = \delta_{\omega_0}$.*

Proof. If $\phi = \delta_{\omega_0}$, then $\phi(fg) = (fg)(\omega_0) = f(\omega_0)g(\omega_0) = \phi(f)\phi(g)$ holds for all $f, g \in C(\Omega)$. That is, ϕ is a multiplicative linear functional.

For the converse, assume that ϕ is a non-zero multiplicative linear functional on $C(\Omega)$. Since $\phi(\mathbf{1}) = \phi(\mathbf{1}^2) = [\phi(\mathbf{1})]^2$ and ϕ is non-zero, it follows that $\phi(\mathbf{1}) = 1$, where $\mathbf{1}$ is the constant function one on Ω. Let $M = \{x \in C(\Omega)\colon\ \phi(x) = 0\}$ be the kernel of the functional ϕ. We claim that there exists a point $\omega_0 \in \Omega$ such that $M \subseteq \{x \in C(\Omega)\colon\ x(\omega_0) = 0\}$.

If there is no such point, then for each $\omega \in \Omega$ we can find a function x_ω in M such that $x_\omega(\omega) \neq 0$. Since x_ω is continuous, there exists a neighborhood V_ω of the point ω such that $x_\omega(t) \neq 0$ for each $t \in V_\omega$. By the compactness of Ω, there exist neighborhoods $V_{\omega_1}, \ldots, V_{\omega_n}$ covering the whole of Ω. Next, consider the function $x = \sum_{i=1}^n x_{\omega_i} \cdot \overline{x}_{\omega_i}$, where, as usual, $\overline{z}$ denotes the complex conjugate of a function $z \in C(\Omega)$. By our choice of the functions, $x(\omega) > 0$ for each $\omega \in \Omega$, and hence $y = \frac{1}{x} \in C(\Omega)$. Now the identity $xy = \mathbf{1}$ coupled with the multiplicativity of ϕ yields $\phi(x)\phi(y) = 1$, and so $\phi(x) \neq 0$. On the other hand, as $\phi(x_\omega) = 0$ for each ω, we certainly have $\phi(x) = 0$, a contradiction. Thus, there exists some $\omega_0 \in \Omega$ such that $M \subseteq \{x \in C(\Omega)\colon\ x(\omega_0) = 0\}$.

Next, we shall show that $\{x \in C(\Omega)\colon\ x(\omega_0) = 0\} \subseteq M$. To this end, let $x \in C(\Omega)$ satisfy $x(\omega_0) = 0$. If $u = \phi(x)\mathbf{1} - x$, then clearly $u \in M$, and so $0 = u(\omega_0) = \phi(x)$. Hence $x \in M$, and thus $M = \{x \in C(\Omega)\colon\ x(\omega_0) = 0\}$.

Finally, if $x \in C(\Omega)$, then $x - \phi(x)\mathbf{1} \in M$, and so $x(\omega_0) - \phi(x) = 0$ or $\phi(x) = x(\omega_0)$, as desired. The uniqueness of ω_0 follows from the separation of points by $C(\Omega)$. ■

We are now ready to characterize the lattice homomorphisms between $C(\Omega)$-spaces.

Theorem 4.25. *A positive operator $T\colon C(\Omega) \to C(Q)$ is a lattice homomorphism if and only if there exist a mapping $\tau\colon Q \to \Omega$ and some weight function $w \in C(Q)$ such that for each $f \in C(\Omega)$ and each $q \in Q$ we have*

$$Tf(q) = w(q)f\bigl(\tau(q)\bigr)\,.$$

Moreover, in this case, $w = T\mathbf{1}_\Omega$ and the mapping τ is uniquely determined and continuous on the set $\{q \in Q\colon\ w(q) > 0\}$.

Proof. If T is of the above form, then clearly T is a lattice homomorphism. For the converse, assume that T is a lattice homomorphism. Then for each

$q \in Q$, we have

$$\big(\delta_q \circ T\big)(f) = \delta_q(Tf) = [Tf](q)\,.$$

Since T is a lattice homomorphism, the linear functional $\delta \circ T\colon C(\Omega) \to \mathbb{R}$ is a lattice homomorphism. Thus, by Lemma 4.23, there exists a unique constant $w(q) \geq 0$ and some (not necessarily unique) point $\tau(q) \in \Omega$ such that

$$[Tf](q) = \big(\delta_q \circ T\big)(q) = w(q)f\big(\tau(q)\big)\,.$$

Letting $f = \mathbf{1}_\Omega$, we obtain $w = T\mathbf{1}_\Omega \in C(Q)$. Moreover, if $w(q) > 0$, then $\tau(q)$ is uniquely determined. We leave it as an exercise to verify that τ is continuous on the set $\{q \in Q\colon\ w(q) > 0\}$. ■

Here is a connection between lattice and algebraic homomorphisms.

Theorem 4.26. *Every algebraic homomorphism between real $C(K)$-spaces is a lattice homomorphism. However, the converse is false.*

Proof. Let $T\colon C(\Omega) \to C(Q)$ be an algebraic homomorphism. Clearly, T is a positive operator. Indeed, if $f \geq 0$, then $Tf = T\big((\sqrt{f})^2\big) = [T(\sqrt{f})]^2 \geq 0$. Now if $f \in C(\Omega)$, then

$$\big|T(f)\big|^2 = \big[T(f)\big]^2 = T(f^2) = T(|f|^2) = \big[T(|f|)\big]^2\,,$$

and so $|T(f)| = T(|f|)$. ■

The algebraic homomorphisms are characterized as follows.

Theorem 4.27. *An operator T between real or complex $C(K)$-spaces is an algebraic homomorphism if and only if there exist a unique clopen subset V of Q and a mapping $\tau\colon Q \to \Omega$ which is continuous on V such that*

$$Tf = \chi_V \cdot (f \circ \tau)$$

for each $f \in C(\Omega)$. The mapping τ is also uniquely determined on V.

Proof. If T has the above form, then T is clearly an algebraic homomorphism. For the converse, assume that T is an algebraic homomorphism. We shall consider the real and the complex cases separately.

CASE I: *Assume that $T\colon C(\Omega) \to C(Q)$ is an algebraic homomorphism between real $C(K)$-spaces.*

By Theorem 4.26, T is a lattice homomorphism. So, from Theorem 4.25, it follows that there exist a mapping $\tau\colon Q \to \Omega$ and some $w \in C(Q)$ such that for each $f \in C(\Omega)$ and each $q \in Q$ we have

$$Tf(q) = w(q)f\big(\tau(q)\big)\,.$$

Moreover, we know that $w = T\mathbf{1}_\Omega$ and that the mapping τ is uniquely determined and continuous on the set $\{q \in Q\colon\ w(q) > 0\}$. Now use the fact that T is an algebraic homomorphism to get

$$w^2 = \big[T\mathbf{1}_\Omega\big]^2 = T\big((\mathbf{1}_\Omega)^2\big) = T(\mathbf{1}_\Omega) = w\,.$$

This implies that $w = \chi_V$ for a unique clopen subset V of Q.

CASE II: *Assume that $T\colon C_{\mathbb{C}}(\Omega) \to C_{\mathbb{C}}(Q)$ is an algebraic homomorphism between complex $C(K)$-spaces.*

For each $q \in Q$, we have

$$\big(\delta_q \circ T\big)(f) = \delta_q(Tf) = [Tf](q)\,.$$

Since T is an algebraic homomorphism, it follows that the linear functional $\delta \circ T\colon C_{\mathbb{C}}(\Omega) \to \mathbb{C}$ is likewise an algebraic homomorphism. Thus, by Lemma 4.24, there exists some point $\tau(q) \in \Omega$ such that

$$[Tf](q) = \big(\delta_q \circ T\big)(q) = f\big(\tau(q)\big)$$

for each $f \in C_{\mathbb{C}}(\Omega)$. From $T(\mathbf{1}_\Omega) = T\big((\mathbf{1}_\Omega)^2\big) = \big[T(\mathbf{1}_\Omega)\big]^2$, it follows that $T(\mathbf{1}_\Omega) = \chi_V$ for a unique clopen subset V of Q. To finish the proof, notice that $Tf = T(\mathbf{1}_\Omega f) = T(\mathbf{1}_\Omega)Tf = \chi_V(f \circ \tau)$. ■

For Markov operators between $C(\Omega)$-spaces, the notions of algebraic homomorphism, lattice homomorphism, and composition operator coincide. Specifically, we have the following result whose proof follows easily from the preceding discussion and is left to the reader.

Theorem 4.28. *For a Markov operator $T\colon C(\Omega) \to C(Q)$ the following statements are equivalent.*

(1) *T is an algebraic homomorphism.*
(2) *T is a lattice homomorphism.*
(3) *T is a composition operator.*

The proof of the next result—which characterizes in terms of complexifications the lattice homomorphisms—is left as an exercise. (See Exercise 5 at the end of the section.)

Theorem 4.29. *A positive operator $T\colon E \to F$ between two Banach lattices is a lattice homomorphism if and only if the operator $T_{\mathbf{C}}\colon E_{\mathbf{C}} \to F_{\mathbf{C}}$ satisfies $|T_{\mathbf{C}}z| = T|z|$ for all $z \in E_{\mathbf{C}}$.*

Our next result in this section deals with the "automatic continuity" of the multiplicative functionals on Banach algebras. In Lemmas 4.23 and 4.24 we did not assume the continuity of the functional ϕ. The continuity of ϕ was a consequence of the lattice or algebraic properties. In the case of a real algebraic homomorphism $\phi\colon C(\Omega) \to \mathbb{R}$ the continuity of ϕ is a consequence of its positivity. Moreover, as we shall see next, the continuity of these functionals is a very special case of the continuity of multiplicative functionals on Banach algebras. The phenomenon of the automatic continuity of multiplicative functionals and multiplicative homomorphisms on $C(\Omega)$ and on many other algebras is well known and very important. We refer to [**95, 199, 304**] for details.

Theorem 4.30. *A multiplicative linear functional on any Banach algebra is continuous and has norm at most one.*

Proof. Let ϕ be a multiplicative functional on a (commutative or not) Banach algebra $\mathcal{A}$. A moment's thought shows that whether ϕ is discontinuous or has norm greater than one, there exists some $a \in \mathcal{A}$ with $\|a\| < 1$ and $\phi(a) = 1$. If we let $b = \sum_{n=1}^{\infty} a^n$, then $ab = b - a$, and the multiplicativity of ϕ yields $\phi(a)\phi(b) = \phi(b) - \phi(a)$. That is, $\phi(b) = \phi(b) - 1$, which is impossible. ∎

A few remarks regarding the existence of non-trivial multiplicative functionals and their automatic continuity are in order. There are two similar instances of automatic continuity in Banach lattices. The first one has already been discussed earlier and it asserts the continuity of *any* positive operator on a Banach lattice (Theorem 1.31). This makes the automatic continuity phenomenon of positive operators different from that of multiplicative homomorphisms, since, as shown by H. Dales and J. Esterle [**96**], for an arbitrary multiplicative homomorphism on a $C(\Omega)$-space the automatic continuity may fail. Surprisingly, band preserving operators on Banach lattices are also automatically continuous; see Theorem 3.29.

The existence of non-trivial multiplicative functionals on Banach algebras is another delicate problem. As noted above each Banach algebra $C(\Omega)$ carries a multitude of multiplicative functionals—for each point $\omega \in \Omega$ there corresponds a continuous multiplicative functional δ_ω. For the Banach algebra $\mathcal{L}(X)$ of all bounded operators on a Banach space X this problem has received special attention. If $1 < \dim X < \infty$, then the zero functional is the only multiplicative functional on $\mathcal{L}(X)$. Moreover, if Y is an arbitrary Banach space and $X = Y \oplus Y \oplus \cdots \oplus Y$, that is, X is the direct sum of $n > 1$ copies of Y, then again only the trivial multiplicative functional exists on $\mathcal{L}(X)$; see Exercise 7 at the end of this section. For a general Banach space X the situation is quite different. To the best of our knowledge it was B. S. Mityagin and I. C. Edelstein [**244**] who were the first to exhibit two types of Banach spaces X with non-trivial multiplicative functionals on the Banach algebra $\mathcal{L}(X)$. For instance, they showed that the famous Banach space J constructed by R. C. James [**159**] can be taken for such an X. This result was generalized by A. Wilansky [**337**]. Nowadays, using the deep results of W. T. Gowers and B. Maurey from [**137**] and [**138**], it has become possible to produce easily a Banach space X such that $\mathcal{L}(X)$ has non-trivial multiplicative functionals; see Exercise 7 in Section 7.1 and [**200**]). All the Banach spaces X with non-trivial multiplicative functionals on $\mathcal{L}(X)$ mentioned above are non-reflexive. It was P. Mankiewicz [**231**] who constructed a reflexive (even a superreflexive) Banach space X with infinitely many multiplicative functionals on $\mathcal{L}(X)$. See also [**97**] for another construction of a Banach space with a similar property.

We shall close the section with an interesting extension property of lattice homomorphisms. In order to describe this property, we need some preliminary discussion regarding extensions of positive operators. Recall that a mapping $p\colon X \to Z$, from a (real) vector space to an ordered vector space, is said to be **sublinear** if:

(a) $p(x+y) \le p(x) + p(y)$ for all $x, y \in X$.

(b) $p(\lambda x) = \lambda p(x)$ for all $\lambda \ge 0$ and all $x \in X$.

An almost verbatim repetition of the proof of the classical Hahn–Banach theorem yields the following result regarding extensions of linear operators.

Theorem 4.31 (Hahn–Banach–Kantorovich). *Let Y be a vector subspace of a (real) vector space X, let F be a Dedekind complete Riesz space, and let $p\colon X \to F$ be a sublinear mapping. If an operator $T\colon Y \to F$ satisfies $Ty \le p(y)$ for all $y \in Y$, then there exists an operator $S\colon X \to F$ that extends T and satisfies $Sx \le p(x)$ for all $x \in X$.*

As an application of this theorem we present the following useful result.

Theorem 4.32 (Kantorovich). *Let Y be a majorizing vector subspace of an ordered vector space X. Then every positive operator from Y to a Dedekind complete Riesz space has a positive linear extension to all of X.*

Proof. Assume that Y is a majorizing vector subspace of an ordered vector space X, and let $T\colon Y \to F$ be a positive operator from Y to a Dedekind complete Riesz space F. Define the mapping $p\colon X \to F$ via the formula

$$p(x) = \inf\{Ty\colon\ y \in Y \text{ and } y \ge x\}.$$

To see that the above infimum exists in F, we argue as follows. Since Y is majorizing, there exists some $z \in Y$ with $z \ge -x$ or $-z \le x$. Similarly, there exists some $y \in Y$ with $y \ge x$ and for any such y the positivity of T implies $-Tz = T(-z) \le Ty$. That is, the subset $\{Ty\colon\ y \in Y \text{ and } y \ge x\}$ of F is non-empty and bounded from below in F. The Dedekind completeness of F guarantees the existence of its infimum.

A straightforward verification shows that p is a sublinear mapping satisfying $Ty = p(y)$ for each $y \in Y$. Theorem 4.31 guarantees that T has a linear extension S to all of X satisfying $Sx \leq p(x)$ for each $x \in X$. Now if $x \in X^+$, then $-x \leq 0$ and so $-Sx = S(-x) \leq p(-x) \leq T(0) = 0$. Thus, $Sx \geq 0$, and hence S is a positive linear extension of T. ■

If Y is a vector subspace of an ordered vector space X and $T\colon Y \to F$ is a positive operator, where F is a Riesz space, let us denote by $\mathcal{E}(T)$ the (possibly empty) collection of all positive linear extensions of T to all of X. It should be clear that $\mathcal{E}(T)$ is a convex subset of the vector space $\mathcal{L}_r(X, F)$. When F is a Dedekind complete Riesz space, the extreme points of the convex set $\mathcal{E}(T)$ have been characterized by Z. Lipecki, D. Plachky, and W. Thomsen [**208**] as follows.

Theorem 4.33 (Lipecki–Plachky–Thomsen). *Let Y be a vector subspace of a Riesz space E, let F be a Dedekind complete Riesz space, and let $T\colon Y \to F$ be a positive operator. Then an operator $S \in \mathcal{E}(T)$ is an extreme point if and only if for each $x \in E$ we have*

$$\inf\big\{S|x-y|\colon\ y \in Y\big\} = 0\,.$$

Proof. Fix $S \in \mathcal{E}(T)$. Assume first that S is an extreme point of $\mathcal{E}(T)$. Define the mapping $p\colon X \to F$ by

$$p(x) = \inf\big\{S|x-y|\colon\ y \in Y\big\}\,.$$

Clearly, p is a sublinear mapping that satisfies $0 \leq p(x) = p(-x) \leq S|x|$ for each $x \in E$ and $p(z) = 0$ for each $z \in Y$.

To see that $p(x) = 0$ holds for each $x \in E$, assume by way of contradiction that $p(u) > 0$ for some $u \in E$. Let $Z = \{\lambda u\colon\ \lambda \in \mathbb{R}\}$ and define the operator $R\colon Z \to F$ by $R(\lambda u) = \lambda p(u)$. From $R(u) = p(u) > 0$, it follows that R is non-zero. Also, it is easy to check that $R(\lambda u) \leq p(\lambda u)$ for all $\lambda \in \mathbb{R}$, and so by Theorem 4.31 there exists a positive linear extension of R to all of E (denoted by R again) such that $Rx \leq p(x)$ for each $x \in E$. Since $y \in Y$ implies $\pm R(z) = R(\pm z) \leq p(\pm z) = p(z)$, it follows that $|Rz| \leq p(z) = 0$ and hence $Rz = 0$ for each $z \in Y$. Now notice that for each $x \in X^+$ we have

$$R(x) \leq p(x) \leq Sx \quad \text{and} \quad -R(x) = R(-x) \leq p(-x) \leq S|-x| = Sx\,.$$

This implies $S - R \geq 0$ and $S + R \geq 0$. Since $Ry = 0$ for each $y \in Y$, we see that $S - R, S + R \in \mathcal{E}(T)$. Now notice that $S + R \neq S$, $S - R \neq S$, and $S = \frac{1}{2}(S+R) + \frac{1}{2}(S-R)$ contradict the assumption that S is an extreme point of $\mathcal{E}(T)$. So, $p(x) = 0$ for each $x \in E$.

For the converse, assume that S satisfies $\inf\big\{S|x-y|\colon\ y \in Y\big\} = 0$ for each $x \in E$, and let two operators $A, B \in \mathcal{E}(T)$ and some $0 < \alpha < 1$ satisfy

$S = \alpha A + (1-\alpha)B$. Then for each $x, y \in E$ we have

$$\big|Ax - Ay\big| \le A|x-y| = \big(\tfrac{1}{\alpha}S - \tfrac{1-\alpha}{\alpha}B\big)|x-y| \le \tfrac{1}{\alpha}S|x-y|\,.$$

In particular, if $y \in Y$, then $Ay = Ty = Sy$, and so for each $x \in E$ we have

$$\big|Sx - Ax\big| \le \big|Sx - Sy\big| + \big|Ay - Ax\big| \le \big(1 + \tfrac{1}{\alpha}\big)S|x-y|\,.$$

From $\inf\big\{S|x-y|\colon\ y \in Y\big\} = 0$, we see that $Ax = Sx$ for each $x \in E$. This implies $A = B = S$, and thus S is an extreme point of $\mathcal{E}(T)$. ■

Corollary 4.34. *Let G be a Riesz subspace of a Riesz space E, let F be a Dedekind complete Riesz space, and let $T\colon G \to F$ be a lattice homomorphism. Then each extreme point of the convex set $\mathcal{E}(T)$ is also a lattice homomorphism.*

Proof. Let S be an extreme point of the convex set $\mathcal{E}(T)$. According to Theorem 4.33 for each $x \in E$ we have

$$\inf\big\{S|x-y|\colon\ y \in G\big\} = 0\,. \qquad (\star)$$

Since T is a lattice homomorphism, $|Sy| = |Ty| = T|y| = S|y|$ for each $y \in G$. Therefore for each $x \in E$ and each $y \in G$ we have

$$\begin{aligned}\big||Sx| - S|x|\big| &\le \big||Sx| - |Sy|\big| + \big||Sy| - S|x|\big| \\ &\le |Sx - Sy| + \big|S|y| - S|x|\big| \le 2S|x-y|\,.\end{aligned}$$

This implies, using $(\star)$, that $|Sx| = S|x|$ for each $x \in E$, that is, S is a lattice homomorphism. ■

The next result of Z. Lipecki [**207**] presents a condition that guarantees the existence of an extreme point in $\mathcal{E}(T)$.

Theorem 4.35 (Lipecki). *If Y is a majorizing vector subspace of a Riesz space E, F is a Dedekind complete Riesz space, and $T\colon Y \to E$ is a positive operator, then the non-empty convex set $\mathcal{E}(T)$ has an extreme point.*

Proof. The proof is a clever application of Zorn's lemma. Observe first that by Theorem 4.32 the convex set $\mathcal{E}(T)$ is non-empty. According to Theorem 4.33 we must establish the existence of an operator $S \in \mathcal{E}(T)$ satisfying

$$\inf\big\{S|x-y|\colon\ y \in Y\big\} = 0$$

for each $x \in X$. Start by considering pairs of the form (Z, A), where Z is a vector subspace majorizing E and $A\colon Z \to F$ is a positive operator. For any such pair (Z, A), we define the mapping $p_{Z,A}\colon Z \to F$ by

$$p_{Z,A}(x) = \inf\big\{Az\colon\ z \in Z \text{ and } z \ge x\big\}\,.$$

A direct verification shows that $p_{Z,A}$ is a sublinear mapping such that $p_{Z,A}(z) = Az$ for each $z \in Z$. Moreover, if two pairs (Z_1, A_1) and (Z_2, A_2) satisfy $Z_1 \subseteq Z_2$ and $A_2 = A_1$ on Z_1, then $p_{Z_2,A_2}(x) \le p_{Z_1,A_1}(x)$ for all $x \in E$.

Now let $\mathcal{C}$ be the collection of all pairs (Z, A) such that:

(a) The vector subspace Z contains Y.

(b) The positive operator $A\colon Z \to F$ coincides with T on Y.

(c) For each $z \in Z$ we have $\inf\{p_{Z,A}(|z-y|)\colon\ y \in Y\} = 0$.

Since $(Y, T) \in \mathcal{C}$, the set $\mathcal{C}$ is non-empty. Next, define an order $\succeq$ on $\mathcal{C}$ by:

$$(Z_2, A_2) \succeq (Z_1, A_1) \quad \text{if} \quad Z_2 \supseteq Z_1 \ \text{and}\ A_2 = A_1 \ \text{on}\ Z_1 .$$

It is easy to check that $\succeq$ is indeed an order on $\mathcal{C}$, and a routine argument shows that every chain in $\mathcal{C}$ has an upper bound. So, by Zorn's lemma, there exists a maximal element, say (V, S). To complete the proof, it suffices to show that $V = E$. If this is established, then it should be clear that $S = p_{V,S}$, and Theorem 4.33 guarantees that S is an extreme point of $\mathcal{E}(T)$.

If $V \neq E$, then there exists some $x \in E$ such that $x \notin V$. Consider the vector subspace $W = \{v + \lambda x\colon\ v \in V \ \text{and}\ \lambda \in \mathbb{R}\}$ and define the operator $B\colon W \to F$ by $B(v+\lambda x) = Sv + \lambda p_{V,S}(x)$. Clearly, V is a proper subspace of W, $B = S$ on V, and we claim that B is a positive operator. To see this, let $0 \le u + \lambda x \in V$. If $\lambda > 0$, then $x \ge -\frac{v}{\lambda}$, and this implies $p_{V,S}(x) \ge -S(\frac{v}{\lambda})$, from which it follows that $B(v + \lambda x) = Sv + \lambda p_{V,S}(x) \ge 0$. The case $\lambda < 0$ is similar and the case $\lambda = 0$ is obvious. To obtain a contradiction, we shall establish that (W, B) satisfies property (c) above.

To this end, notice first that the sublinearity of $p_{W,B}$ implies that the set

$$V_1 = \{z \in X\colon\ \inf\{p_{W,B}(|z-v|)\colon\ v \in V\} = 0\}$$

is a vector subspace of E satisfying $V \subseteq V_1$. We claim that $x \in V_1$. To see this, observe first that $v \in V$ and $v \ge x$ imply $v - x \in W$ and so $p_{W,B}(v - x) = B(v - x) = Sv - p_{V,S}(x)$. Therefore,

$$\begin{aligned} 0 &\le \inf\{p_{W,B}(|x-v|)\colon\ v \in V\} \\ &\le \inf\{p_{W,B}(v-x)\colon\ v \in V \ \text{and}\ v \ge x\} \\ &= \inf\{Sv - p_{V,S}(x)\colon\ v \in V \ \text{and}\ v \ge x\} \\ &= \inf\{Sv\colon\ v \in V \ \text{and}\ v \ge x\} - p_{V,S}(x) = 0 . \end{aligned}$$

This shows that $x \in V_1$, and consequently $W \subseteq V_1$. Now fix $w \in W$ and put $a = \inf\{p_{W,B}(|w-y|)\colon\ y \in Y\}$. Notice that for each $y \in Y$ and each $v \in V$ we have

$$\begin{aligned} 0 \le a &\le p_{W,B}(|w-y|) \le p_{W,B}(|w-v|) + p_{W,B}(|v-y|) \\ &\le p_{W,B}(|w-v|) + p_{V,S}(|v-y|) . \end{aligned}$$

Since $(V,S) \in \mathcal{C}$, we know that $\inf\{p_{V,S}(|v-y|)\colon\ y \in Y\} = 0$, and hence $0 \le a \le p_{W,B}(|w-v|)$ for each $v \in V$. Taking into account that $w \in V_1$, the latter inequality implies $a = \inf\{p_{W,B}(|w-v|)\colon\ v \in V\} = 0$ for each $w \in W$. This establishes $(W,B) \in \mathcal{C}$. However, $(W,B) \succeq (V,S)$ and $(W,B) \ne (V,S)$ contradict the maximality of (V,S), and the proof is finished. ■

The final result on extensions of lattice homomorphisms (due to Z. Lipecki [**207**] and W. A. J. Luxemburg and A. R. Schep [**223**]) is a simple consequence of the preceding discussion.

Corollary 4.36 (Lipecki–Luxemburg–Schep). *If G is a majorizing Riesz subspace of a Riesz space E, then every lattice homomorphism from G to a Dedekind complete Riesz space has a lattice homomorphic extension to all of E.*

Proof. By Theorem 4.32 the set $\mathcal{E}(T)$ is non-empty and by Theorem 4.35 it has an extreme point S. Since T is a lattice homomorphism, Corollary 4.34 guarantees that S is a lattice homomorphic extension of T. ■

For more on extensions of lattice homomorphisms see [**81, 82, 83**].

Exercises

1. By Theorem 4.25, for any given lattice homomorphism $T\colon C(\Omega) \to C(Q)$ there exist a mapping $\tau\colon Q \to \Omega$ and some $w \in C(Q)$ such that for each $f \in C(\Omega)$ and each $q \in Q$ we have
$$Tf(q) = w(q)f(\tau(q))\,.$$
Show that $w = T\mathbf{1}_\Omega$ and that the mapping τ is uniquely determined and continuous on the set $\{q \in Q\colon\ w(q) > 0\}$.
2. Give an example of a lattice homomorphism which is not an algebraic homomorphism.
3. Prove Theorem 4.28.
4. Let Ω and Q be two compact Hausdorff spaces, and consider the convex set $\mathcal{C}$ of all Markov operators from $C(\Omega)$ to $C(Q)$, i.e.,
$$\mathcal{C} = \{T \in \mathcal{L}(C(\Omega), C(Q))\colon\ T \ge 0 \text{ and } T\mathbf{1}_\Omega = \mathbf{1}_Q\}\,.$$
Show that a Markov operator T is a lattice homomorphism if and only if T is an extreme point of $\mathcal{C}$.
5. Prove Theorem 4.29. [HINT: Assume that $T\colon E \to F$ is a lattice homomorphism between real Banach lattices. Let $z = x + \imath y \in E_{\mathbf{c}}$, and put $u = |x| + |y| \in E$ and $v = T|x| + T|y| = Tu \in F$. Then $T(E_u) \subseteq F_v$ and this shows that we can assume without loss of generality that $E = C(\Omega)$

and $F = C(Q)$ and that T is a Markov operator. Now use Theorem 4.28 and view $T: C(\Omega) \to C(Q)$ as a composition operator.]

6. For a multiplicative linear functional ϕ on a Banach algebra $\mathcal{A}$ establish the following.
   (a) If $a \in \mathcal{A}$ is an idempotent element (i.e., $a^2 = a$), then show that $\phi(a) = 0$ or $\phi(a) = 1$.
   (b) Let p and q be two idempotent elements in $\mathcal{A}$. Then show that the linear functional ψ defined on $\mathcal{A}$ by $\psi(a) = \phi(qap)$ is also a multiplicative functional—even though the mapping $a \mapsto qap$ need not be an algebraic homomorphism.

7. Fix a Banach space X and let $Y = X \oplus X \oplus \cdots \oplus X$ be the direct sum of $n > 1$ copies of X; in particular, let $Y = \mathbb{C}^n$ or $Y = \mathbb{R}^n$ where $n > 1$. Show that the only multiplicative linear functional on the Banach algebra $\mathcal{L}(Y)$ is the zero functional. (For an example of a non-trivial multiplicative function on $\mathcal{L}(X)$ see Exercise 7 in Section 4.5.) [HINT: Let ϕ be a multiplicative linear functional on $\mathcal{L}(Y)$. It suffices to show that $\phi(I) = 0$. Define the bounded operator $J: Y \to Y$ by

$$J(x_1, x_2, \ldots, x_n) = (x_n, x_1, x_2, \ldots, x_{n-1})$$

and note that $J^n = I$. Therefore, in order to show that $\phi(I) = 0$, it suffices to establish that $\phi(J) = 0$. Now for each $1 \le i < n$ define the bounded operator $T_i \in \mathcal{L}(Y)$ by $T_i(x_1, x_2, \ldots, x_n) = (0, \ldots, 0, x_i, 0, \ldots, 0)$, where the vector x_i occupies the $(i+1)$-position. Also, define $T_n \in \mathcal{L}(Y)$ by $T_n(x_1, x_2, \ldots, x_n) = (x_n, 0, 0, \ldots, 0)$. Clearly, $T_i^2 = 0$ for all $1 \le i \le n$ and $J = \sum_{i=1}^n T_i$.]

8. For a Banach lattice E, a majorizing vector subspace X of E, and an arbitrary normed Riesz space F establish the following automatic continuity property: *Every positive operator $T: X \to F$ is continuous.* [HINT: By Exercise 21 of Section 1.2 we can assume that F is Dedekind complete. Now, by Kantorovich's Theorem 4.32, the operator T has a positive linear extension $\hat{T}: E \to F$ which is continuous by Theorem 1.31.]

9. (Arendt [**40**]) Let F be an arbitrary Dedekind complete Riesz space and let $R: E \to H$ be a positive operator between two Riesz spaces. Then the positive operator R gives rise to a positive operator $T \mapsto TR$ from $\mathcal{L}_r(H, F)$ to $\mathcal{L}_r(E, F)$. Establish the following.
   (a) If R is a lattice homomorphism, then $T \mapsto TR$ is interval preserving.
   (b) If R is interval preserving, then $T \mapsto TR$ is a lattice homomorphism.

10. (Aliprantis–Burkinshaw–Kranz [**33**]) Let $R: H \to F$ be a positive operator between two Dedekind complete Riesz spaces and let E be an arbitrary Riesz space. For the positive operator $T \mapsto TR$, from $\mathcal{L}_r(H, F)$ to $\mathcal{L}_r(E, F)$, establish the following.
   (a) If R is an order continuous lattice homomorphism, then $T \mapsto RT$ is a lattice homomorphism.
   (b) If the operator $T \mapsto RT$ is a lattice homomorphism and the order dual $F^\sim$ is non-trivial, then R is a lattice homomorphism.

11. Prove Theorem 1.35. [HINT: Use the preceding two exercises by observing that if $T\colon E \to F$ is a bounded operator between two Banach lattices, then $T^*y^* = y^* \circ T$ for each $y^* \in F^*$.]

12. Let $E = \ell_\infty$ and let c denote the Riesz subspace of all convergent sequences. Show that the **limit functional** $\mathcal{L}im\colon c \to \mathbb{R}$, defined by $\mathcal{L}im(x) = \lim_{n\to\infty} x_n$ for each $x = (x_1, x_2, \ldots) \in c$, is a lattice homomorphism that has a lattice homomorphic linear extension to ℓ_∞.

4.4. Fredholm Operators

The objective of this section is to present the basic properties of the class of *Fredholm operators*. They were introduced by I. Fredholm [**127**] in his study of integral equations. These are the operators that have "small deviations" from isomorphisms. To formalize the term "small deviation," we need some preliminary discussion.

Let $T\colon X \to Y$ be an operator between two vector spaces. As usual, we denote by $N(T)$ and $R(T)$ the **null space** and the **range** of T, respectively. That is,

$$N(T) = \{x \in X\colon\ Tx = 0\} \quad \text{and} \quad R(T) = \{Tx\colon\ x \in X\}.$$

Clearly, T is a linear isomorphism from X onto Y if and only if $N(T) = \{0\}$ and $R(T) = Y$ or equivalently $Y/R(T) = \{0\}$. That is, T is an isomorphism if and only if $\dim N(T) = 0$ and $\dim Y/R(T) = 0$.

Assume next that for a bounded operator $T\colon X \to Y$ we can find a closed vector subspace V of X and a closed vector subspace W of Y such that $X = N(T) \oplus V$ and $Y = R(T) \oplus W$. Then $Y/R(T)$ is isomorphic to W and $T\colon V \to R(T)$ is an isomorphism. Moreover, it should be clear that the "measure of deviation" of T from being an isomorphism is associated with the subspaces $N(T)$ and W. The dimension of the null space $N(T)$ is called the **nullity** of T and is denoted $n(T)$, i.e., $n(T) = \dim N(T)$. The dimension of W (or, equivalently, the dimension of the quotient space $Y/R(T)$) is called the **defect** of T (or the **codimension** of T or even the **corank**[3] of T) and is denoted $d(T)$. That is, $d(T) = \dim Y/R(T)$. If $n(T)$ or $d(T)$ is finite, then the extended real number

$$i(T) = n(T) - d(T)$$

is called the **index** of the operator T.

[3] If W is a vector subspace of a vector space V, then the **codimension** of W is the dimension of the quotient vector space V/W. If $T\colon V \to U$ is a linear operator between vector spaces, then the **rank** of T is the dimension of the range $R(T)$ and the **corank** of T is the codimension of the range $R(T)$.

Definition 4.37. *A bounded operator $T\colon X \to Y$ between two Banach spaces is said to be:*

(1) ***semi-Fredholm*** *if it has a closed range and either its nullity or its defect is finite and*

(2) ***Fredholm*** *if its nullity $n(T)$ and defect $d(T)$ are both finite.*

Clearly, every linear isomorphism between two Banach spaces is a Fredholm operator of index zero. Also, any operator between finite dimensional spaces is automatically a Fredholm operator. It is customary to impose in the definition that a Fredholm operator has a closed range. However, this is a redundant assumption.

Lemma 4.38. *Any operator $T\colon X \to Y$ between Banach spaces with a finite defect has a closed range and satisfies*

$$i(T) = \dim N(T) - \dim N(T^*)\,.$$

In particular, every Fredholm operator has a closed range—and hence it is also a semi-Fredholm operator.

Proof. The fact that the operator T has closed range follows from Corollary 2.17. For the index formula notice that according to Theorem 2.13 we have $d(T) = \dim[Y/R(T)] = \dim N(T^*)$. ∎

In our discussion we shall use the following lemma whose proof follows from Exercise 6 in Section 4.1. Many authors use the conclusion of this result as the definition of a Fredholm operator.

Lemma 4.39. *If $T\colon X \to Y$ is a Fredholm operator between Banach spaces, then there exist a closed vector subspace V of X and a finite dimensional vector subspace W of Y such that*

$$X = N(T) \oplus V \quad \text{and} \quad Y = R(T) \oplus W\,.$$

In particular, the surjective operator $T\colon V \to R(T)$ is an isomorphism.

Regarding Fredholm operators between real Banach spaces we have the following result whose straightforward proof is left as an exercise—see Exercise 1 at the end of the section.

Lemma 4.40. *A bounded operator $T\colon X \to Y$ between real Banach spaces is a Fredholm operator if and only if the operator $T_{\mathbf{C}}\colon X_{\mathbf{C}} \to Y_{\mathbf{C}}$ is likewise a Fredholm operator—in which case we have $i(T_{\mathbf{C}}) = i(T)$.*

The sum of two Fredholm operators need not be a Fredholm operator. For instance, on an infinite dimensional Banach space the identity operator

I and $-I$ are Fredholm operators—both of index zero. However, their sum is the zero operator which is not a Fredholm operator.

The classes of compact operators and Fredholm operators are disjoint.

Lemma 4.41. *If X is an infinite dimensional Banach space, then no compact operator with domain X can be a Fredholm operator.*

Proof. Let $T\colon X \to Y$ be a compact operator between two Banach spaces with X infinite dimensional. Assume by way of contradiction that T is also a Fredholm operator. This means that $\dim N(T) < \infty$ and there exist a closed subspace V of X and a finite dimensional subspace W of Y satisfying $X = N(T) \oplus V$ and $Y = R(T) \oplus W$. Clearly, V is infinite dimensional and the operator $S = T|_V\colon V \to R(T)$ is an isomorphism that is also compact. This implies that V is finite dimensional, a contradiction. ∎

Informally speaking, the preceding discussion shows that a compact operator can be thought of as an "almost" finite-rank operator while a Fredholm operator should be perceived as an "almost" isomorphism. As expected, the adjoint of a Fredholm operator is also a Fredholm operator.

Theorem 4.42. *A bounded operator $T\colon X \to Y$ between two Banach spaces is a Fredholm operator if and only if its adjoint operator $T^*\colon Y^* \to X^*$ is also a Fredholm operator. Moreover, if this is the case, then we have $n(T^*) = d(T)$ and $d(T^*) = n(T)$; and so $i(T^*) = -i(T)$.*

Proof. Assume T is a Fredholm operator. Choose a closed subspace V of X and a finite dimensional vector subspace W of Y such that $X = N(T) \oplus V$ and $Y = W \oplus R(T)$. Also, let $S = T|_V$, and note that $S\colon V \to R(T)$ is an isomorphism. Consequently, T can be represented by the direct sum operator $T = 0 \oplus S\colon N(T) \oplus V \to W \oplus R(T)$. Taking adjoints, it follows that $T^* = 0 \oplus S^*\colon W^* \oplus [R(T)]^* \to [N(T)]^* \oplus V^*$. This implies $N(T^*) = W^*$ and $R(T^*) = V^*$. Therefore, T^* is a Fredholm operator satisfying

$$n(T^*) = \dim W^* = \dim W = d(T)$$

and

$$d(T^*) = \dim\big[X^*/R(T^*)\big] = \dim\,[N(T)]^* = \dim N(T) = n(T)\,.$$

For the converse, assume that T^* is a Fredholm operator. Then, by the previous case, $T^{**}\colon X^{**} \to Y^{**}$ is also a Fredholm operator. From $N(T) \subseteq N(T^{**})$, it follows that $\dim N(T) < \infty$. Now recall that $R(T)$ is closed if and only if $R(T^*)$ is closed; see Theorem 2.18. Therefore, by Theorem 2.13 we have $\dim\,[Y/R(T)] = \dim N(T^*) < \infty$. This shows that T is likewise a Fredholm operator. ∎

A composition of Fredholm operators is also a Fredholm operator and the index of the composition is the sum of the indices of the operators. This is known as the **Index Theorem**.

Theorem 4.43 (The Index Theorem). *For a scheme $X \xrightarrow{T} Y \xrightarrow{S} Z$ of bounded operators between Banach spaces we have the following.*

(1) *If S and T are both Fredholm operators, then ST is also a Fredholm operator and $i(ST) = i(S) + i(T)$.*

(2) *If ST is a Fredholm operator, then T is a Fredholm operator if and only if S is a Fredholm operator.*

Proof. The proof is "combinatorial" in nature and is based upon the following simple algebraic fact:

- *If W is a vector subspace of a vector space V, then there exists an algebraic complement of W, i.e., there exists some vector subspace U of V such that $W \oplus U = V$.*

(For a proof see Exercise 3 at the end of the section.)

The key object for our proof is the subspace $N_1 = R(T) \cap N(S)$ of Y. Pick a subspace N_2 of Y such that $N(S) = N_1 \oplus N_2$. So, we have

$$n(S) = \dim N_1 + \dim N_2\,. \tag{$\star$}$$

Next, choose a vector subspace Y_1 of Y such that $R(T) = Y_1 \oplus N_1$. Since $N_2 \cap Y_1 = \{0\}$ (why?), there exists another vector subspace Y_2 of Y such that $Y = Y_1 \oplus N_1 \oplus N_2 \oplus Y_2$. In particular, we have

$$d(T) = \dim N_2 + \dim Y_2\,. \tag{$\star\star$}$$

Since $N(T) \subseteq N(ST)$, there exists a vector subspace X_1 of X such that

$$N(T) \oplus X_1 = N(ST) = T^{-1}\big(N(S)\big) = T^{-1}\big(N(S) \cap R(T)\big) = T^{-1}(N_1)\,.$$

We claim that the operator $T\colon X_1 \to N_1$ is a linear isomorphism. Clearly, T is one-to-one. On the other hand, if $y = Tx \in N_1 \subseteq N(S)$, then $STx = 0$ or $x \in N(ST)$, and so we can write $x = y + z$ with $y \in N(T)$ and z in X_1. Thus, $y = Tx = Tz \in T(X_1)$, and this proves that $T\colon X_1 \to N_1$ is surjective—and hence it is a linear isomorphism. From $N(T) \oplus X_1 = N(ST)$, we see that $n(T) + \dim X_1 = n(ST)$ or

$$n(ST) = n(T) + \dim N_1\,. \tag{$\star\star\star$}$$

Next, observe that S is one-to-one on $Y_1 \oplus Y_2$. This implies

$$\begin{aligned} R(S) &= S(Y_1 \oplus Y_2) = S(Y_1) \oplus S(Y_2) = S(Y_1 \oplus N_1) \oplus S(Y_2) \\ &= S\big(R(T)\big) \oplus S(Y_2) = R(ST) \oplus S(Y_2)\,. \end{aligned}$$

Consequently, if Z_1 is a subspace of Z satisfying $Z = Z_1 \oplus R(S)$, then $Z = Z_1 \oplus R(ST) \oplus S(Y_2)$. So, $d(ST) = \dim Z_1 + \dim S(Y_2) = \dim Z_1 + \dim Y_2$ or

$$d(ST) = d(S) + \dim Y_2 \,. \tag{$\dagger$}$$

With this setup in place we are now ready to prove the theorem.

(1) Assume that S and T are both Fredholm operators. This implies that all subspaces appearing in $(\star)$, $(\star\star)$, $(\star\star\star)$, and $(\dagger)$ are finite dimensional. In particular, ST is a Fredholm operator. Using $(\star\star)$ in connection with $(\star)$ and $(\star\star\star)$, we get

$$\begin{aligned} \dim Y_2 &= d(T) - \dim N_2 = d(T) - n(S) + \dim N_1 \\ &= d(T) - n(S) + \big[n(ST) - n(T)\big] \\ &= n(ST) - n(S) - i(T) \,. \end{aligned}$$

Now a glance at $(\dagger)$ yields

$$d(ST) = d(S) + \dim Y_2 = d(S) + n(ST) - n(S) - i(T)$$

or $n(ST) - d(ST) = i(T) + n(S) - d(S)$. That is, $i(ST) = i(T) + i(S)$.

(2) Assume first that ST and S are Fredholm operators. From the inclusion $N(T) \subseteq N(ST)$, we see that $n(T) < \infty$. Also, from $(\star)$, we get $\dim Y_2 < \infty$, and $(\star\star)$ yields $d(T) < \infty$. Thus, T is a Fredholm operator.

Finally, suppose that ST and T are Fredholm operators. Then $(\star\star)$ yields that $\dim N_2 < \infty$ and $\dim Y_2 < \infty$, and $(\star\star\star)$ implies $\dim N_1 < \infty$. But then from $(\star)$ and $(\dagger)$, we obtain $n(S) < \infty$ and $d(S) < \infty$. That is, S is a Fredholm operator, and the proof is finished. ∎

In order to obtain more properties of Fredholm operators, we need the following useful lemma on the existence of "almost perpendicular" vectors.

Lemma 4.44 (F. Riesz). *If Y is a proper closed subspace of a Banach space X, then for each $0 < \epsilon < 1$ there exists some unit vector $x \in X$ such that $\|y - x\| \geq 1 - \epsilon$ holds for all $y \in Y$.*

Moreover, if Y is finite dimensional, then there exists a unit vector x in X such that the distance from x to Y is one, i.e., $d(x, Y) = 1$.

Proof. By the Hahn–Banach theorem there is a linear functional $x^* \in X^*$ with $\|x^*\| = 1$ satisfying $x^*(y) = 0$ for all $y \in Y$. Pick a unit vector $x \in X$ such that $|x^*(x)| \geq 1 - \epsilon$. Now note that if $y \in Y$, then

$$\|y - x\| \geq \big|x^*(y - x)\big| = \big|x^*(x)\big| \geq 1 - \epsilon \,.$$

For the last part, assume that Y is finite dimensional. Pick another finite dimensional vector subspace Z of X such that Y is a proper subspace of Z. By the preceding case, for each n there exists some unit vector $x_n \in Z$ such that $d(x_n, Y) \geq 1 - \frac{1}{n}$. Since the unit sphere of Z is compact, by passing to a subsequence, we can assume that $x_n \to x \in Z$. Then $\|x\| = 1$ and the continuity of the distance function $z \mapsto d(z, Y)$ yields

$$1 \geq d(x, Y) = \lim_{n\to\infty} d(x_n, Y) \geq \lim_{n\to\infty} \left(1 - \tfrac{1}{n}\right) = 1\,.$$

This implies $d(x, Y) = 1$. ■

We continue with an important property regarding compact perturbations of the identity operator.

Lemma 4.45. *If $K\colon X \to X$ is a compact operator, then the operator $I - K$:*

(1) *has finite ascent and descent, and*

(2) *is a Fredholm operator of index zero.*

Proof. Let $K\colon X \to X$ be a compact operator on a Banach space. We first claim that the vector space $N(I - K)$ is finite dimensional. To see this, notice that if $x \in N(I - K)$, then $Kx = x$. That is, on $N(I - K)$ the compact operator K coincides with the identity operator. This implies that $N(I - K)$ is finite dimensional.

Next, we shall show that $I - K$ has a closed range. To see this, pick a closed complement V of $N(I - K)$, i.e., $X = N(I - K) \oplus V$. (Since $N(I - K)$ is finite dimensional, closed complements of $N(I - K)$ always exist; see Exercise 6 of Section 4.1.) We claim that $I - K$ is bounded below on V. If this is done, then the operator $I - K\colon V \to R(I - K)$ must be an isomorphism, and so $R(I - K)$ must be a closed subspace of X.

To establish that $I - K$ is bounded below on V, assume contrary to our claim that there exists a sequence of unit vectors $\{v_n\} \subseteq V$ such that $v_n - Kv_n \to 0$. Since K is compact, by passing to a subsequence we can suppose that $Kv_n \to v$. This implies that $v_n \to v$ and that $v \neq 0$. The closedness of V guarantees $v \in V$, and the continuity of K yields $v = Kv$. Hence $v - Kv = 0$, that is, $v \in N(I - K)$. However, the latter conclusion contradicts the fact $N(I - K) \cap V = \{0\}$. Thus, $I - K$ must be bounded below on V.

(1) We shall prove only that $\alpha(I - K) < \infty$. The claim $\delta(I - K) < \infty$ can be proven in a similar manner. We already know that the vector space $N(I - K)$ is finite dimensional. Since $(I - K)^n$ can be written as $I - K_1$ for some $K_1 \in \mathcal{K}(X)$, the above conclusion shows that $N\big((I - K)^n\big)$ is also finite dimensional for each n.

Now assume by way of contradiction that $\alpha(I - K) = \infty$. This means that $N\big((I - K)^n\big)$ is a proper subspace of $N\big((I - K)^{n+1}\big)$ for each n. Consequently, by Riesz's Lemma 4.44, there exists a sequence $\{x_n\}$ of unit vectors in X such that $x_n \in N\big((I - K)^{n+1}\big)$ for each n and $\|x_n - x\| \geq 1$ for all

$x \in N\big((I-K)^n\big)$. Since each subspace $N\big((I-K)^n\big)$ is invariant under $I-K$, it follows that $x_m + (I-K)x_n - (I-K)x_m \in N\big((I-K)^n\big)$ for each $1 \le m < n$. Consequently, we have

$$\|Kx_n - Kx_m\| = \big\|x_n - \big[x_m + (I-K)x_n - (I-K)x_m\big]\,\big\| \ge 1$$

for all m and n. This implies that the sequence $\{Kx_n\}$ does not have any convergent subsequence, contradicting the compactness of K. This contradiction establishes that $I-K$ has a finite ascent.

(2) From the preceding conclusions we already know that $N(I-K^*)$ is finite dimensional. Therefore, by Theorem 2.13 we have

$$d(I-K) = \dim\big[X/R(I-K)\big] = \dim N(I-K^*) < \infty\,.$$

This implies that $I-K$ is a Fredholm operator.

To show that the index of $I-K$ equals zero, we proceed as follows. By part (1) the operator $I-K$ has finite ascent and descent—which by Lemma 2.21 coincide. Let $p = \alpha(I-K) = \delta(I-K)$. Then Theorem 2.23 guarantees that $X = N\big((I-K)^p\big) \oplus R\big((I-K)^p\big)$. This identity implies that the Fredholm operator satisfies

$$i\big((I-K)^p\big) = \dim N\big((I-K)^p\big) - \dim N\big((I-K)^p\big) = 0\,.$$

Now applying the Index Theorem, we obtain that $i\big((I-K)^p\big) = p\big[i(I-K)\big]$, and so $i(I-K) = 0$. The proof is finished. ■

The next result, due to F. V. Atkinson [**44**], characterizes the Fredholm operators.

Theorem 4.46 (Atkinson). *For a bounded operator $T\colon X \to Y$ between Banach spaces the following statements are equivalent.*

(1) *T is a Fredholm operator.*

(2) *There exist an operator $S \in \mathcal{L}(Y,X)$ and two finite-rank projections $P_1 \in \mathcal{L}(X)$ and $P_2 \in \mathcal{L}(Y)$ so that $ST = I_X - P_1$ and $TS = I_Y - P_2$.*

(3) *There exist compact operators $K_1 \in \mathcal{L}(X)$ and $K_2 \in \mathcal{L}(Y)$ and an operator $S \in \mathcal{L}(Y,X)$ such that $ST = I_X - K_1$ and $TS = I_Y - K_2$.*

(4) *There exist compact operators $K_1 \in \mathcal{L}(X)$ and $K_2 \in \mathcal{K}(Y)$ and $S_1, S_2 \in \mathcal{L}(Y,X)$ such that $S_1T = I_X - K_1$ and $TS_2 = I_Y - K_2$.*

Proof. (1) $\Longrightarrow$ (2) Assume that T is a Fredholm operator. Then $N(T) < \infty$ and there exist a closed subspace V of X and a finite dimensional subspace W of Y such that $X = N(T) \oplus V$ and $Y = W \oplus R(T)$. Also, let P_N be the projection of X onto $N(T)$ along V and let P_W be the projection of Y onto W along $R(T)$. Clearly, P_N and P_W are bounded finite-rank projections.

Now let $S_1 = T|_V : V \to R(T) = T(V)$, and note that S_1 is continuous. Since the range $R(T)$ is closed, it follows that the operator $S_1^{-1} : T(V) \to V$, defined by $S_1^{-1}(Tv) = v$, is also continuous. Next, consider the operator $S = S_1^{-1}(I_Y - P_W) \in \mathcal{L}(Y, X)$, and note that if $y = w + Tv \in W \oplus T(V) = Y$ with $v \in V$, then

$$Sy = [S_1^{-1}(I_Y - P_W)](w + Tv) = S_1^{-1}(Tv) = v .$$

This implies that for each $x = n + v \in N(T) \oplus V = X$ we have

$$STx = ST(n + v) = S_1^{-1}Tv = v = (I_X - P_N)(n + v) = (I_X - P_N)x .$$

Similarly, for each $y = w + Tv \in W \oplus T(V) = Y$ with $v \in V$ we have

$$TSy = Tv = (I_Y - P_W)y .$$

So, $ST = I_X - P_N$ and $TS = I_Y - P_W$, and the validity of (2) has been established.

(2) $\Longrightarrow$ (3) $\Longrightarrow$ (4) These implications are obvious.

(4) $\Longrightarrow$ (1) Assume that $S_1, S_2 \in \mathcal{L}(Y, X)$ and the compact operators $K_1 \in \mathcal{L}(X)$ and $K_2 \in \mathcal{K}(Y)$ satisfy $S_1T = I_X - K_1$ and $TS_2 = I_Y - K_2$.

Since K_1 is a compact operator, it follows from Lemma 4.45 that $N(S_1T) = N(I_X - K_1)$ is finite dimensional. From $N(T) \subseteq N(S_1T)$, we see that $\dim N(T) < \infty$. The same argument applied to $S_2^*T_2^* = I_{Y^*} - K_2^*$ shows that $N(T^*)$ is also finite dimensional. From Theorem 2.13, it follows that if $R(T)$ is closed, then $\dim \bigl[X/R(T)\bigr] = \dim N(T^*) < \infty$. So, to complete the proof, it remains to be shown that T has a closed range.

To this end, note that Lemma 4.45 implies that $I_Y - K_2$ is a Fredholm operator. So, there exists a finite dimensional subspace Z of Y such that $R(I_Y - K_2) \oplus Z = Y$. Now from $R(I_Y - K_2) = R(TS_2) \subseteq R(T)$, it follows that $R(T) + Z = Y$. This implies that there exists a vector subspace Z_1 of Z (and hence Z_1 is finite dimensional) such that $R(T) \oplus Z_1 = Y$; see Exercise 3 at the end of the section. Now a glance at Theorem 2.16 guarantees that T has a closed range, and the proof is finished. ■

Corollary 4.47. *If $T \in \mathcal{L}(X, Y)$ is a Fredholm operator, then for each $K \in \mathcal{K}(X, Y)$ the operator $T + K$ is also Fredholm and $i(T + K) = i(T)$.*

Proof. Let $T \in \mathcal{L}(X, Y)$ be a Fredholm operator and fix some $K \in \mathcal{K}(X, Y)$. By Theorem 4.46 there exist an operator $S \in \mathcal{L}(Y, X)$ and two compact operators $K_1 \in \mathcal{L}(X)$ and $K_2 \in \mathcal{L}(Y)$ such that $S_1T = I_X - K_1$ and $TS_2 = I_Y - K_2$. It follows that

$$S_1(T + K) = I_X - (K_1 - S_1K) \quad \text{and} \quad (T + K)S_2 = I_Y - (K_2 - KS_2) .$$

Since the operators $K_1 - S_1K$ and $K_2 - KS_2$ are compact, another application of Theorem 4.46 shows that $T + K$ is a Fredholm operator.

From the Index Theorem we know that S_1 is a Fredholm operator, and Lemma 4.45 guarantees that $i(S_1)+i(T)=i(S_1T)=i(I_X-K_1)=0$. Thus, $i(T)=-i(S_1)$. Similarly, $i(T+K)=-i(S_1)$, and so $i(T+K)=i(T)$. ■

The **perturbation** of an operator T by an operator S is $T-S$. In this terminology, Corollary 4.47 states that: *A compact perturbation of a Fredholm operator remains Fredholm and does not change its index.*

If $\mathcal{F}red(X,Y)$ denotes the collection of all Fredholm operators from X to Y (where, as usual, we write $\mathcal{F}red(X)$ instead of $\mathcal{F}red(X,X)$), then the **index function** $i\colon \mathcal{F}red(X,Y)\to\mathbb{Z}$ is defined by $i(T)$ for each $T\in\mathcal{F}red(X,Y)$. It turns out that the index function is continuous when $\mathbb{Z}$ is equipped with its discrete topology. The details are included in the next important result.

Theorem 4.48. *If X and Y are Banach spaces, then $\mathcal{F}red(X,Y)$ is an open subset of $\mathcal{L}(X,Y)$ and the index function $i\colon \mathcal{F}red(X,Y)\to\mathbb{Z}$ is continuous.*

In particular, $\mathcal{F}red(X)$ is an open unital subsemigroup of $\mathcal{L}(X)$ and the index function $i\colon \mathcal{F}red(X)\to\mathbb{Z}$ is a continuous algebraic homomorphism.

Proof. If X is finite dimensional, then $\mathcal{F}red(X,Y)=\emptyset$ provided Y is infinite dimensional and $\mathcal{F}red(X,Y)=\mathcal{L}(X,Y)$ provided Y is finite dimensional. So, we can assume that $T\colon X\to Y$ is a Fredholm operator between two Banach spaces with X infinite dimensional. By Theorem 4.46 there exist an operator $S\in\mathcal{L}(Y,X)$ and two compact operators $K_1\in\mathcal{K}(X)$ and $K_2\in\mathcal{K}(Y)$ such that

$$ST=I_X-K_1 \quad\text{and}\quad TS=I_Y-K_1\,. \tag{$\star$}$$

Since X is infinite dimensional, it follows that $S\neq 0$. By Lemma 4.45 the operator $I_X-K_1\in\mathcal{L}(X)$ is a Fredholm operator of index zero. So, by the Index Theorem 4.43 the operator S is also a Fredholm operator, and from $i(S)+i(T)=i(ST)=i(I_X-K_1)=0$ we get

$$i(T)=-i(S)\,. \tag{$\star\star$}$$

We know that if $A\in\mathcal{L}(X)$ and $B\in\mathcal{L}(Y)$ satisfy $\|I_X-A\|<1$ and $\|I_Y-B\|<1$, then I_X-A and I_Y-B are invertible; and hence they are both isomorphisms. Let $C\in\mathcal{L}(X,Y)$ be any operator such that $\|C\|<\frac{1}{\|S\|}$. Then the operators $SC\in\mathcal{L}(X)$ and $CS\in\mathcal{L}(Y)$ satisfy $\|SC\|<1$ and $\|CS\|<1$. So, the operators $I_X+SC\in\mathcal{L}(X)$ and $I_Y+CS\in\mathcal{L}(Y)$ are isomorphisms, and consequently they are both Fredholm operators of index zero. From Corollary 4.47 it follows that I_X+SC-K_1 and I_Y+CS-K_2 are likewise Fredholm operators of index zero. Moreover, from ($\star$) we get

$$S(T+C)=I_X+SC-K_1 \quad\text{and}\quad (T+C)S=I_Y+CS-K_2\,.$$

Now a glance at the Index Theorem 4.43(2) shows that the operator $T+C$ is Fredholm and satisfies

$$i(S)+i(T+C)=i\big(S(T+C)\big)=i(I_X+SC-K_1)=0\,.$$

Thus, $i(T+C)=-i(S)$, and from $(\star\star)$ we get $i(T+C)=i(T)$.

Therefore, we have established that if $C\in\mathcal{L}(X,Y)$ satisfies $\|C\|<\frac{1}{\|S\|}$, then $T+C$ is a Fredholm operator and $i(T+C)=i(T)$. This shows that $\mathcal{F}red(X,Y)$ is an open subset of $\mathcal{L}(X,Y)$ and that the index function $i\colon \mathcal{F}red(X,Y)\to\mathbb{Z}$ is continuous. The proof is finished. ■

There is a similar result regarding semi-Fredholm operators. To establish it, we need some preliminary discussion.

Lemma 4.49. *For a Banach space X we have the following.*

(1) *If there exist $x_1,x_2,\ldots,x_n\in U_X$ such that for each $x\in U_X$ we can find some x_i such that $\|x-x_i\|<1$, then X is finite dimensional.*

(2) *Let Z be a closed vector subspace of X. Assume that there exist vectors $u_1,u_2,\ldots,u_n\in U_X$ such that for each $x\in U_X$ we can find some u_i satisfying $d(x-u_i,Z)<1$. Then the quotient vector space X/Z is finite dimensional.*

Proof. (1) Let Y be the linear span of $x_1,x_2,\ldots,x_n$. We claim that $X=Y$. If $X\neq Y$, then by Riesz's Lemma 4.44 there exists some unit vector $x\in X$ such that $d(x,Y)=1$. In particular, we have $\|x-x_i\|\geq d(x,Y)=1$ for each i, a contradiction. Hence, $X=Y$.

(2) Let π be the quotient mapping from X to X/Z, and consider the vectors $\pi(u_1),\pi(u_2),\ldots,\pi(u_n)$ in X/Z. This is a finite collection in the closed unit ball of the Banach space X/Z. From our assumption, it follows that for each $x\in U_X$ there exists some i such that $\|\pi(x)-\pi(u_i)\|<1$. By part (1), the vector space X/Z is finite dimensional. ■

The next result deals with the codimension of an operator with a closed range.

Lemma 4.50. *For a bounded operator $T\colon X\to Y$ between Banach spaces with a closed range we have the following.*

(1) *If $d(T)=\infty$ and $N(T)$ is finite dimensional, then every operator in $\mathcal{L}(X,Y)$ close to T also has a closed range, infinite defect, and a finite dimensional null space.*

(2) *If $n(T)=\infty$ and $R(T)$ has a finite codimension, then every operator in $\mathcal{L}(X,Y)$ close to T also has a closed range, infinite nullity, and a finite codimension.*

Proof. (1) The fact that for each operator $S \in \mathcal{L}(X,Y)$ close enough to T the range $R(S)$ is closed has been established in Corollary 2.6. The only thing we need to verify is that $d(S) = \infty$ and $\dim N(S) < \infty$ for all operators $S \in \mathcal{L}(X,Y)$ close to T.

We assume first that T is one-to-one. In this case, since $R(T)$ is closed, $\|Tx\| \geq \gamma\|x\|$ for some $\gamma > 0$ and all $x \in X$. Pick some $\epsilon \in (0,\gamma)$ such that $\frac{3\epsilon}{\gamma-\epsilon} < 1$. We claim that every operator $S \in \mathcal{L}(X,Y)$ with $\|S - T\| < \epsilon$ has a closed range, an infinite defect, and a finite dimensional null space. To establish this, fix some $S \in \mathcal{L}(X,Y)$ satisfying $\|S - T\| < \epsilon$.

From $\|Sx\| \geq (\gamma - \epsilon)\|x\|$ for each $x \in X$, we see that S is bounded below on X. This implies that $N(S) = \{0\}$ and that S has a closed range. Moreover, for each $y \in R(S)$ the vector $x = S^{-1}y$ satisfies the estimate

$$\|x\| \leq \tfrac{1}{\gamma-\epsilon}\,\|y\|\,. \tag{$\star$}$$

To establish that $R(S)$ has infinite codimension, assume contrary to our claim that $\dim Y/R(S) < \infty$. That is, let the Banach space $Y/R(S)$ be finite dimensional. Fix some δ with $0 < \delta < 1 - \frac{3\epsilon}{\gamma-\epsilon}$, and then choose vectors $u_1, u_2, \ldots, u_k$ in the open unit ball of Y such that for each $u \in U_Y$ there exists some u_i satisfying $\|\pi(u) - \pi(u_i)\| < \delta$, where π is the quotient map of Y onto $Y/R(S)$.

We claim that $u_1, u_2, \ldots, u_k$ satisfy the conditions of Lemma 4.49(2) with $Z = R(T)$. To see this, fix any $y \in U_Y$ and consider the vector $\pi(y)$. Pick some i such that $\|\pi(y) - \pi(u_i)\| < \delta$. But then, by the definition of the quotient norm, we can find a vector $r \in R(S)$ such that $\|y - u_i - r\| < \delta$. Clearly, $r = y - u_i - (y - u_i - r)$ and so $\|r\| \leq 2 + \delta < 3$.

Since $r \in R(S)$, there exists some $x \in X$ with $Sx = r$. From ($\star$), we get $\|x\| \leq \frac{1}{\gamma-\epsilon}\|r\| \leq \frac{3}{\gamma-\epsilon}$. Now consider the element $z = Tx \in R(T)$. Clearly, $r - z = Sx - Tx$, and so

$$\|r - z\| = \|Sx - Tx\| = \big\|(S-T)x\big\| \leq \epsilon\|x\| \leq \tfrac{3\epsilon}{\gamma-\epsilon}\,.$$

Keeping in mind that $\|y - u_i - r\| < \delta$, we obtain that

$$d\big(y - u_i, R(T)\big) \leq \|y - u_i - z\| = \big\|y - u_i - r + (r - z)\big\| < \delta + \tfrac{3\epsilon}{\gamma-\epsilon} < 1\,.$$

Therefore, the collection of vectors $u_1, u_2, \ldots, u_k$ satisfies the condition of Lemma 4.49(2). Consequently, the vector space $Y/R(T)$ must be finite dimensional, which is impossible.

For the general case, it remains to notice that the above assumption that T is one-to-one is not essential. Indeed, the null space $N(T)$ is, by hypothesis, finite dimensional and so it is complemented in X. Let V be a closed vector subspace of X such that $X = N(T) \oplus V$. Then the restriction of T to V is one-to-one, and consequently the previous case applies to this restriction. Hence, if $S \in \mathcal{L}(X,Y)$ is close to T, then $S(V)$ has an infinite codimension in Y. Since $S(X) = S(N(T)) + S(V)$ and $S(N(T))$ is finite dimensional, it follows that $S(X)$ also has an infinite codimension in Y.

(2) Assume that a bounded operator $T: X \to Y$ between Banach spaces has a closed range and satisfies $\dim N(T) = \infty$ and $\dim Y/R(T) < \infty$. Consider the adjoint operator $T^*: Y^* \to X^*$ and note that from Theorem 2.13 we have $\dim N(T^*) = \dim [Y/R(T)]^* < \infty$. Moreover, since $R(T)$ is closed, the identity $X^*/R(T^*) = [N(T)]^*$ holds; see Exercise 12 in Section 2.1. From this identity it follows that $\dim X^*/R(T^*) = \infty$. By the preceding case, there exists some $\epsilon > 0$ such that every operator $A \in \mathcal{L}(Y^*, X^*)$ with $\|T^* - A\| < \epsilon$ has a closed range, $\dim N(A) < \infty$ and $\dim X^*/R(A) = \infty$.

Now let an operator $S \in \mathcal{L}(X,Y)$ satisfy $\|T - S\| < \epsilon$. Then we have $\|T^* - S^*\| < \epsilon$. This implies that S^* has a closed range (and hence S likewise has a closed range), $\dim N(S^*) < \infty$ and $\dim X^*/R(S^*) = \infty$. From $\dim [Y/R(S)]^* = \dim N(S^*) < \infty$, we infer that $\dim Y/R(S) < \infty$. Also, from $X^*/R(S^*) = [N(S)]^*$, it follows that $\dim N(S) = \infty$. The proof is finished. ■

Let us denote by $\hat{\mathcal{F}}red(X,Y)$ the collection of all semi-Fredholm operators between the Banach spaces X and Y. That is, $\hat{\mathcal{F}}red(X,Y)$ consists of all bounded operators from X to Y for which the index is a well-defined extended integer. Clearly, we have the inclusion $\mathcal{F}red(X,Y) \subseteq \hat{\mathcal{F}}red(X,Y)$. Combining Theorem 4.48 and Lemma 4.49, we obtain the following generalization of Theorem 4.48.

Theorem 4.51. *If X and Y are Banach spaces, then $\hat{\mathcal{F}}red(X,Y)$ is an open subset of $\mathcal{L}(X,Y)$ and the index function*

$$i\colon \hat{\mathcal{F}}red(X,Y) \longrightarrow \mathbb{Z} \cup \{-\infty, \infty\}$$

is continuous. Moreover, $\mathcal{F}red(X,Y)$ is an open subset of $\hat{\mathcal{F}}red(X,Y)$.

We continue with a useful result regarding invertibility of operators.

Lemma 4.52. *For a bounded operator $T: X \to Y$ between two Banach spaces the following statements are equivalent.*

(1) *There exist continuous projections $P \in \mathcal{L}(X)$ and $Q \in \mathcal{L}(Y)$ such that $R(P) = N(T)$ and $R(Q) = R(T)$.*

(2) *There exist closed subspaces V and W of X and Y, respectively, such that $X = N(T) \oplus V$ and $Y = R(T) \oplus W$.*

(3) *There exists a bounded operator $S: Y \to X$ such that $TST = T$.*

Proof. (1) $\Longrightarrow$ (2) In this case, the closed vector subspaces $V = (I_X - P)(X)$ and $W = (I_Y - Q)(Y)$ satisfy the desired properties.

(2)$\Longrightarrow$(3) Assume that the closed subspaces V and W of X and Y, respectively, satisfy $X = N(T) \oplus V$ and $Y = R(T) \oplus W$. From Theorem 2.16, we know that $R(T)$ is a closed subspace of Y. Consequently, the operator $A = T|_V : V \to R(T)$ is an isomorphism. Next, pick a projection $Q \in \mathcal{L}(Y)$ with range $R(T)$ and consider the operator $S = A^{-1}Q : Y \to X$. Clearly, $S \in \mathcal{L}(Y, X)$. Now, if $x = u + v \in N(T) \oplus V = X$, then

$$TSTx = TA^{-1}QTx = TA^{-1}Tx = TA^{-1}Tv = Tv = Tx\,,$$

and so $TST = T$ holds.

(3)$\Longrightarrow$(1) Assume that $TST = T$ for some operator $S \in \mathcal{L}(Y, X)$. Then $(ST)^2 = S(TST) = ST$, and similarly $(TS)^2 = TS$. That is, the operators $P_1 = ST \in \mathcal{L}(X)$ and $Q = TS \in \mathcal{L}(Y)$ are projections on X and Y, respectively. We claim that $N(P_1) = N(T)$ and $R(Q) = R(T)$.

For the first identity, note that $N(T) \subseteq N(ST)$ and that $x \in N(ST)$ implies $Tx = T(STx) = 0$, i.e., $N(ST) \subseteq N(T)$ is also true. For the second identity, observe that the inclusion $R(TS) \subseteq R(T)$ is obvious and that for each $x \in X$ we have $Tx = TS(Tx) \in R(TS)$. If we let $P = I_X - P_1$, then P is a bounded projection on X and satisfies $P(X) = R(P) = N(T)$. ■

The operators S that satisfy statement (3) of the preceding lemma are referred to as pseudoinverses or as relative inverses of T.

Definition 4.53. *A **pseudoinverse** of a bounded operator $T : X \to Y$ between Banach spaces is any operator $S \in \mathcal{L}(Y, X)$ satisfying $TST = T$.*

Fredholm operators always have pseudoinverses that are also Fredholm operators.

Theorem 4.54. *Every Fredholm operator $T : X \to Y$ between Banach spaces has a pseudoinverse. Moreover, every pseudoinverse $S : Y \to X$ of T is also a Fredholm operator and satisfies $i(S) = -i(T)$.*

Proof. The first part follows immediately from Lemma 4.52. For the second part, let $S : Y \to X$ be a pseudoinverse of T, i.e., $TST = T$. Then, by part (2) of Theorem 4.43, the operator ST is Fredholm. But then, by the same argument, S is a Fredholm operator. Now part (1) of Theorem 4.43 yields $i(T) = i(TST) = i(T) + i(S) + i(T)$. This implies $i(S) = -i(T)$. ■

Finally, we shall close the section with another useful property of Fredholm operators.

Lemma 4.55. *For any Fredholm operator $T \in \mathcal{L}(X)$ we have the following.*

(1) *The vector subspace $Y = \bigcap_{n=1}^{\infty} R(T^n)$ is a T-invariant closed subspace of X satisfying $T(Y) = Y$.*

(2) *If $\lambda \neq 0$, then $N(\lambda - T) = N((\lambda - T)|_Y) = N(\lambda - T|_Y)$.*

(3) *There exists some $\epsilon > 0$ such that:*
   (a) *For each $|\lambda| < \epsilon$ the operator $\lambda - T$ is Fredholm.*
   (b) *The nullity $n(\lambda - T)$ and the defect $d(\lambda - T)$ are constant on the set $\{\lambda \in \mathbb{C}\colon\ 0 < |\lambda| < \epsilon\}$.*

Proof. (a) Let $T\colon X \to X$ be a Fredholm operator. A glance at the Index Theorem 4.43 shows that T^n is also Fredholm for each n. In particular, by Lemma 4.38, the range $R(T^n)$ is closed for each n. This implies that Y is a closed subspace of X. Clearly, Y is T-invariant.

To see that $T(Y) = Y$, observe first that there exists some k such that $R(T^n) \cap N(T) = R(T^k) \cap N(T)$ holds for all $n \geq k$. This should be obvious in view of the fact that $N(T)$ is finite dimensional and $R(T^{n+1}) \subseteq R(T^n)$ for each n. Now fix some $y \in Y$ and let $n \geq k$. Pick some $x \in R(T^k)$ and some $z \in R(T^n) \subseteq R(T^k)$ such that $y = Tx = Tz$. Therefore, $T(x - z) = 0$ and so $x - z \in R(T^k) \cap N(T) \subseteq R(T^n)$. This implies $x \in R(T^n)$ for each $n \geq k$. Consequently, $x \in Y$ and $y = Tx$. This proves that $T(Y) = Y$.

(b) Fix a non-zero scalar λ. Clearly, $N(\lambda - T|_Y) \subseteq N(\lambda - T)$. To see that $N(\lambda - T) \subseteq N(\lambda - T|_Y)$ is also true, let $x \in N(\lambda - T)$. This implies $Tx = \lambda x$, and so $T^n x = \lambda^n x$ for each n. Consequently, $x = \frac{1}{\lambda^n} T^n x \in R(T^n)$ for each n. Thus, $x \in Y$ and so $x \in N(\lambda - T|_Y)$.

(c) According to Theorems 2.10 and 4.48 there exists some $\epsilon > 0$ such that for each $|\lambda| < \epsilon$:

- The operator $\lambda - T \in \mathcal{L}(X)$ is a Fredholm operator and its index satisfies $i(\lambda - T) = i(-T) = i(T)$.
- The operator $(\lambda - T)|_Y = \lambda - T|_Y \in \mathcal{L}(Y)$ is surjective and its index satisfies $i(\lambda - T|_Y) = i(T|_Y)$.

Now fix any $\lambda \in \mathbb{C}$ with $0 < |\lambda| < \epsilon$. Since the operator $\lambda - T|_Y \in \mathcal{L}(Y)$ is surjective, it follows that $n(\lambda - T|_Y) = i(\lambda - T|_Y)$. Now (b) yields

$$\begin{aligned} n(\lambda - T) &= \dim N(\lambda - T) = \dim N(\lambda - T|_Y) \\ &= n(\lambda - T|_Y) = i(\lambda - T|_Y) = i(T|_Y) . \end{aligned}$$

Thus, the function $\lambda \mapsto n(\lambda - T)$ is constant on the set $\{\lambda \in \mathbb{C}\colon\ 0 < |\lambda| < \epsilon\}$. Since the function $\lambda \mapsto i(\lambda - T) = n(\lambda - T) - d(\lambda - T)$ is also constant on the set $\{\lambda \in \mathbb{C}\colon\ 0 < |\lambda| < \epsilon\}$, it follows that the defect function $\lambda \mapsto d(\lambda - T)$ is also constant on $\{\lambda \in \mathbb{C}\colon\ 0 < |\lambda| < \epsilon\}$. The proof is finished. ∎

Exercises

1. Prove Lemma 4.40. [HINT: If $T\colon X \to Y$ is a bounded operator between two real Banach spaces, then note that $N(T_{\mathbf{c}}) = N(T) \oplus \imath N(T) = [N(T)]_{\mathbf{c}}$ and $R(T_{\mathbf{c}}) = R(T) \oplus \imath R(T) = [R(T)]_{\mathbf{c}}$.]

2. Give a direct proof of Lemma 4.44 without using the Hahn–Banach theorem. Also, present an example of an infinite dimensional proper closed subspace Y of a Banach space X such that for each unit vector $x \in X \setminus Y$ we have $d(x, Y) < 1$. [HINT: Fix some $x_0 \in X \setminus Y$ and let $d = d(x_0, Y) > 0$. Next, select some $y_0 \in Y$ with $\|x_0 - y_0\| < \frac{d}{1-\epsilon}$ and let $x = \frac{x_0 - y_0}{\|x_0 - y_0\|}$. Clearly, $\|x\| = 1$ and for any $y \in Y$ we have
$$\|x - y\| = \Big\| \tfrac{x_0 - y_0}{\|x_0 - y_0\|} - y \Big\| = \tfrac{\|x_0 - (y_0 + \|x_0 - y_0\| y)\|}{\|x_0 - y_0\|} \geq \tfrac{d}{\|x_0 - y_0\|} > 1 - \epsilon\,.]$$

3. This exercise presents some elementary algebraic properties that have been used in the proofs of the results in this section. Establish the following.
   (a) If W is a vector subspace of a vector space V, then W has an **algebraic complement**, i.e., there exists a vector subspace U of V satisfying $W \cap V = \{0\}$ and $W + U = V$.
   (b) If U and W are subspaces of a vector space V, then there is a vector subspace W_1 of W satisfying $V \cap W_1 = \{0\}$ and $U + W_1 = U + W$.
   (c) If $T\colon V \to W$ is a linear operator between two vector spaces, then the quotient vector space $V/N(T)$ is linearly isomorphic to $R(T)$.

   [HINT: For (a) and (b) use the fact that any collection of linearly independent vectors can be extended to a Hamel basis. For (c) consider the linear operator $\hat{T}\colon V/N(T) \to R(T)$ defined by $\hat{T}([x]) = Tx$.]

4. Let X be an n-dimensional vector space, let Y be an m-dimensional vector space, and let $T\colon X \to Y$ be an operator. If $r = \dim R(T)$, then show that $n(T) = n - r$ and $d(T) = m - r$; conclude from this that $i(T) = n - m$. [HINT: Use Exercise 9 in Section 8.2.]

5. For a bounded operator $T\colon X \to Y$ between two Banach spaces define $\hat{T}\colon X/N(T) \to Y$ via the formula $\hat{T}[x] = Tx$. Let $\pi\colon X \to X/N(T)$ be the quotient map and $J\colon R(T) \to Y$ be the natural embedding. Now consider the scheme of bounded operators
$$X \xrightarrow{\ \pi\ } X/N(T) \xrightarrow{\ \hat{T}\ } R(T) \xrightarrow{\ J\ } Y\,,$$
and note that $T = J\hat{T}\pi$. If T is a Fredholm operator, then show that all operators in the above scheme are Fredholm operators and compute their indices. Also, verify directly that $i(T) = i(J) + i(\hat{T}) + i(\pi)$.

6. Establish the following property that was left out from the proof of Lemma 4.45: *If $K\colon X \to X$ is a compact operator on a Banach space, then the operator $I - K$ has a finite descent.*

7. (**The Fredholm Alternative**) Let X be a Banach space and $K \in \mathcal{K}(X)$. Show that for each $\lambda \neq 0$ exactly one of the following two exclusive statements is true:

(a) *For each* $y \in X$ *the equation* $x - \lambda Kx = y$ *has a unique solution.*
(b) *The equation* $x - \lambda Kx = 0$ *has a non-zero solution.*
When (b) is true, show also that the equation $x - \lambda Kx = 0$ has a finite number of linearly independent solutions.

8. Let $T \in \mathcal{L}(X, Y)$ and let $S \in \mathcal{L}(Y, X)$ be a pseudoinverse of T. Fix some $y \in Y$ and consider the equation

$$Tx = y\,. \tag{$\star$}$$

Show that equation $(\star)$ has a solution if and only if the vector Sy is a solution of $(\star)$.

9. Show that a bounded operator between Hilbert spaces has a pseudoinverse if and only if it has a closed range.

4.5. Strictly Singular Operators

In this section we shall discuss the properties of the class of strictly singular operators that are related to compact and Fredholm operators. They were introduced by T. Kato [**178**].

Definition 4.56 (Kato). *A bounded operator* $T\colon X \to Y$ *between Banach spaces is said to be* ***strictly singular*** *if there is no infinite dimensional closed subspace* Z *of* X *such that* $T\colon Z \to T(Z)$, *the restriction of* T *to* Z, *is an isomorphism.*

Clearly, the concept of a strictly singular operator is of interest only when the domain space is infinite dimensional. Notice also that a bounded operator $T\colon X \to Y$ between Banach spaces is strictly singular if and only if T is not bounded below on any infinite dimensional closed subspace of X. The next result presents some connections between compact operators and strictly singular operators.

Lemma 4.57. *Let* $T\colon X \to Y$ *be an isomorphism between two Banach spaces. If the restriction of* T *to a closed subspace* Z *of* X *is compact, then* Z *is finite dimensional. In particular, every compact operator is strictly singular.*

Proof. Let Z be a closed subspace of X such that $T\colon Z \to T(Z)$ is compact. Since $T\colon X \to Y$ is an isomorphism, it follows that $T\colon Z \to T(Z)$ is likewise an isomorphism. If $S = (T|_Z)^{-1}$, then $I = TS\colon Z \to Z$ is a compact operator. This implies that Z is finite dimensional. ■

A strictly singular operator need not be a compact operator.

Theorem 4.58. *If* $1 \le p < r < \infty$, *then the natural embedding* $J\colon \ell_p \to \ell_r$ *is a non-compact strictly singular operator.*

Proof. To establish this result we need the following simple property whose proof is left as an exercise.

(P) *If X is an infinite dimensional vector subspace of ℓ_p, then for each m the vector subspace*

$$X_m = \{x \in X\colon\ x_1 = x_2 = \cdots = x_m = 0\}$$

of X is also infinite dimensional.

We denote by P_n the standard projection of ℓ_p onto the first n coordinates of ℓ_p. That is, $P_n y = (y_1, \ldots, y_n, 0, 0, \ldots)$ for each $y \in \ell_p$. We also let $Q_n = I - P_n$.

If e_n denotes the basic n^{th} unit vector in ℓ_p, then $\|e_n\|_p = 1$ for each n and $\|e_n - e_m\|_r = 2^{1/r}$ for $n \neq m$. This shows that J is not a compact operator. To verify that the embedding $J\colon \ell_p \to \ell_r$ is strictly singular, it suffices to show that for each infinite dimensional closed subspace X of ℓ_p there exists a sequence $\{x_n\} \subseteq X$ satisfying $\|x_n\|_p \not\to 0$ and $\|x_n\|_r \to 0$.

To this end, let X be an infinite dimensional closed subspace of ℓ_p. Fix some $0 < \epsilon < 1$ and let $\delta = \frac{1-\epsilon}{2}$. Start by choosing any vector $u_1 \in X$ with $\|u_1\|_p = 1$, and then pick some m_1 such that $\|P_{m_1} u_1\|_p > 1 - \delta$ and $\|Q_{m_1} u_1\|_p < \frac{\delta}{2}$. Let $v_1 = P_{m_1} u_1$ and $w_1 = Q_{m_1} u_1$. By Property (P), the vector subspace X_{m_1} of X is infinite dimensional. Pick some $u_2 \in X_{m_1}$ with $\|u_2\|_p = 1$ and then select $m_2 > m_1$ such that $\|P_{m_2} u_2\|_p > 1 - \delta$ and $\|Q_{m_2} u_2\|_r \leq \|Q_{m_2} u_2\|_p < \frac{\delta}{4}$.

An inductive argument along these lines guarantees the existence of a sequence $\{u_k\} \subseteq X$ of unit vectors and a strictly increasing sequence of natural numbers $\{m_k\}$ such that for each k we have $u_k \in X_{m_{k-1}}$ (where $m_0 = 0$ and $X_0 = X$) and the vectors $v_k = P_{m_k} u_k$ and $w_k = Q_{m_k} u_k$ satisfy:

(a) $\|v_k\|_p > 1 - \delta$, and

(b) $\|w_k\|_r \leq \|w_k\|_p < \frac{\delta}{2^k}$.

Clearly, $\{v_k\}$ is a disjoint sequence and $u_k = v_k + w_k$ for each k.

Now let $x_n = \frac{1}{n^{1/p}} \sum_{k=1}^{n} u_k \in X$. We claim that $\|x_n\|_p \geq \epsilon$ for each n and that $\|x_n\|_r \to 0$. Indeed, using the p-additivity of the norm, we get

$$\begin{aligned} \|x_n\|_p &= \tfrac{1}{n^{1/p}} \Big\| \sum_{k=1}^{n} (v_k + w_k) \Big\|_p \geq \tfrac{1}{n^{1/p}} \Big[\Big\| \sum_{k=1}^{n} v_k \Big\|_p - \Big\| \sum_{k=1}^{n} w_k \Big\|_p \Big] \\ &\geq \tfrac{1}{n^{1/p}} \Big[\Big(\sum_{k=1}^{n} \|v_k\|_p^p \Big)^{1/p} - \sum_{k=1}^{n} \|w_k\|_p \Big] \\ &> \tfrac{1}{n^{1/p}} \Big[\big[n(1-\delta)^p \big]^{1/p} - \delta \Big] = 1 - \delta - \tfrac{\delta}{n^{1/p}} > 1 - 2\delta = \epsilon \,. \end{aligned}$$

On the other hand, the r-additivity of the norm yields

$$\begin{aligned}
\|x_n\|_r &= \frac{1}{n^{1/p}}\Big\|\sum_{k=1}^n (v_k+w_k)\Big\|_r \le \frac{1}{n^{1/p}}\left[\Big\|\sum_{k=1}^n v_k\Big\|_r + \Big\|\sum_{k=1}^n w_k\Big\|_r\right] \\
&\le \frac{1}{n^{1/p}}\left[\Big[\sum_{k=1}^n \|v_k\|_r^r\Big]^{1/r} + \sum_{k=1}^n \|w_k\|_r\right] \\
&\le \frac{1}{n^{1/p}} n^{1/r} + \frac{1}{n^{1/p}}\sum_{k=1}^n \|w_k\|_r \le \frac{1}{n^{1/p-1/r}} + \frac{\delta}{n^{1/p}} \xrightarrow[n\to\infty]{} 0\,.
\end{aligned}$$

This is the desired conclusion.

It is interesting to note that all bounded operators acting in the opposite direction, that is, from ℓ_r to ℓ_p, where $1 \le p < r \le \infty$, are compact and so $\mathcal{L}(\ell_r, \ell_p) = \mathcal{K}(\ell_r, \ell_p)$. This is a well-known theorem due to H. R. Pitt; for a proof see [**205**, Proposition 2.c.3]. For a complete characterization of the pairs $\big(L_{p_1}(\mu), L_{p_2}(\nu)\big)$ for which $\mathcal{L}\big(L_{p_1}(\mu), L_{p_2}(\nu)\big) = \mathcal{K}\big(L_{p_1}(\mu), L_{p_2}(\nu)\big)$ see [**89**, Theorem 4.7]. ■

It should be mentioned that the adjoint of a strictly singular operator need not be strictly singular. For an example see Exercise 6 at the end of the section.

Many properties of strictly singular operators depend on the following technical result.

Lemma 4.59. *For a bounded operator $T\colon X \to Y$ between two Banach spaces with X infinite dimensional we have the following.*

1. *If the restriction of T to any closed subspace of finite codimension is not an isomorphism, then for any closed subspace X_1 of X of finite codimension and each $\epsilon > 0$ there exists an infinite dimensional closed subspace X_2 of X_1 such that $T\colon X_2 \to Y$ is compact and $\|T|_{X_2}\| < \epsilon$.*
2. *If T is strictly singular, then for each infinite dimensional closed subspace X_1 of X there exists an infinite dimensional closed subspace X_2 of X_1 such that $T\colon X_2 \to Y$ is compact and $\|T|_{X_2}\| < \epsilon$.*

Proof. (1) Assume that X_1 is a closed subspace of X of finite codimension and so the restriction of the operator $T \in \mathcal{L}(X,Y)$ to X_1 is not an isomorphism. Replacing X by X_1, we can assume without loss of generality that $X_1 = X$. We claim that in this case the following version of Mazur's Lemma 1.42 is valid:

- *If Z is a finite dimensional vector subspace of X, then for each $\epsilon > 0$ and for each $\delta > 0$ there exists a unit vector $x \in X$ such that $\|Tx\| < \delta$ and $\|z\| \leq (1+\epsilon)\|z + \alpha x\|$ for each $z \in Z$ and each α.*

The proof of this claim is similar to that of Lemma 1.42. Following that proof, we notice that the closed subspace $\bigcap_{i=1}^n \operatorname{Ker} x_i^*$ has finite codimension, and so (in view of our assumptions) the operator T cannot be bounded below on this subspace. Therefore, we can choose a unit vector $x \in \bigcap_{i=1}^n \operatorname{Ker} x_i^*$ satisfying $\|Tx\| < \delta$.

Using the above version of Mazur's lemma and arguing as in the proof of Theorem 1.43, we see that for each $\epsilon > 0$ there exists a normalized basic sequence $\{x_n\}$ in X with basis constant less than 2 such that $\|Tx_n\| < \frac{\epsilon}{8^n}$ for each n. Let X_2 be the infinite dimensional closed subspace generated by $\{x_n\}$. We claim that X_2 satisfies the desired properties.

First, we show that $\|T|_{X_2}\| < \epsilon$. To see this, let $x = \sum_{n=1}^\infty \alpha_n x_n \in X_2$ satisfy $\|x\| \leq 1$. Then $|\alpha_n| = |c_n^*(x)| \leq 4$ for each n; see Corollary 1.40. This implies

$$\|Tx\| = \Big\|\sum_{n=1}^\infty \alpha_n T x_n\Big\| \leq \sum_{n=1}^\infty |\alpha_n| \cdot \|Tx_n\| \leq 4 \sum_{n=1}^\infty \tfrac{\epsilon}{8^n} = \tfrac{4}{7}\epsilon\,,$$

and so $\|T\| = \sup_{\|x\|\leq 1} \|Tx\| \leq \frac{4}{7}\epsilon < \epsilon$.

To see that the restriction $S = T|_{X_2} \colon X_2 \to Y$ is a compact operator, let $P_n \colon X_2 \to X_2$ be the natural projections on the first n coordinates associated with the basis $\{x_n\}$. Since the operator $SP_n \colon X_2 \to Y$ has finite-rank, it is automatically compact. Now notice that if $x = \sum_{i=1}^\infty \alpha_i x_i \in X_2$ satisfies $\|x\| \leq 1$, then

$$\begin{aligned}
\|(S - SP_n)x\| &= \|S(I_{X_2} - P_n)x\| = \Big\|S\Big(\sum_{i=n+1}^\infty \alpha_i x_i\Big)\Big\| = \Big\|\sum_{i=n+1}^\infty \alpha_i T x_i\Big\| \\
&\leq \sum_{i=n+1}^\infty |\alpha_i| \cdot \|Tx_i\| \leq 4 \sum_{i=n+1}^\infty \tfrac{\epsilon}{8^i} = \tfrac{4}{7} \cdot \tfrac{\epsilon}{8^n} < \tfrac{\epsilon}{8^n}\,.
\end{aligned}$$

Therefore, $\|S - SP_n\| \leq \frac{\epsilon}{8^n}$, and so $SP_n \to S$ in $\mathcal{L}(X_2, Y)$. This implies that $S = T|_{X_2}$ is a compact operator, and the proof of the first part is finished.

(2) Assume that T is a strictly singular operator. Then the restriction of T to any infinite dimensional subspace of X is not an isomorphism, and under this condition all arguments of the above proof remain valid for each closed infinite dimensional subspace X_1 of X. ■

Corollary 4.60. *If a bounded operator $T\colon X \to Y$ between two Banach spaces does not have a closed range, then for each closed subspace X_1 of X of finite codimension and for each $\epsilon > 0$ there exists an infinite dimensional closed subspace X_2 of X_1 such that $T\colon X_2 \to Y$ is compact and $\|T|_{X_2}\| < \epsilon$.*

Proof. It is enough to verify that T satisfies the hypotheses of Lemma 4.59, i.e., that the restriction of T to any closed subspace of X of finite codimension is not an isomorphism. To see this, assume by way of contradiction that V and W are closed subspaces such that $X = V \oplus W$, $\dim W < \infty$, and $T\colon V \to T(V)$ is an isomorphism. So, $T(V)$ is a closed subspace of Y and $R(T) = T(V) + T(W)$. Since $T(W)$ is finite dimensional, it follows that $R(T)$ is a closed subspace, which is a contradiction; see Corollary 2.17. ■

We are now ready to present T. Kato's [**178**] characterization of strictly singular operators.

Theorem 4.61 (Kato). *For a bounded operator $T\colon X \to Y$ between two Banach spaces with X infinite dimensional the following statements are equivalent:*

(1) *The operator T is strictly singular.*
(2) *For each infinite dimensional closed subspace X_1 of X and each $\epsilon > 0$ there is an infinite dimensional closed subspace X_2 of X_1 such that the operator $T\colon X_2 \to Y$ is compact and $\|T|_{X_2}\| < \epsilon$.*
(3) *For each infinite dimensional closed subspace X_1 of X there exists an infinite dimensional closed subspace X_2 of X_1 such that the operator $T\colon X_2 \to Y$ is compact.*

Proof. The implication (1) $\Longrightarrow$ (2) follows immediately from part (2) of Lemma 4.59. The implication (2) $\Longrightarrow$ (3) is obvious.

(3) $\Longrightarrow$ (1) Assume that the operator T is not strictly singular. This means that there exists an infinite dimensional closed subspace X_1 of X such that $T\colon X_1 \to T(X_1)$ is an isomorphism. Then, by (3), there exists an infinite dimensional closed subspace X_2 of X_1 such that the operator $T\colon X_2 \to Y$ is compact. However, the latter conclusion contradicts Lemma 4.57 according to which X_2 must be finite dimensional. This establishes that T is strictly singular. ■

A consequence of the preceding theorem is the following result that describes two multiplicative properties of singular operators.

Corollary 4.62. *The collection of all strictly singular operators from a Banach space X to another Banach space Y is a closed subspace of $\mathcal{L}(X,Y)$.*

Moreover, if in a scheme of bounded operators $X \xrightarrow{S} Y \xrightarrow{T} Z$ between Banach spaces either S or T is strictly singular, then TS is strictly singular.

Proof. Let $S, T \in \mathcal{L}(X,Y)$ be two strictly singular operators. It should be clear that αS is also strictly singular for each scalar α. What needs a proof is the fact that $S+T$ is strictly singular.

To see this, let Z be an infinite dimensional closed vector subspace of X. Since S is strictly singular, there exists (by Theorem 4.61) an infinite dimensional closed vector subspace Z_1 of Z such that $S\colon Z_1 \to Y$ is a compact operator. Using Theorem 4.61 once more, we see that there exists an infinite dimensional closed subspace Z_2 of Z_1 such that $T\colon Z_2 \to Y$ is a compact operator. It follows that $S+T\colon Z_2 \to Y$ is a compact operator. Now a glance at Lemma 4.57 guarantees that $(S+T)|_Z$ cannot be an isomorphism. Therefore, the operator $S+T$ is strictly singular.

To see that the vector subspace of all strictly singular operators from X to Y is closed, assume that a sequence $\{T_n\}$ of strictly singular operators from X to Y satisfies $T_n \to T$ in $\mathcal{L}(X,Y)$. Also, let Z be an arbitrary infinite dimensional closed subspace of X. If T is bounded below on Z, then by Lemma 2.4 there exists some n_0 such that T_n is bounded below on Z for all $n \geq n_0$. But this contradicts the fact that no T_n is bounded below on Z. Therefore, T cannot be bounded below on Z, i.e., T is strictly singular.

Now consider a scheme of bounded operators $X \xrightarrow{S} Y \xrightarrow{T} Z$ between Banach spaces. Assume that TS is an isomorphism on an infinite dimensional closed vector subspace V of X. This guarantees that S is bounded below. In particular, S cannot be strictly singular and $S\colon V \to S(V)$ is an isomorphism. Using again the fact that $T\colon S(V) \to TS(V)$ is an isomorphism, we infer that $T\colon S(V) \to Y$ must be bounded below. This shows that T is not a strictly singular operator either. Thus, we have shown that if ST is not strictly singular, then neither S nor T is strictly singular, and the desired conclusion follows. ■

As in the case of compact perturbations of Fredholm operators, a strictly singular perturbation of a Fredholm operator remains Fredholm and has the same index.

Theorem 4.63. *Let $S, T\colon X \to Y$ be two bounded operators between Banach spaces with X infinite dimensional. If T is Fredholm and S is strictly singular, then $T+S$ is a Fredholm operator and $i(T+S) = i(T)$.*

Proof. Since T is a Fredholm operator, there exist closed subspaces V of X and W of Y such that V has a finite defect, W is finite dimensional,

$X = N(T) \oplus V$ and $Y = R(T) \oplus W$. In particular, the surjective operator $T\colon V \to R(T)$ is an isomorphism. Pick some $c > 0$ such that

$$\|Tv\| \geq 2c\|v\| \tag{$\star$}$$

holds for each $v \in V$.

To see that $\dim N(T+S) < \infty$, assume by way of contradiction that $\dim N(T+S) = \infty$. Consequently $\dim [N(T+S) \cap V] = \infty$, since V has a finite defect. Now, for each x in the infinite dimensional closed subspace $X_1 = N(T+S) \cap V$ of X we have $\|Sx\| = \|-Tx\| \geq 2c\|x\|$. This implies that S is invertible on X_1, which is a contradiction. Hence, $\dim N(T+S) < \infty$.

Next, we claim that the range of $T+S$ is closed. To establish this, assume again by way of contradiction that $R(T+S)$ is not closed. Then, by Corollary 4.60, there exists a closed infinite dimensional subspace Z of V such that $\big\|(T+S)|_Z\big\| < c$. Therefore, using $(\star)$, we see that for each $z \in Z$ we have

$$\|Sz\| = \big\|Tz - (T+S)z\big\| \geq \|Tz\| - \|(T+S)z\| \geq 2c\|z\| - c\|z\| = c\|z\|\,.$$

This shows that S restricted to Z is an isomorphism, which is a contradiction. Hence, $R(T+S)$ is closed, and so $T+S$ is a semi-Fredholm operator.

Since λS is a strictly singular operator for each $\lambda \neq 0$, it follows from the last conclusion that $T + \lambda S$ is also a semi-Fredholm operator for each $\lambda \in \mathbb{R}$. Finally, in view of Theorem 4.51, the mapping $\lambda \mapsto i(T+\lambda S)$ from $\mathbb{R}$ to $\mathbb{Z} \cup \{-\infty, \infty\}$ is continuous, and hence it must be constant. This implies that $i(T+\lambda S) = i(T)$ for each $\lambda \in \mathbb{R}$. Letting $\lambda = 1$ yields $i(T+S) = i(T)$. This guarantees that $T+S$ is a Fredholm operator. ■

We conclude the section by mentioning that there exists an interesting analogue, due to F. L. Hernández and B. Rodriguez-Salinas [**150**], of strictly singular operators tailored to utilize the order structure of Banach lattices. The corresponding operators are called **disjointly strictly singular** and their definition is given in Exercise 9 at the end of the section. The reader is referred to [**123, 124**] for some recent developments regarding disjointly strictly singular operators.

Exercises

1. Show that a bounded operator $T\colon X \to Y$ between Banach spaces is strictly singular if and only if T is not bounded below on any infinite dimensional closed subspace of X.

2. Let X_i $(i = 1, 2, \ldots, n)$ and Y_j $(j = 1, 2, \ldots, m)$ be all real (or all complex) Banach spaces and consider a bounded operator

$$\mathcal{T}\colon X_1 \oplus X_2 \oplus \cdots \oplus X_n \to Y_1 \oplus Y_2 \oplus \cdots \oplus Y_m .$$

As in Exercise 11 of Section 1.1, it is easy to see that there exist bounded operators $T_{ji}\colon X_i \to Y_j$ $(i = 1, 2, \ldots, n;\ j = 1, 2, \ldots, m)$ uniquely determined such that $\mathcal{T}$ can be represented by the $m \times n$ matrix

$$\mathcal{T} = \begin{bmatrix} T_{11} & T_{12} & \ldots & T_{1n} \\ T_{21} & T_{22} & \ldots & T_{2n} \\ \vdots & \vdots & \ddots & \vdots \\ T_{m1} & T_{m2} & \ldots & T_{mn} \end{bmatrix}.$$

Prove that $\mathcal{T}$ is strictly singular if and only if each operator T_{ji} is strictly singular.

3. Show that a bounded operator $T\colon X \to Y$ between two real Banach spaces is strictly singular if and only if $T_{\mathbf{c}}\colon X_{\mathbf{c}} \to Y_{\mathbf{c}}$ is strictly singular. [HINT: Use Exercise 2 above.]

4. Prove the following property of ℓ_p-spaces that was used in Theorem 4.58: *If X is an infinite dimensional vector subspace of any ℓ_p-space, then for each $m \in \mathbb{N}$ the vector subspace*

$$X_m = \big\{x = (x_1, x_2, \ldots) \in X\colon\ x_1 = x_2 = \cdots = x_m = 0\big\}$$

is also infinite dimensional.

5. Assume that for a pair of Banach spaces (X, Y) we have the following:
(a) Each infinite dimensional closed subspace of X contains a closed vector subspace that is isomorphic to X.
(b) No closed vector subspace of Y is isomorphic to X.

Show that each operator in $\mathcal{L}(X, Y)$ is strictly singular. (It is known that the pair (ℓ_1, ℓ_p), where $1 < p < \infty$, satisfies properties (a) and (b); see [**205**, Proposition 2.a.2, p. 53].)

6. This exercise presents an example of a strictly singular operator whose adjoint is not strictly singular. Establish the following.
(i) If Y is a separable Banach space, then there exists a bounded surjective operator $T\colon \ell_1 \to Y$ whose adjoint $T^*\colon Y^* \to \ell_\infty$ is a linear isometry.
(ii) If $T\colon \ell_1 \to \ell_p$, where $1 < p < \infty$, is any bounded operator satisfying (i), then its adjoint operator $T^*\colon \ell_q \to \ell_\infty$, where $\frac{1}{p} + \frac{1}{q} = 1$, is not strictly singular.

[HINT: For (i) let $\{y_1, y_2, \ldots\}$ be a countable norm dense subset in the closed unit ball U_Y of Y and then define the operator $T\colon \ell_1 \to Y$ by the formula $T(\lambda_1, \lambda_2, \ldots) = \sum_{n=1}^{\infty} \lambda_n y_n$. For (ii) use Exercise 5 above.]

7. (T. Oikhberg) Show that there exists an infinite dimensional Banach space X with a non-trivial multiplicative linear functional on the Banach algebra $\mathcal{L}(X)$. [HINT: An infinite dimensional Banach space X is called **hereditarily indecomposable**, in brief X is an HI-space, if *no* infinite dimensional subspace of X is isomorphic to a direct sum of two infinite dimensional Banach spaces. The existence of HI-spaces was established by W. T. Gowers and B. Maurey [**137**]. Moreover, it was shown in [**137, 138**] that: *If X is an HI-space and $T \in \mathcal{L}(X)$, then there exist a scalar α and some strictly singular operator $S \in \mathcal{L}(X)$ (both uniquely determined) such that $T = \alpha I + S$, where I is the identity on X.* The linear functional ϕ on $\mathcal{L}(X)$ defined by $\phi(T) = \phi(\alpha I + S) = \alpha$ is non-zero and multiplicative.]

8. Show that if p and q satisfy $1 \le q < p < \infty$, then the natural embedding $J\colon L_p[0,1] \to L_q[0,1]$ is not a strictly singular operator. [HINT: Let X denote the closed vector subspace generated by the standard Rademacher functions on $[0,1]$. It is well known that the two norms $\|\cdot\|_p$ and $\|\cdot\|_2$ are equivalent on X; see [**106**, Theorem 1.10] or [**205**, Theorem 2.b.3]. Now establish that J restricted to X is an isomorphism.]

9. (Hernández–Rodriguez-Salinas [**150**]) A bounded operator $T\colon E \to X$, from a Banach lattice to a Banach space, is said to be **disjointly strictly singular** if for each disjoint sequence $\{x_n\}$ of non-zero vectors in E the restriction of T to the closed vector subspace generated by $\{x_n\}$ is not an isomorphism. Clearly, every strictly singular operator from a Banach lattice to a Banach space is disjointly strictly singular.

Show that if p and q satisfy $1 \le q < p < \infty$, then the natural embedding $J\colon L_p[0,1] \to L_q[0,1]$ is a disjointly strictly singular operator.

Chapter 5

Integral Operators

This chapter studies the classical integral operators in the framework of function spaces. Let E and F be two vector subspaces of $L_0(\mathcal{S}, \Sigma_1, \mu)$ and $L_0(\mathcal{T}, \Sigma_2, \nu)$, respectively. A linear operator $T\colon E \to F$ is said to be an **integral operator** if there exists a jointly measurable function $K(\cdot, \cdot)$ (called the **kernel** of the operator) such that for each $x \in E$ we have

$$Tx(t) = \int_{\mathcal{S}} K(s,t)x(s)\,d\mu(s)$$

for ν-almost all $t \in \mathcal{T}$. As we shall see, the existence of the kernel imposes on the operator several interesting and desirable features that are not present for an arbitrary operator.

The material in this chapter is divided in three sections. In the first section we present the basic properties of integral operators with emphasis on their lattice structure. Among the established order properties, we show that the vector space of all regular integral operators between two Banach function spaces is a band in the Riesz space of all regular operators. Abstract integral operators are introduced in the second section and their close connection with the regular integral operators is thoroughly investigated. It is shown here that the classes of abstract integral operators and regular integral operators coincide. In addition, we present a proof of Bukhvalov's theorem that allows us to recognize when a given operator is an integral operator. As an application of this characterization, we obtain Dunford's famous integral representation theorem for continuous operators from an L_1-space to an L_p-space:

- *If μ and ν are σ-finite measures, then every continuous operator from $L_1(\mu)$ to $L_p(\nu)$ with $1 < p < \infty$ is an integral operator.*

The third section of this chapter studies conditional expectation operators on ideals in $L_1(\mu)$-spaces. The major result here is the characterization of the positive projections that fix the constant functions as conditional expectation operators. The last section of the chapter deals with positive projections and lattice-subspaces.

5.1. The Basics of Integral Operators

In this section, we shall present the basic order-theoretic properties of integral operators. For extensive treatments on integral operators, see the monographs [**78, 107, 118, 149, 174, 175, 188, 191, 343**]. Throughout this section we shall assume that:

- *$(\mathcal{S}, \Sigma_1, \mu)$ and $(\mathcal{T}, \Sigma_2, \nu)$ are two σ-finite measure spaces.*
- *E and F are two ideals in $L_0(\mu)$ and $L_0(\nu)$, respectively.*

We start our discussion with the definition of an integral operator.

Definition 5.1. *Let X and Y be linear subspaces of $L_0(\mu)$ and $L_0(\nu)$, respectively. An operator $T\colon X \to Y$ is said to be an* ***integral operator*** *if there exists a $\mu \times \nu$-measurable function $T(\cdot,\cdot)\colon \mathcal{S} \times \mathcal{T} \to \mathbb{R}$ such that for each $x \in X$ the function $s \mapsto T(s,t)x(s)$ belongs to $L_1(\mu)$ for ν-almost all $t \in \mathcal{T}$ and the equality*

$$Tx(t) = \int_{\mathcal{S}} T(s,t)x(s)\, d\mu(s) \qquad (\star)$$

holds for ν-almost all $t \in \mathcal{T}$.

Before proceeding further, a clarification is needed regarding the definition of an integral operator. The vector spaces X and Y are subspaces of $L_0(\mu)$ and $L_0(\nu)$, respectively. Hence, any operator $R\colon X \to Y$ is in fact an operator between equivalence classes of functions, and so for each equivalence class $[x] \in X$ the element $R[x] \in Y$ is likewise an equivalence class. The definition of the integral operator T tacitly assumes that if $x \in [x]$, then the function $y(t) = \int_{\mathcal{S}} T(s,t)x(s)\, d\mu(s)$ is a representative from the class $T[x]$, i.e., $T[x] = [y]$. In actuality, something stronger is happening here. Notice that we assumed that for a given function $x \in [x]$ there exists a ν-null set B such that the function $s \mapsto T(s,t)x(s)$ is μ-integrable for all $t \notin B$. Now if $x_1 \in [x]$, then it should be clear that for each $t \notin B$ the function $s \mapsto T(s,t)x_1(s)$ is also μ-integrable and

$$\int_{\mathcal{S}} T(s,t)x_1(s)\, d\mu(s) = \int_{\mathcal{S}} T(s,t)x(s)\, d\mu(s)\,.$$

That is, the function

$$y(t) = \int_{\mathcal{S}} T(s,t)x_1(s)\,d\mu(s)\,,\quad t \notin B\,,$$

is the same for any $x_1 \in [x]$. This not only explains why $T[x] = [y]$ but it also shows that we can identify the class $T[x]$ with the function y and write $y = Tx$. In sum, an integral operator $T\colon X \to Y$ gives rise automatically to an operator from equivalence classes to functions, i.e., it is a "lifting" or a "selector."

It should be emphasized that the ν-null set of points outside of which the equality $(\star)$ holds depends upon the function x; see Exercise 6 at the end of the section. The $\mu \times \nu$-measurable function $T(\cdot,\cdot)\colon \mathcal{S} \times \mathcal{T} \to \mathbb{R}$ is referred to as the **kernel** of the integral operator T.[1] It is a common practice to denote the kernel of an integral operator T by the same letter T, and we will follow this practice. This is very efficient and when the reader gets accustomed to this convention, it will cause no problem. We shall also say that a $\mu \times \nu$-measurable function $T(\cdot,\cdot)\colon \mathcal{S} \times \mathcal{T} \to \mathbb{R}$ **defines an integral operator** from X to Y if for each $x \in X$ the function $T(\cdot,t)x(\cdot)$ belongs to $L_1(\mu)$ for ν-almost all $t \in \mathcal{T}$ and the function Tx as given by $(\star)$ in Definition 5.1 belongs to Y.

We shall also adhere to the following agreement: An arbitrary point of $\mathcal{S}$ will be denoted exclusively by the letter s and an arbitrary point of $\mathcal{T}$ by the letter t. Accordingly, $f(s)$ indicates a function $f\colon \mathcal{S} \to \mathbb{R}$ and $g(t)$ denotes a function $g\colon \mathcal{T} \to \mathbb{R}$.

We assumed at the beginning of this section that E and F are ideals in $L_0(\mu)$ and $L_0(\nu)$, respectively. If we consider the supports $\mathcal{S}_1$ and $\mathcal{T}_1$ of E and F, respectively, then it is easy to see that an operator $T\colon E \to F$ is an integral operator if and only if there exists a function $K \in L_0(\mathcal{S}_1 \times \mathcal{T}_1, \mu \times \nu)$ such that for each $x \in E$ the function $s \mapsto K(s,t)x(s)$ belongs to $L_1(\mathcal{S}_1, \mu)$ for ν-almost all $t \in \mathcal{T}$ and the equality

$$Tx(t) = \int_{\mathcal{S}_1} K(s,t)x(s)\,d\mu(s)$$

holds for ν-almost all $t \in \mathcal{T}_1$. In other words, when we deal with integral operators, we can replace $\mathcal{S}$ by $\mathcal{S}_1$ and $\mathcal{T}$ by $\mathcal{T}_1$ and assume without loss of generality that:

- *E and F are order dense ideals in $L_0(\mu)$ and $L_0(\nu)$, respectively.*

Integral operators enjoy the following important continuity property.[2]

---

[1] For this reason, the integral operators are also known as **kernel operators**.

[2] As we shall see in Theorem 5.30, this property characterizes the integral operators.

Lemma 5.2. *Let $T\colon E \to L_0(\nu)$ be an integral operator. If an order bounded sequence $\{x_n\}$ in E satisfies $x_n \xrightarrow{\mu} 0$, then $Tx_n \xrightarrow{p} 0$.*

Proof. Assume $x_n \xrightarrow{\mu} 0$, and let $x \in E^+$ be an order bound for $\{|x_n|\}$. Also, let $T(\cdot,\cdot)$ be the kernel of the operator T. In view of Definition 5.1, we can find some $B \in \Sigma_2$ with $\nu(B) = 0$ such that for each $t \notin B$ and for each n the functions $T(\cdot,t)x(\cdot)$ and $T(\cdot,t)x_n(\cdot)$ belong to $L_1(\mu)$.

By Theorem 1.82 we know that $x_n \xrightarrow{\mu} 0$ if and only if every subsequence of $\{x_n\}$ has a subsequence that converges pointwise μ-a.e. to zero. This implies that for each $t \notin B$ every subsequence of the sequence $\{T(\cdot,t)x_n(\cdot)\}$ has a subsequence that converges pointwise μ-a.e. to zero. Since $|T(\cdot,t)x_n(\cdot)| \leq |T(\cdot,t)x(\cdot)|$ and $Tx_n(t) = \int_{\mathcal{S}} T(s,t)x_n(s)\,d\mu(s)$ holds for all $t \notin B$, the Lebesgue Dominated Convergence Theorem easily implies that $Tx_n(t) \to 0$ for ν-almost all $t \in \mathcal{T}$. ■

Since any sequence that converges pointwise to zero also converges in measure (see Corollary 1.83), the following is an immediate consequence of the preceding result.

Corollary 5.3. *Let $T\colon E \to L_0(\nu)$ be an integral operator. If $\{x_n\}$ is an order bounded sequence in E such that $x_n \xrightarrow{p} 0$, then $Tx_n \xrightarrow{p} 0$. In particular, by Corollary 1.89, the operator T is regular and order continuous.*

The next result is also a consequence of Corollary 1.89.

Corollary 5.4. *Every positive integral operator $T\colon E \to F$ is order continuous.*

As expected, the positive integral operators from E to F are precisely the ones with positive kernels.

Theorem 5.5. *Suppose that E and F are order dense ideals. An integral operator $T\colon E \to F$ is positive if and only if its kernel satisfies $T(s,t) \geq 0$ for $\mu \times \nu$-almost all $(s,t) \in \mathcal{S} \times \mathcal{T}$.*

Proof. Let $T\colon E \to F$ be an integral operator having the $\mu \times \nu$-measurable function $T(\cdot,\cdot)\colon \mathcal{S} \times \mathcal{T} \to \mathbb{R}$ as its kernel. Assume first that $T(s,t) \geq 0$ for all $(s,t) \notin A$, where A is $\mu \times \nu$-null subset of $\mathcal{S} \times \mathcal{T}$. Fix some $x \in E^+$ and then select a μ-null set $B \in \Sigma_1$ such that $x(s) \geq 0$ for all $s \notin B$. Next, choose a ν-null set $C \in \Sigma_2$ such that for all $t \notin C$ we have

$$Tx(t) = \int_{\mathcal{S}} T(s,t)x(s)\,d\mu(s)\,.$$

Put $D = A \cup (B \times \mathcal{T}) \cup (\mathcal{S} \times C)$, and note that $(\mu \times \nu)(D) = 0$. Now, notice that Fubini's Theorem 1.97 guarantees the existence of a ν-null set $C_1 \in \Sigma_2$ such that the set $D_t = \big\{s \in \mathcal{S}\colon\ (s,t) \in D\big\}$ satisfies $\mu^*(D_t) = 0$

for each $t \notin C_1$. Now assume that $t \notin C \cup C_1$ and $s \notin D_t$. Then $(s,t) \notin D$, and so $s \notin B$ (in which case we have $x(s) \geq 0$ and $T(s,t) \geq 0$), and $t \notin C$ (which guarantees that $Tx(t) = \int_{\mathcal{S}} T(s,t)x(s)\,d\mu(s)$). Thus, if t does not belong to the ν-null set $C \cup C_1$, then $T(s,t)x(s) \geq 0$ for each $s \notin D_t$, and so $\mu^*(D_t) = 0$ yields

$$Tx(t) = \int_{\mathcal{S}} T(s,t)x(s)\,d\mu(s) = \int_{\mathcal{S}\setminus D_t} T(s,t)x(s)\,d\mu(s) \geq 0\,.$$

This proves that T is a positive operator. Clearly, this proof remains valid for arbitrary subspaces E and F of $L_0(\mu)$ and $L_0(\nu)$, respectively—and also for arbitrary measures μ and ν.

For the converse, assume that T is a positive operator. We must show that $T(s,t) \geq 0$ for $\mu \times \nu$-almost all $(s,t) \in \mathcal{S} \times \mathcal{T}$. Since $\mathcal{S}$ and $\mathcal{T}$ are both σ-finite measure spaces, we can assume without loss of generality that both measures are finite. Moreover, since E and F are order dense ideals in $L_0(\mu)$ and $L_0(\nu)$, respectively, we can also assume in view of Corollary 1.91 that $\chi_{\mathcal{S}} \in E$ and $\chi_{\mathcal{T}} \in F$. Our last simplifying assumption is that the kernel $T(\cdot,\cdot)$ can be assumed to be $\mu \times \nu$-integrable. To see this, note that since the function $s \mapsto T(s,t) = T(s,t)\chi_{\mathcal{S}}(s)$ belongs to $L_1(\mu)$ for ν-almost all $t \in \mathcal{T}$, by deleting a ν-null set from $\mathcal{T}$ we can suppose that $\int_{\mathcal{S}} |T(s,t)|\,d\mu(s) < \infty$ for all $t \in \mathcal{T}$. So, we can write $\mathcal{T} = \bigcup_{k=1}^{\infty} B_k$, where $B_k = \big\{t \in \mathcal{T}\colon\ \int_{\mathcal{S}} |T(s,t)|\,d\mu(s) \leq k\big\}$. Since the ν-measurable function $g(t) = \int_{\mathcal{S}} |T(s,t)|\,d\mu(s)$ is bounded on B_k, we can conclude from Tonelli's theorem that $T(\cdot,\cdot)$ is integrable over $\mathcal{S} \times B_k$ for each k. Thus, we can assume that the function $T(\cdot,\cdot)$ is $\mu \times \nu$-integrable over $\mathcal{S} \times \mathcal{T}$.

To show that $T(\cdot,\cdot) \geq 0$, it suffices to verify (see Lemma 1.96) that

$$\int_{A\times B} T(s,t)\,d(\mu \times \nu) \geq 0$$

for each $A \in \Sigma_1$ and $B \in \Sigma_2$. Since $\chi_A \in E$ and T is a positive operator, we have $T\chi_A \geq 0$ and consequently $\int_B T\chi_A(t)\,d\nu(t) \geq 0$. By Definition 5.1 the equality

$$T\chi_A(t) = \int_{\mathcal{S}} T(s,t)\chi_A(s)\,d\mu(s) = \int_A T(s,t)\,d\mu(s)$$

holds for ν-almost all t. Therefore, using Fubini's theorem, we obtain

$$\int_{A\times B} T\,d(\mu \times \nu) = \int_B \Big[\int_A T(s,t)\,d\mu(s)\Big]\,d\nu(t) = \int_B T\chi_A(t)\,d\nu(t) \geq 0\,,$$

as desired. ■

Recall that the carrier C_E of an ideal E in $L_0(\mu)$ is the "smallest" μ-measurable set outside of which all functions of E vanish; see Definition 1.93.

Corollary 5.6. *An integral operator $T\colon E \to F$ is positive if and only if its kernel satisfies $T(s,t) \geq 0$ for $\mu \times \nu$-almost all $(s,t) \in C_E \times C_F$.*

Corollary 5.7. *Two integral operators $S,T\colon E \to F$ satisfy $S \geq T$ if and only if $S(s,t) \geq T(s,t)$ for $\mu \times \nu$-almost all $(s,t) \in C_E \times C_F$.*

Another important consequence of Theorem 5.5 is that the kernel of an integral operator is uniquely determined on $C_E \times C_F$.

Corollary 5.8. *Two integral operators $S,T\colon E \to F$ coincide (i.e., $S = T$) if and only if $S(s,t) = T(s,t)$ for $\mu \times \nu$-almost all $(s,t) \in C_E \times C_F$.*

Positive operators that are dominated by integral operators are necessarily integral operators. This basic result is due to A. V. Bukhvalov [**77**] and A. R. Schep [**294**].

Theorem 5.9 (Bukhvalov–Schep)**.** *Every positive operator from E to F that is dominated by an integral operator is itself an integral operator.*

Proof. The proof below—based on the Radon–Nikodým Theorem—is due to A. R. Schep [**294**].

Assume that two operators $S,T\colon E \to F$ satisfy $0 \leq S \leq T$ and that T is an integral operator with kernel $T(\cdot,\cdot)$. Let

$$\Gamma = \big\{A \times B\colon\ \chi_A \in E,\ \ \chi_B \in F,\ \text{ and } T(\cdot,\cdot) \text{ is integrable over } A \times B\big\}\,,$$

and note that Γ is a semiring of subsets of $\mathcal{S} \times \mathcal{T}$. From Corollary 1.91, it follows that the σ-algebra generated by Γ is the product σ-algebra $\Sigma_\mu \otimes \Sigma_\nu$, where for simplicity we write Σ_μ instead of $(\Sigma_1)_\mu$ and Σ_ν instead of $(\Sigma_2)_\nu$.

Using the hypothesis that T is an integral operator, we can define the set function $\pi_T\colon \Gamma \to [0,\infty)$ via the formula

$$\begin{aligned}\pi_T(A \times B) &= \int_{A\times B} T d(\mu \times \nu) = \int_B \Big[\int_A T(s,t)\, d\mu(s)\Big]\, d\nu(t)\\ &= \int_B T\chi_A(t)\, d\nu(t)\,.\end{aligned}$$

A direct verification shows that π_T is a measure on the semiring Γ.

Next, using the operator S, we define a new set function $\pi_S\colon \Gamma \to [0,\infty)$ via the formula

$$\pi_S(A \times B) = \int_B S\chi_A(t)\, d\nu(t) = \int_{\mathcal{T}} \chi_B(t) S\chi_A(t)\, d\nu(t)\,.$$

Clearly, $0 \leq \pi_S \leq \pi_T$. We claim that π_S is a measure. To verify this, it suffices to show (since π_S is dominated by the measure π_T) that π_S is finitely additive; see Exercise 15 in Section 1.6.

To this end, assume that $A\times B=\bigcup_{i=1}^n A_i\times B_i$ is a disjoint union, where $A\times B$ and all the $A_i\times B_i$ are non-empty and lying in Γ. Fix $\tau\in B$ and let $N_\tau=\{i\in\{1,\dots,n\}\colon\ \tau\in B_i\}$. Then for each fixed $\tau\in B$ we have

$$\chi_A(s)=\chi_A(s)\chi_B(\tau)=\sum_{i=1}^n\chi_{A_i}(s)\chi_{B_i}(\tau)=\sum_{i\in N_\tau}\chi_{A_i}(s)$$

for all $s\in\mathcal{S}$. The linearity of S implies that for each $\tau\in B$ we have

$$(S\chi_A)(t)=\chi_B(\tau)(S\chi_A)(t)=\sum_{i=1}^n\chi_{B_i}(\tau)(S\chi_{A_i})(t)=\sum_{i\in N_\tau}(S\chi_{A_i})(t)\qquad(\star)$$

for ν-almost all $t\in\mathcal{T}$. Since the collection $\{N_\tau\colon\ \tau\in B\}$ is finite, there exists a ν-null set C such that $\chi_B(\tau)(S\chi_A)(t)=\sum_{i=1}^n\chi_{B_i}(\tau)(S\chi_{A_i})(t)$ is valid for each $\tau\in\mathcal{T}$ and all $t\notin C$. But then for all $t\notin C$ we have $\chi_B(t)(S\chi_A)(t)=\sum_{i=1}^n\chi_{B_i}(t)(S\chi_{A_i})(t)$. This implies

$$\begin{aligned}\pi_S(A\times B)&=\int_{\mathcal{T}}\chi_B(t)(S\chi_A)(t)\,d\mu(t)=\sum_{i=1}^n\int_{\mathcal{T}}\chi_{B_i}(t)(S\chi_{A_i})(t)\,d\nu(t)\\&=\sum_{i=1}^n\pi_S(A_i\times B_i)\,.\end{aligned}$$

Therefore, π_S is indeed finitely additive—and hence it is a measure. A straightforward verification shows that the outer measures π_S^* and π_T^* satisfy $0\le\pi_S^*\le\pi_T^*$ and that, when restricted to $\Sigma_\mu\otimes\Sigma_\nu$, they are both σ-finite.

Next, observe that the measure π_T^* restricted to the σ-algebra $\Sigma_\mu\otimes\Sigma_\nu$ is absolutely continuous with respect to the σ-finite measure $\mu\times\nu$. From $\pi_S^*\le\pi_T^*$, it follows that π_S^* is also absolutely continuous with respect to $\mu\times\nu$. But then, by the classical Radon–Nikodým theorem, there exists a $\Sigma_\mu\otimes\Sigma_\nu$-measurable function $S(\cdot,\cdot)\colon\mathcal{S}\times\mathcal{T}\to[0,\infty)$ (and hence $S(\cdot,\cdot)$ is also $\mu\times\nu$-measurable) such that $\pi_S^*(U)=\int_U S\,d(\mu\times\nu)$ holds for each $U\in\Sigma_\mu\otimes\Sigma_\nu$. In particular, for each $A\times B\in\Gamma$ we have

$$\begin{aligned}\pi_S(A\times B)&=\int_B(S\chi_A)(t)\,d\nu(t)=\int_{A\times B}S(s,t)\,d(\mu\times\nu)\\&=\int_B\Big[\int_A S(s,t)\,d\mu(s)\Big]\,d\nu(t)\,.\end{aligned}$$

Now fix some $A\in\Sigma_\mu$ with $\mu^*(A)<\infty$ and consider the collection

$$\Lambda=\{B\in\Sigma_\nu\colon\ T(\cdot,\cdot)\text{ is integrable over }A\times B\}\,.$$

Clearly, Λ is a semiring such that the σ-algebra generated by this semiring coincides with Σ_ν; see the proof of Theorem 5.5. From this and Lemma 1.96, it follows that for each $A \in \Sigma_\mu$ with $\mu^*(A) < \infty$ we have

$$S\chi_A(t) = \int_A S(s,t)\,d\mu(s) = \int_S S(s,t)\chi_A(s)\,d\mu(s) \tag{$\star\star$}$$

for ν-almost all $t \in \mathcal{T}$.

To finish the proof, note first that the operator S is order continuous. (This follows from the inequality $0 \le S \le T$ and the fact that T is order continuous by Corollary 5.4.) This and $(\star\star)$ imply immediately that for each $x \in E$ we have $Sx(t) = \int_S S(s,t)x(s)\,d\mu(s)$ for ν-almost all $t \in \mathcal{T}$. That is, S is an integral operator, as desired. ■

From the definition of an integral operator $T\colon E \to F$ with kernel $T(\cdot,\cdot)$, it follows that for each $x \in E$ the function $|T(\cdot,t)|x(\cdot)$ is μ-integrable for ν-almost all t, and therefore we can define (ν-almost everywhere) the function $t \mapsto \int_S |T(s,t)|x(s)\,d\mu(s)$. It is easy to verify that this function is ν-measurable; see Exercise 2 at the end of the section. In other words, every integral operator $T\colon E \to F$ defines automatically via the formula

$$M_T x(t) = \int_S |T(s,t)|x(s)\,d\mu(s) \tag{$\star\star\star$}$$

a new positive integral operator $M_T\colon E \to L_0(\nu)$ with kernel $|T(\cdot,\cdot)|$. It should be emphasized that in general this new integral operator need not map E into F. We will specify below under what conditions $M_T(E) \subseteq F$ and show that in this case M_T coincides with the familiar modulus $|T|$ of the operator T introduced in Section 1.2.

If we consider T as an operator acting from E to $L_0(\nu)$, then it should be clear that $T \le M_T$, and therefore $T\colon E \to L_0(\nu)$ is a regular operator. Since $L_0(\nu)$ is Dedekind complete, T has a modulus given by the familiar Riesz–Kantorovich formula

$$|T|x = \sup\{Tz\colon\; z \in E \;\text{ and }\; |z| \le x\}$$

for each $x \in E^+$ (see Theorem 1.16). As expected, this formula for the modulus of T and the one given by $(\star\star\star)$ coincide.

Theorem 5.10. *If $T\colon E \to F$ is an integral operator, then T as an operator from E to $L_0(\nu)$ is a regular operator and its modulus $|T|\colon E \to L_0(\nu)$ for each $x \in E$ satisfies*

$$|T|x(t) = \int_S |T(s,t)|x(s)\,d\mu(s)$$

for ν-almost all t. That is, $|T| = M_T$.

Proof. As mentioned before, M_T is a positive integral operator satisfying $\pm T \le M_T$. Therefore $|T| \le M_T$. Assume that another positive operator $S\colon E \to L_0(\nu)$ satisfies $\pm T \le S$. Consider the operator $R = M_T \wedge S$. By

Theorem 5.9, the operator R is an integral operator as $0 \le R \le M_T$. Also $\pm T \le R$ and therefore, in view of Corollary 5.7, the kernel $R(\cdot,\cdot)$ of the operator R satisfies $\pm T(s,t) \le R(s,t) \le |T(s,t)|$ for $\mu \times \nu$-almost all (s,t) in $\mathcal{S}\times\mathcal{T}$. This implies that $R(s,t) = T(s,t)$ for $\mu\times\nu$-almost all $(s,t) \in \mathcal{S}\times\mathcal{T}$ or $M_T \wedge S = M_T$. That is, $M_T \le S$, which establishes that $|T| = M_T$. ■

It turns out that the regularity of an integral operator T is precisely the property needed to guarantee that M_T maps E into F. This important result is due to W. A. J. Luxemburg and A. C. Zaanen [**226**].

Theorem 5.11 (Luxemburg–Zaanen). *An integral operator $T\colon E \to F$ is regular if and only if the operator M_T defined by $(\star\star\star)$ has its range in F. In this case M_T coincides with the usual modulus $|T|$ of T in $\mathcal{L}_r(E,F)$.*[3]

Proof. Assume first that T is a regular operator. Since F, as an ideal in $L_0(\nu)$, is a Dedekind complete Riesz space, the modulus $|T|\colon E \to F$ exists in $\mathcal{L}_r(E,F)$ and is given by the Riesz–Kantorovich formula. That is, the equality

$$|T|x = \sup\{Tz\colon\ z \in E \text{ and } |z| \le x\}$$

holds in F for each $x \in E^+$. Moreover, using again that F is an ideal in $L_0(\nu)$, we see that the above supremum remains the same if it is taken in $L_0(\nu)$. That is, the modulus of T calculated in $\mathcal{L}_r(E,F)$ or in $\mathcal{L}_r(E,L_0(\nu))$ is the same. But, by Theorem 5.10, the modulus of T in $\mathcal{L}_r\big(E,L_0(\nu)\big)$ coincides with M_T. Consequently, $M_T(E) \subseteq F$.

For the converse, suppose that the operator M_T maps E to F. Since $T \le M_T$ and M_T is positive, it follows that T is a regular operator, and the proof is finished. ■

From now on, regardless of whether an integral operator $T\colon E \to F$ is regular or not, we will always use the symbol $|T|$ rather than M_T and we will refer to $|T|$ as the **modulus** of the integral operator. It will always be clear from the context between which spaces the modulus acts, and so this will cause no ambiguity.

Corollary 5.12. *Every regular integral operator $T\colon E \to F$ is order continuous. Without the assumption of regularity, the integral operator T is order continuous as an operator from E to $L_0(\nu)$.*

Proof. Let $T\colon E \to F$ be a regular integral operator. Then, by Theorem 5.11, its modulus $|T|\colon E \to F$ is also an integral operator, and so by Corollary 5.4 the operator $|T|$ is order continuous. This implies that T itself is also order continuous.

[3] Due to this, the regular integral operators are also known as **absolute integral operators**.

The last statement of the corollary follows from the previous part since each integral operator $T\colon E \to F$ considered as an operator from E to $L_0(\nu)$ is regular. ∎

Let $T\colon E \to F$ be an integral operator. As stated in Theorem 5.11, the modulus $|T|$ of T maps E to F if and only if T is a regular operator. By Corollary 5.3, we know also that T, as an operator from E to $L_0(\nu)$, is always regular. Below we present an example of an integral operator from E to F whose modulus does not map E to F.

Example 5.13. Consider $[0,1]$ equipped with its Lebesgue measure and let $\mathcal{S} = \mathcal{T} = [0,1]$. Also, for each n let $\Delta_n = \big[1 - \frac{1}{n}, 1 - \frac{1}{n+1}\big)$. The length of Δ_n will be denoted by c_n. For our ideals we take $E = L_1(\mathcal{S})$ and

$$F = \Big\{ y \in L_0(\mathcal{T})\colon\; y\chi_{\Delta_n} \in L_1(\mathcal{T}) \;\text{ and }\; \tfrac{1}{c_n}\int_{\Delta_n} |y(t)|\,dt \longrightarrow 0 \Big\}.$$

The kernel $T(\cdot,\cdot)$ on $\mathcal{S}\times\mathcal{T}$ is defined as follows. For each $s \in [0,1]$ and for each $t \in \Delta_n$, where $n = 1, 2, \dots$, we let

$$T(s,t) = \sin(2\pi n s)\,.$$

It is obvious that $T(\cdot,\cdot)$ is a bounded $\mu\times\nu$-measurable function on $\mathcal{S}\times\mathcal{T}$.

First, we verify that the integral operator T generated by this kernel maps E to F. To see this, note that for each $x \in E$ and each $t \in \Delta_n$ we have $Tx(t) = \int_{\mathcal{S}} x(s)\sin(2\pi n s)\,ds$, and so the Riemann–Lebesgue lemma yields

$$\tfrac{1}{c_n}\int_{\mathcal{T}} |Tx(t)|\chi_{\Delta_n}\,dt = \Big|\int_{\mathcal{S}} x(s)\sin(2\pi n s)\,ds\Big| \longrightarrow 0\,.$$

Thus, $y = Tx \in F$.

Next, we claim that the integral operator $|T|$ with kernel $|T(\cdot,\cdot)|$ does not map E to F. To this end, it suffices to show that the function $v = |T|(\chi_{\mathcal{S}})$ does not belong to F. Indeed, for each $t \in \Delta_n$ we have

$$v(t) = \int_{\mathcal{S}} |\sin(2\pi n s)|\,ds \geq \int_{\mathcal{S}} \sin^2(2\pi n s)\,ds \geq \tfrac{1}{2}\,,$$

and hence $\frac{1}{c_n}\|v\chi_{\Delta_n}\|_1 \geq \frac{1}{2}$. This shows that $v = |T|(\chi_{\mathcal{S}}) \notin F$.

If we equip F with the lattice norm $\|y\| = \sup_n \frac{1}{c_n}\|y\chi_{\Delta_n}\|_1$, then both E and F are Banach lattices with order continuous norms. That is, $|T|$ may fail to map E to F even if both E and E are Banach lattices with order continuous norms. ∎

The vector space of all regular integral operators from E to F will be denoted by $\mathcal{L}_\kappa(E,F)$. Its order structure is described by the following result.

Theorem 5.14 (Bukhvalov–Luxemburg–Schep–Zaanen). *The vector space $\mathcal{L}_\kappa(E,F)$ of all regular integral operators from E to F is a band in $\mathcal{L}_r(E,F)$. Moreover, for each $S,T \in \mathcal{L}_\kappa(E,F)$ and for each $x \in E$ we have*

$$\begin{aligned}\big[(S\vee T)x\big](t) &= \int_{\mathcal{S}} \max\{S(s,t),T(s,t)\}x(s)\,d\mu(s)\,, \quad \textit{and}\\ \big[(S\wedge T)x\big](t) &= \int_{\mathcal{S}} \min\{S(s,t),T(s,t)\}x(s)\,d\mu(s)\end{aligned}$$

for ν-almost all $t \in \mathcal{T}$.

Proof. We can assume without loss of generality that E and F are order dense in $L_0(\mu)$ and $L_0(\nu)$, respectively. Theorems 5.9 and 5.11 guarantee that $\mathcal{L}_\kappa(E,F)$ is an ideal in $\mathcal{L}_r(E,F)$ and that the above formulas for calculating $S\vee T$ and $S\wedge T$ are true. The only thing left to be shown is that $\mathcal{L}_\kappa(E,F)$ is order closed in $\mathcal{L}_r(E,F)$.

To see this, assume that a net $\{T_\alpha\}$ of positive integral operators satisfies $T_\alpha \uparrow S$ in $\mathcal{L}_r(E,F)$. By Corollary 1.91, we can assume without loss of generality that $\mathbf{1}_{\mathcal{S}} \in E$, and so $T_\alpha \mathbf{1}_{\mathcal{S}} \uparrow S\mathbf{1}_{\mathcal{S}}$ in F. Since F has the countable sup property, there is an increasing sequence $\{\alpha_n\}$ of indices such that $T_{\alpha_n}\mathbf{1}_{\mathcal{S}} \uparrow S\mathbf{1}_{\mathcal{S}}$, that is, $\int_{\mathcal{S}} T_{\alpha_n}(s,t)\,d\mu(s) \uparrow S\mathbf{1}_{\mathcal{S}}(t)$ for ν-almost all $t \in \mathcal{T}$. By Theorem 5.5, the kernels $T_{\alpha_n}(s,t)$ satisfy $T_{\alpha_n}(s,t)\uparrow$. This implies that there exists a function $T(\cdot,\cdot)$ such that $T_{\alpha_n}(s,t) \uparrow T(s,t)$ for $\mu\times\nu$-almost all $(s,t) \in \mathcal{S}\times\mathcal{T}$. Without loss of generality, we can suppose that $T_{\alpha_n}(s,t) \uparrow T(s,t)$ for each $(s,t) \in \mathcal{S}\times\mathcal{T}$. Next, we claim that $T(s,t)$ is finite for $\mu\times\nu$-almost all $(s,t) \in \mathcal{S}\times\mathcal{T}$.

To establish this, consider the $\mu\times\nu$-measurable set

$$D = \big\{(s,t) \in \mathcal{S}\times\mathcal{T}\colon\ T(s,t) = \infty\big\}\,,$$

and assume by way of contradiction that $(\mu\times\nu)^*(D) > 0$. Since $\mu\times\nu$ is a σ-finite measure, we can also suppose that $(\mu\times\nu)^*(D) < \infty$. But then, from Fubini's theorem, there exists a ν-measurable subset B of $\mathcal{T}$ with $\nu(B) > 0$ such that for each $t \in B$ the set $A_t = \{s \in \mathcal{S}\colon\ (s,t) \in D\}$ is μ-measurable and satisfies $\mu(A_t) > 0$. Therefore, for each $t \in B$ we have

$$(S\mathbf{1}_{\mathcal{S}})(t) \geq \lim_{n\to\infty} \int_{\mathcal{S}} T_{\alpha_n}(s,t)\chi_{A_t}(s)\,d\mu(s) = \int_{A_t} T(s,t)\,d\mu(s) = \infty\,.$$

This shows that $S\mathbf{1}_{\mathcal{S}} \notin F$, which is impossible. Consequently, the function $T(\cdot,\cdot)$ belongs to $L_0(\mu\times\nu)$.

Now if $x \in E^+$, then $T_{\alpha_n}(s,t)x(s) \uparrow T(s,t)x(s)$ for $\mu\times\nu$-almost all (s,t), and so for ν-almost all $t \in \mathcal{T}$ we have

$$Sx(t) = \lim_{n\to\infty} T_{\alpha_n}x(t) = \lim_{n\to\infty} \int_{\mathcal{S}} T_{\alpha_n}(s,t)x(s)\,d\mu(s) = \int_{\mathcal{S}} T(s,t)x(s)\,d\mu(s)\,.$$

This shows that S is indeed an integral operator, and so $\mathcal{L}_\kappa(E,F)$ is a band in the Dedekind complete Riesz space $\mathcal{L}_r(E,F)$. ■

We conclude the section by providing a convenient criterion for an operator to be an integral operator.

Theorem 5.15. *Let $T\colon E \to F$ be an operator such that $T\colon E \to L_0(\nu)$ is σ-order continuous. If the restriction of T to some order dense ideal in E is an integral operator, then $T\colon E \to F$ itself is an integral operator.*

Proof. Start by observing that the operator $T\colon E \to F$ is an integral operator if and only if $T\colon E \to L_0(\nu)$ is an integral operator. Therefore, without loss of generality, we can suppose that $F = L_0(\nu)$. We can also assume that E is order dense in $L_0(\mu)$.

Suppose first that T is a positive operator and that T is an integral operator on an order dense ideal E_0 in E. Denote by $T_0(\cdot,\cdot)$ the kernel of the operator $T_0 = T|_{E_0}$. By Theorem 5.5, the kernel $T_0(\cdot,\cdot)$ is a non-negative $\mu\times\nu$-measurable function. Consequently, we can assume that $T_0(s,t) \geq 0$ holds for each $(s,t) \in \mathcal{S}\times\mathcal{T}$. We shall finish the proof be proving that T is an integral operator with the same kernel. That is, according to Definition 5.1, we shall show that for each $x \in E$ the function $s \mapsto T_0(s,t)x(s)$ belongs to $L_1(\mu)$ for ν-almost all $t \in \mathcal{T}$ and that the equality

$$Tx(t) = \int_{\mathcal{S}} T_0(s,t)x(s)\,d\mu(s)$$

holds for ν-almost all $t \in \mathcal{T}$.

To this end, it suffices to consider an arbitrary non-negative function $x \in E$. So, fix a function $x \in E^+$ and then pick a sequence $\{x_n\}$ in E_0 satisfying $0 \leq x_n \uparrow x$. Hence, $Tx_n \uparrow Tx$ since T is σ-order continuous. In particular, $Tx_n(t) \uparrow Tx(t)$ for ν-almost all $t \in \mathcal{T}$. Let D be the set of points at which there is no convergence or where $Tx(t) = \infty$. Thus, $\nu(D) = 0$.

Since $x_n \in E_0$, there exists a ν-null set B_n such that for each $t \notin B_n$:

(1) The function $s \mapsto T_0(s,t)x_n(s)$ belongs to $L_1(\mu)$.
(2) $Tx_n(t) = \int_{\mathcal{S}} T_0(s,t)x_n(s)\,d\mu(s)$.

Let $B = D \cup \bigcup_{n=1}^\infty B_n$, and observe that $\nu(B) = 0$. For each $t \notin B$ we have $T_0(s,t)x_n(s) \uparrow T_0(s,t)x(s)$ for μ-almost all $s \in \mathcal{S}$, and

$$\int_{\mathcal{S}} T_0(s,t)x_n(s)\,d\mu(s) = Tx_n(t) \uparrow Tx(t) < \infty\,. \qquad (\star)$$

So, by B. Levi's theorem, $\int_{\mathcal{S}} T_0(s,t)x(s)\,d\mu(s) \leq Tx(t) < \infty$, that is, the function $s \mapsto T_0(s,t)x(s)$ belongs to $L_1(\mu)$ for ν-almost all $t \in \mathcal{T}$, as required.

Furthermore, using once more that $T_0(s,t)x_n(s) \uparrow T_0(s,t)x(s)$, we see that

$$\int_{\mathcal{S}} T_0(s,t)x_n(s)\,d\mu(s) \uparrow \int_{\mathcal{S}} T_0(s,t)x(s)\,d\mu(s)\,.$$

The latter and ($\star$) yield $Tx(t) = \int_{\mathcal{S}} T_0(s,t)x(s)\,d\mu(s)$ for each $t \notin B$. That is, T is indeed a kernel operator with the same kernel as T_0.

To show that the assumption that T is a positive operator is not essential, we proceed as follows. Since $T\colon E \to L_0(\nu)$ is σ-order continuous, T is regular (see Corollary 1.89(a)), and hence the operator $T^+\colon E \to L_0(\nu)$ exists and is σ-order continuous. By Theorem 5.10, the integral operator $T_0\colon E_0 \to L_0(\nu)$ is regular and $T_0^+\colon E_0 \to L_0(\nu)$ is also an integral operator with kernel $T_0^+(\cdot,\cdot)$. It remains to note that the restriction of T^+ to E_0 coincides with T_0^+. Thus, instead of T we can consider the positive operators T^+ and T^-, and the desired conclusion follows. ■

Exercises

1. Assume that μ and ν are σ-finite measures. Show in detail that if the kernels of two integral operators $S,T\colon E \to F$ satisfy $S(s,t) = T(s,t)$ for $\mu \times \nu$-almost all $(s,t) \in \mathcal{S} \times \mathcal{T}$, then $Sx = Tx$ for each $x \in E$.

2. Let $(\mathcal{S}, \Sigma_1, \mu)$ and $(\mathcal{T}, \Sigma_2, \nu)$ be two σ-finite measure spaces, and consider a $\mu \times \nu$-measurable function $K\colon \mathcal{S} \times \mathcal{T} \to \mathbb{R}$. Assume that a function $f \in L_0(\mu)$ satisfies $K(\cdot,t)f(\cdot) \in L_1(\mu)$ for ν-almost all $t \in \mathcal{T}$. Show that the function $g\colon \mathcal{T} \to \mathbb{R}$ defined ν-almost everywhere by the formula

$$g(t) = \int_{\mathcal{S}} K(s,t)f(s)\,d\mu(s)$$

is ν-measurable, i.e., $g \in L_0(\nu)$. [HINT: By Lemma 1.90 there exists a sequence $\{K_n\}$ of $\mu \times \nu$-integrable functions satisfying $|K_n(s,t)| \le |K(s,t)|$ for $\mu \times \nu$-almost all $(s,t) \in \mathcal{S} \times \mathcal{T}$, and $K_n(s,t) \to K(s,t)$ for $\mu \times \nu$-almost all (s,t). Since each K_n is $\mu \times \nu$-integrable, Fubini's Theorem 1.97 guarantees that the formula $g_n(t) = \int_{\mathcal{S}} K_n(s,t)f(s)\,d\mu(s)$ defines a ν-integrable function. Now a straightforward application of the Lebesgue Dominated Convergence Theorem yields $g_n(t) \to g(t)$ for ν-almost all $t \in \mathcal{T}$.]

3. Let $T(\cdot,\cdot)$ be a $\mu \times \nu$-measurable function such that:
   (a) $\int_{\mathcal{S}} |T(s,t)|\,d\mu(s) \le 1$ for ν-almost all $t \in \mathcal{T}$.
   (b) $\int_{\mathcal{T}} |T(s,t)|\,d\nu(t) \le 1$ for μ-almost all $s \in \mathcal{S}$.
   Prove that the integral operator T with kernel $T(\cdot,\cdot)$ is a bounded operator from $L_2(\mu)$ to $L_2(\nu)$.

4. ([**78**, Section 4.1]) Let $T\colon L_2(\mu) \to L_2(\nu)$ be an integral operator. Assume that there exists a subset $B \in \Sigma_2$ such that $\nu(B) = 0$ and for each function $x \in L_2(\mu)$ we have $\int_{\mathcal{S}} |T(s,t)x(s)|\,d\mu(s) < \infty$ for each $t \notin B$. That is, the

exceptional set in the definition of the integral operator T is independent of the "input" function x.

Prove that $\int_{\mathcal{S}} |T(s,t)|^2 \, d\mu(s) < \infty$ for ν-almost all $t \in \mathcal{T}$. (Operators that satisfy this condition are known as **Carleman operators**.)

5. Assume that $(\mathcal{S}, \Sigma_1, \mu)$ and $(\mathcal{T}, \Sigma_2, \nu)$ are two σ-finite measure spaces, and let $1 < p, q < \infty$ satisfy $\frac{1}{p} + \frac{1}{q} = 1$. Show that an integral operator $T\colon L_p(\mu) \to L_0(\nu)$ maps the closed unit ball of $L_p(\mu)$ to an order bounded subset of $L_0(\nu)$ (i.e., there exists a function $g \in L_0(\nu)$ such that $|Tx| \le g$ for each $x \in L_p(\mu)$ with $\|x\|_p \le 1$) if and only if there exists a ν-null set $B \in \Sigma_2$ such that

$$\int_{\mathcal{S}} |T(s,t)|^q \, d\mu(s) < \infty$$

for each $t \in \mathcal{T} \setminus B$.

In particular, an integral operator $T\colon L_2(\mu) \to L_2(\nu)$ is a Carleman operator if and only if T maps the closed unit ball of $L_2(\mu)$ into an order bounded subset of $L_0(\nu)$.

6. Give an example of an integral operator on $L_2[0,1]$ which is not a Carleman operator—where $[0,1]$ is equipped with its Lebesgue measure. In view of Exercise 4, this demonstrates that in general the exceptional set in the definition of an integral operator depends on the function the operator is acting upon. [HINT: Use Exercise 3 above.]

7. Let $\{T_n\}$ be an increasing sequence of regular integral operators and let T be another regular integral operator. Show that $T_n \uparrow T$ holds in $\mathcal{L}_r(E,F)$ if and only if $T_n(s,t) \uparrow T(s,t)$ for $\mu \times \nu$-almost all (s,t) in $C_E \times C_F$.

8. Let E be an ideal in $L_0(\mu)$, where $(\mathcal{S}, \Sigma, \mu)$ is a non-atomic σ-finite measure space. If $T\colon E \to E$ is a regular integral operator, then show that T is disjoint from the identity operator, i.e., $|T| \wedge I = 0$.

9. (Abramovich–Aliprantis–Zame [**20**]) Show that for a Dedekind complete AM-space E with unit e the following two statements are equivalent.

(a) There exists a probability measure space (Ω, Σ, π) and a surjective lattice isometry $T\colon E \to L_\infty(\pi)$ with $Te = \mathbf{1}$.

(b) E admits a strictly positive order continuous linear functional.

[HINT: Assume that $\phi\colon E \to \mathbb{R}$ is a strictly positive order continuous linear functional. We can assume that $\phi(e) = 1$. Clearly, the formula $|||x||| = \phi(|x|)$ defines an L-norm on E. If $\hat{E}$ is the norm completion of $(E, |||\cdot|||)$, then $\hat{E}$ is an AL-space (see Exercise 6 of Section 3.4) having e as a weak unit. Now, according to the Kakutani–Bohnenblust–Nakano Representation Theorem 3.5, $\hat{E}$ is lattice isometric to some $L_1(\pi)$-space, where e corresponds to the constant function $\mathbf{1}$. Show that π is a probability measure and that E is lattice isometric to $L_\infty(\pi)$.]

10. Let $(\mathcal{S}, \Sigma_1, \mu)$ and $(\mathcal{T}, \Sigma_2, \nu)$ be two arbitrary measure spaces, and assume that $1 < p, q < \infty$ satisfy $\frac{1}{p} + \frac{1}{q} = 1$. If a function $T(\cdot,\cdot) \in L_p(\mu \times \nu)$, then show that the formula

$$Tx(t) = \int_{\mathcal{S}} T(s,t)x(s) \, d\mu(s)$$

defines a regular integral operator from $L_q(\mu)$ to $L_p(\nu)$ such that its norm satisfies $\|T\| \le \|T(\cdot,\cdot)\|_p$. (When $p = q = 2$, such an operator T is called a **Hilbert–Schmidt operator**.)

11. Show that each Hilbert–Schmidt operator is a Carleman operator.

12. Assume that $(\mathcal{S}, \Sigma_1, \mu)$ are $(\mathcal{T}, \Sigma_2, \nu)$ are two σ-finite measure spaces, and assume that $(\mathcal{S}, \Sigma_1, \mu)$ is **atomic**—which in this case means that $\mathcal{S}$ is a union of at most countably many atoms.

Show that a linear operator $T \colon E \to L_0(\nu)$, where E is an ideal in $L_0(\mu)$, is an integral operator if and only if T is order continuous.

5.2. Abstract Integral Operators

This section is a continuation of the previous one. The main objective here is to present two important characterizations of integral operators (Theorems 5.28 and 5.30). Also an integral representation of order continuous linear functionals (Theorem 5.26) will be obtained that is very useful in applications.

Recall that if L is a Riesz space, then $L_n^\sim$ denotes the **order continuous dual** of L, i.e., the band in $L^\sim$ consisting of all order continuous linear functionals. If L is a normed Riesz space, then we shall denote by L_n^* the ideal in L^* of all order continuous and norm bounded linear functionals on L, that is, $L_n^* = L^* \cap L_n^\sim$. If L is a Banach lattice, then of course $L^* = L^\sim$ and $L_n^* = L_n^\sim$. Riesz spaces with "small" order continuous duals should be viewed as "pathological" objects from the point of view of the theory of Riesz spaces.

Definition 5.16. *A Riesz space L is said to be a* ***normal Riesz space*** *if its order continuous dual $L_n^\sim$ separates the points of L, that is, if for each non-zero $x \in L$ there exists some $\phi \in L_n^\sim$ satisfying $\phi(x) \neq 0$.*

Let L and M be two Riesz spaces with M Dedekind complete. If Φ is a vector subspace of $L^\sim$, then the **Φ-integral operators** from L to M are the operators in the band generated in $\mathcal{L}_r(L, M)$ by $\Phi \otimes M$. The $L_n^\sim$-integral operators (i.e., when $\Phi = L_n^\sim$) are called abstract integral operators. They were introduced by G. Ya. Lozanovsky in [**219**] under the name "almost integral operators." Here is their exact definition.

Definition 5.17 (Lozanovsky). *Let L and M be two Riesz spaces with M Dedekind complete. An order bounded operator $T \colon L \to M$ is said to be an* ***abstract integral operator*** *(or an* ***almost integral operator****) if it is an $L_n^\sim$-integral operator, i.e., if T belongs to the band $(L_n^\sim \otimes M)^{dd}$ generated by $L_n^\sim \otimes M$ in $\mathcal{L}_r(L, M)$. The band $(L_n^\sim \otimes M)^{dd}$ is called the* ***band of abstract integral operators****.*

Thus, a positive operator $T: L \to M$ is an abstract integral operator if and only if there exists a net $\{T_\alpha\}$ of positive operators such that $T_\alpha \uparrow T$ in $\mathcal{L}_r(L, M)$ and for each α there exist $0 \le \phi_\alpha \in L_n^\sim$ and $0 \le v_\alpha \in M$ satisfying $T_\alpha \le \phi_\alpha \otimes v_\alpha$. Non-trivial abstract integral operators exist provided that $L_n^\sim \ne \{0\}$. For instance, when L is a normal Riesz space, then there are many abstract integral operators.

From the preceding definition and Theorem 1.29, the following result should be immediate.

Lemma 5.18. *Every abstract integral operator is order continuous.*

As in the previous section, (Ω, Σ, π) denotes our "generic" measure space, $(\mathcal{S}, \Sigma_1, \mu)$ and $(\mathcal{T}, \Sigma_2, \nu)$ denote σ-finite measure spaces. A **function space** E is simply an ideal in some $L_0(\pi)$.[4] In this case we shall also say that E is **associated with** $L_0(\pi)$ or with the measure π.

Definition 5.19. *A function space is said to be:*

(1) *A* ***normed function space*** *if G carries a lattice norm. If G is complete in this norm, then G is a* ***Banach function space***.

(2) *A* ***normal function space*** *if G is normal, i.e., if $G_n^\sim$ separates the points of G.*

Not every function space is normal. For instance, we have the following result that is essentially due to M. M. Day [**101**].

Lemma 5.20 (Day). *If (Ω, Σ, π) is a non-atomic σ-finite measure space, then for each $0 \le p < 1$ we have $L_p^\sim(\pi) = \{0\}$.*

Proof. For the proof of the $0 < p < 1$ case, see [**25**, Theorem 12.31]. To see that $L_0^\sim(\pi) = \{0\}$, let $0 \le \phi \in L_0^\sim(\pi)$. We claim that ϕ is order continuous. To see this, assume that $x_n \downarrow 0$ in $L_0(\pi)$. By Corollary 1.86, the sequence $\{x_n\}$ is relatively uniformly convergent to zero. So, there exists a null sequence $\{\epsilon_n\}$ of real numbers and some $x \in L_0^+(\pi)$ satisfying $0 \le x_n \le \epsilon_n x$ for each n. This implies $0 \le \phi(x_n) \le \epsilon_n \phi(x)$ for each n, and so $\phi(x_n) \to 0$. Thus, ϕ is order continuous.

Now, by the first part, $\phi = 0$ on each ideal $L_p(\mu)$ with $0 < p < 1$. Since each of these ideals is order dense in $L_0(\pi)$ and ϕ is order continuous, it follows that $\phi = 0$ on $L_0(\pi)$, as desired. ∎

If an integral operator $T: E \to F$ between Banach function spaces is positive, then it is automatically norm continuous. As mentioned earlier, if T is not positive, then the modulus of T does not necessarily act from E to F; and so the norm continuity of T does not follow from the automatic

[4] Many authors define a function space to be simply a vector subspace of some $L_0(\pi)$-space.

continuity of positive operators. Nevertheless, integral operators are always norm continuous.

Lemma 5.21. *Let G be a Banach function space associated with some $L_0(\pi)$. If $\|x_n - x\| \to 0$ holds in G, then $\{x_n\}$ has a subsequence which is relatively uniformly convergent to x (and hence $\{x_n\}$ also has a subsequence which is pointwise convergent to x).*

In particular, if $x_n \geq 0$ for each n and the series $x = \sum_{n=1}^{\infty} x_n$ is norm convergent, then $x(\omega) = \sum_{n=1}^{\infty} x_n(\omega)$ for π-almost all $\omega \in \Omega$.

Proof. Assume that $\|x_n - x\| \to 0$. Then, by a general property of Banach lattices, $\{x_n\}$ has a subsequence which is relatively uniformly convergent to x; see Exercise 13 in Section 1.2. This subsequence automatically converges π-a.e. pointwise to x. ■

Corollary 5.22. *Every Banach function space G associated with a σ-finite measure space (Ω, Σ, π) has a weak order unit, i.e., it contains a function that is strictly positive on the carrier of G.*

In particular, if G is order dense in $L_0(\pi)$, then there exists some $u \in G^+$ satisfying $u(\omega) > 0$ for π-almost all $\omega \in \Omega$.

Proof. Clearly, we can assume without loss of generality that the ideal G is order dense in $L_0(\pi)$. By Corollary 1.91 there exists a disjoint sequence $\{\Omega_n\} \subseteq \Sigma$ such that $\Omega = \bigcup_{n=1}^{\infty} \Omega_n$ and the function χ_{Ω_n} belongs to G for each $n \in \mathbb{N}$. Then the series $u = \sum_{n=1}^{\infty} \frac{1}{2^n(1+\|\chi_{\Omega_n}\|)} \chi_{\Omega_n}$, where $\|\cdot\|$ denotes the norm of G, is norm convergent in G. By Lemma 5.21, we have $u(\omega) = \sum_{n=1}^{\infty} \frac{1}{2^n(1+\|\chi_{\Omega_n}\|)} \chi_{\Omega_n}(\omega)$ for π-almost all $\omega \in \Omega$. This implies that u is a weak unit. ■

Theorem 5.23. *Every integral operator between Banach function spaces is norm continuous.*

Proof. Let $T\colon E \to F$ be an integral operator, where E and F are Banach function spaces. It suffices to show that T has a closed graph; see Theorem 1.5. To this end, assume that $\|x_n - x\| \to 0$ in E and $\|Tx_n - y\| \to 0$ in F. By passing to appropriate subsequences (using Lemma 5.21), we can assume without loss of generality that $\{x_n\}$ is order bounded and converges μ-almost everywhere to x and that $\{Tx_n\}$ converges ν-almost everywhere to y. Now, Corollary 5.3 implies $Tx_n \xrightarrow{p} Tx$. This yields $Tx = y$. ■

Our next goal is to establish the important fact (due to G. Ya. Lozanovsky) that each normed function space is necessarily normal. To prove this, we need the following technical result due to W. A. J. Luxemburg and A. C. Zaanen [**224**, Lemma 31.1].

Lemma 5.24 (Luxemburg–Zaanen). *Let ϕ be a positive linear functional on a normed Riesz space L. Assume that $\{u_n\}$ is an order bounded sequence in L^+ satisfying $\sum_{n=1}^{\infty}\|u_n\|<\infty$. Then for each $\epsilon>0$ there exist two sequences $\{v_n\}$ and $\{w_n\}$ in L^+ such that $v_n \xrightarrow{o} 0$, $\phi(w_n)<\epsilon$ for each n, and $u_n=v_n+w_n$ for each n.*

Proof. By hypothesis $0\le u_n\le u$ holds for some $u\in L$ and for each n. Fix $\epsilon>0$, and for each n and m let

$$u_{n,m}=u_n\vee u_{n+1}\vee\cdots\vee u_{n+m}\,.$$

From $u_{n,m}\uparrow_m\le u$ and the positivity of ϕ, we see that $\phi(u_{n,m})\uparrow_m\le\phi(u)$. So, there is a strictly increasing sequence $\{m_n\}$ of natural numbers such that

$$0\le\phi(u_{n,m})-\phi(u_{n,m_n})<\tfrac{\epsilon}{2^n} \qquad (\star)$$

for all $m\ge m_n$. Next, for each n let

$$z_n=u_{1,m_1}\wedge u_{2,m_2}\wedge\cdots\wedge u_{n,m_n}\,.$$

From $0\le z_n\le u_{n,m_n}\le\sum_{i=n}^{n+m_n}u_i$, we get $\|z_n\|\le\sum_{i=n}^{\infty}\|u_i\|\to 0$, and so $\|z_n\|\to 0$. Obviously $z_n\downarrow$, and so $z_n\downarrow 0$. Now for each n let

$$y_n=\sum_{i=1}^{n}\big(u_{i,m_n}-u_{i,m_i}\big)\ge 0\,.$$

Since $1\le i\le n$ implies $m_i\le m_n$, it follows from $(\star)$ that

$$\phi(y_n)=\sum_{i=1}^{n}\phi(u_{i,m_n}-u_{i,m_i})<\sum_{i=1}^{n}\tfrac{\epsilon}{2^i}<\epsilon\,.$$

We claim that $u_n\le z_n+y_n$ for each n. To establish this inequality, it suffices to prove that $u_n\le u_{j,m_j}+y_n$ for each $j=1,\ldots,n$. For this, note that if $1\le j\le n\le m_n$, then

$$\begin{aligned}u_{j,m_j}+y_n &= u_{j,m_j}+\sum_{i=1}^{n}(u_{i,m_n}-u_{i,m_i})\\ &\ge u_{j,m_j}+\big(u_{j,m_n}-u_{j,m_j}\big)=u_{j,m_n}\ge u_n\,.\end{aligned}$$

Finally, using the Riesz Decomposition Property, we can write each u_n as $u_n=v_n+w_n$, where $0\le v_n\le z_n$ and $0\le w_n\le y_n$. Clearly, the sequences $\{v_n\}$ and $\{w_n\}$ satisfy the desired properties. ■

Theorem 5.25 (Lozanovsky). *If G is a normed function space, then $G_{\rm n}^*$ separates the points of G. In particular, G is a normal Riesz space.*

Proof. Let $H = G \cap L_1(\pi)$ and let $\phi(x) = \int_\Omega x\, d\pi$ for $x \in L_1(\pi)$. We define the function $\rho\colon H^+ \to [0, \infty)$ by the formula

$$\rho(x) = \inf\{\|y\| + \phi(x - y)\colon\ y \in H \text{ and } 0 \le y \le x\},$$

where $\|\cdot\|$ denotes the lattice norm of G.

We claim that $\rho(x) > 0$ for each $0 < x \in H$. To see this, assume by way of contradiction that $\rho(x) = 0$ for some $0 < x \in H$. Consequently, there exists a sequence $\{u_n\} \subseteq [0, x]$ satisfying $\|u_n\| + \phi(x - u_n) < \frac{1}{2^n}$ for each n. From $\sum_{n=1}^\infty \|u_n\| < \infty$ and Lemma 5.24 (applied to ϕ), it follows that given any $\epsilon > 0$ there exist two sequences $\{v_n\}$ and $\{w_n\}$ such that $v_n \xrightarrow{o} 0$, $\phi(w_n) < \epsilon$ and $u_n = v_n + w_n$ for each n. Then for each n we have

$$\begin{aligned} 0 < \phi(x) &= \phi(x - u_n) + \phi(u_n) = \phi(x - u_n) + \phi(v_n) + \phi(w_n) \\ &< \tfrac{1}{2^n} + \phi(v_n) + \epsilon\,. \end{aligned}$$

Letting $n \to \infty$, we get $0 < \phi(x) \le \epsilon$ for each $\epsilon > 0$, which is impossible. Hence, $\rho(x) > 0$ for each $0 < x \in H$.

A direct verification shows that the function $x \mapsto \rho(|x|)$ defines a lattice norm on H. Clearly, $\rho(|x|) \le \|x\|$ and $\rho(|x|) \le \phi(|x|)$ for each $x \in H$.

To prove that G^*_{n} separates the points of G, it suffices to show (see Exercise 4 at the end of the section) that for each $0 < u \in G$ there exists some $0 < \phi \in G^*_{\mathrm{n}}$ with $\phi(u) > 0$. So, fix some $u > 0$. Since $L_1(\pi)$ is an order dense ideal in $L_0(\pi)$ (this is the only place in the proof where we use the assumption that π is σ-finite), it follows that H is an order dense ideal in G. Therefore, we can assume without loss of generality that $u \in H$. In particular, u also belongs to $L_1(\pi)$.

By the Hahn–Banach theorem, there exists a positive linear functional ψ on H such that $\psi(u) = \rho(u) > 0$ and $|\psi(x)| \le \rho(|x|)$ for each $x \in H$. In particular, $\psi(x) \le \rho(x) \le \phi(x)$ for each $x \in H^+$ and hence ψ is order continuous on H. Notice next that the formula

$$\psi_H(x) = \sup\{\psi(y)\colon\ y \in H \text{ and } 0 \le y \le x\},\quad x \in H^+,$$

defines an additive function from H^+ to $\mathbb{R}$. Therefore, by Theorem 1.15, the formula $\psi_H(x) = \psi_H(x^+) - \psi_H(x^-)$ extends ψ_H to a positive linear functional on G. Clearly, $|\psi_H(x)| \le \psi_H(|x|) \le \rho(|x|) \le \|x\|$ for each $x \in G$ and so ψ_H is norm continuous on G. Moreover, it is easy to see that $\psi_H \le \phi$ and hence the order continuity of ϕ implies that ψ_H is also order continuous, i.e., $\psi_H \in G^*_{\mathrm{n}}$. To finish the proof, note that $\psi_H(u) = \psi(u) > 0$. ■

The importance of order continuous linear functionals has been already demonstrated many times in our discussions. Our next result adds a new feature to this story by providing an extremely useful integral representation for these functionals.[5] In the special case of spaces of bounded functions this representation was obtained by B. Z. Vulikh and in general by G. Ya. Lozanovsky. We refer to the paper by B. Z. Vulikh and G. Ya. Lozanovsky [**329**] for additional historic details.

Theorem 5.26 (Lozanovsky). *Assume that G is a function space associated with a σ-finite measure π. Then for each $\phi \in G_n^\sim$ there exists a function $g_\phi \in L_0(\pi)$ such that for each $x \in G$ we have the integral representation*

$$\phi(x) = \int_\Omega g_\phi(\omega)x(\omega)\,d\pi(\omega)\,.$$

The function g_ϕ is uniquely determined on the carrier of G.

If G is an order dense normal function space in $L_0(\pi)$, then the function g_ϕ is unique and the mapping $\phi \mapsto g_\phi$, from $G_n^\sim$ to $L_0(\pi)$, is a lattice isomorphism whose image G' is the order dense normal function space in $L_0(\pi)$ described by

$$G' = \big\{g \in L_0(\pi)\colon\ gx \in L_1(\pi)\ \ \textit{for all}\ \ x \in G\big\}\,.^6$$

Moreover, G' can be identified with $G_n^\sim$, where each function $g \in G'$ acts on G as an order continuous linear functional via the formula

$$\langle x, g\rangle = g(x) = \int_\Omega g(\omega)x(\omega)\,d\pi(\omega)\,.$$

Proof. Consider the collection $\Gamma = \{A \in \Sigma\colon\ \chi_A \in G\}$. Clearly Γ is a semiring of subsets of Ω and, by Corollary 1.91, the σ-algebra generated by Γ coincides with Σ.

Fix some $0 \le \phi \in G_n^\sim$, and then define the set function $\xi\colon \Gamma \to [0,\infty)$ by $\xi(A) = \phi(\chi_A)$. The order continuity of ϕ implies easily that ξ is a measure. Consequently, the usual Carathéodory extension ξ^* of ξ, restricted to Σ, is a measure on Σ. Next, observe that the measure ξ^* is obviously σ-finite (since ξ is σ-finite) and that ξ^* is absolutely continuous with respect to the σ-finite measure π. Indeed, if $A \in \Sigma$ satisfies $\pi(A) = 0$, then $\chi_A = 0 \in G$, and so $\xi^*(A) = \xi(A) = \phi(\chi_A) = 0$. Therefore, by the classical Radon–Nikodým theorem (here we use again that π is σ-finite), there exists some function $0 \le g_\phi \in L_0(\pi)$ such that $\xi^*(A) = \int_A g_\phi(\omega)\,d\pi(\omega)$ holds for all $A \in \Sigma$.

Consequently, if a function $f \in G^+$ is a Σ-step function (that is, of the form $f = \sum_{i=1}^n \alpha_i\chi_{A_i}$ with each $A_i \in \Sigma$), then $\phi(f) = \int_\Omega g_\phi(\omega)f(\omega)\,d\pi(\omega)$. Now let $x \in G^+$. Then there exists a sequence of Σ-step functions $\{f_n\}$ such

[5] For this reason the order continuous linear functionals are also called **normal integrals**.

[6] The ideal G' is often referred to as the **Köthe dual** of G.

that $0 \le f_n \uparrow x$ holds in G. The order continuity of ϕ and Beppo Levi's theorem imply that

$$\phi(x) = \lim_{n\to\infty} \phi(f_n) = \lim_{n\to\infty} \int_\Omega g_\phi(\omega) f_n(\omega)\, d\pi(\omega) = \int_\Omega g_\phi(\omega) x(\omega)\, d\pi(\omega)\,.$$

This establishes the representation of ϕ. The verification of the uniqueness of g_ϕ on the carrier of G is straightforward.

Assume now that G is normal and order dense. The uniqueness of g_ϕ follows easily from the fact that in this case the carrier of G is all of Ω. Moreover, it should be clear that the mapping $\phi \mapsto g_\phi$ is linear and that its image G' is an ideal in $L_0(\pi)$. To see that $\phi \mapsto g_\phi$ is a lattice isomorphism, fix $\phi \in G_n^\sim$. Then for each $x \in G^+$ we have

$$\phi^+(x) = \sup_{0\le y\le x} \phi(y) = \sup_{0\le y\le x} \int_\Omega g_\phi(\omega) y(\omega)\, d\pi(\omega) = \int_\Omega [g_\phi]^+(\omega) x(\omega)\, d\pi(\omega)\,.$$

This implies that $g_{\phi^+} = [g_\phi]^+$, and so $\phi \mapsto g_\phi$ is a lattice isomorphism.

For the last part, we need to verify that $G_n^\sim$ is order dense in $L_0(\pi)$. To this end, take any $0 < y \in L_0(\pi)$, and then choose some $x \in E^+$ satisfying $0 < x \le y$. Since the order continuous dual $G_n^\sim$ separates the points of G, there exists some $0 < \phi \in G_n^\sim$ (with the corresponding function $0 < g_\phi \in L_0(\pi)$) such that $\phi(x) = \int_\Omega g_\phi(\omega) x(\omega)\, d\pi(\omega) > 0$. This implies $0 < g_\phi \wedge x \le x \le y$, and from $g_\phi \wedge x \in G_n^\sim$, it follows that $G_n^\sim$ is order dense in $L_0(\pi)$. The proof is finished. ■

From the preceding theorem and Corollary 5.22 we get the following.

Corollary 5.27. *If G is a Banach function space associated with a σ-finite measure π, then the order continuous dual G_n^* has a weak unit—and so it is a principal band in G^*.*

We are now ready to present a characterization of integral operators as abstract integral operators. This basic result is due to G. Ya. Lozanovsky [**219**].

Theorem 5.28 (Lozanovsky). *If E and F are function spaces and E is normal, then the bands of abstract integral operators and regular integral operators from E to F coincide, i.e., $\mathcal{L}_\kappa(E,F) = (E_n^\sim \otimes F)^{\rm dd}$.*

Proof. By Lemma 1.94, we can assume without loss of generality that E and F are order dense ideals in $L_0(\mu)$ and $L_0(\nu)$, respectively. For simplicity we denote the band $(E_n^\sim \otimes F)^{\rm dd}$ by $\mathcal{B}$. Take an arbitrary $\phi\otimes e \in E_n^\sim \otimes F$. By Theorem 5.26 there is some $g \in L_0(\mu)$ such that $\phi(x) = \int_S g(s)x(s)\, d\mu(s)$ holds for all $x \in E$. Therefore, for each $x \in E$ we have

$$[(\phi\otimes e)x](t) = \Big[\int_S g(s)x(s)\, d\mu(s)\Big] e(t) = \int_S [g(s)e(t)]\, x(s)\, d\mu(s)$$

for ν-almost all $t \in \mathcal{T}$. Consequently, $\phi \otimes e$ is a regular integral operator with kernel $g(s)e(t)$, i.e., $\phi \otimes e \in \mathcal{L}_\kappa(E,F)$. Since, by Theorem 5.14, the regular integral operators form a band in $\mathcal{L}_r(E,F)$, it follows that $\mathcal{B} \subseteq \mathcal{L}_\kappa(E,F)$.

For the converse we must establish that $\mathcal{L}_\kappa(E,F) \subseteq \mathcal{B}$. Take a $\mu \times \nu$-measurable subset U of $\mathcal{S} \times \mathcal{T}$ with $(\mu \times \nu)^*(U) < \infty$ and assume that the $\mu \times \nu$-measurable function χ_U defines an integral operator T acting from E to F; so, in particular, $Tx(t) = \int_{\mathcal{S}} \chi_U(s,t)x(s)\,d\mu(s) \in F$ for each $x \in E$. Since $(\mu \times \nu)^*(U) < \infty$, there exists a sequence $\{A_n\}$ of μ-measurable subsets of $\mathcal{S}$ and a sequence $\{B_n\}$ of ν-measurable subsets of $\mathcal{T}$ such that

$$U \subseteq \bigcup_{n=1}^{\infty} A_n \times B_n\,. \tag{$\dagger$}$$

Now recall that the order denseness of E implies (according to Theorem 5.26) that E_n^* is also order dense. In particular, it follows from Corollary 1.91 that for each n there exist μ-measurable sets $A_1^n, A_2^n, A_3^n, \ldots$ with $\chi_{A_m^n} \in E_{\rm n}^\sim$ and ν-measurable sets $B_1^n, B_2^n, B_3^n, \ldots$ with $\chi_{B_m^n} \in F$ such that $A_n = \bigcup_{m=1}^\infty A_m^n$ and $B_n = \bigcup_{k=1}^\infty B_k^n$. This shows that in the union in ($\dagger$), we can suppose that $\chi_{A_n} \in E_{\rm n}^\sim$ and $\chi_{B_n} \in F$ for each n. Next, put $U_n = U \cap \big(\bigcup_{k=1}^n A_k \times B_k\big)$ and note that the integral operator T_n with kernel χ_{U_n} satisfies the inequalities $0 \le T_n \le \sum_{k=1}^n \chi_{A_k} \otimes \chi_{B_k}$. Since $\chi_{A_k} \otimes \chi_{B_k} \in \mathcal{B}$ for each k, it follows that $T_n \in \mathcal{B}$. From $\chi_{U_n} \uparrow \chi_U$ we get $T_n \uparrow T$ and so $T \in \mathcal{B}$; see Exercise 7 in Section 5.1.

Finally, let $T\colon E \to F$ be a positive integral operator with kernel $T(\cdot,\cdot)$. In view of Theorem 5.5, we can assume without loss of generality that $T(s,t) \ge 0$ for all $(s,t) \in \mathcal{S} \times \mathcal{T}$. Since the measure $\mu \times \nu$ is σ-finite, there exists a sequence $\{\Psi_n\}$ of $\mu \times \nu$-step functions such that $0 \le \Psi_n(s,t) \uparrow T(s,t)$ for each $(s,t) \in \mathcal{S} \times \mathcal{T}$. Now notice that if $\Psi = \sum_{j=1}^m \alpha_j \chi_{U_j}$ is a $\mu \times \nu$-step function satisfying $0 \le \Psi \le T$, then each $\mu \times \nu$-measurable set U_j satisfies the conditions of the previous part. Consequently, the integral operator generated by χ_{U_j} belongs to $\mathcal{B}$, and so the integral operator with kernel Ψ also belongs to $\mathcal{B}$. Thus, the integral operator T_n with kernel Ψ_n belongs to $\mathcal{B}$. Since $\mathcal{B}$ is a band, it follows (again by Exercise 7 in Section 5.1) that $T \in \mathcal{B}$. Therefore, $\mathcal{L}_\kappa(E,F) \subseteq \mathcal{B}$ is also true, and the proof is finished. ■

Recall that if $T\colon L \to M$ is an order continuous regular operator between two Riesz spaces with M Dedekind complete, then its **order adjoint operator** $T^\sim\colon M^\sim \to L^\sim$, defined for each $x \in L$ and each $y \in M^\sim$ via the duality identity

$$\langle Tx, y\rangle = \langle x, T^\sim y\rangle = T^\sim y(x)\,,$$

carries order continuous linear functionals to order continuous linear functionals, i.e., $T^\sim(M_{\rm n}^\sim) \subseteq L_{\rm n}^\sim$ holds. To see this, let $y \in M_{\rm n}^\sim$ and assume $x_\alpha \xrightarrow{o} 0$ holds in L. Then the order continuity of T implies $Tx_\alpha \xrightarrow{o} 0$ in M, and so $T^\sim y(x_\alpha) = y(Tx_\alpha) \to 0$ in $\mathbb{R}$. This shows that $T^\sim y$ is order continuous. (If L and M are Banach function spaces, then of course $L^\sim = L^*$, $M^\sim = M^*$, and $T^\sim = T^*$.)

Now let $T\colon E \to F$ be a regular integral operator between two normal function spaces. Then, by Corollary 5.12, T is order continuous, and so $T^\sim(F_{\rm n}^\sim) \subseteq E_{\rm n}^\sim$. That is, if we consider the restriction of $T^\sim\colon F^\sim \to E^\sim$ to $F_{\rm n}^\sim$, then we have another order continuous regular operator $T^\sim\colon F_{\rm n}^\sim \to E_{\rm n}^\sim$ between normal function spaces. It turns out that this operator is also an integral operator.

Theorem 5.29. *If $T\colon E \to F$ is a regular integral operator between two normal function spaces, then its adjoint operator $T^\sim\colon F_{\rm n}^\sim \to E_{\rm n}^\sim$ is also a regular integral operator, and its kernel $T^\sim(t,s)$ satisfies $T^\sim(t,s) = T(s,t)$ for $\nu \times \mu$-almost all $(t,s) \in \mathcal{T} \times \mathcal{S}$. That is, if $y \in F_{\rm n}^\sim$, then*

$$T^\sim y(s) = \int_{\mathcal{T}} T(s,t)y(t)\,d\nu(t)$$

holds for μ-almost all $s \in \mathcal{S}$.

Proof. Let $T\colon E \to F$ be a regular integral operator between two normal function spaces. Without loss of generality, we can assume that E and F are order dense ideals in $L_0(\mu)$ and $L_0(\nu)$, respectively, and that T is a positive operator. Fix some $0 \le y \in F_{\rm n}^\sim$. Then for each $x \in F^+$ we have

$$\begin{aligned}\langle Tx, y\rangle &= \int_{\mathcal{T}}\Big[\int_{\mathcal{S}} T(s,t)x(s)\,d\mu(s)\Big]y(t)\,d\nu(t)\\ &= \int_{\mathcal{T}}\Big[\int_{\mathcal{S}} T(s,t)x(s)y(t)\,d\mu(s)\Big]\,d\nu(t)\,.\end{aligned}$$

So, by Tonelli's Theorem 1.98, the function $(s,t) \mapsto T(s,t)x(s)y(t)$ is $\mu \times \nu$-integrable. In particular, for each fixed $x \in E^+$ the function $T(s,\cdot)x(s)y(\cdot)$ is integrable over $\mathcal{T}$ for μ-almost all $s \in \mathcal{S}$, and the function

$$g(s) = \int_{\mathcal{T}} T(s,t)y(t)x(s)\,d\nu(t) = \Big[\int_{\mathcal{T}} T(s,t)y(t)\,d\nu(t)\Big]x(s)$$

is μ-integrable. Since E is order dense in $L_0(\mu)$, it easily follows that the function $h(s) = \int_{\mathcal{T}} T(s,t)y(t)\,d\nu(t)$ is μ-measurable. Moreover, for each $x \in E$ and each $y \in F_{\rm n}^\sim$ it follows from Fubini's Theorem 1.97 that

$$\begin{aligned}\langle Tx, y\rangle &= \langle x, T^\sim y\rangle = \int_{\mathcal{S}}\Big[\int_{\mathcal{T}} T(s,t)y(t)x(s)\,d\nu(t)\Big]d\mu(s)\\ &= \int_{\mathcal{S}}\Big[\int_{\mathcal{T}} T(s,t)y(t)\,d\nu(t)\Big]x(s)\,d\mu(s) = \int_{\mathcal{S}} h(s)x(s)\,d\mu(s)\,.\end{aligned}$$

It now readily follows from Theorem 5.26 that $h \in E_{\rm n}^\sim$ and that $T^\sim y(s) = h(s) = \int_{\mathcal{T}} T(s,t)y(t)\,d\nu(t)$ holds for μ-almost all $s \in \mathcal{S}$. That is, $T^\sim$ is an integral operator whose kernel satisfies $T^\sim(t,s) = T(s,t)$ for $\nu \times \mu$-almost all $(t,s) \in \mathcal{T} \times \mathcal{S}$. ■

We are now ready to state and prove another important characterization of integral operators that is due to A. V. Bukhvalov [**77**].

Theorem 5.30 (Bukhvalov). *For an operator $T\colon E \to F$ between two function spaces the following four statements are equivalent.*

(1) *T is an integral operator.*

(2) *If $\{x_n\} \subseteq E$ is order bounded and satisfies $x_n \xrightarrow{\mu} 0$, then $Tx_n \xrightarrow{\nu} 0$.*

(3) *The operator T satisfies the two conditions:*
   (a) *If $\{x_n\} \subseteq E$ is order bounded and $x_n \xrightarrow{\mu} 0$, then $Tx_n \xrightarrow{\nu} 0$.*
   (b) *If $\chi_A \in E$ for some measurable set A of finite measure, then for every sequence $\{B_n\}$ of measurable subsets of A with $\mu(B_n) \to 0$ we have $T\chi_{B_n} \xrightarrow{\nu} 0$.*

(4) *The operator $T\colon E \to L_0(\nu)$ is σ-order continuous and satisfies condition* 3(b) *above.*

Proof. (1) $\Longrightarrow$ (2) This follows from Lemma 5.2.

(2) $\Longrightarrow$ (3) Condition 3(b) is a special case of (2) and condition 3(a) follows immediately from Corollary 1.83.

(3) $\Longrightarrow$ (4) This follows from Corollary 1.89.

(4) $\Longrightarrow$ (1) This proof is quite involved. Before proceeding with it, we need to make a few simplifications. First of all, note that we can assume without loss of generality that $F = L_0(\nu)$ and (in view of Lemma 1.94) that E is order dense in $L_0(\mu)$. Hence, by Corollary 1.89, the σ-order continuous operator $T\colon E \to L_0(\nu)$ is regular and order continuous.

Let us show that we can also assume that the operator T is positive. Indeed, observe that since T is order continuous, T^+ and T^- are both order continuous. So, our claim will be justified if we show that T^+ satisfies 3(b) (in which case $T^- = T^+ - T$ satisfies 3(b) automatically as a difference of two operators satisfying this condition).

To see this, assume that $\{B_n\}$ is a sequence of μ-measurable sets such that $\mu(B_n) \to 0$ and $B_n \subseteq A$ for some measurable subset $A \subseteq \mathcal{S}$ such that $\mu(A) < \infty$ and $\chi_A \in E$. Now fix a μ-measurable subset C of A and note that the obvious identity $C \setminus C \cap (A \setminus B_n) = C \cap B_n$ implies $\mu[C \setminus C \cap (A \setminus B_n)] \to 0$.

Therefore, using condition 3(b), we get $T(\chi_{C\setminus C\cap(A\setminus B_n)}) \xrightarrow{p} 0$. Consequently,

$$\begin{aligned} T\chi_C &= \lim_{n\to\infty} T\big(\chi_{C\cap(A\setminus B_n)}\big) \le \liminf_{n\to\infty} T^+\big(\chi_{C\cap(A\setminus B_n)}\big) \\ &\le \liminf_{n\to\infty} T^+\big(\chi_{A\setminus B_n}\big) \\ &= T^+\chi_A - \limsup_{n\to\infty} T^+(\chi_{B_n})\,. \end{aligned}$$

Since, by Theorem 1.17, $T^+\chi_A = \sup\{T\chi_C\colon\ C \in \Sigma_1 \text{ and } C \subseteq A\}$, the previous inequality yields

$$T^+\chi_A \le T^+\chi_A - \limsup_{n\to\infty} T^+(\chi_{B_n})\,,$$

whence $\limsup T^+\chi_{B_n} \le 0$. Since the inequality $0 \le \liminf T^+\chi_{B_n}$ is trivially true, it follows that $\limsup T^+\chi_{B_n} = \liminf T^+\chi_{B_n} = 0$, that is, $T^+\chi_{B_n} \xrightarrow{p} 0$ indeed holds, and so T^+ satisfies 3(b). Therefore, we can assume without loss of generality that T is also a positive operator.

Next, we claim that E can be assumed to be a normal function space. To see this, consider the order dense ideal $E_0 = E \cap L_1(\mu)$. Clearly, E_0 is a normal function space since the order continuous linear functionals on $L_1(\mu)$ are order continuous on E_0 as well. Now the operator $T_0 = T|_{E_0}$, the restriction of the operator T to E_0, obviously is order continuous and satisfies 3(b) because the operator T satisfies it. If we can show that T_0 is an integral operator, then Theorem 5.15 will guarantee that T is also an integral operator. Thus, replacing E with E_0 and T with T_0, we can also assume without loss of generality that E is a normal function space.

We now proceed with the rest of the proof. From Theorem 5.28 we know that $\mathcal{L}_k(E,F) = (E_{\rm n}^\sim \otimes F)^{\rm dd}$. Consequently our operator T can be written as $T = R + S$ with $R \in \mathcal{L}_k(E,F)$ and $S \perp (E_{\rm n}^\sim \otimes F)$. This shows that T is an integral operator if and only if $S = 0$. From $0 \le S \le T$ and the order continuity of T, it easily follows that S is also order continuous and satisfies 3(b). Therefore, to finish the proof, we need to establish that the zero operator is the only positive operator S that satisfies condition 3(b), is order continuous and is disjoint from $E_{\rm n}^\sim \otimes F$. Assume by way of contradiction that $S > 0$ is such an operator.

Since (in view of Theorem 5.26) $E_{\rm n}^\sim$ is order dense in $L_0(\mu)$, it follows that $E \cap E_{\rm n}^\sim$ is an order dense ideal. In particular, $E \cap E_{\rm n}^\sim$ is order dense in E. Therefore, the order continuity of S implies that there exists a measurable set A of finite μ-measure such that $\chi_A \in E \cap E_{\rm n}^\sim$ and $e = S\chi_A > 0$. So, there is some $\epsilon > 0$ and some ν-measurable subset D of $\mathcal{T}$ with $\nu(D) > 0$ such that $e(t) > 2\epsilon$ for each $t \in D$, i.e., $e = S\chi_A \ge 2\epsilon\chi_D$.

Consider $\phi \otimes e \in E_n^\sim \otimes F$, where we let $\phi = \chi_A$. By our assumption $S \wedge (\phi \otimes e) = 0$, and so (by Theorem 1.17) we obtain

$$\begin{aligned} 0 &= \big[S \wedge (\phi \otimes e)\big](\chi_A) \\ &= \inf\Big\{S\chi_C + \Big(\int_{A\setminus C} \phi\, d\mu\Big)e\colon\ C \in \Sigma_1 \text{ and } C \subseteq A\Big\} \\ &= \inf\Big\{S\chi_C + \mu(A \setminus C)e\colon\ C \in \Sigma_1 \text{ and } C \subseteq A\Big\}. \end{aligned}$$

Notice that for each measurable subset C of A we have

$$S\chi_C + \mu(A \setminus C)e \le S\chi_A + \mu(A)e = [1 + \mu(A)]e\,. \tag{$\dagger$}$$

Since F has the countable sup property, there is a sequence $\{C_n\} \subseteq \Sigma_1$ such that $C_n \subseteq A$ for each n and

$$\inf_{n\in\mathbb{N}}\Big\{S\chi_{C_n} + \mu(A \setminus C_n)e\colon\ C_n \in \Sigma_1 \text{ and } C_n \subseteq A\Big\} = 0\,. \tag{$\star$}$$

We claim now that there exists a subsequence $\{C_{n_m}\}$ of $\{C_n\}$ such that $\mu(A \setminus C_{n_m}) \to 0$ and

$$\inf_{m\in\mathbb{N}}\Big\{S\chi_{C_{n_m}} + \mu(A \setminus C_{n_m})e\Big\} = 0\,.$$

If this claim is established, then we get a contradiction as follows. Since $\mu(A \setminus C_{n_m}) \to 0$, it follows from 3(b) that $S\chi_{A\setminus C_{n_m}} \xrightarrow{\ p\ } 0$ or, equivalently, that $S\chi_{C_{n_m}} \xrightarrow{\ p\ } S\chi_A = e$. Consequently, by Egorov's theorem, there exists a ν-measurable subset D_1 of D with $\nu(D_1) > 0$ and some m_1 such that $e(t) - S\chi_{C_{n_m}}(t) < \epsilon$ holds for all $m \ge m_1$ and all $t \in D_1$. Therefore (since $e(t) \ge 2\epsilon$ for each $t \in D$), we can conclude that

$$S\chi_{C_{n_m}} \ge \epsilon\chi_{D_1} \tag{$\star\star$}$$

for all $m \ge m_1$. Notice now that no matter how large the integer m_1 might be, there exists some $\epsilon_1 > 0$ such that

$$\inf_{m<m_1}\Big\{S\chi_{C_{n_m}} + \mu(A \setminus C_{n_m})e\Big\} \ge \epsilon_1\chi_D\,. \tag{$\star\star\star$}$$

To see this, fix $1 \le m < m_1$. If $\mu(A \setminus C_{n_m}) = 0$, then $C_{n_m} = A$ μ-a.e., and so

$$S\chi_{C_{n_m}} + \mu(A \setminus C_{n_m})e = S\chi_{C_{n_m}} = S\chi_A = e \ge 2\epsilon\chi_D\,.$$

Otherwise, if $\mu(A \setminus C_{n_m}) > 0$, then

$$S\chi_{C_{n_m}} + \mu(A \setminus C_{n_m})e \ge \mu(A \setminus C_{n_m})e\,.$$

Since we have only a finite number of terms, the existence of a desired $\epsilon_1 > 0$ is established. From ($\star\star$) and ($\star\star\star$), it follows that

$$\inf_{m\in\mathbb{N}}\Big\{S\chi_{C_{n_m}} + \mu(A \setminus C_{n_m})e\Big\} \ge (\epsilon \wedge \epsilon_1)\chi_{D_1} > 0\,,$$

which is a contradiction.

It remains to prove the existence of a subsequence $\{C_{n_m}\}$ of $\{C_n\}$ with the required properties. For each $k \in \mathbb{N}$ let $\mathbb{N}_k = \{n \geq k\colon\ \mu(A \setminus C_n) < \frac{1}{k}\}$ and note that for each $n \notin \mathbb{N}_k$ we have

$$S\chi_{C_n} + \mu(A \setminus C_n)e \geq \tfrac{1}{k}e\,.$$

In view of (†) and (⋆), this implies

$$\inf_{n\in\mathbb{N}_k} \Big\{S\chi_{C_n} + \mu(A \setminus C_n)e\colon\ C_n \in \Sigma_1 \text{ and } C_n \subseteq A\Big\} = 0\,.$$

In particular, we have that $\inf_{n\in\mathbb{N}_k} S\chi_{C_n} = 0$, $\inf_{n\in\mathbb{N}_k} \mu(A \setminus C_n) = 0$, and that each set $\mathbb{N}_k$ is infinite. Let $\mathbb{N}_k = \{m_{k,1}, m_{k,2}, \ldots\}$, where $m_{k,n} < m_{k,n+1}$ for each n.

Now for each k and r in $\mathbb{N}$ put

$$y_{k,r} = \bigwedge_{i=1}^{r} S\chi_{C_{m_{k,i}}}\,.$$

From $\inf_{n\in\mathbb{N}_k} S\chi_{C_n} = 0$, it follows that for each k we have $y_{k,r} \downarrow_r 0$. Consequently, by the diagonal property (see Theorem 1.87) there exists a sequence $\{y_k\}$ in $L_0(\nu)$ with $y_k \downarrow 0$ and a strictly increasing sequence $\{r_k\}$ of natural numbers such that $y_{k,r_k} \leq y_k$ for each k. Let $M = \bigcup_{k=1}^{\infty}\{m_{k,1}, \ldots, m_{k,r_k}\}$, and note that M is a countable subset on $\mathbb{N}$. Therefore, $M = \{n_1, n_2, \ldots\}$, where $n_1 < n_2 < \cdots$, and clearly $\lim_{m\to\infty} \mu(A \setminus C_{n_m}) = 0$.

We claim that $\{C_{m_n}\}$ is the desired subsequence, i.e, that

$$g = \inf_{m\in\mathbb{N}} \Big\{S\chi_{C_{n_m}} + \mu(A \setminus C_{n_m})e\Big\} = 0\,.$$

To see this, fix $\delta > 0$, and then pick some m_0 such that $\mu(A \setminus C_{n_m}) < \delta$ for each $m \geq m_0$. Then for each k we have:

$$g \leq \inf_{m\geq m_0} \Big\{S\chi_{C_{n_m}} + \mu(A \setminus C_{n_m})e\Big\} \leq \inf_{m\geq m_0} S\chi_{C_{n_m}} + \delta e \leq y_k + \delta e\,,$$

where the last inequality follows from the definition of functions $y_{k,r}$ and the fact that $y_{k,r_k} \leq y_k$. But then $y_k \downarrow 0$ yields $g \leq \delta e$. Since $\delta > 0$ is arbitrary, we get $g = 0$. This completes the proof of the theorem. ■

Some historical comments regarding Bukhvalov's Theorem 5.30 are in order. It was proven by A. V. Bukhvalov in [**77**] under the assumption that the domain space is normal. The basic ideas of the proof are due to H. Nakano [**253**], where he established that a positive bilinear form lies in the band generated by the finite-rank bilinear forms if and only if it satisfies a continuity property similar to the one of statement (2). Bukhvalov's theorem was also reproved independently by A. R. Schep [**294**]. Removing the assumption that $E_{\mathrm{n}}^{\sim}$ is total on E is not particularly important since the function spaces with trivial order dual may carry only trivial integral operators. For instance, it follows from a well known theorem by S. Kwapien [**196**] (see also [**170**, Theorem 8.4]) that the function space $L_0[0,1]$ admits only the trivial integral operator.

The first simple application of Theorem 5.30 describes an important class of Banach function spaces on which every continuous operator is an integral operator.

Corollary 5.31. *Let E be a Banach function space with order continuous norm. Then every continuous operator $T\colon E \to L_\infty(\nu)$ is a regular integral operator. That is, $\mathcal{L}(E, L_\infty(\nu)) = \mathcal{L}_\kappa(E, L_\infty(\nu))$.*

Proof. Since $L_\infty(\nu)$ is a Dedekind complete AM-space with unit, we know from Theorem 3.9 that $\mathcal{L}(E, L_\infty(\nu)) = \mathcal{L}_r(E, L_\infty(\nu))$. So, it remains to show that every positive operator $T\colon E \to L_\infty(\nu)$ is an integral operator.

Assume that E is associated with $L_0(\mu)$, and let an order bounded sequence $\{x_n\} \subseteq E$ satisfy $x_n \xrightarrow{\mu} 0$. The order continuity of the norm in E easily implies (see Exercise 8 at the end of the section) that $\|x_n\| \to 0$, and so $\|Tx_n\|_\infty \to 0$. In particular, we have $Tx_n \xrightarrow{\nu} 0$. Hence T satisfies condition (2) of Theorem 5.30, and so T is an integral operator. ■

Corollary 5.32. *If F is a reflexive Banach function space associated with $L_0(\nu)$, then every continuous operator $T\colon L_1(\mu) \to F$ is a regular integral operator. That is, $\mathcal{L}(L_1(\mu), F) = \mathcal{L}_\kappa(L_1(\mu), F)$.*

Proof. Let $T\colon L_1(\mu) \to F$ be a continuous operator. By Theorem 3.9 we know that the operator T is regular. Also, T is order continuous since $L_1(\mu)$ has order continuous norm. Taking the adjoint operator, we see that $T^*\colon F^* \to L_1(\mu)^* = L_\infty(\mu)$. Since the Banach function space F is reflexive, both F and F^* have order continuous norms, and hence $F^* = F^*_{\rm n}$. So, by Corollary 5.31, the operator $T^*\colon F^* \to L_1(\mu)^* = L_\infty(\mu)$ is a regular integral operator.

Since $L_\infty(\mu)^*_{\rm n} = L_1(\mu)$ (see Exercise 6 at the end of the section), Theorem 5.29 implies that $T = T^{**}\colon L_\infty(\mu)^*_{\rm n} = L_1(\mu) \to (F^*_{\rm n})^* = F$ must be a regular integral operator. ■

The two well-known classical representation theorems of N. Dunford [**117**] are special cases of the preceding results.

Corollary 5.33 (Dunford's Theorem). *We have the following.*

(1) *If $1 \le p < \infty$, then every continuous operator $T\colon L_p(\mu) \to L_\infty(\nu)$ is a regular integral operator.*

(2) *If $1 < p < \infty$, then every continuous operator $T\colon L_1(\mu) \to L_p(\nu)$ is a regular integral operator.*

We shall close the section with one more characterization of integral operators between Banach function spaces when the measures μ and ν are both finite. To state the result, we need some preliminary discussion.

If D is any subset of $L_0(\pi)$ and $u \in L_0(\pi)$, then let $uD = \{ux\colon\ x \in D\}$. Any subset of the form uD with u satisfying $u(\omega) > 0$ for π-almost all $\omega \in \Omega$ is called a **twist** of D. It is not difficult to see that a twist of an ideal is an ideal and a twist of an order dense ideal is also order dense. Moreover, if G is a Banach function space, then every twist uG of G is a Banach function space. (To see this, define the lattice norm on uG by $|||ux||| = \|x\|$ for each $x \in G$.)

Regarding twists, we have the following remarkable result of G. Ya. Lozanovsky [**220**, Theorem 7] whose lengthy and technical proof is omitted.

Theorem 5.34 (Lozanovsky's Twist Theorem). *Assume that G is an order dense Banach function space associated with a finite measure π. Then there exists a twist uG of G such that $L_\infty(\pi) \subseteq uG \subseteq L_1(\pi)$.*

By means of Lozanovsky's Twist Theorem we can now establish the following representation theorem for integral operators. This representation theorem is also due to G. Ya. Lozanovsky and it can be found in [**329**].

Theorem 5.35 (Lozanovsky). *Assume that $(\mathcal{S}, \Sigma_1, \mu)$ and $(\mathcal{T}, \Sigma_2, \nu)$ are two finite measure spaces, and let E and F be two order dense Banach function spaces associated with $L_0(\mu)$ and $L_0(\nu)$, respectively. If $T\colon E \to F$ is a regular integral operator, then there exist a strictly positive function $u \in L_0(\mu)$, a strictly positive function $v \in L_0(\nu)$, and a $\mu \times \nu$-integrable function K such that for each $x \in E$ we have*

$$Tx(t) = v(t) \int_{\mathcal{S}} K(s,t)u(s)x(s)\, d\mu(s)$$

for ν-almost all $t \in \mathcal{T}$.

Proof. Without loss of generality, we can assume that T is a positive integral operator. Suppose first that $L_\infty(\mu) \subseteq E$ and $F \subseteq L_1(\nu)$. Then $\mathbf{1}_{\mathcal{S}} \in E$, and so $T\mathbf{1}_{\mathcal{S}}(\cdot) = \int_{\mathcal{S}} T(s,\cdot)\, d\mu(s) \in L_1(\nu)$. But then Tonelli's theorem implies that $T(\cdot,\cdot) \in L_1(\mu \times \nu)$, and our conclusion holds true with $u = \mathbf{1}_{\mathcal{S}}$, $v = \mathbf{1}_{\mathcal{T}}$, and $K = T(\cdot,\cdot)$.

Now consider the general case. By Lozanovsky's Twist Theorem 5.34, there exist two strictly positive measurable functions $u \in L_0(\mu)$ and w in $L_0(\nu)$ such that $L_\infty(\mu) \subseteq uE \subseteq L_1(\mu)$ and $L_\infty(\nu) \subseteq wF \subseteq L_1(\nu)$. Next, define the positive operator $K\colon uE \to wF$ via the formula $K(ux) = wTx$.

From Theorem 5.30 it follows immediately that K is an integrable operator and, by the preceding case, its kernel $K(\cdot,\cdot) \in L_1(\mu \times \nu)$. Now if we let $v = \frac{1}{w}$, then v is a strictly positive function in $L_0(\nu)$, and for each $x \in E$ we have

$$Tx(t) = v(t) \int_S K(s,t)u(s)x(s)\,d\mu(s)$$

for ν-almost all $t \in T$. ■

We conclude the section with one comment regarding the σ-finiteness assumption that has been used throughout this and the previous sections. There exists a considerably weaker condition (the so-called direct sum property) that is usually sufficient instead of the σ-finiteness. Here is its definition: *A measure space (Ω, Σ, π) satisfies the* ***direct sum property*** *if there exists a family $\{\Omega_\lambda\}_{\lambda\in\Lambda}$ of pairwise disjoint sets in Σ such that $0 < \pi(\Omega_\lambda) < \infty$ for all $\lambda \in \Lambda$, and for each $A \in \Sigma$ with finite measure we can find an at most countable family $\Lambda_0 \subseteq \Lambda$ and a μ-null set $N \in \Sigma$ such that $A = \bigcup_{\lambda\in\Lambda_0}(A \cap \Omega_\lambda) \cup N$.*

Exercises

1. Let E and G be two Banach function spaces associated with $L_0(\pi)$. Show that $E \cap G$ is also a Banach function space—which is order dense if both E and F are order dense ideals. [HINT: If $\|\cdot\|_E$ and $\|\cdot\|_G$ denote the norms on E and G, respectively, then consider the lattice norm $\|\cdot\|$ on $E \cap G$ defined by $\|x\| = \|x\|_E + \|x\|_G$.]

2. Let G be an ideal in a Riesz space E, let F be a Dedekind complete Riesz space, and let $T\colon G \to F$ be a positive operator. Assuming that T has a regular extension to all of E, establish the following.

(a) The operator T has a minimal positive extension to all of E. That is, there exists a positive extension $T_G\colon E \to F$ of T such that if $S\colon E \to F$ is any positive extension of T, then $T_G \le S$.

(b) The minimal extension T_G of T is given by the formula

$$T_G(x) = \sup\{Ty\colon\ y \in G \text{ and } 0 \le y \le x\}$$

for each $x \in E^+$. In particular, $T_G(x) = 0$ for each $x \in G^{\mathrm{d}}$.

(c) If T is order continuous, then T_G is also order continuous.

3. By means of Nakano's theorem (see Exercise 22 of Section 1.2), show directly that if E is a Banach function space associated with a σ-finite measure π, then its order continuous dual E_{n}^* has weak units.

4. Let E be a normed Riesz space. Show that $E^*_{\rm n}$ separates the points of E if and only if for each $0 < x \in E$ there exists some $0 < \phi \in E^*_{\rm n}$ satisfying $\phi(x) > 0$.

5. Verify the uniqueness of the function g in the Representation Theorem 5.26.

6. Let (Ω, Σ, π) be a σ-finite measure space. Show that $L_\infty(\pi)^*_{\rm n} = L_1(\pi)$. [HINT: Let $E = L_\infty(\pi)^*_{\rm n}$. By Theorem 5.26, E is a Banach function space associated with $L_0(\pi)$. Clearly, $L_1(\pi) \subseteq E$. Now assume that $f \in E$. This means that $\int_\Omega f(\omega)x(\omega)\,d\pi(\omega) \in \mathbb{R}$ for each x in $L_\infty(\pi)$. In particular, if $x = \operatorname{Sign} f$, then $x \in L_\infty(\pi)$ and $\int_\Omega |f|\,d\pi = \int_\Omega f(\omega)x(\omega)\,d\pi(\omega) \in \mathbb{R}$. This shows that $f \in L_1(\pi)$ and $L_\infty(\pi)^*_{\rm n} = L_1(\pi)$.]

7. Let (Ω, Σ, π) be a σ-finite measure space. If E is a Banach function space in $L_0(\pi)$, then show that the center of E coincides with $L_\infty(\pi)$, that is, $\mathcal{Z}(E) = L_\infty(\pi)$.

8. For an order bounded sequence $\{x_n\}$ in a Banach function space with order continuous norm consider the following convergence properties:

$$\text{(a) } x_n \xrightarrow{o} x \qquad \text{(b) } x_n \xrightarrow{p} x \qquad \text{(c) } x_n \xrightarrow{\mu} x \qquad \text{(d) } \|x_n - x\| \longrightarrow 0\,.$$

Show that (a) $\Longleftrightarrow$ (b), (c) $\Longleftrightarrow$ (d), and (b) $\Longrightarrow$ (c). Does (d) $\Longrightarrow$ (b)? [HINT: Use Exercise 4 in Section 5.1 and Lemma 1.30(c).]

9. If $\{x_n\}$ is an order bounded positive sequence in some Banach function space E in $L_0(\pi)$, then prove that $x_n \xrightarrow{\pi} 0$ is equivalent to $x_n \xrightarrow{\sigma(E, E^*_{\rm n})} 0$, i.e., to the property: $\int_\Omega x_n(\omega)y(\omega)\,d\pi(\omega) \to 0$ for each $y \in E^*_{\rm n}$.

Is this equivalence true if $\{x_n\}$ is not assumed to be a positive sequence?

10. If E and F are Banach function spaces, then show that $(E^*_{\rm n} \otimes F)^{\rm dd}$ is a principal band in $\mathcal{L}_r(E, F)$. [HINT: We can assume that E and F are order dense ideals. If e and ϕ are weak units in F and $E^*_{\rm n}$, respectively, then show that $\mathcal{B}_{\phi \otimes e} = (E^*_{\rm n} \otimes F)^{\rm dd}$.]

11. Prove Corollary 5.31 using Theorem 5.28 instead of Theorem 5.30.

12. Show that Dunford's Theorem (see Corollary 5.33) is not valid for continuous operators between arbitrary L_1-spaces. [HINT: The identity operator $I\colon L_1[0,1] \to L_1[0,1]$ is disjoint from every regular integral operator; see Exercise 8 in Section 5.1.]

13. Assume that $(\mathcal{S}, \Sigma_1, \mu)$, $(\mathcal{T}, \Sigma_2, \nu)$, and $(\mathcal{Q}, \Sigma_3, \theta)$ are three σ-finite measure spaces and that E, F, and G are ideals in $L_0(\mu)$, $L_0(\nu)$, and $L_0(\pi)$, respectively. Assume also that in the scheme $E \xrightarrow{S} F \xrightarrow{T} G$ the operators S and T are both regular and order continuous.

If either S or T is an integral operator, then show that TS is likewise an integral operator.

14. Assume that $(\mathcal{S}, \Sigma_1, \mu)$ and $(\mathcal{T}, \Sigma_2, \nu)$ are two σ-finite measure spaces with $(\mathcal{S}, \Sigma_1, \mu)$ non-atomic. Also, suppose that E and F are ideals in $L_0(\mathcal{S}, \Sigma_1, \mu)$ and $L_0(\mathcal{T}, \Sigma_2, \nu)$, respectively, and let $S\colon E \to F$ be a regular integral operator.

Let F_0 denote the order ideal in F generated by the range space of the operator S and assume that there is an operator $T\colon F_0 \to E$ such that $TS = I_E$, the identity operator on E (in a sense T is a right inverse of the operator S). Show that T cannot be an integral operator.

15. Assume that $(\mathcal{S}, \Sigma_1, \mu)$, $(\mathcal{T}, \Sigma_2, \nu)$, and $(\mathcal{Q}, \Sigma_3, \theta)$ are three σ-finite measure spaces and that E, F, and G are ideals in $L_0(\mu)$, $L_0(\nu)$, and $L_0(\theta)$, respectively. Suppose also that in the scheme $E \xrightarrow{S} F \xrightarrow{T} G$ both S and T are regular integral operators with kernels $S(\cdot,\cdot)$ and $T(\cdot,\cdot)$, respectively.

Show that the composition operator $R = TS$ is also an integral operator whose kernel is given by the **convolution** of the two kernels $S(\cdot,\cdot)$ and $T(\cdot,\cdot)$, that is,

$$R(s,q) = \int_{\mathcal{T}} S(s,t)T(t,q)\,d\nu(t)\,.$$

16. Let $(\mathcal{S}, \Sigma_1, \mu)$ and $(\mathcal{T}, \Sigma_2, \nu)$ be two σ-finite measure spaces with $(\mathcal{S}, \Sigma_1, \mu)$ non-atomic, and let $0 < K(\cdot,\cdot) \in L_0(\mu \times \nu)$. Do there exist a function $0 < f \in L_0(\mu)$ and a subset $B \in \Sigma_2$ such that:

(a) $\nu(B) > 0$.

(b) For each $t \in B$ we have $K(s,t) \geq f(s)$ for μ-almost all $s \in \mathcal{S}$?

17. (Benedek–Panzone [**54**]) Let $(\mathcal{S}, \Sigma_1, \mu)$ and $(\mathcal{T}, \Sigma_2, \nu)$ be two σ-finite measure spaces. For each $p \in [1, \infty)$ we denote by $L_{p,\infty}(\mu \times \nu)$ (or simply by $L_{p,\infty}$) the collection of all functions $f \in L_0(\mu \times \nu)$ such that

$$\|f\|_{p,\infty} = \operatorname*{ess\,sup}_{s \in \mathcal{S}} \Big[\int_{\mathcal{T}} |f(s,t)|^p \, d\nu(t) \Big]^{\frac{1}{p}} < \infty\,.$$

As usual, the equivalent functions are identified. Similarly, we denote by $L_{p,1}(\mu\times\nu)$ (or simply by $L_{p,1}$) the collection of all functions $f \in L_0(\mu\times\nu)$ such that

$$\|f\|_{p,1} = \int_{\mathcal{S}} \Big[\int_{\mathcal{T}} |f(s,t)|^p \, d\nu(t) \Big]^{\frac{1}{p}} d\mu(s) < \infty\,.$$

The non-negative real numbers $\|f\|_{p,\infty}$ and $\|f\|_{p,1}$ are called the **mixed norms** of f. Show that:

(a) The function $\|\cdot\|_{p,\infty}$ is well defined, i.e., if $f, g \in L_{p,\infty}$ satisfy $f = g$ $\mu \times \nu$-a.e., then $\|f\|_{p,\infty} = \|g\|_{p,\infty}$.

(b) The collections $L_{p,\infty}$ and $L_{p,1}$ (with their mixed norms) are both Banach function spaces associated with $L_0(\mu \times \nu)$.

(c) If $1 < p, q < \infty$ satisfy $\frac{1}{p} + \frac{1}{q} = 1$, then $L^*_{p,1} = L_{q,\infty}$.

18. The result of this exercise can be found in many monographs; see for instance [**107, 118, 174, 175**]. For the notation used here see Exercise 17 above.

Assume that $(\mathcal{S}, \Sigma_1, \mu)$ and $(\mathcal{T}, \Sigma_2, \nu)$ are two σ-finite measure spaces. Show that a $\mu \times \nu$-measurable function $T(\cdot,\cdot)\colon \mathcal{S} \times \mathcal{T} \to \mathbb{R}$ defines, via the usual formula

$$Tx(t) = \int_{\mathcal{S}} T(s,t)x(s)\,d\mu(s)\,,$$

an integral operator from $L_1(\mu)$ to $L_p(\nu)$ for some $1 < p < \infty$ if and only if $T(\cdot,\cdot)$ belongs to $L_{p,\infty}$. Moreover, prove that in this case the integral operator T determined by the kernel $T(\cdot,\cdot)$ is a regular operator and its norm is given by $\|T\| = \|T(\cdot,\cdot)\|_{p,\infty}$.

19. (Lozanovsky [**221**]) Dunford's theorem stated in part (2) of Corollary 5.33 asserts that: *If $(\mathcal{S},\Sigma_1,\mu)$ and $(\mathcal{T},\Sigma_2,\nu)$ are σ-finite measure spaces and $1 < p < \infty$, then every bounded operator $T\colon L_1(\mu) \to L_p(\nu)$ is a regular integral operator.* Present an alternate proof of Dunford's theorem using Exercise 18 above.

20. Let E and F be two Banach function spaces with F having order continuous norm. Show that every regular integral operator from E to F carries order bounded subsets of E to norm totally bounded subsets of F. That is, show that if $x \in E^+$, then for each regular integral operator $T\colon E \to F$ the set $T[0,x]$ is norm totally bounded. [HINT: Use Exercise 9 of Section 2.3.]

5.3. Conditional Expectations and Positive Projections

In this section (Ω,Σ,μ) will denote a fixed probability measure space. The constant function one on Ω will be denoted by $\mathbf{1}$. For each $1 \le p \le \infty$ the familiar Dedekind complete (real) Banach lattice $L_p(\Omega,\Sigma,\mu)$ will also be denoted by $L_p(\Sigma)$. As usual, we view the equivalence classes as functions.

If $\mathcal{A}$ is a σ-subalgebra of Σ and $1 \le p \le \infty$, then by $L_p(\mathcal{A})$ we shall denote the Dedekind complete Banach lattice $L_p(\Omega,\mathcal{A},\mu)$. This Banach lattice consists of the equivalences classes of all $\mathcal{A}$-measurable functions whose absolute p^{th} powers are μ-integrable. Also, $L_\infty(\mathcal{A})$ consists of all essentially bounded $\mathcal{A}$-measurable functions. That is, an $\mathcal{A}$-measurable function $f\colon \Omega \to \mathbb{R}$ belongs to $L_\infty(\mathcal{A})$ if and only if $\|f\|_{\mathcal{A},\infty} < \infty$, where

$$\|f\|_{\mathcal{A},\infty} = \inf\big\{M > 0\colon\ \mu\big(\{\omega \in \Omega\colon\ |f(\omega)| > M\}\big) = 0\big\}.$$

Whenever the σ-subalgebra under consideration is clear from the context, we will simply write $\|f\|_\infty$ instead of $\|f\|_{\mathcal{A},\infty}$ or $\|f\|_{\Sigma,\infty}$. Clearly, each $L_p(\mathcal{A})$ is a vector sublattice of $L_p(\Sigma)$. If $1 \le p < \infty$ and $f \in L_p(\mathcal{A})$, then it should be obvious that the $\|\cdot\|_p$-norm of f in $L_p(\mathcal{A})$ coincides with the $\|\cdot\|_p$-norm of f in $L_p(\Sigma)$. The same conclusion is true for $p = \infty$. That is, $\|f\|_{\mathcal{A},\infty} = \|f\|_{\Sigma,\infty}$ for each function $f \in L_\infty(\mathcal{A})$. Thus, for each $1 \le p \le \infty$ the Banach lattice $L_p(\mathcal{A})$ is a closed vector sublattice of $L_p(\Sigma)$ containing the constant functions. We shall see later in this section that every closed vector sublattice of an $L_p(\Sigma)$-space that includes the function $\mathbf{1}$ is of this type.

Before proceeding further, let us make one more cautionary comment about the Banach lattice $L_p(\mathcal{A})$, where $\mathcal{A}$ is a σ-subalgebra of Σ. If we consider the probability measure space $(\Omega,\mathcal{A},\mu)$ in its own right, then we can think of $L_p(\mathcal{A})$ as the usual L_p-space associated with the measure space $(\Omega, \mathcal{A}, \mu)$. On the other hand, if we consider $L_p(\mathcal{A})$ as a vector sublattice of $L_p(\Sigma)$, then we view the classes of $L_p(\mathcal{A})$ as classes in $L_p(\Sigma)$. That is, we tacitly identify $L_p(\mathcal{A})$ with its image in $L_p(\Sigma)$ under the lattice isometry $[f]_{\mathcal{A}} \mapsto [f]_\Sigma$. The following example clarifies the situation. Let $\Omega = [0,1]$, let Σ be the Σ-algebra of all Lebesgue measurable subsets of $[0,1]$, and let μ be the Lebesgue measure. Consider also the σ-subalgebra of Σ given by $\mathcal{A} = \{\emptyset, [0,1], [0,\frac{1}{2}], (\frac{1}{2},1]\}$. Then it is easy to see that

$$L_p(\mathcal{A}) = \{c_1\chi_{[0,\frac{1}{2}]} + c_2\chi_{(\frac{1}{2},1]} \colon \ c_1, c_2 \in \mathbb{R}\},$$

and so each class of $L_p(\mathcal{A})$ consists of one function. However, if we consider $L_p(\mathcal{A})$ as a vector sublattice of $L_p(\Sigma)$, then we can view the classes in $L_p(\mathcal{A})$ as classes in $L_p(\Sigma)$. To clarify it better, consider the function $f = \chi_{[0,\frac{1}{2}]}$. If we consider $L_p(\mathcal{A})$ as an L_p-space in its own right, then the function f is identified with the class $[f]_{\mathcal{A}} = \{f\}$ in $L_p(\mathcal{A})$. On the other hand, if we consider $L_p(\mathcal{A})$ as a vector sublattice of $L_p(\Sigma)$, then the function f is identified with the class $[f]_\Sigma = \{g \in L_p(\Sigma) \colon \ g = f \ \ \mu\text{-a.e.}\}$ in $L_p(\Sigma)$.

Now fix a σ-subalgebra $\mathcal{A}$ of Σ. By the classical Radon–Nikodym theorem for each $f \in L_1(\Sigma)$ there exists a unique function $h \in L_1(\mathcal{A})$ such that $\int_A h\, d\mu = \int_A f\, d\mu$ for each $A \in \mathcal{A}$. The function h is called the **conditional expectation** of f with respect to $\mathcal{A}$ and is denoted $\mathcal{E}(f|\mathcal{A})$. That is, $\mathcal{E}(f|\mathcal{A})$ is a μ-a.e. uniquely determined $\mathcal{A}$-measurable and μ-integrable function satisfying the identity

$$\int_A f\, d\mu = \int_A \mathcal{E}(f|\mathcal{A})\, d\mu$$

for each $A \in \mathcal{A}$. Clearly, if $f, g \in L_1(\Sigma)$, then for all $\alpha, \beta \in \mathbb{R}$ we have

$$\mathcal{E}(\alpha f + \beta g|\mathcal{A}) = \alpha\mathcal{E}(f|\mathcal{A}) + \beta\mathcal{E}(g|\mathcal{A}).$$

That is, $\mathcal{E}(\cdot|\mathcal{A})$ defines an operator on $L_1(\Sigma)$ whose range is $L_1(\mathcal{A})$.

Definition 5.36 (Kolmogorov [**187**]). *The operator $\mathcal{E}(\cdot|\mathcal{A}) \colon L_1(\Sigma) \to L_1(\Sigma)$ is called the **conditional expectation operator** induced by $\mathcal{A}$.*

The proof of the next result is straightforward and is omitted.

Lemma 5.37. *For every σ-subalgebra $\mathcal{A}$ of Σ the conditional expectation operator $\mathcal{E}(\cdot|\mathcal{A}) \colon L_1(\Sigma) \to L_1(\Sigma)$ is a strictly positive order continuous contractive projection with range $L_1(\mathcal{A})$ that leaves $L_\infty(\Sigma)$ invariant. That is:*

(1) *If $f \in L_1(\Sigma)$ and $f > 0$, then $\mathcal{E}(f|\mathcal{A}) > 0$ in $L_1(\mathcal{A})$.*

(2) *If $f_n \downarrow 0$ in $L_1(\Sigma)$, then $\mathcal{E}(f_n|\mathcal{A}) \downarrow 0$ in $L_1(\mathcal{A})$.*[7]

---

[7] Recall (see Theorem 1.80 and the paragraph thereafter) that since we deal here with a finite measure space, each ideal E in $L_0(\Sigma)$ satisfies the countable sup property. Accordingly, a positive operator $T \colon E \to E$ is order continuous if and only if $f_n \downarrow 0$ implies $Tf_n \downarrow 0$.

(3) *The operator* $\mathcal{E}(\cdot|\mathcal{A})$ *has norm one, i.e.,* $\|\mathcal{E}(\cdot|\mathcal{A})\|_1 = 1$.

(4) *If* $g \in L_1(\mathcal{A})$, *then* $\mathcal{E}(g|\mathcal{A}) = g$, *and so*

$$\mathcal{E}(\cdot|\mathcal{A})^2 = \mathcal{E}(\cdot|\mathcal{A}) \quad \textit{and} \quad \mathcal{E}(\mathbf{1}|\mathcal{A}) = \mathbf{1}.$$ [8]

(5) *The operator* $\mathcal{E}(\cdot|\mathcal{A})$ *leaves* $L_\infty(\Sigma)$ *invariant and* $\|\mathcal{E}(\cdot|\mathcal{A})\|_\infty = 1$.

Not only do the conditional expectation operators leave $L_\infty(\Sigma)$ invariant, but they also leave invariant all the $L_p(\Sigma)$-spaces.

Lemma 5.38. *For each* σ*-subalgebra* $\mathcal{A}$ *of* Σ *and for each* $1 \le p \le \infty$ *the conditional expectation operator* $\mathcal{E}(\cdot|\mathcal{A})$ *leaves* $L_p(\Sigma)$ *invariant and is contractive on this space.*

Proof. From the definition of the conditional expectation operator and from Lemma 5.37 it follows that we need to consider only $1 < p < \infty$. Let $\mathcal{A}$ be a σ-subalgebra of Σ. Take any non-negative function $f \in L_p(\Sigma)$ and apply to it Jensen's inequality for conditional expectation operators with the convex function $\psi(t) = |t|^p$; see Exercise 5(b) at the end of the section. This gives

$$0 \le [\mathcal{E}(f|\mathcal{A})]^p \le \mathcal{E}(f^p|\mathcal{A}). \qquad (\dagger)$$

Since $f^p \in L_1(\Sigma)$, the function $\mathcal{E}(f^p|\mathcal{A})$ belongs to $L_1(\mathcal{A}) \subseteq L_1(\Sigma)$. Therefore, from ($\dagger$) we get $\big[\mathcal{E}(f|\mathcal{A})\big]^p \in L_1(\Sigma)$, that is, $\mathcal{E}(f|\mathcal{A}) \in L_p(\Sigma)$. This proves that $\mathcal{E}(\cdot|\mathcal{A})$ leaves $L_p(\Sigma)$ invariant.

Now, according to the definition of the conditional expectation we have $\int_\Omega \mathcal{E}(f^p|\mathcal{A})\, d\mu = \int_\Omega f^p\, d\mu$, and so

$$\begin{aligned} \|\mathcal{E}(f|\mathcal{A})\|_p &= \Big[\int_\Omega [\mathcal{E}(f|\mathcal{A})]^p\, d\mu\Big]^{\frac{1}{p}} \\ &\le \Big[\int_\Omega \mathcal{E}(f^p|\mathcal{A})\, d\mu\Big]^{\frac{1}{p}} \\ &= \Big[\int_\Omega f^p\, d\mu\Big]^{\frac{1}{p}} = \|f\|_p\,. \end{aligned}$$

Therefore, $\mathcal{E}(\cdot|\mathcal{A})\colon L_p(\Sigma) \to L_p(\Sigma)$ is a contraction, as claimed. ■

Following the work of G. Birkhoff [**63**], we shall say that a bounded operator $T\colon L_1(\Sigma) \to L_1(\Sigma)$ satisfies the **averaging property** (or that the operator T is an **averaging operator**) if for each pair $h, f \in L_1(\Sigma)$ such that $hTf \in L_1(\Sigma)$ [in particular, this is so whenever $h \in L_\infty(\Sigma)$ and $f \in L_1(\Sigma)$], we have

$$T(hTf) = ThTf\,.$$

Conditional expectations enjoy the averaging property.

[8] Positive projections leaving the function **1** fixed are sometimes called **Markov projections**. This is in agreement with our Definition 3.15 of a Markov operator.

Theorem 5.39. *Every conditional expectation operator is an averaging operator. That is, if $\mathcal{A}$ is any σ-subalgebra of Σ and $h, f \in L_1(\Sigma)$ satisfy $h\mathcal{E}(f|\mathcal{A}) \in L_1(\Sigma)$, then*

$$\mathcal{E}(h\mathcal{E}(f|\mathcal{A})|\mathcal{A}) = \mathcal{E}(h|\mathcal{A})\mathcal{E}(f|\mathcal{A})\,.$$

Proof. In view of the order (or norm) continuity of $\mathcal{E}(\cdot|\mathcal{A})$ it is enough to verify that

$$\mathcal{E}(\chi_B \mathcal{E}(\chi_C|\mathcal{A})|\mathcal{A}) = \mathcal{E}(\chi_B|\mathcal{A})\mathcal{E}(\chi_C|\mathcal{A}) \qquad (\star)$$

for each $B, C \in \Sigma$. Since $\mathcal{E}(\chi_C|\mathcal{A})$ is an essentially bounded non-negative $\mathcal{A}$-measurable function, there exists a sequence $\{h_n\}$ of non-negative $\mathcal{A}$-step functions satisfying $h_n \uparrow \mathcal{E}(\chi_C|\mathcal{A})$. This and the fact that $\mathcal{E}(h_n|\mathcal{A}) = h_n$ for each n imply at once that in order to prove $(\star)$, it suffices to establish that

$$\mathcal{E}(\chi_B\chi_D|\mathcal{A}) = \mathcal{E}(\chi_B|\mathcal{A})\chi_D$$

for each $B \in \Sigma$ and each $D \in \mathcal{A}$. To do so, we just have to prove (by the uniqueness of the Radon–Nikodym derivative) that

$$\int_A \chi_B\chi_D\,d\mu = \int_A \mathcal{E}(\chi_B|\mathcal{A})\chi_D\,d\mu$$

for each $A \in \mathcal{A}$ or, equivalently,

$$\int_{A\cap D} \chi_B\,d\mu = \int_{A\cap D} \mathcal{E}(\chi_B|\mathcal{A})\,d\mu\,.$$

Since $A \cap D \in \mathcal{A}$, the last equality holds true automatically by virtue of the definition of $\mathcal{E}(\cdot|\mathcal{A})$, and the proof is finished. ■

Our next objective is to establish that the properties of the conditional expectation operators in Lemma 5.37 basically characterize the Markov projections not only on $L_1(\Sigma)$-spaces but also on a much broader class of subspaces of $L_1(\Sigma)$-spaces.

We begin with a characterization of the order closed vector sublattices of $L_0(\mu)$ containing the constant function $\mathbf{1}$.

Lemma 5.40. *If E is an ideal in $L_0(\Sigma)$ containing the constants, then for a vector sublattice Y of E with $\mathbf{1} \in Y$ the following statements are equivalent.*

(1) *$Y = E \cap L_0(\mathcal{A})$ for some σ-subalgebra $\mathcal{A}$ of Σ.*

(2) *Y is order closed in E.*

(3) *Y is closed under μ-a.e. pointwise monotone sequential limits in E.*

Proof. The implications (1) $\Longrightarrow$ (2) $\Longrightarrow$ (3) are trivial. Hence, only the implication (3) $\Longrightarrow$ (1) needs a proof. So, assume that (3) is true. Consider the collection of sets

$$\mathcal{A} = \big\{A \in \Sigma\colon\ \chi_A \in Y\big\}\,.$$

We claim that $\mathcal{A}$ is a σ-subalgebra of Σ. Using the identities

$$\chi_\Omega = \mathbf{1} \in Y, \qquad \chi_{A^c} = \mathbf{1} - \chi_A, \qquad \chi_{A \cap B} = \chi_A \wedge \chi_B$$

we conclude at once that $\mathcal{A}$ is a subalgebra of Σ. To see that $\mathcal{A}$ is a σ-algebra, assume that some sequence $\{A_n\} \subseteq \mathcal{A}$ satisfies $A_n \uparrow A$. This certainly implies that $A \in \Sigma$ and $\chi_{A_n} \uparrow \chi_A$. From $\chi_A \in L_\infty(\Sigma) \subseteq E$, our condition (3) guarantees that $\chi_A \in Y$. Thus, $A \in \mathcal{A}$, and so $\mathcal{A}$ is a σ-subalgebra of Σ.

Since the vector sublattice of all $\mathcal{A}$-step functions is included in Y and is order dense in $E \cap L_0(\mathcal{A})$, it follows from (3) that $E \cap L_0(\mathcal{A}) \subseteq Y$. Now let $f \in Y$ and consider the set $A = \{\omega \in \Omega\colon\ f(\omega) > 0\}$. Since Y is a vector sublattice and $\mathbf{1} \in Y$, it follows that $\{nf^+ \wedge \mathbf{1}\} \subseteq Y$ and moreover $nf^+ \wedge \mathbf{1} \uparrow \chi_A$. Using once more (3), we see that $\chi_A \in Y$ and so $A \in \mathcal{A}$. Finally, note that since $f - c\mathbf{1} \in Y$ for each scalar c, the preceding case implies

$$\{\omega \in \Omega\colon\ f(\omega) > c\} = \{\omega \in \Omega\colon\ [f - c\mathbf{1}](\omega) > 0\} \in \mathcal{A}.$$

So, f is $\mathcal{A}$-measurable and hence $f \in L_0(\mathcal{A})$. This implies $Y \subseteq E \cap L_0(\mathcal{A})$. Consequently, $Y = E \cap L_0(\mathcal{A})$, as desired. ■

In a Banach lattice with order continuous norm, a vector sublattice is order closed if and only if it is norm closed. In the absence of the order continuity of the norm, the assumption of the order closedness of Y in Lemma 5.40 cannot be replaced with the norm closedness of Y; see Exercise 6 at the end of the section.

An algebraic analogue of Lemma 5.40 for $L_\infty(\Sigma)$ is as follows.

Corollary 5.41. *For a subalgebra Y of the Banach algebra $L_\infty(\Sigma)$ the following statements are equivalent.*

(1) *$Y = L_\infty(\mathcal{A})$ for some σ-subalgebra $\mathcal{A}$ of Σ.*

(2) *Y contains the constants and is order closed in $L_\infty(\Sigma)$.*

Proof. It should be clear that (1) implies (2). Now assume that (2) is true. Since norm convergence in $L_\infty(\Sigma)$ implies order convergence, it easily follows that Y is a normed closed subalgebra of $L_\infty(\Sigma)$. But then it is well known from the proof of the Stone–Weierstrass theorem (see for instance [**31**, Theorem 11.5, p. 89]) that Y is a vector sublattice. The validity of (1) now follows from Lemma 5.40. ■

The next result presents two basic order properties of contractive operators on L_p-spaces.

Lemma 5.42. *For a contractive operator $T\colon L_p(\Sigma) \to L_p(\Sigma)$, $1 \le p \le \infty$, satisfying $T\mathbf{1} = \mathbf{1}$ we have the following:*

(a) *If $1 \le p < \infty$, then $T^*\mathbf{1} = \mathbf{1}$.*

(b) *If $p = 1$ or $p = \infty$, then T is a positive operator.*

Proof. (a) Let q be the conjugate index of p, i.e., $\frac{1}{p} + \frac{1}{q} = 1$. Since T is contractive, T^* is also contractive, and consequently $\|T^*\mathbf{1}\|_q \le 1$. From $T\mathbf{1} = \mathbf{1}$ and Hölder's inequality, we obtain

$$1 = \langle \mathbf{1}, \mathbf{1} \rangle = \langle T\mathbf{1}, \mathbf{1} \rangle = \langle \mathbf{1}, T^*\mathbf{1} \rangle \le \|\mathbf{1}\|_p \cdot \|T^*\mathbf{1}\|_q \le 1 .$$

So, in this case, Hölder's inequality is the equality

$$\int_\Omega \mathbf{1} T^*\mathbf{1}\, d\mu = \|\mathbf{1}\|_p \cdot \|T^*\mathbf{1}\|_q = \|T^*\mathbf{1}\|_q = 1 . \qquad (\star\star)$$

This guarantees the existence of some constant $c > 0$ such that $|T^*\mathbf{1}| = c\mathbf{1}$; see for instance [**32**, Problem 31.4]. Clearly, $c = 1$, and from this and $(\star\star)$ we get $T^*\mathbf{1} = \mathbf{1}$.

(b) Consider first the case $p = \infty$. Pick any $0 \le x \in L_\infty(\Sigma)$. Dividing by $\|x\|_\infty$ if necessary, we can assume that $0 \le x \le \mathbf{1}$, and so $0 \le \mathbf{1} - x \le \mathbf{1}$. Taking into account that $\|T\| = 1$ and that the closed unit ball of $L_\infty(\Sigma)$ is the order interval $[-\mathbf{1}, \mathbf{1}]$, it follows that $\mathbf{1} - Tx = T(\mathbf{1} - x) \in [-\mathbf{1}, \mathbf{1}]$. Therefore, $\mathbf{1} - Tx \le \mathbf{1}$ and so $Tx \ge 0$. This shows that T is positive.

Let now $p = 1$. It follows from (a) that $T^*\mathbf{1} = \mathbf{1}$. Therefore, by the previous case the adjoint operator $T^*\colon L_\infty(\Sigma) \to L_\infty(\Sigma)$ is positive. This implies that the second adjoint T^{**} of T is also a positive operator. Since T^{**} agrees with T on $L_1(\Sigma)$, it follows that T itself is positive too. ■

Exercise 7 at the end of the section shows that for other values of p the conclusions of the preceding lemma do not hold in general. Recall that a Banach lattice E is said to have **strictly monotone norm** if $|x| < |y|$ implies $\|x\| < \|y\|$. The L_p-spaces with $1 \le p < \infty$ have strictly monotone norms.

Lemma 5.43. *Let $T\colon E \to E$ be an operator on a vector lattice and assume that, either*

(a) *T is a strictly positive projection, or*

(b) *E is a Banach lattice with strictly monotone norm and T is a contractive positive operator.*

Then the fixed space $\mathcal{F}(T) = \{x \in E\colon\ Tx = x\}$ of the operator T is a vector sublattice of E (which, in case (b)*, is also norm closed).*

Proof. (a) To begin with, notice that since T is a projection, the fixed space $\mathcal{F}(T)$ of T simply coincides with the range $T(E)$ of T.

Now take any $x \in T(E)$. Then $|x| = |Tx| \le T|x|$, and so $T|x| - |x| \ge 0$. Since $T^2 = T$, it follows that $T\big(T|x| - |x|\big) = 0$, and so the strict positivity of T implies $T|x| = |x|$, that is, $|x| \in \mathcal{F}(T)$. Therefore, $\mathcal{F}(T) = T(E)$ is a vector sublattice of E.

(b) Clearly, $\mathcal{F}(T)$ is a closed vector subspace of E. To see that $\mathcal{F}(T)$ is a vector sublattice, pick any $x \in \mathcal{F}(T)$. Then $|x| = |Tx| \le T|x|$. If $|x| < T|x|$ were true, then the strict monotonicity of the norm would imply $\big\||x|\big\| < \big\|T|x|\big\|$ and so $\|T\| > 1$, a contradiction. Hence $|x| = T|x|$, i.e., $|x| \in \mathcal{F}(T)$. This shows that $\mathcal{F}(T)$ is a vector sublattice of E. ■

Our next result is essentially due to J. L. Kelley [**180**], G. L. Seever [**299**], and D. E. Wulbert [**341**]. It asserts that any contractive projection on $L_\infty(\Sigma)$ with range $L_\infty(\mathcal{A})$ has the averaging property.

Lemma 5.44 (Kelley–Seever–Wulbert). *If $\mathcal{A}$ is a σ-subalgebra of Σ, then any contractive projection $P\colon L_\infty(\Sigma) \to L_\infty(\Sigma)$ with range $L_\infty(\mathcal{A})$ is an averaging operator, i.e., it satisfies*

$$P\big(fP(g)\big) = P(f)P(g)$$

for each $f, g \in L_\infty(\Sigma)$.

Proof. Since $\mathbf{1} \in L_\infty(\mathcal{A})$, we have $P\mathbf{1} = \mathbf{1}$ and hence $\|P\| = 1$. By Lemma 5.42(b) the operator P is positive.

Since $Px = x$ for each $x \in L_\infty(\mathcal{A})$ and since the $\mathcal{A}$-measurable step functions are norm dense in $L_\infty(\mathcal{A})$, to prove the desired averaging property it suffices to show that $P(f\chi_A) = (Pf)\chi_A$ for all $f \in L_\infty(\Sigma)$ and $A \in \mathcal{A}$.

So, let $f \in L_\infty(\Sigma)$ and $A \in \mathcal{A}$. Without loss of generality we can assume that the function f satisfies $0 \le f \le \mathbf{1}$. Then $0 \le f\chi_A \le \chi_A$ and hence $0 \le P(f\chi_A) \le P\chi_A = \chi_A$. Similarly, $0 \le P(f\chi_{A^c}) \le \chi_{A^c}$. Therefore,

$$\operatorname{Supp} P(f\chi_A) \subseteq A \quad \text{and} \quad \operatorname{Supp} P(f\chi_{A^c}) \subseteq A^c\,. \tag{$\dagger$}$$

These inclusions should, of course, be understood to be valid μ-a.e. with respect to $\mathcal{A}$. For instance, the inclusion $\operatorname{Supp} P(f\chi_A) \subseteq A$ means that the set $D = \big\{\omega \in \operatorname{Supp} P(f\chi_A)\colon\ \omega \notin A\big\}$ belongs to $\mathcal{A}$ and $\mu(A) = 0$. Since $f\chi_A + f\chi_{A^c} = f$, it follows that

$$P(f\chi_A) + P(f\chi_{A^c}) = Pf = (Pf)\chi_A + (Pf)\chi_{A^c}\,,$$

and from this and ($\dagger$) we get $P(f\chi_A) = (Pf)\chi_A$, as desired. ■

It is worthwhile pointing out that Lemma 5.44 combined with the order continuity of the conditional expectation operator provides an alternate proof of Theorem 5.39.

We are now ready to present a characterization due to A. A. Mekler [**236**] of the order continuous positive projections that preserve **1**. The characterization was also proven independently by P. G. Dodds, C. B. Huijsmans, and B. de Pagter in [**110**]. Recall that (by Theorem 5.26) for any ideal E in $L_0(\Sigma)$ its order continuous dual $E^\sim_{\rm n}$ can be identified with an ideal in $L_0(\Sigma)$ as follows:

$$E^\sim_{\rm n} = \{\phi \in L_0(\Sigma)\colon\ \phi x \in L_1(\Sigma) \text{ for all } x \in E\}.$$

In particular, notice that:

- *If* $\mathbf{1} \in E$, *i.e., if* $L_\infty(\Sigma) \subseteq E$, *then* $E^\sim_{\rm n} \subseteq L_1(\Sigma)$.
- *If* $L_\infty(\Sigma) \subseteq E \subseteq L_1(\Sigma)$, *then* $L_\infty(\Sigma) \subseteq E^\sim_{\rm n} \subseteq L_1(\Sigma)$.

In the sequel, keep in mind that an "ideal E in $L_1(\Sigma)$ containing the constants" is simply any ideal E in $L_0(\Sigma)$ satisfying $L_\infty(\Sigma) \subseteq E \subseteq L_1(\Sigma)$.

Theorem 5.45 (Mekler). *If an ideal E in $L_1(\Sigma)$ contains the constants, then for an operator $P\colon E \to E$ the following statements are equivalent.*

(1) *P is an order continuous positive projection, its range is a vector sublattice of E, and $P\mathbf{1} = \mathbf{1}$.*

(2) *There exist a σ-subalgebra $\mathcal{A}$ of Σ and a non-negative function ϕ in $E^\sim_{\rm n}$ such that $\mathcal{E}(\phi|\mathcal{A}) = \mathbf{1}$ and $Px = \mathcal{E}(\phi x|\mathcal{A})$ for each $x \in E$.*

Proof. (1) $\Longrightarrow$ (2) The order continuity of the projection P implies immediately that the vector sublattice $Y = P(E)$ of E satisfies statement (3) of Lemma 5.40. So, according to this lemma, there exists a σ-subalgebra $\mathcal{A}$ of Σ such that $Y = E \cap L_0(\mathcal{A})$. Since $E \subseteq L_1(\Sigma)$, we have $Y = E \cap L_1(\mathcal{A})$. Next, we shall consider two cases.

CASE I: $E = L_\infty(\Sigma)$.

In this case we have $Y = L_\infty(\Sigma) \cap L_1(\mathcal{A}) = L_\infty(\mathcal{A})$. (Why?) By Lemma 5.44 the operator P has the averaging property on $L_\infty(\Sigma)$, i.e., $P(x_1 P(x_2)) = P(x_1)P(x_2)$ for each $x_1, x_2 \in L_\infty(\Sigma)$.

Now consider the positive linear functional $\Phi\colon L_\infty(\Sigma) \to \mathbb{R}$ defined by $\Phi(x) = \int P(x)\,d\mu$. The order continuity of the operator P implies that Φ is order continuous, that is, $\Phi \in L_\infty(\Sigma)^*_{\rm n} = L_1(\Sigma)$. Therefore, there exists a non-negative function $\phi \in L_1(\Sigma)$ such that $\Phi(x) = \int \phi x\,d\mu$ for each $x \in E$. In particular, for each $A \in \mathcal{A}$ and each $x \in E$ we have

$$\int_A \phi x\,d\mu = \int \phi x \chi_A\,d\mu = \Phi(x\chi_A) = \int P(x\chi_A)\,d\mu. \qquad (\star)$$

Using that P satisfies the averaging property and that $P\chi_A = \chi_A$, we get

$$P(x\chi_A) = P(xP\chi_A) = P(x)P(\chi_A) = P(x)\chi_A$$

for each $x \in E$. From this and $(\star)$, we obtain

$$\int_A \phi x\, d\mu = \int P(x)\chi_A\, d\mu = \int_A P(x)\, d\mu\,.$$

Since Px is $\mathcal{A}$-measurable and μ-integrable for each $x \in E$, it follows from the definition of the conditional expectation that $\mathcal{E}(\phi x|\mathcal{A}) = Px$ for each $x \in E$. In particular, $\mathcal{E}(\phi|\mathcal{A}) = \mathcal{E}(\phi\mathbf{1}|\mathcal{A}) = P\mathbf{1} = \mathbf{1}$. This establishes the desired representation of P in the special case when $E = L_\infty(\Sigma)$.

CASE II: *E is an ideal such that $L_\infty(\Sigma) \subseteq E \subseteq L_1(\Sigma)$.*

The hypotheses that P is positive and $P\mathbf{1} = \mathbf{1}$ guarantee that P leaves $L_\infty(\Sigma)$ invariant and $\|P\|_\infty = 1$. So, if P_∞ denotes the restriction of P to $L_\infty(\Sigma)$, then P_∞ is a contractive projection on $L_\infty(\Sigma)$ satisfying $P_\infty\mathbf{1} = \mathbf{1}$.

We claim that the range of P_∞ is a vector sublattice of $L_\infty(\Sigma)$. To see this, let $x, y \in L_\infty(\Sigma)$, and put $z = Px \vee Py \in L_\infty(\Sigma)$. Since the range of the operator $P\colon E \to E$ is a vector sublattice of E, there exists some $u \in E$ such that $Pu = z$. But then the function $z \in L_\infty(\Sigma)$ satisfies $z = Pz = P_\infty z$. This shows that the range of P_∞ is a vector sublattice of $L_\infty(\Sigma)$.

According to Case I, there exist a σ-subalgebra $\mathcal{A}$ of Σ and a nonnegative function $\phi \in L_1(\Sigma)$ such that $P_\infty x = \mathcal{E}(\phi x|\mathcal{A})$ for each $x \in L_\infty(\Sigma)$. This means that the desired representation

$$Px = \mathcal{E}(\phi x|\mathcal{A})$$

is valid for each $x \in L_\infty(\Sigma)$. Since $L_\infty(\Sigma)$ is an order dense ideal in $L_1(\Sigma)$, it follows that $L_\infty(\Sigma)$ is also an order dense ideal in E. Now use the fact that P and $\mathcal{E}(\cdot|\mathcal{A})$ are both order continuous on E to infer that $Px = \mathcal{E}(\phi x|\mathcal{A})$ for each $x \in E$.

The obtained representation implies, in particular, that for each $x \in E$ the function ϕx is integrable. This means that $\phi \in E_{\mathrm{n}}^\sim$.

(2) $\Longrightarrow$ (1) Assume that there exist a σ-subalgebra $\mathcal{A}$ of Σ and a nonnegative function $\phi \in E_{\mathrm{n}}^\sim$ such that $\mathcal{E}(\phi|\mathcal{A}) = \mathbf{1}$ and $Px = \mathcal{E}(\phi x|\mathcal{A})$ for each $x \in E$. In particular, notice that $\phi = \phi\mathbf{1} \in L_1(\Sigma)$. Lemma 5.37 guarantees that P is a positive order continuous operator. We also have $P\mathbf{1} = \mathcal{E}(\phi\mathbf{1}|\mathcal{A}) = \mathcal{E}(\phi|\mathcal{A}) = \mathbf{1}$ by the condition on ϕ.

Let us verify that P is a projection. Keeping in mind that $\mathcal{E}(\cdot|\mathcal{A})$ satisfies the averaging property, for each $x \in E$ we obtain

$$\begin{aligned} P^2x &= P(\mathcal{E}(\phi x|\mathcal{A})) = \mathcal{E}(\phi\mathcal{E}(\phi x|\mathcal{A})|\mathcal{A}) \\ &= \mathcal{E}(\phi|\mathcal{A})\mathcal{E}(\phi x|\mathcal{A}) = \mathcal{E}(\phi x|\mathcal{A}) = Px\,. \end{aligned}$$

It remains to verify that $Y = P(E)$, the range of P, is a vector sublattice of E. To this end, let $y \in Y$ and then choose a vector $x \in E$ such that $y = Px = \mathcal{E}(\phi x|\mathcal{A})$. Let $A = \{\omega \in \Omega : y(\omega) \geq 0\}$. Then $A \in \mathcal{A}$ because y is $\mathcal{A}$-measurable, and so $\mathcal{E}(\chi_A|\mathcal{A}) = \chi_A$. Notice also that $y^+ = y\chi_A$. Using this and the averaging property of the operator $\mathcal{E}(\cdot|\mathcal{A})$ once more, we obtain

$$\begin{aligned} y^+ &= y\chi_A = \mathcal{E}(\phi x|\mathcal{A})\mathcal{E}(\chi_A|\mathcal{A}) = \mathcal{E}(\phi x\mathcal{E}(\chi_A|\mathcal{A})|\mathcal{A}) \\ &= \mathcal{E}(\phi x\chi_A|\mathcal{A}) = P(x\chi_A)\,. \end{aligned}$$

Since the function $x\chi_A$ belongs to E, it follows that $y^+ \in Y$. This shows that Y is a vector sublattice of E, and the proof is finished. ■

Theorem 5.45 has a number of important consequences.

Corollary 5.46. *Let E be an ideal in $L_1(\Sigma)$ containing the constants. For an operator $P\colon E \to E$ the following conditions are equivalent.*

(1) *P is an order continuous strictly positive projection and $P\mathbf{1} = \mathbf{1}$.*

(2) *There exist a σ-subalgebra $\mathcal{A}$ of Σ and a strictly positive function $\phi \in E_n^\sim$ such that $\mathcal{E}(\phi|\mathcal{A}) = \mathbf{1}$ and $Px = \mathcal{E}(\phi x|\mathcal{A})$ for each $x \in E$.*

Proof. Only the implication (1)$\Longrightarrow$(2) needs a proof. Since P is strictly positive, it follows from Lemma 5.43(a) that the range $Y = P(E)$ is a vector sublattice of E. Therefore, Theorem 5.45 applies and guarantees the existence of a σ-subalgebra $\mathcal{A}$ of Σ and a non-negative function $\phi \in E_n^\sim$ such that $\mathcal{E}(\phi|\mathcal{A}) = \mathbf{1}$ and $Px = \mathcal{E}(\phi x|\mathcal{A})$ for each $x \in E$. To see that ϕ is strictly positive, let $A \in \Sigma$ with $\mu(A) > 0$. Then $0 < \chi_A \in E$, and by the strict positivity of P we get $P\chi_A > 0$. Therefore,

$$\int_A \phi\, d\mu = \int_\Omega \phi\chi_A\, d\mu = \int_\Omega \mathcal{E}(\phi\chi_A|\mathcal{A})\, d\mu = \int_\Omega P\chi_A\, d\mu > 0$$

holds for all $A \in \Sigma$ with $\mu(A) > 0$. This guarantees that ϕ is a strictly positive function. ■

Corollary 5.47. *For an operator $P\colon L_p(\Sigma) \to L_p(\Sigma)$, where $1 \leq p < \infty$, the following statements are equivalent.*

(1) *P is a strictly positive projection such that $P\mathbf{1} = \mathbf{1}$.*

(2) *There exist a σ-subalgebra $\mathcal{A}$ of Σ and a strictly positive function ϕ in $L_q(\Sigma)$, where $\frac{1}{p}+\frac{1}{q} = 1$, such that $\mathcal{E}(\phi|\mathcal{A}) = \mathbf{1}$ and $Px = \mathcal{E}(\phi x|\mathcal{A})$ for each $x \in L_p(\Sigma)$.*

Proof. The equivalence of these two statements follows immediately from Corollary 5.46 by observing that the order continuity of the norm implies that every positive operator on any $L_p(\Sigma)$ with $1 \leq p < \infty$ is automatically order continuous. ■

Let E be an ideal in $L_1(\Sigma)$ containing the constants. Then we know that $L_\infty(\Sigma) \subseteq E_n^\sim \subseteq L_1(\Sigma)$. This means that we can consider $L_\infty(\Sigma)$ as an ideal in the order dual $E^\sim$. With this observation in mind, we can state the following result.

Lemma 5.48. *For an ideal E in $L_1(\Sigma)$ containing the constants and a positive operator $T\colon E \to E$ we have the following.*

(1) *The order adjoint of T fixes $\mathbf{1}$, i.e., $T^\sim \mathbf{1} = \mathbf{1}$, if and only if for each $x \in E$ we have $\int Tx\, d\mu = \int x\, d\mu$.*

(2) *If $T^\sim \mathbf{1} = \mathbf{1}$, then T is strictly positive and order continuous.*

Proof. (1) Assume first that $T^\sim \mathbf{1} = \mathbf{1}$. Then for each $x \in E$ we have

$$\int x\, d\mu = \langle x, \mathbf{1} \rangle = \langle x, T^\sim \mathbf{1} \rangle = \langle Tx, \mathbf{1} \rangle = \int Tx\, d\mu\,.$$

For the converse, assume that $\int x\, d\mu = \int Tx\, d\mu$ holds for each $x \in E$. Rewriting this identity in duality form and taking into account that $\mathbf{1} \in E^\sim$, we see that

$$\langle x, \mathbf{1} \rangle = \langle Tx, \mathbf{1} \rangle = \langle x, T^\sim \mathbf{1} \rangle\,,$$

or $\langle x, T^\sim \mathbf{1} - \mathbf{1} \rangle = 0$ for each $x \in E$. This implies $T^\sim \mathbf{1} = \mathbf{1}$.

(2) We show first that T is strictly positive. To this end, let $0 < x \in E$. From part (1), we have $\int Tx\, d\mu = \int x\, d\mu > 0$. This implies $Tx > 0$, and so T is strictly positive.

For the order continuity of T, assume $x_n \downarrow 0$ in E. This implies $x_n \downarrow 0$ in $L_1(\Sigma)$, and so from part (1) we get

$$\int_\Omega Tx_n\, d\mu = \int_\Omega x_n\, d\mu \downarrow 0\,.$$

Since $Tx_n \downarrow$, this implies that $Tx_n \downarrow 0$ μ-a.e. Consequently, $Tx_n \downarrow 0$ holds in E, and so T is order continuous. ■

We now present the definition of a bistochastic projection.

Definition 5.49. *Let E be an ideal in $L_1(\Sigma)$ containing the constants. A positive projection $P\colon E \to E$ is said to be* ***bistochastic*** *if both P and its order adjoint operator $P^\sim\colon E^\sim \to E^\sim$ are Markov, that is, $P\mathbf{1} = \mathbf{1}$ and $P^\sim \mathbf{1} = \mathbf{1}$.*

From Lemma 5.48 we have the following.

Corollary 5.50. *Every bistochastic projection is strictly positive and order continuous.*

We continue with a few more representations of projections my means of conditional expectation operators.

Corollary 5.51. *Let E be an ideal in $L_1(\Sigma)$ containing the constants. For an operator $P\colon E \to E$ the following statements are equivalent.*

(1) *P is a bistochastic projection.*

(2) *There exists a unique σ-subalgebra $\mathcal{A}$ of Σ such that $Px = \mathcal{E}(x|\mathcal{A})$ for each $x \in E$.*

Proof. Again only the implication (1)$\Longrightarrow$(2) is non-trivial. Assume that P is a bistochastic projection. By Corollary 5.50, we know that P is strictly positive and order continuous. Therefore, in view of Theorem 5.45, there exist a σ-subalgebra $\mathcal{A}$ of Σ and a non-negative function $\phi \in E_{\rm n}^{\sim} \subseteq L_1(\Sigma)$ such that $\mathcal{E}(\phi|\mathcal{A}) = \mathbf{1}$ and $Px = \mathcal{E}(\phi x|\mathcal{A})$ for each $x \in E$. The only thing left to verify is that $\phi = \mathbf{1}$.

Assume to the contrary that $\phi \neq \mathbf{1}$. Since $\int_\Omega \phi\, d\mu = 1$, it follows that there exist a constant $\gamma > 1$ and a set $B \in \Sigma$ with $\mu(B) > 0$ such that $\phi\chi_B > \gamma\chi_B$. Using that P is a bistochastic operator and the definition of the conditional expectation operator $\mathcal{E}(\cdot|\mathcal{A})$, we obtain that

$$\begin{aligned}\mu(B) &= \int_\Omega \chi_B\, d\mu = \int_\Omega P\chi_B\, d\mu = \int_\Omega \mathcal{E}(\phi\chi_B|\mathcal{A})\, d\mu \\ &= \int_\Omega \phi\chi_B\, d\mu > \int_\Omega \gamma\chi_B\, d\mu = \gamma\mu(B)\,,\end{aligned}$$

which is a contradiction. Therefore, $\phi = \mathbf{1}$ and so $P = \mathcal{E}(\cdot|\mathcal{A})$. ■

If P is a contractive projection on $L_1(\Sigma)$ satisfying $P\mathbf{1} = \mathbf{1}$, then by Lemma 5.42(a) the projection P is bistochastic, and so the preceding corollary immediately yields the classical result of R. G. Douglas [**112**].

Corollary 5.52 (Douglas)**.** *For an operator $P\colon L_1(\Sigma) \to L_1(\Sigma)$ the following statements are equivalent.*

(1) *P is a contractive projection satisfying $P\mathbf{1} = \mathbf{1}$.*

(2) *P is a conditional expectation operator. That is, there exists a unique σ-subalgebra $\mathcal{A}$ of Σ such that $P = \mathcal{E}(\cdot|\mathcal{A})$.*

Corollary 5.52 was generalized to $L_p(\Sigma)$-spaces for $1 < p < \infty$ with $p \neq 2$ by T. Andô [**34**]. Under the assumption that the contractive projection is positive the restriction $p \neq 2$ is redundant. We present below a very simple proof of this result.

Corollary 5.53 (Andô)**.** *For a positive operator $P\colon L_p(\Sigma) \to L_p(\Sigma)$, where $1 \le p < \infty$, the following statements are equivalent.*

(1) *P is a contractive projection satisfying $P\mathbf{1} = \mathbf{1}$.*

(2) *P is a conditional expectation operator. That is, there exists a σ-subalgebra $\mathcal{A}$ of Σ satisfying $Px = \mathcal{E}(x|\mathcal{A})$ for each $x \in L_p(\Sigma)$.*

Proof. Notice that for $p = 1$ the result is simply Douglas' result (Corollary 5.52). So, we can assume that $1 < p < \infty$.

(1) $\Longrightarrow$ (2) The projection P is certainly order continuous since the norm on $L_p(\Sigma)$ is order continuous. In view of Lemma 5.43(b), the range of P is a vector sublattice of $L_p(\Sigma)$, since the operator P is contractive and the norm on $L_p(\Sigma)$ is strictly monotone. Hence, by Theorem 5.45, the representation $Px = \mathcal{E}(\phi x|\mathcal{A})$ holds. We want to show that $\phi = \mathbf{1}$. For this to be true it suffices to know that $P^*\mathbf{1} = \mathbf{1}$. But this is guaranteed by Lemma 5.42(a).

(2) $\Longrightarrow$ (1) The implication was established in Lemma 5.38. ■

Regarding the contractive projections on L_p-spaces some historical comments are in order. A version of Douglas' theorem (Corollary 5.52) for positive operators appeared in the book by J. Neveu [**256**, Exercise IV.3.1, p. 123] that was published in 1963. Several other versions of Corollary 5.52 (with different assumptions about the operator) go back even earlier, namely, to the papers by S-T. C. Moy [**246**] and G. C. Rota [**285**]. As we have already mentioned, Douglas' theorem was generalized to L_p-spaces by T. Andô [**34**]. Subsequently, the description of positive projections on Banach function spaces via conditional expectations was studied by many authors; for details see [**59, 110, 236, 237**] and the references therein. The proofs presented here are adaptations and refinements of those in [**9, 110, 236, 237**]. It seems that S-T. C. Moy [**246**] was the first who used the conditional expectations for representations of some operators that were not assumed to be projections. Some considerable generalizations of his results were obtained by P. G. Dodds, C. B. Huijsmans and B. de Pagter [**110**]; see for instance Exercise 10 at the end of the section. For a current survey on contractive projections on Banach spaces see the article by B. Randrianantoanina [**274**].

Our next comment relates some of the results obtained in this sections with the interpolation of operators on Banach function spaces. For this comment we need some terminology from the theory of Banach function spaces.

- *Two functions $x, y \in L_0(\mu)$ are said to be* ***equimeasurable*** *if they generate the same distribution function, i.e., if the equality*

$$\mu\big(\{\omega\in\Omega\colon\ x(\omega)\le t\}\big) = \mu\big(\{\omega\in\Omega\colon\ y(\omega)\le t\}\big)$$

  holds for each $t \in \mathbb{R}$.

- *A Banach function space E associated with some $L_0(\mu)$-space is called* ***rearrangement invariant*** *if every $y \in L_0(\mu)$ that is equimeasurable to some $x \in E$ belongs to E and $\|y\| = \|x\|$.*

- *A rearrangement invariant Banach function space E is called an* ***interpolation space*** *if each bounded operator on $L_1(\Sigma)$ that leaves $L_\infty(\Sigma)$ invariant and is bounded on $L_\infty(\Sigma)$ also leaves E invariant and is bounded on E.*

Regarding, interpolation spaces we have the following classical result known as the Riesz–Thorin theorem [**56**, Theorem 1.1.1].

Theorem 5.54 (Riesz–Thorin). *Each $L_p(\Sigma)$ is an interpolation space.*

The Riesz–Thorin theorem provides, of course, another proof of Lemma 5.38, according to which the conditional expectation operator $\mathcal{E}(\cdot|\mathcal{A})$ leaves each $L_p(\Sigma)$-space invariant, that is, $\mathcal{E}(L_p(\Sigma)|\mathcal{A}) \subseteq L_p(\Sigma)$ for each σ-subalgebra $\mathcal{A}$ of Σ. However, it is important to point out that this is not true for an arbitrary ideal E in $L_1(\Sigma)$. For some algebras $\mathcal{A}$ the inclusion $\mathcal{E}(E|\mathcal{A}) \subseteq E$ may hold and for some it may not. Remarkably, as was proved by A. A. Mekler [**238**], if $\mathcal{E}(E|\mathcal{A}) \subseteq E$ for each σ-subalgebra $\mathcal{A}$ of Σ, then E is an interpolation space.

We continue with two examples showing that all conditions in the preceding corollaries are essential. Our first example shows that without the assumption $P\mathbf{1} = \mathbf{1}$ the conclusion of Theorem 5.45 is no longer true.

Example 5.55. Let $E = L_1[0,1]$, where $[0,1]$ is equipped with the Lebesgue measure. We define an operator $P\colon L_1[0,1] \to L_1[0,1]$ by letting

$$Px = \Big[x + \int_0^{\frac{1}{2}} x(s)\,ds\Big]\chi_{[\frac{1}{2},1]}\,.$$

A direct verification shows that P is a strictly positive projection on $L_1[0,1]$ with $P\mathbf{1} = \frac{3}{2}\chi_{[\frac{1}{2},1]}$ and range $L_1[\frac{1}{2},1]$. We claim that this projection P does not have a representation of the form $Px = \mathcal{E}(\phi x|\mathcal{A})$ for some ϕ in $E_n^\sim = L_\infty[0,1]$ and some σ-subalgebra $\mathcal{A} \subseteq \Sigma$.

To see this, assume contrary to our claim that $Px = \mathcal{E}(\phi x|\mathcal{A})$ for some $\phi \in L_\infty[0,1]$ and some σ-subalgebra $\mathcal{A} \subseteq \Sigma$. This would imply (see Exercise 9 at the end of the section) that $\mathcal{E}(\phi|\mathcal{A})$ were a characteristic function. However, $\mathcal{E}(\phi|\mathcal{A}) = P\mathbf{1} = \frac{3}{2}\chi_{[\frac{1}{2},1]}$, a contradiction. Therefore, P cannot have a representation of the form $Px = \mathcal{E}(\phi x|\mathcal{A})$.

On the other hand, if we define $Q\colon L_1[0,1] \to L_1[0,1]$ by $Qx = x\chi_{[\frac{1}{2},1]}$, then Q is also a positive projection with range $L_1[\frac{1}{2},1]$ satisfying $Q\mathbf{1} = \chi_{[\frac{1}{2},1]}$. Moreover, it is easy to see that if

$$\mathcal{A} = \Big\{ \big[0,\tfrac{1}{2}\big] \cup A\colon\ A \text{ is a Lebesgue measurable subset of } \big[\tfrac{1}{2},1\big] \Big\}$$

and $\phi = \chi_{[\frac{1}{2},1]}$, then $Qx = \mathcal{E}(\phi x|\mathcal{A})$ for each $x \in E$. ∎

Our next example shows that in general a Markov projection can easily fail to be contractive and its range can fail to be a vector sublattice.

Example 5.56. We shall establish the existence of a positive projection $P\colon L_1[0,1] \to L_1[0,1]$ such that:

(1) $P\mathbf{1} = \mathbf{1}$.

(2) $\|P\| > 1$.

(3) The range of P is not a vector sublattice of $L_1[0,1]$.

Consider the sets $A = \big[0, \frac{1}{3}\big)$, $B = \big[\frac{1}{3}, \frac{2}{3}\big)$, and $C = \big[\frac{2}{3}, 1\big]$, and let $\mathcal{A}$ be the σ-subalgebra of the Lebesgue σ-algebra Σ of $[0,1]$ generated by these three sets. It follows that

$$L_1(\mathcal{A}) = \big\{c_1\chi_A + c_2\chi_B + c_3\chi_C\colon\; c_1, c_2, c_3 \in \mathbb{R}\big\},$$

and the positive operator $\mathcal{E}(\cdot|\mathcal{A})\colon L_1(\Sigma) \to L_1(\Sigma)$ is given by

$$\mathcal{E}(x|\mathcal{A}) = 3\Big[\int_0^{\frac{1}{3}} x(s)\,ds\Big]\chi_A + 3\Big[\int_{\frac{1}{3}}^{\frac{2}{3}} x(s)\,ds\Big]\chi_B + 3\Big[\int_{\frac{2}{3}}^{1} x(s)\,ds\Big]\chi_C$$

and has $L_1(\mathcal{A})$ as its range. (See Exercise 4 at the end of the section.)

Next, define $Q\colon L_1(\mathcal{A}) \to L_1(\mathcal{A})$ by

$$Q\big(c_1\chi_A + c_2\chi_B + c_3\chi_C\big) = c_1\chi_A + c_2\chi_B + \tfrac{c_1+c_2}{2}\chi_C,$$

and note that Q is a positive projection satisfying $Q\mathbf{1} = \mathbf{1}$.

If we define $P\colon L_1[0,1] \to L_1[0,1]$ by $P = Q\mathcal{E}(\cdot|\mathcal{A})$, then P is a positive projection satisfying $P\mathbf{1} = \mathbf{1}$. In particular, P is a Markov projection. Now notice that if $x = 3\chi_A$, then $\|x\|_1 = 1$, $\mathcal{E}(x|\mathcal{A}) = 3\chi_A$ and $Px = 3\chi_A + \frac{3}{2}\chi_C$. This implies $\|Px\|_1 = \frac{3}{2} > 1$, and so $\|P\| > 1$ holds.

To see that the range of P is not a vector sublattice, notice that the two functions $y_1 = 3\chi_A$ and $y_2 = \chi_B$ satisfy $Py_1 = 3\chi_A + \frac{3}{2}\chi_B$ and $Py_2 = \chi_B + \frac{1}{2}\chi_C$. This implies $Py_1 \vee Py_2 = 3\chi_A + \chi_B + \frac{3}{2}\chi_B \notin R(P)$, and so the range of P is not a vector sublattice of $L_1[0,1]$. ∎

It is worth noting that for a Markov projection on $L_1[0,1]$ condition (3) in the preceding example is in general stronger than condition (2); see Exercise 13 at the end of the section.

We conclude the section with one more important property of the conditional expectation operator asserting that the operator $\mathcal{E}(\cdot|\mathcal{A})$ is basically self-adjoint.

Theorem 5.57. *For each σ-subalgebra $\mathcal{A}$ of the σ-algebra Σ the adjoint operator $\mathcal{E}(\cdot|\mathcal{A})^*\colon L_\infty(\Sigma) \to L_\infty(\Sigma)$ satisfies $\mathcal{E}(y|\mathcal{A})^* = \mathcal{E}(y|\mathcal{A})$ for each $y \in L_\infty(\Sigma)$. That is, $\mathcal{E}(\cdot|\mathcal{A})^*$ coincides with the restriction of the conditional expectation operator $\mathcal{E}(\cdot|\mathcal{A})$ to the space $L_\infty(\Sigma)$.*

Proof. In this proof the projection $\mathcal{E}(\cdot|\mathcal{A})\colon L_1(\Sigma) \to L_1(\Sigma)$ will be denoted for brevity by P. Recall that P is bistochastic, that is, $P^*\mathbf{1} = \mathbf{1}$.

The operator $P^*\colon L_\infty(\Sigma) \to L_\infty(\Sigma)$ is also a bistochastic projection on $L_\infty(\Sigma)$, and therefore by Corollary 5.51 there exists a σ-subalgebra $\mathcal{B}$ of Σ such that for each $y \in L_\infty(\Sigma)$ we have $P^*y = \mathcal{E}(y|\mathcal{B})$. That is, on the space $L_\infty(\Sigma)$ we have the two (positive order continuous) bistochastic projections $P = \mathcal{E}(\cdot|\mathcal{A})$ and $P^* = \mathcal{E}(\cdot|\mathcal{B})$.

Obviously, the projections $\mathcal{E}(\cdot|\mathcal{A})$ and $\mathcal{E}(\cdot|\mathcal{B})$ can be assumed acting on $L_2(\Sigma)$. To complete the proof, it suffices to show that $\mathcal{E}(y|\mathcal{A}) = \mathcal{E}(y|\mathcal{B})$ for each $y \in L_2(\Sigma)$. Since $\mathcal{E}(\cdot|\mathcal{A})$ and $\mathcal{E}(\cdot|\mathcal{B})$ are projections on $L_2(\Sigma)$, it is enough to prove that $\mathcal{E}(\cdot|\mathcal{A})$ and $\mathcal{E}(\cdot|\mathcal{B})$ have the same range and the same kernel. Since (by Lemma 5.38) both projections $\mathcal{E}(\cdot|\mathcal{A})$ and $\mathcal{E}(\cdot|\mathcal{B})$ are contractions on $L_2(\Sigma)$, it follows that $\mathcal{E}(\cdot|\mathcal{A})$ and $\mathcal{E}(\cdot|\mathcal{B})$ are orthogonal projections; see Exercise 14 at the end of the section. In particular, if $\mathcal{E}(\cdot|\mathcal{A})$ and $\mathcal{E}(\cdot|\mathcal{B})$ have the same range, then they automatically have the same kernel—which is the orthogonal complement of the common range. Consequently, $\mathcal{E}(\cdot|\mathcal{A}) = \mathcal{E}(\cdot|\mathcal{B})$ if and only if $\mathcal{E}(\cdot|\mathcal{A})$ and $\mathcal{E}(\cdot|\mathcal{B})$ have the same range as projections on $L_2(\Sigma)$. To verify the latter, it suffices to establish that $\chi_A \in L_2(\mathcal{B})$ for each $A \in \mathcal{A}$ and $\chi_B \in L_2(\mathcal{A})$ for each $B \in \mathcal{B}$.

Let $A \in \mathcal{A}$. Using the definition of the adjoint operator and the fact that $P\chi_A = \chi_A$, we have:

$$\langle \mathcal{E}(\chi_A|\mathcal{B}), \chi_A\rangle = \langle P^*\chi_A, \chi_A\rangle = \langle \chi_A, P\chi_A\rangle = \langle \chi_A, \chi_A\rangle = \mu(A)\,.$$

At the same time, using Hölder's inequality, we obtain

$$\langle \mathcal{E}(\chi_A|\mathcal{B}), \chi_A\rangle = \langle P^*\chi_A, \chi_A\rangle \le \|P^*\chi_A\|_\infty \cdot \|\chi_A\|_1 \le \|\chi_A\|_1 = \mu(A)\,.$$

That is, we have obtained the equality in the Hölder inequality, and therefore the function $\mathcal{E}(\chi_A|\mathcal{B})$ must be equal to 1 on the set A. Similarly, we can show that $\mathcal{E}(\chi_{A^c}|\mathcal{B})$ must be equal to 1 on the set A^c. Since both functions $\mathcal{E}(\chi_A|\mathcal{B})$ and $\mathcal{E}(\chi_{A^c}|\mathcal{B})$ are non-negative and $\mathcal{E}(\chi_A|\mathcal{B}) + \mathcal{E}(\chi_{A^c}|\mathcal{B}) = \mathcal{E}(\mathbf{1}|\mathcal{B}) = \mathbf{1}$, we can conclude that $\chi_A = \mathcal{E}(\chi_A|\mathcal{B}) \in L_2(\mathcal{B})$.

Next, take any $B \in \mathcal{B}$. We have

$$\langle P\chi_B, \chi_B\rangle = \langle \chi_B, P^*\chi_B\rangle = \langle \chi_B, \mathcal{E}(\chi_B|\mathcal{B})\rangle = \langle \chi_B, \chi_B\rangle = \mu(B)\,.$$

Using Hölder's inequality once again, we obtain

$$\langle P\chi_B, \chi_B\rangle \le \|P\chi_B\|_1 \cdot \|\chi_B\|_\infty = \|P\chi_B\|_1 \le \|\chi_B\|_1 = \mu(B)\,.$$

As above, this implies $\chi_B = P\chi_B \in L_2(\mathcal{A})$. The proof is finished. ∎

Exercises

1. Let (Ω, Σ, μ) be a probability space, and assume that $\mathcal{A}$ is a σ-subalgebra of Σ. Show that for each function $f \in L_\infty(\mathcal{A})$ we have $\|f\|_{\mathcal{A},\infty} = \|f\|_{\Sigma,\infty}$.

2. This exercise shows that every Banach lattice with order continuous norm and a weak unit can be represented by a Banach function space that "lives" between L_∞ and L_1.

Let E be a Banach lattice with order continuous norm and a weak unit e. Show that there exists a probability measure space (Ω, Σ, μ) and an ideal F in $L_1(\Sigma)$ containing the constants, i.e., $L_\infty(\Sigma) \subseteq F \subseteq L_1(\mu)$, and a surjective lattice isomorphism $T\colon E \to F$ such that $T(e) = \mathbf{1}$. [HINT: Let $\phi\colon E \to \mathbb{R}$ be a strictly positive linear functional satisfying $\phi(e) = 1$. (Since E has order continuous norm, such a functional always exists; see [**30**, Theorem 12.14, p. 183].) Now consider the L-norm $|||\cdot|||$ on E defined by $|||x||| = \phi(|x|)$. Then E is an ideal in the norm completion $\hat{E}$ of $(E, |||\cdot|||)$ and e is a weak unit in $\hat{E}$. Finally, use the Kakutani–Bohnenblust–Nakano Representation Theorem 3.5 to identify $\hat{E}$ with an $L_1(\mu)$-space for some probability measure space (Ω, Σ, μ) with e corresponding to the function $\mathbf{1}$. This implies that $L_\infty(\mu) \subseteq E \subseteq L_1(\mu)$.]

3. Prove Lemma 5.37.

4. Assume that $\{\Omega_i\}_{i\in I}$ is an at most countable **partition** of a set Ω. That is, I is at most countable, $\Omega_i \neq \emptyset$ for each $i \in I$, $\Omega_i \cap \Omega_j = \emptyset$ for $i \neq j$, and $\bigcup_{i\in I} \Omega_i = \Omega$. Establish the following.

(a) The σ-algebra generated by the family $\{\Omega_i\}_{i\in I}$ is given by
$$\mathcal{A} = \left\{\textstyle\bigcup_{i\in J} \Omega_i\colon\ J \subseteq I\right\}.$$

(b) A function $f\colon \Omega \to \mathbb{R}$ is $\mathcal{A}$-measurable if and only if it is of the form $f = \sum_{i\in I} \alpha_i \chi_{\Omega_i}$.

(c) If Ω is at most countable, then every σ-algebra of subsets of Ω is generated by a unique at most countable partition of Ω.

(d) Now suppose that (Ω, Σ, μ) is a probability measure space and that $\{\Omega_i\}_{i\in I}$ is an at most countable measurable partition of Ω, i.e, besides $\{\Omega_i\}_{i\in I}$ being a partition of Ω, we also have $\Omega_i \in \Sigma$ for each $i \in I$. If $\mu(\Omega_i) > 0$ for each $i \in I$, then for any $f \in L_1(\Omega, \Sigma, \mu)$ we have
$$\mathcal{E}(f|\mathcal{A}) = \sum_{i\in I} \left[\tfrac{1}{\mu(\Omega_i)} \textstyle\int_{\Omega_i} f\, d\mu\right] \chi_{\Omega_i}\,.$$

5. For an arbitrary convex function $\psi\colon \mathbb{R} \to \mathbb{R}$, any function $f \in L_1(\Sigma)$ for which $\psi \circ f \in L_1(\Sigma)$, and any σ-subalgebra $\mathcal{A}$ of Σ establish the following inequalities—known as **Jensen's inequalities**:

(a) $\psi\big(\int f\, d\mu\big) \leq \int \psi \circ f\, d\mu$.

(b) $\psi\big(\mathcal{E}(f|\mathcal{A})\big) \leq \mathcal{E}(\psi \circ f|\mathcal{A})$ in $L_1(\mathcal{A})$.

6. Give an example of an ideal E in $L_1(\Sigma)$ such that E is a Banach lattice and contains a non-trivial norm closed order dense ideal in E.

7. Show that the conclusions of Lemma 5.42 do not hold in general for the remaining values of the indices.

8. Show that the conditional expectation operator $\mathcal{E}(\cdot|\mathcal{A})$ satisfies the following property: *If a function* $f \in L_1(\Sigma)$ *has its support in some set* $A \in \mathcal{A}$, *then the function* $\mathcal{E}(f|\mathcal{A})$ *also has its support in* A.

9. (Dodds–Huijsmans–de Pagter [**110**]) Suppose that E is an ideal in $L_1(\Sigma)$ containing the constants, and let $0 \le \phi \in E^\sim_{\mathrm{n}}$. Assume also that for some σ-subalgebra $\mathcal{A}$ of Σ the operator $P\colon L_1(\Sigma) \to L_1(\Sigma)$, defined by $Px = \mathcal{E}(\phi x|\mathcal{A})$, leaves E invariant.

Show that P is a projection if and only if there exists a subset $A \in \mathcal{A}$ such that $\operatorname{Supp}\phi \subseteq A$ and $\mathcal{E}(\phi|\mathcal{A}) = \chi_A$ in $L_1(\mathcal{A})$.

10. (Moy [**246**]; Dodds–Huijsmans–de Pagter [**110**]) This exercise is a modification of Theorem 5.45 for linear operators that are not necessarily projections. Let E be an ideal in $L_1(\Sigma)$ containing the constants, and let $T\colon E \to E$ be a linear operator such that:

(a) If $x_n \xrightarrow{o} 0$ in E, then $\|Tx_n\|_1 \to 0$.

(b) T leaves $L_\infty(\Sigma)$ invariant.

(c) For each $h \in L_\infty(\Sigma)$ and each $g \in E$ we have $T(hTg) = ThTg$.

Then there exists a σ-subalgebra $\mathcal{A}$ of Σ and a function $\phi \in E^\sim_{\mathrm{n}}$ so that $Tx = \mathcal{E}(\phi x|\mathcal{A})$ for each $x \in E$.

11. Give an example of a Markov projection on $L_1[0,1]$ that is not bistochastic. Show that such a projection cannot be contractive.

12. For a vector subspace Y of some $L_p(\Sigma)$ with $1 \le p < \infty$ establish the following.

(a) If Y is a closed vector sublattice, then Y has a weak unit.

(b) Y is a closed vector sublattice if and only if it is the range of a contractive positive projection on $L_p(\Sigma)$.

13. According to the previous exercise, if a Markov projection P on some $L_p(\Sigma)$ with $1 \le p < \infty$ is contractive, then its range is a vector sublattice of $L_p(\Sigma)$. Show that the converse is not true. That is, give an example of a Markov projection P on $L_p(\Sigma)$ such that its range is a vector sublattice of $L_p(\Sigma)$ but $\|P\| > 1$.

14. Show that a projection on a Hilbert space is an orthogonal projection if and only if it is a contraction.

15. Use Theorem 5.57 to obtain an alternative proof of Lemma 5.38. That is, show that for each σ-subalgebra $\mathcal{A}$ of Σ and for each $1 \le p \le \infty$ the operator $\mathcal{E}(\cdot|\mathcal{A})$ leaves the space $L_p(\Sigma)$ invariant and is contractive on this space.

5.4. Positive Projections and Lattice-subspaces

In spite of the somewhat "negative" nature of Examples 5.55 and 5.56, there is a considerable amount of interesting information that can be obtained about general positive projections and their ranges. This will be done in this section. As our starting point, we shall use Example 5.56. Though the range Y of the projection Q is not a vector sublattice of X, nevertheless, Y

considered in its own right is a vector lattice! This special feature possessed by certain subspaces is formalized in the next definition.

Definition 5.58. *A vector subspace Y of an ordered vector space X is said to be a **lattice-subspace** if Y equipped with the ordering induced by X is a vector lattice. Equivalently, Y is a lattice-subspace of X if Y equipped with the cone $Y_+ = Y \cap X_+$ is a vector lattice.*

If Y is a lattice-subspace of an ordered vector space X, then the lattice operations in the vector lattice (Y, Y_+) will be denoted by $\curlyvee$ and $\curlywedge$ (or by $\vee_Y$ and $\wedge_Y$, respectively). That is, if $y_1, y_2 \in Y$, then the supremum of the set $\{y_1, y_2\}$ in the vector lattice (Y, Y_+) is denoted by $y_1 \curlyvee y_2$ or $y_1 \vee_Y y_2$. Similarly, the infimum of the set $\{y_1, y_2\}$ in the vector lattice (Y, Y_+) is denoted by $y_1 \curlywedge y_2$ or $y_1 \wedge_Y y_2$. The absolute value of any $y \in Y$ will be denoted by $|y|_Y$.

It should be obvious that a vector sublattice of a vector lattice is automatically a lattice-subspace. The converse is not true in general; namely, the class of lattice-subspaces is considerably larger than the class of vector sublattices, and so, an arbitrary lattice-subspace need not be a vector sublattice. For instance, if we consider the Riesz space $E = C[0,1]$ and let Y be the vector subspace of E consisting of all linear functions, i.e., Y consists of all functions of the form $y(t) = mt + b$, then Y is a lattice-subspace of E that fails to be a vector sublattice.

The next result is essentially due to H. H. Schaefer [**290**, p. 214] and it reveals several order properties of the range of a positive projection.

Theorem 5.59 (Schaefer). *The range F of a positive projection $P\colon E \to E$ on a vector lattice satisfies the following properties.*

(1) *F is a lattice-subspace of E and its lattice operations are given by*
$$x \curlyvee y = P(x \vee y), \quad x \curlywedge y = P(x \wedge y), \quad \text{and} \quad |x|_F = P(|x|).$$
(2) *If e is a strong unit in E, then Pe is a strong unit in F.*
(3) *If E is Dedekind complete, then F is Dedekind complete.*
(4) *If P is strictly positive, then F is a vector sublattice of E.*
(5) *If E is a normed lattice, then the norm*
$$|||x||| = \big\||x|_F\big\| = \big\|P|x|\big\|, \quad x \in F,$$
is a lattice norm on F and $\|x\| \leq |||x|||$ for each $x \in F$.
(6) *If E is a Banach lattice, then the restriction of $\|\cdot\|$ to F is equivalent to $|||\cdot|||$; and consequently $(F, |||\cdot|||)$ is a Banach lattice.*

Proof. (1) Clearly, $F_+ = E_+ \cap F$ is a cone. We must show that this cone induces a lattice ordering on F. We shall prove only the supremum formula.

To this end, pick any $x, y \in F$. Then from $x \le x \vee y$ and $y \le x \vee y$ we see that $x = Px \le P(x \vee y)$ and $y \le P(x \vee y)$. That is, $P(x \vee y)$ is an upper bound in F for the set $\{x, y\}$. To see that this is the least upper bound, assume $x \le z$ and $y \le z$ for some $z \in F$. Then $x \vee y \le z$, and the positivity of P implies $P(x \vee y) \le Pz = z$. Hence, $x \veebar y = P(x \vee y)$ in F.

(2) Assume that some element $e \in E$ is a strong unit, and take any $x \in F$. There exists $\lambda > 0$ such that $x \le \lambda e$. The positivity of P implies that $x = Px \le \lambda Pe$. This shows that Pe is a strong unit in F.

(3) Assume that E is Dedekind complete and let $0 \le x_\alpha \downarrow$ in F. Then there exists some $x \in E$ such that $x_\alpha \downarrow x$ in E. Clearly, $Px \le Px_\alpha = x_\alpha$ for each α, that is, Px in a lower bound of $\{x_\alpha\}$ in F. On the other hand, if $x_\alpha \ge y$ holds in F for some $y \in F$ and each α, then $x \ge y$ in E, and so $Px \ge Py = y$. Therefore, $x_\alpha \downarrow Px$ in F and this shows that F is Dedekind complete.

(4) This follows from Lemma 5.43(b).

(5) Clearly, $|x|_F \le |y|_F$ in F implies $|||x||| = \||x|_F\| \le \||y|_F\| = |||y|||$. Consequently, $|||\cdot|||$ is a lattice norm on F. Now if $x \in F$, then we have $|x| = |Px| \le P|x| = |x|_F$, and so $\|x\| \le \||x|_F\| = |||x|||$.

(6) If E is a Banach lattice, then the positivity of P guarantees that P is norm continuous, and hence the projection P has closed range. From (5), for each $x \in F$ we have

$$\|x\| \le |||x||| = \|P(|x|)\| \le \|P\| \cdot \|x\| ,$$

which means that the two norms $|||\cdot|||$ and $\|\cdot\|$ are equivalent on F. In particular, $\bigl(F, |||\cdot|||\bigr)$ is a Banach lattice. ■

A vector subspace Y of an ordered vector space E is said to be **positively complemented** if there exists a positive projection on E with range Y. In this terminology, the main conclusion of Theorem 5.59 states that:

- *Every positively complemented subspace of a vector lattice is a lattice-subspace.*

The next result, due to T. Andô (personal communication) and independently to N. Ghoussoub [**135**, Lemma III.3], shows that the range of a positive projection under some mild conditions is lattice isomorphic to a vector sublattice.

Theorem 5.60. *If $P\colon E \to E$ is an order continuous positive projection on a Dedekind complete vector lattice, then there exists a positively complemented vector sublattice F of E that is order isomorphic to the lattice-subspace $P(E)$.*

If E is also a Banach lattice, then the vector sublattice F is closed.

Proof. We know that the range $Y = P(E)$ is a lattice-subspace of E and we claim that Y is order isomorphic to a positively complemented vector sublattice of E. Consider first $N = \{x \in E\colon\ P|x| = 0\}$, the null ideal of P. Since P is order continuous, N is a band in E. Hence, $E = N \oplus N^d$. We denote by Q the band projection from E onto N^d. Notice that if $x = x_1 + x_2 \in N \oplus N^d$, then $Px = Px_2 = PQx$, and so $PQ = P$.

Now let $P_1 = QP$. Clearly, $P_1 \geq 0$ and

$$P_1^2 = (QP)(QP) = Q(PQ)P = QPP = QP = P_1 .$$

That is, P_1 is a positive projection on E. Obviously, P_1 leaves N^d invariant. We claim that P_1 is strictly positive on the disjoint complement N^d. Indeed, if $0 < x \in N^d$ and $P_1x = QPx = 0$, then $Px \in N$, and so $Px = P(Px) = 0$, contrary to the assumption that $x \in N^d$. That is, $0 < x \in N^d$ implies $P_1x > 0$. Clearly, $P_1(E) = P_1\big(N^d\big)$. Therefore, letting $F = P_1(E)$ and using Theorem 5.59(4), we see that F is a positively complemented vector sublattice of N^d and hence of E. In particular, notice that F is a closed vector sublattice if E is a Banach lattice.

Finally, we claim that $P_1\colon Y \to F$ is an onto order isomorphism. To see this, note first that $x \in Y$ and $P_1x = 0$ imply $Qx = QPx = P_1x = 0$ or $x \in N$, from which it follows that $x = Px = 0$. That is, P_1 is one-to-one. Also, it should be obvious that P_1 is positive and onto. To finish the proof, it remains to be shown that $P_1^{-1}\colon F \to Y$ is also positive. To this end, fix $y \in F^+$ and put $x = P_1^{-1}y \in Y$. This implies $0 \leq P_1x = QPx = Qx$. Now write $x = x_1 + x_2 \in N \oplus N^d$ and note that $x_2 = Qx \geq 0$. Consequently, $P_1^{-1}y = x = Px = Px_1 + Px_2 = Px_2 \geq 0$, and the proof is finished. ∎

Theorem 5.59 does not characterize lattice-subspaces. That is, not every lattice-subspace of a vector lattice is the range of a positive projection. This is demonstrated by the next example taken from [**18**].

Example 5.61. Consider the positive functions $x_1, x_2 \in C[0,1]$ defined by

$$x_1(t) = \begin{cases} (t-\frac{1}{2})^2 & \text{if } 0 \leq t \leq \frac{1}{2} \\ t - \frac{1}{2} & \text{if } \frac{1}{2} < t \leq 1 \end{cases} \quad \text{and} \quad x_2(t) = \begin{cases} \frac{1}{2} - t & \text{if } 0 \leq t \leq \frac{1}{2} \\ (t-\frac{1}{2})^2 & \text{if } \frac{1}{2} < t \leq 1 \,. \end{cases}$$

Let F be the vector space generated in $C[0,1]$ by the linearly independent functions x_1 and x_2. We claim that F is a lattice-subspace of $C[0,1]$.

To prove this, it clearly suffices to show that if $\lambda_1x_1 + \lambda_2x_2 \geq 0$, then $\lambda_1 \geq 0$ and $\lambda_2 \geq 0$. So, assume that $\lambda_1x_1 + \lambda_2x_2 \geq 0$. This implies that for $0 \leq t < \frac{1}{2}$, we have $\lambda_1\big(t-\frac{1}{2}\big)^2 + \lambda_2\big(\frac{1}{2}-t\big) \geq 0$, or $\lambda_1\big(\frac{1}{2}-t\big) + \lambda_2 \geq 0$. Letting

$t \to \frac{1}{2}-$, we obtain $\lambda_2 \geq 0$. Similarly, from $\lambda_1(t-\frac{1}{2})+\lambda_2(t-\frac{1}{2})^2 \geq 0$ for each $\frac{1}{2} < t \leq 1$, we infer that $\lambda_1 \geq 0$.

Next, we claim that there is no positive projection on $C[0,1]$ whose range is F. To see this, assume by way of contradiction that there exists a positive projection $P\colon C[0,1] \to C[0,1]$ such that $P\big(C[0,1]\big) = F$. Consider the positive linear functional $f\colon F \to \mathbb{R}$ defined by

$$f(\lambda_1 x_1 + \lambda_2 x_2) = \lambda_1 .$$

Then $f \circ P$ is a positive linear functional on $C[0,1]$. So, by the Riesz Representation Theorem, there exists a (unique non-zero) regular Borel measure ν on $[0,1]$ such that $[f \circ P](x) = \int_{[0,1]} x\,d\nu$ for each $x \in C[0,1]$. Letting $x = x_2$, we get $0 = [f \circ P](x_2) = \int_{[0,1]} x_2\,d\nu$, from which it follows that $\operatorname{Supp}\nu = \{\frac{1}{2}\}$. Finally, letting $x = x_1$, we obtain

$$1 = [f \circ P](x_1) = \int_{[0,1]} x_1\,d\nu = x_1\big(\tfrac{1}{2}\big)\nu\big(\{\tfrac{1}{2}\}\big) = 0 ,$$

a contradiction. So, F cannot be the range of a positive projection. ∎

In spite of the non-existence of positive projections from a vector lattice E onto its lattice-subspaces, there is an important projection associated with any lattice-subspace Y of a vector lattice E. Namely, there is a positive projection defined on the vector sublattice $\mathcal{R}(Y)$ of E generated by Y and having Y as its range. This result is not trivial since even for a two dimensional Y the vector sublattice $\mathcal{R}(Y)$ can be infinite dimensional; see Exercise 6 at the end of the section. To introduce these special projections, we need to describe the vector sublattice generated by a vector subspace.

For every non-empty subset A of a vector lattice, the symbol $A^\wedge$ will denote the collection of all vectors that can be written as infima of finite subsets of A. That is, a vector a belongs to $A^\wedge$ if there exist vectors $a_1, a_2, \ldots, a_k \in A$ such that $a = \bigwedge_{i=1}^k a_i$. Similarly, $A^\vee$ is the set consisting of all suprema of the finite subsets of A. We write $A^{\vee\wedge}$ for $(A^\vee)^\wedge$ and $A^{\wedge\vee}$ for $(A^\wedge)^\vee$. It always turns out that $A^{\vee\wedge} = A^{\wedge\vee}$. This is a consequence of the following simple result whose proof is based upon the distributive law of the lattice operations in vector lattices and is left as an exercise.

Lemma 5.62. *Let $\{x_{ij}\colon\ i \in I = \{1,2,\ldots,n\};\ j \in J = \{1,2,\ldots,m\}\}$ be a finite subset of vectors in a vector lattice. Then*

$$\bigwedge_{\tau \in J^I} \bigvee_{i=1}^{n} x_{i,\tau(i)} = \bigvee_{i=1}^{n} \bigwedge_{j=1}^{m} x_{ij} \quad \text{and} \quad \bigvee_{\tau \in I^J} \bigwedge_{j=1}^{m} x_{\tau(j),j} = \bigwedge_{i=1}^{n} \bigvee_{j=1}^{m} x_{ij} ,$$

where J^I denotes the (finite) collection of all mappings from I to J.

If A is a non-empty subset of a vector sublattice E, then the collection of all vector sublattices of E that include A is non-empty. The intersection of this collection of vector sublattices is the smallest vector sublattice that includes A. This vector sublattice is called the **vector sublattice** (or the **Riesz subspace**) **generated by** A and will be denoted $\mathcal{R}(A)$. The next result presents a description of $\mathcal{R}(A)$.

Lemma 5.63. *The vector sublattice $\mathcal{R}(Y)$ generated by a vector subspace Y of a vector lattice coincides with $Y^{\wedge\vee}$ and also with $Y^{\vee\wedge}$. That is,*

$$\mathcal{R}(Y) = Y^{\wedge\vee} = Y^{\vee\wedge}.$$

Consequently, the vector sublattice generated by a non-empty subset A of a vector lattice is $\mathcal{R}(A) = [A]^{\wedge\vee}$, where $[A]$ denotes the linear span of A.

Proof. The verification of this identity follows easily from Lemma 5.62. We leave the details as an exercise. ■

The relationship between lattice-subspaces and positive projections is completely revealed by the following result of S. Miyajima [**245**].

Theorem 5.64 (Miyajima). *A vector subspace Y of a vector lattice E is a lattice-subspace of E if and only if there exists a (unique) positive projection $P\colon \mathcal{R}(Y) \to \mathcal{R}(Y)$ whose range is Y. This projection P is defined by*

$$P\Big(\bigvee_{i=1}^{n} \bigwedge_{j=1}^{m} x_{ij}\Big) = \boldsymbol{\bigvee}_{i=1}^{n} \boldsymbol{\bigwedge}_{j=1}^{m} x_{ij}\,.$$

Moreover, the operator $P\colon \mathcal{R}(Y) \to Y$ is a surjective interval preserving lattice homomorphism.

Proof. Let Y be a vector subspace of a vector lattice E such that there exists a positive projection $P\colon \mathcal{R}(Y) \to \mathcal{R}(Y)$ whose range is Y. By Theorem 5.59(1) the vector space Y is a lattice-subspace of $\mathcal{R}(Y)$, and consequently it is also a lattice-subspace of E.

For the converse, assume that Y is a lattice-subspace of E. We start by establishing that the mapping as described in the theorem is well defined. That is, we shall prove that

$$\bigvee_{i=1}^{n} \bigwedge_{j=1}^{m} x_{ij} = \bigvee_{p=1}^{k} \bigwedge_{q=1}^{\ell} y_{pq} \Longrightarrow \boldsymbol{\bigvee}_{i=1}^{n} \boldsymbol{\bigwedge}_{j=1}^{m} x_{ij} = \boldsymbol{\bigvee}_{p=1}^{k} \boldsymbol{\bigwedge}_{q=1}^{\ell} y_{pq}\,, \qquad (\star)$$

where the vectors x_{ij} and y_{pq} belong to Y. To establish this, we need to show first that if $\bigvee_{i=1}^{n} \bigwedge_{j=1}^{m} x_{ij} \geq 0$, then $\boldsymbol{\bigvee}_{i=1}^{n} \boldsymbol{\bigwedge}_{j=1}^{m} x_{ij} \geq 0$.

So, assume that $\bigvee_{i=1}^{n} \bigwedge_{j=1}^{m} x_{ij} \geq 0$. From Lemma 5.62 it follows that $\bigvee_{i=1}^{n} x_{i,\tau(i)} \geq 0$ for each $\tau \in J^I$. Since $\boldsymbol{\bigvee}_{i=1}^{n} x_{i,\tau(i)} \geq \bigvee_{i=1}^{n} x_{i,\tau(i)} \geq 0$, applying Lemma 5.62 again (but this time to the vector lattice (F, F_+)), we obtain that $\boldsymbol{\bigvee}_{i=1}^{n} \boldsymbol{\bigwedge}_{j=1}^{m} x_{ij} = \boldsymbol{\bigwedge}_{\tau \in J^I} \boldsymbol{\bigvee}_{i=1}^{n} x_{i,\tau(i)} \geq 0$.

Now assume that $\bigvee_{i=1}^{n}\bigwedge_{j=1}^{m} x_{ij} = \bigvee_{p=1}^{k}\bigwedge_{q=1}^{\ell} y_{pq}$. From the inequalities $\bigvee_{p=1}^{k}\bigwedge_{q=1}^{\ell} y_{pq} \geq \bigwedge_{q=1}^{\ell} y_{pq} \geq \boldsymbol{\bigwedge}_{q=1}^{\ell} y_{pq}$ for each p, it follows that

$$\bigvee_{i=1}^{n}\bigwedge_{j=1}^{m}\big[x_{ij} - \boldsymbol{\bigwedge}_{q=1}^{\ell} y_{pq}\big] = \bigvee_{p=1}^{k}\bigwedge_{q=1}^{\ell} y_{pq} - \boldsymbol{\bigwedge}_{q=1}^{\ell} y_{pq} \geq 0\,,$$

and hence $\boldsymbol{\bigvee}_{i=1}^{n}\boldsymbol{\bigwedge}_{j=1}^{m}\big[x_{ij} - \boldsymbol{\bigwedge}_{q=1}^{\ell} y_{pq}\big] = \boldsymbol{\bigvee}_{i=1}^{n}\boldsymbol{\bigwedge}_{j=1}^{m} x_{ij} - \boldsymbol{\bigwedge}_{q=1}^{\ell} y_{pq} \geq 0$. This implies $\boldsymbol{\bigvee}_{i=1}^{n}\boldsymbol{\bigwedge}_{j=1}^{m} x_{ij} \geq \boldsymbol{\bigwedge}_{q=1}^{\ell} y_{pq}$ for each p, and thus

$$\boldsymbol{\bigvee}_{i=1}^{n}\boldsymbol{\bigwedge}_{j=1}^{m} x_{ij} \geq \boldsymbol{\bigvee}_{p=1}^{k}\boldsymbol{\bigwedge}_{q=1}^{\ell} y_{pq}\,.$$

By symmetry, the reverse inequality is also true, and so the validity of $(\star)$ has been established. In other words, the mapping P is well defined.

Next, observe that if $x = \bigvee_{i=1}^{n}\bigwedge_{j=1}^{m} x_{ij} = \bigwedge_{p=1}^{k}\bigvee_{q=1}^{\ell} y_{pq}$, then

$$Px = \boldsymbol{\bigvee}_{i=1}^{n}\boldsymbol{\bigwedge}_{j=1}^{m} x_{ij} = \boldsymbol{\bigwedge}_{p=1}^{k}\boldsymbol{\bigvee}_{q=1}^{\ell} y_{pq}\,.$$

This follows easily from Lemma 5.62. We are now ready to establish the basic properties of the mapping P.

- *The mapping P is additive.*

To see this, let $x = \bigvee_{i=1}^{n}\bigwedge_{j=1}^{m} x_{ij}$ and $y = \bigvee_{p=1}^{k}\bigwedge_{q=1}^{\ell} y_{pq}$, and note that

$$\begin{aligned} P(x+y) &= P\big(\bigvee_{i=1}^{n}\bigwedge_{j=1}^{m} x_{ij} + \bigvee_{p=1}^{k}\bigwedge_{q=1}^{\ell} y_{pq}\big) \\ &= P\big(\bigvee_{i=1}^{n}\bigvee_{p=1}^{k}\bigwedge_{j=1}^{n}\bigwedge_{q=1}^{\ell}(x_{ij}+y_{pq})\big) \\ &= \boldsymbol{\bigvee}_{i=1}^{n}\boldsymbol{\bigvee}_{p=1}^{k}\boldsymbol{\bigwedge}_{j=1}^{n}\boldsymbol{\bigwedge}_{q=1}^{\ell}(x_{ij}+y_{pq}) \\ &= \boldsymbol{\bigvee}_{i=1}^{n}\boldsymbol{\bigwedge}_{j=1}^{m} x_{ij} + \boldsymbol{\bigvee}_{p=1}^{k}\boldsymbol{\bigwedge}_{q=1}^{\ell} y_{pq} \\ &= P(x) + P(y)\,. \end{aligned}$$

- *The mapping P is homogeneous.*

Let $x = \bigvee_{i=1}^{n}\bigwedge_{j=1}^{m} x_{ij}$ and fix some scalar α. If $\alpha \geq 0$, then it should be clear that $P(\alpha x) = \alpha P(x)$. If $\alpha < 0$, then note that

$$\begin{aligned} P\big(\alpha\big[\bigvee_{i=1}^{n}\bigwedge_{j=1}^{m} x_{ij}\big]\big) &= P\big(\big[\bigwedge_{i=1}^{n}\bigvee_{j=1}^{m} \alpha x_{ij}\big]\big) = \boldsymbol{\bigwedge}_{i=1}^{n}\boldsymbol{\bigvee}_{j=1}^{m} \alpha x_{ij} \\ &= \alpha\big[\boldsymbol{\bigvee}_{i=1}^{n}\boldsymbol{\bigwedge}_{j=1}^{m} x_{ij}\big] = \alpha P(x)\,. \end{aligned}$$

- *The mapping $P\colon \mathcal{R}(Y) \to Y$ is a surjective interval preserving lattice homomorphism.*

If $x, y \in \mathcal{R}(Y)$, then it should be clear that $P(x \vee y) = P(x) \boldsymbol{\vee} P(y)$, and therefore P is also a lattice homomorphism. To see that P is an interval preserving operator, assume that two vectors $0 \leq x \in \mathcal{R}(Y)$ and $y \in Y$ satisfy $0 \leq y \leq Px$. If we consider the vector $z = x \wedge y \in \mathcal{R}(Y)$, then $0 \leq z \leq x$ holds and $P(z) = P(x \wedge y) = P(x) \wedge P(y) = P(x) \wedge y = y$. Thus, $P[0, x] = [0, Px]$.

- *The positive projection $P: \mathcal{R}(Y) \to \mathcal{R}(Y)$ is the only positive projection on $\mathcal{R}(Y)$ having Y as its range.*

To establish this claim, let $Q: \mathcal{R}(Y) \to \mathcal{R}(Y)$ be another positive projection with range Y. In view of Theorem 5.59 we know that for each finite collection of vectors $x_1, x_2, \ldots, x_n$ in Y we have

$$Q\big(\bigvee_{i=1}^n x_i\big) = \mathbb{V}_{i=1}^n x_i \quad \text{and} \quad Q\big(\bigwedge_{i=1}^n x_i\big) = \mathbb{\Lambda}_{i=1}^n x_i\,.$$

Now fix some vector $x = \bigvee_{i=1}^n \bigwedge_{j=1}^m x_{ij}$ in $\mathcal{R}(Y)$. Then for each i we have $x \geq \bigwedge_{j=1}^m x_{ij}$, and the positivity of Q yields $Q(x) \geq Q\big(\bigwedge_{j=1}^m x_{ij}\big) = \mathbb{\Lambda}_{j=1}^m x_{ij}$ for each i. This implies $Q(x) \geq \mathbb{V}_{i=1}^n \mathbb{\Lambda}_{j=1}^m x_{ij} = P(x)$. Since the vector x can also be written in the form $x = \bigvee_{p=1}^k \bigwedge_{q=1}^\ell y_{pq}$, a similar argument shows that $Q(x) \leq P(x)$. Therefore, $Q(x) = P(x)$ for each $x \in \mathcal{R}(Y)$, and the proof is finished. ■

If Y is a lattice-subspace of a vector lattice, then the unique positive projection $P: \mathcal{R}(Y) \to \mathcal{R}(Y)$ with range Y determined in Theorem 5.64 will be referred to as the **Miyajima projection**. The reader can find more about lattice-subspaces in [**18, 268, 269, 270**].

Exercises

1. If P is a positive projection on an $L_p(\mu)$-space for some $1 \leq p < \infty$, then show that its range is again order isomorphic (up to an equivalent norm) to an $L_p(\nu)$-space for some measure space $(\Omega_1, \Sigma_1, \nu)$.
2. Prove Lemma 5.62.
3. Prove Lemma 5.63.
4. Show that a vector subspace of a finite dimensional vector lattice is a lattice-subspace if and only if it is the range of a positive projection.
5. For a vector subspace Y of some $C(K)$-space establish the following.
   (a) Y separates the points of K if and only if $\mathcal{R}(Y)$ separates the points of K.
   (b) Assume that the induced cone $Y_+ = Y \cap [C(K)]^+$ is generating in Y, i.e., $Y = Y_+ - Y_+$. Then Y majorizes $C(K)$ if and only if $\mathcal{R}(Y)$ majorizes $C(K)$.
   (c) Assume that:
      (i) Y is a Dedekind complete lattice-subspace. (In particular, this is so if Y is a finite dimensional lattice-subspace.)
      (ii) Y separates the points of K and $\mathbf{1} \in Y$.

   Then the Miyajima projection $P: \mathcal{R}(Y) \to \mathcal{R}(Y)$ extends uniquely to a positive projection on $C(K)$ with range Y.

6. Let Y be the vector subspace of $C[0,1]$ generated by the functions $\mathbf{1}$ and s, where $s(t) = t$ for each $t \in [0,1]$. That is, Y consists of all linear functions. Establish the following.

(a) Y is a lattice-subspace of E.

(b) $\mathcal{R}(Y)$ consists of all piecewise linear continuous functions.

(c) The Miyajima projection $P\colon \mathcal{R}(Y) \to \mathcal{R}(Y)$ is given by

$$Px(t) = tx(1) + (1-t)x(0)$$

for each $x \in \mathcal{R}(Y)$ and each $t \in [0,1]$.

(d) The operator $Q\colon C[0,1] \to C[0,1]$ defined by

$$Qx(t) = tx(1) + (1-t)x(0)$$

for each $x \in C[0,1]$ and each $t \in [0,1]$ is the only extension of the Miyajima projection to a positive projection on all of $C[0,1]$.

(e) There are infinitely many bounded projections on $C[0,1]$ with range Y.

7. If Y is a majorizing lattice-subspace of a Banach lattice, then show that the Miyajima projection $P\colon \mathcal{R}(Y) \to \mathcal{R}(Y)$ is continuous.

8. This exercise presents an application of lattice-subspaces to harmonic analysis. Recall that a C^2-function $u\colon \mathcal{O} \to \mathbb{R}$, where $\mathcal{O}$ is an open subset of $\mathbb{R}^2$, is said to be **harmonic** if it satisfies **Laplace's equation** $\nabla^2 u = \frac{\partial^2 u}{\partial x^2} + \frac{\partial^2 u}{\partial y^2} = 0$ at every point of $\mathcal{O}$.

Now let $D = \big\{(x,y)\colon\ x^2+y^2 < 1\big\}$ be the open unit disk. We shall denote by $\mathcal{H}$ the vector space of all continuous functions defined on the closed unit disk $\overline{D} = \big\{(x,y)\colon\ x^2+y^2 \le 1\big\}$ and harmonic on D. The classical *Dirichlet theorem* states that if f is a continuous function on the unit circle $\partial D = \big\{(x,y)\colon\ x^2+y^2 = 1\big\}$, then there exists a unique continuous extension $\widehat{f}$ of f to all of $\overline{D}$ which is harmonic on D. The harmonic extension $\widehat{f}$ in polar coordinates is given by Poisson's classical formula

$$\widehat{f}(r,\theta) = \tfrac{1}{2\pi}\int_0^{2\pi} \tfrac{f(\cos\phi,\,\sin\phi)(1-r^2)}{1-2r\sin(\theta-\phi)+r^2}\,d\phi,\ \ (r,\theta) \in [0,1) \times [0,2\pi]\,.$$

Establish the following.

(a) A function $f \in C\big(\overline{D}\big)$ satisfies $\widehat{f_{\partial D}} \ge 0$ if and only if $f_{\partial D} \ge 0$, where $f_{\partial D}$ denotes the restriction of the function f to ∂D.

(b) The mapping $Q\colon C\big(\overline{D}\big) \to C\big(\overline{D}\big)$, defined by $Qf = \widehat{f_{\partial D}}$, is a positive projection whose range is $\mathcal{H}$.

(c) $\mathcal{H}$ is a lattice-subspace of $C\big(\overline{D}\big)$ and Q is the unique extension of the Miyajima projection to $C\big(\overline{D}\big)$.

(d) $\mathcal{H}$ is an AM-space with unit $\mathbf{1}$, the constant function one, and the mapping $f \mapsto \widehat{f}$, from $C(\partial D)$ to $\mathcal{H}$, is a surjective lattice isometry.

Chapter 6

Spectral Properties

This chapter introduces the spectrum of a bounded operator and studies its basic properties. We start with a general discussion of the spectrum, of the resolvent set, and of the spectral radius. Subsequently, we introduce and study the resolvent function. In particular, we pay special attention to the resolvent function of a positive operator.

The various important parts of the spectrum (the point spectrum, the residual spectrum, the continuous spectrum and the approximate point spectrum) are all introduced and some of their interrelationships are obtained. The chapter culminates with an extensive treatment of the "Functional Calculus" with a special emphasis on the Spectral Mapping Theorem and some of its applications.

The Spectral Mapping Theorem is used in conjunction with the notion of a spectral set to construct reducing pairs of subspaces for a given operator. We also study the isolated points in the spectrum that are poles of the resolvent function.

6.1. The Spectrum of an Operator

In this section, unless otherwise stated, X will denote a non-trivial (i.e., $X \neq \{0\}$) complex Banach space. If a real Banach space X is given, then X should be replaced by its complexification $X_{\mathbf{C}}$. As mentioned earlier, the symbol $\mathcal{L}(X, Y)$ will denote the Banach space of all bounded (linear) operators from a Banach space X into another Banach space Y and $\mathcal{L}(X)$ stands for $\mathcal{L}(X, X)$. Unless otherwise stated, the letter I will denote the identity operator on X. Identifying each complex number λ with λI, we can consider λ as an operator on X.

Definition 6.1. *The* ***spectrum*** $\sigma(T)$ *of a bounded operator* $T\colon X \to X$ *is the set of all complex numbers* λ *such that the operator* $\lambda - T$ *is not invertible on* X*, that is,*

$$\sigma(T) = \big\{\lambda \in \mathbb{C}\colon\ (\lambda - T)^{-1} \text{ does not exist}\big\}.$$

The complement of the spectrum is called the ***resolvent set*** *of* T *and is denoted by* $\rho(T)$*, that is,*

$$\rho(T) = \mathbb{C}\backslash\sigma(T) = \big\{\lambda \in \mathbb{C}\colon\ \lambda - T \text{ is an invertible operator from } X \text{ onto } X\big\}.$$

It should be noticed immediately from the above definition that both the spectrum $\sigma(T)$ and the resolvent set $\rho(T)$ depend only on the topology of X and not on any particular norm generating the topology.

Let us make a comment regarding the resolvent and the spectrum of an operator defined on a real Banach space. So, let $T\colon X \to X$ be a bounded operator on a real Banach space X. Then, identifying T with $T + \imath 0$, we can consider T as a bounded operator on the complexification $X_{\mathbf{C}}$ of X, i.e., $T + \imath 0 \in \mathcal{L}(X_{\mathbf{C}}) = \mathcal{L}(X) \oplus \imath\mathcal{L}(X)$. Now if $\lambda = \alpha + \imath\beta$ is a complex number, then the operator $\lambda - T$ can be identified with the operator $(\alpha - T) + \imath\beta I$ on $X_{\mathbf{C}}$, where I is the identity operator on X. This means that $\lambda - T$ is an invertible operator (on $X_{\mathbf{C}}$) if and only if there exist two operators $A, B \in \mathcal{L}(X)$ satisfying

$$[(\alpha - T) + \imath\beta I](A + \imath B) = (A + \imath B)[(\alpha - T) + \imath\beta I] = I + \imath 0$$

or $[(\alpha - T)A - \beta B] + \imath[(\alpha - T)B + \beta A] = [A(\alpha - T) - \beta B] + \imath[B(\alpha - T) + \beta A] = I + \imath 0$. Thus, we have the following result regarding the spectrum of a bounded operator on a real Banach space.

Lemma 6.2. *Let* X *be a real Banach space and let* $T \in \mathcal{L}(X)$*. A complex number* $\lambda = \alpha + \imath\beta$ *belongs to the resolvent of* T*, i.e.,* $\lambda \in \rho(T)$*, if and only if there exist two operators* $A, B \in \mathcal{L}(X)$ *satisfying*

$$A(\alpha - T) = (\alpha - T)A = I + \beta B \quad \text{and} \quad (\alpha - T)B = B(\alpha - T) = -\beta A.$$

When $\lambda = \alpha + \imath\beta$ is a real number (i.e., when $\beta = 0$), the preceding lemma yields $A(\alpha - T) = (\alpha - T)A = I$. In view of Definition 6.1, there exists a function $\lambda \mapsto (\lambda - T)^{-1}$ from $\rho(T)$ into $\mathcal{L}(X)$. This function is called the **resolvent function** (or simply the **resolvent**) of T and is denoted by $R(\cdot, T)$, i.e.,

$$R(\lambda, T) = (\lambda - T)^{-1}, \quad \lambda \in \rho(T).$$

If there is no need to indicate the operator T, then we shall simply write $R(\lambda)$ instead of $R(\lambda, T)$. Notice that the assumption $X \neq \{0\}$ guarantees that $R(\lambda, T) \neq 0$ for each $\lambda \in \rho(T)$.

Theorem 6.3. *If $T \in \mathcal{L}(X)$ and $|\lambda| > \|T\|$, then $\lambda \in \rho(T)$ and the resolvent $R(\lambda, T)$ is given by the Laurent series*

$$R(\lambda, T) = \sum_{n=0}^{\infty} \lambda^{-(n+1)} T^n,$$

where the series converges in the operator norm in the Banach space $\mathcal{L}(X)$. Moreover, for each $|\lambda| > \|T\|$ we have

$$\|R(\lambda, T)\| \le \tfrac{1}{|\lambda| - \|T\|}$$

and so, in particular, $\lim_{\lambda \to \infty} R(\lambda, T) = 0$.

Proof. The norm convergence of the series follows from $\frac{\|T\|}{|\lambda|} < 1$. We need to verify that the sum of the series, i.e., the bounded operator $S = \sum_{n=0}^{\infty} \lambda^{-(n+1)} T^n$, coincides with $R(\lambda, T)$. To see this, note that

$$\begin{aligned} (\lambda - T)S &= (\lambda - T) \lim_{m \to \infty} \sum_{n=0}^{m} \lambda^{-(n+1)} T^n \\ &= \lim_{m \to \infty} \Big[(\lambda - T) \sum_{n=0}^{m} \lambda^{-(n+1)} T^n \Big] \\ &= \lim_{m \to \infty} \Big[\sum_{n=0}^{m} \lambda^{-n} T^n - \sum_{n=0}^{m} \lambda^{-(n+1)} T^{n+1} \Big] \\ &= \lim_{m \to \infty} \Big[I - \big(\tfrac{T}{\lambda}\big)^{m+1} \Big] = I. \end{aligned}$$

Similarly, $S(\lambda - T) = I$, and so $S = R(\lambda, T)$. For the norm inequality note that

$$\|R(\lambda, T)\| \le \tfrac{1}{|\lambda|} \sum_{n=0}^{\infty} \big\| \tfrac{T^n}{\lambda^n} \big\| \le \tfrac{1}{|\lambda|} \sum_{n=0}^{\infty} \big(\tfrac{\|T\|}{|\lambda|} \big)^n = \frac{1}{|\lambda|(1 - \frac{\|T\|}{|\lambda|})} = \tfrac{1}{|\lambda| - \|T\|},$$

and the proof is finished. ■

The series expansion $R(\lambda, T) = \sum_{n=0}^{\infty} \lambda^{-(n+1)} T^n$ is usually referred to as the **Neumann series**. It appeared for the first time in the work of C. Neumann [**254**] devoted to potential theory.

Lemma 6.4. *For an operator $T \in \mathcal{L}(X)$ and any λ, $\mu \in \rho(T)$ we have:*

(1) *$R(\lambda)R(\mu) = R(\mu)R(\lambda)$.*
(2) *If $S \in \mathcal{L}(X)$ satisfies $ST = TS$, then $R(\lambda, T)S = SR(\lambda, T)$ for each $\lambda \in \rho(T)$. In other words, if T belongs to the commutant of an operator S, then $R(\lambda, T)$ belongs to it too. In particular, T and $R(\lambda, T)$ commute.*
(3) *$R(\lambda) - R(\mu) = -(\lambda - \mu)R(\lambda)R(\mu)$.*

Proof. (1) Notice that $(\mu - T)(\lambda - T) = (\lambda - T)(\mu - T)$. So, by taking inverses on both sides, we get $R(\lambda)R(\mu) = R(\mu)R(\lambda)$.

(2) Fix $\lambda \in \rho(T)$. Then we have

$$R(\lambda, T)S(\lambda - T) = R(\lambda, T)(\lambda - T)S = S = SR(\lambda, T)(\lambda - T),$$

and so by "canceling" $\lambda - T$, we get $R(\lambda, T)S = SR(\lambda, T)$.

(3) Using part (2) we get

$$\begin{aligned}(\lambda - T)(\mu - T)\big[R(\lambda) - R(\mu)\big] &= (\lambda - T)(\mu - T)\big[(\lambda - T)^{-1} - (\mu - T)^{-1}\big]\\ &= (\mu - T) - (\lambda - T) = -(\lambda - \mu),\end{aligned}$$

and the desired identity follows. ■

Definition 6.5. *For a given operator $T \in \mathcal{L}(X)$, the identity*

$$R(\lambda) - R(\mu) = -(\lambda - \mu)R(\lambda)R(\mu)$$

is called the ***resolvent identity*** *(of T).*

A glance at Corollary 2.11 guarantees that the resolvent set of an operator is always open. Our next theorem shows that in actuality much more is true.

Theorem 6.6. *If $T \in \mathcal{L}(X)$ and $\lambda_0 \in \rho(T)$, then the open disk $D\big(\lambda_0, \frac{1}{\|R(\lambda_0)\|}\big)$ is included in $\rho(T)$ and for each $\lambda \in D\big(\lambda_0, \frac{1}{\|R(\lambda_0)\|}\big)$ we have*

$$R(\lambda) = \sum_{n=0}^{\infty}(\lambda_0 - \lambda)^n R(\lambda_0)^{n+1} = \sum_{n=0}^{\infty}(-1)^n(\lambda - \lambda_0)^n R(\lambda_0)^{n+1}.$$

Moreover, letting $\gamma = |\lambda - \lambda_0| \cdot \|R(\lambda_0)\| < 1$, we have

$$\|R(\lambda)\| \le \tfrac{1}{1-\gamma}\|R(\lambda_0)\| \qquad (\star)$$

and

$$\|R(\lambda) - R(\lambda_0)\| \le \tfrac{\gamma}{1-\gamma}\|R(\lambda_0)\|. \qquad (\star\star)$$

Proof. Let $|\lambda - \lambda_0| < \frac{1}{\|R(\lambda_0)\|}$. Then $\gamma = |\lambda - \lambda_0| \cdot \|R(\lambda_0)\| < 1$ and so the series $\sum_{n=0}^{\infty}(\lambda_0 - \lambda)^n R(\lambda_0)^{n+1}$ is norm convergent. Now note that

$$\begin{aligned}
&(\lambda - T)\sum_{n=0}^{\infty}(\lambda_0 - \lambda)^n(\lambda_0 - T)^{-(n+1)} \\
&= \big[-(\lambda_0 - \lambda) + (\lambda_0 - T)\big]\sum_{n=0}^{\infty}(\lambda_0 - \lambda)^n(\lambda_0 - T)^{-(n+1)} \\
&= \sum_{n=0}^{\infty}\big[(\lambda_0 - \lambda)^n(\lambda_0 - T)^{-n} - (\lambda_0 - \lambda)^{n+1}(\lambda_0 - T)^{-(n+1)}\big] \\
&= \lim_{m\to\infty}\sum_{n=0}^{m}\big[(\lambda_0 - \lambda)^n(\lambda_0 - T)^{-n} - (\lambda_0 - \lambda)^{n+1}(\lambda_0 - T)^{-(n+1)}\big] \\
&= \lim_{m\to\infty}\big[I - (\lambda_0 - \lambda)^{m+1}R(\lambda_0)^{m+1}\big] = I.
\end{aligned}$$

Similarly, $\big[\sum_{n=0}^{\infty}(\lambda_0 - \lambda)^n(\lambda_0 - T)^{-(n+1)}\big](\lambda - T) = I$. This establishes the representation for $R(\lambda)$. In view of Theorem 1.72, the function $R(\cdot)$ is analytic in $\rho(T)$.

To establish the inequality $(\star)$, observe that

$$\begin{aligned}
\|R(\lambda)\| &= \Big\|\sum_{n=0}^{\infty}(\lambda_0 - \lambda)^n R(\lambda_0)^{n+1}\Big\| \\
&\le \sum_{n=0}^{\infty}|\lambda_0 - \lambda|^n \|R(\lambda_0)\|^{n+1} \\
&= \|R(\lambda_0)\|\sum_{n=0}^{\infty}\gamma^n \\
&= \tfrac{1}{1-\gamma}\cdot\|R(\lambda_0)\|.
\end{aligned}$$

Finally, to establish $(\star\star)$, use the identity

$$R(\lambda) - R(\lambda_0) = \sum_{n=1}^{\infty}(\lambda_0 - \lambda)^n R(\lambda_0)^{n+1},$$

in connection with the triangle inequality. ■

Corollary 6.7. *For each operator $T \in \mathcal{L}(X)$ the resolvent set $\rho(T)$ is an open subset of the complex plane and the operator-valued function*

$$\lambda \longmapsto R(\lambda, T) = R(\lambda),$$

from $\rho(T)$ to $\mathcal{L}(X)$, is analytic.

In particular, the resolvent function $\lambda \mapsto R(\lambda)$ is continuous.

Corollary 6.8. *For every operator $T \in \mathcal{L}(X)$ the derivative of its resolvent function is given by*

$$\frac{dR(\lambda,T)}{d\lambda} = -R^2(\lambda,T).$$

Proof. By Corollary 6.7 the resolvent function is continuous. So, using the resolvent identity, we see that

$$\frac{dR(\lambda,T)}{d\lambda} = \lim_{\mu\to\lambda} \frac{R(\mu,T)-R(\lambda,T)}{\mu-\lambda} = -\lim_{\mu\to\lambda} R(\mu,T)R(\lambda,T) = -R^2(\lambda,T),$$

as claimed. ■

Here is another important consequence of Theorem 6.6.

Corollary 6.9. *For an operator $T \in \mathcal{L}(X)$ we have:*

(1) *If $\lambda \in \rho(T)$, then $\|R(\lambda,T)\| \geq \frac{1}{d(\lambda,\sigma(T))}$, where $d\big(\lambda,\sigma(T)\big)$ is the distance from λ to the spectrum $\sigma(T)$.*

(2) *Assume that a sequence $\{\lambda_n\} \subseteq \rho(T)$ satisfies $\lambda_n \to \lambda_0 \in \mathbb{C}$. Then $\lambda_0 \in \sigma(T)$ if and only if $\|R(\lambda_n)\| \to \infty$.*

Proof. (1) Fix $\lambda \in \rho(T)$. If $\mu \in \sigma(T)$, then according to Theorem 6.6, we see that $|\lambda-\mu| \geq \frac{1}{\|R(\lambda)\|}$. (Otherwise, $|\lambda-\mu| < \frac{1}{\|R(\lambda)\|}$ implies $\mu \in \rho(T)$, a contradiction.) So,

$$d\big(\lambda,\sigma(T)\big) = \inf\big\{|\lambda-\mu|\colon\ \mu \in \sigma(T)\big\} \geq \frac{1}{\|R(\lambda)\|}.$$

(2) If $\lambda_0 \in \sigma(T)$, then the conclusion follows from the relation

$$\big\|R(\lambda_n,T)\big\| \geq \frac{1}{d(\lambda_n,\sigma(T))} \xrightarrow[n\to\infty]{} \infty.$$

Conversely, if $\big\|R(\lambda_n,T)\big\| \to \infty$, then $\lambda_0 \in \sigma(T)$ must be true. Otherwise, if $\lambda_0 \in \rho(T)$, then by the continuity of the resolvent function $R(\cdot,T)\colon \rho(T) \to X$ we would have $\big\|R(\lambda_n,T)\big\| \to \big\|R(\lambda_0,T)\big\| \in \mathbb{R}$. ■

It turns out that the spectrum of a bounded operator is always nonempty and compact. This is a basic result.

Theorem 6.10. *The spectrum of a bounded operator on a (non-trivial) complex Banach space is a non-empty compact subset of the complex plane.*

Proof. Let X be a complex Banach space and let $T \in \mathcal{L}(X)$. Assume by way of contradiction that $\sigma(T) = \emptyset$ or equivalently that $\rho(T) = \mathbb{C}$. This means that the resolvent function $R(\cdot)$ is an entire function. The estimate $\big\|R(\lambda)\big\| \leq \frac{1}{|\lambda|-\|T\|}$ (for $|\lambda| > \|T\|$), established in Theorem 6.3, implies $\lim_{\lambda\to\infty} R(\lambda) = 0$. The latter shows that $R(\cdot)$ is bounded at infinity. Since $R(\cdot)$ is analytic, it follows that $R(\cdot)$ is a bounded entire function.

Consequently, by Liouville's Theorem 1.74, $R(\cdot)$ is a constant function. In particular, notice that

$$(-T)^{-1} = R(0) = \lim_{\lambda\to\infty} R(\lambda) = 0\,,$$

which implies that $X = \{0\}$, a contradiction. Hence, $\sigma(T) \neq \emptyset$.

From $\sigma(T) = \mathbb{C} \setminus \rho(T)$, we see that $\sigma(T)$ is a closed set. On the other hand, since $|\lambda| > \|T\|$ implies $\lambda \in \rho(T)$, we also see that $\sigma(T)$ is a bounded subset of $\mathbb{C}$. So, $\sigma(T)$ is a non-empty compact subset of $\mathbb{C}$. ■

It is interesting to observe that every non-empty compact subset K of the complex plane $\mathbb{C}$ is the spectrum of a bounded operator. To see this, choose a countable dense subset $\{\lambda_1, \lambda_2, \ldots\}$ of K and then consider the linear operator $T\colon \ell_2 \to \ell_2$ defined by

$$T(z_1, z_2, \ldots) = (\lambda_1 z_1, \lambda_2 z_2, \ldots)\,.$$

If e_n is the sequence whose n^{th} component is one and every other is zero, then we have $(\lambda_n - T)(e_n) = 0$ and so $\lambda_n \in \sigma(T)$ for each n. Now an easy verification shows that if $\lambda \notin K$, then $(\lambda - T)^{-1}$ exists, and from this it follows that $\sigma(T) = K$. For another example see Exercise 15 at the end of the section.

Definition 6.11. *The* ***spectral radius*** $r(T)$ *of an arbitrary operator* T *in* $\mathcal{L}(X)$ *is the smallest non-negative real number* r *for which the closed disk* $\{\lambda \in \mathbb{C}\colon\ |\lambda| \le r\}$ *contains the spectrum* $\sigma(T)$. *That is,*

$$r(T) = \sup\{|\lambda|\colon\ \lambda \in \sigma(T)\} = \max\{|\lambda|\colon\ \lambda \in \sigma(T)\}\,.$$

Clearly, $r(T) \le \|T\|$, and strict inequality is possible. As noticed before, the spectrum of an operator $T \in \mathcal{L}(X)$ does not depend on any particular equivalent norm on X. Therefore, the spectral radius $r(T)$ is also independent of any particular equivalent norm on X. Nevertheless, there is a very important formula due to I. M. Gelfand that allows one to calculate the spectral radius of an operator in terms of the norm.

Theorem 6.12 (Gelfand [**133**]). *If* $T \in \mathcal{L}(X)$, *then*

$$r(T) = \lim_{n\to\infty} \|T^n\|^{\frac{1}{n}} = \inf_n \|T^n\|^{\frac{1}{n}}\,.$$

In addition, for each $\lambda \in \mathbb{C}$ *satisfying* $|\lambda| > r(T)$ *the resolvent* $R(\lambda, T)$ *is given by the Neumann series*

$$R(\lambda, T) = \sum_{n=0}^{\infty} \lambda^{-(n+1)} T^n\,.$$

Proof. Let $T \in \mathcal{L}(X)$. If $\lambda \in \sigma(T)$, then we claim that $\lambda^n \in \sigma(T^n)$. Indeed, if $\lambda^n \notin \sigma(T^n)$, then there exists some bounded operator $S\colon X \to X$ such that $(\lambda^n - T^n)S = S(\lambda^n - T^n) = I$. Write

$$\lambda^n - T^n = (\lambda - T)\big(\lambda^{n-1} + \lambda^{n-2}T + \cdots + \lambda T^{n-2} + T^{n-1}\big),$$

and note that the operators $A = \big(\lambda^{n-1}+\lambda^{n-2}T+\cdots+\lambda T^{n-2}+T^{n-1}\big)S$ and $B = S\big(\lambda^{n-1}+\lambda^{n-2}T+\cdots+\lambda T^{n-2}+T^{n-1}\big)$ satisfy $(\lambda-T)A = B(\lambda-T) = I$. This implies $\lambda \in \rho(T)$, which is a contradiction.[1]

Now choose some $\alpha \in \sigma(T)$ with $|\alpha| = r(T)$. From $\alpha^n \in \sigma(T^n)$, we see that $|\alpha|^n \le \|T^n\|$, and so $|\alpha| = r(T) \le \|T^n\|^{\frac{1}{n}}$ holds for each n. It follows that

$$r(T) \le \inf_n \|T^n\|^{\frac{1}{n}} \le \liminf_{n\to\infty} \|T^n\|^{\frac{1}{n}}. \qquad (\star)$$

Note that the resolvent function $R(\lambda, T)$ is analytic on the open annulus $A = \big\{\lambda \in \mathbb{C}\colon\ |\lambda| > r(T)\big\}$ and that A is the largest open annulus centered at zero that is included in $\rho(T)$. By Laurent's Theorem 1.78 we must have $R(\lambda,T) = \sum_{n=0}^{\infty} \lambda^{-(n+1)}T^n$ for each $|\lambda| > r(T)$, and from Theorem 1.77 we see that $r(T) = \limsup_{n\to\infty} \|T^n\|^{\frac{1}{n}}$. A glance at $(\star)$ guarantees that $r(T) = \lim_{n\to\infty} \|T^n\|^{\frac{1}{n}} = \inf_n \|T^n\|^{\frac{1}{n}}$. ∎

We refer to Exercise 6 at the end of this section for an elementary proof of the existence of the limit $\lim_{n\to\infty} \|T^n\|^{\frac{1}{n}}$.

Definition 6.13. *An operator $T \in \mathcal{L}(X)$ is said to be:*

(1) ***nilpotent*** *if $T^k = 0$ for some positive integer k and*

(2) ***quasinilpotent*** *if $r(T) = \lim_{n\to\infty} \|T^n\|^{\frac{1}{n}} = 0$.*

Clearly, every nilpotent operator is quasinilpotent. Also notice that an operator $T \in \mathcal{L}(X)$ is quasinilpotent if and only if $\sigma(T) = \{0\}$.

Recall that if $T\colon X \to Y$ is a bounded operator between Banach spaces, then its **adjoint operator** $T^*\colon Y^* \to X^*$ is defined via the duality formula

$$\langle x, T^*y^*\rangle = \langle Tx, y^*\rangle$$

for all $x \in X$ and all $y^* \in Y^*$. It is well known that the mapping $T \mapsto T^*$, from $\mathcal{L}(X,Y)$ to $\mathcal{L}(Y^*,X^*)$, is a linear isometry.

Theorem 6.14. *The spectrum of an operator $T \in \mathcal{L}(X)$ coincides with the spectrum of its adjoint, i.e.,*

$$\sigma(T) = \sigma(T^*).$$

In particular, we have $r(T) = r(T^*)$.

[1] Here, we use the fact that if $A, B, C \in \mathcal{L}(X)$ satisfy $CA = BC = I$, then $A = B = C^{-1}$. To see this, note that $A = IA = (BC)A = B(CA) = BI = B$.

Proof. The proof is a consequence of the following two facts:

(1) $(\lambda - T)^* = \lambda - T^*$ for all $\lambda \in \mathbb{C}$ and all $T \in \mathcal{L}(X)$.

(2) An operator $S \in \mathcal{L}(X)$ is invertible if and only if its adjoint S^* is invertible.

To verify (2), notice first that $SA = AS = I$ implies $S^*A^* = A^*S^* = I^*$. Conversely, if $S^*\colon X^* \to X^*$ is invertible, then by the previous remark the operator $S^{**}\colon X^{**} \to X^{**}$ is invertible and so S (which is the restriction of S^{**} to X) is one-to-one and carries X onto a closed subspace $S(X)$ of X. If $S(X) \neq X$, then by the Separation Theorem there exists some non-zero $x^* \in X^*$ which vanishes on $S(X)$. That is,

$$\langle Sx, x^* \rangle = \langle x, S^*x^* \rangle = 0$$

holds for all $x \in X$, and so $S^*x^* = 0$. Since S^* is invertible, we infer that $x^* = 0$, which is a contradiction. Hence, $S(X) = X$ and so S is invertible if and only if S^* is invertible. ∎

We close the section with the definition of the peripheral spectrum of an operator.

Definition 6.15. *For an operator $T \in \mathcal{L}(X)$, the non-empty compact set*

$$\sigma_{\text{per}}(T) = \{\lambda \in \sigma(T)\colon \ |\lambda| = r(T)\} = \sigma(T) \cap \{\lambda \in \mathbb{C}\colon \ |\lambda| = r(T)\}$$

is known as the ***peripheral spectrum*** *of the operator. That is, the* peripheral spectrum *of an operator is the non-empty part of the spectrum that lies on the circle centered at zero with radius $r(T)$.*

Finally, let us indicate that the preceding definition and results regarding operators can be extended verbatim to complex unital Banach algebras. To see this, let $\mathcal{A}$ be a complex Banach algebra with unit e. If x is a vector in $\mathcal{A}$, then the **spectrum** of x is defined by

$$\sigma(x) = \{\lambda \in \mathbb{C}\colon \ \lambda e - x \text{ is not invertible in } \mathcal{A}\}.$$

The **resolvent** $\rho(x)$ of x is the complement of $\sigma(x)$, i.e., $\rho(x) = \mathbb{C} \setminus \sigma(x)$. The **spectral radius** of x is defined as usual by $r(x) = \sup_{\lambda \in \sigma(x)} |\lambda|$. Similarly, the **resolvent function** $R(\cdot, x)\colon \rho(x) \to \mathcal{A}$ of x is defined by $R(\lambda, x) = (\lambda e - x)^{-1}$.

It is easy to see that all results concerning operators that were obtained in this section are also true for vectors in complex unital Banach algebras. For instance, we have the following theorem.

Theorem 6.16. *If x is a vector in a complex unital Banach algebra, then:*

(a) *Its spectrum $\sigma(x)$ is a non-empty compact subset of $\mathbb{C}$.*

(b) *Its spectral radius is given by $r(x) = \lim_{n\to\infty} \|x^n\|^{\frac{1}{n}} = \inf_n \|x^n\|^{\frac{1}{n}}$.*

Exercises

1. Establish the following identity—known as the **second resolvent identity**. If $S, T \in \mathcal{L}(X)$ and $\lambda \in \rho(T) \cap \rho(S)$, then
$$R(\lambda, S) - R(\lambda, T) = R(\lambda, S)(S - T)R(\lambda, T).$$

2. Generalize Corollary 6.8 by proving that the derivatives of the resolvent function of an operator $T \in \mathcal{L}(X)$ are given by the formulas
$$\frac{d^n R(\lambda, T)}{d\lambda^n} = (-1)^n n! \big[R(\lambda, T)\big]^{n+1}$$
for $n = 1, 2, \ldots$. [HINT: Use Theorems 1.73(3) and 6.6.]

3. If X is a Banach space and $T \in \mathcal{L}(X)$, then show that $\lim_{\lambda\to\infty} \lambda R(\lambda, T) = I$. [HINT: Use either Exercise 11 of Section 2.1 or Exercise 6 of Section 1.5.]

4. For any operator $T \in \mathcal{L}(X)$ show that $[R(\lambda, T)]^* = R(\lambda, T^*)$ holds for each $\lambda \in \rho(T)$.

5. Let $T \in \mathcal{L}(X)$. If $\lambda_0 \in \rho(T)$, then show that $R(\lambda_0, T)$ cannot be quasinilpotent. [HINT: If $S \in \mathcal{L}(X)$ is invertible, then $0 \notin \sigma(T)$.]

6. Show directly that if $T\colon X \to X$ is a bounded operator on a (real or complex) Banach space, then $\lim_{n\to\infty} \|T^n\|^{\frac{1}{n}}$ always exists in $\mathbb{R}$ and
$$\lim_{n\to\infty} \|T^n\|^{\frac{1}{n}} = \inf_n \|T^n\|^{\frac{1}{n}}.$$
[HINT: Let $\alpha = \inf_n \|T^n\|^{\frac{1}{n}} \geq 0$. From $\alpha \leq \|T^n\|^{\frac{1}{n}}$, we see that $\liminf_{n\to\infty} \|T^n\|^{\frac{1}{n}} \geq \alpha$. Next, let $\epsilon > 0$ and then fix some k such that $\|T^k\|^{\frac{1}{k}} < \alpha + \epsilon$. Now for any positive integer n write $n = km + r$, where $m \geq 0$ and $0 \leq r < k$. Note that
$$\begin{aligned} \|T^n\| &= \|T^r(T^k)^m\| \leq \|T\|^r \cdot \|T^k\|^m = \|T\|^r \cdot (\|T^k\|^{\frac{1}{k}})^{km} \\ &= \|T\|^r \cdot (\alpha + \epsilon)^{n-r} \leq M(\alpha + \epsilon)^n, \end{aligned}$$
where $M = \max_{0\leq r<k}(\frac{\|T\|}{\alpha+\epsilon})^r$. Thus, $\|T^n\|^{\frac{1}{n}} \leq M^{\frac{1}{n}}(\alpha+\epsilon)$, from which we get $\limsup_{n\to\infty} \|T^n\|^{\frac{1}{n}} \leq \alpha + \epsilon$ for all $\epsilon > 0$, or $\limsup_{n\to\infty} \|T^n\|^{\frac{1}{n}} \leq \alpha$.]

7. For any operator $T \in \mathcal{L}(X)$ show that
$$r(T) = \inf\big\{\alpha > 0\colon \tfrac{\|T^n\|}{\alpha^n} \to 0\big\}.$$

8. If $S, T\colon X \to X$ are two commuting bounded operators on an arbitrary Banach space, then show that $r(ST) \leq r(S)r(T)$. [HINT: If $ST = TS$, then $(ST)^n = S^n T^n$ for each n.]

9. Let $T\colon X \to X$ be a bounded operator and assume that for some non-zero $\alpha \in \mathbb{C}$ the series $\sum_{n=0}^{\infty} \alpha^{-(n+1)} T^n x$ converges in norm for each $x \in X$. Show that $\alpha \in \rho(T)$. [HINT: Use the Uniform Boundedness Principle.]

10. For an arbitrary operator $T \in \mathcal{L}(X)$ show that $r(T) = \lim_{n\to\infty} \|T^n\|^{\frac{1}{n}} = 0$ if and only if $\lim_{n\to\infty} \|T^n x\|^{\frac{1}{n}} = 0$ holds for each $x \in X$. [HINT: Use the previous exercise.]

11. If X is a Banach space and $T \in \mathcal{L}(X)$, then show that

$$\sigma(T^k) = \{\lambda^k\colon\ \lambda \in \sigma(T)\}$$

for each k (and conclude from this that $r(T^k) = [r(T)]^k$ holds).

Use the above conclusion to establish that an operator $T \in \mathcal{L}(X)$ satisfies $\|T^n\| \to 0$ if and only if $r(T) < 1$. [HINT: For the second part, note that if $\|T^n\| \to 0$, then $\|T^k\| < 1$ for some k and therefore we have $r(T) = \inf_n \|T^n\|^{\frac{1}{n}} < 1$. On the other hand, if $r(T) = \lim_{n\to\infty} \|T^n\|^{\frac{1}{n}} < 1$, then fix $r(T) < \delta < 1$ and note that $\|T^n\| \le \delta^n$ for all $n \ge n_0$.]

12. If $T\colon X \to X$ is an invertible bounded operator on a Banach space, then show that $\sigma(T^{-1}) = \{\lambda^{-1}\colon\ \lambda \in \sigma(T)\}$.

13. If $T \in \mathcal{L}(X)$ and $\|T^k\| < 1$ for some k, then show that $\lim_{n\to\infty} \|T^n\| = 0$. [HINT: Note that $r(T) = \inf_n \|T^n\|^{\frac{1}{n}} < 1$ and use Exercise11 above.]

14. Let Ω be a compact Hausdorff space and let μ be a Borel measure on Ω. Assume that either $E = C(\Omega)$ or E is a Banach function space that is order dense in $L_0(\mu)$. If $\phi\colon \Omega \to \mathbb{C}$ is a continuous function, then show that the spectrum of the multiplication operator $M_\phi\colon E \to E$ coincides with the range of ϕ, i.e., $\sigma(M_\phi) = R_\phi = \{\phi(\omega)\colon\ \omega \in \Omega\}$.

15. For a non-empty compact subset K of $\mathbb{C}$ establish the following.
(a) There exists a bounded operator $T\colon \ell_2 \to \ell_2$ such that $\sigma(T) = K$.
(b) If $S\colon C_{\mathbb{C}}(K) \to C_{\mathbb{C}}(K)$ is the multiplication operator defined by $(Sx)(z) = zx(z)$, then $\sigma(S) = K$.

(See also Exercise 11 in Section 7.1.)

16. If an invertible operator $T \in \mathcal{L}(X)$ satisfies $\|T^n\| \le M < \infty$ for all integers n (such an operator is called a **Cesaro operator** or a **doubly power bounded operator**), then show that $\sigma(T) \subseteq \{\lambda \in \mathbb{C}\colon\ |\lambda| = 1\}$. [HINT: If $|\lambda| < 1$, consider the operator $S = -\sum_{n=0}^{\infty} \lambda^n T^{-(n+1)}$ and note that $S(\lambda - T) = (\lambda - T)S = I$.]

17. Let $T_i\colon X_i \to X_i$ $(i = 1, \ldots, n)$ be bounded operators on Banach spaces. If $X = X_1 \oplus \cdots \oplus X_n$ is the direct sum Banach space, then show that:
(a) The **direct sum operator** $T = T_1 \oplus \cdots \oplus T_n\colon X \to X$, defined by

$$T(x_1 \oplus \cdots \oplus x_n) = T_1 x_1 \oplus \cdots \oplus T_n x_n\,,$$

is invertible if and only if each T_i is invertible—in which case the inverse operator T^{-1} satisfies $T^{-1} = T_1^{-1} \oplus T_2^{-1} \oplus \cdots \oplus T_n^{-1}$.
(b) The spectrum of the direct sum operator T is the union of the spectra of the operators T_i. That is,

$$\sigma(T_1 \oplus \cdots \oplus T_n) = \bigcup_{i=1}^{n} \sigma(T_i)\,.$$

18. Show that an operator $T\colon X \to X$ on a complex Banach space is nilpotent if and only if it is quasinilpotent and has finite descent.

19. If $S, T \in \mathcal{L}(X)$, then show that $\sigma(ST) \cup \{0\} = \sigma(TS) \cup \{0\}$. Conclude from this that:
(a) $\sigma(ST) = \{0\}$ if and only if $\sigma(TS) = \{0\}$.
(b) $r(ST) = r(TS)$.
Give an example of two operators $S, T \in \mathcal{L}(X)$ such that $\sigma(ST) \neq \sigma(TS)$. [HINT: If $A, B, C \in \mathcal{L}(X)$ satisfy the identity $(I - AB)C = C(I - AB) = I$, then notice that $(I + BCA)(I - BA) = (I - BA)(I + BCA) = I$. For a counterexample, notice that if S and T denote the backward and forward shift operators on ℓ_1, then $\sigma(ST) = \{1\}$ while $\sigma(TS) = \{0, 1\}$.]

20. Let $T \in \mathcal{L}(X)$ be an operator and let $z_0 \in \mathbb{C}$ be a point such that every $\lambda \in \sigma(T) \setminus \{z_0\}$ is an isolated point of $\sigma(T)$. Show that $\sigma(T)$ is at most countable and that if $\sigma(T)$ is countable, say $\sigma(T) = \{\lambda_1, \lambda_2, \ldots\}$, then $\lambda_n \to z_0$. [HINT: For each $\epsilon > 0$ let $C_\epsilon = \big\{\lambda \in \sigma(T)\colon\ |\lambda - z_0| \geq \epsilon\big\}$. Establish that each C_ϵ is a finite set.]

6.2. Special Points of the Spectrum

As before, X will denote a non-trivial complex Banach space and $T \in \mathcal{L}(X)$ a fixed operator.

A complex number λ is an **eigenvalue** of T whenever there exists a non-zero vector x (called an **eigenvector** corresponding to λ) such that $Tx = \lambda x$. Notice that the eigenvalues of T are exactly the complex numbers λ for which $\lambda - T$ is not one-to-one, and so they belong to the spectrum of T. The set of all eigenvalues of T is known as the **point spectrum** of T and is denoted by $\sigma_p(T)$. That is,

$$\begin{aligned}\sigma_p(T) &= \big\{\lambda \in \sigma(T)\colon\ \lambda - T \ \text{ is not one-to-one}\big\}\\ &= \big\{\lambda \in \sigma(T)\colon\ Tx = \lambda x \text{ for some } x \neq 0\big\}.\end{aligned}$$

Recall that if an operator is one-to-one, then (in view of the Open Mapping Theorem) it fails to be invertible if and only if it is not surjective. Accordingly, when a non-invertible operator $\lambda - T$ is one-to-one, we distinguish two cases:

(1) The range of the operator $\lambda - T$ is not dense, i.e., $\overline{(\lambda - T)(X)} \neq X$.

(2) The operator $\lambda - T$ has dense range, i.e., $\overline{(\lambda - T)(X)} = X$.

In case (1) we say that λ belongs to the residual spectrum $\sigma_r(T)$ of T. That is, the **residual spectrum** $\sigma_r(T)$ is defined by

$$\sigma_r(T) = \big\{\lambda \in \sigma(T) \setminus \sigma_p(T)\colon\ \overline{(\lambda - T)(X)} \neq X\big\}.$$

In case (2) we say that λ belongs to the continuous spectrum $\sigma_c(T)$ of T. That is, the **continuous spectrum** $\sigma_c(T)$ is defined by

$$\sigma_c(T) = \big\{\lambda \in \sigma(T) \setminus \sigma_p(T)\colon\ \overline{(\lambda - T)(X)} = X\,\big\}\,.$$

Clearly, $\sigma_p(T)$, $\sigma_c(T)$, and $\sigma_r(T)$ are pairwise disjoint and

$$\sigma(T) = \sigma_p(T) \cup \sigma_r(T) \cup \sigma_c(T)\,.$$

Another important part of the spectrum is the **approximate point spectrum** $\sigma_a(T)$ that is defined as follows.

$$\sigma_a(T) = \big\{\lambda \in \mathbb{C}\colon\ \forall\, \epsilon > 0\ \exists\ x \text{ with } \|x\| = 1 \text{ and } \|\lambda x - Tx\| < \epsilon\big\}\,.$$

The reader should be able to verify easily the following equivalent ways of defining the approximate point spectrum.

$$\begin{aligned}
\sigma_a(T) &= \big\{\lambda \in \mathbb{C}\colon\ \lambda - T \ \text{ is not bounded below}\,\big\}\\
&= \big\{\lambda \in \mathbb{C}\colon\ \exists\ \{x_n\} \subseteq X \text{ with } \|x_n\| = 1\ \forall\ n \text{ and } \lambda x_n - Tx_n \to 0\,\big\}\\
&= \big\{\lambda \in \mathbb{C}\colon\ \exists\ \{x_n\} \subseteq X \text{ satisfying } x_n \not\to 0 \text{ and } \lambda x_n - Tx_n \to 0\,\big\}\\
&= \big\{\lambda \in \mathbb{C}\colon\ \exists\ \{\lambda_n\} \subseteq \mathbb{C} \text{ with } \lambda_n \to \lambda \text{ and } \{x_n\} \subseteq X \text{ with}\\
&\qquad\qquad \|x_n\| = 1 \text{ for all } n \text{ such that } \lambda_n x_n - Tx_n \to 0\,\big\}\,.
\end{aligned}$$

The points of the approximate point spectrum are also known as **approximate eigenvalues**. Clearly, $\sigma_p(T) \subseteq \sigma_a(T) \subseteq \sigma(T)$.

Theorem 6.17. *The continuous spectrum of an operator $T \in \mathcal{L}(X)$ is included in the approximate point spectrum of T, i.e., $\sigma_c(T) \subseteq \sigma_a(T)$. In particular, we have $\sigma(T) = \sigma_a(T) \cup \sigma_r(T)$.*

Proof. Take an arbitrary $\lambda \in \sigma_c(T)$. By the definition of the continuous spectrum, the range of $\lambda - T$ is dense and consequently (by Lemma 2.8) the operator $\lambda - T$ is not bounded below. Therefore, $\lambda \in \sigma_a(T)$. ■

The important topological properties of the approximate spectrum are included in the next theorem.

Theorem 6.18. *For an operator $T \in \mathcal{L}(X)$ we have the following:*

(1) *The approximate point spectrum $\sigma_a(T)$ is a closed set.*

(2) *Every boundary point of the spectrum belongs to the approximate point spectrum. That is, $\partial\sigma(T) \subseteq \sigma_a(T)$ (and so $\sigma_a(T) \neq \emptyset$).*

(3) *The complement of $\sigma_a(T)$ in $\sigma(T)$, i.e., the set*

$$\sigma(T) \setminus \sigma_a(T) = \big\{\lambda \in \mathbb{C}\colon\ \lambda - T \ \textit{is bounded below and } R(\lambda - T) \neq X\big\}\,,$$

is an open subset of the complex plane.

Proof. (1) Take any $\alpha \notin \sigma_a(T)$. Then the operator $\alpha - T$ is bounded below. For each $\lambda \in \mathbb{C}$ we have $\|(\lambda - T) - (\alpha - T)\| = |\lambda - \alpha|$. So, if $\lambda \in \mathbb{C}$ is close enough to α, then by Lemma 2.4(2) the operator $\lambda - T$ is also bounded below. That is, for all λ sufficiently close to α we have $\lambda \notin \sigma_a(T)$ proving that $\sigma_a(T)$ is closed.

(2) Let $\lambda \in \partial\sigma(T)$. Then there exists a sequence $\{\lambda_n\} \subseteq \rho(T)$ satisfying $\lambda_n \to \lambda$. Clearly, $\lambda_n - T \to \lambda - T$. If $\lambda - T$ were bounded below, then Theorem 2.9 would imply that $\lambda - T$ is invertible, contrary to $\lambda \in \sigma(T)$. Hence $\lambda \in \sigma_a(T)$, and so $\partial\sigma(T) \subseteq \sigma_a(T)$.

(3) Let $\alpha \in \sigma(T) \setminus \sigma_a(T)$. Since $\alpha \notin \sigma_a(T)$ and $\sigma_a(T)$ is closed, there exists some $\epsilon > 0$ such that $D_\epsilon(\alpha) \cap \sigma_a(T) = \emptyset$. Also, from $\partial\sigma(T) \subseteq \sigma_a(T)$, we know that $\alpha \notin \partial\sigma(T)$, and hence α is an interior point of $\sigma(T)$. It follows that there exists some $0 < r < \epsilon$ such that $D_r(\alpha) \subseteq \sigma(T) \setminus \sigma_a(T)$. That is, α is an interior point of $\sigma(T) \setminus \sigma_a(T)$ and so $\sigma(T) \setminus \sigma_a(T)$ is an open subset of the complex plane $\mathbb{C}$. ■

It is important to know that each point of the residual spectrum of an operator is an eigenvalue of its adjoint.

Theorem 6.19. *If $T \in \mathcal{L}(X)$, then*

$$\sigma_p(T^*) = \left\{\lambda \in \mathbb{C}\colon\ \overline{(\lambda - T)(X)} \neq X\right\}.$$

In particular, we have $\sigma_r(T) \subseteq \sigma_p(T^*)$.

Proof. The proof will be based on the duality identity

$$\big\langle(\lambda - T)x, x^*\big\rangle = \big\langle x, (\lambda - T^*)x^*\big\rangle. \tag{$\star$}$$

If $\lambda - T$ does not have dense range, then by the Separation Theorem there exists some non-zero $x^* \in X^*$ satisfying $\big\langle(\lambda - T)x, x^*\big\rangle = 0$ for all $x \in X$. But then ($\star$) implies $(\lambda - T^*)x^* = 0$ or $T^*x^* = \lambda x^*$, i.e., λ is an eigenvalue of T^*. Conversely, if $T^*x^* = \lambda x^*$ holds for some non-zero $x^* \in X^*$, then ($\star$) implies $\big\langle(\lambda - T)x, x^*\big\rangle = 0$ for all $x \in X$, and so the range of $\lambda - T$ cannot be dense in X. ■

Regarding eigenvalues we have the following interesting result.

Theorem 6.20. *If an operator $T\colon X \to X$ on a complex Banach space is surjective but not invertible, then there exists some $\epsilon > 0$ such that every λ in the open disk $D(0, \epsilon)$ is an eigenvalue of T.*

Proof. According to Corollary 2.12 there exists some $\epsilon > 0$ such that every operator $S \in \mathcal{L}(X)$ that satisfies $\|T - S\| < \epsilon$ is surjective and non-invertible. In particular, for each $\lambda \in D(0, \epsilon)$ the operator $T - \lambda$ is surjective but not invertible. This implies that each $\lambda \in D(0, \epsilon)$ is an eigenvalue of T. ■

Let us use the shift operator as an example to illustrate the various parts of the spectrum. The **(forward) shift operator** $T\colon \ell_p \to \ell_p$ $(1 \le p \le \infty)$ is defined by

$$T(z_1, z_2, \ldots) = (0, z_1, z_2, \ldots)\,.$$

For $1 \le p < \infty$ we have $(\ell_p)^* = \ell_q$, where $\frac{1}{p} + \frac{1}{q} = 1$, and the duality $\langle \ell_p, \ell_q \rangle$ is defined by

$$\langle z, \xi \rangle = \sum_{i=1}^{\infty} z_i \xi_i$$

for all $z = (z_1, z_2, \ldots) \in \ell_p$ and all $\xi = (\xi_1, \xi_2, \ldots) \in \ell_q$. It is easy to verify that the adjoint $T^*\colon \ell_q \to \ell_q$ of the shift operator is given by

$$T^*(z_1, z_2, \ldots) = (z_2, z_3, \ldots)\,,$$

which is also known as the **backward shift operator**.

Example 6.21. Let $1 < p < \infty$ and let $T\colon \ell_p \to \ell_p$ be the shift operator. Notice that $\|T^n z\|_p = \|z\|_p$ holds for all $z \in \ell_p$, and so $\|T^n\| = 1$ for each n. Therefore,

$$r(T) = \lim_{n \to \infty} \|T^n\|^{\frac{1}{n}} = 1 = \|T\|\,.$$

We shall establish that $\sigma(T) = \sigma(T^*) = \big\{\lambda \in \mathbb{C}\colon\ |\lambda| \le 1\big\}$.

1. *The eigenvalues of the forward and backward shift operators.*

Obviously, $\sigma_p(T) = \emptyset$. We claim that $\sigma_p(T^*) = \{\lambda \in \mathbb{C}\colon\ |\lambda| < 1\}$. Indeed, if $|\lambda| < 1$, then the vector $z_\lambda = (1, \lambda, \lambda^2, \ldots) \in \ell_q$ satisfies

$$T^* z_\lambda = (\lambda, \lambda^2, \ldots) = \lambda(1, \lambda, \lambda^2, \ldots) = \lambda z_\lambda\,.$$

(Compare this conclusion with Theorem 6.20.) An easy argument shows that no $\lambda \in \mathbb{C}$ with $|\lambda| = 1$ can be an eigenvalue of T^*. Taking into account that $\|T\| = 1$, we conclude that

$$\sigma(T) = \sigma(T^*) = \big\{\lambda \in \mathbb{C}\colon\ |\lambda| \le 1\big\}\,.$$

2. *The residual spectra of the forward and backward shift operators.*

Since $\sigma_p(T) = \emptyset$, it follows that $\sigma_r(T) = \sigma_p(T^*) = \{\lambda \in \mathbb{C}\colon\ |\lambda| < 1\}$; see Theorem 6.19. Also, we have $\sigma_r(T^*) = \emptyset$. Indeed, Theorem 6.19 applied to the adjoint operator $T^*\colon \ell_q \to \ell_q$ implies

$$\sigma_r(T^*) \subseteq \sigma_p(T^{**}) = \sigma_p(T) = \emptyset\,.$$

3. *The continuous spectra of the forward and backward shift operators.*

We have

$$\sigma_c(T) = \sigma(T) \setminus \big[\sigma_p(T) \cup \sigma_r(T)\big] = \big\{\lambda \in \mathbb{C}\colon\ |\lambda| = 1\big\}$$

and

$$\sigma_c(T^*) = \sigma(T^*) \setminus \big[\sigma_p(T^*) \cup \sigma_r(T^*)\big] = \big\{\lambda \in \mathbb{C}\colon\ |\lambda| = 1\big\}.$$

4. *The approximate point spectra of the forward and backward shift operators.*

Here we claim

$$\sigma_a(T) = \big\{\lambda \in \mathbb{C}\colon\ |\lambda| = 1\big\} \text{ and } \sigma_a(T^*) = \sigma(T^*) = \big\{\lambda \in \mathbb{C}\colon\ |\lambda| \le 1\big\}.$$

To see that $\sigma_a(T) = \big\{\lambda \in \mathbb{C}\colon\ |\lambda| = 1\big\}$, note first that

$$\partial\sigma(T) = \big\{\lambda \in \mathbb{C}\colon\ |\lambda| = 1\big\} \subseteq \sigma_a(T).$$

To establish that equality holds, now observe that for $|\alpha| < 1$ and $x \in \ell_p$ with $\|x\|_p = 1$, we have

$$\|\alpha x - Tx\|_p \ge \|Tx\|_p - |\alpha|\|x\|_p = \|x\|_p - |\alpha|\|x\|_p = 1 - |\alpha| > 0,$$

which means that $\alpha \notin \sigma_a(T)$. ∎

Exercises

1. Show that an operator $T \in \mathcal{L}(X)$ is not one-to-one if and only if 0 is an eigenvalue of T.
2. Let $T\colon V \to V$ be a linear operator on a vector space. Show that a collection of eigenvectors of T any two of which correspond to distinct eigenvalues is a linearly independent set.
3. Let $T\colon X \to X$ be a bounded operator. Show that if for some $\lambda \in \mathbb{C}$ there exists a sequence $\{x_n\}$ of unit vectors such that $\|\lambda x_n - Tx_n\| \to 0$, then $\lambda \in \sigma(T)$, i.e., show that $\sigma_a(T) \subseteq \sigma(T)$.
4. Show that a complex number is an approximate eigenvalue for an operator $T \in \mathcal{L}(X)$ if and only if there exists a sequence $\{x_n\}$ in X satisfying $x_n \not\to 0$ and $\lambda x_n - Tx_n \to 0$.
5. Let $T\colon X \to X$ be a bounded below operator on a Banach space. If $\gamma > 0$ satisfies $\|Tx\| \ge \gamma\|x\|$ for each $x \in X$, then show that

$$\sigma_a(T) \cap \{\lambda \in \mathbb{C}\colon\ |\lambda| < \gamma\} = \emptyset.$$

In particular, if the operator $T\colon X \to X$ is an isometry on a Banach space, then show that $\sigma_a(T) \cap \{\lambda \in \mathbb{C}\colon\ |\lambda| < 1\} = \emptyset$. [HINT: If $|\lambda| < \gamma$ and $x \in X$ satisfies $\|x\| = 1$, then $\|Tx - \lambda x\| \ge \|Tx\| - |\lambda|\|x\| \ge (\gamma - |\lambda|) > 0$.]
6. Let $T\colon X \to X$ be a non-invertible bounded operator on a Banach space. If some $\gamma > 0$ satisfies $\|Tx\| \ge \gamma\|x\|$ for each $x \in X$, then show that

$$D(0,\gamma) = \{\lambda \in \mathbb{C}\colon\ |\lambda| < \gamma\} \subseteq \sigma_r(T).$$

[HINT: Let $\beta = \sup\{\alpha \ge 0\colon\ D(0,\alpha) \subseteq \sigma_r(T)\}$. Clearly, $D(0,\beta) \subseteq \sigma_r(T)$. We must show that $\beta \ge \gamma$. To establish this claim, assume by way of contradiction that $\beta < \gamma$. This implies the existence of a sequence of complex numbers $\{\lambda_n\}$ satisfying $\beta < |\lambda_n| < \beta + \frac{\gamma-\beta}{2n}$ and $\lambda_n \notin \sigma_r(T)$ for each n and $\lambda_n \to \lambda$. From $|\lambda| = \beta$, it follows that $\lambda \in \sigma(T)$. Now notice that

$$\|(\lambda_n - T)x\| \geq \|Tx\| - \|\lambda_n x\| \geq \gamma\|x\| - |\lambda_n| \cdot \|x\| \geq \tfrac{\gamma-\beta}{2}\|x\|$$

holds for each $x \in X$. Since each operator $\lambda_n - T$ has dense range, it follows from Lemma 2.8 that each $\lambda_n - T$ is an invertible operator. But then, by Theorem 2.5, the operator $\lambda - T$ is itself invertible, a contradiction.]

7. If $P \in \mathcal{L}(X)$ is a non-trivial projection (i.e., if $P^2 = P$ and $P \neq 0, I$), then show that $\sigma(P) = \sigma_p(P) = \{0, 1\}$.

8. Let $S \in \mathcal{L}(X)$ and $T \in \mathcal{L}(Y)$ be two bounded operators, where X and Y are two complex Banach spaces. The operators S and T are said to be **similar** whenever there exists an invertible operator $R \in \mathcal{L}(X, Y)$ satisfying $T = RSR^{-1}$. In other words, S and T are similar if and only if there exists an isomorphism (i.e., a surjective, one-to-one, and continuous operator) $R\colon X \to Y$ such the diagram shown in Figure 1 is commutative.

If S and T are similar operators, then show that

$$\sigma(S) = \sigma(T), \quad \sigma_p(S) = \sigma_p(T), \quad \sigma_a(S) = \sigma_a(T), \quad \text{and} \quad \sigma_r(S) = \sigma_r(T)\,.$$

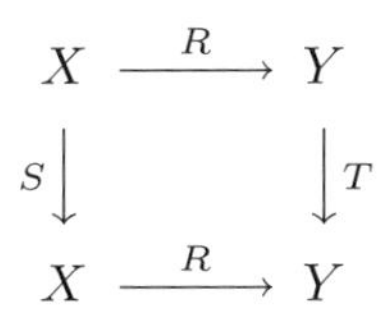

Figure 1

9. For two operators $S, T \in \mathcal{L}(X)$ establish the following.

(a) A non-zero complex number is an eigenvalue of ST if and only if it is an eigenvalue of TS.

(b) If X is finite dimensional, then show that $\sigma(ST) = \sigma(TS)$; compare this with Exercise 19 of Section 6.1 and Exercise 7 of Section 8.2.

[HINT: $(ST)x = \lambda x$ implies $(TS)Tx = \lambda Tx$.]

6.3. The Resolvent of a Positive Operator

In this section we shall exhibit a few additional properties of the resolvent function when our Banach space X is the complexification of a Banach lattice. So, let E be a (real) Banach lattice. As before, if $T \in \mathcal{L}(E)$, then by the spectrum, the resolvent, the resolvent function, the spectral radius, etc. of T, we shall always understand the corresponding notions for the extension $T_{\mathbf{c}}$ of T to $E_{\mathbf{c}}$. In particular, the spectrum of an operator $T \in \mathcal{L}(E)$ is the set

$$\sigma(T) = \big\{\lambda \in \mathbb{C}\colon\ \lambda - T_{\mathbf{c}} \ \text{ is not invertible on } E_{\mathbf{c}}\big\}\,.$$

Since the two norms $\|\cdot\|$ and $\|\cdot\|_{\mathbf{C}}$ on $E_{\mathbf{C}}$, defined by

$$\|z\| = \sup_{\theta\in[0,2\pi]} \|x\cos\theta + y\sin\theta\| \quad \text{and} \quad \|z\|_{\mathbf{C}} = \|\,|z|\,\|\,,$$

are equivalent, the following result should be immediate from Lemma 1.7.

Lemma 6.22. *If $T\colon E \to E$ is a bounded operator on a real Banach lattice, then*

$$r(T) = \max_{\lambda\in\sigma(T)} |\lambda| = \lim_{n\to\infty} \big[\|T^n\|_{\mathbf{C}}\big]^{\frac{1}{n}} = \lim_{n\to\infty} \|T^n\|^{\frac{1}{n}}\,.$$

Concerning the resolvent of a positive operator, we have the following.

Lemma 6.23. *For a positive operator $T\colon E \to E$ on a real Banach lattice the following statements hold.*

(1) *If $\lambda \in \mathbb{C}$ satisfies $|\lambda| > r(T)$, then*

$$|R(\lambda,T)z| \le R(|\lambda|,T)|z| \ \text{ for all } z \in E_{\mathbf{C}}\,.$$

(2) *If λ_1 and λ_2 are real numbers and $\lambda_1 > \lambda_2 > r(T)$, then*

$$0 \le R(\lambda_1,T) \le R(\lambda_2,T)\,.$$

(3) *If λ is a real number satisfying $\lambda > r(T)$, then*

$$0 \le TR(\lambda,T) \le \lambda R(\lambda,T)\,.$$

Proof. (1) Using Lemma 3.22 and Exercise 1 at the end of the section, we see that

$$\begin{aligned} \big|R(\lambda,T)z\big| &= \Big|\sum_{n=0}^{\infty} \lambda^{-(n+1)} T^n z\Big| \le \sum_{n=0}^{\infty} |\lambda|^{-(n+1)} |T^n z| \\ &\le \sum_{n=0}^{\infty} |\lambda|^{-(n+1)} T^n |z| = R\big(|\lambda|,T\big)|z|\,. \end{aligned}$$

(2) Assume that λ is real and satisfies $\lambda > r(T)$. Then each operator $\lambda^{-(k+1)}T^k$ is positive, and so the operator $\sum_{k=0}^{n} \lambda^{-(k+1)}T^k$ is positive for each n. Therefore, if $x \in E^+$, it follows from the closedness of the positive cone E^+ that

$$R(\lambda,T)x = \Big[\sum_{k=0}^{\infty} \lambda^{-(k+1)} T^k\Big] x = \lim_{n\to\infty} \Big[\sum_{k=0}^{n} \lambda^{-(k+1)} T^k\Big] x \ge 0\,.$$

So, $R(\lambda,T)$ is a positive operator. The other inequality now follows from the resolvent identity by observing that

$$R(\lambda_2,T) - R(\lambda_1,T) = (\lambda_1-\lambda_2)R(\lambda_1,T)R(\lambda_2,T) \ge 0\,.$$

(3) Here we have

$$\begin{aligned} 0 \le TR(\lambda, T) &= \sum_{n=0}^{\infty} \lambda^{-(n+1)} T^{n+1} = \lambda \sum_{n=0}^{\infty} \lambda^{-(n+2)} T^{n+1} \\ &= \lambda \sum_{n=1}^{\infty} \lambda^{-(n+1)} T^{n} \le \lambda R(\lambda, T), \end{aligned}$$

and the proof is finished. ■

We close the section with an important property of certain positive operators.

Lemma 6.24. *Let $T\colon E \to E$ be a positive operator on a Banach lattice. Assume $\alpha v \le Tv$ for some $0 < v \in E$ and some scalar $\alpha > 0$. Then for each real number $\lambda > r(T)$ there exists some $0 < u \in E$ such that $v \in E_u$ and $\alpha u \le Tu \le \lambda u$. In particular, the principal ideal E_u is T-invariant.*

Proof. Consider the resolvent operator $R(\lambda, T)$. Since $\lambda > r(T)$, we know that $R(\lambda, T) = \sum_{n=0}^{\infty} \frac{T^n}{\lambda^{n+1}}$. We claim that the vector $u = R(\lambda, T)v$ has the desired properties. Indeed,

$$Tu = T\Big(\sum_{n=0}^{\infty} \frac{T^n v}{\lambda^{n+1}}\Big) = \sum_{n=0}^{\infty} \frac{T^n(Tv)}{\lambda^{n+1}} \ge \sum_{n=0}^{\infty} \frac{T^n(\alpha v)}{\lambda^{n+1}} = \alpha \sum_{n=0}^{\infty} \frac{T^n v}{\lambda^{n+1}} = \alpha u\,.$$

Applying Lemma 6.23(3), we get

$$Tu = TR(\lambda, T)v \le \lambda R(\lambda, T)v = \lambda u\,.$$

Finally, from $R(\lambda, T) \ge \frac{1}{\lambda} I$ we obtain $u = R(\lambda, T)v \ge \frac{1}{\lambda} v$. This implies $u > 0$ and $v \in E_u$. ■

Exercises

1. Assume that $\sum_{n=1}^{\infty} |z_n|$ exists in E, where $z_n = x_n + \imath y_n \in E_{\mathbf{c}}$ for each n. Show that $\sum_{n=1}^{\infty} z_n$ exists in $E_{\mathbf{c}}$ and $|\sum_{n=1}^{\infty} z_n| \le \sum_{n=1}^{\infty} |z_n|$, where the series are assumed to converge in norm. [HINT: If k and θ are fixed, then

$$\begin{aligned} \Big(\sum_{n=1}^{k} x_n\Big) \cos\theta + \Big(\sum_{n=1}^{k} y_n\Big) \cos\theta &= \sum_{n=1}^{k} (x_n \cos\theta + y_n \sin\theta) \\ &\le \sum_{n=1}^{k} |z_n| \le \sum_{n=1}^{\infty} |z_n|\,. \end{aligned}$$

Hence, $(\sum_{n=1}^{\infty} x_n) \cos\theta + (\sum_{n=1}^{\infty} y_n) \sin\theta \le \sum_{n=1}^{\infty} |z_n|$ holds for all θ, and from this the desired inequality follows.]

2. If $S, T\colon E \to E$ are positive operators, then show that

$$r(S+T) \geq \max\{r(S), r(T)\}.$$

3. Let $T\colon E \to E$ be a positive operator on a Banach lattice. If T is σ-order continuous (resp. order continuous), then show that $R(\lambda, T)$ is also σ-order continuous (resp. order continuous) for each $\lambda > r(T)$.

4. Let $T\colon E \to E$ be a positive operator on a Banach lattice. If a scalar $\alpha > 0$ satisfies $\alpha x \leq Tx$ for some $x > 0$ or $\alpha x^* \leq T^* x^*$ for some $0 < x^* \in E^*$, then show that $\alpha \leq r(T)$.

5. Let $T\colon X \to X$ be a bounded operator on a complex Banach space and let $\rho_\infty(T)$ denote the unbounded (open) component in $\mathbb{C}$ of the resolvent $\rho(T)$. Establish the following.

(a) If $\sigma(T)$ is at most countable (in particular, if every non-zero point in $\sigma(T)$ is an isolated point), then $\rho_\infty(T) = \rho(T)$.

(b) A closed subspace Y of the Banach space X is T-invariant if and only if Y is $R(\lambda, T)$-invariant for each $\lambda \in \rho_\infty(T)$.

(c) If a closed subspace Y of X is T-invariant and $T|_Y$ is the restriction of T to Y, then $\rho_\infty(T) \subseteq \rho(T|_Y)$ and $R(\lambda, T|_Y) = R(\lambda, T)|_Y$ for each $\lambda \in \rho_\infty(T)$.

(d) If $\rho_\infty(T) = \rho(T)$, then for every T-invariant closed subspace Y of X we have $\sigma(T|_Y) \subseteq \sigma(T)$.

(e) Assume that an open annulus $A_{R_1,R_2}(\lambda_0)$ lies entirely in $\rho_\infty(T)$ and that $R(\lambda, T) = \sum_{n=-\infty}^{\infty} (\lambda - \lambda_0)^n A_n$ is the Laurent series expansion of the resolvent in $A_{R_1,R_2}(\lambda_0)$. If Y is a T-invariant closed subspace of X, then Y is also A_n-invariant for each $n \in \mathbb{Z}$.

6. Give an example of a positive operator $T\colon E \to E$ on a Banach lattice and some T-invariant closed ideal J in E such that:

(a) There is some $\lambda \in \rho(T)$ for which J is not $R(\lambda, T)$-invariant.

(b) The spectrum $\sigma(T|_J)$ is not included in $\sigma(T)$, where $T|_J$ denotes the restriction of T to J.

(Compare these conclusions with Exercise 5 above.)

6.4. Functional Calculus

In the course of the proof of Theorem 6.12 we have established that if T is an operator on a complex Banach space and $\lambda \in \sigma(T)$, then $\lambda^n \in \sigma(T^n)$ for each n. More generally, if $p(z) = a_n z^n + \cdots + a_1 z + a_0$ is a polynomial and $p(T)$ denotes the operator obtained by the formal substitution of an operator $T \in \mathcal{L}(X)$ for z in $p(z)$, then one can show easily (see Exercise 1 at the end of the section) that $\sigma(p(T)) = p(\sigma(T))$. This result is a very special case of a powerful theory that allows one to associate with each operator $T \in \mathcal{L}(X)$ a broad class $\mathcal{F}(T)$ of analytic functions (that includes all polynomials) such that for each $f \in \mathcal{F}(T)$ an operator $f(T)$ can be defined whose spectrum $\sigma(f(T))$ coincides with $f(\sigma(T))$.

The present section is an introduction to this "functional calculus." A rigorous treatment of this theory depends heavily on the theory of Banach-valued functions of a complex variable, some elementary properties of which were presented in Section 1.5. Detailed presentations of functional calculus can be found in the classic monographs by N. Dunford and J. T. Schwartz [**118**] and E. Hille and R. S. Phillips [**152**]. See also [**113**] and [**317**].

As before, X will denote a non-trivial complex Banach space and $T\colon X \to X$ a bounded operator. Recall that a **neighborhood** of a subset A of $\mathbb{C}$ is any open set G such that $A \subseteq G$. We shall denote by $\mathcal{F}(T)$ the collection of all complex-valued functions that are analytic on neighborhoods of the spectrum $\sigma(T)$ of T. That is, $f \in \mathcal{F}(T)$ if and only if there exists a neighborhood G (depending on f) of $\sigma(T)$ such that $f\colon G \to \mathbb{C}$ is analytic—the neighborhood G can always be assumed to be bounded.

To be more precise, $\mathcal{F}(T)$ consists of pairs (f, V), where V is an open bounded set including $\sigma(T)$ and $f\colon V \to \mathbb{C}$ is an analytic function. If we define the relation $\sim$ on $\mathcal{F}(T)$ by letting $(f, V) \sim (g, W)$ whenever $f = g$ on some neighborhood U of $\sigma(T)$ satisfying $U \subseteq V \cap W$, then $\sim$ is an equivalence relation on $\mathcal{F}(T)$. With this equivalence relation in mind, we shall interpret the elements in $\mathcal{F}(T)$ as equivalence classes rather than as individual functions. The proof of the next result is straightforward and is omitted.

Lemma 6.25. *For every operator $T \in \mathcal{L}(T)$ the collection $\mathcal{F}(T)$ of all equivalence classes of functions equipped with the pointwise operations of addition and multiplication is a unital algebra.*

It is a very important fact that to each function $f \in \mathcal{F}(T)$ we can associate an operator $f(T) \in \mathcal{L}(X)$. The definition of the operator $f(T)$ will be described next. But first, let us say that a subset $\mathcal{C}$ of $\mathbb{C}$ is a **Jordan contour** if there exists a finite number of pairwise disjoint (closed simple) rectifiable Jordan curves $\mathcal{C}_1, \mathcal{C}_2, \ldots, \mathcal{C}_m$ in $\mathbb{C}$ such that $\mathcal{C} = \bigcup_{n=1}^{m} \mathcal{C}_m$.

Definition 6.26. *Let V be a neighborhood of a subset A of $\mathbb{C}$. A Jordan contour $\mathcal{C}$ in V **surrounds** A if there exists an open subset W of V such that $A \subseteq W \subseteq \overline{W} \subseteq V$ and $\mathcal{C} = \partial W$.*

The Jordan contour $\mathcal{C}$, viewed as the boundary of W, is always assumed to be positively oriented.

The following is a very useful result. Although it is intuitively clear, its proof is cumbersome and requires some lengthy technical arguments; for details see [**317**, p. 310].

Lemma 6.27. *Every neighborhood V of a compact set A admits a Jordan contour in V surrounding A.*

Now let $T \in \mathcal{L}(X)$ and $f \in \mathcal{F}(T)$. Pick any neighborhood V of $\sigma(T)$ on which f is analytic, and let $\mathcal{C}$ be a Jordan contour in V surrounding $\sigma(T)$. Then the operator-valued function $\lambda \mapsto f(\lambda)R(\lambda, T)$ is continuous on $\mathcal{C}$, and hence it is Riemann integrable over $\mathcal{C}$. Moreover, Cauchy's Integral Theorem 1.70 shows that if $\mathcal{C}_1$ is another Jordan contour surrounding $\sigma(T)$ in the **domain of analyticity**[2] of f, then

$$\int_{\mathcal{C}} f(\lambda)R(\lambda, T)\, d\lambda = \int_{\mathcal{C}_1} f(\lambda)R(\lambda, T)\, d\lambda .$$

Thus, the integral $\int_{\mathcal{C}} f(\lambda)R(\lambda, T)\, d\lambda$ is well defined, i.e., it is independent of the Jordan contour $\mathcal{C}$ surrounding $\sigma(T)$ in the domain of analyticity of f. This observation gives rise to the operator $f(T)$ as follows.

Definition 6.28. *Let $T \in \mathcal{L}(X)$ and fix $f \in \mathcal{F}(T)$. If $\mathcal{C}$ is any Jordan contour surrounding $\sigma(T)$ in the domain of analyticity of f, then the operator $f(T)$ in $\mathcal{L}(X)$ is defined by*

$$f(T) = \tfrac{1}{2\pi i} \int_{\mathcal{C}} f(\lambda)R(\lambda, T)\, d\lambda .$$

The algebraic properties of the operators $f(T)$ are included in the next result. These properties are referred to collectively as "functional calculus."

Theorem 6.29 (The Functional Calculus). *The mapping $f \mapsto f(T)$, from $\mathcal{F}(T)$ to $\mathcal{L}(X)$, is an algebraic homomorphism. That is, for each pair f, g in $\mathcal{F}(T)$ and all scalars α and β in $\mathbb{C}$ we have*

$$(\alpha f + \beta g)(T) = \alpha f(T) + \beta g(T) \quad \text{and} \quad (fg)(T) = f(T)g(T) .$$

Moreover:

(1) *If $S \in \mathcal{L}(X)$ commutes with T, then S commutes with $f(T)$. In particular, $Tf(T) = f(T)T$ for each $f \in \mathcal{F}(T)$.*

(2) *If $f \in \mathcal{F}(T)$, then $f \in \mathcal{F}(T^*)$ and $f(T^*) = f(T)^*$.*

(3) *If a function f satisfies $f(\lambda) = \sum_{n=0}^{\infty} a_n \lambda^n$ for all λ in a neighborhood of $\sigma(T)$, then $f \in \mathcal{F}(T)$ and $f(T) = \sum_{n=0}^{\infty} a_n T^n$.*

Proof. Fix $T \in \mathcal{L}(X)$ and $f, g \in \mathcal{F}(T)$. Pick a neighborhood V of $\sigma(T)$ such that both f and g are analytic on V, and let $\mathcal{C}_1$ be a Jordan contour

[2] This means that there exists some neighborhood U of $\sigma(T)$ such that $\mathcal{C}_1 \subseteq U$ and f is analytic on U.

in V surrounding $\sigma(T)$. Then for each $\alpha,\beta\in\mathbb{C}$ we have

$$\begin{aligned}\alpha f(T)+\beta g(T) &= \alpha\Big[\tfrac{1}{2\pi\imath}\int_{\mathcal{C}_1} f(\lambda)R(\lambda,T)\,d\lambda\Big]+\beta\Big[\tfrac{1}{2\pi\imath}\int_{\mathcal{C}_1} g(\lambda)R(\lambda,T)\,d\lambda\Big]\\ &= \tfrac{1}{2\pi\imath}\int_{\mathcal{C}_1}\big[\alpha f(\lambda)+\beta g(\lambda)\big]R(\lambda,T)\,d\lambda=(\alpha f+\beta g)(T)\,.\end{aligned}$$

Let W be the neighborhood of $\sigma(T)$ such that $\sigma(T)\subseteq W\subseteq\overline{W}\subseteq V$ and $\mathcal{C}_1=\partial W$. Now pick another Jordan contour $\mathcal{C}_2$ in W surrounding $\sigma(T)$. By Cauchy's Integral Theorem 1.70 we have $\frac{1}{2\pi\imath}\int_{\mathcal{C}_1}\frac{g(\mu)}{\mu-\lambda}\,d\mu=g(\lambda)$ for each $\lambda\in\mathcal{C}_2$ and $\frac{1}{2\pi\imath}\int_{\mathcal{C}_2}\frac{f(\lambda)}{\lambda-\mu}\,d\lambda=0$ for each $\mu\in\mathcal{C}_1$. Therefore, using Fubini's theorem and the Resolvent Identity (see Definition 6.5), we see that

$$\begin{aligned}f(T)g(T) &= \Big[\tfrac{1}{2\pi\imath}\int_{\mathcal{C}_2} f(\lambda)R(\lambda,T)\,d\lambda\Big]\cdot\Big[\tfrac{1}{2\pi\imath}\int_{\mathcal{C}_1} g(\mu)R(\mu,T)\,d\mu\Big]\\ &= -\tfrac{1}{4\pi^2}\int_{\mathcal{C}_2}\int_{\mathcal{C}_1} f(\lambda)g(\mu)R(\lambda,T)R(\mu,T)\,d\mu d\lambda\\ &= \tfrac{1}{4\pi^2}\int_{\mathcal{C}_2}\int_{\mathcal{C}_1}\tfrac{f(\lambda)g(\mu)}{\lambda-\mu}\big[R(\lambda,T)-R(\mu,T)\big]\,d\mu d\lambda\\ &= \tfrac{1}{2\pi\imath}\int_{\mathcal{C}_2}\Big[\tfrac{1}{2\pi\imath}\int_{\mathcal{C}_1}\tfrac{g(\mu)}{\mu-\lambda}\,d\mu\Big]f(\lambda)R(\lambda,T)\,d\lambda\\ &\quad -\tfrac{1}{4\pi^2}\int_{\mathcal{C}_1}\Big[\int_{\mathcal{C}_2}\tfrac{f(\lambda)}{\lambda-\mu}\,d\lambda\Big]g(\mu)R(\mu,T)\,d\mu\\ &= \tfrac{1}{2\pi\imath}\int_{\mathcal{C}_2} g(\lambda)f(\lambda)R(\lambda,T)\,d\lambda=(fg)(T)\,.\end{aligned}$$

The above establish that the mapping $f\mapsto f(T)$, from $\mathcal{F}(T)$ to $\mathcal{L}(X)$, is an algebraic homomorphism. We shall finish the proof by verifying the remaining properties.

(1) From $ST=TS$ and Lemma 6.4 we see that $SR(\lambda,T)=R(\lambda,T)S$ for each $\lambda\in\rho(T)$. Now the commutativity of S and $f(T)$ follows immediately from the definition of the Riemann integral of a Banach-valued function.

(2) Theorem 6.14 asserts that $\sigma(T)=\sigma(T^*)$, and so $\mathcal{F}(T)=\mathcal{F}(T^*)$. Also, we know that $R(\lambda,T^*)=[R(\lambda,T)]^*$ for all $\lambda\in\rho(T)$; see Exercise 4 of Section 6.1. Now the desired conclusion follows from the definition of the Riemann integral by observing that:

$$\begin{aligned}f(T^*) &= \tfrac{1}{2\pi\imath}\int_{\mathcal{C}_1} f(\lambda)R(\lambda,T^*)\,d\lambda=\tfrac{1}{2\pi\imath}\int_{\mathcal{C}_1} f(\lambda)[R(\lambda,T)]^*\,d\lambda\\ &= \Big[\tfrac{1}{2\pi\imath}\int_{\mathcal{C}_1} f(\lambda)R(\lambda,T)\,d\lambda\Big]^*=f(T)^*\,.\end{aligned}$$

(3) Consider first an entire function of the form $h(\lambda) = \lambda^k$, where $k \in \mathbb{N}$, and fix any positively oriented circle $\mathcal{C}$ centered at zero with radius greater than $\|T\|$. Using that the Neumann series (see Theorem 6.3) converges uniformly on any compact subset outside of the closed disk centered at zero with radius $\|T\|$, we see that

$$\begin{aligned} h(T) &= \tfrac{1}{2\pi\imath} \int_{\mathcal{C}} \lambda^k R(\lambda, T)\, d\lambda = \tfrac{1}{2\pi\imath} \int_{\mathcal{C}} \lambda^k \Big[\sum_{n=0}^{\infty} \lambda^{-(n+1)} T^n\Big]\, d\lambda \\ &= \sum_{n=0}^{\infty} \Big[\tfrac{1}{2\pi\imath} \int_{\mathcal{C}} \tfrac{d\lambda}{\lambda^{n+1-k}}\Big] T^n = T^k . \end{aligned}$$

This conclusion combined with the first part shows that for every polynomial $p(\lambda) = \sum_{n=0}^{k} a_n \lambda^n$ we have $p(T) = \sum_{n=0}^{k} a_n T^n$.

We now consider the general case. Since the series $f(\lambda) = \sum_{n=0}^{\infty} a_n \lambda^n$ converges in a neighborhood of $\sigma(T)$, there exist $0 < r_1 < r_2$ such that $\sigma(T) \subseteq B(0, r_1)$ and the series $\sum_{n=0}^{\infty} a_n \lambda^n$ converges for each $\lambda \in B(0, r_2)$. In particular, the series $f(\lambda) = \sum_{n=0}^{\infty} a_n \lambda^n$ converges uniformly on $B(0, r_1)$. That is, if $p_k(\lambda) = \sum_{n=0}^{k} a_n \lambda^n$, then the sequence of polynomials $\{p_k\}$ converges uniformly to f on $B(0, r_1)$. This implies (see Exercise 5 at the end of the section) that

$$f(T) = \lim_{k \to \infty} p_k(T) = \lim_{k \to \infty} \sum_{n=0}^{k} a_n T^n = \sum_{n=0}^{\infty} a_n T^n$$

holds in $\mathcal{L}(X)$, and the proof is finished. ■

From part (3) of the preceding theorem we get the following.

Corollary 6.30. *If $T \in \mathcal{L}(X)$ and V is a neighborhood of $\sigma(T)$, then for any Jordan contour $\mathcal{C}$ in V surrounding $\sigma(T)$ we have*

$$\tfrac{1}{2\pi\imath} \int_{\mathcal{C}} R(\lambda, T)\, d\lambda = I .$$

We are now ready to state and prove the Spectral Mapping Theorem.

Theorem 6.31 (The Spectral Mapping Theorem). *If $T \in \mathcal{L}(X)$, then for every function $f \in \mathcal{F}(T)$ we have $\sigma\big(f(T)\big) = f\big(\sigma(T)\big)$. That is,*

$$\sigma\big(f(T)\big) = \big\{f(\lambda)\colon\ \lambda \in \sigma(T)\big\} .$$

Proof. Let $f \in \mathcal{F}(T)$ and choose a neighborhood V of $\sigma(T)$ on which f is analytic. Fix $\xi \in \sigma(T)$ and define the function $g\colon V \to \mathbb{C}$ by

$$g(\lambda) = \begin{cases} \frac{f(\lambda) - f(\xi)}{\lambda - \xi} & \text{if } \lambda \neq \xi \\ f'(\xi) & \text{if } \lambda = \xi\ . \end{cases}$$

Clearly, the function g is analytic on V and satisfies $f(\xi)-f(\lambda)=(\xi-\lambda)g(\lambda)$ for all $\lambda\in V$. From Theorem 6.29, we get $f(\xi)-f(T)=(\xi-T)g(T)$. Since $\xi\in\sigma(T)$ and $\xi-T$ commutes with $g(T)$, it follows from the previous identity that the operator $f(\xi)-f(T)$ cannot be invertible, i.e., $f(\xi)\in\sigma\big(f(T)\big)$. Hence, $f\big(\sigma(T)\big)\subseteq\sigma\big(f(T)\big)$.

For the reverse inclusion, let $\xi\in\sigma\big(f(T)\big)$ and assume by way of contradiction that $\xi\notin f\big(\sigma(T)\big)$. Consider the function $h(\lambda)=\frac{1}{\xi-f(\lambda)}$ and note that $h\in\mathcal{F}(T)$. From $h(\lambda)\big(\xi-f(\lambda)\big)=1$ for each $\lambda\in V$ and Theorem 6.29, it follows that $h(T)\big[\xi-f(T)\big]=\big[\xi-f(T)\big]h(T)=I$, contrary to the non-invertibility of $\xi-f(T)$. Therefore, $\xi\in f\big(\sigma(T)\big)$, and so $\sigma\big(f(T)\big)\subseteq f\big(\sigma(T)\big)$. Consequently, $f\big(\sigma(T)\big)=\sigma\big(f(T)\big)$. ■

The rest of the section deals with properties of spectral sets. To illustrate their role, we begin with a simple example. Let $T_i\colon X_i\to X_i$ $(i=1,2)$ be two bounded operators defined on complex Banach spaces. Also let $\sigma_i=\sigma(T_i)$ be the spectrum of the operator T_i. Consider now the direct sum Banach space $X=X_1\oplus X_2$ and the direct sum (continuous) operator $T\colon X\to X$ defined by $T(x_1,x_2)=T_1x_1\oplus T_2x_2$. It is easy to check that $\sigma(T)=\sigma_1\cup\sigma_2$ (see also Exercise 17 of Section 6.1) and that X_1 and X_2 are complemented T-invariant closed subspaces of X. These decomposition properties are the basic features behind the concept of a spectral set of an operator. As we shall see, whenever the spectrum of an operator T can be represented in the form $\sigma_1\cup\sigma_2=\sigma(T)$, where σ_1 and σ_2 are closed and disjoint, then there exists a *unique* pair (Y,Z) of T-reducing subspaces such that $\sigma(T|_Y)=\sigma_1$ and $\sigma(T|_Z)=\sigma_2$. Therefore, in such a case, the operator $T\colon X\to X$ can be reduced to two "simpler" parts.

The above discussion can be formalized by introducing the concept of a spectral set of an operator.

Definition 6.32. *Let $T\colon X\to X$ be a continuous operator on a Banach space. A subset σ of $\sigma(T)$ is called a* **spectral set** *of T if σ is a closed and open subset of $\sigma(T)$ in the relative topology of $\sigma(T)$.*

Equivalently, a closed subset σ of $\sigma(T)$ is a spectral set whenever its relative complement $\sigma'=\sigma(T)\setminus\sigma$ is also a closed set.

Now let σ be a non-empty spectral set of an operator $T\in\mathcal{L}(X)$ such that $\sigma'=\sigma(T)\setminus\sigma$ is also non-empty. If we let $2\delta=d(\sigma,\sigma')>0$, then the two non-empty open sets $V_\sigma=\bigcup_{\lambda\in\sigma}B(\lambda,\delta)$ and $V_{\sigma'}=\bigcup_{\lambda\in\sigma'}B(\lambda,\delta)$ satisfy

$$\sigma\subseteq V_\sigma,\quad \sigma'\subseteq V_{\sigma'},\quad\text{and}\quad V_\sigma\cap V_{\sigma'}=\emptyset\,.$$

That is, there exist disjoint neighborhoods V_σ and $V_{\sigma'}$ of σ and σ', respectively. In particular, $V_\sigma\cup V_{\sigma'}$ is a neighborhood of $\sigma(T)$. Next, consider the functions $f_\sigma, f_{\sigma'}\colon V_\sigma \cup V_{\sigma'} \to \mathbb{C}$ defined by

$$f_\sigma(\lambda) = \begin{cases} 1 & \text{if } \lambda \in V_\sigma \\ 0 & \text{if } \lambda \in V_{\sigma'} \end{cases} \quad \text{and} \quad f_{\sigma'}(\lambda) = \begin{cases} 0 & \text{if } \lambda \in V_\sigma \\ 1 & \text{if } \lambda \in V_{\sigma'} \,. \end{cases}$$

Clearly, $f_\sigma, f_{\sigma'} \in \mathcal{F}(T)$, and

$$(f_\sigma)^2 = f_\sigma\,, \quad (f_{\sigma'})^2 = f_{\sigma'}\,, \quad f_\sigma f_{\sigma'} = 0\,, \quad \text{and} \quad f_\sigma + f_{\sigma'} = \mathbf{1}\,.$$

Consequently, according to Theorem 6.29 we have

$$\begin{aligned} \big[f_\sigma(T)\big]^2 = f_\sigma(T)\,, \quad \big[f_{\sigma'}(T)\big]^2 &= f_{\sigma'}(T)\,, \quad f_\sigma(T) f_{\sigma'}(T) = 0\,, \quad \text{and} \\ f_\sigma(T) + f_{\sigma'}(T) &= I\,. \end{aligned}$$

It follows that $f_\sigma(T)$ and $f_{\sigma'}(T)$ are both bounded projections on X satisfying $X = R\big(f_\sigma(T)\big) \oplus R\big(f_{\sigma'}(T)\big)$. Moreover, the Spectral Mapping Theorem yields

$$\sigma\big(f_\sigma(T)\big) = f_\sigma\big(\sigma(T)\big) = \{0,1\} \quad \text{and} \quad \sigma\big(f_{\sigma'}(T)\big) = f_{\sigma'}\big(\sigma(T)\big) = \{0,1\}\,.$$

This shows that $f_\sigma(T)$ and $f_{\sigma'}(T)$ are both non-trivial bounded projections. Since $f_\sigma(T)$ and $f_{\sigma'}(T)$ commute with T, we see that the pair $\big(R\big(f_\sigma(T)\big), R\big(f_{\sigma'}(T)\big)\big)$ of complemented closed subspaces reduces T.

We emphasize that the projection $f_\sigma(T)$ is independent of the open set V_σ. This projection plays an important role in operator theory and has a special name.

Definition 6.33. *The projection $f_\sigma(T)$ is called the* ***spectral projection*** *associated with the spectral set σ and is denoted $P_\sigma(T)$, i.e., $P_\sigma(T) = f_\sigma(T)$.*

We can also consider the extreme cases $\sigma = \sigma(T)$ and $\sigma = \emptyset$. In this cases, we have $P_{\sigma(T)}(T) = I$ and $P_{\emptyset} = 0$.

A fundamental property of the spectral projection $P_\sigma(T)$ is included in the next result.

Theorem 6.34. *Assume that a bounded operator $T\colon X \to X$ on a Banach space has a non-trivial spectral set σ. Then T admits a unique (non-trivial) reducing pair (Y_σ, Z_σ) such that*

$$\sigma(T|_{Y_\sigma}) = \sigma \quad \text{and} \quad \sigma(T|_{Z_\sigma}) = \sigma(T) \setminus \sigma\,,$$

where $T|_{Y_\sigma}$ and $T|_{Z_\sigma}$ denote the restrictions of T to Y_σ and Z_σ, respectively.

Moreover, the spectral projection $P_\sigma(T)$ is the projection of X onto Y_σ along Z_σ.

Proof. As above, let V_σ and $V_{\sigma'}$ be two disjoint non-empty neighborhoods of σ and $\sigma' = \sigma(T) \setminus \sigma$, respectively, and let f_σ and $f_{\sigma'}$ also be defined as above. If Y_σ and Z_σ are the ranges of the non-trivial projections $f_\sigma(T)$ and $f_{\sigma'}(T)$, then the pair (Y_σ, Z_σ) of non-trivial closed subspaces reduces T; see

Theorem 2.22. First, we shall verify that $\sigma(T|_{Y_\sigma}) = \sigma$ and $\sigma(T|_{Z_\sigma}) = \sigma'$. By symmetry, it suffices to establish that $\sigma(T|_{Y_\sigma}) = \sigma$. For simplicity, for each $R \in \mathcal{L}(X)$ we shall denote the restriction of R to Y_σ by R_σ, i.e., $R_\sigma = T|_{Y_\sigma}$.

Let $\lambda \in \sigma(T_\sigma)$, and assume by way of contradiction that $\lambda \notin \sigma$. We can suppose that $\lambda \notin V_\sigma$; otherwise we replace V_σ by $V_\sigma \cap (\mathbb{C} \setminus \{\lambda\})$. Define the function $g\colon V_\sigma \cup V_{\sigma'} \to \mathbb{C}$ by $g(\mu) = \frac{1}{\lambda-\mu}$ if $\mu \in V_\sigma$ and $g(\mu) = 0$ if $\mu \in V_{\sigma'}$. Clearly, $g \in \mathcal{F}(T)$, $g(\mu)(\lambda-\mu)f_\sigma(\mu) = f_\sigma(\mu)$ and $g(\mu)f_\sigma(\mu) = g(\mu)$ for each $\mu \in V_\sigma \cup V_{\sigma'}$. Consequently, from Theorem 6.29 we get

$$g(T)(\lambda - T)f_\sigma(T) = (\lambda - T)g(T)f_\sigma(T) = f_\sigma(T)\,, \quad \text{and} \tag{$\star$}$$

$$g(T)f_\sigma(T) = f_\sigma(T)g(T) = g(T)\,. \tag{$\star\star$}$$

Now $(\star\star)$ shows that $g(T)$ leaves Y_σ invariant and a glance at $(\star)$ guarantees that $g(T)_\sigma(\lambda - T_\sigma) = (\lambda - T_\sigma)g(T)_\sigma = I_\sigma$. Therefore, $\lambda - T_\sigma$ is invertible on Y_σ, and hence $\lambda \notin \sigma(T_\sigma)$, a contradiction. This contradiction implies $\lambda \in \sigma$, and so $\sigma(T_\sigma) \subseteq \sigma$.

For the reverse inclusion, let $\lambda \in \sigma$ and assume that $\lambda \notin \sigma(T_\sigma)$. So, there exists some $S \in \mathcal{L}(Y_\sigma)$ such that $S(\lambda - T_\sigma) = (\lambda - T_\sigma)S = I_\sigma$. Next, consider the analytic function $h\colon V_\sigma \cup V_{\sigma'} \to \mathbb{C}$ defined by $h(\mu) = 0$ if $\mu \in V_\sigma$ and $h(\mu) = \frac{1}{\lambda-\mu}$ if $\mu \in V_{\sigma'}$. Clearly, $h(\mu)(\lambda - \mu) = f_{\sigma'}(\mu)$ for each $\mu \in V_\sigma \cup V_{\sigma'}$. So, Theorem 6.29 implies $h(T)(\lambda - T) = (\lambda - T)h(T) = f_{\sigma'}(T)$. Consequently,

$$\begin{aligned}\big[Sf_\sigma(T) + h(T)\big](\lambda - T) &= (\lambda - T)\big[Sf_\sigma(T) + h(T)\big] \\ &= f_\sigma(T) + f_{\sigma'}(T) = I\,.\end{aligned}$$

This shows that the operator $\lambda - T$ is invertible on X or $\lambda \notin \sigma(T)$, contrary to the assumption that $\lambda \in \sigma \subseteq \sigma(T)$. Thus, $\lambda \in \sigma(T_\sigma)$, and so $\sigma(T_\sigma) = \sigma$.

Finally, we shall prove the uniqueness of the reducing pair. To this end, assume that (Y, Z) is another reducing pair for T satisfying

$$\sigma(T|_Y) = \sigma \quad \text{and} \quad \sigma(T|_Z) = \sigma(T) \setminus \sigma\,.$$

Let $\mathcal{C}_1$ be a Jordan contour in V_σ surrounding σ, and let $\mathcal{C}_2$ be a Jordan contour in $V_{\sigma'}$ surrounding σ'. The union $\mathcal{C}$ of $\mathcal{C}_1$ and $\mathcal{C}_2$ is a Jordan contour in $V_\sigma \cup V_{\sigma'}$ surrounding $\sigma(T)$. Now note that if $y \in Y$, then we have

$$\begin{aligned} y &= \Big[\tfrac{1}{2\pi\imath}\int_{\mathcal{C}_1} R(\lambda, T|_Y)\,d\lambda\Big]y = \Big[\tfrac{1}{2\pi\imath}\int_{\mathcal{C}_1} R(\lambda, T)\,d\lambda\Big]y \\ &= \Big[\tfrac{1}{2\pi\imath}\int_{\mathcal{C}} f_\sigma(\lambda)R(\lambda, T)\,d\lambda\Big]y = f_\sigma(T)y\,.\end{aligned}$$

This implies $y \in Y_\sigma$, and so $Y \subseteq Y_\sigma$. Similarly, $Z \subseteq Z_\sigma$. If $v \in Y_\sigma$, then write $v = y + z \in Y \oplus Z = X$, and note that $z = v - y \in Y_\sigma \cap Z_\sigma = \{0\}$. That is, $z = 0$ or $v = y \in Y$. Thus, $Y_\sigma \subseteq Y$, and hence $Y = Y_\sigma$. Similarly, $Z = Z_\sigma$, and the uniqueness of the reducing pair has been established. ■

It should be noticed that in the extreme cases $\sigma = \sigma(T)$ and $\sigma = \emptyset$ we have $\big(Y_{\sigma(T)}, Z_{\sigma(T)}\big) = \big(X, \{0\}\big)$ and $\big(Y_{\emptyset}, Z_{\emptyset}\big) = \big(\{0\}, X\big)$.

A special class of spectral sets is provided by the isolated points of the spectrum of an operator $T \in \mathcal{L}(X)$. If λ_0 is an isolated point in the spectrum $\sigma(T)$ of an operator $T \in \mathcal{L}(X)$, then there exists some $r > 0$ such that $B(\lambda_0, r) \cap \sigma(T) = \{\lambda_0\}$. In this case, we know that $R(\cdot, T)$ has a unique Laurent series expansion on the open annulus $\{\lambda \in \mathbb{C}\colon\ 0 < |\lambda - \lambda_0| < r\}$. The unique series expansion of $R(\cdot, T)$ in such an annulus will be referred to as the Laurent series expansion of the resolvent **around** λ_0.

Adopting the standard complex-variable terminology, we shall say that an isolated point λ_0 in the spectrum $\sigma(T)$ of a bounded operator $T \in \mathcal{L}(X)$ is a **pole** of order $k \geq 1$ of the resolvent $R(\cdot, T)$ if in the Laurent expansion $R(\lambda, T) = \sum_{n=-\infty}^{\infty} (\lambda - \lambda_0)^n A_n$ around λ_0 we have $A_{-k} \neq 0$ and $A_{-n} = 0$ for all $n > k$. If $k = 1$, then λ_0 is called a **simple pole**. If $A_{-n} \neq 0$ for infinitely many $n \in \mathbb{N}$, then we shall say that the resolvent $R(\cdot, T)$ has an **essential singularity** at λ_0. Regarding isolated points, we have the following results.

Lemma 6.35. *If $T \in \mathcal{L}(X)$ and λ_0 is an isolated point in $\sigma(T)$, then λ_0 is either a pole or an essential singularity of the resolvent $R(\lambda, T)$.*

Proof. Consider the Laurent series $R(\lambda, T) = \sum_{n=-\infty}^{\infty} (\lambda - \lambda_0)^n A_n$ around the point λ_0. If λ_0 were not a singularity, then $A_{-n} = 0$ for each $n \geq 1$ and so $\lim_{\lambda \to \lambda_0} R(\lambda, T) = A_0 \in \mathcal{L}(X)$. But $(\lambda - T)R(\lambda, T) = R(\lambda, T)(\lambda - T) = I$, and so letting $\lambda \to \lambda_0$, we get $(\lambda_0 - T)A_0 = A_0(\lambda_0 - T) = I$, contrary to the hypothesis that $\lambda_0 \in \sigma(T)$. ∎

Lemma 6.36. *Let $T \in \mathcal{L}(X)$ and let $\lambda_0 \in \sigma(T)$. If the operator $\lambda_0 - T$ has finite ascent and descent, then λ_0 is an isolated point of the spectrum of T.*

Proof. Assume that $\lambda_0 \in \sigma(T)$ is such that $\alpha(\lambda_0 - T) = \delta(\lambda_0 - T) = p < \infty$. Let $Y = N\big((\lambda_0 - T)^p\big)$ and $Z = R\big((\lambda_0 - T)^p\big)$. By Theorem 2.23 the T-invariant closed subspaces Y and Z satisfy $X = Y \oplus Z$ and the operator $\lambda_0 - T$ is quasinilpotent on Y and invertible on Z.

The fact that $\lambda_0 - T$ is quasinilpotent on Y implies that the spectrum of the operator $\lambda_0 - T|_Y$ in $\mathcal{L}(Y)$ is $\{0\}$. Therefore, the spectrum of T restricted to Y is $\{\lambda_0\}$. That is, $\sigma(T|_Y) = \{\lambda_0\}$.

Since the operator $\lambda_0 - T$ is invertible on Z, it follows that there exists a neighborhood V of λ_0 such that the operator $\lambda - T|_Z$ is invertible on Z for each $\lambda \in V$; see Corollary 2.11. This implies that for each $\lambda \in V \setminus \{\lambda_0\}$ the operators $\lambda - T|_Y$ and $\lambda - T|_Z$ are invertible on Y and Z, respectively. But then for each $\lambda \in V \setminus \{\lambda_0\}$ the operator $\lambda - T = (\lambda - T|_Y) \oplus (\lambda - T|_Z)$ is also invertible on X; see Exercise 17 in Section 6.1. Therefore, λ_0 is an isolated point in $\sigma(T)$. ∎

The coefficients of the Laurent series expansion of the resolvent around an isolated point of the spectrum of an operator satisfy some useful identities that are included in the next lemma.

Lemma 6.37. *Let $T\colon X \to X$ be a bounded operator on a Banach space and let λ_0 be an isolated point of the spectrum of $\sigma(T)$. If*

$$R(\lambda, T) = \sum_{n=-\infty}^{\infty} (\lambda - \lambda_0)^n A_n$$

is the Laurent series expansion of the resolvent function $R(\cdot, T)$ around λ_0, then the coefficients of the expansion satisfy the identities:

$$\begin{aligned} A_{-1} &= I + (T - \lambda_0)A_0\,, \\ A_{n-1} &= (T - \lambda_0)A_n \ \ \textit{for each} \ \ n \neq 0\,, \quad \textit{and} \\ A_n T &= T A_n \ \ \textit{for each} \ \ n\,. \end{aligned}$$

In particular, for each $n \geq 1$ we have

$$(T - \lambda_0)^n A_n = A_0 \qquad \textit{and} \qquad A_{-n} = (T - \lambda_0)^{n-1} A_{-1}\,.$$

Proof. Multiplying $R(\lambda, T) = \sum_{n=-\infty}^{\infty} (\lambda - \lambda_0)^n A_n$ on the left by $\lambda - T$ and using that $(\lambda - T)R(\lambda, T) = I$, we obtain

$$\begin{aligned} I &= \sum_{n=-\infty}^{\infty} (\lambda - T)(\lambda - \lambda_0)^n A_n \\ &= \sum_{n=-\infty}^{\infty} \big[(\lambda - \lambda_0) + (\lambda_0 - T)\big](\lambda - \lambda_0)^n A_n \\ &= \sum_{n=-\infty}^{\infty} (\lambda - \lambda_0)^{n+1} A_n + \sum_{n=-\infty}^{\infty} (\lambda_0 - T)(\lambda - \lambda_0)^n A_n \\ &= \sum_{n=-\infty}^{\infty} (\lambda - \lambda_0)^n \big[A_{n-1} + (\lambda_0 - T)A_n\big]\,. \end{aligned}$$

The uniqueness of the Laurent series expansion yields $I = A_{-1} + (\lambda_0 - T)A_0$ and $A_{n-1} + (\lambda_0 - T)A_n = 0$ for each $n \neq 0$. These are the first two desired identities.

If we multiply $R(\lambda, T) = \sum_{n=-\infty}^{\infty} (\lambda - \lambda_0)^n A_n$ on the right by $\lambda - T$, then as above we can obtain that $A_n T = T A_n$ for each n. The last two identities are straightforward. ■

If λ_0 is an isolated point of the spectrum, then the set $\sigma = \{\lambda_0\}$ is clearly a spectral set. For simplicity we shall write $P_{\lambda_0}(T)$ rather than $P_{\{\lambda_0\}}(T)$. In this case, we shall also write Y_{λ_0} and Z_{λ_0} instead of the more laborious symbols $Y_{\{\lambda_0\}}$ and $Z_{\{\lambda_0\}}$.

Following the standard terminology from the theory of complex variables, we shall call the operator A_{-1} in the Laurent series expansion $R(\lambda, T) = \sum_{n=-\infty}^{\infty} (\lambda - \lambda_0)^n A_n$ the **residue operator** at λ_0. Remarkably, when λ_0 is an isolated point, the residue operator coincides with the spectral projection $P_{\lambda_0}(T)$. The details are included in the next result.

Lemma 6.38. *If λ_0 is an isolated point in the spectrum of an operator T in $\mathcal{L}(X)$, then in the Laurent series expansion $R(\lambda, T) = \sum_{n=-\infty}^{\infty} (\lambda - \lambda_0)^n A_n$ around λ_0 the residue operator A_{-1} is a projection that coincides with the spectral projection associated with $\{\lambda_0\}$. That is, $A_{-1} = P_{\lambda_0}(T)$.*

Proof. Since λ_0 is an isolated point of the spectrum $\sigma(T)$, there exists some $r > 0$ such that $\{\lambda \in \mathbb{C}\colon\ 0 < |\lambda - \lambda_0| \leq 3r\} \cap \sigma(T) = \emptyset$. Let C be any positively oriented circle with radius less than $3r$ centered at λ_0. By Theorem 1.78 the coefficients of the Laurent series expansion of the resolvent function $R(\lambda, T)$ around λ_0 are given by

$$A_n = \tfrac{1}{2\pi i} \int_C \tfrac{R(\lambda,T)}{(\lambda-\lambda_0)^{n+1}}\, d\lambda \quad \text{for } n = 0, \pm 1, \pm 2, \ldots\ .$$

In particular, we have

$$A_{-1} = \tfrac{1}{2\pi i} \int_C R(\lambda, T)\, d\lambda\,.$$

For the rest of the proof the symbol C_δ will denote the positively oriented circle centered at λ_0 and having radius $\delta > 0$. Now consider the open set $V = D(\lambda_0, r) \cup \{z \in \mathbb{C}:\ 2r < |z - \lambda_0| < \rho\}$, where $\rho > 0$ is large enough so that $\sigma(T) \subseteq V$. Clearly, $\partial V = C_r \cup C_\rho \cup (-C_{2r})$, where $-C_{2r}$ denotes the circle C_{2r} oriented negatively.

Let f be the complex-valued function defined by

$$f(\lambda) = \begin{cases} 0 & \text{if } |\lambda - \lambda_0| > \frac{3}{2}r \\ 1 & \text{if } |\lambda - \lambda_0| < \frac{5}{4}r\,. \end{cases}$$

Clearly, f belongs to $\mathcal{F}(T)$ and is identically 1 on the open disk $D(\lambda_0, r)$. So, by the definition of the spectral projection $P_{\lambda_0}(T)$ associated with the spectral set $\{\lambda_0\}$, we see that $f(T) = P_{\lambda_0}(T)$. It remains to notice that

$$A_{-1} = \tfrac{1}{2\pi i} \int_{C_r} R(\lambda, T)\, d\lambda = \tfrac{1}{2\pi i} \int_{\partial V} f(\lambda) R(\lambda, T)\, d\lambda = f(T) = P_{\lambda_0}(T)\,.$$

The proof is finished. ∎

We are now ready to characterize the isolated points of the spectrum that are poles of the resolvent.

Theorem 6.39. *Let λ_0 be an isolated point in the spectrum of an operator $T \in \mathcal{L}(X)$ and let $R(\lambda, T) = \sum_{n=-\infty}^{\infty}(\lambda - \lambda_0)^n A_n$ be the Laurent series expansion of the resolvent around λ_0. Then the following statements are equivalent.*

(1) *The point λ_0 is a pole of the resolvent of order k.*

(2) *We have $(T - \lambda_0)^{k-1}A_{-1} \neq 0$ and $(T - \lambda_0)^k A_{-1} = 0$.*

(3) *The ascent and descent of the operator $\lambda_0 - T$ are both finite—and in this case they are equal to the order of the pole λ_0.*

Proof. (1) $\Longrightarrow$ (2) We have $A_{-k} \neq 0$ and $A_{-n} = 0$ for all $n > k$. But then it follows from Lemma 6.37 that $(T - \lambda_0)^{k-1}A_{-1} = A_{-k} \neq 0$ and $(T - \lambda_0)^k A_{-1} = A_{-(k+1)} = 0$.

(2) $\Longrightarrow$ (3) From Lemma 6.38, we know that $A_{-1} = P_{\lambda_0}(T)$. So, we have $(T - \lambda_0)^{k-1}P_{\lambda_0}(T) \neq 0$ and $(T - \lambda_0)^k P_{\lambda_0}(T) = 0$.

We shall verify first that $N\big((\lambda_0 - T)^k\big) = N\big((\lambda_0 - T)^{k+1}\big)$. We already know that $N\big((\lambda_0 - T)^k\big) \subseteq N\big((\lambda_0 - T)^{k+1}\big)$. To prove the reverse inclusion, let $x \in N\big((\lambda_0 - T)^{k+1}\big)$, and assume by way of contradiction that x does not belong to $N\big((\lambda_0 - T)^k\big)$, that is, the vector $y = (\lambda_0 - T)^k x \neq 0$. It follows that $(\lambda_0 - T)y = (\lambda_0 - T)^{k+1}x = 0$ or $Ty = \lambda_0 y$. This implies (see Exercise 3 at the end of this section) that $f(T)y = f(\lambda_0)y$ for each $f \in \mathcal{F}(T)$. In particular, letting $f = f_\sigma$, where f_σ is the function defined before Theorem 6.34, we obtain that $P_{\lambda_0}(T)y = f_\sigma(\lambda_0)y = y$. Consequently,

$$0 = (T - \lambda_0)^k P_{\lambda_0}(T)x = P_{\lambda_0}(T)(T - \lambda_0)^k x = P_{\lambda_0}(T)y = y \neq 0\,,$$

which is a contradiction. Hence, $N\big((\lambda_0 - T)^k\big) = N\big((\lambda_0 - T)^{k+1}\big)$. This shows that the operator $\lambda_0 - T$ has finite ascent and that $p = \alpha(\lambda_0 - T) \leq k$. In particular, we have $N\big((\lambda_0 - T)^p\big) = N\big((\lambda_0 - T)^n\big)$ for all $n \geq p$.

Next, notice that $(T - \lambda_0)^{k-1}P_{\lambda_0}(T) \neq 0$ guarantees the existence of some vector x in the range of the projection $P_{\lambda_0}(T)$ such that

$$(T - \lambda_0)^{k-1}x = (T - \lambda_0)^{k-1}P_{\lambda_0}(T)x \neq 0\,.$$

From $(T - \lambda_0)^k x = (T - \lambda_0)^k P_{\lambda_0}(T)x = 0$, it follows that

$$N\big((\lambda_0 - T)^{k-1}\big) \neq N\big((\lambda_0 - T)^k\big)\,.$$

This shows that $p \geq k$ is also true. Thus, $p = k$.

Next, we consider the unique decomposition $X = Y_{\lambda_0} \oplus Z_{\lambda_0}$ described in Theorem 6.34, where $(Y_{\lambda_0}, Z_{\lambda_0})$ is a reducing pair for T, $\sigma(T|_{Y_{\lambda_0}}) = \{\lambda_0\}$ and $\sigma(T|_{Z_{\lambda_0}}) = \sigma(T) \setminus \{\lambda_0\}$. In particular, the operator $\lambda_0 - T$ is invertible on Z_{λ_0}. This implies that for each $n \in \mathbb{N}$ the operator $(\lambda_0 - T)^n$ is also invertible on Z_{λ_0}.

Since $P_{\lambda_0}(T)(X) = Y_{\lambda_0}$, the identity $(T - \lambda_0)^k P_{\lambda_0}(T) = 0$, implies that $(\lambda_0 - T)^k = 0$ on Y_{λ_0}. Consequently,

$$R\big((\lambda_0 - T)^{k+1}\big) = R\big((\lambda_0 - T)^k\big) = Z_{\lambda_0}\,.$$

Thus, $\lambda_0 - T$ has finite descent, and Lemma 2.21 guarantees that the descent $\delta(\lambda_0 - T) = k$.

(3) $\Longrightarrow$ (1) Assume that $\alpha(\lambda_0 - T) = \delta(\lambda_0 - T) = p < \infty$. If we let $Y = N\big((\lambda_0 - T)^p\big)$ and $Z = R\big((\lambda_0 - T)^p\big)$, then by Theorem 2.23 both Y and Z are T-invariant closed subspaces satisfying $X = Y \oplus Z$, and the operator $\lambda_0 - T$ is quasinilpotent on Y and invertible on Z.

From the identity $T = T|_Y \oplus T|_Z$, we get $\sigma(T) = \sigma(T|_Y) \cup \sigma(T|_Z)$; see Exercise 17 of Section 6.1. Since $\lambda_0 \notin \sigma(T|_Z)$ and $\sigma(T|_Y) = \{\lambda_0\}$, we see that $\sigma(T|_Z) = \sigma(T) \setminus \{\lambda_0\}$. This conclusion implies that $Y = Y_{\lambda_0}$ and $Z = Z_{\lambda_0}$; see Theorem 6.34. In particular, $P_{\lambda_0}(T)$ is the projection of X on Y along Z. So, if we take an arbitrary $n > p$ and $x \in X$, then from $P_{\lambda_0}(T)x \in Y = N\big((\lambda_0 - T)^p\big) = N\big((\lambda_0 - T)^{n-1}\big)$ it follows that

$$A_{-n}x = (T - \lambda_0)^{n-1} P_{\lambda_0}(T)x = 0\,.$$

That is, $A_{-n} = 0$ holds for all $n > p$, and so λ_0 is a pole of the resolvent $R(\cdot, T)$ whose order k is less than or equal to p. But then from the implication (2) $\Rightarrow$ (3), we see that necessarily $k = p$. ■

Corollary 6.40. *If $T \in \mathcal{L}(X)$ and $\lambda_0 \in \sigma(T)$ is a pole of order k of the resolvent function $R(\cdot, T)$ around λ_0, then:*

(1) *λ_0 is an eigenvalue of T.*

(2) *$Y_{\lambda_0} = N\big((\lambda_0 - T)^k\big)$ and $Z_{\lambda_0} = R\big((\lambda_0 - T)^k\big)$.*

Proof. Let $T \in \mathcal{L}(X)$ and assume that $\lambda_0 \in \sigma(T)$ is a pole of order k of the resolvent function $R(\cdot, T)$ around λ_0. According to Theorem 6.39 we have $(T - \lambda_0)^{k-1} A_{-1} \neq 0$ and $(T - \lambda_0)^k A_{-1} = 0$. If we choose any vector $x \in X$ such that $y = (T - \lambda_0)^{k-1} A_{-1}x \neq 0$, then clearly the vector y satisfies $Ty = \lambda_0 y$. Therefore, λ_0 is an eigenvalue.

The identities $Y_{\lambda_0} = N\big((\lambda_0 - T)^k\big)$ and $Z_{\lambda_0} = R\big((\lambda_0 - T)^k\big)$ follow immediately from Theorems 6.34 and 2.23. ■

Corollary 6.41. *If X is a finite dimensional complex Banach space and $T \in \mathcal{L}(X)$, then every point in the spectrum of T is a pole of the resolvent function $R(\lambda, T)$.*

Proof. Fix $T \in \mathcal{L}(X)$. From $\dim X < \infty$ it follows that the spectrum of T is finite, it consists of eigenvalues, and obviously for each $\lambda \in \mathbb{C}$ the operator $\lambda - T$ has finite ascent and descent. The conclusion now follows immediately from Theorem 6.39. ■

Exercises

1. Show directly (i.e., by avoiding the Spectral Mapping Theorem) that if $p(z) = \sum_{k=0}^{n} a_k z^k$ is a polynomial, then for every bounded operator $T\colon X \to X$ on a Banach space we have

$$\sigma(p(T)) = p(\sigma(T)) = \{p(\lambda)\colon\ \lambda \in \sigma(T)\}.$$

[HINT: Fix $\mu \in \mathbb{C}$ and write $\mu - p(z) = -a_n(z - z_1)(z - z_2)\cdots(z - z_n)$, where $z_1, \ldots, z_n$ are the n solutions of the equation $p(z) = \mu$. Then

$$\mu - p(T) = -a_n(T - z_1)(T - z_2)\cdots(T - z_n).$$

Now notice that $\mu \in \sigma(p(T))$ if and only if $z_i \in \sigma(T)$ for some i.]

2. Assume that f is an analytic function defined on a neighborhood V of ∞, that is, the function f is defined and is analytic on an open set of the form $V = \{\lambda \in \mathbb{C}\colon\ |\lambda| > R\}$ for some $R > 0$. Suppose also that f vanishes at infinity. If $\mathcal{C}$ is a Jordan contour in V surrounding the spectrum of an operator $T \in \mathcal{L}(X)$, then show that $\int_{\mathcal{C}} f(\lambda) R(\lambda, T)\, d\lambda = 0$.

3. Let $T\colon X \to X$ be a bounded operator on a Banach space. Fix any function $f \in \mathcal{F}(T)$ and let $\mathcal{C}$ be a Jordan contour surrounding $\sigma(T)$ in the domain of analyticity of f. Show that

$$f(T)x = \tfrac{1}{2\pi\imath} \int_{\mathcal{C}} f(\lambda) R(\lambda, T) x\, d\lambda$$

holds for each $x \in X$. In particular, if $Tx = \alpha x$ for some $x \in X$ and a scalar $\alpha \in \mathbb{C}$, then show that $f(T)x = f(\alpha)x$.

4. Let $T \in \mathcal{L}(X)$, let $f \in \mathcal{F}(T)$, and let $g \in \mathcal{F}(f(T))$. Establish that the composition $g \circ f$ belongs to $\mathcal{F}(T)$ and that $(g \circ f)(T) = g(f(T))$.

5. Let $T\colon X \to X$ be a bounded operator on a Banach space, and let V be an open neighborhood of $\sigma(T)$. Assume that a sequence $\{f_n\}$ of analytic functions defined on V converges uniformly to some function $f\colon V \to \mathbb{C}$. Show that $f \in \mathcal{F}(T)$ and that $\|f_n(T) - f(T)\| \to 0$.

6. Let $T\colon X \to X$ be a bounded operator on a Banach space, and let σ be a spectral set of T. Also, let (Y, Z) be the pair of reducing subspaces as determined in Theorem 6.34. If $h \in \mathcal{F}(T)$, then show that

$$h(T)|_Y = h(T|_Y).$$

7. Let X be a Banach space. Show that there do not exist operators S, T in $\mathcal{L}(X)$ such that $ST - TS = I$, the identity operator on X. [HINT: Use Exercise 19 in Section 6.1.]

8. Let $T\colon X \to X$ be a bounded operator and let σ be a spectral set of T. Also, let (Y_σ, Z_σ) be the unique T-reducing pair of subspaces such that $X = Y_\sigma \oplus Z_\sigma$, $\sigma(T|_{Y_\sigma}) = \sigma$, and $\sigma(T|_{Z_\sigma}) = \sigma(T) \setminus \sigma$; see Theorem 6.34. If we let $T_\sigma = T|_{Y_\sigma}$, then show that

$$\sigma_p(T_\sigma) = \sigma \cap \sigma_p(T), \quad \sigma_r(T_\sigma) = \sigma \cap \sigma_r(T), \quad \text{and} \quad \sigma_c(T_\sigma) = \sigma \cap \sigma_c(T).$$

9. Let λ_0 be an isolated point in the spectrum of an operator $T \in \mathcal{L}(X)$. Show that the resolvent of T has an essential singularity at λ_0 if and only if either the ascent or the descent of $\lambda_0 - T$ is infinite.

10. Prove Corollary 6.40.

11. Let $T \in \mathcal{L}(X)$ and let λ_0 be an isolated point in the spectrum $\sigma(T)$ such that every neighborhood of λ_0 contains points that are not eigenvalues of T. Show that λ_0 is a pole of the resolvent $R(\cdot, T)$ if and only if $\lambda_0 - T$ has finite descent.

12. If a quasinilpotent operator $T \in \mathcal{L}(X)$ satisfies $p = \delta(T) < \infty$, then show that $T^p = 0$. [HINT: Use the preceding exercise.]

13. For a spectral set σ of an operator $T \in \mathcal{L}(X)$ establish the following.

(a) If $0 \notin \sigma$, then there exists an operator $S \in \mathcal{L}(X)$ such that

$$P_\sigma(T) = TS = ST\,.$$

(b) If $0 \in \sigma$, then there exists an operator $S \in \mathcal{L}(X)$ such that

$$P_\sigma(T) = I + TS = I + ST\,.$$

[HINT: Let (Y_σ, Z_σ) be the unique T-reducing pair such that $Y_\sigma \oplus Z_\sigma = X$, and let $P_\sigma(T)$ be the projection of X onto Y_σ along Z_σ. Assume $0 \notin \sigma$. This means that the operator $T\colon Y_\sigma \to Y_\sigma$ is invertible. Let $R \in \mathcal{L}(Y_\sigma)$ be the inverse of $T|_{Y_\sigma}$ and put $S = RP_\sigma(T)$.]

14. Let σ be a spectral set of an operator $T \in \mathcal{L}(X)$. If there exists some $r > 0$ such that $\sigma \subseteq \{\lambda \in \mathbb{C}\colon\ |\lambda| < r\}$ and $\sigma(T) \setminus \sigma \subseteq \{\lambda \in \mathbb{C}\colon\ |\lambda| > r\}$, then show that the range of the spectral projection $P_\sigma(T)$ is given by

$$R\big(P_\sigma(T)\big) = \big\{x \in X\colon\ \tfrac{T^n x}{r^n} \to 0\big\}\,.$$

15. (R. Melton) For an entire function $f\colon \mathbb{C} \to \mathbb{C}$ having the series expansion $f(\lambda) = \sum_{n=0}^{\infty} a_n \lambda^n$ the following statements are equivalent.

(a) Each coefficient a_n of the expansion is non-negative.

(b) For each positive operator T on any Banach lattice E the operator $f(T) = \sum_{n=1}^{\infty} a_n T^n$ is positive.

(c) There exists an infinite dimensional Banach lattice E such that $f(K)$ is positive for each positive compact operator K on E.

Chapter 7

Some Special Spectra

This chapter is a continuation of the previous one. Our objective here is to investigate the structure of the spectra of several classes of operators. The first section provides a complete description of the spectrum of compact and strictly singular operators. We establish here that each strictly singular (in particular, each compact) operator has an at most countable spectrum whose non-zero points are isolated eigenvalues. If the spectrum of such an operator is countable, then it forms a sequence that converges to zero. These results are illustrated with several examples of compact integral operators.

The second section of this chapter presents a useful method of converting approximate eigenvalues of an operator into eigenvalues by enlarging the Banach space upon which the operator is acting. The third section studies the structure of the spectrum of lattice homomorphisms. The spectrum of these operators exhibits a remarkable degree of symmetry, and the main results of this section show that the spectrum of each lattice homomorphism and of each interval preserving positive operator is cyclic. The fourth section introduces the order spectrum of an order bounded operator and studies some basic properties the order spectrum.

The fifth and final section of this chapter is devoted to the essential spectrum of an operator. The essential spectrum of an operator $T \in \mathcal{L}(X)$ is the non-empty compact subset of the spectrum of T consisting of all complex numbers λ for which the operator $\lambda - T$ fails to be a Fredholm operator. This is the same as the spectrum of T in the Calkin algebra $\mathfrak{C}(X) = \mathcal{L}(X)/\mathcal{K}(X)$ and, therefore, there exists a possibility of studying the essential spectrum of T in the context of Banach algebras. A discussion regarding connections between the essential spectrum and measures of non-compactness closes the section.

7.1. The Spectrum of a Compact Operator

Recall that an operator $T\colon X \to Y$ between two Banach spaces is said to be **compact** whenever the image under T of the unit ball of X is a norm totally bounded subset of Y (or, equivalently, whenever the image under T of each bounded sequence in X has a convergent subsequence in Y). A compact operator is, of course, automatically continuous.

The objective of this section is to establish that the spectrum of a compact operator is an at most countable set. As before, X denotes a complex Banach space.

An operator $T \in \mathcal{L}(X)$ is said to be **power compact** if there exists some $k \in \mathbb{N}$ such that T^k is a compact operator.

Theorem 7.1. *If $T \in \mathcal{L}(X)$ is a power compact operator (in particular, if T is compact) and X is infinite dimensional, then $0 \in \sigma(T)$.*

Proof. Assume that some power of T, say $S = T^k$, is a compact operator. If $0 \notin \sigma(T)$, then T is invertible and so $S = T^k$ is also invertible, i.e., S^{-1} exists. But then the identity operator $I = SS^{-1}\colon X \to X$ is compact. Hence, X must be finite dimensional, a contradiction. ■

The next result is crucial for the description of the spectrum of a compact operator.

Lemma 7.2. *If $T \in \mathcal{L}(X)$ is a compact operator, then for each $\epsilon > 0$ the set of all eigenvalues of T with modulus greater than ϵ, i.e., the set*

$$\{\lambda \in \sigma_p(T)\colon\ |\lambda| > \epsilon\},$$

is finite (and possibly empty).

Proof. Assume by way of contradiction that there exists a sequence $\{\lambda_n\}$ of distinct eigenvalues satisfying $|\lambda_n| > \epsilon > 0$ for all n. For each n pick a unit vector x_n such that $Tx_n = \lambda_n x_n$. Clearly, the set of eigenvectors $\{x_1, x_2, \ldots\}$ is a linearly independent set; see Exercise 2 at the end of Section 6.2.

Let V_n be the linear span of $\{x_1, \ldots, x_n\}$. Then:

- Each V_n is invariant under T, i.e., $T(V_n) \subseteq V_n$.
- V_{n-1} is a proper closed subspace of V_n for each $n \geq 2$.

By Lemma 4.44, for each n there exists a unit vector $y_n \in V_n$ such that

$$\|y_n - y\| \geq \tfrac{1}{2} \ \text{ for all } \ y \in V_{n-1}\,.$$

We claim that $Ty_n - \lambda_n y_n \in V_{n-1}$. Indeed, if $y_n = \sum_{i=1}^{n} \alpha_i x_i$, then

$$Ty_n - \lambda_n y_n = \sum_{i=1}^{n} \alpha_i \lambda_i x_i - \sum_{i=1}^{n} \alpha_i \lambda_n x_i = \sum_{i=1}^{n-1} \alpha_i(\lambda_i - \lambda_n) x_i \in V_{n-1}\,.$$

Therefore, for $n > m$ we have $Ty_m + \lambda_n y_n - Ty_n \in V_{n-1}$, and hence

$$\begin{aligned}\|Ty_n - Ty_m\| &= \big\|\lambda_n y_n + (Ty_n - \lambda_n y_n - Ty_m)\big\| \\ &= |\lambda_n|\big\|y_n - \tfrac{1}{\lambda_n}(Ty_m + \lambda_n y_n - Ty_n)\big\| \\ &\geq \tfrac{\epsilon}{2}\,.\end{aligned}$$

This shows that $\{Ty_n\}$ cannot have any convergent subsequence, contrary to the compactness of T. This completes the proof. ■

And now we come to a complete description of the spectrum of a compact operator.

Theorem 7.3. *For any compact operator $T \in \mathcal{L}(X)$ we have:*

(1) *The spectrum of T is at most countable.*

(2) *Every non-zero point of the spectrum is an eigenvalue.*

(3) *If the spectrum of T is countable and $\{\lambda_1, \lambda_2, \ldots\}$ is any enumeration of $\sigma(T)$, then $\lambda_n \to 0$. In particular, every non-zero point in the spectrum of T is an isolated point.*

Proof. Let $T \in \mathcal{L}(X)$ be a compact operator. From Lemma 7.2 it follows that T has at most countably many eigenvalues and that if $\{\lambda_1, \lambda_2, \ldots\}$ is any sequence of distinct eigenvalues, then $\lambda_n \to 0$.

Our next goal is to show that each non-zero boundary point of the spectrum is an eigenvalue, i.e., each $\lambda \in \partial\sigma(T) \setminus \{0\}$ belongs to $\sigma_p(T)$. So, let $\lambda \in \partial\sigma(T) \setminus \{0\}$. By Theorem 6.18(2), we know that λ belongs to the approximate point spectrum of T. So, there exists a sequence $\{x_n\}$ of unit vectors such that $Tx_n - \lambda x_n \to 0$. The compactness of T guarantees that some subsequence of $\{Tx_n\}$ (which we shall denote by $\{Tx_n\}$ again) converges. From $x_n = \frac{1}{\lambda}Tx_n - \frac{1}{\lambda}(Tx_n - \lambda x_n)$, we see that the sequence $\{x_n\}$ is also norm convergent, say $x_n \to x$. Since $\|x_n\| = 1$ for each n, we infer that $\|x\| = 1$. Using once again that $Tx_n - \lambda x_n \to 0$, we obtain $Tx - \lambda x = 0$, or $Tx = \lambda x$. Thus, every non-zero boundary point of $\partial\sigma(T)$ is an eigenvalue of T, and hence $\partial\sigma(T)$ is at most countable.

To finish the proof, we shall show that $\sigma(T)$ has no interior points. Indeed, if $\sigma(T)$ had an interior point λ_0, then it is easy to see that $\sigma(T)$ would contain a boundary point in every direction from λ_0, i.e., $\partial\sigma(T)$ would be uncountable. So, $\sigma(T) = \partial\sigma(T)$, and the proof is finished. ■

If $T\colon X \to X$ is a compact operator and X is infinite dimensional, then zero is always a boundary point of the spectrum of T. This implies that when X is infinite dimensional, zero belongs to the approximate point spectrum of T.

Corollary 7.4. *For a power compact operator $T \in \mathcal{L}(X)$ we have:*

(1) *Its spectrum is at most countable.*

(2) *If the spectrum of T is a countable set and $\{\lambda_1, \lambda_2, \dots\}$ is an arbitrary enumeration of $\sigma(T)$, then $\lambda_n \to 0$.*

Proof. The conclusions follow immediately from Theorem 7.3 and the fact that $\sigma(T^k) = \{\lambda^k\colon \lambda \in \sigma(T)\}$. (Use here either the Spectral Mapping Theorem 6.31 or Exercise 11 in Section 6.1.) ∎

Corollary 7.5. *If $T \in \mathcal{L}(X)$ is a compact operator, then each non-zero $\lambda_0 \in \sigma(T)$ is a pole of the resolvent of T.*

Proof. Assume that $T\colon X \to X$ is a compact operator on a Banach space and let λ_0 be a non-zero point in the spectrum of T. By Theorem 7.3 we know that λ_0 is an isolated point of $\sigma(T)$. Also, by Lemma 4.45, the operator $\lambda_0 - T$ has finite ascent and descent. Now a glance at Theorem 6.39 shows that λ_0 is a pole of the resolvent of T. ∎

As expected every finite-rank operator has a finite spectrum.

Theorem 7.6. *The spectrum of a finite-rank operator $T\colon X \to X$ on a Banach space is a finite set and consists of eigenvalues.*

If X is infinite dimensional and $T = \sum_{i=1}^{k} x_i^* \otimes u_i$ is the unique representation of T with respect to a basis $u_1, \dots, u_k$ of the range of T, then the spectrum of T consists of zero together with the eigenvalues of the $k \times k$ square matrix $A = \big[x_i^*(u_j)\big]$.

Proof. If X is finite dimensional, then the result is obvious. So, assume that X is infinite dimensional. Clearly, T is not one-to-one and so $Tx = 0$ for some $x \neq 0$, which shows that 0 is an eigenvalue of T.

Assume that $Tx = \lambda x$ for some $\lambda \neq 0$ and $x \neq 0$ and let $T = \sum_{i=1}^{k} x_i^* \otimes u_i$ be the representation of T with respect to a basis $\{u_1, \dots, u_k\}$ of $T(X)$. It follows that

$$Tu_j = \sum_{i=1}^{k} x_i^*(u_j) u_i\,, \quad j = 1, \dots, k\,.$$

From $x = \frac{1}{\lambda} Tx$, we get $x \in T(X)$. Thus, $x = \sum_{j=1}^{k} \alpha_j u_j$ for appropriate (uniquely determined) scalars α_j. Now if α denotes the column vector with coordinates $\alpha_1, \dots, \alpha_k$ and $A = \big[x_i^*(u_j)\big]$, then it is easy to see that the equality $Tx = \lambda x$ is equivalent to $A\alpha = \lambda\alpha$. This shows that the non-zero eigenvalues of T are precisely the non-zero eigenvalues of the $k \times k$ square matrix A. ∎

The next result describes a large class of compact integral operators.

Theorem 7.7. *Assume that $(\mathcal{S}, d)$ is a compact metric space and μ is a Borel measure on $\mathcal{S}$. If $K\colon \mathcal{S}\times\mathcal{S}\to\mathbb{R}$ is a continuous function, then the integral operator $T\colon L_p(\mu)\to L_p(\mu)$, defined by*

$$Tx(t) = \int_{\mathcal{S}} K(s,t)x(s)\,d\mu(s)\,,$$

is a compact operator for each $1\le p\le\infty$.

Proof. Assume $1\le p\le\infty$. Fix some $t_0\in\mathcal{S}$ and let $\epsilon>0$. By the uniform continuity of K on $\mathcal{S}\times\mathcal{S}$ there exists some $\delta>0$ such that $d(t_0,t)<\delta$ implies $|K(s,t)-K(s,t_0)|<\epsilon$ for each $s\in\mathcal{S}$. So, if $d(t_0,t)<\delta$ and $f\in L_p(\mu)$ satisfies $\|f\|_p\le 1$, then Hölder's inequality implies

$$\begin{aligned}\big|Tf(t)-Tf(t_0)\big| &= \Big|\int_{\mathcal{S}}\big[K(s,t)-K(s,t_0)\big]f(s)\,d\mu(s)\Big|\\ &= \epsilon\Big[\int_{\mathcal{S}}|f(s)|\,d\mu(s)\Big]\\ &\le \epsilon\big[\mu(\mathcal{S})\big]^{\frac{1}{q}}\|f\|_p\le\epsilon\big[\mu(\mathcal{S})\big]^{\frac{1}{q}}\,,\end{aligned}$$

where $\frac{1}{p}+\frac{1}{q}=1$. Hence, $\{Tf\colon\ \|f\|_p\le 1\}$ is a $\|\cdot\|_\infty$-bounded and equicontinuous subset of $C(\mathcal{S})$. From the classical Ascoli–Arzelà Theorem (see [**31**, Theorem 7.9, p. 60]), we obtain that $\{Tf\colon\ \|f\|_p\le 1\}$ is a $\|\cdot\|_\infty$-totally bounded subset of $C(\mathcal{S})$, and consequently a $\|\cdot\|_p$-totally bounded subset of $L_p(\mu)$. In other words, the integral operator T is compact. ■

Here is another important example of a compact integral operator.

Example 7.8 (The Volterra Operator)**.** Consider the jointly measurable function $K\colon [0,1]\times[0,1]\to\mathbb{R}$ defined by

$$K(s,t)=\begin{cases}0 & \text{if } 0\le t<s\le 1\\ 1 & \text{if } 0\le s\le t\le 1\,,\end{cases}$$

and let $V\colon L_p[0,1]\to L_p[0,1]$ $(1\le p\le\infty)$ be the integral operator with kernel K. That is, for each $x\in L_p[0,1]$ we have

$$Vx(t)=\int_0^1 K(s,t)x(s)\,ds=\int_0^t x(s)\,ds\,.$$

This integral operator is known as the **Volterra operator**. According to Theorem 5.29 its adjoint operator $V^*\colon L_q[0,1]\to L_q[0,1]$ is given by

$$V^*y(s)=\int_0^1 K(s,t)y(t)\,dt=\int_s^1 y(t)\,dt\,,$$

where $1\le q\le\infty$ satisfies $\frac{1}{p}+\frac{1}{q}=1$.

The Volterra operator is compact. Indeed, for each x in $L_p[0,1]$ it follows from Hölder's inequality that

$$|Vx(t_1) - Vx(t_2)| = \left| \int_{t_1}^{t_2} x(s)\, ds \right| \le |t_1 - t_2|^{\frac{1}{q}} \|x\|_p$$

for all $t_1, t_2 \in [0,1]$. This implies that $\{Vx\colon \ \|x\|_p \le 1\}$ is a bounded and equicontinuous subset of $C[0,1]$, and from this we infer that V is a compact operator.

The spectrum of V consists of the zero element alone, i.e., $\sigma(V) = \{0\}$. The fact that $0 \in \sigma(V)$ follows immediately from Theorem 7.1. On the other hand, if $\lambda \ne 0$, then for every $y \in L_p[0,1]$, it is easy to check that the equation

$$(\lambda - V)x = y$$

has a unique solution x (see Exercise 3 at the end of the section) given by

$$x(s) = \tfrac{1}{\lambda^2} e^{\frac{s}{\lambda}} \int_0^s y(\xi) e^{-\frac{\xi}{\lambda}}\, d\xi + \tfrac{1}{\lambda} y(s)\,.$$

This shows that $(\lambda - V)^{-1}$ exists for each $\lambda \ne 0$, and therefore $\sigma(V) = \{0\}$, i.e., V is a quasinilpotent operator. Finally, it should be noticed that 0 is not an eigenvalue of V. ∎

We continue with an important result (known as the Krein–Rutman theorem) regarding the spectral radius of a compact positive operator. Since the peripheral spectrum of an operator is a non-empty compact set, there always exists a point in the spectrum that lies on the circle centered at zero with radius $r(T)$. In the case of a positive operator, we get some extra information about the peripheral spectrum. Namely, as we shall see next, the spectral radius of a positive operator always belongs to the spectrum of the operator.

Theorem 7.9. *The spectral radius of a positive operator on a Banach lattice belongs to the spectrum of the operator.*

Proof. Let $T\colon E \to E$ be a positive operator on a Banach lattice. For each n put $\lambda_n = r(T) + \frac{1}{n}$. Clearly, $\{\lambda_n\} \subseteq \rho(T)$ and $\lambda_n \to r(T)$. To prove that $r(T) \in \sigma(T)$, it suffices to show (in view of Corollary 6.9) that $\lim_{n\to\infty} \|R(\lambda_n, T)\|_{\mathbf{c}} = \infty$.

To see this, start by fixing some $\alpha \in \sigma(T)$ with $|\alpha| = r(T)$ and let $\mu_n = \frac{\alpha\lambda_n}{|\alpha|}$ for each n. Then $\mu_n \in \rho(T)$, $|\mu_n| = \lambda_n$ for each n, and $\mu_n \to \alpha$. By Corollary 6.9, we have $\lim_{n\to\infty} \|R(\mu_n, T)\|_{\mathbf{c}} = \infty$. Next, for each n pick a unit vector $z_n \in E_{\mathbf{c}}$ such that

$$\big\| R(\mu_n, T) z_n \big\| \ge \tfrac{1}{2} \|R(\mu_n, T)\|_{\mathbf{c}}\,.$$

From Lemma 6.23, we have $|R(\mu_n,T)z_n| \le R(\lambda_n,T)|z_n|$. Hence

$$\|R(\lambda_n,T)\|_{\mathbf{c}} \ge \big\|\, R(\lambda_n,T)|z_n|\,\big\| \ge \tfrac{1}{2}\|R(\mu_n,T)\|_{\mathbf{c}}\,,$$

and so $\lim_{n\to\infty}\|R(\lambda_n,T)\|_{\mathbf{c}} = \infty$, as desired. ∎

And now we come to a classical result due to M. G. Krein and M. A. Rutman [**192**].

Theorem 7.10 (Krein–Rutman). *If $T\colon E\to E$ is a compact positive operator on a Banach lattice with $r(T)>0$, then the spectral radius $r(T)$ is an eigenvalue having a positive eigenvector. That is, there exists some $x>0$ such that $Tx = r(T)x$.*

Proof. Let $T\colon E\to E$ be a compact positive operator on a Banach lattice with $r(T)>0$. From Theorems 7.9 and 7.3 it follows that $r(T)$ is an eigenvalue of T. What we do not know yet is that this eigenvalue has a positive eigenvector. We shall accomplish this next. The elegant proof below is due to F. F. Bonsall [**68**].

Let $\lambda_n = r(T)+\frac{1}{n}$; and so $\lambda_n \downarrow r(T)$. Now Theorem 7.9 and Corollary 6.9 guarantee that $\lim_{n\to\infty}\|R(\lambda_n,T)\|_{\mathbf{c}} = \infty$. Since $R(\lambda_n,T)\ge 0$ holds for each n, it follows from Corollary 3.23 that $\|R(\lambda_n,T)\| = \|R(\lambda_n,T)\|_{\mathbf{c}}$. Consequently, for each n there exists a positive unit vector $y_n\in E$ such that $\|R(\lambda_n,T)y_n\| \ge \frac{1}{2}\|R(\lambda_n,T)\| = \frac{1}{2}\|R(\lambda_n,T)\|_{\mathbf{c}} > 0$.

Put $x_n = \frac{R(\lambda_n,T)y_n}{\|R(\lambda_n,T)y_n\|}$ and note that x_n is a positive unit vector. From

$$Tx_n - r(T)x_n = \big[\lambda_n - r(T)\big]x_n - (\lambda_n - T)x_n = \tfrac{1}{n}x_n - \tfrac{y_n}{\|R(\lambda_n,T)y_n\|}\,,$$

we obtain that

$$\big\|Tx_n - r(T)x_n\big\| \le \tfrac{1}{n} + \tfrac{2}{\|R(\lambda_n,T)\|} \xrightarrow[n\to\infty]{} 0\,. \qquad (\star)$$

By the compactness of T, the sequence $\{Tx_n\}$ has a convergent subsequence, which we shall denote by $\{Tx_n\}$ again. From $r(T)>0$ and $(\star)$, it follows that $x_n\to x$ for some positive unit vector x. Using $(\star)$ once again, we obtain $Tx = r(T)x$, as desired. ∎

We close the section by proving that the structure of the spectrum of a strictly singular operator is identical to that of a compact operator. However, the proof for strictly singular operators is much more involved and for this reason we have chosen to keep the much simpler proof of Theorem 7.3 for compact operators as well.

Theorem 7.11. *If $T \in \mathcal{L}(X)$ is a strictly singular operator on an infinite dimensional Banach space, then:*

(1) *Its spectrum $\sigma(T)$ is at most countable.*

(2) *The point zero belongs to the spectrum and is the only possible accumulation point of $\sigma(T)$.*

(3) *Each non-zero point of the spectrum is an eigenvalue.*

Proof. We begin by observing that $\lambda = 0$ belongs to $\sigma(T)$. Indeed, if $0 \notin \sigma(T)$, then the operator T would be invertible, which is impossible since T is strictly singular.

From Theorem 4.63 we know that for each non-zero $\lambda \in \mathbb{C}$ the operator $\lambda - T$ is Fredholm and that $i(\lambda - T) = i(\lambda I) = 0$. By virtue of part (c) of Lemma 4.55, there exists some open disk $D(\lambda)$ centered at λ and such that for each $\mu \in D(\lambda)$ the operator $\mu - T$ is Fredholm and its nullity $n(\mu - T)$ and defect $d(\mu - T)$ are constant on the set $D(\lambda) \setminus \{\lambda\}$. Moreover, we claim that $n(\mu - T) = d(\mu - T) = 0$ on $D(\lambda)$ with the possible exception of the point λ.

To establish this claim, fix some $\lambda \neq 0$ and then choose any point λ_1 with $|\lambda_1| > \|T\|$ and such that the closed line segment $L = [\lambda, \lambda_1]$ joining λ and λ_1 does not include zero. Since the operator $\mu - T$ is invertible whenever $|\mu| > \|T\|$, it follows that $n(\mu - T) = d(\mu - T) = 0$ for all μ close to λ_1, and consequently for all $\mu \in D(\lambda_1) \setminus \{\lambda_1\}$ by our choice of the disk $D(\lambda_1)$.

The compactness and connectedness of the segment L guarantee the existence of a finite number of disks $D(\xi_1), \ldots, D(\xi_k)$ such that they cover L and $D(\xi_j) \cap D(\xi_{j+1}) \neq \emptyset$ for $1 \leq j \leq k-1$. (See also Exercise 7 in Section 1.5.) Since the functions $n(\mu - T)$ and $d(\mu - T)$ are constant on each set $D(\xi_i) \setminus \{\xi_i\}$, it follows that they are constant on the union of the disks $D(\xi_j)$ except possibly for the points $\xi_1, \ldots, \xi_k$. Moreover, since on $D(\lambda_1) \setminus \{\lambda_1\}$ both functions $n(\mu - T)$ and $d(\mu - T)$ are equal to zero, it follows that $n(\mu - T) = d(\mu - T) = 0$ for all $\mu \in \bigcup_{j=1}^{k} D(\xi_j) \setminus \{\xi_j\}$. In particular, this implies that $n(\mu - T) = d(\mu - T) = 0$ for all $\mu \in D(\lambda) \setminus \{\lambda\}$. In other words, the operator $\mu - T$ is invertible for all μ in $D(\lambda)$ with a possible exception of the center λ. Since $\lambda \neq 0$ is arbitrary, this proves that each non-zero point in the spectrum $\sigma(T)$ is an isolated point.

Next, we shall show that the set $\sigma(T)$ is at most countable. For this, it suffices to verify that for every $\epsilon > 0$ the set $A_\epsilon = \{\lambda \in \sigma(T)\colon\ |\lambda| \geq \epsilon\}$ is finite (and possibly empty). Since A_ϵ is compact and does not contain zero, there exists a finite collection of disks $D(\xi_1), \ldots, D(\xi_n)$ as described earlier that cover A_ϵ. As shown above, only the centers $\xi_1, \ldots, \xi_n$ of these disks can belong to the spectrum, and so $A_\epsilon \subseteq \{\xi_1, \ldots, \xi_n\}$.

Finally, let us prove that each non-zero $\lambda \in \sigma(T)$ is an eigenvalue. Recall that λ is an isolated point in the spectrum, and consequently we can consider the spectral subspace X_λ corresponding to λ. By Theorem 6.34 the closed subspace X_λ is T-invariant and $\sigma(T_\lambda) = \{\lambda\}$, where T_λ denotes as usual the restriction of T to X_λ.

We claim that $\dim X_\lambda < \infty$. To see this, assume by way of contradiction that $\dim X_\lambda = \infty$. Since T is strictly singular on X, it follows that the operator T_λ is strictly singular on X_λ. But then the conclusion at the very beginning of the proof implies $0 \in \sigma(T_\lambda) = \{\lambda\}$, contrary to $\lambda \neq 0$. Hence, the Banach space X_λ is finite dimensional.

Now notice that the operator $T_\lambda \colon X_\lambda \to X_\lambda$ necessarily has an eigenvalue. As λ is the only point in the spectrum $\sigma(T_\lambda)$, we conclude that λ is an eigenvalue of T_λ, and therefore of T as well. ■

Exercises

1. Show that the only compact operators with closed range are the finite-rank operators.

2. Consider the interval $[0,1]$ equipped with its Lebesgue measure and for each $1 \leq p \leq \infty$ define the positive integral operator $T \colon L_p[0,1] \to L_p[0,1]$ by

$$Tf(t) = \int_0^1 (s^2 + t^2) f(s)\, ds\,.$$

Show that T is a finite-rank integral operator and find its spectrum.

3. Consider the Volterra operator $V \colon L_p[0,1] \to L_p[0,1]$ $(1 \leq p < \infty)$ defined by

$$Vf(t) = \int_0^t f(s)\, ds\,.$$

Show that V is quasinilpotent, i.e., $\sigma(V) = \{0\}$, by proving that for each $\lambda \neq 0$ and each $g \in L_p[0,1]$ the equation $(\lambda - V)f = g$ has a unique solution given by

$$f(s) = \tfrac{1}{\lambda^2} e^{\frac{s}{\lambda}} \int_0^s g(\xi) e^{-\frac{\xi}{\lambda}}\, d\xi + \tfrac{1}{\lambda} g(s)\,.$$

[HINT: Consider the operator $S \colon L_p[0,1] \to L_p[0,1]$ defined by

$$Sf(t) = \tfrac{1}{\lambda^2} e^{\frac{t}{\lambda}} \int_0^t f(\xi) e^{-\frac{\xi}{\lambda}}\, d\xi + \tfrac{1}{\lambda} f(t)\,.$$

Prove that S is a bounded operator satisfying $(\lambda - V)Sf = S(\lambda - V)f = f$ for each continuous function f. Use the fact that $C[0,1]$ is norm dense in $L_p[0,1]$ to conclude that $S = (\lambda - V)^{-1}$.]

4. Show directly (i.e., without using the previous exercise) that the Volterra operator has no eigenvalues—and conclude from this that the Volterra operator is quasinilpotent.

5. Let $T\colon E \to E$ be a positive contraction on a Banach lattice. Show that 1 belongs to the approximate point spectrum of T if and only if $r(T) = 1$.

6. Show that the hypothesis $r(T) > 0$ cannot be dropped from the Krein–Rutman Theorem 7.10.

7. Show that the compactness hypothesis of the operator in the Krein–Rutman Theorem 7.10 cannot be dropped. [HINT: Use the shift operator in Example 6.21.]

8. Let $T\colon E \to E$ be a positive operator on a Banach lattice and let p be a polynomial with non-negative coefficients. Show that $r(p(T)) = p(r(T))$. [HINT: Use Exercise 1 of Section 6.4.]

9. Prove the following generalization of the Krein–Rutman Theorem 7.10. If a power compact positive operator $T\colon E \to E$ satisfies $r(T) > 0$, then there exists some vector $x > 0$ such that $Tx = r(T)x$. [HINT: Assume that $r = r(T) > 0$ and that T^k is compact for some $k > 1$. According to Theorem 7.10 there exists some $x > 0$ such that $T^k x = r^k x$. Now notice that if $S = \sum_{i=0}^{k-1} r^i T^{k-1-i}$, then $y = Sx > 0$ and

$$Ty - ry = TSx - rSx = (T - r)Sx = (T^k - r^k)x = 0\,,$$

proving that $Ty = ry$.]

10. Prove Lemma 4.18 using the Krein–Rutman theorem. That is, show by using the Krein–Rutman theorem that if Ω is a compact Hausdorff space without isolated points, then the only weakly compact multiplication operator on $C(\Omega)$ is the zero operator. [HINT: Assume that there exists a non-zero weakly compact multiplication operator M_ϕ on $C(\Omega)$. Then $M_\phi M_\phi = M_{\phi^2}$ is a non-zero compact positive multiplication operator; see [**30**, p. 337]. So, replacing ϕ by ϕ^2, we can suppose that $\phi > 0$ and that M_ϕ is compact. Clearly,

$$\begin{aligned} r &= r(M_\phi) = \lim_{n\to\infty} \|M_\phi^n\|^{\frac{1}{n}} = \lim_{n\to\infty} \|M_{\phi^n}\|^{\frac{1}{n}} \\ &= \lim_{n\to\infty} (\|\phi^n\|_\infty)^{\frac{1}{n}} = \|\phi\|_\infty > 0\,, \end{aligned}$$

and so by the Krein–Rutman Theorem 7.10 there exists some $0 < x \in C(\Omega)$ such that $M_\phi x = \phi x = rx$, or $(\phi - r)x = 0$. The latter guarantees that $\phi(\omega) = r$ for all ω in a non-empty open subset V of Ω. It follows that the closed unit ball of $C(\overline{V})$ is compact. This implies that $\overline{V}$ (and hence V) must be a finite set. But then this guarantees that Ω has isolated points.]

11. Show that there exist a Banach space X and a non-empty compact subset K of $\mathbb{C}$ such that $\sigma(T) \neq K$ for each $T \in \mathcal{L}(X)$. (See also Exercise 15 in Section 6.1.)

12. Establish the following generalization of Theorem 7.11: *If $S \in \mathcal{L}(X)$ is strictly singular and an operator $T \in \mathcal{L}(X)$ has an at most countable spectrum, then the spectrum of $S+T$ is at most countable and the points of $\sigma(T)$ and zero are the only possible accumulation points of $\sigma(S+T)$.*

13. (T. Oikhberg and V. Troitsky) Let ϕ be a non-zero multiplicative functional on the Banach algebra $\mathcal{L}(X)$, where X is an arbitrary Banach space with $\dim X > 1$. Show that:
   (a) For each $T \in \mathcal{L}(X)$ we have $\phi(T) \in \sigma(T)$.
   (b) If for $T \in \mathcal{L}(X)$ there is a reducing pair of closed subspaces (V, W) such that $\dim V < \infty$ and T vanishes on W, then $\phi(T) = 0$.
   (c) If $T \in \mathcal{L}(X)$ is a strictly singular operator (in particular, if T is compact), then $\phi(T) = 0$.

7.2. Turning Approximate Eigenvalues into Eigenvalues

The purpose of this section is to present a method that allows us to treat the approximate eigenvalues of an operator as eigenvalues. The method is based on the following result that will be proven in this section: *If X is a (real or complex) Banach space, then X can be enlarged to a new Banach space $\widehat{X}$ such that every bounded operator $T\colon X \to X$ can be extended to a bounded operator $\widehat{T}\colon \widehat{X} \to \widehat{X}$ satisfying $\sigma(T) = \sigma(\widehat{T})$ and $\sigma_a(T) = \sigma_p(\widehat{T})$.*

We start with some preliminary discussion regarding sums of sequences of Banach spaces. Let $X_1, X_2, \ldots$ be a sequence of Banach spaces all over the same field ($\mathbb{R}$ or $\mathbb{C}$). For each $1 \le p < \infty$ we define the **ℓ_p-sum** of the sequence $\{X_n\}$ by

$$(X_1 \oplus X_2 \oplus \cdots)_p = \Big\{ x = (x_1, x_2, \ldots) \in \prod_{n=1}^{\infty} X_n\colon\ \|x\|_p = \Big(\sum_{n=1}^{\infty} \|x_n\|^p\Big)^{\frac{1}{p}} < \infty \Big\}.$$

We also define

$$(X_1 \oplus X_2 \oplus \cdots)_\infty = \Big\{ x = (x_1, x_2, \ldots) \in \prod_{n=1}^{\infty} X_n\colon\ \|x\|_\infty = \sup_{n \in \mathbb{N}} \|x_n\| < \infty \Big\},$$

and

$$(X_1 \oplus X_2 \oplus \cdots)_0 = \Big\{ x = (x_1, x_2, \ldots) \in \prod_{n=1}^{\infty} X_n\colon\ \lim_{n\to\infty} \|x_n\| = 0 \Big\}.$$

It is also customary to denote these Banach spaces by $\ell_p(X_1 \oplus X_2 \oplus \cdots)$ and $c_0(X_1 \oplus X_2 \oplus \cdots)$. In the case $X = X_1 = X_2 = \cdots$ we shall write

$$\ell_p(X) = \ell_p(X \oplus X \oplus \cdots) \quad \text{and} \quad c_0(X) = c_0(X \oplus X \oplus \cdots).$$

It is a routine matter to verify that under the coordinatewise operations of addition and scalar multiplication, each ℓ_p-sum is a Banach space. If each X_n is also a Banach lattice, then under the coordinatewise ordering and lattice operations the ℓ_p-sums are likewise Banach lattices.

Clearly, the c_0-sum of the sequence $\{X_n\}$ is a closed subspace of its ℓ_∞-sum. If, in addition, each X_n is a Banach lattice, then the c_0-sum of the sequence $\{X_n\}$ is a closed ideal in the ℓ_∞-sum of the sequence. Moreover, if each X_n is an AM-space with unit e_n, then the ℓ_∞-sum of the sequence $\{X_n\}$ is also an AM-space with unit $e = (e_1, e_2, \ldots)$.

Definition 7.12. *If X is a Banach space, then we let*

$$\widehat{X} = \ell_\infty(X)/c_0(X),$$

the quotient Banach space of $\ell_\infty(X)$ over the closed subspace $c_0(X)$.

Clearly, $\widehat{X}$ is a Banach lattice if X is a Banach lattice. The vectors of $\widehat{X}$ are equivalence classes and, as usual, for each $x = (x_1, x_2, \ldots) \in \ell_\infty(X)$ its equivalence class will be denoted by $[x]$. It should be obvious that two vectors $x = (x_1, x_2, \ldots)$ and $y = (y_1, y_2, \ldots)$ in $\ell_\infty(X)$ are equivalent (i.e., $[x] = [y]$) if and only if $\lim_{n\to\infty} \|x_n - y_n\| = 0$. This implies that for each $x = (x_1, x_2, \ldots) \in \ell_\infty(X)$ the norm of the vector $[x] \in \widehat{X}$ is given by

$$\big\|[x]\big\| = \inf_{y\in[x]} \|y\|_\infty = \limsup_{n\to\infty} \|x_n\|.$$

Our next objective is to show that the Banach space $\widehat{X}$ has the properties stated at the beginning of the section.

First of all, notice that there is a natural mapping $x \mapsto \dot{x}$ from X into $\widehat{X}$, defined by $\dot{x} = (x, x, x, \ldots)$. That is, $\dot{x}$ is the equivalence class of the constant sequence $(x, x, x, \ldots)$. It is a routine matter to verify that the mapping $x \mapsto \dot{x}$ is a linear isometry—which is also a lattice isometry if X is a Banach lattice. Thus, identifying X with its image $\dot{X}$, we can assume without loss of generality that X is a closed subspace of the Banach space $\widehat{X}$. In case X is a Banach lattice, $\widehat{X}$ is also a Banach lattice and $\dot{X}$ is a closed Riesz subspace of $\widehat{X}$.

Next, assume that $T\colon X \to X$ is a bounded operator. Let $[x] \in \widehat{X}$, and pick an arbitrary sequence $(x_1, x_2, \ldots) \in [x]$. Since $(x_1, x_2, \ldots)$ is a bounded sequence in X and T is continuous, the sequence $(Tx_1, Tx_2, \ldots)$ is also bounded. Moreover, if $y = (y_1, y_2, \ldots) \in [x]$, then $x_n - y_n \to 0$, and so by the continuity of T we get $Tx_n - Ty_n = T(x_n - y_n) \to 0$. This implies that the two sequences $(Tx_1, Tx_2, \ldots)$ and $(Ty_1, Ty_2, \ldots)$ belong to the same equivalence class, which we shall denote by $\widehat{T}([x])$. That is, the bounded operator T gives rise to a mapping $\widehat{T}\colon \widehat{X} \to \widehat{X}$ via the formula

$$\widehat{T}([(x_1, x_2, \ldots)]) = [(Tx_1, Tx_2, \ldots)].$$

A direct verification shows that $\widehat{T}\colon \widehat{X} \to \widehat{X}$ is a linear operator, which is also bounded. As a matter of fact, $\|\widehat{T}\| = \|T\|$.

To see this, start by observing that for each $x = (x_1, x_2, \ldots)$ we have

$$\begin{aligned} \big\|\widehat{T}[x]\big\| &= \big\|[(Tx_1, Tx_2, \ldots)]\big\| = \limsup_{n\to\infty} \|Tx_n\| \\ &\le \limsup_{n\to\infty} \|T\| \cdot \|x_n\| = \|T\| \cdot \limsup_{n\to\infty} \|x_n\| \\ &= \|T\| \cdot \big\|[x]\big\| \,. \end{aligned}$$

This implies that $\widehat{T}$ is a bounded operator and that $\|\widehat{T}\| \le \|T\|$. Now if $x \in X$ satisfies $\|x\| \le 1$, then $\|\dot{x}\| = \|x\| \le 1$, and so $\|\widehat{T}\| \ge \|\widehat{T}\dot{x}\| = \|Tx\|$. This implies

$$\|\widehat{T}\| \ge \sup_{\|x\|\le 1} \|Tx\| = \|T\| \,,$$

and hence $\|\widehat{T}\| = \|T\|$.

Next, notice that if $x \in X$ and, as before, $\dot{x}$ denotes the equivalence class with representative $(x, x, \ldots)$, then the constant sequence $(Tx, Tx, Tx, \ldots)$ is a representative of the class $\widehat{T}\dot{x}$. This shows that for each $x \in X$ we have $\widehat{T}\dot{x} = [(Tx, Tx, Tx, \ldots)]$. That is, $\widehat{T}$ is a bounded linear extension of T from X to all of $\widehat{X}$. In other words, we have established the following theorem.

Theorem 7.13. *If X is a Banach space and $T\colon X \to X$ is a bounded operator, then $\widehat{T}\colon \widehat{X} \to \widehat{X}$ is a linear extension of T to all of $\widehat{X}$ satisfying $\|\widehat{T}\| = \|T\|$. Moreover, when X is a Banach lattice, T is positive if and only if $\widehat{T}$ is positive.*

More properties of the mapping $T \mapsto \widehat{T}$ are included in the next results.

Theorem 7.14. *If X is a Banach space, then the mapping $T \mapsto \widehat{T}$, from $\mathcal{L}(X)$ to $\mathcal{L}(\widehat{X})$, is a linear isometry that is also an algebraic homomorphism, i.e., $\widehat{ST} = \widehat{S}\widehat{T}$ for all $S, T \in \mathcal{L}(X)$.*

Proof. It should be clear that the mapping $T \mapsto \widehat{T}$ is linear and (in view of Theorem 7.13) it is also an isometry. To see that this mapping is an algebraic homomorphism, note that if $S, T \in \mathcal{L}(X)$ and $x = (x_1, x_2, \ldots) \in \ell_\infty(X)$, then

$$\begin{aligned} \widehat{ST}[x] &= [(STx_1, STx_2, \ldots)] = \widehat{S}[(Tx_1, Tx_2, \ldots)] \\ &= \widehat{S}(\widehat{T}[(x_1, x_2, \ldots)]) = \widehat{S}\widehat{T}[x] \,. \end{aligned}$$

Therefore, $\widehat{ST} = \widehat{S}\widehat{T}$, and so $T \mapsto \widehat{T}$ is an algebraic homomorphism. ■

Theorem 7.15. *A bounded operator $T\colon X \to X$ on a Banach space is invertible if and only if $\widehat{T}\colon \widehat{X} \to \widehat{X}$ is invertible. Moreover, if T is invertible, then $(\widehat{T})^{-1} = \widehat{(T^{-1})}$.*

Proof. Assume first that T is invertible, and so $TT^{-1} = T^{-1}T = I$. Since $S \mapsto \widehat{S}$ is an algebraic homomorphism, we have $\widehat{T}\widehat{T^{-1}} = \widehat{T^{-1}}\widehat{T} = \widehat{I} = I$. This shows that $\widehat{T}$ is invertible and that $(\widehat{T})^{-1} = \widehat{(T^{-1})}$.

For the converse, suppose that $\widehat{T}$ is invertible. Since T is an extension of $\widehat{T}$, it follows that T is one-to-one. To see that T is surjective, fix some $y \in X$. Pick some $x = (x_1, x_2, \ldots) \in \ell_\infty(X)$ such that $\widehat{T}[x] = \dot{y} = [(y, y, \ldots)]$. This means that $Tx_n \to y$, and consequently $\widehat{T}\dot{x}_n \to \dot{y}$ in $\widehat{X}$. Since $\widehat{T}$ is invertible, we see that $\{\dot{x}_n\}$ is a Cauchy sequence in $\widehat{X}$, and so $\{x_n\}$ is a Cauchy sequence in X. Since X is a Banach space, we have $x_n \to x$ in X, and consequently $\widehat{T}\dot{x}_n \to \widehat{T}\dot{x}$ in $\widehat{X}$. This implies $\dot{y} = \widehat{T}\dot{x}$, or $y = Tx$. So, T is also surjective, and hence it is an invertible operator. ■

We are now ready to establish the desired the relationships between the spectra of the operators T and $\widehat{T}$. We start with the complex version of this result.

Theorem 7.16. *If $T\colon X \to X$ is a bounded operator on a complex Banach space, then:*

(1) $\rho(T) = \rho(\widehat{T})$ *and* $\sigma(T) = \sigma(\widehat{T})$.

(2) $\sigma_a(T) = \sigma_a(\widehat{T}) = \sigma_p(\widehat{T})$.

(3) $\widehat{R(\lambda, T)} = R(\lambda, \widehat{T})$ *holds for each* $\lambda \in \rho(T)$.

Proof. (1) If $\lambda \in \mathbb{C}$, then it follows from Theorems 7.14 and 7.15 that the operator $\lambda - T$ is invertible in $\mathcal{L}(X)$ if and only if $\lambda - \widehat{T}$ is invertible in $\mathcal{L}(\widehat{X})$. This shows that $\rho(T) = \rho(\widehat{T})$, and so $\sigma(T) = \mathbb{C} \setminus \rho(T) = \mathbb{C} \setminus \rho(\widehat{T}) = \sigma(\widehat{T})$.

(2) To prove that $\sigma_a(T) = \sigma_p(\widehat{T})$ notice that:

$$\begin{aligned} \lambda \in \sigma_a(T) \quad &\Longleftrightarrow \exists\, \{z_n\} \subseteq X \text{ such that } z_n \not\to 0 \text{ and } \|\lambda z_n - Tz_n\| \to 0 \\ &\Longleftrightarrow \exists\, [z] = [(z_1, z_2, \ldots)] \neq 0 \ \text{ such that } \ \widehat{T}[z] = \lambda[z] \\ &\Longleftrightarrow \lambda \in \sigma_p(\widehat{T})\,. \end{aligned}$$

Now fix $\lambda \in \sigma_a(\widehat{T})$, and let $\epsilon > 0$. This implies that there exists some vector $[z] = [(z_1, z_2, \ldots)] \in \widehat{X}$ satisfying

$$\|[z]\| = \limsup_{n\to\infty} \|z_n\| = 1 \quad \text{ and } \quad \|\widehat{T}[z] - \lambda[z]\| = \limsup_{n\to\infty} \|Tz_n - \lambda z_n\| < \epsilon\,.$$

Then there exists some k such that $\|Tz_n - \lambda z_n\| < \epsilon$ holds for all $n \geq k$. Since $\limsup_{n\to\infty} \|z_n\| = 1$, there are infinitely many $n \geq k$ such that $\|z_n\| > \frac{1}{2}$. This shows that $\lambda \in \sigma_a(T)$, and so $\sigma_a(\widehat{T}) \subseteq \sigma_a(T)$. Since $\sigma_a(T) \subseteq \sigma_a(\widehat{T})$ is obviously true, we see that $\sigma_a(T) = \sigma_a(\widehat{T})$.

(3) If $\lambda \in \rho(T)$, then a glance at Theorem 7.15 shows that

$$\big[R(\lambda,T)\big]^{\wedge} = \big[(\lambda - T)^{-1}\big]^{\wedge} = (\lambda - \widehat{T})^{-1} = R(\lambda, \widehat{T}),$$

and the proof is finished. ■

The real version of Theorem 7.16 can be formulated as follows.

Corollary 7.17. *Let $T\colon X \to X$ be a bounded operator on a real Banach space and let $T_{\mathbf{c}}$ be the extension of T to the complexification $X_{\mathbf{c}}$ of X given by the formula $T_{\mathbf{c}}(x + \imath y) = Tx + \imath Ty$. Then we have:*

(1) $\rho(T_{\mathbf{c}}) = \rho(\widehat{T_{\mathbf{c}}})$ *and* $\sigma(T_{\mathbf{c}}) = \sigma(\widehat{T_{\mathbf{c}}})$.

(2) $\sigma_a(T_{\mathbf{c}}) = \sigma_a(\widehat{T_{\mathbf{c}}}) = \sigma_p(\widehat{T_{\mathbf{c}}})$.

(3) $\big[R(\lambda, T_{\mathbf{c}})\big]^{\wedge} = R(\lambda, \widehat{T_{\mathbf{c}}})$ *holds for each* $\lambda \in \rho(T_{\mathbf{c}})$.

We close the section by mentioning that the Banach space $\widehat{X}$ satisfying the properties of Theorem 7.16 is not unique. Instead of $\widehat{X}$ we can use any ultrapower of X. In this case, the corresponding analogue of Theorem 7.16 takes the following form (whose proof is left for the reader).

Theorem 7.18. *If $T\colon X \to X$ is a bounded operator on a Banach space, then for every ultrafilter $\mathcal{U}$ on $\mathbb{N}$ we have:*

(1) $\sigma(T) = \sigma(T_{\mathcal{U}})$.

(2) $\sigma_a(T) = \sigma_a(T_{\mathcal{U}}) = \sigma_p(T_{\mathcal{U}})$.

Exercises

1. Show that if X is a Banach space, then for each $x = (x_1, x_2, \ldots) \in \ell_\infty(X)$ the norm of the vector $[x] \in \widehat{X}$ is given by

$$\big\|[x]\big\| = \inf_{y \in [x]} \|y\|_\infty = \limsup_{n \to \infty} \|x_n\|.$$

2. Assume that $\{X_n\}$ is a sequence of real Banach spaces. Show that (with the obvious identifications) for each $1 \le p \le \infty$ we have

$$\ell_p((X_1)_{\mathbf{c}} \oplus (X_2)_{\mathbf{c}} \oplus \cdots) = \ell_p(X_1 \oplus X_2 \oplus \cdots) \oplus \imath\ell_p(X_1 \oplus X_2 \oplus \cdots).$$

Also show that

$$c_0((X_1)_{\mathbf{c}} \oplus (X_2)_{\mathbf{c}} \oplus \cdots) = c_0(X_1 \oplus X_2 \oplus \cdots) \oplus \imath c_0(X_1 \oplus X_2 \oplus \cdots).$$

3. Let $\{X_n\}$ be a sequence of real (or complex) Banach spaces, and assume that $1 < p, q < \infty$ satisfy $\frac{1}{p} + \frac{1}{q} = 1$. Show that

$$(X_1 \oplus X_2 \oplus \cdots)_p^* = (X_1^* \oplus X_2^* \oplus \cdots)_q\,,$$

where the equality is understood subject to the natural duality

$$\langle x, x^* \rangle = \sum_{n=1}^{\infty} \langle x_n, x_n^* \rangle$$

for all $x = (x_1, x_2, \ldots)$ in $(X_1 \oplus X_2 \oplus \cdots)_p$ and all $x^* = (x_1^*, x_2^*, \ldots)$ in $(X_1^* \oplus X_2^* \oplus \cdots)_q$. Likewise show that

$$\begin{aligned}(X_1 \oplus X_2 \oplus \cdots)_1^* &= (X_1^* \oplus X_2^* \oplus \cdots)_\infty\,, \quad \text{and}\\ (X_1 \oplus X_2 \oplus \cdots)_0^* &= (X_1^* \oplus X_2^* \oplus \cdots)_1\,.\end{aligned}$$

4. Let $\{X_n\}$ be a sequence of all real (or complex) Banach spaces. Fix any $1 \le p < \infty$ and let $X = (X_1 \oplus X_2 \oplus \cdots)_p$. For each $\mathcal{T} = (T_1, T_2, \ldots)$ in $(\mathcal{L}(X_1) \oplus \mathcal{L}(X_2) \oplus \cdots)_\infty$ define the mapping $\mathcal{F}_\mathcal{T}\colon X \to X$ by

$$\mathcal{F}_\mathcal{T}(x_1, x_2, \ldots) = (T_1x_1, T_2x_2, \ldots)\,.$$

Establish the following.

(a) $\mathcal{F}_\mathcal{T}$ is a bounded operator on X, i.e., $\mathcal{F}_\mathcal{T} \in \mathcal{L}(X)$.

(b) The mapping $\mathcal{T} \mapsto \mathcal{F}_\mathcal{T}$, from $(\mathcal{L}(X_1) \oplus \mathcal{L}(X_2) \oplus \cdots)_\infty$ to $\mathcal{L}(X)$, is a linear isometry—and so for each $1 \le p < \infty$ the Banach space $(\mathcal{L}(X_1) \oplus \mathcal{L}(X_2) \oplus \cdots)_\infty$ is a closed subspace of the Banach space $\mathcal{L}((X_1 \oplus X_2 \oplus \cdots)_p)$.

5. Verify that the mapping $x \mapsto \dot{x}$, from X to $\widehat{X}$ defined by $\dot{x} = [(x, x, x, \ldots)]$, is a linear isometry—which is also a lattice isometry if X is in addition a Banach lattice.

6. If X is a Banach lattice, then show that a bounded operator $T\colon X \to X$ is positive if and only if the operator $\widehat{T}\colon \widehat{X} \to \widehat{X}$ is positive.

7. If $T\colon X \to X$ is a surjective operator on a Banach space, then show that $\widehat{T}\colon \widehat{X} \to \widehat{X}$ is also surjective. Conversely, if T has a closed range and $\widehat{T}$ is surjective, then show that T is also surjective.

8. Show that a positive operator $T\colon E \to E$ on a real Banach lattice is a lattice homomorphism if and only if $\widehat{T}\colon \widehat{E} \to \widehat{E}$ is likewise a lattice homomorphism.

9. Let X be a real Banach space with complexification $X_\mathbf{c} = X \oplus \imath X$. Establish the following:

(a) A sequence $\{x_n + \imath y_n\} \subseteq X_\mathbf{c}$ is bounded, that is,

$$z = (x_1 + \imath y_1, x_2 + \imath y_2, \ldots) \in \ell_\infty(X_\mathbf{c})\,,$$

if and only if the sequences $x = (x_1, x_2, \ldots)$ and $y = (y_1, y_2, \ldots)$ in X are both bounded, i.e., $x, y \in \ell_\infty(X)$. In this case, write $z = x + \imath y$.

(b) Show that the mapping $\mathcal{T}\colon \widehat{X}_\mathbf{c} \to \widehat{X} \oplus \imath\widehat{X}$, defined by

$$\mathcal{T}([x + \imath y]) = [x] + \imath[y]$$

for all $x = (x_1, x_2, \ldots)$ and $y = (y_1, y_2, \ldots)$ in $\ell_\infty(X)$, is a surjective (topological) isomorphism—which is a lattice isomorphism if X is also a Banach lattice.

[HINT: For the continuity of $\mathcal{T}$ note that

$$\begin{aligned}
\|\mathcal{T}([x + \imath y])\| &= \|[x] + \imath[y]\| = \sup_{\theta\in\mathbb{R}} \|[x]\cos\theta + [y]\sin\theta\| \\
&= \sup_{\theta\in\mathbb{R}} \|[(x_1\cos\theta + y_1\sin\theta, x_2\cos\theta + y_2\sin\theta, \ldots)]\| \\
&= \sup_{\theta\in\mathbb{R}} \big[\limsup_{n\to\infty} \|x_n\cos\theta + y_n\sin\theta\|\big] \\
&\le \limsup_{n\to\infty} \|x_n + \imath y_n\| = \|[x + \imath y]\| \,.\big]
\end{aligned}$$

$$\begin{array}{ccc}
\widehat{X_c} & \xrightarrow{\ \mathcal{T}\ } & (\widehat{X})_c \\
{\scriptstyle \widehat{T_c}}\big\downarrow & & \big\downarrow{\scriptstyle (\widehat{T})_c} \\
\widehat{X_c} & \xrightarrow{\ \mathcal{T}\ } & (\widehat{X})_c
\end{array}$$

Figure 1

10. Let $T\colon X \to X$ be a bounded operator on a real Banach space, and let $\widehat{T}\colon \widehat{X} \to \widehat{X}$ be its extension to the Banach space $\widehat{X}$. If $X_\mathbf{c} = X \oplus \imath X$ is the complexification of X, then in the preceding exercise we defined the mapping $\mathcal{T}\colon \widehat{X_\mathbf{c}} \to \widehat{X} \oplus \imath\widehat{X}$ by

$$\mathcal{T}([x + \imath y]) = [x] + \imath[y]$$

for all $x = (x_1, x_2, \ldots)$ and $y = (y_1, y_2, \ldots)$ in $\ell_\infty(X)$ and claimed that it was a surjective (topological) isomorphism. Now establish the following properties.

(a) The diagram shown in Figure 1 is commutative.

(b) The spectra of the operators T, $\widehat{T_\mathbf{c}}$, and $\widehat{T} + \imath\widehat{T}$ coincide. That is,

$$\sigma(T) = \sigma(\widehat{T_\mathbf{c}}) = \sigma(\widehat{T} + \imath\widehat{T}), \quad \sigma_p(T) = \sigma_p(\widehat{T_\mathbf{c}}) = \sigma_p(\widehat{T} + \imath\widehat{T}), \quad \text{and}$$
$$\sigma_a(T) = \sigma_a(\widehat{T_\mathbf{c}}) = \sigma_a(\widehat{T} + \imath\widehat{T}) \,.$$

(c) If X is a Banach lattice and T is a lattice homomorphism, then

$$|(\widehat{T})_\mathbf{c}(z)| = \widehat{T}|z|$$

holds for all $z \in (\widehat{X})_\mathbf{c}$.

11. Prove Theorem 7.18.

7.3. The Spectrum of a Lattice Homomorphism

The objective of this section is to establish that the spectrum of a lattice homomorphism has a nice symmetric property known as *cyclicity.* In our discussion here we shall use the letter Γ to denote the unit circle, i.e.,

$$\Gamma = \big\{\lambda \in \mathbb{C}\colon\ |\lambda| = 1\big\} \,.$$

Keep in mind that the unit circle is a multiplicative group.

Definition 7.19. *Let A be a subset of complex numbers. We shall say that a complex number $\alpha = |\alpha|e^{\imath\theta}$ is a* **cyclic point** *of A if $|\alpha|e^{\imath k\theta} \in A$ for each integer $k \in \mathbb{Z}$. The set A is called* **cyclic** *if every point of A is cyclic.*

All open and closed discs centered at zero are cyclic sets and likewise so is any set consisting of roots of unity for some n, that is, any set of the form $\{z \in \mathbb{C}\colon\ z^n = 1\} = \{e^{\frac{2p\pi}{n}\imath}\colon\ p = 0, 1, \ldots, n-1\}$. More generally, any subgroup of Γ is a cyclic set. Also, according to this definition, every subset of non-negative real numbers considered as a set of complex numbers is a cyclic set.

H. H. Schaefer [**290**, Proposition 4.2, p. 324] has shown that the peripheral spectrum of a Markov operator enjoys the following "cyclicity" property.

Theorem 7.20 (Schaefer). *Let E be an AM-space with unit e, and let $T\colon E \to E$ be a Markov operator. Then the set of eigenvalues*

$$\mathcal{E} = \{\lambda \in \Gamma\colon\ T_{\mathbf{C}}z = \lambda z \ \text{for some}\ z \in E_{\mathbf{C}}\ \text{with}\ |z| = e\}$$

is a subgroup of the unit circle C (and hence $\mathcal{E}$ is a cyclic set).

Proof. Since $Te = e$, we have $1 \in \mathcal{E}$. By Theorem 3.20, we can assume that $E = C(\Omega)$ and $E_{\mathbf{C}} = C_{\mathbb{C}}(\Omega)$ for some compact Hausdorff space Ω, where the unit e corresponds to the constant function $\mathbf{1}$ on Ω. Also, we know that $E^* = \text{ca}\,(\Omega)$ and $E^*_{\mathbf{C}} = \text{ca}\,(\Omega) \oplus \imath\text{ca}\,(\Omega)$, where $\text{ca}\,(\Omega)$ is the AL-space of all regular (finite) Borel measures on Ω. Since T is a Markov operator, T^* carries regular Borel probability measures to regular Borel probability measures. For simplicity, we shall denote $T_{\mathbf{C}}$ by T.

Now fix $\omega \in \Omega$ and let $\mu_\omega = T^*\delta_\omega$. Assume that there exist $z \in C_{\mathbb{C}}(\Omega)$ and $\lambda \in \Gamma$ such that $|z| = \mathbf{1}$ and $Tz = \lambda z$. Then we have

$$\lambda z(\omega) - (Tz)(\omega) = \langle Tz, \delta_\omega\rangle = \langle z, T^*\delta_\omega\rangle = \langle z, \mu_\omega\rangle = \int_\Omega z(t)\, d\mu_\omega(t)\,,$$

and so $\int_\Omega \lambda^{-1}[z(\omega)]^{-1} z(t)\, d\mu_\omega(t) = 1$. Since $|\lambda^{-1}[z(\omega)]^{-1}z(t)| = 1$ for each $t \in \Omega$, it follows (how?) that $\lambda^{-1}z(t)[z(\omega)]^{-1} = 1$ for each t lying in the support of μ_ω. This implies $z(t) = \lambda z(\omega)$ for all t in the support of μ_ω.

Similarly, if another $\lambda_1 \in \Gamma$ satisfies $Tz_1 = \lambda_1 z_1$ for some $z_1 \in C_{\mathbb{C}}(\Omega)$ with $|z_1| = \mathbf{1}$, then $z_1(t) = \lambda_1 z_1(\omega)$ for all t in the support of μ_ω, and so $\lambda\lambda_1 z(\omega)z_1(\omega) = z(t)z_1(t)$ for all t in the support of μ_ω. Hence,

$$\begin{aligned} \lambda\lambda_1 z(\omega)z_1(\omega) &= \int_\Omega z(t)z_1(t)\, d\mu_\omega(t) = \langle zz_1, \mu_\omega\rangle = \langle zz_1, T^*\delta_\omega\rangle \\ &= \langle Tzz_1, \delta_\omega\rangle = (Tzz_1)(\omega)\,. \end{aligned}$$

Thus, $T(zz_1) = \lambda\lambda_1 zz_1$, so that $\lambda\lambda_1 \in \mathcal{E}$. A similar argument shows that if $\lambda \in \mathcal{E}$, then $\overline{\lambda} \in \mathcal{E}$, where $\overline{\lambda}$ is the complex conjugate of λ. Hence, $\mathcal{E}$ is a subgroup of Γ. ■

Corollary 7.21. *Let $T: E \to E$ be a positive operator on a Banach lattice having a positive eigenvector u corresponding to an eigenvalue $r > 0$, i.e., $Tu = ru$. If T has another eigenvalue of the form $\lambda = re^{\imath\theta}$ corresponding to an eigenvector $z \in E_{\mathbf{C}}$ with $|z| = u$, then λ is a cyclic point of the point spectrum $\sigma_p(T)$ of T.*

Proof. Let $S = \frac{T}{r}$, and note that the positive operator $S: E \to E$ satisfies $Su = u$. This implies that $S: E_u \to E_u$ is a Markov operator and $z \in (E_u)_{\mathbf{C}}$. Since $Sz = e^{\imath\theta}z$ and $|z| = u$, it follows from Theorem 7.20 that $e^{\imath k\theta}$ is an eigenvalue of S for all $k \in \mathbb{Z}$. Therefore, the complex number $\lambda = re^{\imath\theta}$ is a cyclic point in the point spectrum of the operator T. ■

Corollary 7.22. *The point spectrum of a lattice homomorphism is cyclic.*

Proof. Let $T: E \to E$ be a lattice homomorphism, and let $Tz = \lambda z$ for some non-zero $z \in E_{\mathbf{C}}$ and some non-zero $\lambda = re^{\imath\theta} \in \mathbb{C}$. Then a glance at Theorem 4.29 shows that

$$T|z| = |Tz| = |\lambda z| = |\lambda| \cdot |z| = r|z| .$$

Now use Corollary 7.21 to conclude that $re^{\imath k\theta} \in \sigma_p(T)$ for each $k \in \mathbb{Z}$. ■

The next remarkable result is due to E. Scheffold [**293**].

Theorem 7.23 (Scheffold). *The spectrum of a lattice homomorphism is cyclic.*

Proof. Let $T: E \to E$ be a lattice homomorphism. Combining Exercise 8 of Section 7.2 with Corollary 7.17, we can assume without loss of generality that $\sigma_a(T) = \sigma_p(T)$. Also, from Corollary 7.22, we know that $\sigma_p(T)$ is cyclic. So, what needs to be shown is that if $\lambda = |\lambda|e^{\imath\theta} \in \sigma_r(T)$, then $|\lambda|e^{\imath k\theta} \in \sigma(T)$ for each $k \in \mathbb{Z}$.

To this end, let $\lambda = |\lambda|e^{\imath\theta} \in \sigma_r(T)$. Then $\lambda \in \sigma_p(T^*)$; see Theorem 6.19. Therefore, there exists some non-zero vector $z^* \in E^*_{\mathbf{C}} = E^* \oplus \imath E^*$ satisfying $T^*z^* = \lambda z^*$. This implies

$$|\lambda| \cdot |z^*| = |\lambda z^*| = |T^*z^*| \le T^*|z^*| .$$

Now if we fix some real number $\lambda_0 > r(T)$, then according to Lemma 6.24 there exists some $0 < \phi \in E^*$ such that the principal ideal E^*_ϕ generated by ϕ in E^* is T^*-invariant, $|z^*| \in E^*_\phi$, and $|\lambda|\phi \le T^*\phi \le \lambda_0\phi$. By Theorem 3.39, there exists a unique lattice homomorphism $T_\phi: E(\phi) \to E(\phi)$ whose adjoint makes the diagram in Figure 2 commutative. (The construction of the *AL*-space $E(\phi)$ is described in Section 3.4.)

$$\begin{array}{ccc} E_\phi^* & \xrightarrow{\ \Phi\ } & E(\phi)^* \\ {\scriptstyle T^*}\big\downarrow & & \big\downarrow{\scriptstyle T_\phi^*} \\ E_\phi^* & \xrightarrow{\ \Phi\ } & E(\phi)^* \end{array}$$

Figure 2

From the construction of the space $E(\phi)$ and the fact that T is a lattice homomorphism, for each $x \in E$ we have

$$\begin{aligned} |\lambda| \cdot \big\|[x]\big\|_\phi &= |\lambda|\phi(|x|) \\ &\le T^*\phi(|x|) = \phi(T|x|) \\ &= \phi(|Tx|) = \big\|[Tx]\big\|_\phi \\ &= \|T_\phi([x])\|_\phi \,. \end{aligned}$$

This implies $|\lambda| \cdot \|z\| \le \|T_\phi(z)\|_\phi$ for each $z \in E(\phi)$, i.e., T_ϕ is a bounded below operator.

Now observe that if $0 \notin \sigma(T_\phi)$, then T is a surjective lattice isomorphism, and since Φ is a lattice isometry, $T^*\colon E_\phi^* \to E_\phi^*$ is also a lattice isomorphism. Since $\lambda \in \sigma_p(T^*|_{E_\phi^*})$, it follows from Corollary 7.22 that

$$|\lambda|e^{\imath k\theta} \in \sigma_p\big(T^*|_{E_\phi^*}\big) \subseteq \sigma_p(T^*) \subseteq \sigma(T^*) = \sigma(T)$$

for each $k \in \mathbb{Z}$. So, in this case, λ is a cyclic point of the spectrum $\sigma(T)$.

Next, assume that $0 \in \sigma(T_\phi)$. Since T_ϕ is bounded below, it follows (see Exercises 5, 6, and 8 of Section 6.2) that

$$D(0,|\lambda|) \subseteq \sigma_r(T_\phi) \subseteq \sigma_p\big(T_\phi^*\big) = \sigma_p\big(T^*|_{E_\phi^*}\big) \subseteq \sigma_p(T^*) \subseteq \sigma(T^*) = \sigma(T)\,.$$

That is, the open disk centered at zero with radius $|\lambda|$ lies in $\sigma(T)$, and so λ is automatically a cyclic point of $\sigma(T)$, and the proof is finished. ■

Corollary 7.24. *The spectrum of an interval preserving positive operator is cyclic.*

Proof. If $T\colon E \to E$ is an interval preserving operator, then $T^*\colon E^* \to E^*$ is a lattice homomorphism; see Theorem 1.35. Therefore, by Theorem 7.23, the set $\sigma(T) = \sigma(T^*)$ is cyclic. ■

We conclude the section with a brief comment on some recent developments regarding the symmetry of the spectrum of an operator. Recall that the spectrum of an operator is said to be **rotation invariant** if for each point λ in the spectrum the circle centered at zero with radius $|\lambda|$ lies entirely in the spectrum.

If $T\colon E \to E$ is a disjointness preserving operator on a Banach lattice, then there are several general results asserting rotational invariance of $\sigma(T)$. For example, this is so if E is a Banach function space and $|T|^n \wedge |T|^m = 0$ whenever $n \neq m$; for details see Chapters 12 and 13 in [**21**].

Exercises

1. Let $G = \{z_1, \ldots, z_p\}$ be a subset of the unit circle Γ consisting of p distinct numbers. If G is closed under multiplication, then show that

$$G = \{z \in \mathbb{C}\colon\ z^p = 1\} = \left\{e^{\frac{2\pi k}{p}\imath}\colon\ k = 0, 1, \ldots, p-1\right\}.$$

[HINT: Fix some $z \in G$. Since each of the $p+1$ elements $z, z^2, \ldots, z^{p+1}$ is in G, necessarily $z^m = z^\ell$ for some $1 \leq m < \ell \leq p+1$. This implies $z^n = 1$ for some $1 \leq n \leq p$. If we let $\nu = \min\{n \in \mathbb{N}\colon\ 1 \leq n \leq k \text{ and } z^n = 1\}$, then $H = \{1, z, \ldots, z^{\nu-1}\}$ is a subgroup of G and so p is divisible by ν. This guarantees that $z^p = 1$.]

2. Fix an angle $\theta \in \mathbb{R}$ and let $G = \{e^{\imath n\theta}\colon\ n \in \mathbb{N}\}$. Establish the following.
   (a) If θ is a rational multiple of 2π, then G is a finite subgroup of the unit circle Γ (and so G consists of roots of unity).
   (b) If θ is not a rational multiple of 2π, then G is dense in Γ.

3. (Kronecker's theorem) Let G be an arbitrary subgroup of the unit circle Γ. Show that either G is dense in Γ or else it consists of roots of unity, i.e., $G = \{e^{\frac{2\pi k}{p}\imath}\colon\ k = 0, 1, \ldots, p-1\}$ for some $p \in \mathbb{N}$.

4. Establish the following result that was used in the proof of Theorem 7.20. *If μ is a regular Borel probability measure on a compact Hausdorff space Ω and some $z \in C_{\mathbb{C}}(\Omega)$ satisfies $|z| = \mathbf{1}$ and $\int_\Omega z(t)\,d\mu(t) = 1$, then $z(t) = 1$ for each t in the support of μ.*

5. Let $T\colon E \to E$ be an interval preserving positive operator. If $\lambda = |\lambda|e^{\imath\theta}$ belongs to the residual spectrum of T, then show that for each integer n the complex number $|\lambda|e^{\imath n\theta}$ is an eigenvalue of the adjoint operator T^*.

6. Assume that for some continuous function $\phi\colon \Omega \to \mathbb{R}$ on a compact Hausdorff space the multiplication operator $M_\phi\colon C(\Omega) \to C(\Omega)$ is a lattice homomorphism.

   Show directly (i.e., without using Theorem 7.23) that the spectrum of the multiplication operator M_ϕ is cyclic.

7. Let $T\colon X \to X$ be a bounded operator on a Banach space. If the approximate point spectrum of T is rotation invariant, then show that the spectrum of T is also rotation invariant.

7.4. The Order Spectrum of an Order Bounded Operator

The lattice structure of Banach lattices can be used to define a new concept of a spectrum for a bounded operator that allows a deeper look at the order properties of the inverse operator. This is the order spectrum which is defined as follows.

Definition 7.25. *Let $T\colon E \to E$ be a bounded operator on a (real) Banach lattice. The* **order spectrum** *$\sigma_o(T)$ of T consists of all complex numbers λ for which the operator $\lambda - T$ does not have an order bounded inverse on the complexification $E_{\mathbf{C}}$ of E. That is,*

$$\sigma_o(T) = \{\lambda \in \mathbb{C}\colon\ \lambda - T \text{ does not have an order bounded inverse on } E_{\mathbf{C}}\}.$$

Notice that if E is a Dedekind complete Banach lattice and $T\colon E \to E$ is an order bounded operator, then T belongs to the complex unital Banach algebra $\mathcal{L}_r(E_{\mathbf{C}})$ and its spectrum in $\mathcal{L}_r(E_{\mathbf{C}})$ obviously coincides with the order spectrum of T. So, when E is Dedekind complete and $T\colon E \to E$ is order bounded, the order spectrum $\sigma_o(T)$ is a non-empty compact subset of $\mathbb{C}$. Letting $r_o(T) = \max_{\lambda \in \sigma_o(T)} |\lambda|$, we have

$$r_o(T) = \lim_{n\to\infty} \left(\|T^n\|_r\right)^{\frac{1}{n}}.$$

(See Theorem 6.16 and the discussion preceding it.) It should be clear that for any bounded operator T we have the inclusion $\sigma(T) \subseteq \sigma_o(T)$.

In what follows, we will restrict our attention to order bounded operators. Let us mention at once that in general the order spectrum $\sigma_o(T)$ can properly include the usual spectrum $\sigma(T)$ (see Example 7.36 below). Nevertheless, there exists a method that allows us to reduce the notion of the order spectrum to the usual spectrum. More precisely, each Dedekind complete Banach lattice E can be embedded into a larger Dedekind complete Banach lattice $\check{E}$ in such a way that each order bounded operator T on E will have an order bounded extension $\check{T}$ to this enlargement satisfying $\sigma_o(T) = \sigma(\check{T})$. The purpose of this section is to describe this method.

We begin by assuming that E is a Riesz space and that A is an arbitrary fixed index set. Only this is necessary to define the extension $\check{E}$.

It is well known, and easy to check, that the Cartesian product E^A under the pointwise operations is also a Riesz space. That is, E^A is a Riesz space under the operations defined by

$$[f+g](\alpha) = f(\alpha)+g(\alpha), \quad [\lambda f](\alpha) = \lambda f(\alpha), \quad \text{and} \quad [f \vee g](\alpha) = f(\alpha) \vee g(\alpha)$$

for all $f, g \in E^A$, all $\lambda \in \mathbb{R}$, and all $\alpha \in A$. The positive cone of E^A consists of all positive functions, that is, functions $f \in E^A$ with $f(\alpha) \in E^+$ for each $\alpha \in A$.

There is a natural embedding of E into E^A. This embedding simply assigns to each $x \in E$ the constant function $\check{x}\colon A \to E$ with value x, i.e, $\check{x}(\alpha) = x$ for each $\alpha \in A$. It should be clear that $x \mapsto \check{x}$ is indeed a lattice isomorphism and so E (identified with its image in E^A) can be viewed as a Riesz subspace of E^A. To avoid introducing unnecessary notation, from now on, every vector $x \in E$ will play a double role: On the one hand, it can be viewed as an element in E and, on the other hand, it can be considered as the constant function x on A; that is, if x is considered as a function on A, then $x(\alpha) = x$ for each $\alpha \in A$. If clarity requires to differentiate between the vector $x \in E$ and the constant function x on A, then we shall use the symbol $\check{x}$. The next result is left as an exercise.

Lemma 7.26. *A Riesz space E is Dedekind complete if and only if the Riesz space E^A is Dedekind complete.*

The ideal generated by E in E^A will be our "enlargement" space.

Definition 7.27. *The ideal generated by E in E^A will be denoted $\check{E}$, i.e.,*

$$\check{E} = \big\{ f \in E^A \colon \ \exists \ x \in E \ \text{ such that } |f(\alpha)| \le x \ \text{ for each } \alpha \in A \big\}.$$

Assume now that $\|\cdot\|$ is a lattice norm on the Riesz space E. There is a natural way to extend $\|\cdot\|$ to a lattice norm on $\check{E}$. Namely, for each $f \in \check{E}$ we let

$$\|f\| = \inf\big\{ \|x\| \colon \ x \in E \ \text{ and } \ |f| \le x \big\}. \tag{$\star$}$$

It is a routine matter to verify that $(\star)$ indeed defines a lattice norm on $\check{E}$ that extends the original norm of E. In actuality, as the next result shows, more is true.

Lemma 7.28. *If E is a Banach lattice, then $\check{E}$ equipped with the lattice norm given by $(\star)$ is a Banach lattice containing E as a closed vector sublattice.*

For any mapping $T \colon E \to E$ there is a natural extension $\check{T} \colon E^A \to E^A$ given by the formula

$$\big(\check{T}f\big)(\alpha) = T\big(f(\alpha)\big),$$

where $f \in E^A$ and $\alpha \in A$. Notice that $\check{T}$ is linear if and only if T is linear. The reader should verify also that if $S, T \colon E \to E$ are two operators and λ is an arbitrary scalar, then

$$(S+T)\check{} = \check{S} + \check{T}, \quad (\lambda T)\check{} = \lambda \check{T}, \quad \text{and} \quad (ST)\check{} = \check{S}\check{T}.$$

Regarding invertibility of operators we have the following.

Lemma 7.29. *Let $T \colon E \to E$ be an operator on a Riesz space. Then:*

(1) *T is one-to-one if and only if $\check{T} \colon \check{E} \to \check{E}$ is one-to-one.*

(2) *T is surjective if and only if $\check{T} \colon E^A \to E^A$ is surjective.*

Proof. (1) Assume that the operator T is one-to-one and let $\check{T}f = 0$. This implies $\check{T}f(\alpha) = T\big(f(\alpha)\big) = 0$ for each $\alpha \in A$. Since T is one-to-one, it follows that $f(\alpha) = 0$ for each $\alpha \in A$, i.e., $f = 0$. Hence, $\check{T}$ is one-to-one.

Now suppose that $\check{T}$ is one-to-one on $\check{E}$, and let $Tx = 0$. So, $\check{T}\check{x} = 0$, and thus $\check{x} = 0$. This implies $x = 0$, and hence T is one-to-one.

(2) Assume that $\check{T}\colon E^A \to E^A$ is surjective, and let $y \in E$. Pick some $f \in E^A$ such that $\check{T}f = \check{y}$, and then note that for each index $\alpha \in A$ we have $T\big(f(\alpha)\big) = \check{T}f(\alpha) = \check{y}(\alpha) = y$. This shows that T is surjective.

For the converse, assume that T is surjective, and let $f \in E^A$. For each $\alpha \in A$ pick some $g(\alpha) \in E$ such that $T\big(g(\alpha)\big) = f(\alpha)$. Then the function $g \in E^A$ satisfies $\check{T}g = f$, so that $\check{T}$ is surjective. ■

Several important order properties of operators transfer to their extensions.

Lemma 7.30. *An operator $T\colon E \to E$ on a Riesz space is:*

(1) *positive if and only if $\check{T}$ is positive,*

(2) *a lattice homomorphism if and only if $\check{T}$ is a lattice homomorphism,*

(3) *interval preserving if and only if $\check{T}$ is interval preserving.*

Proof. (1) Assume that T is positive, and let $0 \le f \in E^A$. Then for each $\alpha \in A$ we have $\check{T}f(\alpha) = T\big(f(\alpha)\big) \ge 0$ and so $\check{T}f \ge 0$ in E^A. Therefore, $\check{T}$ is a positive operator. Conversely, if $\check{T}$ is positive and $x \in E^+$, then for each index α we have $0 \le (\check{T}\check{x})(\alpha) = Tx$, proving that T is a positive operator.

(2) Assume first that T is a lattice homomorphism, and let $f, g \in E^A$. Then we have

$$\begin{aligned}[\check{T}(f \vee g)](\alpha) &= T([f \vee g](\alpha)) = T\big(f(\alpha) \vee g(\alpha)\big) = T\big(f(\alpha)\big) \vee T\big(g(\alpha)\big) \\ &= \check{T}f(\alpha) \vee \check{T}g(\alpha) = [\check{T}f \vee \check{T}g](\alpha)\end{aligned}$$

for each $\alpha \in A$. Thus, $\check{T}(f \vee g) = \check{T}f \vee \check{T}g$, and so $\check{T}$ is a lattice homomorphism.

For the converse, assume that $\check{T}$ is a lattice homomorphism, and let $x, y \in E$. If α is any index, then

$$\begin{aligned}Tx \vee Ty &= (\check{T}\check{x})(\alpha) \vee (\check{T}\check{y})(\alpha) = [\check{T}\check{x} \vee \check{T}\check{y}](\alpha) = \big[\check{T}(\check{x} \vee \check{y})\big](\alpha) \\ &= T\big([\check{x} \vee \check{y}](\alpha)\big) = T\big(\check{x}(\alpha) \vee \check{y}(\alpha)\big) = T(x \vee y)\,.\end{aligned}$$

This shows that T is a lattice homomorphism.

(3) Assume that T is interval preserving, and let $f \ge 0$ and $g \in E^A$ satisfy $0 \le g \le \check{T}f$. That is, $0 \le g(\alpha) \le \check{T}f(\alpha) = T\big(f(\alpha)\big)$ holds for each $\alpha \in A$. Since T is interval preserving, for each $\alpha \in A$ there exists some

$0 \leq h(\alpha) \leq f(\alpha)$ such that $T(h(\alpha)) = g(\alpha)$. Then the function $h \in E^A$ satisfies $0 \leq h \leq f$ and $\check{T}h = g$. This shows that $\check{T}$ is interval preserving.

Finally, assume that $\check{T}$ is interval preserving, and let $x, y \in E^+$ satisfy $0 \leq x \leq Ty$. This implies $0 \leq \check{x} \leq \check{T}\check{y}$, and so there exists some $f \in E^A$ satisfying $0 \leq f \leq \check{y}$ and $\check{T}f = \check{x}$. Now take any index α and note that $0 \leq f(\alpha) \leq y$ and $T\big(f(\alpha)\big) = \check{T}f(\alpha) = \check{x}(\alpha) = x$. This shows that T is interval preserving. ■

For the remaining results in this section, we need to add a technical condition ensuring that the index set A is not too small. Namely, we shall assume that:

- *The cardinality of the index set A is at least as large as the cardinality of the Riesz space E.*

The next result characterizes the order bounded operators on E as those whose extensions to E^A leave $\check{E}$ invariant.

Theorem 7.31. *An operator $T\colon E \to E$ on a Riesz space is order bounded if and only if the operator $\check{T}\colon E^A \to E^A$ leaves $\check{E}$ invariant. In this case $\check{T}$ is also order bounded.*

Proof. Assume first that T is order bounded, and let $f \in \check{E}$. Pick some $x \in E^+$ satisfying $|f(\alpha)| \leq x$ for each $\alpha \in A$. Since T is order bounded, there exists some $y \in E^+$ such that $T[-x, x] \subseteq [-y, y]$. Now note that $\big|\check{T}f(\alpha)\big| = \big|T\big(f(\alpha)\big)\big| \leq y$ for each $\alpha \in A$. Therefore, $\check{T}f \in \check{E}$, and so $\check{T}$ leaves $\check{E}$ invariant. Moreover, notice that for each $g \in \check{E}$ satisfying $|g| \leq f$ we have $-\check{y} \leq \check{T}g \leq \check{y}$. This shows that $\check{T}\colon \check{E} \to \check{E}$ is indeed an order bounded operator.

For the converse, assume that $\check{T}$ leaves $\check{E}$ invariant. Fix some $x \in E^+$. Since the cardinality of the index set A is at least as large as the cardinality of E, there exists a function $f\colon A \to E$ whose range is the entire order interval $[-x, x]$. In view of our hypothesis, $\check{T}f \in \check{E}$ and so there exists an element $y \in E^+$ such that $|\check{T}f| \leq \check{y}$. Therefore $T[-x, x] \subseteq [-y, y]$. This proves that T is order bounded, and the proof is finished. ■

We are now ready for the basic results of this section.

Theorem 7.32. *Let $T\colon E \to E$ be an order bounded operator on a Riesz space. If $\check{T}\colon \check{E} \to \check{E}$ is invertible, then T is also invertible and its inverse satisfies $(\check{T})^{-1} = (T^{-1})\check{}$ and is order bounded.*

Proof. Let $S\colon \check{E} \to \check{E}$ be the inverse of $\check{T}\colon \check{E} \to \check{E}$, i.e., $S\check{T} = \check{T}S = I_{\check{E}}$. Part (1) of Lemma 7.29 guarantees that T is one-to-one and the proof of part (2) of the same lemma shows that T is also surjective. Consequently, T^{-1} exists—and, of course, is a linear operator.

From $TT^{-1} = T^{-1}T = I$, it follows that $\check{T}(T^{-1})\check{} = (T^{-1})\check{}\check{T} = \check{I}$ on E^A. Since $\check{T}$ leaves $\check{E}$ invariant, it follows that $S = (\check{T})^{-1} = (T^{-1})\check{}$; see Exercise 1 at the end of this section. Finally, observing that $(\check{T})^{-1}$ leaves $\check{E}$ invariant, we get from Theorem 7.31 that T^{-1} is also an order bounded operator. ■

Recall that every linear operator on $E_{\mathbf{C}} = E \oplus \imath E$ is of the form $S + \imath T$, where S and T are linear operators on E. Let us say that an operator $S+\imath T$ is **order bounded** if both S and T are order bounded. Also, if $S+\imath T$ is an operator on $E_{\mathbf{C}}$, let us write $(S+\imath T)\check{} = \check{S} + \imath\check{T}$; clearly, this operator acts on $(E^A)_{\mathbf{C}}$. Notice that if S and T are order bounded, then $\check{S}+\imath\check{T}$ leaves the vector subspace $(\check{E})_{\mathbf{C}} = \check{E} \oplus \imath\check{E}$ of $(E^A)_{\mathbf{C}}$ invariant. The complex version of Theorem 7.32 is now immediate.

Theorem 7.33. *Let $S, T\colon E \to E$ be two order bounded operators on a Riesz space. If $(S+\imath T)\check{}\colon \check{E}_{\mathbf{C}} \to \check{E}_{\mathbf{C}}$ is invertible, then $S+\imath T\colon E_{\mathbf{C}} \to E_{\mathbf{C}}$ is also invertible, and its inverse is order bounded and satisfies*

$$[(S+\imath T)^{-1}]\check{} = [(S+\imath T)\check{}]^{-1} = (\check{S}+\imath\check{T})^{-1}.$$

Theorem 7.33 implies at once the result announced at the beginning of the section about reducing the order spectrum to the standard spectrum.

Theorem 7.34. *If T is an order bounded operator on a Dedekind complete Banach lattice, then $\sigma_o(T) = \sigma_o(\check{T}) = \sigma(\check{T})$.*

Corollary 7.35. *Lattice homomorphisms and interval preserving positive operators have cyclic order spectra.*

Proof. Let $T\colon E \to E$ be a positive operator which is either interval preserving or a lattice homomorphism. According to Theorem 7.34 we have $\sigma_o(T) = \sigma_o(\check{T}) = \sigma(\check{T})$. Also, from Lemma 7.30, we know that $\check{T}\colon \check{E} \to \check{E}$ is either interval preserving or a lattice homomorphism, respectively. To complete the proof, it remains to notice that by Theorem 7.23 and Corollary 7.24 these types of positive operators have cyclic spectra. ■

We conclude the section with the promised example, due to T. Andô (see [**291**]), of an order bounded operator for which the order spectrum is larger than the usual spectrum.

Example 7.36 (Andô). For each k consider the finite dimensional real Hilbert space $E_k = \mathbb{R}^2 \oplus \mathbb{R}^2 \oplus \cdots \oplus \mathbb{R}^2$ of k factors of $\mathbb{R}^2$ with the standard coordinatewise inner product. With the usual coordinatewise order E_k becomes a Dedekind complete Banach lattice.

Consider now the ℓ_2-sum $E = (E_1 \oplus E_2 \oplus \cdots)_2$ of the sequence $\{E_k\}$. It is a Dedekind complete Banach lattice which is a (real) Hilbert space under the inner product defined by

$$(x, y) = \sum_{k=1}^{\infty} (x_k, y_k)$$

for all $x = (x_1, x_2, \ldots)$ and $y = (y_1, y_2, \ldots)$ in E.

Next, for each $k \in \mathbb{N}$ we consider the 2×2 orthogonal matrix

$$M_k = \begin{bmatrix} \cos \frac{1}{k} & -\sin \frac{1}{k} \\ \sin \frac{1}{k} & \cos \frac{1}{k} \end{bmatrix}.$$

A direct computation shows that $M_k^n = \begin{bmatrix} \cos \frac{n}{k} & -\sin \frac{n}{k} \\ \sin \frac{n}{k} & \cos \frac{n}{k} \end{bmatrix}$ for each $n \in \mathbb{N}$. Therefore, the modulus of M_k^n (considered as an operator on $\mathbb{R}^2$) is given by the 2×2 symmetric matrix

$$|M_k^n| = \begin{bmatrix} |\cos \frac{n}{k}| & |\sin \frac{n}{k}| \\ |\sin \frac{n}{k}| & |\cos \frac{n}{k}| \end{bmatrix}.$$

It is easy to see that the eigenvalues of $|M_k^n|$ are $|\cos \frac{n}{k}| \pm |\sin \frac{n}{k}|$. So, the Euclidean norm of $|M_k^n|$ is $\big\||M_k^n|\big\| = |\cos \frac{n}{k}| + |\sin \frac{n}{k}|$.

Next, for each $k \in \mathbb{N}$ we define the operator $T_k \colon E_k \to E_k$ via the formula $T_k = M_k \oplus M_k \oplus \cdots \oplus M_k$. Clearly, T_k is a unitary operator on E_k which is also order bounded. This implies $\|T_k\| = 1$. Moreover, the modulus of T_k^n is given by $|T_k^n| = |M_k^n| \oplus |M_k^n| \oplus \cdots \oplus |M_k^n|$.

Now consider the operator $\mathcal{T} = (T_1, T_2, \ldots) \in \mathcal{L}_r(E)$; see Exercise 4 in Section 7.2. It should be noticed that $\mathcal{T}$ is indeed an order bounded operator which is also unitary. This implies that $\mathcal{T}$ as an operator on $E_{\mathbf{C}}$ (the complexification of E) is also unitary, and so $r(\mathcal{T}) = 1$.

Fix any $0 < \epsilon < 1$ and then pick some n_0 such that $\cos \frac{1}{n} > \epsilon$ and $\sin \frac{1}{n} > \epsilon \cdot \frac{1}{n}$ for each $n \geq n_0$. This and Exercise 4 in Section 7.2 imply that for all $n \geq n_0$ we have

$$\begin{aligned} \|\mathcal{T}^n\|_r &= \sup_{k \in \mathbb{N}} \big\||T_k^n|\big\| \\ &\geq \big\||T_{n^2}^n|\big\| \\ &\geq \big\||M_{n^2}^n|\big\| \\ &= \cos \tfrac{1}{n} + \sin \tfrac{1}{n} \\ &> \epsilon + \epsilon \cdot \tfrac{1}{n} \\ &= \epsilon\big(1 + \tfrac{1}{n}\big). \end{aligned}$$

Consequently, we have

$$r_o(T) = \lim_{n\to\infty} \left(\|T^n\|_r\right)^{\frac{1}{n}} \geq \lim_{n\to\infty} \left[\epsilon\left(1+\tfrac{1}{n}\right)\right]^{\frac{1}{n}} = e > 1 = r(T)\,.$$

So, the spectrum of T is a proper subset of the order spectrum of T. ∎

Exercises

1. Let $T\colon V \to V$ be an invertible operator on a vector space, and let $T^{-1}\colon V \to V$ be its inverse. Assume that W is a T-invariant subspace of V such that $T\colon W \to W$ is also invertible. Then show that T^{-1} leaves W invariant and that the inverse of $T|_W$ (the restriction of T to W) is the restriction of T^{-1} to W, i.e., $(T|_W)^{-1} = T^{-1}|_W$.

2. Describe the Banach lattice $\check{E}$ when $E = \mathbb{R}$.

3. Prove Lemma 7.26.

4. Prove Lemma 7.28.

5. Prove Theorem 7.33.

6. Prove Theorem 7.34.

7. Establish the following elementary properties that were used in Example 7.36. Consider the matrix

$$M = \begin{bmatrix} \cos\phi & -\sin\phi \\ \sin\phi & \cos\phi \end{bmatrix}$$

acting on $\mathbb{R}^2$, where ϕ is an arbitrary angle. Then:

(a) For each $n \in \mathbb{N}$ we have $M^n = \begin{bmatrix} \cos n\phi & -\sin n\phi \\ \sin n\phi & \cos n\phi \end{bmatrix}$.

(b) The modulus of M^n is the matrix

$$|M^n| = \begin{bmatrix} |\cos n\phi| & |\sin n\phi| \\ |\sin n\phi| & |\cos n\phi| \end{bmatrix}.$$

(c) The eigenvalues of the symmetric matrix $|M^n|$ are

$$|\cos n\phi| \pm |\sin n\phi|\,.$$

(d) The Euclidean norm of the matrix $|M^n|$ is

$$\||M^n|\| = |\cos n\phi| + |\sin n\phi|\,.$$

7.5. The Essential Spectrum of a Bounded Operator

In this section, unless otherwise stated, X will denote an infinite dimensional complex Banach space. As we already know, the vector space $\mathcal{K}(X)$ of all compact operators on X is a closed algebraic ideal in $\mathcal{L}(X)$. It is easy to see that the quotient vector space $\mathcal{L}(X)/\mathcal{K}(X)$ under the algebraic operations

- $[S] + [T] = [S + T]$,
- $\lambda[T] = [\lambda T]$, and
- $[S][T] = [ST]$

is a unital algebra with the element $[I]$ as its unit. Moreover, the quotient vector space $\mathcal{L}(X)/\mathcal{K}(X)$ under the quotient norm

$$\big\|[T]\big\| = \inf\big\{\|S\|:\ S \in [T]\big\} = \inf\big\{\|S\|:\ S - T \in \mathcal{K}(X)\big\}$$

is a Banach space. Since the quotient norm also satisfies the properties

$$\big\|[S][T]\big\| \le \big\|[S]\big\| \cdot \big\|[T]\big\| \quad \text{and} \quad \big\|[I]\big\| = 1\,,$$

$\mathcal{L}(X)/\mathcal{K}(X)$ is a unital Banach algebra. This Banach algebra is called the **Calkin algebra**[1] of X and is denoted $\mathfrak{C}(X)$, i.e., $\mathfrak{C}(X) = \mathcal{L}(X)/\mathcal{K}(X)$.

The quotient map (also called the **canonical projection**) of $\mathcal{L}(X)$ onto $\mathfrak{C}(X)$ will be denoted by π. That is, $\pi\colon \mathcal{L}(X) \to \mathfrak{C}(X)$ is defined by

$$\pi(T) = [T] = T + \mathcal{K}(X)\,.$$

From the inequality $\big\|\pi(T)\big\| \le \|T\|$, it follows that π is a contraction. The above properties are summarized in the following result.

Theorem 7.37. *If X is an infinite dimensional Banach space, then the Calkin algebra $\mathfrak{C}(X)$ is a unital Banach algebra. The canonical projection $\pi\colon \mathcal{L}(X) \to \mathfrak{C}(X)$ is a contraction and an algebraic homomorphism.*

There is standard terminology regarding properties of operators as elements in the Calkin algebra. It is customary to say that an operator $T \in \mathcal{L}(X)$ possess a property (P) **essentially** if its canonical projection $\pi(T)$ satisfies property (P) in the Calkin algebra $\mathfrak{C}(X)$.

For instance, we say that an operator $T \in \mathcal{L}(X)$ is **essentially invertible** if $\pi(T)$ is invertible in the Calkin algebra $\mathfrak{C}(X)$. Incidentally, the essentially invertible operators are precisely the Fredholm operators. This is simply a restatement of Theorem 4.46.

Theorem 7.38. *An operator $T \in \mathcal{L}(X)$ is essentially invertible if and only if it is a Fredholm operator.*

We now introduce the essential spectrum of an operator.

Definition 7.39. *The* ***essential*** *(or the* ***Fredholm****)* ***spectrum*** *of an operator $T \in \mathcal{L}(X)$ is the spectrum of $\pi(T)$ in the algebra $\mathfrak{C}(X)$. That is,*

$$\begin{aligned}\sigma_{\text{ess}}(T) &= \big\{\lambda \in \mathbb{C}:\ \lambda - \pi(T) \ \textit{ is not invertible in } \mathfrak{C}(X)\big\}\\ &= \big\{\lambda \in \mathbb{C}:\ \lambda - T \ \textit{ is not a Fredholm operator}\big\}\,.\end{aligned}$$

[1] It was introduced by J. W. Calkin in [**84**].

An immediate consequence of Theorem 6.16 is the following.

Lemma 7.40. *The essential spectrum of an operator $T \in \mathcal{L}(X)$ is a non-empty compact subset of $\sigma(T)$.*

Since for each $T \in \mathcal{L}(X)$ and each $\lambda \in \mathbb{C}$ we have $\lambda - T^* = (\lambda - T)^*$, the following result is an immediate consequence of Theorem 4.42.

Lemma 7.41. *The essential spectrum of any operator $T \in \mathcal{L}(X)$ coincides with the essential spectrum of its adjoint, i.e., $\sigma_{\text{ess}}(T) = \sigma_{\text{ess}}(T^*)$.*

The next result presents some operators whose essential spectrum coincides with their spectrum.

Theorem 7.42. *If an operator $T \in \mathcal{L}(X)$ does not have any non-trivial closed invariant subspace, then the essential spectrum and the spectrum of T coincide, i.e., $\sigma_{\text{ess}}(T) = \sigma(T)$.*

Proof. Assume that T has no non-trivial closed invariant subspaces and let $\lambda \in \sigma(T)$. Clearly, the closed vector subspace $N(\lambda - T)$ is T-invariant and hence either $N(\lambda - T) = X$ or $N(\lambda - T) = \{0\}$. If $N(\lambda - T) = X$, then $T = \lambda I$, which contradicts our hypothesis. Therefore, $N(\lambda - T) = \{0\}$.

Now the assumption $\lambda \in \sigma(T)$ guarantees that $R(\lambda - T) \neq X$, and clearly $R(\lambda - T) \neq \{0\}$. Since $R(\lambda - T)$ is T-invariant, it follows that $R(\lambda - T)$ is dense in X. Therefore, $\lambda - T$ does not have a closed range, and hence it fails to be a Fredholm operator. This implies that the operator $\lambda - T$ is not essentially invertible. That is, $\lambda \in \sigma_{\text{ess}}(T)$. Consequently, we have $\sigma(T) \subseteq \sigma_{\text{ess}}(T) \subseteq \sigma(T)$ or $\sigma_{\text{ess}}(T) = \sigma(T)$. ∎

The next result identifies a special class of isolated points in the spectrum of a bounded operator.

Lemma 7.43. *Let $T \in \mathcal{L}(X)$ and $\lambda_0 \in \sigma(T) \setminus \sigma_{\text{ess}}(T)$. If there is a path lying outside of $\sigma_{\text{ess}}(T)$ and joining λ_0 with a point in the resolvent set $\rho(T)$, then λ_0 is an isolated point of the spectrum $\sigma(T)$.*

Proof. Fix $\lambda_0 \in \sigma(T) \setminus \sigma_{\text{ess}}(T)$, and let C be a path joining λ_0 with a point $\mu \in \rho(T)$ such that $C \cap \sigma_{\text{ess}}(T) = \emptyset$. So, for each $z \in C$ the operator $z - T$ is a Fredholm operator. Now, according to Lemma 4.55, for each $z \in C$ there exists an open disk $D(z)$ centered at z such that the nullity $n(\zeta - T)$ and the defect $d(\zeta - T)$ are constant on $D(z) \setminus \{z\}$.

Since the curve C is a compact set, there exists a finite subcollection $\{D(z_1), D(z_2), \ldots, D(z_n)\}$ from these disks that covers C. By Exercise 7 in Section 1.5, we can assume without loss of generality that this collection is also a chain such that $z_1 = \lambda_0$ and $z_n = \mu$. Since for each $\zeta \in D(z_n) \setminus \{z_n\}$ we have $n(\zeta - T) = d(\zeta - T) = 0$, it easily follows that $n(\zeta - T) = d(\zeta - T) = 0$ in the open disk $D(z_1)$ with the exception of $\zeta = \lambda_0$. This implies that $\zeta - T$ is invertible for all $\zeta \in D(\lambda_0) \setminus \{\lambda_0\}$. So, λ_0 is an isolated point of $\sigma(T)$. ∎

Theorem 7.44. *Let λ_0 be an isolated point in the spectrum of an operator $T \in \mathcal{L}(X)$. If $\lambda_0 \notin \sigma_{\text{ess}}(T)$, then:*

(1) *The spectral projection $P_{\lambda_0}(T)$ is of finite-rank.*

(2) *The operator $\lambda_0 - T$ has finite ascent and descent.*

(3) *The point λ_0 is an eigenvalue of T and is a pole of the resolvent of T whose order equals the ascent (or the descent) of $\lambda_0 - T$.*

Proof. Let λ_0 be an isolated point in the spectrum of an operator $T \in \mathcal{L}(X)$ such that $\lambda_0 \notin \sigma_{\text{ess}}(T)$. So, $\lambda_0 - T$ is a Fredholm operator.

(1) Let $P_{\lambda_0}(T)$ be the spectral projection corresponding to the spectral set $\{\lambda_0\}$, and let $X_1 = R\big(P_{\lambda_0}(T)\big)$ and $X_2 = N\big(P_{\lambda_0}(T)\big)$. Then $X = X_1 \oplus X_2$ and (in view of Theorem 6.34) both X_1 and X_2 are invariant under T. Moreover, letting $T_1 = T|_{X_1}$ and $T_2 = T|_{X_2}$, we know (by Theorem 6.34 again) that $\sigma(T_1) = \{\lambda_0\}$ and $\sigma(T_2) = \sigma(T) \setminus \{\lambda_0\}$. Since $\lambda_0 I_{X_2} - T_2 \in \mathcal{L}(X_2)$ is one-to-one and onto, for each $n \geq 1$ we have

$$N\big((\lambda_0 - T)^n\big) = N\big((\lambda_0 I_{X_1} - T_1)^n\big) \oplus \{0\}\,, \quad \text{and} \tag{$\star$}$$

$$R\big((\lambda_0 - T)^n\big) = R\big((\lambda_0 I_{X_1} - T_1)^n\big) \oplus X_2\,. \tag{$\star\star$}$$

Now notice that there exists a (not necessarily closed) vector subspace Y of X_1 such that $R(\lambda_0 I_{X_1} - T_1) \oplus Y = X_1$. Clearly,

$$X = X_1 \oplus X_2 = R(\lambda_0 I_{X_1} - T_1) \oplus Y \oplus X_2 = R(\lambda_0 - T) \oplus Y\,.$$

Since $\lambda_0 - T$ is Fredholm, Y is finite dimensional, and so from

$$\dim N(\lambda_0 I_{X_1} - T_1) = \dim N(\lambda_0 - T) < \infty\,,$$

we see that the operator $\lambda_0 I_{X_1} - T_1 \in \mathcal{L}(X_1)$ is Fredholm, i.e., $\lambda_0 \notin \sigma_{\text{ess}}(T_1)$. But then from $\sigma_{\text{ess}}(T_1) \subseteq \sigma(T_1) = \{\lambda_0\}$ we obtain $\sigma_{\text{ess}}(T_1) = \emptyset$. This implies that X_1 is finite dimensional. Thus, $P_{\lambda_0}(T)$ is of finite-rank.

(2) and (3) Since X_1 is finite dimensional, it follows that $\lambda_0 I_{X_1} - T_1$ has finite ascent and descent. This, in conjunction with ($\star$) and ($\star\star$), guarantees that $\lambda_0 - T$ likewise has finite ascent and descent. Now a glance at Theorem 6.39 shows that λ_0 is a pole of the resolvent of T.

Since $\sigma(T_1) = \{\lambda_0\}$ and $\dim X_1 < \infty$, there exists some non-zero vector $x \in X_1$ such that $Tx = T_1 x = \lambda_0 x$. Thus, λ_0 is an eigenvalue of T. The last conclusion follows immediately from Theorem 6.39. ∎

Corollary 7.45. *For an isolated point λ_0 in the spectrum of an operator $T \in \mathcal{L}(X)$ the following statements are equivalent.*

(a) *The spectral projection $P_{\lambda_0}(T)$ is of finite-rank.*

(b) *The operator $\lambda_0 - T$ is Fredholm, i.e., $\lambda_0 \notin \sigma_{\rm ess}(T)$.*

Proof. The implication (b) $\Longrightarrow$ (a) has been established in Theorem 7.44. Now assume that (a) is true.

Let $X_1 = R\big(P_{\lambda_0}(T)\big)$ and $X_2 = N(P_{\lambda_0}(T))$. According to our hypothesis X_1 is finite dimensional, $X = X_1 \oplus X_2$, and the closed subspaces X_1 and X_2 of X are both invariant under T. If T_i is the restriction of T to X_i, then $\sigma(T_1) = \{\lambda_0\}$ and $\sigma(T_2) = \sigma(T) \setminus \{\lambda_0\}$. Since $\lambda_0 \notin \sigma(T_2)$, it follows that $\lambda_0 - T_2 \in \mathcal{L}(X_2)$ is an isomorphism.

This implies, on the one hand, $N(\lambda_0 - T) = N(\lambda_0 - T_1) \subseteq X_1$, and so $n(T) = \dim N(\lambda_0 - T) < \infty$. On the other hand, from $X_2 \subseteq R(\lambda_0 - T)$, it easily follows that $d(T) = \dim X/R(\lambda_0 - T) < \infty$. Therefore, $\lambda_0 - T$ is a Fredholm operator. ∎

We now introduce the essential spectral radius of a bounded operator.

Definition 7.46. *The* ***essential spectral radius*** *of an operator $T \in \mathcal{L}(X)$ is the spectral radius of the element $\pi(T)$ in the Calkin algebra $\mathfrak{C}(X)$, i.e,*

$$\begin{aligned} r_{\rm ess}(T) &= r\big(\pi(T)\big) \\ &= \max\big\{|\lambda|\colon\ \lambda \in \sigma_{\rm ess}(T)\big\} \\ &= \lim_{n\to\infty} \big\|\pi(T^n)\big\|^{\frac{1}{n}} \\ &= \inf_n \big\|\pi(T^n)\big\|^{\frac{1}{n}} \\ &= \lim_{n\to\infty} \Big[\inf_{K\in\mathcal{K}(X)} \|T^n - K\| \Big]^{\frac{1}{n}} . \end{aligned}$$

The essentially quasinilpotent operators are defined as follows.

Definition 7.47. *An operator $T \in \mathcal{L}(X)$ is said to be* ***essentially quasinilpotent*** *(or a* ***Riesz operator****) if $\sigma_{\rm ess}(T) = \{0\}$, or equivalently, if its essential spectral radius is zero.*

The next result presents a useful characterization of the essentially quasinilpotent operators.

Theorem 7.48. *An operator $T \in \mathcal{L}(X)$ is essentially quasinilpotent if and only if every $\lambda \in \sigma(T) \setminus \{0\}$ is:*

(1) *an isolated point of $\sigma(T)$, and*

(2) *its spectral projection $P_\lambda(T)$ has finite-rank.*

Proof. Assume that T is essentially quasinilpotent, i.e., $\sigma_{\text{ess}}(T) = \{0\}$, and let $\lambda \in \sigma(T) \setminus \{0\}$. Clearly, λ can be joined with any point of $\rho(T)$ by a path lying outside $\sigma_{\text{ess}}(T) = \{0\}$. So, according to Lemma 7.43, λ is an isolated point of $\sigma(T)$. By Theorem 7.44 the spectral projection $P_\lambda(T)$ has finite-rank.

For the converse, assume that every point $\lambda \in \sigma(T) \setminus \{0\}$ satisfies (1) and (2). Now fix $\lambda \in \sigma(T) \setminus \{0\}$. By (1), we know that λ is an isolated point of $\sigma(T)$. Therefore, Corollary 7.45 applies and shows that $\lambda \notin \sigma_{\text{ess}}(T)$. Since $\sigma_{\text{ess}}(T) \subseteq \sigma(T)$ and $\sigma_{\text{ess}}(T) \neq \emptyset$, it follows that $\sigma_{\text{ess}}(T) = \{0\}$, i.e., T is essentially quasinilpotent. ■

There are several interesting consequences of the preceding result. We list some of them next.

Corollary 7.49. *If an operator $T \in \mathcal{L}(X)$ is essentially quasinilpotent, then for each $\lambda \in \sigma(T) \setminus \{0\}$:*

(a) *λ is a pole of the resolvent of T,*

(b) *λ is an eigenvalue of T,*

(c) *the operator $\lambda - T$ has finite ascent and descent, and*

(d) *the order of the pole λ equals the ascent (or the descent) of the operator $\lambda - T$.*

Proof. Assume that an operator $T \in \mathcal{L}(X)$ satisfies $\sigma_{\text{ess}}(T) = \{0\}$, and let $\lambda_0 \in \sigma(T) \setminus \{0\}$. By Theorem 7.48, we know that λ_0 is an isolated point of $\sigma(T)$. Theorem 7.44 implies that λ_0 is an eigenvalue of T and that the operator $\lambda_0 - T$ has finite ascent and descent and, consequently, from part (3) of the same theorem λ_0 is a pole of the resolvent of T whose order equals the ascent (or the descent) of the operator $\lambda_0 - T$. ■

Corollary 7.50. *The spectrum of an essentially quasinilpotent operator is at most countable and if it is countable, say $\{\lambda_1, \lambda_2, \ldots\}$, then $\lambda_n \to 0$.*

Proof. By Theorem 7.48, we know that every non-zero point in the spectrum of T is an isolated point of $\sigma(T)$. This implies that $\sigma(T)$ is at most countable. The rest follows immediately from Exercise 20 of Section 6.1. ■

Corollary 7.51. *If $T \in \mathcal{L}(X)$ is essentially quasinilpotent, then its resolvent set $\rho(T)$ is always a connected set.*

Proof. This follows from Theorem 7.48 in conjunction with Exercise 5(a) of Section 6.3. ■

Corollary 7.52. *The restriction of an essentially quasinilpotent operator to any invariant closed subspace is again essentially quasinilpotent.*

Proof. Suppose that $T \in \mathcal{L}(X)$ is essentially quasinilpotent and that Y is a closed T-invariant subspace of X. According to part (d) of Exercise 5 in Section 6.3, we have $\sigma(T|_Y) \subseteq \sigma(T)$.

Now let $\lambda_0 \in \sigma(T|_Y) \setminus \{0\}$. By Theorem 7.48, λ_0 is an isolated point of $\sigma(T)$. Let $R(\lambda, T) = \sum_{n=-\infty}^{\infty} (\lambda - \lambda_0)^n A_n$ be the Laurent series expansion of $R(\lambda, T)$ around λ_0. By Exercise 5(e) in Section 6.3, we know that the subspace Y is A_n-invariant for each n. In particular, we have $R(\lambda, T|_Y) = \sum_{n=-\infty}^{\infty} (\lambda - \lambda_0)^n A_n|_Y$, and so $P_{\lambda_0}(T|_Y) = A_{-1}|_Y = P_{\lambda_0}(T)|_Y$. Since $P_{\lambda_0}(T)$ has finite-rank, it follows that $P_{\lambda_0}(T|_Y)$ also has finite-rank. By Theorem 7.44, $T|_Y$ is essentially quasinilpotent. ■

The next result characterizes the essentially quasinilpotent operators in terms of a decomposition property of their resolvent functions.

Theorem 7.53. *An operator $T \in \mathcal{L}(X)$ is essentially quasinilpotent if and only if its resolvent can be written in the form*

$$R(\lambda, T) = B(\lambda) + K(\lambda),$$

where the functions $B\colon \mathbb{C} \setminus \{0\} \to \mathcal{L}(X)$ and $K\colon \rho(T) \to \mathcal{K}(X)$ are both analytic and $\lim_{\lambda \to \infty} B(\lambda) = 0$.

Proof. Assume first that $R(\lambda, T)$ has the stated decomposition. Then the analytic function B has the Laurent series expansion $B(\lambda) = \sum_{n=0}^{\infty} \frac{B_n}{\lambda^{n+1}}$ (see Exercise 6 of Section 1.5) and $\lim_{n\to\infty} \sqrt[n]{\|B_n\|} = 0$ (see Theorem 1.77). From the Neumann series expansion of the resolvent, it follows that for each $|\lambda| > r(T)$ we have $K(\lambda) = \sum_{n=0}^{\infty} \frac{T^n - B_n}{\lambda^{n+1}} = \sum_{n=0}^{\infty} \frac{K_n}{\lambda^{n+1}}$, where we let $K_n = T^n - B_n \in \mathcal{L}(X)$ for each $n \geq 0$. Now if we take any $r > r(T)$ and consider the positively oriented circle C_r with center at zero and radius r, then from the Cauchy formula

$$K_n = \frac{1}{2\pi i} \int_{C_r} \frac{K(\lambda)}{\lambda^{n+1}}\, d\lambda$$

and the definition of the Riemann integral, we infer that each K_n is a compact operator. In particular, we have $\|[T^n]\|^{\frac{1}{n}} \leq \|T^n - K_n\|^{\frac{1}{n}} = \sqrt[n]{\|B_n\|}$ for each $n \geq 0$. From $\lim_{n\to\infty} \sqrt[n]{\|B_n\|} = 0$, we get

$$r_{\text{ess}}(T) = \lim_{n\to\infty} \|[T^n]\|^{\frac{1}{n}} = 0.$$

That is, T is essentially quasinilpotent.

For the converse, assume that the operator T is essentially quasinilpotent. This guarantees the existence of compact operators $K_0, K_1, K_2, \ldots$ such that $\sqrt[n]{\|T^n - K_n\|} \to 0$. Put $B_n = T^n - K_n$ for each $n \geq 0$ and note that the series $B(\lambda) = \sum_{n=0}^{\infty} \frac{B_n}{\lambda^{n+1}}$ converges for each $\lambda \in \mathbb{C} \setminus \{0\}$. So, B defines an analytic function on $\mathbb{C} \setminus \{0\}$. Since $R(\lambda, T) = \sum_{n=0}^{\infty} \frac{T^n}{\lambda^{n+1}}$ holds for each $|\lambda| > r(T)$, it follows that the series $K(\lambda) = \sum_{n=0}^{\infty} \frac{K_n}{\lambda^{n+1}}$ converges for each $|\lambda| > r(T)$ to a (necessarily) compact operator. In particular, we have $R(\lambda, T) = B(\lambda) + K(\lambda)$ for each $|\lambda| > r(T)$.

The function $\widehat{K}\colon \rho(T) \to \mathcal{L}(X)$ defined by $\widehat{K}(\lambda) = R(\lambda, T) - B(\lambda)$ is clearly analytic. Since $\widehat{K}(\lambda) = K(\lambda)$ for each $|\lambda| > r(T)$ and $\rho(T)$ is connected, it follows that $\widehat{K}$ is the unique analytic extension of K to $\rho(T)$. To finish the proof, it remains to be shown that $\widehat{K}(\lambda)$ is a compact operator for each $\lambda \in \rho(T)$.

To see this, notice that if $\pi\colon \mathcal{L}(X) \to \mathfrak{C}(X)$ is the canonical projection of $\mathcal{L}(X)$ onto the Calkin algebra $\mathfrak{C}(X)$, then the function $f\colon \rho(T) \to \mathfrak{C}(X)$, defined by $f(\lambda) = \pi\big(R(\lambda, T)\big) - \pi\big(\widehat{K}(\lambda)\big)$, is clearly analytic. Moreover, the function $g\colon \rho(T) \to \mathfrak{C}(X)$ defined by $g(\lambda) = \pi\big(R(\lambda, T)\big)$ is also analytic and $f(\lambda) = g(\lambda)$ holds for each $|\lambda| > r(T)$. Since $\rho(T)$ is connected, it follows that $f(\lambda) = g(\lambda)$ for each $\lambda \in \rho(T)$. This implies $\pi\big(\widehat{K}(\lambda)\big) = 0$ for each $\lambda \in \rho(T)$. That is, $\widehat{K}(\lambda)$ is a compact operator for each $\lambda \in \rho(T)$, and the proof is finished. ■

Our next goal is to establish that the Calkin algebra can be represented as a subalgebra of bounded operators on an appropriate Banach space. To do this, we need some preliminary discussion.

Let $\mathcal{U}$ be an ultrafilter on $\mathbb{N}$ and let $X_{\mathcal{U}}$ be the ultrapower of X determined by the ultrafilter $\mathcal{U}$; see Section 1.4. If $T \in \mathcal{L}(X)$, then there exists a natural extension of T to an operator $T_{\mathcal{U}}$ on $X_{\mathcal{U}}$ defined by

$$T_{\mathcal{U}}[(x_1, x_2, \ldots)] = [(Tx_1, Tx_2, \ldots)] = (Tx_1, Tx_2, \ldots) + \mathcal{N}_{\mathcal{U}}$$

for each $(x_1, x_2, \ldots) \in \ell_\infty(X)$. Because X is a closed vector subspace of $X_{\mathcal{U}}$, the quotient space $\widetilde{X} = X_{\mathcal{U}}/X$ is a Banach space.

Next, we shall indicate how each operator $T \in \mathcal{L}(X)$ gives rise to an operator $\widetilde{T} \in \mathcal{L}(\widetilde{X})$. To do this, we need to introduce some notation. First of all, if $x = (x_1, x_2, \ldots) \in \ell_\infty(X)$, let us write $x_{\mathcal{U}}$ to denote the equivalence class of x in $X_{\mathcal{U}}$. Second, we shall denote the equivalence class of $x_{\mathcal{U}}$ in $\widetilde{X}$ by $\widetilde{x}$ or by $[x_{\mathcal{U}}]$. With this notation in place, a sequence $x = (x_1, x_2, \ldots)$ in $\ell_\infty(X)$ satisfies

$$\widetilde{x} = 0 \iff \lim\nolimits_{\mathcal{U}} x_n \text{ exists in } X\,.$$

In particular, notice that for a sequence $y = (y_1, y_2, \ldots) \in \ell_\infty(X)$ we have $y_{\mathcal{U}} \in [x_{\mathcal{U}}]$ if and only if there exists some $v \in X$ such that $x_n - y_n - v \xrightarrow{\mathcal{U}} 0$.

Each operator $T \in \mathcal{L}(X)$ gives rise to a bounded operator $\widetilde{T}\colon \widetilde{X} \to \widetilde{X}$ defined via the formula

$$\widetilde{T}\widetilde{x} = (T_{\mathcal{U}}x_{\mathcal{U}})\widetilde{\ } = [(Tx_1, Tx_2, \ldots)_{\mathcal{U}}]$$

for each $(x_1, x_2, \ldots) \in \ell_\infty(X)$.

Let us verify that the above formula is well defined and that it defines a bounded operator on $\widetilde{X}$. Fix $y_{\mathcal{U}} \in [x_{\mathcal{U}}]$. This means that there exists some $u \in X$ such that $x_n - y_n - u \xrightarrow{\mathcal{U}} 0$. Consequently, $Tx_n - Ty_n - Tu \xrightarrow{\mathcal{U}} 0$, and so $(Ty_1, Ty_2, \ldots)_{\mathcal{U}} \in [(Ty_1, Ty_2, \ldots)_{\mathcal{U}}]$. This shows that $\widetilde{T}$ is well defined, and clearly it is linear. Moreover, for each $y_{\mathcal{U}} \in [x_{\mathcal{U}}]$ we have

$$\begin{aligned} \|\widetilde{T}[x_{\mathcal{U}}]\| &= \|\widetilde{T}[y_{\mathcal{U}}]\| \\ &= \|[(Ty_1, Ty_2, \ldots)_{\mathcal{U}}]\| \\ &\leq \|(Ty_1, Ty_2, \ldots)_{\mathcal{U}}\| \\ &\leq \|T_{\mathcal{U}}\| \cdot \|(y_1, y_2, \ldots)_{\mathcal{U}}\|, \end{aligned}$$

and from this we easily get

$$\|\widetilde{T}[x_{\mathcal{U}}]\| \leq \|T\| \cdot \|[x_{\mathcal{U}}]\|.$$

That is, $\widetilde{T}\colon \widetilde{X} \to \widetilde{X}$ is indeed a bounded operator.

The basic algebraic properties of the operators $\widetilde{T}$ are included in the next result. These properties (as well as many others) have been obtained by J. J. Buoni, R. Harte, and A. W. Wickstead [**79**], J. J. M. Chadwick and A. W. Wickstead [**87**], and M. P. W. Wolff [**340**].

Lemma 7.54. *For an operator $T \in \mathcal{L}(X)$ we have the following.*

(1) *$\widetilde{T} = 0$ if and only if T is a compact operator.*

(2) *$\widetilde{T}$ is one-to-one if and only if T is a semi-Fredholm operator with finite nullity, i.e., $n(T) < \infty$.*

(3) *$\widetilde{T}$ is surjective and T has closed range if and only if T is a semi-Fredholm operator with finite defect, i.e., $d(T) < \infty$.*

(4) *$\widetilde{T}$ is invertible if and only if T is a Fredholm operator.*

Proof. (1) Fix an operator $T \in \mathcal{L}(X)$. Assume first that $\widetilde{T} = 0$ and let $\{x_n\}$ be a sequence in U_X. If we let $x = (x_1, x_2, \ldots) \in \ell_\infty(X)$, then from $\widetilde{T}[x_{\mathcal{U}}] = [(Tx_1, Tx_2, \ldots)_{\mathcal{U}}] = 0$, it follows that there exists some $z \in X$ such that $Tx_n \xrightarrow{\mathcal{U}} z$. In particular, it follows (see part (3) of Lemma 1.59) that there exists a subsequence $\{x_{k_n}\}$ of $\{x_n\}$ such that $Tx_{k_n} \to z$ in X. This implies that T is a compact operator.

For the converse, suppose that T is a compact operator and pick any vector $x = (x_1, x_2, \ldots) \in \ell_\infty(X)$. Since the sequence $\{Tx_n\}$ lies in a compact set, it follows from part (4) of Lemma 1.59 that there exists some $z \in X$ such that $Tx_n \xrightarrow{\mathcal{U}} z$ or $Tx_n - z \xrightarrow{\mathcal{U}} 0$. This implies that $0 \in [(Tx_1, Tx_2, \ldots)_{\mathcal{U}}]$, and so $\widetilde{T}[x_{\mathcal{U}}] = [(Tx_1, Tx_2, \ldots)_{\mathcal{U}}] = 0$. Consequently, $\widetilde{T} = 0$.

(2) Suppose that T is a semi-Fredholm operator with $\dim N(T) < \infty$. By the definition of semi-Fredholmness, we know that the range $R(T)$ of

T is closed. Since $N(T)$ is finite dimensional, there exists a closed vector subspace V of X such that $X = N(T) \oplus V$. In particular, the operator $S = T|_Y : V \to R(T)$ is an isomorphism.

Now assume that a sequence $x = (x_1, x_2, \ldots) \in \ell_\infty(X)$ satisfies $\widetilde{T}\widetilde{x} = 0$. Clearly, this is equivalent to saying that there exists some $z \in X$ such that $Tx_n \xrightarrow{\mathcal{U}} z$. Now for each n write $x_n = y_n + v_n$ with $y_n \in N(T)$ and $v_n \in V$, and note that both sequences $\{y_n\}$ and $\{v_n\}$ are bounded. Since $N(T)$ is finite dimensional, the sequence $\{y_n\}$ lies in a compact set, and so there exists some $y \in N(T)$ such that $y_n \xrightarrow{\mathcal{U}} y$. From $Tv_n = Tx_n \xrightarrow{\mathcal{U}} z$ in X, it follows that $Tv_n \xrightarrow{\mathcal{U}} z$ in $R(T)$, and hence $v_n = S^{-1}(Tv_n) \xrightarrow{\mathcal{U}} S^{-1}(z)$ in X. Consequently, $x_n = y_n + v_n \xrightarrow{\mathcal{U}} y + S^{-1}(z)$ holds in X. This shows that $\widetilde{x} = 0$, and so $\widetilde{T}$ is one-to-one.

For the converse, assume that the operator $\widetilde{T}$ is one-to-one. Let a sequence $\{x_n\}$ in $N(T)$ satisfy $\|x_n\| \leq 1$ for each n. Now consider the sequence $x = (x_1, x_2, \ldots) \in \ell_\infty(X)$ and note that $\widetilde{T}\widetilde{x} = [(Tx_1, Tx_2, \ldots)_{\mathcal{U}}] = 0$. Since the operator $\widetilde{T}$ is one-to-one, we get $\widetilde{x} = 0$. This means that there exists some $y \in X$ such that $x_n \xrightarrow{\mathcal{U}} y$. By Lemma 1.59(3), there exists a subsequence of $\{x_n\}$ that is norm convergent to y. This shows that the closed unit ball of $N(T)$ is a compact set, and hence the vector space $N(T)$ is finite dimensional.

Next, pick a closed vector subspace W of X such that $X = N(T) \oplus W$. To establish that T has closed range, it suffices to show that T is bounded below on W. If this is not the case, there exists a sequence $\{w_n\}$ of unit vectors in W such that $\|Tw_n\| \to 0$. In particular, we have $Tw_n \xrightarrow{\mathcal{U}} 0$, and from this we infer that the vector $w = (w_1, w_2, \ldots) \in \ell_\infty(X)$ satisfies $\widetilde{T}\widetilde{w} = 0$. Using again that $\widetilde{T}$ is one-to-one, we obtain $\widetilde{w} = 0$. This means that there exists some $u \in X$ such that $w_n \xrightarrow{\mathcal{U}} u$. By Lemma 1.59(3), we can assume (by passing to a subsequence if necessary) that $\|w_n - u\| \to 0$. Thus, $u \in W$ and $\|u\| = 1$. Now the continuity of T implies $\|Tw_n - Tu\| \to 0$, and so $Tu = 0$. Finally, using that T is one-to-one on W, we get $u = 0$, contrary to $\|u\| = 1$. This contradiction establishes that T is bounded below on W, and so $R(T)$ is closed.

(3) Assume first that $\widetilde{T}$ is surjective and that T has closed range. Also, suppose by way of contradiction that $X/R(T)$ is infinite dimensional. Then, using Riesz's Lemma 4.44 in conjunction with Exercise 1 of Section 3.4, we see that there exists a sequence $\{y_n\}$ of unit vectors such that for each n and all $y \in R(T) \oplus \mathrm{Span}\{y_1, y_2, \ldots, y_n\}$ we have

$$\|y_{n+1} - y\| > \tfrac{1}{2}. \qquad (\star)$$

Put $y = (y_1, y_2, \ldots) \in \ell_\infty(X)$. Since $\widetilde{T}$ is surjective, there exists some $x = (x_1, x_2, \ldots) \in \ell_\infty(X)$ such that $\widetilde{T}\widetilde{x} = \widetilde{y}$. This means that for some vector $a \in X$ we have $Tx_n - y_n - a \xrightarrow{\mathcal{U}} 0$. Consequently, by Lemma 1.59(3), there exists a strictly increasing sequence $\{k_n\}$ of natural numbers such that $\|Tx_{k_n} - y_{k_n} - a\| \to 0$. By passing to another subsequence if necessary we can also assume that

$$\|Tx_{k_n} - y_{k_n} - a\| < \tfrac{1}{6} \tag{$\star\star$}$$

for all n. Now using ($\star$) and ($\star\star$), we see that

$$\begin{aligned} \tfrac{1}{2} &< \|y_{k_{n+1}} - (Tx_{k_{n+1}} + y_{k_n} - Tx_{k_n})\| \\ &= \|(Tx_{k_n} - y_{k_n}) - (Tx_{k_{n+1}} - y_{k_{n+1}})\| \\ &\le \|Tx_{k_n} - y_{k_n} - a\| + \|Tx_{k_{n+1}} - y_{k_{n+1}} - a\| \\ &< \tfrac{1}{6} + \tfrac{1}{6} = \tfrac{1}{3}, \end{aligned}$$

which is impossible. Therefore, $X/R(T)$ is finite dimensional.

For the converse, assume that T is a semi-Fredholm operator such that $X/R(T)$ is finite dimensional. By the definition of semi-Fredholmness the range $R(T)$ is closed and so there exists a finite dimensional vector subspace Z of X such that $X = R(T) \oplus Z$. Let $y = (y_1, y_2, \ldots) \in \ell_\infty(X)$. To finish the proof, we must show that there exists some $x = (x_1, x_2, \ldots) \in \ell_\infty(X)$ such that $\widetilde{T}\widetilde{x} = \widetilde{y}$.

To this end, for each n let $y_n = r_n + z_n$, where $r_n \in R(T)$ and $z_n \in Z$. Since $R(T)$ is closed, it easily follows that both sequences $r = (r_1, r_2, \ldots)$ and $z = (z_1, z_2, \ldots)$ belong to $\ell_\infty(X)$. Since $\{z_n\}$ lies in Z (which is finite dimensional) and is bounded, it also lies in a compact set, and so according to Lemma 1.59(4) there exists some $b \in X$ such that $z_n \xrightarrow{\mathcal{U}} b$ or, equivalently, $\widetilde{z} = 0$. Next, notice that (in view of Corollary 2.15) there exists a sequence $x = (x_1, x_2, \ldots) \in \ell_\infty(X)$ such that $r_n = Tx_n$ for each n. Finally, notice that the identities $y_n = r_n + z_n$ are equivalent to $y_{\mathcal{U}} = T_{\mathcal{U}}x_{\mathcal{U}} + z_{\mathcal{U}}$. This implies

$$\widetilde{y} = [y_{\mathcal{U}}] = [T_{\mathcal{U}}x_{\mathcal{U}}] + [z_{\mathcal{U}}] = \widetilde{T}\widetilde{x} + \widetilde{z} = \widetilde{T}\widetilde{x},$$

and the proof of this part is finished.

(4) This follows immediately from properties (2) and (3) above. ■

To continue our discussion, we now need to introduce the notion of the measure of non-compactness for sets and operators.

Definition 7.55. *If A is a bounded subset of a metric space, then the* ***measure of non-compactness*** *$\chi(A)$ of A is the infimum of all $r > 0$ such that A can be covered by a finite number of open balls of radius r. That is,*

$$\chi(A) = \inf\left\{r > 0\colon\ \exists\ x_1, \ldots, x_n \text{ such that } A \subseteq \textstyle\bigcup_{i=1}^{n} B(x_i, r)\right\}.$$

For an unbounded subset A of a metric space we let $\chi(A) = \infty$.

If X is a Banach space and $T \in \mathcal{L}(X)$, then the ***measure of non-compactness*** *$\chi(T)$ of T is defined by*

$$\chi(T) = \chi\big(T(U_X)\big).$$

The next result contains several elementary properties of the measure of non-compactness that follow easily from the preceding definition.

Lemma 7.56. *For the measure of non-compactness we have:*

(1) *For each subset A of a metric space we have $\chi(A) = \chi(\overline{A})$.*

(2) *A subset A of a metric space is relatively compact if and only if $\chi(A) = 0$.*

(3) *A bounded operator $T \in \mathcal{L}(X)$ is compact if and only if $\chi(T) = 0$.*

(4) *If X, Y and Z are Banach spaces, $T \in \mathcal{L}(X, Y)$ and $S \in \mathcal{L}(Y, Z)$, then $\chi(ST) \le \chi(S)\chi(T)$.*

We also have the following formula, due to V. G. Troitsky [**322**], regarding the measure of non-compactness for bounded subsets of a separable Banach space.

Lemma 7.57 (Troitsky)**.** *If A is a bounded subset of a separable Banach space, then*

$$\chi(A) = \max\left\{\|\widetilde{x}\|\colon\ x = (x_1, x_2, \ldots) \in A^{\mathbb{N}}\right\}.$$

Proof. Let A be a bounded subset of a Banach space X. Fix $\rho > \chi(A)$, and then pick $a_1, a_2, \ldots, a_m \in A$ such that $A \subseteq \bigcup_{k=1}^{m} B(a_k, \rho)$. Now let $x = (x_1, x_2, \ldots) \in A^{\mathbb{N}}$ and for each $1 \le k \le m$ put $\mathbb{N}_k = \big\{n \in \mathbb{N}\colon\ x_n \in B(a_k, \rho)\big\}$. Clearly, $\mathbb{N} = \bigcup_{k=1}^{m} \mathbb{N}_k$. So, according to Lemma 1.55, for some $1 \le k \le m$ we have $\mathbb{N}_k \in \mathcal{U}$. If we let $y = (x_1 - a_k, x_2 - a_k, \ldots) \in \ell(X)$, then $y_{\mathcal{U}} \in \widetilde{x}$, and so (by Exercise 5 of Section 1.4) we have $\|\widetilde{x}\| \le \|y_{\mathcal{U}}\| = \lim_{\mathcal{U}} \|x_n - a_k\| \le \rho$. This implies

$$\|\widetilde{x}\| \le \chi(A).$$

If $\chi(A) = 0$, i.e., if A is a norm totally bounded set, then $\|\widetilde{x}\| = 0$ for all $x = (x_1, x_2, \ldots) \in A^{\mathbb{N}}$. So, we can assume that $\chi(A) > 0$, i.e., that A is not norm totally bounded.

Since X is a separable Banach space, there exists a countable dense subset $Z = \{z_1, z_2, \ldots\}$ of X. Fix r such that $0 < r < \chi(A)$. This implies that for each n the set A is not a subset of $\bigcup_{i=1}^{n} B(z_i, r)$. So, for each n there exists some $x_n \in A \setminus \bigcup_{i=1}^{n} B(z_i, r)$. Put $x = (x_1, x_2, \ldots) \in A^{\mathbb{N}}$. We shall finish the proof by showing that $\|\widetilde{x}\| = \chi(A)$.

To see this, start by observing that from $\|x_n - z_k\| \geq r$ for each $n > k$ it follows that $\|x_{\mathcal{U}} - z_k\| = \lim_{\mathcal{U}} \|x_n - z_k\| \geq r$, and so $\|x_{\mathcal{U}} - z_k\| \geq \chi(A)$ for each k. Since Z is dense in X, we have $\|x_{\mathcal{U}} - z\| \geq \chi(A)$ for each $z \in X$. But then we have

$$\chi(A) \leq d(\widetilde{x}, X) = \|\widetilde{x}\| \leq \chi(A).$$

That is, $\|\widetilde{x}\| = \chi(A)$, as desired. ■

The importance of the measure of non-compactness of an operator is highlighted in the next result.

Corollary 7.58 (Troitsky [**322**]). *If X is a separable Banach space, then for each operator $T \in \mathcal{L}(X)$ we have $\|\widetilde{T}\| = \chi(T)$.*

Proof. Assume that X is a separable Banach space and let $T \in \mathcal{L}(X)$. Using Lemma 7.57, we see that

$$\begin{aligned}
\chi(T) &= \chi\big(T(U_X)\big) \\
&= \max\big\{\|\widetilde{y}\|:\ y = (y_1, y_2, \ldots) \in \big[T(U_X)\big]^{\mathbb{N}}\big\} \\
&= \max\big\{\big\|\widetilde{T}\widetilde{x}\big\|:\ x = (x_1, x_2, \ldots) \in \big(U_X\big)^{\mathbb{N}}\big\} \\
&= \max\big\{\big\|\widetilde{T}\widetilde{x}\big\|:\ \widetilde{x} \in U_{\widetilde{X}}\big\} \\
&= \big\|\widetilde{T}\big\|,
\end{aligned}$$

and the proof is finished. ■

The preceding corollary is also valid for non-separable Banach spaces, provided that the chosen ultrafilter $\mathcal{U}$ satisfies a certain "saturation" condition. For details see [**322**].

The proof of the next lemma is a direct consequence of the preceding results and is omitted.

Lemma 7.59. *If X is a separable Banach space, then the measure of non-compactness $\chi\colon \mathcal{L}(X) \to \mathbb{R}$ is a seminorm. This seminorm induces a norm on the Calkin algebra $\mathfrak{C}(X)$.*

And now we come to the identification of the Banach algebra $\mathfrak{C}(X)$ as an algebra of bounded operators on a Banach space.

Theorem 7.60 (Troitsky [**322**]). *If X is a separable Banach space, then the mapping $[T] \mapsto \widetilde{T}$, from $\mathfrak{C}(X)$ to $\mathcal{L}(\widetilde{X})$, is a well-defined algebraic isomorphism—and so $\mathfrak{C}(X)$ can be considered as an algebra of operators on the Banach space $\widetilde{X}$.*

In addition, if $\mathfrak{C}(X)$ is equipped with the norm $\chi(\cdot)$, then the mapping $[T] \mapsto \widetilde{T}$ is a linear isometry from $\mathfrak{C}(X)$ into $\mathcal{L}(\widetilde{X})$.

Proof. From part (1) of Lemma 7.54 it follows immediately that the mapping $[T] \mapsto \widetilde{T}$, from $\mathfrak{C}(X)$ to $\mathcal{L}(\widetilde{X})$, is well defined. Clearly, this mapping is also linear.

To see that it is an algebraic homomorphism, let $S, T \in \mathcal{L}(X)$ and fix $x = (x_1, x_2, \ldots) \in \ell_\infty(X)$. From

$$\begin{aligned} \widetilde{ST}[x_{\mathcal{U}}] &= [((ST)x_1, (ST)x_2, \ldots)_{\mathcal{U}}] \\ &= [(S(Tx_1), S(Tx_2), \ldots)_{\mathcal{U}}] \\ &= \widetilde{S}[(Tx_1, Tx_2, \ldots)_{\mathcal{U}}] \\ &= (\widetilde{S}\widetilde{T})[x_{\mathcal{U}}], \end{aligned}$$

we obtain that $\widetilde{ST} = \widetilde{S}\widetilde{T}$ so that the mapping $[T] \mapsto \widetilde{T}$ is also an algebraic homomorphism.

The fact that $[T] \mapsto \widetilde{T}$ is a linear isometry when $\mathfrak{C}(X)$ is equipped with the norm $\chi(\cdot)$ follows from Corollary 7.58. This guarantees that the operator $[T] \mapsto \widetilde{T}$ is also an algebraic isomorphism. ■

Next we will establish an interesting formula, due to R. D. Nussbaum [**258**], for calculating the essential spectral radius of a bounded operator.

Theorem 7.61 (Nussbaum). *For a bounded operator $T\colon X \to X$ on a Banach space we have the following.*

1. *The spectrum of $\widetilde{T}$ coincides with the essential spectrum of T, i.e., $\sigma(\widetilde{T}) = \sigma_{\text{ess}}(T)$.*
2. *If X is separable, then the essential spectral radius of T satisfies*

$$r_{\text{ess}}(T) = r(\widetilde{T}) = \lim_{n\to\infty} \sqrt[n]{\|\widetilde{T}^n\|} = \lim_{n\to\infty} \sqrt[n]{\chi(T^n)}\,.$$

Proof. (1) Using part (4) of Lemma 7.54 and the definition of the essential spectrum, we see that:

$$\begin{aligned} \lambda \in \sigma(\widetilde{T}) &\iff \lambda - \widetilde{T} \text{ is not invertible} \\ &\iff \lambda - T \text{ is not Fredholm} \\ &\iff \lambda \in \sigma_{\text{ess}}(T)\,. \end{aligned}$$

This implies $\sigma(\widetilde{T}) = \sigma_{\text{ess}}(T)$.

(2) This follows immediately from part (1) and Corollary 7.58. ■

We close the chapter with a discussion regarding a decomposition property of the essentially quasinilpotent operators. If $T = Q + K$ for some $K \in \mathcal{K}(X)$ and some quasinilpotent operator $Q \in \mathcal{L}(X)$, then we have

$$\|\pi(T)^n\|^{\frac{1}{n}} = \|\pi(Q)^n\|^{\frac{1}{n}} \le \|Q^n\|^{\frac{1}{n}} \longrightarrow 0.$$

This implies that $\pi(T)$ is quasinilpotent in the Calkin algebra $\mathfrak{C}(X)$, i.e., T is essentially quasinilpotent.

T. T. West [**333**] has proven that the converse of the preceding conclusion is true for Hilbert spaces. We state this result below. Its proof is beyond the scope of this monograph and is omitted.

Theorem 7.62 (West). *A bounded operator $T: H \to H$ on a Hilbert space is essentially quasinilpotent if and only if there exist a quasinilpotent operator $Q \in \mathcal{L}(H)$ and a compact operator $K \in \mathcal{K}(H)$ such that $T = Q + K$.*

We shall say that an operator $T \in \mathcal{L}(X)$ has a **West Decomposition** if T can be written in the form $T = Q + K$, where Q is quasinilpotent and K is compact. As noticed before, if an operator has a West decomposition, then it is essentially quasinilpotent.

We shall also say that a Banach space X has the **West Decomposition Property** if every essentially quasinilpotent operator on X has a West decomposition. In this terminology, Theorem 7.62 states that:

- *Every Hilbert space has the West Decomposition Property.*

The following question arises here:

- *Does every Banach space have the West Decomposition Property?*

The answer to this question for general Banach spaces is unknown. K. R. Davidson and D. A. Herrero [**99**] have shown that the Banach spaces c_0 and ℓ_p for $1 \le p < \infty$ have the West decomposition property.

There is another special part of the spectrum of an operator—known as the Weyl spectrum—that is related to the essential spectrum and to the West decomposition property. Its definition is as follows.

Definition 7.63. *The **Weyl spectrum** $\sigma_w(T)$ of an operator $T \in \mathcal{L}(X)$ is the intersection of the spectra of all compact perturbations of T, i.e.,*

$$\sigma_w(T) = \bigcap_{K \in \mathcal{K}(X)} \sigma(T - K).$$

Clearly, $\sigma_{\text{ess}}(T) \subseteq \sigma_w(T) \subseteq \sigma(T)$. It is not known if there is a particular compact operator K such that $\sigma(T-K) = \sigma_w(T)$. That is:

- *Is the Weyl spectrum of a bounded operator equal to the spectrum of a compact perturbation of the operator?*

A positive answer was obtained by J. G. Stampfli [**311**] for Hilbert spaces and by K. R. Davidson and D. A. Herrero [**99**] for c_0 and ℓ_p-spaces. For other results regarding the West Decomposition Property and the Weyl spectrum in general Banach spaces see [**99, 163, 346, 347, 348**]. For more about the Calkin algebra see the monograph by S. R. Caradus, W. E. Pfaffenberger, and B. Yood [**85**].

Exercises

1. For the identity operator I on a Banach space X establish the following:
(a) X is infinite dimensional if and only if $\|[I]\| = 1$.
(b) X is finite dimensional if and only if $\|[I]\| = 0$.

2. Let $\mathcal{A}$ be a unital Banach algebra, and let $G(\mathcal{A})$ denote the non-empty collection of all invertible elements of $\mathcal{A}$. Show that $G(\mathcal{A})$ is an open subset of $\mathcal{A}$ and that the mapping $x \mapsto x^{-1}$, from $G(\mathcal{A})$ to $G(\mathcal{A})$, is continuous. In particular, $G(\mathcal{A})$ is a topological group. [HINT: Note that if e is the unit element of the algebra $\mathcal{A}$ and if $\|a\| < 1$, then the element $e-a$ is invertible and $(e-a)^{-1} = \sum_{n=0}^{\infty} a^n$. Now assume that an element $a \in \mathcal{A}$ is invertible and that some $x \in \mathcal{A}$ satisfies $\|x-a\| < \frac{1}{\|a^{-1}\|}$. Then from $\|e - xa^{-1}\| = \|(a-x)a^{-1}\| \leq \|x-a\| \cdot \|a^{-1}\| < 1$, it follows that $xa^{-1} = e - (e - xa^{-1})$ is an invertible element. This implies that x is invertible. For the continuity of $x \mapsto x^{-1}$ see Exercise 11 of Section 2.1.]

3. Let $S\colon \ell_p \to \ell_p$, where $1 \leq p < \infty$, be the forward shift defined by

$$S(x_1, x_2, x_3, \ldots) = (0, x_1, x_2, x_3, \ldots).$$

Show that the essential spectrum of S coincides with the unit circle, that is,

$$\sigma_{\text{ess}}(S) = \Gamma = \{\lambda \in \mathbb{C}\colon\ |\lambda| = 1\}.$$

4. Show that if an operator $T\colon X \to X$ on an infinite dimensional Banach space satisfies $\sigma(T) = \{\lambda_0\}$, then $\sigma_{\text{ess}}(T) = \{\lambda_0\}$.

5. Show that the power compact operators on a Banach space are precisely the essentially nilpotent operators.

Deduce from this that every bounded operator on a Banach lattice that is dominated by a compact positive operator is essentially nilpotent.

6. Let λ_0 be an isolated point in the spectrum of an operator $T \in \mathcal{L}(X)$. Show without using Theorem 7.44 that if the spectral projection $P_{\lambda_0}(T)$ is of finite-rank, then λ_0 is a pole of the resolvent of T.

7. Give an example of an operator $T \in \mathcal{L}(X)$ and some $\lambda \in \sigma(T) \setminus \sigma_{\text{ess}}(T)$ that is not an eigenvalue of T. [HINT: Consider the forward shift operator $S: \ell_2 \to \ell_2$, i.e, $S(x_1, x_2, \ldots) = (0, x_1, x_2, \ldots)$, and note that 0 is not an eigenvalue of S while $0 \in \sigma(S) \setminus \sigma_{\text{ess}}(S)$.]

8. Show that every strictly singular operator on an infinite dimensional Banach space is essentially quasinilpotent. [HINT: Assume that $T \in \mathcal{L}(X)$ is strictly singular and that $\lambda \in \sigma(T) \setminus \{0\}$. From Theorem 7.11 we know that λ is an isolated point of $\sigma(T)$. Now according to Exercise 13 of Section 6.4 there exists an operator $S \in \mathcal{L}(X)$ such that $P_{\lambda_0}(T) = TS = ST$. This implies that the operator $P_{\lambda_0}(T) \in \mathcal{L}(X)$ is strictly singular. Since $P_{\lambda_0}(T)$ is the identity operator on its range, it follows that $P_{\lambda_0}(T)$ has finite-rank. Now use Theorem 7.48.]

9. Prove Lemma 7.56.

10. Prove Lemma 7.59.

Chapter 8

Positive Matrices

It is well known that operators on $\mathbb{R}^n$ or $\mathbb{C}^n$ can be represented by square matrices with real or complex entries. The spectrum of an operator on $\mathbb{R}^n$ or $\mathbb{C}^n$ coincides, of course, with the spectrum of the matrix representing the operator and therefore it consists of the eigenvalues of the matrix. Matrices with non-negative entries enjoy many special spectral properties. For instance, the spectral radius of a matrix with non-negative entries is always an eigenvalue corresponding to a positive eigenvector—although, of course, the spectral radius of an arbitrary matrix need not be an eigenvalue.

The purpose of this chapter is to present a general discussion of the spectral properties possessed by square matrices with non-negative entries. Matrices with non-negative entries are also known as **positive** (or **non-negative**) **matrices**. These are precisely the matrices corresponding to positive operators on $\mathbb{R}^n$ or $\mathbb{C}^n$. Because of their importance and wide range of applications, positive matrices have been studied extensively by many people in various disciplines. For instance, in engineering positive matrices appear quite often in control systems of the form

$$\dot{x}(t) = Ax(t) + Bu(t)\,,$$

where A and B are appropriate matrices, $x(\cdot)$ is the **state function** and $u(\cdot)$ is the **control function**. In economics, positive matrices are associated with production. If $a_{ij} \geq 0$ is the amount of the jth good needed to produce one unit of the ith good (where $i, j = 1, \ldots, n$), then the non-negative matrix $A = [a_{ij}]$ is called the **production matrix** or the **input-output matrix** of the economy. Now if $x \geq 0$ is the production vector (where the component x_i denotes the amount of the ith good produced) and $c \geq 0$ designates the consumption vector (where the component c_j is the amount of the jth good used for consumption), then the "supply = demand" equilibrium condition yields

$$x = Ax + c \qquad \text{or} \qquad (I - A)x = c\,.$$

Predicting the consumption vector c (input), the above equation allows us to find the needed production vector x (output). For extensive treatments on positive matrices and their applications, we refer the reader to the monographs [**57, 58, 157, 198, 243**]. For a rigorous study of finite dimensional vector spaces the reader can consult [**145**].

When positivity is combined with the notion of irreducibility, it results in several remarkable spectral properties the most important of which are described by the celebrated Perron–Frobenius theorem. As usual, a square matrix is called **irreducible** if (considered as an operator) it does not leave invariant any non-trivial ideal. For positive matrices this is equivalent to saying that for each i and j there exists some natural number k such that the $(i,j)^{\text{th}}$ entry of the matrix A^k is positive. The Perron–Frobenius theorem now can be summarized as follows.

- *Every positive irreducible matrix has a positive spectral radius and a peripheral spectrum that consists of a finite number of complex numbers that are equally spaced along the spectral circle. Moreover, the spectral radius is an eigenvalue with algebraic and geometric multiplicity one.*

8.1. The Banach Lattices $M_n(\mathbb{R})$ and $M_n(\mathbb{C})$

The inner product of two vectors $x, y \in \mathbb{C}^n$ will be denoted, as usual, by (x, y), i.e.,

$$(x, y) = \sum_{i=1}^{n} x_i \overline{y}_i\,.$$

The unit sphere of $\mathbb{C}^n$ under the **Euclidean norm** will be denoted by $\mathbf{S}$, i.e., $\mathbf{S} = \big\{x \in \mathbb{C}^n\colon\ \|x\| = \sqrt{(x,x)} = 1\big\}$. The positive part of $\mathbf{S}$ will be denoted by $\mathbf{S}^+$, i.e. $\mathbf{S}^+ = \{x \in \mathbf{S}\colon\ x \geq 0\}$. Both $\mathbf{S}$ and $\mathbf{S}^+$ are compact subsets of $\mathbb{C}^n$. Whenever necessary, $\mathbf{S}^+$ will be considered as a compact subset of $\mathbb{R}^n$. Recall that a positive vector $x > 0$ in $\mathbb{C}^n$ is said to be **strictly positive**, in symbols $x \gg 0$, whenever each component of x is positive.

We shall consider any given $n \times n$ matrix (with complex or real entries) as an operator acting on $\mathbb{C}^n$ via the usual matrix multiplication formula. Recall that throughout this book the vectors in $\mathbb{C}^n$ are usually denoted as row vectors; for instance, we write $x = (x_1, x_2, \ldots, x_n)$. However, whenever we encounter the expression Ax, where $A = [a_{ij}]$ is an $n \times n$ matrix, then it will be understood that the vector x is now considered as a column vector, i.e., $x = \begin{bmatrix} x_1 \\ \vdots \\ x_n \end{bmatrix}$, and the vector Ax is computed via standard matrix multiplication. That is,

$$Ax = \begin{bmatrix} \sum_{j=1}^n a_{1j}x_j \\ \sum_{j=1}^n a_{2j}x_j \\ \vdots \\ \sum_{j=1}^n a_{mj}x_j \end{bmatrix} = \Big(\sum_{j=1}^n a_{1j}x_j, \sum_{j=1}^n a_{2j}x_j, \ldots, \sum_{j=1}^n a_{mj}x_j \Big).$$

If $A = [a_{ij}]$ is an $n \times n$ matrix with complex entries, then $\overline{A} = \big[\overline{a}_{ij}\big]$ is the **conjugate** of A and $A^t = [b_{ij}]$, where $b_{ij} = a_{ji}$, is the **transpose** of A. Clearly, $\overline{A^t} = \overline{A}^t$. We have the fundamental duality identities:

$$(Ax, y) = (x, \overline{A}^t y) \qquad \text{and} \qquad (x, Ay) = (\overline{A}^t x, y)$$

for all $x, y \in \mathbb{C}^n$.

An $n \times n$ matrix $A = [a_{ij}]$ is said to be **Hermitian** if $\overline{A}^t = A$. A Hermitian matrix with real entries is known as a **symmetric matrix**. Notice that for every $n \times n$ matrix A the matrix $\overline{A}^t A$ is always a Hermitian matrix. The proofs of the next few results are left as exercises.

Lemma 8.1. *An $n \times n$ matrix A is Hermitian if and only if $(Ax, y) = (x, Ay)$ for all $x, y \in \mathbb{C}^n$.*

An $n \times n$ matrix U is called **unitary** if $U^{-1} = \overline{U}^t$. A unitary matrix with real entries is referred to as an **orthogonal matrix**. The unitary matrices are characterized as follows.

Lemma 8.2. *For an $n \times n$ matrix A the following are equivalent.*

(a) *A is unitary.*

(b) *The rows of A form an orthonormal basis in $\mathbb{C}^n$.*

(c) *The columns of A form an orthonormal basis in $\mathbb{C}^n$.*

(d) *A is **inner product preserving**, i.e., $(Ax, Ay) = (x, y)$ for $x, y \in \mathbb{C}^n$.*

(e) *A is Euclidean **norm preserving**, i.e., $\|Ax\| = \|x\|$ for $x, y \in \mathbb{C}^n$.*

(f) *A carries an orthonormal basis to an orthonormal basis.*

Recall that the **characteristic polynomial** of an $n \times n$ matrix A is the polynomial of degree n defined by

$$p_A(\lambda) = \det(\lambda I - A).$$

The spectrum of A consists precisely of the roots of the polynomial p_A, i.e.,

$$\sigma(A) = \big\{\lambda \in \mathbb{C}\colon\ p_A(\lambda) = 0\big\}.$$

It is well known that each $\lambda \in \sigma(A)$ is an eigenvalue of the matrix A, that is, $\sigma(A) = \sigma_p(A)$. If $\lambda_1, \lambda_2, \ldots, \lambda_n$ are the roots of the characteristic polynomial p_A (counted with their multiplicities), then it is easy to see that

$$\sum_{i=1}^{n} \lambda_i = \sum_{i=1}^{n} a_{ii} = \operatorname{tr}(A) \qquad \text{and} \qquad \prod_{i=1}^{n} \lambda_i = \det(A)\,.$$

Recall that the sum of the diagonal entries $\sum_{i=1}^{n} a_{ii}$ of A is known as the **trace** of A.

Each matrix "satisfies" its characteristic polynomial. This important result is known as the Cayley–Hamilton theorem.[1]

Theorem 8.3 (Cayley–Hamilton). *If an $n \times n$ matrix A has characteristic polynomial $p_A(\lambda) = \lambda^n + a_{n-1}\lambda^{n-1} + a_{n-2}\lambda^{n-2} + \cdots + a_1\lambda + a_0$, then*

$$A^n + a_{n-1}A^{n-1} + a_{n-2}A^{n-2} + \cdots + a_1 A + a_0 I = 0\,.$$

The **operator** (or the **Euclidean**) **norm** of an $n \times n$ matrix A will be denoted by $\|A\|$, i.e.,

$$\|A\| = \sup\bigl\{\|Ax\|\colon\ x \in \mathbb{C}^n \ \text{ and } \|x\| \le 1\bigr\}\,.$$

If λ_1 is the largest eigenvalue of the Hermitian matrix $\overline{A}^t A$, then $\|A\| = \sqrt{\lambda_1}$; see Exercise 9 at the end of this section.

The vector space of all $n \times n$ matrices with real (resp. complex) entries will be denoted by $M_n(\mathbb{R})$ (resp. $M_n(\mathbb{C})$). Clearly, $M_n(\mathbb{R})$ is a Banach lattice and $M_n(\mathbb{C})$ is the complexification of $M_n(\mathbb{R})$, i.e., $M_n(\mathbb{C}) = M_n(\mathbb{R}) \oplus \imath M_n(\mathbb{R})$. The Banach lattice $M_n(\mathbb{R})$ is lattice isomorphic to $\mathbb{R}^{n^2}$ and the complex Banach lattice $M_n(\mathbb{C})$ to $\mathbb{C}^{n^2}$.

Exercises

1. If A is an $n \times n$ matrix with real or complex entries, then show that
$$(Ax, y) = (x, \overline{A}^t y)$$
for all $x, y \in \mathbb{C}^n$. In particular, show that A is Hermitian if and only if $(Ax, y) = (x, Ay)$ for all $x, y \in \mathbb{C}^n$.
2. Prove Lemma 8.2.
3. As in the case of operators, two $n \times n$ matrices A and B are said to be **similar** if there exists an invertible matrix C such that $A = C^{-1}BC$. Recall also that the **trace** $\operatorname{tr}(A)$ of an $n \times n$ matrix $A = [a_{ij}]$ is the sum of its diagonal entries, i.e., $\operatorname{tr}(A) = \sum_{i=1}^{n} a_{ii}$. Establish the following.

[1] This famous theorem was proved in 1853 for some special matrices by William Rowan Hamilton (1805–1865). A few years later, Arthur Cayley (1821–1895) announced (without providing a proof) the validity of the theorem for all matrices.

(i) Two similar matrices have the same characteristic polynomial—and hence they have the same eigenvalues counted with their multiplicities.

(ii) This part deals with the trace of matrices.

(a) Show that if A and B are two arbitrary $n \times n$ matrices, then $\operatorname{tr}(AB) = \operatorname{tr}(BA)$.

(b) Show that the trace of an $n \times n$ matrix A coincides with the sum of its eigenvalues. That is, if $\lambda_1, \ldots, \lambda_n$ are the eigenvalues of A (counted with their multiplicities), then $\operatorname{tr}(A) = \sum_{i=1}^n \lambda_i$.

(c) Show that two similar $n \times n$ matrices have the same trace, i.e., the trace of an $n \times n$ matrix is *similarity invariant.*

(d) Consider an $n \times n$ matrix A with column vectors $c_1, \ldots, c_n$. Also, let e_i denote the ith unit vector and consider e_i as a linear functional acting on $\mathbb{C}^n$ (or on $\mathbb{R}^n$) via the formula $e_i(x) = (x, e_i) = x_i$. Show that A, as an operator on $\mathbb{C}^n$ (or on $\mathbb{R}^n$), can be identified with the finite-rank operator $T = \sum_{i=1}^n e_i \otimes c_i$.

(e) Using the notation of part (b), show that $\operatorname{tr}(A) = \sum_{i=1}^n e_i(c_i)$. (Compare this with Exercise 4 of Section 4.1.)

4. Show that each eigenvalue of a unitary matrix has modulus one. [HINT: If U is a unitary matrix and a unit vector $z \in \mathbb{C}^n$ satisfies $Uz = \lambda z$, then

$$|\lambda|^2 = \lambda\overline{\lambda}(z, z) = (\lambda z, \lambda z) = (Uz, Uz) = (\overline{U}^t Uz, z) = (z, z) = 1\,.$$

See also Exercise 5 of Section 6.2.]

5. Show that every Hermitian matrix has real eigenvalues and that eigenvectors corresponding to distinct eigenvalues are orthogonal. [HINT: If A is a Hermitian matrix and a unit vector $z \in \mathbb{C}^n$ satisfies $Az = \lambda z$, then

$$\begin{aligned} \overline{\lambda} &= \overline{\lambda}(z, z) = (z, \lambda z) = (z, Az) = (\overline{A}^t z, z) \\ &= (Az, z) = (\lambda z, z) = \lambda(z, z) = \lambda\,.] \end{aligned}$$

6. Recall that a square matrix A is called **diagonalizable** if there exists an invertible $n \times n$ matrix R (called a **diagonalizing matrix** for A) such that the matrix $R^{-1}AR$ is diagonal. Establish the following.

(a) A matrix A is diagonalizable if and only if A has n linearly independent eigenvectors.

(b) Every Hermitian matrix is diagonalizable by a unitary matrix.

7. An $n \times n$ matrix A is said to be **positive semidefinite** if $(Az, z) \geq 0$ for all $z \in \mathbb{C}^n$. Establish the following.

(a) If A is a positive semidefinite matrix, then A has non-negative eigenvalues.

(b) A Hermitian matrix A is positive semidefinite if and only if it has non-negative eigenvalues.

[HINT: For (b) use the fact that a Hermitian matrix is diagonalizable by a unitary matrix.]

8. For each k let $A_k = [a_{ij}^{(k)}]$ be an $n \times n$ matrix. It is easy to see that the sequence of matrices $\{A_k\}$ converges to a matrix $A = [a_{ij}]$ in the Banach lattice $M_n(\mathbb{C})$ if and only if $\lim_{k\to\infty} a_{ij}^{(k)} = a_{ij}$ for each i and j.

Show that an $n \times n$ matrix A satisfies $A^k \to 0$ in $M_n(\mathbb{C})$ if and only if every eigenvalue of A has modulus less than one. [HINT: See Exercise 11 of Section 6.1.]

9. Consider a matrix $A \in M_n(\mathbb{C})$. If λ_1 is the (non-negative) largest eigenvalue of the Hermitian matrix $\overline{A}^t A$, then show that $\|A\| = \sqrt{\lambda_1}$. In particular, if A is Hermitian, then show that its Euclidean norm coincides with its spectral radius. [HINT: Let $x \in \mathbb{C}^n$ satisfy $\|x\| = 1$. From the identity $(x, \overline{A}^t Ax) = (Ax, Ax) = \|Ax\|^2 \geq 0$, we infer that the Hermitian matrix $\overline{A}^t A$ is also positive semidefinite and so it has non-negative eigenvalues, say $\lambda_1 \geq \lambda_2 \geq \cdots \geq \lambda_n \geq 0$. If U is a unitary matrix that diagonalizes $\overline{A}^t A$ and $x = Uy$, then $\|y\| = 1$ and

$$\|Ax\|^2 = (x, \overline{A}^t Ax) = (y, \overline{U}^t \overline{A}^t AUy) = \sum_{i=1}^n \lambda_i |y_i|^2 \leq \lambda_1 .$$

This implies $\|A\| = \sqrt{\lambda_1}$.]

10. Show that a square matrix is quasinilpotent if and only if it is nilpotent. (Compare this with Exercise 18 of Section 6.1.)

11. Let $A = [a_{ij}] \in M_n(\mathbb{C})$. If we let $a_{ij} = b_{ij} + \imath c_{ij}$ and consider the real matrices $B = [b_{ij}]$ and $C = [c_{ij}]$, then we can write $A = B + \imath C$ and view A as a vector in the complexification $M_n(\mathbb{R}) \oplus \imath M_n(\mathbb{R}) = M_n(\mathbb{C})$ of the real Banach lattice $M_n(\mathbb{R})$.

Show that the modulus of $A = B + \imath C$ in $M_n(\mathbb{R}) \oplus \imath M_n(\mathbb{R})$ is the positive matrix $|A| = [|a_{ij}|]$.

12. Prove the Cayley–Hamilton Theorem 8.3.

8.2. Operators on Finite Dimensional Spaces

In this section, we shall discuss the structure of operators on finite dimensional spaces and their connections with square matrices. Unless otherwise stated, X will denote a finite dimensional complex vector space of dimension n. We start our discussion by mentioning (without proofs) a few standard facts that will be used throughout this book regarding finite dimensional vector spaces and operators acting on them.

- *For an operator $T\colon X \to X$ on a finite dimensional vector space the following statements are equivalent.*
  (a) *T is invertible.*
  (b) *T is one-to-one.*
  (c) *T is surjective.*
- *If $\{e_1, \ldots, e_n\}$ is a basis of X, then $(\lambda_1, \lambda_2, \ldots, \lambda_n) \mapsto \sum_{i=1}^n \lambda_i e_i$ is a surjective linear isomorphism between $\mathbb{C}^n$ and X. Under this mapping,*

we can identify X with $\mathbb{C}^n$. Moreover, we can define the ***Euclidean norm*** *on X relative to the basis $\{e_1, \ldots, e_n\}$ by*

$$\|x\| = \Big\|\sum_{i=1}^{n} \lambda_i e_i\Big\| = \Big(\sum_{i=1}^{n} |\lambda_i|^2\Big)^{\frac{1}{2}}.$$

- *There exists a unique Hausdorff linear topology on X, called the* ***Euclidean topology*** *on X. In particular, all norms defined on X are equivalent.*
- *Every linear operator $T\colon X \to X$ is continuous.*
- *Let $T\colon X \to X$ be an operator, and fix a basis $\{e_1, \ldots, e_n\}$ in X. If for each j we write $Te_j = \sum_{i=1}^{n} a_{ij} e_i$ and consider the $n \times n$ matrix $A = [a_{ij}]$, i.e., the jth column of A is the column vector $\begin{bmatrix} a_{1j} \\ a_{2j} \\ \vdots \\ a_{nj} \end{bmatrix}$, then for each vector $x = \sum_{j=1}^{n} \lambda_j e_j = \begin{bmatrix} \lambda_1 \\ \vdots \\ \lambda_n \end{bmatrix} = \lambda$, we have*

$$\begin{aligned} Tx &= \sum_{j=1}^{n} \lambda_j Te_j = \sum_{j=1}^{n} \lambda_j \Big(\sum_{i=1}^{n} a_{ij} e_i\Big) = \sum_{i=1}^{n}\Big(\sum_{j=1}^{n} a_{ij}\lambda_j\Big) e_i \\ &= \Big(\sum_{j=1}^{n} a_{1j}\lambda_j, \sum_{j=1}^{n} a_{2j}\lambda_j, \ldots, \sum_{j=1}^{n} a_{nj}\lambda_j\Big) = A\lambda, \end{aligned}$$

  where $A\lambda$ is the usual matrix multiplication of the matrix A with the column vector λ. The matrix A is called the ***representing matrix*** *of T with respect to the basis $\{e_1, \ldots, e_n\}$.*

  In other words, for each fixed basis, the operator T can be identified with an $n \times n$ matrix. In particular, we can identify $\mathcal{L}(X)$ with $M_n(\mathbb{C})$. This means that once a basis of X has been chosen, then every $n \times n$ matrix defines—as above—a unique operator on X.
- *If we consider another basis $\{f_1, f_2, \ldots, f_n\}$ in X and let B be the representing matrix of the operator $T\colon X \to X$ relative to this new basis, then B is similar to A. That is, there exists an invertible matrix C such that $B = C^{-1}AC$. This implies that A and B have the same characteristic polynomial. Therefore, all matrices representing T have the same characteristic polynomial. This common characteristic polynomial is called the* ***characteristic polynomial*** *of the operator T and is denoted by $p_T(\lambda)$.*

  It should be clear that the characteristic polynomial of T has degree n, the dimension of the vector space X.
- *The spectrum of an operator $T\colon X \to X$ consists of eigenvalues. These eigenvalues are the roots of the characteristic polynomial of T. That is, $\sigma(T) = \{\lambda \in \mathbb{C}\colon\ p_T(\lambda) = 0\}$.*

  In particular, every operator on a finite dimensional complex vector space has an eigenvalue—and hence an eigenvector.

- *Recall that a pair (Y, Z) of (closed) subspaces of X is* ***reducing*** *for an operator $T: X \to X$, or that T is* ***reduced*** *by the pair (Y, Z), if Y and Z are both T-invariant and $X = Y \oplus Z$.*

  If (Y, Z) is a reducing pair of subspaces for an operator $T: X \to X$, then

$$p_T(\lambda) = p_{T|_Y}(\lambda) p_{T|_Z}(\lambda) .$$

- *If $T: X \to X$ is an operator on a finite dimensional vector space, then*

$$\dim(\mathrm{Ker}(T)) + \dim(\mathrm{R}(T)) = \dim(X) .$$

Having listed all these basic properties above, we now proceed to describe the structure of linear operators on finite dimensional vector spaces. We start with the following result.

Theorem 8.4. *If $T: X \to X$ is an operator on an n-dimensional complex vector space, then there exist n subspaces $V_1, V_2, \ldots, V_n$ of X such that:*

(a) $V_1 \subseteq V_2 \subseteq \cdots \subseteq V_n = X$.

(b) *Each V_i has dimension i.*

(c) *Each V_i is T-invariant, i.e., $T(V_i) \subseteq V_i$.*

Proof. The proof is by induction on the dimension n of the complex vector space X. For $n = 1$ the conclusion is obviously true. So, for the induction step, assume that our claim is true for some n and let $T: X \to X$ be an operator on an $(n+1)$-dimensional complex vector space. Then the topological dual X^* of X is also an $(n+1)$-dimensional complex vector space, and so its adjoint operator $T^*: X^* \to X^*$ has an eigenvector. That is, there exists some non-zero vector $x^* \in X^*$ and some complex number λ such that $T^*x^* = \lambda x^*$. Fix some vector $e \in X$ with $x^*(e) = 1$ and put $V = \mathrm{Ker}\, x^* = \{x \in X\colon\ x^*(x) = 0\}$. We claim that $V \oplus \mathbb{C}e = X$.

To see this, note first that if $x \in X$, then the vector $y = x - x^*(x)e$ belongs to V and so $x = y + x^*(x)e \in V + \mathbb{C}e$. So, $V + \mathbb{C}e = X$. On the other hand, if $x \in V \cap \mathbb{C}e$, then $x = \mu e$ and so $\mu = x^*(\mu e) = 0$. This implies $x = 0$, and therefore $V \oplus \mathbb{C}e = X$. Clearly, V is an n-dimensional complex vector space.

Next, we claim that V is T-invariant. To see this, let $v \in V$ and note that from

$$\langle Tv, x^* \rangle = \langle v, T^*x^* \rangle = \langle v, \lambda x^* \rangle = \lambda \langle v, x^* \rangle = \lambda x^*(v) = 0 ,$$

it follows that $Tv \in V$. So, by our induction hypothesis, there exist subspaces $V_1, V_2, \ldots, V_n$ of V that satisfy properties (a), (b), and (c) for the operator $T: V \to V$. It follows that the subspaces $V_1, V_2, \ldots, V_n, V_{n+1} = X$ satisfy properties (a), (b), and (c) for the operator $T: X \to X$, and the proof is finished. ■

The structure of operators on finite dimensional vector spaces is closely related to the structure of nilpotent operators. Recall that an operator $T\colon X \to X$ on a vector space is said to be **nilpotent** if $T^k = 0$ for some k. The natural number $q = \min\{k \in \mathbb{N}\colon\ T^k = 0\}$ is called the **index of nilpotence** of T.

The concepts of nilpotence and invertibility of operators on finite dimensional spaces are in a way "opposite" notions. This is immediate from our next result that is the finite dimensional version of Theorem 2.23. (Keep in mind that every operator on a finite dimensional vector space automatically has finite ascent and descent.)

Theorem 8.5. *If $T\colon X \to X$ is an operator on a finite dimensional vector space, then there exists a unique T-reducing pair (Y, Z) of vector subspaces such that $T\colon Y \to Y$ is nilpotent and $T\colon Z \to Z$ is invertible.*[2]

Therefore, in order to describe the structure of operators on finite dimensional vector spaces, we begin by studying the structure of nilpotent operators. We start with the following result.

Lemma 8.6. *Let $T\colon X \to X$ be a linear operator. If for some $q \geq 1$ a vector $u \in X$ satisfies $T^{q-1}u \neq 0$ and $T^q u = 0$, then the q vectors*

$$u, Tu, T^2u, \ldots, T^{q-1}u$$

are linearly independent. In particular, if X is finite dimensional and T is nilpotent with index q, then $q \leq \dim(X)$.

Proof. To see that the q vectors $u, Tu, T^2u, \ldots, T^{q-1}u$ are linearly independent, let

$$\alpha_1 u + \alpha_2 Tu + \alpha_3 T^2 u + \cdots + \alpha_q T^{q-1} u = 0\,. \tag{1}$$

Now suppose by way of contradiction that $\alpha_i \neq 0$ for some i, and consider the natural number $j = \min\{1 \leq i \leq q\colon\ \alpha_i \neq 0\}$. Then, applying the operator T^{q-j} to (**1**), we get $\alpha_j T^{q-1}u = 0$, which is a contradiction. Hence, $\alpha_i = 0$ for each i, and so the vectors $u, Tu, T^2u, \ldots, T^{q-1}u$ are linearly independent.

Now assume that X is finite dimensional and that T is nilpotent with index of nilpotence q. Hence, there exists at least one vector $u \in X$ such that $T^{q-1}u \neq 0$. This implies that $q \leq \dim(X)$. ■

The next result describes a key "reducibility" property of nilpotent operators.

---

[2] This is also expressed by saying that: *Every operator on a finite dimensional vector space is the direct sum of a nilpotent operator and an invertible operator.*

Theorem 8.7. *Let $T\colon X \to X$ be a nilpotent operator on a finite dimensional vector space with index of nilpotence q, and let a vector $u \in X$ satisfy $T^{q-1}u \neq 0$. If U is the T-invariant subspace generated by the linearly independent vectors $u, Tu, \ldots, T^{q-1}u$, then there exists a T-invariant complement W to U, i.e., $U \oplus W = X$ and $T(W) \subseteq W$.*

Proof. The proof is by induction on the index of nilpotence q. If T is nilpotent with index $q = 1$ (i.e., $T = 0$), then the conclusion should be obvious. So, for the induction step, assume that the claim is true for nilpotent operators of index $q \geq 1$.

Now let X be a finite dimensional vector space and let $T\colon X \to X$ be a nilpotent operator with index $q + 1$. Also, let u be a vector of X satisfying $T^q u \neq 0$, and let U be the $(q+1)$-dimensional T-invariant subspace generated by the vectors $u, Tu, \ldots, T^q u$. We denote by V the q-dimensional T-invariant subspace generated by the linearly independent vectors $Tu, T^2u, \ldots, T^q u$. Clearly,

$$T(U) = V\,.$$

The range $R(T)$ of T is a T-invariant subspace and the operator $T\colon R(T) \to R(T)$ is nilpotent with index of nilpotence q. So, applying our induction hypothesis to the subspace V of $R(T)$, we see that there exists a T-invariant vector subspace Y of $R(T)$ such that

$$V \oplus Y = R(T)\,. \tag{$\star$}$$

Now consider the vector subspace $Z = T^{-1}(Y)$ of X. Since $T(Y) \subseteq Y$, it follows that $Y \subseteq Z$ and that Z is T-invariant. We claim that

$$U + Z = X\,. \tag{$\star\star$}$$

To see this, let $x \in X$. Then, by $(\star)$, there exist some $u' \in U$ and some $y \in Y$ such that $Tx = Tu' + y$ or $T(x - u') = y \in Y$. Therefore, the vector $z = x - u'$ belongs to Z. This implies $x = u' + z \in U + Z$, and the validity of $(\star\star)$ has been established.

Next, we claim that $Y \cap (U \cap Z) = Y \cap U = \{0\}$. To see this, let $x \in Y \cap U$ and choose scalars β_i such that $x = \sum_{i=0}^{q} \beta_i T^i u$. Then $Tx \in Y$ and $Tx \in T(U) = V$, and so from $(\star)$ we infer that $Tx \in V \cap Y = \{0\}$, i.e., $Tx = \sum_{i=0}^{q} \beta_i T^{i+1} u = 0$. This implies $\beta_i = 0$ for all $0 \leq i \leq q-1$, and so $x = \beta_q T^q u \in V$. In turn, this yields $x \in Y \cap V = \{0\}$, i.e., $x = 0$.

Finally, consider the vector subspace $Y \oplus (U \cap Z)$ of Z and let Y_1 be an arbitrary complement of $Y \oplus (U \cap Z)$ in Z, that is,

$$Y \oplus (U \cap Z) \oplus Y_1 = Z\,.$$

To finish the proof, we shall show that if $W = Y \oplus Y_1$, then the pair of subspaces (U, W) reduces the operator T. To establish this, note first that

$x \in U \cap (Y \oplus Y_1)$ implies $x \in (U \cap Z) \cap (Y \oplus Y_1) = \{0\}$, i.e., $x = 0$. Next, from $U \oplus Y \oplus Y_1 \supseteq Y \oplus (U \cap Z) \oplus Y_1 = Z$, $U \oplus Y \oplus Y_1 \supseteq U$, and $(\star\star)$, it follows that $U \oplus Y \oplus Y_1 = X$. Moreover, $T(Z) \subseteq Y$ implies $T(Y \oplus Y_1) \subseteq Y \subseteq Y \oplus Y_1$. This shows that the pair (U, W) reduces the operator T, and the proof is finished. ■

An easy consequence of the preceding result is the following.

Corollary 8.8. *Let $T: X \to X$ be a nilpotent operator on a finite dimensional vector space. Then there exist natural numbers $q_1 \geq q_2 \geq \cdots \geq q_k$ and vectors $u_1, u_2, \ldots, u_k$ satisfying:*

(1) *$T^{q_i - 1} u_i \neq 0$ and $T^{q_i} u_i = 0$ for each i.*

(2) *The collection of vectors*

$$\begin{gathered} T^{q_1-1}u_1, T^{q_1-2}u_1, \ldots, Tu_1, u_1\,, \\ T^{q_2-1}u_2, T^{q_2-2}u_2, \ldots, Tu_2, u_2\,, \\ \vdots \\ T^{q_k-1}u_k, T^{q_k-2}u_k, \ldots, Tu_k, u_k \end{gathered}$$

is a basis of X.

In particular, the matrix A representing T with respect to this basis has the following form: Every entry of A outside the diagonal above the main diagonal of A is zero. On the diagonal above the main diagonal, we have (starting from the top) a string of $q_1 - 1$ ones followed by a zero if $k \neq 1$. Then there is a string of $q_2 - 1$ ones followed by a zero if $k \neq 2$. This pattern follows until we exhaust all the q_i.

And now we come to the basic structural property of operators on finite dimensional complex vector spaces.

Theorem 8.9. *Let $T: X \to X$ be an operator on a finite dimensional complex vector space. Assume that $\lambda_1, \ldots, \lambda_k$ are the distinct roots of the characteristic polynomial of T with multiplicities $m_1, \ldots, m_k$, respectively. Then there exist T-invariant subspaces $X_1, \ldots, X_k$ of X such that:*

(1) *$\dim(X_i) = m_i$ for each $i = 1, 2, \ldots, k$.*

(2) *$X = X_1 \oplus X_2 \oplus \cdots \oplus X_k$.*

(3) *The operator $\lambda_i - T: X_i \to X_i$ is nilpotent for each $i = 1, 2, \ldots, k$.*

Proof. According to Theorem 8.5, for each $i = 1, \ldots, k$ there exists a pair of subspaces (X_i, Y_i) that reduces the operator $T - \lambda_i$ such that $T - \lambda_i: X_i \to X_i$ is nilpotent and $T - \lambda_i: Y_i \to Y_i$ is invertible. Clearly, each pair (X_i, Y_i) also reduces T. Indeed, if $x \in X_i$, then $Tx = (T - \lambda_i)x + \lambda_i x \in X_i$, and similarly $y \in Y_i$ yields $Ty \in Y_i$. This implies that $p_T(\lambda) = p_{T|_{X_i}}(\lambda) p_{T|_{Y_i}}(\lambda)$.

We claim that $p_{T|_{Y_i}}(\lambda_i) \neq 0$. Otherwise, there would exist some non-zero vector $y \in Y_i$ such that $(T - \lambda_i)y = 0$, contradicting the invertibility of $T - \lambda_i \colon Y_i \to Y_i$. Now this conclusion, coupled with the identity $\sigma(T|_{X_i}) = \sigma((T - \lambda_i)|_{X_i}) + \lambda_i = \{\lambda_i\}$, implies that $p_{T|_{X_i}}(\lambda) = (\lambda - \lambda_i)^{m_i}$. This guarantees that X_i has dimension m_i.

Next, we claim that if $i \neq j$, then $X_i \cap X_j = \{0\}$. To see this, assume that for some $i \neq j$ we have $X_i \cap X_j \neq \{0\}$. Since $X_i \cap X_j$ is T-invariant, it follows that there exists some $\lambda \in \mathbb{C}$ and some non-zero $z \in X_i \cap X_j$ such that $Tz = \lambda z$. This implies $\lambda = \lambda_i = \lambda_j$, contrary to $\lambda_i \neq \lambda_j$. Hence, $X_i \cap X_j = \{0\}$ holds true for $i \neq j$. Finally, a simple dimensionality argument shows that $X_1 \oplus X_2 \oplus \cdots \oplus X_k = X$, and the proof is finished. ■

Now let A be an $n \times n$ matrix with distinct eigenvalues $\lambda_1, \ldots, \lambda_k$ counted with their multiplicities $m_1, \ldots, m_k$, respectively. Then Theorem 8.9 coupled with Corollary 8.8 informs us that A is similar to a matrix B with the following properties:

(a) The entries on the main diagonal of the matrix B are the eigenvalues of A repeated according to their multiplicities. Starting from the position $(1, 1)$, we have the eigenvalue λ_1 repeated m_1-times followed by the eigenvalue λ_2 repeated m_2-times, and so on.

(b) All entries of B which are not on the main diagonal or on the diagonal above the main one are zero.

(c) On the diagonal above the main diagonal the entries are only zeros and ones and have the following pattern: On the part above the string of the m_i-eigenvalues λ_i, the entries start with a zero followed by a string of $(m_i - 1)$-ones, and so on.

The matrix B is known as the **Jordan form** of the matrix A.

Exercises

1. Show that for an operator $T \colon X \to X$ on a finite dimensional vector space the following statements are equivalent.
   (a) T is invertible.
   (b) T is one-to-one.
   (c) T is surjective.

2. A square matrix A is said to be **upper triangularizable** if there exists an invertible matrix C (called an **upper triangularizing matrix** for A) such that $C^{-1}AC$ is an upper triangular matrix.

Show that every matrix A is upper triangularizable and that if C is an upper triangularizing matrix for A, then the diagonal entries of $C^{-1}AC$ consist of the eigenvalues of A (counted with their multiplicities). [HINT: Use Theorem 8.4.]

3. Let A be an $n \times n$ matrix having distinct eigenvalues $\lambda_1, \ldots, \lambda_k$ with multiplicities $m_1, \ldots, m_k$, respectively. If $q(\cdot)$ is a polynomial, then show that the characteristic polynomial of the $n \times n$ matrix $q(A)$ is

$$p_{q(A)}(\lambda) = [\lambda - q(\lambda_1)]^{m_1}[\lambda - q(\lambda_2)]^{m_2} \cdots [\lambda - q(\lambda_k)]^{m_k}.$$

[HINT: By Exercise 2 above, we know that every matrix is similar to an upper triangular matrix, and so we can suppose that A is itself an upper triangular matrix. In particular, the diagonal entries of A are the eigenvalues λ_i; each appearing m_i times—the multiplicity of the eigenvalue λ_i. It follows that $q(A)$ is also an upper triangular matrix having diagonal entries $q(\lambda_1), \ldots, q(\lambda_k)$ with each $q(\lambda_i)$ appearing m_i times.]

4. This exercise describes the *minimal polynomial* of an operator $T\colon X \to X$ on an n-dimensional complex Banach space.

(a) Show that the spectrum of T is finite and consists of eigenvalues. Moreover, if we let $\sigma(T) = \{\lambda_1, \ldots, \lambda_k\}$, then show that $k \le n$.

(b) Show that there exists a non-zero polynomial $p(\lambda)$ such that $p(T)=0$.

(c) Prove that some root of the polynomial $p(\lambda)$ in (b) must belong to the spectrum of T; and so the spectrum of T is non-empty.

(d) Assume that $\pi(\lambda)$ is a polynomial of minimal degree and leading coefficient one satisfying $\pi(T) = 0$. Show that

$$\pi(\lambda) = (\lambda - \lambda_1)^{\alpha_1} \cdots (\lambda - \lambda_k)^{\alpha_k},$$

where, as in (a), $\sigma(T) = \{\lambda_1, \ldots, \lambda_k\}$ and each α_i is the ascent of the operator $\lambda_i - T$. (This establishes that the polynomial $\pi(\lambda)$ is uniquely determined; it is referred to as the **minimal polynomial** of the operator T.)

(e) Show that a polynomial $q(\lambda)$ satisfies $q(T) = 0$ if and only if $q(\lambda)$ is divisible by the minimal polynomial $\pi(\lambda)$ of T. In particular, prove that two polynomials $\psi(\lambda)$ and $\phi(\lambda)$ satisfy $\psi(T) = \phi(T)$ if and only if there exists a polynomial $\theta(\lambda)$ such that $\phi(\lambda) = \psi(\lambda) + \theta(\lambda)\pi(\lambda)$.

[HINT: Let $\{x_1, \ldots, x_n\}$ be a basis of X. Since for each i the $n+1$ vectors $x_i, Tx_i, \ldots, T^n x_i$ must be linearly dependent, there exists a non-zero polynomial $p_i(\lambda)$ such that $p_i(T)x_i = 0$. If $p(\lambda) = p_1(\lambda)p_2(\lambda)\cdots p_n(\lambda)$, then $p(\lambda)$ is a non-zero polynomial satisfying $p(T) = 0$.]

5. Let $T\colon X \to X$ be an operator on a finite dimensional vector space. If a pair of subspaces (Y, Z) reduces T, then show that $p_T(\lambda) = p_{T|_Y}(\lambda)p_{T|_Z}(\lambda)$.

6. Let $T\colon X \to X$ be an operator on an n-dimensional vector space. If two $n \times n$ matrices A and B represent T with respect to two different bases of X, then show that A and B are similar.

7. If $S, T\colon X \to X$ are two operators on a finite dimensional vector space, then show that ST and TS have the same characteristic polynomial—and conclude from this that if A and B are two $n \times n$ square matrices, then AB and BA have the same characteristic polynomial. [HINT: Prove first the result for the special case when S is a projection, i.e, $S^2 = S$. For the general case show that there exists an invertible operator $R\colon X \to X$ such that RS is a projection, and then apply the preceding conclusion to the pair of operators RS and TR^{-1}.]

8. Let V be a finite dimensional vector space. If A is a non-empty subset of V, then its **annihilator** $A^\perp$ is the vector subspace of V^* defined by

$$A^\perp = \{v^* \in V^*\colon\ v^*(a) = 0 \text{ for all } a \in A\}.$$

Similarly, the **annihilator** of any non-empty subset B of V^* is the vector subspace of V defined by $B^\perp = \{v \in V\colon\ b^*(v) = 0 \text{ for all } b^* \in B\}$.

Assume that $T\colon X \to Y$ is an operator between two finite dimensional vector spaces. As usual, we define its adjoint operator $T^*\colon Y^* \to X^*$ via the duality identity

$$\langle Tx, y^*\rangle = \langle x, T^*y^*\rangle = y^*(Tx)$$

for all $x \in X$ and all $y^* \in Y^*$. Establish the following:

(a) $[R(T)]^\perp = \operatorname{Ker}(T^*)$.

(b) $[\operatorname{Ker}(T)]^\perp = R(T^*)$

(c) $\dim R(T^*) = \dim R(T)$.

9. If $T\colon X \to Y$ is an operator between finite dimensional vector spaces, then show that

$$\dim \operatorname{Ker}(T) + \dim \mathrm{R}(T) = \dim X.$$

10. Use Theorem 8.9 to present an alternate proof of the Cayley–Hamilton Theorem 8.3.

11. Show that every square matrix is similar to its transpose.

12. Let $T\colon Y \to Y$ be a bounded operator on a Banach space. A vector $y \in Y$ is said to be **cyclic** (or more precisely T-**cyclic**) if the vector subspace generated by $\{y, Ty, T^2y, T^3y, \dots\}$ is dense in the Banach space Y.

For an operator $T\colon X \to X$ on an n-dimensional vector space establish the following.

(a) A vector $x \in X$ is T-cyclic if and only if $\{x, Tx, \dots, T^{n-1}x\}$ is a basis of X.

(b) The operator T has a cyclic vector if and only if T has a matrix representation of the form:

$$\begin{bmatrix} 0 & 0 & 0 & \cdots & 0 & 0 & -a_0 \\ 1 & 0 & 0 & \cdots & 0 & 0 & -a_1 \\ 0 & 1 & 0 & \cdots & 0 & 0 & -a_2 \\ \vdots & \vdots & \vdots & \ddots & \vdots & \vdots & \vdots \\ 0 & 0 & 0 & \cdots & 0 & 0 & -a_{n-3} \\ 0 & 0 & 0 & \cdots & 1 & 0 & -a_{n-2} \\ 0 & 0 & 0 & \cdots & 0 & 1 & -a_{n-1} \end{bmatrix} \qquad (\star)$$

(c) If x is a T-cyclic vector and $T^n x = -a_0x - a_1Tx - \cdots - a_{n-1}T^{n-1}x$, then the characteristic polynomial of T is given by

$$p_T(\lambda) = \lambda^n + a_{n-1}\lambda^{n-1} + a_{n-2}\lambda^{n-2} + \cdots + a_1\lambda + a_0.$$

8.3. Matrices with Non-negative Entries

Here we shall take a look at square matrices with non-negative entries. We start by classifying various types of such matrices.

Definition 8.10. *A square $n \times n$ matrix $A = [a_{ij}]$ is said to be:*

(1) ***positive*** *(**non-negative**) if $a_{ij} \geq 0$ for all i and j (equivalently, if $Ax \geq 0$ for each $x > 0$),*

(2) ***strictly positive*** *if $a_{ij} \geq 0$ for all i and j and every column has a positive entry (equivalently, if $Ax > 0$ for each $x > 0$), and*

(3) ***strongly positive*** *if $a_{ij} > 0$ for all i and j (equivalently, if $Ax \gg 0$ for each $x > 0$).*

It should be clear that a strongly positive matrix is strictly positive and that a strictly positive matrix is positive. It should also be noticed that our terminology regarding matrices is in complete agreement with our terminology regarding operators introduced in Definition 1.13. That is, a matrix A is positive if and only if A, as an operator on $\mathbb{R}^n$ (or $\mathbb{C}^n$), is a positive operator—in which case we write $A \geq 0$ or $0 \leq A$. The operator notation $B \leq A$ (or $A \geq B$) for matrices with real entries is equivalent, in this case, to $b_{ij} \leq a_{ij}$ for all i and j. Similarly, a matrix A is a strictly positive matrix if and only if A, as an operator on $\mathbb{R}^n$ (or $\mathbb{C}^n$), is a strictly positive operator. It is easy to see that every $n \times n$ matrix $A = [a_{ij}]$ has a modulus which is given by the matrix $|A| = [|a_{ij}|]$.

Theorem 8.11. *The spectral radius of a positive matrix is an eigenvalue having a positive eigenvector.*

Proof. Let A be a positive matrix. If $r(A) > 0$, then the conclusion follows immediately from the Krein–Rutman Theorem 7.10. However, in the finite dimensional case, the proof of the Krein–Rutman theorem can be used to cover the case $r(A) = 0$ as well. Indeed, in the finite dimensional case one has to notice that the sequence of positive unit vectors $\{x_n\}$ always has a convergent subsequence to some positive vector. This positive vector is an eigenvector corresponding to the spectral radius of the matrix. ∎

In spite of the fact that the proof of Theorem 8.11 mimics that of the Krein–Rutman Theorem 7.10, there is a difference between these two results that is worth pointing out. In the Krein–Rutman theorem, we assume that the spectral radius of the compact operator is strictly positive, while in Theorem 8.11 we do not exclude that the spectral radius can be zero. In Theorem 8.11 the conclusion about the existence of a positive eigenvector remains true even if the spectral radius is zero. We refer to Exercise 10 at the end of this section for an independent proof of this fact.

Definition 8.12. *With each non-negative $n \times n$ matrix A, we associate the following two non-empty sets*

$$\begin{aligned} \mathcal{L}_A &= \{\lambda \geq 0\colon\ \exists\ x > 0 \ \textit{with}\ Ax \geq \lambda x\} \\ &= \{\lambda \geq 0\colon\ \exists\ x \in \boldsymbol{S}^+ \ \textit{with}\ Ax \geq \lambda x\} \end{aligned}$$

and

$$\begin{aligned} \mathcal{U}_A &= \{\lambda \geq 0\colon\ \exists\ x \gg 0 \ \textit{with}\ \lambda x \gg Ax\} \\ &= \{\lambda \geq 0\colon\ \exists\ 0 \ll x \in \boldsymbol{S}^+ \ \textit{with}\ \lambda x \gg Ax\}\,. \end{aligned}$$

Notice that if $0 \leq \lambda_1 \leq \lambda$ and $\lambda \in \mathcal{L}_A$, then $\lambda_1 \in \mathcal{L}_A$ and, similarly, $\lambda_1 \geq \lambda$ and $\lambda \in \mathcal{U}_A$ imply $\lambda_1 \in \mathcal{U}_A$. Notice also that if $\lambda \geq 0$ and $x > 0$ satisfy $Ax \geq \lambda x$, then

$$\lambda\|x\| = \|\lambda x\| \leq \|Ax\| \leq \|A\| \cdot \|x\|\,,$$

and so $\lambda \leq \|A\|$. That is, $\mathcal{L}_A \subseteq \big[0, \|A\|\big]$. Since $0 \in \mathcal{L}_A$, we see that $\mathcal{L}_A$ is a non-empty interval in $\mathbb{R}$ that is bounded from above. The set $\mathcal{L}_A$ is also closed. To see this, assume that a sequence $\{\lambda_n\} \subseteq \mathcal{L}_A$ satisfies $\lambda_n \to \lambda$. For each n pick some $x_n \in \mathbf{S}^+$ with $Ax_n \geq \lambda_n x_n$. Since $\mathbf{S}^+$ is a compact set, by passing to a subsequence (if necessary), we can assume that $x_n \to x \in \mathbf{S}^+$. Hence, $Ax = \lim_{n\to\infty} Ax_n \geq \lim_{n\to\infty} \lambda_n x_n = \lambda x$. This shows $\lambda \in \mathcal{L}_A$ and so $\mathcal{L}_A$ is closed. Consequently, $\mathcal{L}_A = [0, \lambda_A]$, where

$$\lambda_A = \sup \mathcal{L}_A = \max \mathcal{L}_A\,.$$

The set $\mathcal{U}_A$ will be described in Corollary 8.23. The properties of the non-negative number λ_A will be discussed in detail below. The first property follows immediately from the definition of λ_A.

Lemma 8.13. *If two $n \times n$ positive matrices A and B satisfy $A \leq B$, then $\lambda_A \leq \lambda_B$.*

More properties of λ_A are included in the next result.

Lemma 8.14. *For an $n \times n$ positive matrix A we have the following:*

(1) $\lambda_A = r(A)$, *the spectral radius of A.*

(2) $\lambda_A = \lambda_{A^t}$.

(3) *If A is strictly positive, then $\lambda_A > 0$.*

(4) *λ_A is the only possible eigenvalue of A having a strictly positive eigenvector.*

Proof. (1) By Theorem 8.11 there exists some $x > 0$ with $Ax = r(A)x$ and so $r(A) \leq \lambda_A$. On the other hand, if $x \in \mathbf{S}^+$ and $\lambda \geq 0$ satisfy $\lambda x \leq Ax$, then $\lambda^k = \|\lambda^k x\| \leq \|A^k x\| \leq \|A^k\|$ for each k. Consequently, $\lambda \leq \|A^k\|^{\frac{1}{k}}$

for each k and so $\lambda \leq \lim_{k\to\infty} \|A^k\|^{\frac{1}{k}} = r(A)$. This implies $\lambda_A \leq r(A)$, and therefore $\lambda_A = r(A)$.

(2) This follows from (1) and the fact that A and A^t have the same spectrum.

(3) Choose some $x > 0$ with $Ax = \lambda_A x$. Since A is strictly positive, we have $Ax > 0$ and this implies $\lambda_A > 0$.

(4) Assume that $Ax = \alpha x$ holds for some $\alpha \in \mathbb{C}$ and some $x \gg 0$. From (1) and (2) we know that there is some $y > 0$ such that $A^t y = r(A)y$. Since $x \gg 0$, we see that $(x, y) = \sum_{i=1}^n x_i y_i > 0$. Now note that

$$\begin{aligned} \big[\alpha - r(A)\big](x, y) &= \alpha(x, y) - r(A)(x, y) = (\alpha x, y) - (x, r(A)y) \\ &= (Ax, y) - (x, A^t y) = (Ax, y) - (Ax, y) = 0\,, \end{aligned}$$

from which it follows that $\alpha = r(A)$. ■

The following result (whose proof is left as an exercise) characterizes the matrices in $M_n(\mathbb{R})$ that are lattice homomorphisms or interval preserving operators on $\mathbb{R}^n$.

Theorem 8.15. *Let A be an $n \times n$ non-negative matrix. Then:*

(1) *A defines a lattice homomorphism on $\mathbb{R}^n$ if and only if each row has at most one non-zero entry.*

(2) *A is interval preserving if and only if every column of A has at most one non-zero entry.*

Exercises

1. Let $A = [a_{ij}]$ be an $n \times n$ matrix with real entries. Show that A (as a regular operator on $\mathbb{R}^n$) has a modulus and that $|A| = [|a_{ij}|]$.
2. Show (by means of an example) that although the spectral radius of a (non-zero) non-negative matrix is an eigenvalue, it need not have a strictly positive eigenvector.
3. Assume that a non-zero positive matrix A commutes with a strongly positive matrix. Show that if $\lambda_A > 0$ and $Ax \geq \lambda_A x$ for some $x > 0$, then $Ax = \lambda_A x$. [HINT: Argue as in the proof of Theorem 8.22.]
4. (Frobenius [**130**]) Let $A = [a_{ij}]$ be an $n \times n$ positive matrix. For each i let $r_i = \sum_{j=1}^n a_{ij}$ (the sum of all elements in row i) and put $\rho = \min_i r_i$ and $R = \max_i r_i$. Show that $\rho \leq r(A) \leq R$. [HINT: Pick some $x > 0$ such that $Ax = r(A)x$. From $r(A)x = Ax \leq Rx$, we get $r(A) \leq R$. On the other hand, if $e = (1, 1, \ldots, 1)$, then $\rho e \leq Ae$.]

5. (Gershgorin [**134**]) This exercise presents a simple but very useful "approximation" of the spectrum of a square matrix. Let $A = [a_{ij}] \in M_n(\mathbb{C})$ and for each i let $r_i = \sum_{j\neq i}^n |a_{ij}|$ and $C_i = \{z \in \mathbb{C}\colon\ |z - a_{ii}| \leq r_i\}$. Show that $\sigma(A) \subseteq \bigcup_{i=1}^n C_i$. [HINT: Let λ be an eigenvalue of the matrix A and pick $x = (x_1, \ldots, x_n) \neq 0$ satisfying $Ax = \lambda x$. Choose some k such that $|x_k| = \max_{1\leq i\leq n} |x_i| > 0$. Now use the equation $(\lambda - a_{kk})x_k = \sum_{j\neq k} a_{kj}x_j$ to get $|\lambda - a_{kk}| \leq \sum_{j\neq k} |a_{kj}||\frac{x_j}{x_k}| \leq r_k$.]

6. If a matrix $A = [a_{ij}] \in M_n(\mathbb{C})$ satisfies $|a_{ii}| > \sum_{j\neq i} |a_{ij}|$ for each i, then show that A is invertible. [HINT: Use the preceding exercise.]

7. A non-negative $n \times n$ matrix $A = [a_{ij}]$ is said to be a **Markov** matrix whenever the sum of the elements in each row equals one, i.e., whenever $\sum_{j=1}^n a_{ij} = 1$ for each i. (This is equivalent, of course, to saying that A defines a Markov operator on $\mathbb{R}^n$.) Show in four different ways that every Markov matrix A satisfies $r(A) = 1$ by using:
(a) Exercise 4 above,
(b) Lemma 8.14(4),
(c) Gershgorin's theorem (see Exercise 5 above), and
(d) the spectral radius formula.
[HINT: If $x = (1, 1, \ldots, 1)$, then $Ax = x$, so that 1 is an eigenvalue of A. By Lemma 8.14(4), we see that $r(A) = 1$. Gershgorin's theorem also informs us that if λ is an eigenvalue of A, then for some i we have

$$|\lambda| - a_{ii} \leq |\lambda - a_{ii}| \leq \sum_{j\neq i} a_{ij} = 1 - a_{ii}\,.$$

Consequently, $|\lambda| \leq (1 - a_{ii}) + a_{ii} = 1$, which shows that the spectrum of A lies entirely inside the closed unit disk.]

8. Prove Theorem 8.15.

9. If an $n \times n$ positive matrix $A = [a_{ij}]$ satisfies $\sigma(A) = \{1\}$, then show that $a_{ii} = 1$ for all i. Use this and Theorem 8.15 to show that if a positive matrix A with $\sigma(A) = \{1\}$ defines either a lattice homomorphism or an interval preserving operator, then $A = I$.

10. Let A be an $n \times n$ positive matrix. If A is nilpotent (i.e., $r(A) = 0$), then show that there exists some vector $x > 0$ such that $Ax = 0$.

8.4. Irreducible Matrices

It is easy to see that a vector subspace J of $\mathbb{C}^n$ (or $\mathbb{R}^n$) is an ideal if and only if it is a band—and also if and only if it is of the form

$$J = \big\{(z_1, \ldots, z_n) \in \mathbb{C}^n\colon\ z_i = 0 \ \text{ for all } \ i \in \mathcal{I}\big\}\,,$$

for some subset $\mathcal{I}$ of $\{1, 2, \ldots, n\}$. We are now ready to define the notion of ideal irreducibility for matrices. In Chapter 9, this definition will be extended to arbitrary operators; see Definition 9.2.

Definition 8.16. *An $n \times n$ matrix A is said to be* ***reducible*** *whenever A leaves invariant a non-trivial ideal in $\mathbb{C}^n$. That is, A is reducible if there exists a vector subspace J of $\mathbb{C}^n$ of the form*

$$J = \{(z_1, \ldots, z_n) \in \mathbb{C}^n \colon \ z_i = 0 \ \text{ for all } \ i \in \mathcal{I}\},$$

where $\mathcal{I}$ is a non-empty proper subset of $\{1, 2, \ldots, n\}$, such that $A(J) \subseteq J$. A square matrix that is not reducible is called an ***irreducible matrix.***[3]

Here is a simple but very useful property of irreducible matrices.

Lemma 8.17. *Irreducible matrices have no zero rows or columns. In particular, every irreducible matrix carries strictly positive vectors to strictly positive vectors.*

Proof. Let A be an irreducible matrix. If A has a zero row, say the ith, then for each $x \in \mathbb{C}^n$ we have $(Ax)_i = 0$, i.e., the ith component of Ax is zero. This implies that the proper ideal $J = \{x \in \mathbb{C}^n \colon \ x_i = 0\}$ is A-invariant, a contradiction.

On the other hand, if A has a zero column, say the jth, and we consider the proper ideal $J = \{x \in \mathbb{C}^n \colon \ x_i = 0 \ \text{for all} \ i \neq j\}$, then $A(J) = \{0\} \subseteq J$, which is also impossible. ■

If $A = [a_{ij}]$ is an $n \times n$ matrix, then for each k we shall write $A^k = [a_{ij}^{(k)}]$, i.e., $a_{ij}^{(k)}$ denotes the $(i,j)^{\text{th}}$ element of the matrix A^k. Also, let us recall that an $n \times n$ matrix is said to be a **permutation matrix** if its rows consist of a permutation of the basic unit row vectors. That is, a permutation matrix is a matrix which can be obtained by permuting the rows (or, equivalently, the columns) of the identity matrix. Three important properties that are satisfied by any $n \times n$ permutation matrix P are the following.

(1) $P^tP = PP^t = I$, that is, $P^{-1} = P^t$.

(2) If A is an $n \times m$ matrix, then PA is obtained from A by permuting the rows of A in the same manner we permuted the rows of the identity matrix to get P. For instance, if P is the 3×3 permutation matrix with rows e_2, e_3, e_1 and A has rows r_1, r_2, r_3, then PA is the matrix with rows r_2, r_3, r_1.

(3) If A is an $m \times n$ matrix, then AP is obtained from A by permuting the columns of A in the same manner we permuted the columns of the identity matrix to get P.

We are now ready to characterize the irreducible positive matrices.

---

[3] Many authors refer to an irreducible matrix as an **indecomposable matrix**.

Theorem 8.18. *For an $n \times n$ positive matrix A the following statements are equivalent.*

1. *The matrix A is irreducible.*
2. *The matrix $I + A$ is irreducible.*
3. *The matrix $(I + A)^{n-1}$ is strongly positive.*
4. *For each position (i, j) there exists some integer k such that the $(i, j)^{\text{th}}$ entry $a_{ij}^{(k)}$ of A^k is positive, i.e., $a_{ij}^{(k)} > 0$.*
5. *There is no permutation matrix P such that*
$$P^t AP = \begin{bmatrix} B & C \\ 0 & D \end{bmatrix},$$
where B and D are square matrices.

Proof. (1) $\iff$ (2) This follows from the fact that for a vector subspace V of $\mathbb{C}^n$ we have $A(V) \subseteq V$ if and only if $(I + A)(V) \subseteq V$.

(1) $\Longrightarrow$ (3) We claim that:

- *If A is irreducible and a positive vector x has $1 \leq k \leq n - 1$ positive components, then the positive vector $(I + A)x$ has at least $k + 1$ positive components.*

If the claim is true, then from the identity
$$(I + A)^{n-1}x = \underbrace{(I + A)(I + A) \cdots (I + A)}_{(n-1)-\text{factors}} x$$
and an easy inductive argument, we easily see that $(I + A)^{n-1}x \gg 0$ for each $x > 0$, and this guarantees that $(I + A)^{n-1}$ is strongly positive.

To establish the above claim, fix an arbitrary $x > 0$ and assume that the set $\mathcal{I} = \{t \in \{1, \ldots, n\}\colon\ x_t > 0\}$ has k elements, where $1 \leq k \leq n - 1$. From $(I + A)x = x + Ax$, we see that $[(I + A)x]_t \geq x_t > 0$ for each $t \in \mathcal{I}$. To see that $[(I + A)x]_t = x_t + (Ax)_t > 0$ for some $t \notin \mathcal{I}$, we must show that $(Ax)_t > 0$ for some $t \notin \mathcal{I}$. To this end, assume by way of contradiction that $(Ax)_t = 0$ for all $t \notin \mathcal{I}$ and consider the non-trivial ideal
$$J = \{z \in \mathbb{C}^n\colon\ z_i = 0 \ \text{ for all } \ i \notin \mathcal{I}\}.$$
If $z \in J$, then clearly $|z| \leq \lambda x$ holds for some $\lambda > 0$, and therefore from $|Az| \leq A|z| \leq \lambda Ax$ it follows that $(Az)_t = 0$ for all $t \notin \mathcal{I}$, i.e., $Az \in J$. In other words, $A(J) \subseteq J$, contrary to the irreducibility of A, and the validity of our claim follows.

(3) $\Longrightarrow$ (4) Assume $(I + A)^{n-1}$ is strongly positive. First, we observe that A cannot have a zero row. For if a row of A is zero, say the ith row, then the ith row of $I + A$ is equal to the basic unit vector e_i. But then $I + A$

carries each vector whose i^{th} component equals zero to a vector whose i^{th} component is also zero. Consequently, $(I+A)^{n-1}$ carries vectors with i^{th} component equal to zero to vectors with i^{th} component also zero. However, the latter conclusion contradicts the strong positivity of $(I+A)^{n-1}$.

Since no row of A is zero, it follows that $A(I+A)^{n-1}$ is strongly positive. Hence, from

$$A(I+A)^{n-1} = A\Big[\sum_{k=0}^{n-1}\tbinom{n-1}{k}A^k\Big] = \sum_{k=0}^{n-1}\tbinom{n-1}{k}A^{k+1},$$

we see that for each (i,j) there exists some integer k (in fact $1 \le k \le n$) such that $a_{ij}^{(k)} > 0$.

(4) $\Longrightarrow$ (5) Assume by way of contradiction that there exists some permutation matrix P such that $P^tAP = \begin{bmatrix} B & C \\ 0 & D \end{bmatrix}$, where B and D are square matrices. Then for each k we have

$$A^k = P\begin{bmatrix} B_k & C_k \\ 0 & D_k \end{bmatrix}P^t,$$

where B_k and D_k are also square matrices having the same sizes as B and D, respectively. From the above representation of A^k, we infer that there is at least one position (i,j) such that the $(i,j)^{\text{th}}$ entry of A^k is zero for each k. However, the latter contradicts our hypothesis and the validity of (5) is established.

(5) $\Longrightarrow$ (1) Assume by way of contradiction that A leaves invariant a non-trivial ideal $J = \{z \in \mathbb{C}^n \colon\ z_i = 0 \text{ for all } i \in \mathcal{I}\}$. Let $\mathcal{I} = \{i_1, \dots, i_k\}$, where $1 \le i_1 < \cdots < i_k \le n$ and $1 \le k \le n-1$. The inclusion $A(J) \subseteq J$ means that if some $z \in \mathbb{C}^n$ satisfies $z_{i_j} = 0$ for each $j = 1, \dots, k$, then $(Az)_{i_j} = 0$ also holds for each $j = 1, \dots, k$.

Let Q be a permutation matrix whose last k rows are the (row) unit vectors $e_{i_1}, \dots, e_{i_k}$. Then for each $z \in J$ the column vector Qz has its last k components all equal to zero, i.e., $(Qz)_i = 0$ for each $i = n-k+1, \dots, n$. In particular, for each $z \in J$, the column vector QAz also has its last k components all equal to zero.

Now let us look at the permutation matrix $P = Q^t$, the transpose of Q. Clearly, the last k columns of Q^t are the column unit vectors $e_{i_1}, \dots, e_{i_k}$. In particular, the components of the first $n-k$ column vectors of Q^t are all zero at the coordinates $i_1, \dots, i_k$. So, if the columns of AQ^t are $c_1, \dots, c_{n-k}, \dots, c_n$, then the column vectors $c_1, \dots, c_{n-k}$ all have zero components at the coordinates $i_1, \dots, i_k$. But then for each $i = 1, \dots, n-k$ the column vector Pc_i must have (by the choice of Q) its last k components equal to zero. In other words, the matrix $P^tAP = QAQ^t$ is of the form

$$P^tAP = \begin{bmatrix} B & C \\ 0 & D \end{bmatrix},$$

where 0 is the $k \times (n-k)$-zero matrix, B is an $(n-k) \times (n-k)$ square matrix and D is a $k \times k$ square matrix. To complete the proof, notice that the last conclusion contradicts our assumption. ■

Corollary 8.19. *Every strongly positive matrix is irreducible.*

An irreducible positive matrix need not be strongly positive. For instance, it is easy to verify that the non-negative matrix $A = \begin{bmatrix} 0 & 1 \\ 1 & 0 \end{bmatrix}$ is irreducible but fails to be strongly positive. The same matrix also demonstrates two other properties:

- *A power of an irreducible matrix need not be an irreducible matrix.*
- *For a given irreducible positive matrix A there may be no power A^m which is strongly positive.*

Corollary 8.20. *Every irreducible positive matrix commutes with some strongly positive matrix.*

Proof. If A is a irreducible positive matrix, then A commutes with the strongly positive matrix $(I + A)^{n-1}$. ■

Corollary 8.21. *Every irreducible positive matrix A is strictly positive; and hence its spectral radius is positive, i.e., $r(A) = \lambda_A > 0$.*

Proof. Let A be a positive irreducible matrix and let $x > 0$. If $Ax = 0$, then $A^kx = 0$ for each k contradicting statement (4) of Theorem 8.18. ■

More interesting spectral properties of positive irreducible matrices are included in the next result.

Theorem 8.22. *Assume that A is a positive irreducible matrix. If some vector $x > 0$ satisfies $Ax \geq \lambda_A x$, then:*

(1) $Ax = \lambda_A x$.

(2) *The positive vector x is strictly positive, i.e., $x \gg 0$.*

(3) *The eigenspace $N_{\lambda_A} = \{z \in \mathbb{C}^n\colon\ Az = \lambda_A z\}$ is one dimensional and is generated by the strictly positive vector x.*

Proof. (1) and (2) By Corollary 8.20, there exists a strongly positive matrix B such that $AB = BA$. Clearly, this matrix AB is also strongly positive. Indeed, if $z > 0$, then $Bz \gg 0$, and (since A has no zero rows) $ABz \gg 0$.

Now let $Ax \geq \lambda_A x$ for some $x > 0$ and assume by way of contradiction that $Ax > \lambda_A x$. By Corollary 8.21 we have $\lambda_A > 0$. Now consider the vector $y = \frac{1}{\lambda_A}(Ax - \lambda_A x)$ and note that $y > 0$ and $Ax = \lambda_A(x + y)$. Hence, $ABy \gg 0$ and $B(x + y) \gg 0$. It follows that

$$AB(x + y) = ABx + ABy \gg ABx = BAx = \lambda_A B(x + y)\,.$$

This implies $AB(x + y) \geq (\lambda_A + \epsilon)B(x + y)$ for some $\epsilon > 0$, proving that $\lambda_A + \epsilon \in \mathcal{L}_A$ contrary to the definition of λ_A. Thus, $Ax = \lambda_A x$.

To see that $x = (x_1, \ldots, x_n) \gg 0$, notice that $A^k x = \lambda_A^k x$ holds for each k. Next, fix some j with $x_j > 0$. Then, if i is given, pick some k with $a_{ij}^{(k)} > 0$. It follows that $\lambda_A^k x_i = (A^k x)_i \geq a_{ij}^{(k)} x_j > 0$, and so $x_i > 0$ for each i, i.e., $x \gg 0$.

(3) Assume first that some non-zero vector $z \in \mathbb{C}^n$ satisfies $Az = \lambda_A z$. Then $|z| > 0$ and $\lambda_A|z| = |\lambda_A z| = |Az| \leq A|z|$. By (1) and (2), we have $|z| \gg 0$, and so each component of z is non-zero. Now consider the vector $y = x_1 z - z_1 x$. Clearly, $Ay = \lambda_A y$ and since the first component of y is zero, it follows from the above discussion that $y = 0$. This implies $z = \frac{z_1}{x_1}x$. ■

Corollary 8.23. *Every positive irreducible matrix A satisfies the following properties.*

(1) *The set $\mathcal{U}_A = \{\lambda > 0\colon\ \exists\ x \gg 0\ \text{with}\ \lambda x \gg Ax\}$ coincides with the open interval (λ_A, ∞), i.e., $\mathcal{U}_A = (\lambda_A, \infty)$.*

(2) *If an eigenvalue λ_0 of A has an eigenvector $y > 0$, then $\lambda_0 = \lambda_A$ and $y \gg 0$.*

(3) *For a matrix B satisfying $0 \leq B \leq A$ we have $\lambda_B = \lambda_A$ if and only if $B = A$ (and so if B is a non-negative matrix satisfying $B < A$, then $\lambda_B < \lambda_A$).*

Proof. (1) By Theorem 8.22, there exists $x \gg 0$ such that $Ax = \lambda_A x$. This implies $\mu x \gg Ax$ for each $\mu > \lambda_A$ and so $(\lambda_A, \infty) \subseteq \mathcal{U}_A$.

Conversely, assume by way of contradiction that some $\lambda \in \mathcal{U}_A$ satisfies $\lambda \leq \lambda_A$. Pick some $y \gg 0$ with $\lambda y \gg Ay$. Also, by Theorem 8.11 and Theorem 8.22, there exists some $z \gg 0$ such that $A^t z = \lambda_A z$. Now note that

$$\begin{aligned}(Ay, A^t z) &= (Ay, \lambda_A z) = \lambda_A (Ay, z) = \lambda_A (y, A^t z) \\ &\geq \lambda (y, A^t z) = (\lambda y, A^t z) \\ &> (Ay, A^t z)\,,\end{aligned}$$

which is impossible. Hence, $\lambda > \lambda_A$ and so $\mathcal{U}_A = (\lambda_A, \infty)$.

(2) Assume that $Ay = \lambda_0 y$ for some $y > 0$. By Corollary 8.20 there exists a strongly positive matrix B such that $AB = BA$. Consequently, we have $ABy = BAy = \lambda_0 By$ and the vector $z = By$ is strictly positive. This, coupled with the equality $Az = \lambda_0 z$, implies that $\lambda_0 > 0$. On the other hand, it follows that $Az \ll (\lambda_0 + \epsilon)z$ for each $\epsilon > 0$. Hence, $\lambda_0 + \epsilon \in \mathcal{U}_A$ and so (by the previous part) $\lambda_A \leq \lambda_0 + \epsilon$ for all $\epsilon > 0$. This implies $\lambda_A \leq \lambda_0$. Since the inequality $\lambda_0 \leq r(A) = \lambda_A$ is always true, we conclude that $\lambda_0 = \lambda_A$. Now invoke Theorem 8.22(2) to get $y \gg 0$.

(3) Assume $0 \leq B \leq A$. If $B = A$, then clearly $\lambda_B = \lambda_A$. For the converse, assume $\lambda_B = \lambda_A$. Since B is a non-negative matrix, there exists some $x > 0$ such that $\lambda_A x = \lambda_B x = Bx \leq Ax$. But then Theorem 8.22(2) implies $x \gg 0$ and $\lambda_A x = Ax$. Hence $(A - B)x = 0$, and using that $A - B \geq 0$ and $x \gg 0$, we easily get $A = B$. ■

Exercises

1. For a vector subspace J of $\mathbb{C}^n$ show that the following statements are equivalent.
   (a) J is an ideal, i.e., $J = J_0 \oplus \imath J_0$, where J_0 is an ideal in $\mathbb{R}^n$.
   (b) J is a band, i.e., $J = B \oplus \imath B$, where B is a band in $\mathbb{R}^n$.
   (c) There exists some non-empty subset $\mathcal{I}$ of $\{1, 2, \ldots, n\}$ such that

$$J = \{(z_1, \ldots, z_n) \in \mathbb{C}^n\colon\ z_i = 0 \ \text{ for all } \ i \in \mathcal{I}\}\,.$$

2. Show that an $n \times n$ positive matrix A is irreducible if and only if A^t is irreducible. [HINT: Use the identity $(I + A^t)^{n-1} = [(I + A)^{n-1}]^t$ in conjunction with Theorem 8.18 .]

3. For positive $n \times n$ matrices A and B establish the following.
   (a) If A^k is irreducible for some $k \geq 1$, then A is irreducible.
   (b) If A is irreducible, then $A + B$ is irreducible.
   (c) If A and B are irreducible, then AB and BA need not be irreducible.

4. Show that a non-negative matrix A is irreducible if and only if the matrix $B = I + A$ defines a Krein operator, that is, for each $x > 0$ we have $B^k x \gg 0$ for some k.

5. Use the preceding exercise and Theorem 9.42 to present an alternate proof of the fact that the spectral radius of a positive irreducible matrix has a unique (up to scalar multiples) strictly positive eigenvector.

6. Let $A = [a_{ij}]$ be a positive irreducible matrix and let $A_t = [a_{ij}(t)]$ be a positive matrix-valued function defined for each t in some subinterval (a, b) of $\mathbb{R}$ such that $\lim_{t \to t_0} a_{ij}(t) = a_{ij}$ for each (i, j) and some $t_0 \in (a, b)$. Show that $\lim_{t \to t_0} r(A_t) = r(A)$. [HINT: Fix a sequence $\{t_n\} \subseteq (a, b)$ satisfying $t_n \to t_0$. By Theorem 8.11 for each n there exists a positive unit vector x_n with $A_{t_n} x_n = r(A_{t_n}) x_n$. By passing to a subsequence, we can assume that $x_n \to x > 0$. It follows that

$$\lim_{n \to \infty} r(A_{t_n}) x_n = \lim_{n \to \infty} A_{t_n} x_n = Ax > 0 ,$$

where the last conclusion holds true by virtue of the strict positivity of A (Corollary 8.21). This yields $\lim_{n \to \infty} r(A_{t_n}) = \lambda_0$ and so $Ax = \lambda_0 x$. By Corollary 8.23, $\lambda_0 = \lambda_A = r(A)$ and this implies $\lim_{t \to t_0} r(A_t) = r(A)$.]

8.5. The Perron–Frobenius Theorem

We start by characterizing the eigenvalues in the peripheral spectrum of a positive irreducible matrix.

Lemma 8.24. *Let A be a positive irreducible $n \times n$ matrix with spectral radius $r = r(A)$. For an angle ϕ the following statements are equivalent.*

(1) *The complex number $re^{\imath\phi}$ is an eigenvalue of A.*

(2) *There exists a diagonal matrix D such that $|D| = I$ and*

$$A = e^{\imath\phi} DAD^{-1} .$$

Proof. (1) $\Longrightarrow$ (2) Assume that $re^{\imath\phi}$ is an eigenvalue of A. Therefore, there exists some non-zero vector $x = (x_1, \ldots, x_n) \in \mathbb{C}^n$ such that $Ax = re^{\imath\phi} x$. This implies $r|x| = |re^{\imath\phi} x| = |Ax| \le A|x|$ and Theorem 8.22 guarantees that $|x| \gg 0$ and

$$A|x| = r|x| .$$

In particular, $x_j \neq 0$ for each j and so the diagonal matrix

$$D = \operatorname{diag}\left(\tfrac{x_1}{|x_1|}, \ldots, \tfrac{x_n}{|x_n|}\right) ,$$

is well defined and satisfies $|D| = I$. Obviously, $D|x| = x$. Next, consider the matrix

$$B = [b_{kj}] = e^{-\imath\phi} D^{-1} AD ,$$

and notice that

$$\begin{aligned} B|x| &= e^{-\imath\phi}D^{-1}AD|x| = e^{-\imath\phi}D^{-1}Ax \\ &= e^{-\imath\phi}D^{-1}re^{\imath\phi}x = rD^{-1}x = r|x| \\ &= A|x|\,. \end{aligned}$$

At the same time, from

$$\begin{aligned} |B||x| &= \left|e^{-\imath\phi}D^{-1}AD\right||x| \le |D^{-1}||A||D||x| \\ &= A|x| = B|x| \le |B||x|\,, \end{aligned}$$

we see that $B|x| = |B||x|$, that is

$$\sum_{j=1}^{n} b_{kj}|x_j| = \sum_{j=1}^{n} |b_{kj}||x_j| \qquad (\star)$$

for each $k = 1, \ldots, n$. This implies that for each k there exists some angle θ_k such that $b_{kj} = |b_{kj}|e^{\imath\theta_k}$ for $j = 1, \ldots, n$; see Exercise 1 at the end of this section. But then $(\star)$ yields $\bigl(\sum_{j=1}^{n} |b_{kj}||x_j|\bigr)e^{\imath\theta_k} = \sum_{j=1}^{n} |b_{kj}||x_j|$, and so for each fixed k either $b_{kj} = 0$ for each j or else $e^{\imath\theta_k} = 1$. In either case, $b_{kj} = |b_{kj}| \ge 0$ for all k and j.

From the definition of B it follows that

$$B = |B| = \left|e^{-\imath\phi}D^{-1}AD\right| \le |D^{-1}|\,A\,|D| = A\,,$$

and therefore $a_{kj} - b_{kj} \ge 0$ for all k and j. Now taking into account that $A|x| = B|x|$, we see that $\sum_{j=1}^{n}(a_{kj} - b_{kj})|x_j| = 0$ for all k. Since $|x_j| > 0$ for each j, the latter implies $a_{kj} = b_{kj}$ for all k and j, i.e., $A = B$, and our conclusion follows. (Another way of proving that $A = B$ is by using Corollary 8.23(3) in conjunction with the inequality $0 \le B \le A$.)

(2) $\Longrightarrow$ (1) Assume that $A = e^{\imath\phi}DAD^{-1} = D(e^{\imath\phi}A)D^{-1}$. This means that A is similar to $e^{\imath\phi}A$ and so the matrices $e^{\imath\phi}A$ and A have the same spectrum. This implies $e^{\imath\phi}\sigma(A) = \sigma(e^{\imath\phi}A) = \sigma(A)$. Since $r \in \sigma(A)$, we conclude that $re^{\imath\phi} \in \sigma(A)$. ∎

Let us say that a subset Δ of complex numbers is **rotation invariant under an angle** θ if $e^{\imath\theta}\Delta = \Delta$.

The spectrum of an irreducible positive matrix exhibits the following rotational invariance.

Corollary 8.25. *Let A be a positive irreducible matrix with spectral radius $r = r(A)$. The spectrum of A is rotation invariant under an angle θ if and only if $re^{\imath\theta}$ is an eigenvalue of A.*

Proof. If $re^{\imath\theta}$ is an eigenvalue of A, then, by Lemma 8.24, A is similar to $e^{\imath\theta}A$. This implies $e^{\imath\theta}\sigma(A) = \sigma(e^{\imath\theta}A) = \sigma(A)$.

Conversely, if $\sigma(A)$ is rotation invariant under an angle θ, then (given that r is an eigenvalue of A) it follows that $re^{\imath\theta} \in \sigma(A)$. ■

We are now ready to state the classical Perron–Frobenius theorem regarding the spectral properties of an irreducible positive matrix. The theorem was proven first for strongly positive matrices by O. Perron [**265**] and was generalized later to positive irreducible matrices by G. Frobenius [**131**].

Theorem 8.26 (Perron–Frobenius). *If A is a positive irreducible matrix, then:*

(1) *Its spectral radius $r(A)$ is an eigenvalue and satisfies*

$$\begin{aligned} r(A) &= \sup\{\lambda \geq 0 \colon \ \exists \ x > 0 \ \text{ with } Ax \geq \lambda x\} \\ &= \inf\{\lambda > 0 \colon \ \exists \ x \gg 0 \ \text{ with } \lambda x \gg Ax\} > 0\,. \end{aligned}$$

(2) *The eigenspace of $r(A)$ is one dimensional and is generated by a strictly positive eigenvector.*

(3) *The eigenvalue $r(A)$ is a simple root of the characteristic polynomial of A.*

(4) *The peripheral spectrum of A is of the form*

$$\sigma_{\text{per}}(A) = \Big\{ r(A) e^{\frac{2\pi k}{p}\imath} \colon \ k = 0, 1, \ldots, p-1 \Big\}$$

and consists of simple roots of the characteristic polynomial. Moreover, if the characteristic polynomial of A is

$$p_A(\lambda) = \lambda^n + a_{n_1}\lambda^{n_1} + \cdots + a_{n_k}\lambda^{n_k}\,,$$

where $n > n_1 > \cdots > n_k \geq 0$ and $a_{n_j} \neq 0$ for each j, then

$$\begin{aligned} p &= \gcd\{n - n_1, n - n_2, \ldots, n - n_k\} \\ &= \gcd\{n - n_1, n_1 - n_2, \ldots, n_{k-1} - n_k\}\,, \end{aligned}$$

where gcd *is an abbreviation for the "greatest common divisor."*

(5) *The spectrum of A is rotation invariant under the angle $\frac{2\pi}{p}$.*

(6) *If $\mu \in \sigma(A)$ is non-zero, then the p distinct complex numbers*

$$\mu, \mu e^{\frac{2\pi}{p}\imath}, \mu e^{\frac{4\pi}{p}\imath}, \ldots, \mu e^{\frac{2(p-1)\pi}{p}\imath}$$

in $\sigma(A)$ are roots of the characteristic polynomial of A having the same multiplicity.

Proof. (1) This is part (1) of Lemma 8.14 and part (1) of Corollary 8.23.

(2) This follows immediately from Theorem 8.22(3).

(3) The polynomial $p_A(\lambda) = \det(\lambda I - A)$ is the characteristic polynomial of A. Let us denote by A_k the $(n-1) \times (n-1)$ matrix which is obtained

from A by deleting its kth row and k^{th} column. Also, let $p_k(\cdot)$ denote the characteristic polynomial of A_k, i.e., $p_k(\lambda) = \det(\lambda I - A_k)$. An easy computation (using the familiar formula for differentiating a determinant whose entries are functions) shows that

$$p'_A(\lambda) = \sum_{k=1}^{n} \det(\lambda I - A_k) = \sum_{k=1}^{n} p_k(\lambda)\,. \qquad (\star\star)$$

Now let B_k denote the $n \times n$ matrix obtained from A by making all entries in the kth row and k^{th} column of A zero. Then $0 \le B_k \le A$ and $B_k \ne A$, and from Corollary 8.23(3), we know that $\lambda_{B_k} < \lambda_A = r(A)$. Since the characteristic polynomial p_{B_k} of B_k satisfies $p_{B_k}(\lambda) = \lambda p_k(\lambda)$, it easily follows that λ_A is greater than any real root of any one of the polynomials p_k. Since each p_k has leading term λ^{n-1}, it follows that $p_k(\lambda_A) > 0$ for each k. A glance at $(\star\star)$ shows that $p'_A(\lambda_A) > 0$ and this implies that λ_A is a simple root of the characteristic polynomial $p_A(\lambda)$ of A.

(4) and (5) Let $\sigma_{\text{per}}(A) = \{r(A)e^{\imath\phi_k}\colon\ k = 0, \ldots, p-1\}$ be the peripheral spectrum of A. By Corollary 8.25, the set $\sigma_{\text{per}}(A)$ is rotation invariant under $e^{\imath\phi_k}$. This guarantees that the subset of complex numbers

$$G = \{e^{\imath\phi_k}\colon\ k = 0, \ldots, p-1\}$$

is a group, and consequently $G = \{z \in \mathbb{C}\colon\ z^p = 1\}$, i.e., G has the desired form; see Exercise 1 in Section 7.3. Now invoke Corollary 8.25 once again to obtain that the whole spectrum $\sigma(A)$ is rotation invariant under the angle $\frac{2\pi}{p}$. The simplicity of the roots can be established either directly or by employing the conclusion of part (6) below.

To see that p can be computed by the described formula, note first that it follows from Lemma 8.24 that the two matrices A and $e^{\frac{2\pi}{p}\imath}A$ are similar, and so they have the same characteristic polynomial. Computing the characteristic polynomials yields

$$\begin{aligned} p_A(\lambda) &= \lambda^n + a_{n_1}\lambda^{n_1} + \cdots + a_{n_k}\lambda^{n_k} \\ &= \lambda^n + a_{n_1}e^{\frac{2\pi(n-n_1)}{p}\imath}\lambda^{n_1} + \cdots + a_{n_k}e^{\frac{2\pi(n-n_k)}{p}\imath}\lambda^{n_k}\,. \end{aligned}$$

This implies $a_{n_j}e^{\frac{2\pi(n-n_j)}{p}\imath} = a_{n_j}$, or $e^{\frac{2\pi(n-n_j)}{p}\imath} = 1$ for each j. Therefore, p divides $n - n_j$ for each j.

To establish that p is the *gcd* of the numbers $n - n_j$, we distinguish two cases: $p = 1$ and $p > 1$. Assume first that $p = 1$, i.e., $\sigma_{\text{per}}(A) = \{r(A)\}$. In this case, if $m > 1$ is a divisor of each $n - n_j$, then (as above) the matrix $e^{\frac{2\pi}{m}\imath}A$ has the same characteristic polynomial as A. This guarantees that $r(A)e^{\frac{2\pi}{m}\imath}$ is a point of the peripheral spectrum of A which is different from $r(A)$, a contradiction. Now consider $p > 1$. If $m > 1$ is a divisor of each

$n - n_j$, then (as in the previous case) the spectrum of A is rotation invariant under the angle $\frac{2\pi}{m}$. By Corollary 8.25, $r(A)e^{\frac{2\pi}{m}\imath}$ is an eigenvalue lying in the peripheral spectrum of A. In particular, $\frac{2\pi}{m} \geq \frac{2\pi}{p}$ or $p \geq m$, and the formula for p has been established.

(6) Assume that $\lambda_1, \ldots, \lambda_k$ are the distinct eigenvalues of A with multiplicities $m_1, \ldots, m_k$, respectively. It follows from Exercise 3 of Section 8.2 that the matrix $e^{\frac{2\pi}{p}\imath}A$ has the distinct eigenvalues $\lambda_1 e^{\frac{2\pi}{p}\imath}, \ldots, \lambda_k e^{\frac{2\pi}{p}\imath}$ with multiplicities $m_1, \ldots, m_k$, respectively. But since the matrices A and $e^{\frac{2\pi}{p}\imath}A$ are similar, $\lambda_1 e^{\frac{2\pi}{p}\imath}, \ldots, \lambda_k e^{\frac{2\pi}{p}\imath}$ are also the distinct eigenvalues of A with multiplicities $m_1, \ldots, m_k$, respectively. In other words, a multiplication of an eigenvalue by $e^{\frac{2\pi}{p}\imath}$ does not change its multiplicity, and our final conclusion follows. ∎

Corollary 8.27. *Let $A = [a_{ij}]$ be an $n \times n$ irreducible positive matrix. If $a_{ii} \neq 0$ for some i, then $\sigma_{\text{per}}(A) = \{r(A)\}$.*

Proof. In this case $\text{tr}(A) = \sum_{i=1}^{n} a_{ii} > 0$ and the characteristic polynomial of A is given by $p_A(\lambda) = \lambda^n - \text{tr}(A)\lambda^{n-1} + \cdots + \det(A)$. Therefore, by part (4) of Theorem 8.26, the peripheral spectrum is the singleton $\{r_A\}$. ∎

A special class of irreducible positive matrices is provided by the class of primitive matrices.

Definition 8.28. *An irreducible $n \times n$ positive matrix A is said to be* ***primitive*** *if its peripheral spectrum is the singleton $\{r(A)\}$.*

The primitive matrices are precisely the irreducible positive matrices which have a strongly positive power. The details are included in the next result.

Theorem 8.29. *For an $n \times n$ irreducible positive matrix A with spectral radius $r = r(A)$ the following statements are equivalent.*

1. *The matrix A is primitive, i.e., $\sigma_{\text{per}}(A) = \{r\}$.*
2. *For some m the matrix A^m is strongly positive.*
3. *There is a strongly positive matrix S such that $\left(\frac{1}{r}A\right)^n \longrightarrow S$.*

Proof. (3) $\Longrightarrow$ (2) Since S is strongly positive and $\left(\frac{1}{r}A\right)^n \longrightarrow S$, it follows that there exists some m such that each entry of $\left(\frac{1}{r}A\right)^n$ is greater than zero for all $n \geq m$. Hence, A^m is strongly positive.

(2) $\Longrightarrow$ (1) Assume that A^m is strongly positive for some m. By Corollary 8.19, A^m is irreducible and since it is also strongly positive, it follows from Corollary 8.27 that r^m is the only point in the peripheral spectrum

of A^m and, by part (3) of Theorem 8.26, r^m is a simple root of the characteristic polynomial of A^m. This implies that $\sigma_{\text{per}}(A) = \{r\}$. Indeed, if the peripheral spectrum of A were not a singleton, then (by Exercise 3 of Section 8.2) either the peripheral spectrum of A^m would fail to be a singleton or else r^m would be an eigenvalue of A^m of multiplicity greater than one, which contradicts in both cases our hypothesis. Hence, A is a primitive matrix.

(1) $\Longrightarrow$ (3) Replacing A by $\frac{1}{r}A$, we can assume that $r = 1$. The Jordan form of A guarantees the existence of an invertible matrix B such that

$$B^{-1}AB = \begin{bmatrix} 1 & 0 \\ 0 & C \end{bmatrix}, \tag{$\star$}$$

where C is an $(n-1)\times(n-1)$ upper triangular matrix having along its main diagonal only eigenvalues of A with moduli strictly less than one. The latter implies $C^n \to 0$ (see Exercise 8 of Section 8.1). Since $A^n = B\begin{bmatrix} 1 & 0 \\ 0 & C^n \end{bmatrix}B^{-1}$ for each n, we see that

$$A^n \longrightarrow B\begin{bmatrix} 1 & 0 \\ 0 & 0 \end{bmatrix}B^{-1} = S \geq 0\,.$$

To see that S is strictly positive, let $b = (b_1, b_2, \ldots, b_n)$ denote the first row of B^{-1} and $c^t = (c_1, c_2, \ldots, c_n)$ be the transpose of the first column c of B. Clearly, both b and c are non-zero vectors. From ($\star$), we have

$$AB = B\begin{bmatrix} 1 & 0 \\ 0 & C \end{bmatrix} \quad \text{and} \quad A^t(B^{-1})^t = (B^{-1})^t\begin{bmatrix} 1 & 0 \\ 0 & C^t \end{bmatrix},$$

and from this it easily follows that $Ac = c$ and $A^t b^t = b^t$. Since A and A^t are irreducible and both have spectral radius 1, we infer that all components of the vectors b and c are non-zero; see Theorem 8.22(2). Now notice that the $(i,j)^{\text{th}}$ entry s_{ij} of S satisfies $s_{ij} = c_i b_j \neq 0$ and this guarantees that indeed $s_{ij} > 0$ for all i and j. ■

Corollary 8.30. *If A is a primitive matrix with $r = r(A)$, then for each $x > 0$ the limit $y = \lim_{n\to\infty}\big(\frac{1}{r}A\big)^n x$ exists, is strictly positive, and satisfies $Ay = ry$.*

Proof. Fix $x > 0$. By Theorem 8.29, we have $\big(\frac{1}{r}A\big)^n x \to Sx = y \gg 0$. Moreover, note that $Ay = A\big(\lim\limits_{n\to\infty}\big[\frac{1}{r}A\big]^n x\big) = r\lim\limits_{n\to\infty}\big(\frac{1}{r}A\big)^{n+1}x = ry$. ■

In view of the preceding results one might think that it would be easy to figure out which finite subsets of complex numbers are the spectra of irreducible positive matrices. Surprisingly, this is a very difficult problem and a complete answer is not yet known. Even when we place a bound on the number of points in a subset σ of $\mathbb{C}$, the irreducible positive matrices that have the set σ as their spectrum can be of very large sizes. For more detailed information see [**72**] and [**184**].

We close the section by mentioning that there is extensive literature regarding extensions of the Perron–Frobenius theory to matrices that leave a cone invariant. For details we refer the reader to the survey article by B.-S. Tam [**316**] and the references therein.

Exercises

1. Show that n complex numbers $z_1, \ldots, z_n$ satisfy $\left|\sum_{i=1}^n z_i\right| = \sum_{i=1}^n |z_i|$ if and only if there exists some angle θ such that $z_i = |z_i|e^{\imath\theta}$ for each $i = 1, \ldots, n$.

2. Consider the polynomials:

$$\text{(i) } \lambda^3 - 1\,, \quad \text{(ii) } \lambda^3 - \lambda\,, \quad \text{and} \quad \text{(iii) } \lambda^4 + \lambda^2 - 1\,.$$

Which of the above polynomials are the characteristic polynomials of positive irreducible matrices? [HINT: Consider the matrices

$$\begin{bmatrix} 0 & 1 & 0 \\ 0 & 0 & 1 \\ 1 & 0 & 0 \end{bmatrix} \quad \text{and} \quad \begin{bmatrix} 0 & \frac{\sqrt{2}}{2} & 0 \\ \frac{\sqrt{2}}{2} & 0 & \frac{\sqrt{2}}{2} \\ 0 & \frac{\sqrt{2}}{2} & 0 \end{bmatrix}.]$$

3. Consider the following Markov matrix

$$A = \begin{bmatrix} 0 & 0 & \frac{1}{2} & \frac{1}{2} \\ 1 & 0 & 0 & 0 \\ 0 & 1 & 0 & 0 \\ 0 & 0 & 1 & 0 \end{bmatrix}.$$

Show that A is an irreducible positive matrix satisfying $\sigma_{\text{per}}(A) = \{1\}$. [HINT: Note that $p_A(\lambda) = \lambda^4 - \frac{\lambda}{2} - \frac{1}{2}$ and that A^{10} is strongly positive.]

4. This exercise illustrates the Perron–Frobenius theorem. Consider the positive matrix

$$A = \begin{bmatrix} 0 & 0 & 0 & 0 & 1 & 0 & 0 & 0 \\ 0 & 0 & 0 & 0 & 0 & 1 & 0 & 0 \\ 0 & 0 & 0 & 0 & 0 & 0 & 1 & 0 \\ 0 & 0 & 0 & 0 & 0 & 0 & 0 & 1 \\ 1 & 1 & 1 & 0 & 0 & 0 & 0 & 0 \\ 1 & 1 & 0 & 1 & 0 & 0 & 0 & 0 \\ 1 & 0 & 1 & 1 & 0 & 0 & 0 & 0 \\ 0 & 1 & 1 & 1 & 0 & 0 & 0 & 0 \end{bmatrix}.$$

(a) Show that A is irreducible.

(b) Show that the characteristic polynomial of A is

$$p_A(\lambda) = (\lambda^2 - 3)(\lambda - 1)^2(\lambda + 1)^2(\lambda^2 + 1).$$

Conclude from this that on the unit circle there are six eigenvalues: 1 with multiplicity two, -1 with multiplicity two, $\imath$ and $-\imath$.

(c) Show that the spectrum of A is rotation invariant under 180°.

5. Let A be a primitive matrix and let $(\frac{1}{r}A)^n \to S$, where $r = r(A)$. Show that each column of S is an eigenvector corresponding to the eigenvalue r.

6. Show that an $n \times n$ positive matrix is primitive if and only if it defines a Krein operator.

7. We have two glasses; call them 1 and 2. Glass 1 contains 100 grams of water and glass 2 contains 100 grams of wine. We transfer 10 grams from glass 1 into glass 2 and stir the mixture very well. Subsequently, we transfer 10 grams from glass 2 to glass 1 and again stir the mixtures in both glasses very well. We repeat this process ad infinitum.

Assume that w_n^1 and w_n^2 denote the amounts (in grams) of wine in glasses 1 and 2, respectively, after the nth stage of this process. Show that $\lim_{n\to\infty} w_n^1 = \lim_{n\to\infty} w_n^2 = 50$.

[HINT: Note that $\begin{bmatrix} w_{n+1}^1 \\ w_{n+1}^2 \end{bmatrix} = \begin{bmatrix} \frac{10}{11} & \frac{1}{11} \\ \frac{1}{11} & \frac{10}{11} \end{bmatrix} \begin{bmatrix} w_n^1 \\ w_n^2 \end{bmatrix}$. This, in conjunction with Corollary 8.30, implies $\begin{bmatrix} w_n^1 \\ w_n^2 \end{bmatrix} = \begin{bmatrix} \frac{10}{11} & \frac{1}{11} \\ \frac{1}{11} & \frac{10}{11} \end{bmatrix}^n \begin{bmatrix} 0 \\ 100 \end{bmatrix} \to c \begin{bmatrix} 1 \\ 1 \end{bmatrix}$. Now use that $w_n^1 + w_n^2 = 100$.]

Chapter 9

Irreducible Operators

This chapter extends the discussion about irreducible matrices to infinite dimensional spaces. Among other things, we shall present conditions which guarantee that the spectral radius of a positive operator is non-zero. The conditions are known as ideal and band irreducibility. An operator T on a Banach lattice is said to be **ideal** (resp. **band**) **irreducible** if there is no non-trivial closed ideal (resp. band) which is T-invariant. The following list contains samples of several results that will be obtained in this chapter.

- Every ideal irreducible compact positive operator has a positive spectral radius (B. de Pagter).
- If two positive operators commute with each other and one is ideal irreducible and the other is compact, then both have positive spectral radii (Abramovich–Aliprantis–Burkinshaw).
- A band irreducible positive kernel operator has a positive spectral radius (Andô–Krieger).
- If a positive operator on a $C(\Omega)$-space (where Ω is a compact Hausdorff space) has a positive eigenvalue λ having an order unit as an eigenvector, then its adjoint also has λ as an eigenvalue (M. G. Krein).

In the process of establishing these results, we will also introduce several new classes of positive operators and study their spectral properties.

9.1. Irreducible and Expanding Operators

Before presenting some general results regarding the spectral radius of positive operators, we shall pause for a while to discuss various classes of positive operators and their relationships. We shall group these operators into three categories according to their special properties. The first two categories of operators will be defined in terms of certain invariant subspaces—the ideals and bands.

Let $T\colon X \to X$ be a bounded operator on a Banach space. Recall that a vector subspace V of X is **invariant under** T (or simply T**-invariant**) if $T(V) \subseteq V$. A vector subspace V of X is said to be **hyperinvariant for** T (or simply T**-hyperinvariant**) whenever V is invariant under every bounded operator on X that commutes with T, i.e., whenever $S \in \mathcal{L}(X)$ and $ST = TS$ imply $S(V) \subseteq V$. If X is a Banach lattice and T is a positive operator, then a vector subspace V of X is called **lattice hyperinvariant for** T (or simply ℓ**-hyperinvariant**) whenever V is invariant under every positive operator on X that commutes with T. As usual, a vector subspace V of a vector space X is **non-trivial** if $V \neq \{0\}$ and $V \neq X$.

The **null ideal** N_T of a positive operator $T\colon E \to F$ between two Riesz spaces is the ideal in E defined by

$$N_T = \{x \in E\colon\ T|x| = 0\}\,.$$

Clearly, $T = 0$ if and only if $N_T = E$. It should also be obvious that if E and F are Banach lattices, then the null ideal N_T is closed in E. Moreover, observe that a positive operator $T\colon E \to F$ is strictly positive (see Definition 1.13) if and only if $N_T = \{0\}$. If $T\colon E \to F$ is a positive operator between two Riesz spaces, then the positive operator $\hat{T}\colon E/N_T \to F$, defined by $\hat{T}[x] = Tx$, is strictly positive.

Lemma 9.1. *If $T\colon E \to E$ is an arbitrary positive operator on a Banach lattice, then its null ideal and the closed ideal generated by its range are both ℓ-hyperinvariant.*

Proof. Let $T\colon E \to E$ be a positive operator on a Banach lattice and let $S\colon E \to E$ be a positive operator commuting with T. If $x \in N_T$, then from $0 \leq T|Sx| \leq T(S|x|) = S(T|x|) = 0$, it follows that $T|Sx| = 0$, and so $Sx \in N_T$. That is, $S(N_T) \subseteq N_T$ and this proves that N_T is a closed ℓ-hyperinvariant ideal for T.

Next, notice that the closed ideal J_T generated by the range of T is described by

$$J_T = \big\{x \in E\colon\ \exists\, y \geq 0 \text{ such that } |x| \leq Ty\big\}^{-}.$$

Again, assume that $S \in \mathcal{L}(E)$ is a positive operator that commutes with T. If $|x| \leq Ty$ for some $y \geq 0$, then $|Sx| \leq S|x| \leq S(Ty) = T(Sy)$, which shows that $Sx \in J_T$. Thus, J_T is S-invariant, and so J_T is a closed ℓ-hyperinvariant ideal for T. ■

We are now ready to define the classes of ideal and band irreducible operators. They extend the corresponding notions for matrices; see Definition 8.16.

Definition 9.2. *A bounded operator $T\colon E \to E$ on a Banach lattice is:*

(1) ***ideal irreducible*** *if T has no invariant non-trivial closed ideals, that is, whenever a closed ideal J satisfies $T(J) \subseteq J$, then either $J = \{0\}$ or $J = E$, and*

(2) ***band irreducible*** *if T has no invariant non-trivial bands, i.e., whenever a band B satisfies $T(B) \subseteq B$, then either $B = \{0\}$ or $B = E$.*

The next result indicates that "ideal irreducibility" is a relatively strong property. In particular, as we shall see, only Banach lattices with quasi-interior points admit ideal irreducible positive operators.

Theorem 9.3. *If $T\colon E \to E$ is an ideal irreducible positive operator on a Banach lattice, then:*

(1) *T is band irreducible.*

(2) *T is strictly positive, i.e., $x > 0$ implies $Tx > 0$.*

(3) *E has quasi-interior points.*

(4) *T carries quasi-interior points to quasi-interior points.*

Proof. (1) The conclusion follows from the fact that every band is a closed ideal.

(2) Let $T\colon E \to E$ be an ideal irreducible positive operator on a Banach lattice. The null ideal N_T of T is closed and T-invariant. Since T is ideal irreducible, it follows that $N_T = \{0\}$, which means that T is strictly positive.

(3) By Theorem 4.21 there exists some $u > 0$ such that $T(E_u) \subseteq E_u$. This implies $T(\overline{E}_u) \subseteq \overline{E}_u$, i.e., $\overline{E}_u$ is T-invariant. Since T is ideal irreducible and $E_u \neq \{0\}$, we see that $\overline{E}_u = E$. So, u is a quasi-interior point.

(4) Let $u > 0$ be a quasi-interior point of E and let J_T denote the closed ideal generated by the range of T. That is,

$$J_T = \big\{x \in E\colon \ \exists\, y \geq 0 \text{ such that } |x| \leq Ty\big\}^{-}.$$

By Lemma 9.1, J_T is a closed ℓ-hyperinvariant ideal for T, and so $J_T = E$. Since $u > 0$ is a quasi-interior point, the principal ideal E_u generated by u is norm dense in E. Clearly, $T(E_u) \subseteq E_{Tu} \subseteq J_T$, and so

$$T(E) = T\big(\overline{E}_u\big) \subseteq \overline{T(E_u)} \subseteq \overline{E}_{Tu} \subseteq J_T\,.$$

This implies $\overline{E}_{Tu} = J_T = E$, i.e., Tu is also a quasi-interior point of E. ■

A similar result holds true for band irreducible positive operators; see Exercise 10 at the end of the section. We now introduce the classes of "expanding" operators.

Definition 9.4. *A positive operator $T\colon E \to E$ on a Banach lattice is said to be:*

(1) ***strongly expanding*** *if for each $x > 0$ the vector Tx is a quasi-interior point,*

(2) ***expanding*** *if for each $x > 0$ the vector Tx is a weak unit, and*

(3) ***weakly expanding,*** *if for each $x > 0$ there exists some n (depending upon x) such that the vector $T^n x$ is a weak unit.*

There are some other classes of operators that are closely related to the classes of operators introduced above. These are the classes of Krein and strong Krein operators that were introduced in [**8**]. Their precise definition is as follows.

Definition 9.5. *A positive operator $T\colon E \to E$ on an arbitrary AM-space with unit is said to be:*

(1) *a **strong Krein operator** whenever T maps each non-zero positive vector to a (strong) unit, and*

(2) *a **Krein operator** whenever for each $x > 0$ there exists some n (depending upon x) such that the vector $T^n x$ is a (strong) unit.*

The properties introduced in Definitions 9.2, 9.4, and 9.5 are quite strong, and positive operators with such properties need not exist on an arbitrary Banach lattice. For instance, Krein and strongly expanding operators require the presence of strong units and of quasi-interior points, respectively, while expanding operators can possibly exist only on spaces with weak units, and—as we shall see—ideal irreducible positive operators can exist only on Banach lattices with the countable sup property. Some interesting examples of compact integral Krein operators on $C(\Omega)$-spaces will be presented in Example 9.39.

The next result shows schematically the relationships between the various classes of positive operators introduced above.

Theorem 9.6. *We have the following implication scheme for positive operators on a Banach lattice:*

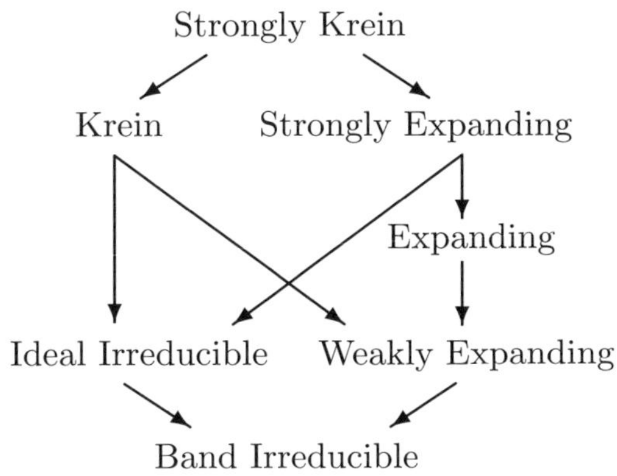

In general, no other implication is true.

Proof. We leave the details of proving the indicated implications as an exercise. Several examples below demonstrate that in general no other implication holds. ■

Example 9.7. *An order continuous band irreducible positive operator that is neither ideal irreducible nor weakly expanding.*

Consider the positive operator $T\colon \ell_\infty \to \ell_\infty$ defined by

$$T(x_1, x_2, \ldots) = \Big(\sum_{n=1}^{\infty} \tfrac{x_n}{2^n}, x_1, x_2, \ldots \Big).$$

Clearly, $T(c_0) \subseteq c_0$ and so T is not ideal irreducible. Also, it is easy to see that T is order continuous. Now, if $x = (1, 0, 0, \ldots) > 0$, then it easily follows that $T^k x$ is not a weak unit for each k, and hence T is not weakly expanding.

To see that T is band irreducible, assume that $T(B) \subseteq B$ for some band $B \neq \{0\}$. If $0 < x \in B$, then the element $T^k x \in B$ has its first k components non-zero. This means that for each k the band B contains a positive vector whose k^{th} component is non-zero. So, if some $v \in \ell_\infty$ satisfies $v \perp B$, then $v = 0$, which shows that $B = \ell_\infty$, i.e., T is a band irreducible operator. ■

Example 9.8. *A Krein operator which is neither a strong Krein nor an expanding operator.*

The operator $T\colon \ell_\infty \to \ell_\infty$ defined by

$$T(x_1, x_2, x_3, \ldots) = \big(x_1 + x_2, x_1 + x_3, x_1 + x_4, \ldots\big)$$

is a Krein operator which is neither strong Krein nor expanding. ■

Example 9.9. *An expanding operator which is not strongly expanding.*

Consider the operator $T\colon C[0,1] \to C[0,1]$ defined by the formula

$$Tx(t) = \int_0^1 \sin(st)x(s)\,ds\,.$$

It is easy to verify that T is an expanding compact operator. Since $Tx(0) = 0$ for each $x \in C[0,1]$, this operator fails to be strongly expanding. Also, notice that if $J = \{x \in C[0,1]\colon\ x(0) = 0\}$, then J is a non-trivial closed ideal and $T(J) \subseteq J$ holds. This shows that T is not ideal irreducible; for more properties see Exercise 8 of Section 9.2. ∎

Example 9.10. *An ideal irreducible operator which is not weakly expanding.*

Consider the operator T in Example 9.7 restricted to c_0. As noticed there, $T(c_0) \subseteq c_0$. Since c_0 has order continuous norm, the notions of ideal irreducibility and band irreducibility coincide, and so $T\colon c_0 \to c_0$ is ideal irreducible. The final argument in Example 9.7 shows that $T\colon c_0 \to c_0$ is not weakly expanding. ∎

Example 9.11. *A weakly expanding operator which is neither expanding nor ideal irreducible.*

Consider first the positive operator $S\colon c_0 \to c_0$ defined by

$$S(x_1, x_2, \ldots) = \big(x_2 + x_1, x_3 + \tfrac{x_1}{2}, x_4 + \tfrac{x_1}{3}, \ldots\big)\,.$$

It should be immediate that S is weakly expanding but fails to be expanding. Arguing as in the last part of Example 9.7, we see that S is ideal irreducible. Moreover, since c_0 has order continuous norm, it follows that S is also order continuous.

Next, extend S to the Banach lattice c of all convergent sequences by letting

$$T(x_1, x_2, \ldots) = S(x_1 - x_\infty, x_2 - x_\infty, \ldots) + 2x_\infty e\,,$$

where $x_\infty = \lim_{n\to\infty} x_n$ and $e = (1, 1, 1, \ldots)$. The reader can verify that $T\colon c \to c$ is a strictly positive operator, it is weakly expanding, and (by virtue of $T(c_0) \subseteq c_0$) it fails to be ideal irreducible. In addition, it should be noted that T is not σ-order continuous. ∎

Example 9.12. *A rank-one expanding order continuous positive operator which fails to be ideal irreducible.*

Consider the sequence $u = \big(1, \frac{1}{2}, \frac{1}{3}, \ldots\big)$ and note that u is a weak unit of ℓ_∞. Also, let $\phi\colon \ell_\infty \to \mathbb{R}$ be a strictly positive order continuous linear functional; for instance we can define ϕ by $\phi(x) = \sum_{n=1}^\infty \frac{x_n}{2^n}$. Now consider the rank-one order continuous positive operator $T = \phi \otimes u\colon \ell_\infty \to \ell_\infty$. Clearly, $T(c_0) \subseteq c_0$, and so T is not ideal irreducible. However, if $0 < x \in \ell_\infty$, then $T(x) = \phi(x)u$ is a weak unit, and so the compact positive operator T is expanding. Finally, it is interesting to note that if we replace u by a strong unit (for instance, let $u = (1, 1, 1, \ldots)$), then $\phi \otimes u$ becomes a strong Krein operator on ℓ_∞ which is not ideal irreducible. ∎

It is clear that each expanding operator is also strictly positive. The next lemma presents two useful sufficient conditions for the converse to be true.

Lemma 9.13. *Let $S\colon E \to E$ be a strictly positive operator. If an ideal irreducible (resp. a σ-order continuous band irreducible) positive operator $T\colon E \to E$ satisfies $TS \leq \lambda S$ for some $\lambda > 0$, then $T^k S$ and ST^k are strongly expanding (resp. expanding) for each $k = 0, 1, 2, \ldots$.*

Proof. We shall consider only the case when T is positive and ideal irreducible. The other case is similar.

The proof is by induction on k. Assume first $k = 0$. Let $x > 0$ and put $u = Sx > 0$. From $T(Sx) \leq \lambda Sx$ (i.e., $Tu \leq \lambda u$), we see that $T(\overline{E}_u) \subseteq \overline{E}_u$. By the ideal irreducibility of T, we get $\overline{E}_u = E$. That is, $u = Sx$ is a quasi-interior point, and so S is strongly expanding. Now assume that $T^{k-1}S$ is strongly expanding. Since $T^k S = T(T^{k-1}S) = T^{k-1}(TS) \leq \lambda T^{k-1}S$ and (by the induction hypothesis) $T^{k-1}S$ is strictly positive, the above argument shows that $T^k S$ is strongly expanding.

Finally, since S is strongly expanding and T^k is strictly positive, it should be immediate that ST^k is also strongly expanding. ∎

It is natural to ask here if the order of operators in Lemma 9.13 is essential, i.e., if the assumption $TS \leq \lambda S$ can be replaced by $ST \leq \lambda S$. We refer to Exercise 17 at the end of this section for examples demonstrating that the order is essential.

Corollary 9.14. *If $T\colon E \to E$ is an ideal irreducible positive operator, then for each $\lambda > r(T)$ and each $k = 0, 1, 2, \ldots$ the operator $R(\lambda, T)T^k$ is strongly expanding (and hence ideal irreducible).*

Similarly, if $T\colon E \to E$ is a σ-order continuous band irreducible positive operator, then for each $\lambda > r(T)$ and each $k = 0, 1, 2, \ldots$ the operator $R(\lambda, T)T^k$ is expanding.

Proof. Note that for each $\lambda > r(T)$ the operator $R(\lambda, T)$ is strictly positive and satisfies $TR(\lambda, T) \leq \lambda R(\lambda, T)$; see Lemma 6.23(3). ∎

The existence of an ideal irreducible positive operator on a Banach lattice E imposes a severe restriction on E.

Lemma 9.15. *If a Banach lattice E admits a compact strictly positive operator, then E also admits a strictly positive linear functional. In particular, in this case, E has the countable sup property.*[1]

Proof. Let $T\colon E \to E$ be a compact strictly positive operator. Fix a countable set $\{\phi_1, \phi_2, \ldots\}$ of continuous non-zero linear functionals in the range $R(T^*)$ of T^* that is norm dense in $R(T^*)$ and put $\phi = \sum_{n=1}^{\infty} \frac{|\phi_n|}{2^n \|\phi_n\|} \geq 0$. (This is possible since $T^*\colon E^* \to E^*$ is a compact operator.) We claim that ϕ is strictly positive on E.

To this end, let $x > 0$. If $\phi(x) = 0$, then $\phi_n(x) = 0$ for each n and so $\psi(Tx) = T^*\psi(x) = 0$ for each $\psi \in E^*$. Hence, $Tx = 0$, contrary to the strict positivity of T. Thus, $\phi(x) > 0$ and so ϕ is strictly positive. This implies that E has the countable sup property. ■

Corollary 9.16. *If a Banach lattice E admits a compact strictly positive operator, then for positive operators on E the notions of σ-order continuity and order continuity coincide.*

Corollary 9.17. *If a Banach lattice E admits an ideal irreducible positive operator, then E has the countable sup property.*

Proof. Let $T\colon E \to E$ be an ideal irreducible positive operator. Fix some scalar $\lambda > r(T)$ and note that (by Corollary 9.14) the positive operator $Q = R(\lambda, T)$ is strongly expanding. Now fix $0 < \phi \in E^*$ and $0 < u \in E$, and let $K = \phi \otimes u$. Notice that for each quasi-interior vector $x > 0$, it follows that $Kx = \phi(x)u > 0$. Hence, $KQ\colon E \to E$ is a compact strictly positive operator and so, by Lemma 9.15, E has the countable sup property. ■

Exercises

1. Verify the "Implication Scheme" of Theorem 9.6.
2. Show that every weakly expanding operator is strictly positive.
3. Let $S, T\colon E \to E$ be two arbitrary positive operators on a Banach lattice. If $0 \leq S \leq T$ and S is band (resp. ideal) irreducible, then show that T is likewise band (resp. ideal) irreducible.
4. If S and T are two positive operators on a Banach lattice and one of them is band (resp. ideal) irreducible, then show that $S+T$ is band (resp. ideal) irreducible.

[1] Recall that a Riesz space E is said to have the **countable sup property** whenever for every subset D of E having a supremum there exists an at most countable subset C of D satisfying $\sup C = \sup D$. Any Riesz space that admits a strictly positive linear functional has the countable sup property; see [**29**, Theorem 2.6, p. 12].

5. Give an example of a band irreducible positive operator T that is not a supremum of a family of ideal irreducible positive operators dominated by T.

6. We know that each central operator leaves every ideal invariant while an ideal irreducible operator does not leave invariant any non-trivial closed ideal. So, the following conjecture seems plausible: *If S is a central positive operator and T is an ideal irreducible positive operator, then $S \wedge T = 0$.* Disprove this conjecture.

7. If a positive operator T on a Banach lattice is either strongly expanding, or expanding, or strongly Krein and S is a strictly positive operator, then show that TS also possesses the corresponding property of T.

8. If a Banach lattice E has order continuous norm, then show that a positive operator on E is expanding if and only if it is strongly expanding.

9. Show that a positive operator on a Banach lattice with order continuous norm is band irreducible if and only if it is ideal irreducible.

10. If $T\colon E \to E$ is a σ-order continuous band irreducible positive operator, then show that:

(a) T is strictly positive.

(b) T carries weak units to weak units.

(c) E has weak units.

[HINT: For (a) if for some $x > 0$ we have $Tx = 0$, then $T(B_x) = \{0\} \subseteq B_x$ (where B_x denotes the band generated by x), and so $B_x = E$ which implies $T = 0$. For (b) fix a weak unit $u > 0$, and note that $Tu > 0$ implies $B_{Tu} \neq \{0\}$. Now if $x > 0$, then $x \wedge nu \uparrow x$ implies $Tx \wedge nTu \uparrow Tx$, and so $Tx \in B_{Tu}$. So, B_{Tu} is T-invariant, and this implies $B_{Tu} = E$. For (c) mimic the proof of Theorem 9.3(4).]

11. For a positive operator $T\colon E \to E$ on a Banach lattice establish the following properties.

(a) If T^k is ideal (resp. band) irreducible for some $k > 1$, then T is also ideal (resp. band) irreducible.

(b) If T is ideal irreducible, then T^2 need not be ideal irreducible.

[HINT: For (b) consider $T\colon \mathbb{R}^2 \to \mathbb{R}^2$ defined by $T(x, y) = (y, x)$.]

12. For a Markov operator $T\colon E \to E$ on an AM-space with unit e establish the following.

(a) There exists some $0 < \phi \in E^*$ satisfying $T^*\phi = \phi$. (Note that if $E = C(\Omega)$ and $e = \mathbf{1}$, then ϕ is a regular Borel measure on Ω, known as an **invariant measure** for T.)

(b) If $0 < \phi \in E^*$ satisfies $T^*\phi = \phi$, then the rank-one positive operator $S = \phi \otimes e$ satisfies $ST = TS = S$.

[HINT: For (a) consider the w^*-compact convex subset

$$\mathcal{C} = \big\{\psi \in E^*\colon\ \psi \geq 0 \text{ and } \psi(e) = 1\big\}$$

of E^*. Note that the adjoint operator $T^*\colon E^* \to E^*$ is w^*-continuous and satisfies $T^*(\mathcal{C}) \subseteq \mathcal{C}$. Thus, by the classical Tychonoff Fixed Point Theorem, there exists some $\phi \in \mathcal{C}$ such that $T^*\phi = \phi$.]

13. Consider the positive operator $T\colon C[0,1] \to C[0,1]$ defined by

$$Tf(t) = \tfrac{1}{t}\int_0^t f(s)\,ds\,, \qquad (\star)$$

for all $0 < t \le 1$ and $Tf(0) = f(0)$. Show that:

(a) T defines indeed a positive operator on $C[0,1]$.

(b) T is one-to-one—and hence strictly positive.

(c) T is a Markov operator.

(d) T is not band irreducible—and hence not ideal irreducible either.

(e) The point spectrum of T is the set

$$\sigma_p(T) = \{\lambda \in \mathbb{C}\colon\ |\lambda - \tfrac{1}{2}| < \tfrac{1}{2}\} \cup \{1\}\,.$$

[HINT: Notice that if $|\lambda - \frac{1}{2}| < \frac{1}{2}$, then the continuous complex-valued function $f \in C_{\mathbb{C}}[0,1]$, given by $f(s) = s^{\frac{1}{\lambda}-1}$, satisfies $Tf = \lambda f$.]

(f) T is not a compact operator.

14. Let $1 < p < \infty$. We consider the operator T of the preceding exercise defined on $L_p[0,\infty)$ via the same formula

$$Tf(t) = \tfrac{1}{t}\int_0^t f(s)\,ds\,, \qquad (\star\star)$$

for all $s > 0$ and $Tf(0) = f(0)$. Show that $(\star\star)$ defines a one-to-one strictly positive operator from $L_p[0,\infty)$ to $L_p[0,\infty)$ such that $\|T\|_p = \frac{p}{p-1}$.

15. Prove the converse of Corollary 9.14. That is, show that if $T\colon E \to E$ is a positive operator on a Banach lattice such that $R(\lambda, T)$ is strongly expanding for some $\lambda > r(T)$, then T is ideal irreducible. [HINT: If a closed ideal J is T-invariant, then J is also $R(\lambda, T)$-invariant.]

16. For a positive operator $T\colon E \to E$ on a Banach lattice show the following.

(a) The operator T is ideal irreducible if and only if $\{0\}$ and E^* are the only w^*-closed bands in E^* that are invariant under the adjoint operator $T^*\colon E^* \to E^*$.

(b) If the adjoint operator T^* is band irreducible, then T is ideal irreducible. Give an example of an ideal irreducible operator whose adjoint is not band irreducible.

(c) If E has order continuous norm, then T is ideal irreducible if and only if T^* is band irreducible.

[HINT: For each ideal J of E let

$$J^{\mathrm{o}} = \{x^* \in E^*\colon\ \langle x, x^*\rangle = 0 \text{ for all } x \in J\}$$

and for each band B of E^* let $B^{\mathrm{o}} = \{x \in E\colon\ \langle x, x^*\rangle = 0 \text{ for all } x^* \in B\}$. For (b) notice that the adjoint of any positive operator on $C[0,1]$ cannot be ideal irreducible since the norm dual of $C[0,1]$ does not have any quasi-interior points. To establish (c), use (a) in conjunction with the fact that a Banach lattice has order continuous norm if and only if every band in E^* is w^*-closed; see [**29**, Theorem 9.1, p. 60].]

17. Give examples of two positive operators S and T on atomic and non-atomic Banach lattices such that T is ideal irreducible, $ST \leq S$, and S is a non-expanding strictly positive operator. (This shows that the inequality $TS \leq \lambda S$ in Lemma 9.13 cannot be replaced by $ST \leq \lambda S$.)

9.2. Ideal Irreducibility and the Spectral Radius

The Krein–Rutman Theorem 7.10 asserts that when the spectral radius of a compact positive operator is positive, then it is an eigenvalue corresponding to a positive eigenvector. However, the Krein–Rutman theorem does not tell us when the spectral radius of a compact positive operator is positive! A glance at the Volterra operator shows that the spectral radius can be zero. The purpose of this section is to present conditions that guarantee the positivity of the spectral radius for compact positive operators. To avoid trivialities, we shall assume that *all operators appearing in the statements of the results below are non-zero and that the dimensions of the Banach lattices are greater than one.*

We start our discussion with the notion of local quasinilpotence which was introduced first in [**10**].

Definition 9.18. *A bounded operator* $T\colon X \to X$ *on a Banach space is said to be* (***locally***) ***quasinilpotent*** *at a vector* x_0 *if* $\lim_{n\to\infty} \|T^n x_0\|^{\frac{1}{n}} = 0$.

It is well known that an operator $T\colon X \to X$ on a Banach space is quasinilpotent if and only if it is locally quasinilpotent at each point of X; see Exercise 10 in Section 6.1. On the other hand, a one-to-one operator can easily be locally quasinilpotent at some points without being quasinilpotent; see Exercise 4 at the end of this section.

As we shall see later, non-zero compact operators on infinite dimensional Banach spaces always have non-trivial closed hyperinvariant subspaces; see Theorem 10.15. The next basic result, due to B. de Pagter, guarantees the existence of non-trivial closed ℓ-hyperinvariant ideals for locally quasinilpotent compact positive operators. B. de Pagter's proof is a skillful adaptation to the Banach lattice setting of H. M. Hilden's proof of Lomonosov's Invariant Subspace Theorem 10.19 as presented by A. J. Michaels in [**241**]; see Theorem 10.15. The fact that quasinilpotence can be replaced by local quasinilpotence has been observed in [**7**].

Recall that if $T\colon X \to X$ is a bounded operator on a Banach space, then its **commutant** $\{T\}'$ is the subalgebra of $\mathcal{L}(X)$ consisting of all bounded operators on X that commute with T, i.e.,

$$\{T\}' = \{S \in \mathcal{L}(X)\colon\ ST = TS\}.$$

Theorem 9.19 (de Pagter [**260**]). *Every non-zero compact positive operator which is locally quasinilpotent at a non-zero positive vector has a non-trivial closed ℓ-hyperinvariant ideal.*

Proof. Let $T\colon E \to E$ be a (non-zero) compact positive operator on a Banach lattice. We can assume that T is strictly positive; otherwise (according to Lemma 9.1) the null ideal N_T of T is a non-trivial closed ℓ-hyperinvariant ideal. By Lemma 9.1 the non-zero closed ideal J_T generated by the range of T,

$$J_T = \big\{x \in E\colon\ \exists\ y \geq 0 \text{ such that } |x| \leq Ty\big\}^{-},$$

is lattice hyperinvariant for T. If $J_T \neq E$, then J_T is the desired closed ℓ-hyperinvariant ideal for T. So, assume that $J_T = E$.

The compactness of T implies that the range of T is separable; see Exercise 6 in Section 2.4. So, there exists a countable subset $\{w_1, w_2, \ldots\}$ of the range $T(E)$ consisting of non-zero vectors that is norm dense in $T(E)$. Let $w = \sum_{n=1}^{\infty} \frac{|w_n|}{2^n\|w_n\|} > 0$. The obvious inclusion $\{w_1, w_2, \ldots\} \subseteq E_w$ implies $T(E) \subseteq \overline{E}_w$. Hence, $E = J_T \subseteq \overline{E}_w$, and so $\overline{E}_w = E$. In other words, in this case, we can also suppose that E has quasi-interior points. Next, consider the collection of operators

$$\mathcal{A} = \big\{A \in \mathcal{L}(E)\colon\ \exists\ 0 \leq B \in \{T\}' \text{ such that } |Ax| \leq B|x| \text{ for all } x \in E\big\}.$$

Clearly, $\mathcal{A}$ is a subalgebra of $\mathcal{L}(E)$ containing the center of E, i.e., $\mathcal{Z}(E) \subseteq \mathcal{A}$. Now for each $x \in E$, let

$$\mathcal{A}x = \big\{Ax\colon\ A \in \mathcal{A}\big\}.$$

We claim that $\mathcal{A}x$ is an ℓ-hyperinvariant vector subspace of E. To see this, let $0 \leq S \in \mathcal{L}(E)$ satisfy $ST = TS$. Choose any $A \in \mathcal{A}$ and consider the vector $y = Ax \in \mathcal{A}x$. Pick $0 \leq B \in \{T\}'$ satisfying $|Az| \leq B|z|$ for all $z \in E$, and note that $|SAz| \leq S|Az| \leq (SB)|z|$ for each $z \in E$ and that $SB \in \{T\}'$. This implies $SA \in \mathcal{A}$ and so $Sy = SAx \in \mathcal{A}x$. In other words, $\mathcal{A}x$ is ℓ-hyperinvariant for each $x \in E$. Consequently, the norm closure $\overline{\mathcal{A}x}$ is likewise an ℓ-hyperinvariant subspace of E for each x.

Since $I \in \mathcal{A}$, we have $x = Ix \in \mathcal{A}x$, and so $\mathcal{A}x$ is non-zero for each non-zero vector $x \in E$. Next, we claim that $\overline{\mathcal{A}x}$ is, in fact, a closed ideal in E for each x. To see this, let $y \in E$ satisfy $|y| \leq \sum_{i=1}^{k} |A_i x|$, where each $A_i \in \mathcal{A}$. Since (in view of the Riesz Decomposition Property [**29**, Theorem 1.2, p. 3]) we can write $y = \sum_{i=1}^{k} y_i$ with $|y_i| \leq |A_i x|$ for each i, we can assume that $|y| \leq |Ax|$ for some $A \in \mathcal{A}$. Now a glance at Lemma 4.17 guarantees the existence of a sequence $\{T_n\} \subseteq \mathcal{Z}(E) \subseteq \mathcal{A}$ such that $T_n Ax \to y$. Since $T_n A \in \mathcal{A}$ for each n, we immediately get that $y \in \overline{\mathcal{A}x}$, and so $\overline{\mathcal{A}x}$ is an ideal.

To finish the proof, it suffices to show that $\overline{\mathcal{A}x} \neq E$ for some $x \neq 0$. To this end, assume by way of contradiction that $\overline{\mathcal{A}x} = E$ for each $x \neq 0$. Let $u > 0$ be a vector of local quasinilpotence of T, i.e., $\lim_{n\to\infty} \|T^n u\|^{\frac{1}{n}} = 0$.

The strict positivity of T implies $Tu > 0$. So, by the continuity of T, there exists an open ball U centered at u such that

$$0 \notin \overline{U} \quad \text{and} \quad 0 \notin \overline{T(U)}\,. \tag{$\star$}$$

From our assumption that $\overline{\mathcal{A}x} = E$ for each $x \in \overline{T(U)}$, we see that $u \in \overline{\mathcal{A}x}$ for each $x \in \overline{T(U)}$.

Now for each $x \in \overline{T(U)}$ pick some operator $A_x \in \mathcal{A}$ such that $A_x(x) \in U$, and then choose some open ball B_x centered at x such that $A_x(B_x) \subseteq U$. The compactness of T implies that $\overline{T(U)}$ is a norm compact set. Since the open balls $\{B_x\colon\ x \in \overline{T(U)}\}$ cover $\overline{T(U)}$, there exist $x_1, \ldots, x_k \in \overline{T(U)}$ such that $\overline{T(U)} \subseteq \bigcup_{i=1}^k B_{x_i}$. For simplicity let us write B_i instead of B_{x_i} and A_i instead of A_{x_i}. Next, notice that there exists some operator $0 \leq R \in \{T\}'$ satisfying $|A_i z| \leq R|z|$ for all $i = 1, \ldots, k$ and all $z \in E$.

Since $Tu \in \overline{T(U)}$, there exists some index $j_1 \in \{1, \ldots, k\}$ such that $Tu \in B_{j_1}$. This implies $A_{j_1}Tu \in U$ and so $TA_{j_1}Tu \in \overline{T(U)}$. Repeating this argument inductively, we obtain a sequence $\{j_n\}$ of indices from $\{1, \ldots, k\}$ such that

$$z_n = (A_{j_n}T)(A_{j_{n-1}}T)\cdots(A_{j_1}T)u \in U$$

for each n. From $|A_i z| \leq R|z|$ and $RT = TR$ we get $|z_n| \leq R^n T^n u$ for each n. The latter implies $\|z_n\| \leq \|R\|^n \cdot \|T^n u\|$. Hence,

$$\|z_n\|^{\frac{1}{n}} \leq \|R\| \cdot \|T^n u\|^{\frac{1}{n}} \to 0\,,$$

and so $z_n \to 0$. Since $z_n \in U$ for each n, it follows that $0 \in \overline{U}$, contrary to $(\star)$. This contradiction completes the proof of the theorem. ■

Corollary 9.20 (de Pagter [**260**]). *An ideal irreducible compact positive operator on a Banach lattice has a positive spectral radius.*

It should be pointed out that the compactness assumption in the above corollary is essential; see the comments at the end of Section 9.3. We are now in a position to sharpen the previous result by proving the positivity of the spectral radii of two commuting positive operators.

Corollary 9.21 (Abramovich–Aliprantis–Burkinshaw [**7**]). *If an ideal irreducible positive operator $T\colon E \to E$ commutes with a compact positive operator $S\colon E \to E$, then $r(T) > 0$ and $r(S) > 0$.*

Proof. Let $S, T\colon E \to E$ be two (non-zero) positive operators on a Banach lattice. Assume that S is compact, T is ideal irreducible, and $ST = TS$. From Theorem 9.3, it follows that $ST = TS > 0$.

Assume that $r(ST) = 0$. Then, by Theorem 9.19, the compact positive operator ST has a non-trivial closed ℓ-hyperinvariant ideal J. Since the operator T commutes with ST, we see that $T(J) \subseteq J$, which contradicts the ideal irreducibility of T. Hence, $r(ST) > 0$. It remains to observe that (in view of the commutativity of S and T) we have $0 < r(ST) \leq r(S)r(T)$; see Exercise 8 of Section 6.1. This implies $r(T) > 0$ and $r(S) > 0$. ■

The next results show that the compactness assumption in Corollary 9.20 can be considerably relaxed.

Corollary 9.22 (Abramovich–Aliprantis–Burkinshaw [**7**]). *If T is a compactly dominated ideal irreducible positive operator on a Banach lattice, then every positive operator S in the commutant of T has a positive spectral radius, i.e., $r(S) > 0$.*

Proof. Let $0 \leq T \leq K$ for some compact operator K and let S be a positive operator commuting with the ideal irreducible operator T. By Theorem 9.3 the operator T is strictly positive. Hence, by Corollary 2.35, the strictly positive operator T^3 is compact. Theorem 9.3 shows that E has quasi-interior points and that T carries quasi-interior points to quasi-interior points. This implies that the positive operator ST^3 is non-zero. Furthermore, ST^3 is compact and commutes with T. Finally, Corollary 9.21 implies $r(ST^3) > 0$, and so from $r(ST^3) \leq r(S)r(T^3)$ we obtain $r(S) > 0$. ■

Recall that a Banach space X is said to have the **Dunford–Pettis Property** if $x_n \xrightarrow{w} 0$ in X and $x_n^* \xrightarrow{w} 0$ in X^* imply $x_n^*(x_n) \to 0$. It is well known that AL- and AM-spaces have the Dunford–Pettis Property; see [**30**, Section 19].

Lemma 9.23. *If $X \xrightarrow{S} Y \xrightarrow{T} Z$ are weakly compact operators between Banach spaces and Y has the Dunford–Pettis property, then TS is a compact operator.*

Proof. See [**30**, Theorem 19.8, p. 337]. ■

Corollary 9.24. *Let $T\colon E \to E$ be an ideal irreducible weakly compact positive operator on a Banach lattice. If E has the Dunford–Pettis property (in particular, if E is an AL- or an AM-space), then T has a positive spectral radius.*

Proof. In this case T^2 is a strictly positive compact operator which commutes with the ideal irreducible operator T. ■

B. de Pagter's Theorem 9.19 guarantees that a compact quasinilpotent positive operator $T\colon E \to E$ on a Banach lattice has a non-trivial ℓ-hyperinvariant ideal J. What this result does not guarantee is that $T(J) \neq \{0\}$. Here is a simple example.

Example 9.25. Consider the positive operator $T\colon \mathbb{R}^2 \to \mathbb{R}^2$ defined by the formula $T(x,y) = (y,0)$. Clearly, $T^2 = 0$. It is easy to verify that the ideal (in fact, the band) $J = \{(x,0)\colon\ x \in \mathbb{R}\}$ is the only non-trivial closed hyperinvariant ideal for T. However, $T(J) = \{0\}$. ∎

Surprisingly, if we exclude the case $T^2 = 0$, then the lattice hyperinvariant ideal J guaranteed by Theorem 9.19 can be chosen to satisfy $T(J) \neq \{0\}$.

Theorem 9.26. *If a compact quasinilpotent positive operator $T\colon E \to E$ on a Banach lattice satisfies $T^2 \neq 0$, then there exists a non-trivial closed ℓ-hyperinvariant ideal J for T such that $T(J) \neq \{0\}$.*

Proof. Let $T\colon E \to E$ be a compact quasinilpotent positive operator on a Banach lattice such that $T^2 \neq 0$. Let $N = \{x \in E\colon\ T|x| = 0\}$, the null ideal of T, and define the operator $\hat{T}\colon E/N \to E/N$ by $\hat{T}[x] = [Tx]$. Clearly, $\hat{T}$ is a quasinilpotent compact positive operator. We also claim that $\hat{T}$ is non-zero. Indeed, if we pick some $x > 0$ satisfying $T^2x > 0$, then $Tx \notin N$, and hence $\hat{T}[x] = [Tx] > 0$ in E/N, which shows that $\hat{T} > 0$.

By Theorem 9.19 there exists a non-trivial closed ℓ-hyperinvariant ideal $\hat{J}$ for $\hat{T}$. Let $J = \{x \in E\colon\ [x] \in \hat{J}\}$. Since the quotient map $x \mapsto [x]$ is a lattice homomorphism, it follows that J is a non-trivial closed ideal in E that obviously is invariant under T. We claim that $T(J) \neq \{0\}$. To see this, pick some $0 < x \in J$ satisfying $0 < [x] \in \hat{J}$ and notice that $Tx > 0$. Otherwise, $Tx = 0$ implies $x \in N$, that is, $[x] = 0$, which is a contradiction.

Finally, notice that J is an ℓ-hyperinvariant ideal for T. To verify this, take a positive operator $S\colon E \to E$ that commutes with T. If $[x] = [y]$, i.e., if $x - y \in N$, then from $0 \le T|Sx - Sy| \le TS|x-y| = ST|x-y| = 0$, we see that $Sx - Sy \in N$, that is, $[Sx] = [Sy]$. Therefore, S defines an operator $\hat{S} \ge 0$ on E/N via the formula $\hat{S}[x] = [Sx]$. As $\hat{S}$ commutes with $\hat{T}$, the inclusion $\hat{S}(\hat{J}) \subseteq \hat{J}$ holds and from this we infer that $S(J) \subseteq J$. ∎

We conclude the section with a general result concerning strongly expanding operators.

Theorem 9.27. *Let $Q, S\colon E \to E$ be two positive operators on a Banach lattice such that:*

(1) *Q is a strongly expanding operator.*
(2) *S dominates a compactly dominated positive operator—in particular, this is so if S itself is dominated by a compact operator.*

Then $r(QS) > 0$, i.e., the spectral radius of the operator QS is positive.

Proof. Let Q and S satisfy the properties stated in the theorem. Pick a positive operator L and a compact operator K satisfying $0 < L \le S$ and $L \le K$.

Consider the positive operator $C = QL$ and note that C is non-zero since Q is strongly expanding. Note also that if $C^k x > 0$ for some $x > 0$ and some k, then $C^k x$ is a quasi-interior point. (This follows from the fact that C^k can be written in the form $C^k = QR$ for some $R > 0$.) This guarantees that $C^k > 0$ for each k. Since $0 < C = QL \le QK$, Corollary 2.35 implies that C^3 is a compact operator.

Assume that $r(C) = 0$. Then $r(C^3) = [r(C)]^3 = 0$ and taking into account that $(C^3)^2 = C^6 > 0$, Theorem 9.26 guarantees the existence of a non-trivial closed ℓ-hyperinvariant ideal J for C^3 with $C^3(J) \ne \{0\}$. Hence, for some $x > 0$ the quasi-interior point $C^3 x$ belongs to the closed ideal J, which implies that $J = E$, contradicting our assumption that $r(C) = 0$. So, $r(QS) \ge r(C) > 0$. ■

Corollary 9.28. *If the commutant of a positive operator T contains an ideal irreducible positive operator S and a compact positive operator K, then $r(T) > 0$.*

Proof. Let $TS = ST$ and $TK = KT$, where $S > 0$ is ideal irreducible and $K > 0$ is compact. Fix $\lambda > r(S)$ and note that the operator $R(\lambda, S)$ is strongly expanding by Corollary 9.14. This implies $TR(\lambda, S)K > 0$. Since both $R(\lambda, S)$ and K commute with T, we get $T\big[R(\lambda, S)K\big] = \big[R(\lambda, S)K\big]T > 0$. Consequently, $TK = KT > 0$. Next, notice that the operators $R(\lambda, S)$ and TK satisfy the hypotheses of Theorem 9.27, and so

$$0 < r\big(R(\lambda, S)TK\big) = r\big(T[R(\lambda, S)K]\big) \le r(T)r\big(R(\lambda, S)K\big).$$

Hence, $r(T) > 0$. ■

Corollary 9.29. *Let $S, T\colon E \to E$ be two commuting positive operators on a Banach lattice. If T is ideal irreducible and some power of S dominates a compactly dominated positive operator, then $r(S) > 0$ and $r(T) > 0$.*

Proof. Assume S^m dominates some compactly dominated positive operator. Fix some $\lambda > r(T)$ and let $Q = R(\lambda, T)T$. By Corollary 9.14, Q is strongly expanding. Now from Theorem 9.27, it follows that

$$0 < r\big(QS^m\big) = r\big(R(\lambda,T)TS^m\big) \le r\big(R(\lambda,T)\big)r(T)\big[r(S)\big]^m,$$

and so $r(S) > 0$ and $r(T) > 0$ both hold. ■

The last four results are due to the authors and O. Burkinshaw [**7**].

Exercises

1. Let E be a Banach lattice and let $T\colon E \to E$ be a positive operator having a non-trivial closed invariant ideal. Show that there exists some $0 < u \in E$ satisfying $\overline{E}_u \neq E$ and $T(\overline{E}_u) \subseteq \overline{E}_u$. [HINT: Let J be a non-trivial closed T-invariant ideal and fix some $0 < x \in J$. Next consider the element $u = \sum_{n=0}^{\infty} \frac{1}{2^n} \cdot \frac{T^n x}{\|T\|^n}$.]
2. Let $T\colon X \to X$ be a bounded operator on a Banach space. If λ is an eigenvalue of T, then show that its eigenspace $N_T(\lambda) = \{x \in X\colon Tx = \lambda x\}$ is a closed T-hyperinvariant subspace.
3. (Ercan–Onal [**122**]) Let $T\colon X \to X$ be a bounded operator on a Banach space and let x be a vector in X. Show that T is locally quasinilpotent at x if and only if $\big|x^*(T^n x)\big|^{\frac{1}{n}} \to 0$ for each $x^* \in X^*$.

   If X is a Banach lattice, then show that T is locally quasinilpotent at x if and only if $\big|x^*(T^n x)\big|^{\frac{1}{n}} \to 0$ holds for each $0 \le x^* \in X^*$.
4. This exercise presents a one-to-one positive operator which is locally quasinilpotent at some positive vector but fails to be quasinilpotent. Consider the positive operator $T\colon \ell_2 \to \ell_2$ defined by

   $$T(x_1, x_2, x_3, \ldots) = \big(x_1, x_1, \tfrac{x_2}{2}, \tfrac{x_3}{3}, \ldots\big).$$

   Show that T is one-to-one, is locally quasinilpotent at e_2, and fails to be quasinilpotent. [HINT: From $Te_n = \frac{1}{n} e_{n+1}$ for all $n \ge 2$, we see that $T^n e_2 = \frac{1}{(n+1)!} e_{n+2}$. Hence $\|T^n e_2\| = \frac{1}{(n+1)!}$, from which it follows that $\lim_{n\to\infty} \|T^n e_2\|^{\frac{1}{n}} = 0$. On the other hand, $\|T^n\| \ge \|T^n e_1\| \ge 1$ implies $\|T^n\|^{\frac{1}{n}} \ge 1$ for each n.]
5. Let $T\colon X \to X$ be a continuous operator on a Banach space and let $\mathcal{Q}_T$ denote the set of all points at which T is locally quasinilpotent, i.e.,

   $$\mathcal{Q}_T = \Big\{x \in X\colon \lim_{n\to\infty} \|T^n x\|^{\frac{1}{n}} = 0\Big\}.$$

   Show that $\mathcal{Q}_T$ is a T-hyperinvariant vector subspace.[2] [HINT: Let x, y in $\mathcal{Q}_T$ and fix $\epsilon > 0$. Pick n_0 such that $\|T^n x\| < \epsilon^n$ and $\|T^n y\| < \epsilon^n$ for

[2] For some applications and historical comments regarding $\mathcal{Q}_T$ see [**233, 234**].

all $n \geq n_0$. So, $\|T^n(x+y)\|^{\frac{1}{n}} \leq (\|T^n x\| + \|T^n y\|)^{\frac{1}{n}} < 2\epsilon$ for all $n \geq n_0$. Now let $x_0 \in \mathcal{Q}_T$, let $S \in \mathcal{L}(X)$ satisfy $ST = TS$, and note that

$$\|T^n(Sx_0)\|^{\frac{1}{n}} = \|S(T^n x_0)\|^{\frac{1}{n}} \leq \|S\|^{\frac{1}{n}} \cdot \|T^n x_0\|^{\frac{1}{n}} \xrightarrow[n\to\infty]{} 0\,.]$$

6. For the notation used here refer to Exercise 5 above. For a bounded operator $T\colon X \to X$ on a Banach space establish the following properties.
(a) $\mathcal{Q}_T = \{0\}$ is possible.
(b) $\mathcal{Q}_T$ can be dense without being equal to X.
[HINT: For (a) notice that every isometry T satisfies $\mathcal{Q}_T = \{0\}$. For (b) look at the backward shift.]

7. This exercise presents a pair of positive operators that satisfy the hypotheses of Corollary 9.21. Consider the positive operator $T\colon \ell_\infty \to \ell_\infty$ defined by

$$T(x_1, x_2, x_3, \ldots) = \left(\tfrac{x_1+x_2}{2}, \tfrac{x_1+x_3}{2}, \tfrac{x_1+x_4}{2}, \ldots\right).$$

Establish the following:
(a) The operator T is a Krein (and hence an ideal irreducible) non-compact positive operator.
(b) The spectral radius of T equals one, i.e., $r(T) = 1$.
(c) There exists some $0 \leq \phi \in \ell_\infty^*$ satisfying $\|\phi\| = \phi(e) = 1$ and $T^*\phi = \phi$, where e denotes the constant one sequence, i.e., $e = (1, 1, 1, \ldots)$.
(d) The rank-one positive operator $S = \phi \otimes e$ on ℓ_∞ satisfies $TS = ST = S$.
[HINT: Note that T is a Markov operator and then use part (b) of Exercise 12 of Section 9.1.]

8. Consider the operator $T\colon C[0,1] \to C[0,1]$ defined by the formula

$$Tx(t) = \int_0^1 \sin(st)x(s)\,ds\,.$$

Show the following:
(a) T is one-to-one and strictly positive.
(b) T is a compact operator.
(c) If $J = \{x \in C[0,1]\colon\ x(0) = 0\}$, then J is a non-trivial closed ideal and $T(J) \subseteq J$ holds—and so T is not ideal irreducible.
(d) For each $x > 0$ the function Tx is a quasi-interior point of J. In particular, the operator $T\colon J \to J$ (the restriction of T to J) is strongly expanding—and hence ideal irreducible.
(e) The spectral radius of the operator $T\colon C[0,1] \to C[0,1]$ is positive (i.e., $r(T) > 0$) and coincides with the spectral radius of the operator $T\colon J \to J$ (the restriction of T to J).
[HINT: For (a) notice that if $Tx = 0$, then $\int_0^1 s^{2n} \sin(st)x(s)\,ds = 0$ for each $n = 0, 1, \ldots$. Now observe that the algebra generated by $\{\mathbf{1}, s^2\}$ is (by the Stone–Weierstrass theorem) norm dense in $C[0,1]$.]

9. Generalize Corollary 9.21 as follows: If an ideal irreducible positive operator $T\colon E \to E$ commutes with a compact positive operator $S\colon E \to E$, then neither T nor S is locally quasinilpotent at any positive vector.

10. Generalize Corollary 9.22 as follows: If $T\colon E \to E$ is a compactly dominated ideal irreducible positive operator on a Banach lattice, then every positive operator on E that commutes with T (in particular, T itself) cannot be locally quasinilpotent at any positive vector of E.

11. Show that in Corollary 9.22 the hypothesis "S commutes with T" cannot be replaced by "S commutes with some power of T" even when T is compact. [HINT: Consider the pair of positive operators $S, T\colon \mathbb{R}^2 \to \mathbb{R}^2$ defined by $S(x, y) = (y, 0)$ and $T(x, y) = (y, x)$. Then $T^2 = I$ and so $ST^2 = T^2 S$. However, $r(S) = 0$.]

12. Show that if an ideal irreducible positive operator $T\colon E \to E$ on a Banach lattice has a compact power (in particular, if T is dominated by a compact operator), then $r(T) > 0$. [HINT: Use Corollary 9.21.]

13. (de Pagter [**260**]) Show that if a power of an ideal irreducible positive operator T dominates a compactly dominated positive operator, then $r(T) > 0$. [HINT: Use Corollary 9.29.]

9.3. Band Irreducibility and the Spectral Radius

In this section, we shall discuss the spectral radius of band irreducible positive operators. As we shall see, the results are analogous to the ones of the previous section but not identical. When we deal with band irreducible positive operators, quite often we must assume that they are also σ-order continuous. As in the previous section, we shall suppose that all operators appearing in the statements of the results are non-zero and they act on Banach lattices of dimension greater than one.

Lemma 9.30. *Let $T\colon E \to E$ be a σ-order continuous, compact, and strictly positive operator on a Banach lattice. Then every positive operator on E that commutes with T is order continuous.*

Proof. By Corollary 9.16, we know that every σ-order continuous positive operator on E is automatically order continuous. Now assume that a positive operator $S\colon E \to E$ satisfies $ST = TS$. Let $x_n \downarrow 0$. Then $Tx_n \downarrow 0$ and by the compactness of T, we infer[3] that $\|Tx_n\| \downarrow 0$. So, $\|STx_n\| \downarrow 0$. On the other hand, if $Sx_n \geq x \geq 0$ holds for each n, then from $STx_n = TSx_n \geq Tx \geq 0$ and $\|STx_n\| \downarrow 0$, we get $Tx = 0$. Since T is strictly positive, the latter shows that $x = 0$, and so $Sx_n \downarrow 0$, i.e., S is σ-order continuous—and consequently S is order continuous. ■

Let us mention that without the compactness of the operator T Lemma 9.30 is not true even if T is in addition ideal irreducible; see Exercise 1 at the end of the section.

[3] Here is the argument: The compactness of T implies that the sequence $\{Tx_n\}$ has a norm convergent subsequence, say $\{Tx_{k_n}\}$. Since $Tx_{k_n} \downarrow 0$, Lemma 1.30(c) guarantees $\|Tx_{k_n}\| \downarrow 0$, and from this we obtain $\|Tx_n\| \downarrow 0$.

The analogue of Theorems 9.26 and 9.19 for lattice hyperinvariant bands is as follows.

Theorem 9.31. *If $T\colon E \to E$ is an order continuous compact quasinilpotent positive operator on a Banach lattice, then there exists a non-trivial ℓ-hyperinvariant band B for T. Moreover, if $T^2 \neq 0$, then the band B can be chosen to satisfy $T(B) \neq \{0\}$.*

Proof. If the null ideal $N = \{x \in E\colon\ T|x| = 0\}$ is non-zero, then N is (by Lemma 9.1) a non-trivial T-hyperinvariant band. So, we can assume that $N_T = \{0\}$, and hence T is strictly positive. Let J_T denote the closed ideal generated by the range of T in E, i.e.,

$$J_T = \{x \in E\colon\ \exists\ y \geq 0 \text{ such that } |x| \leq Ty\}^- .$$

By Lemma 9.1, J_T is a closed lattice hyperinvariant ideal for T. Since the operator $T\colon J_T \to J_T$ is strictly positive and quasinilpotent, it follows from Theorem 9.26 that T, as an operator on J_T, has a non-trivial closed ℓ-hyperinvariant ideal $J \subseteq J_T$. Since J_T itself is ℓ-hyperinvariant, it follows that J, considered as an ideal of E, is also ℓ-hyperinvariant. Now let B be the band generated by J in E. By Lemma 9.30, every positive operator commuting with T is order continuous. This implies that B is an ℓ-hyperinvariant band for T. To finish the proof, it suffices to show that $B \neq E$. To see this, assume by way of contradiction that $B = E$. Hence, for each $x > 0$ there exists a net $\{x_\alpha\}$ in J such that $0 \leq x_\alpha \uparrow x$. By the order continuity of T, we get $Tx_\alpha \uparrow Tx$, and by the compactness of T we see that the increasing net $\{Tx_\alpha\}$ is norm convergent in E. It follows that $\|Tx_\alpha - Tx\| \to 0$. Since J is T-invariant, $\{Tx_\alpha\} \subseteq J$, and thus $Tx \in J$ for all $x \in E$. This implies $J = J_T$, contradicting our assumption $J \neq J_T$.

To establish the second part, assume now that $T^2 \neq 0$. In this case, we consider the operator $\hat{T}\colon E/N \to E/N$ defined by $\hat{T}[x] = [Tx]$, where N is again the null ideal of T. Since T is order continuous, N is a band. We claim that $\hat{T} > 0$. Indeed, if we pick $x > 0$ such that $T^2x > 0$, then $Tx \notin N$ and so $\hat{T}[x] > 0$. Clearly, $\hat{T}$ is quasinilpotent and compact. Furthermore, we claim that $\hat{T}$ is order continuous.

To see this, let $[x_\alpha] \downarrow 0$ in E/N. Replacing $\{x_\alpha\}$ by the net of all finite infima of the net $\{x_\alpha^+\}$, we can assume that $0 \leq x_\alpha \downarrow$ holds in E. Since E is Archimedean, there exists a net $\{y_\lambda\}$ of E satisfying $0 \leq y_\lambda \leq x_\alpha$ for each α and each λ and $x_\alpha - y_\lambda \downarrow_{(\alpha,\lambda)} 0$; see [**225**, Theorem 22.5, p. 115]. From $[x_\alpha] \downarrow 0$, we see that $[y_\lambda] = 0$, i.e., $Ty_\lambda = 0$ for each λ. Therefore, replacing $\{x_\alpha\}$ by $\{x_\alpha - y_\lambda\}$, we can assume without loss of generality that $x_\alpha \downarrow 0$ in E. By the order continuity of T, it follows that $Tx_\alpha \downarrow 0$. So, by the order continuity of the quotient map $x \mapsto [x]$ (see Exercise 2 at the end of this section), we obtain $\hat{T}[x_\alpha] = [Tx_\alpha] \downarrow 0$. Thus, $\hat{T}$ is order continuous.

By the first part of the proof, there exists a non-trivial ℓ-hyperinvariant band $\hat{D}$ in E/N. Let $D = \{x \in E\colon\ [x] \in \hat{D}\}$. The order continuity of the quotient map $x \mapsto [x]$ implies that D is a non-trivial ℓ-hyperinvariant band. To complete the proof, it suffices to show that $T(D) \neq \{0\}$. To see this, pick any $0 < x \in D$ satisfying $0 < [x] \in \hat{D}$. We claim that $Tx > 0$. Otherwise, $Tx = 0$ would imply that $x \in N$, i.e., $[x] = 0$, which is a contradiction. ■

Using the preceding theorem we will show now that every order continuous compact positive operator lying in the commutant of a band irreducible positive operator has a non-zero spectral radius.

Corollary 9.32. *Every order continuous compact positive operator T that commutes with a band irreducible positive operator has a positive spectral radius, i.e., $r(T) > 0$.*

Proof. Assume that T commutes with a band irreducible positive operator S. If $r(T) = 0$, then by Theorem 9.31 the operator T has a non-trivial hyperinvariant band B. In particular, $S(B) \subseteq B$ must hold, which is impossible. So, $r(T) > 0$. ■

Corollary 9.33 (Schaefer–Grobler)**.** *Every σ-order continuous band irreducible compact positive operator on a Banach lattice has a positive spectral radius.*

Proof. Let $T\colon E \to E$ be a σ-order continuous band irreducible compact positive operator. Then T is strictly positive (see Exercise 10 in Section 9.1), and so (by Lemma 9.30) T is also order continuous. By Corollary 9.32, $r(T) > 0$. ■

Under the assumptions of Corollary 9.33 not only T but also every positive operator in the commutant of T has positive spectral radius.

Corollary 9.34. *Let T be a σ-order continuous band irreducible compact positive operator. Then every positive operator S in the commutant of T has positive spectral radius, i.e., $r(S) > 0$.*

Proof. Let $0 < S \in \{T\}'$. Observe first that ST is a σ-order continuous positive operator. To see this, note that since T is strictly positive, we have $ST = TS > 0$. Next, let $x_n \downarrow 0$. Then $Tx_n \downarrow 0$ and by the compactness of T, we get $\|Tx_n\| \downarrow 0$. So, from $0 \le STx_n \downarrow$ and $\|STx_n\| \downarrow 0$, we obtain $STx_n \downarrow 0$, which means that ST is σ-order continuous. Since ST commutes with T, Lemma 9.30 implies that ST is also order continuous.

Now notice that since the compact positive operator ST commutes with the band irreducible positive operator T, Corollary 9.32 guarantees that $0 < r(ST) \le r(S)r(T)$. Consequently, $r(S) > 0$. ■

The analogue of Theorem 9.27 for bands is as follows.

Theorem 9.35. *Let $Q, S\colon E \to E$ be two positive operators such that:*

(1) *Q is σ-order continuous and expanding.*

(2) *S dominates a compactly dominated σ-order continuous positive operator—in particular, this is the case if S itself is σ-order continuous and dominated by a compact operator.*

Then $r(QS) > 0$, i.e., the spectral radius of the operator QS is positive.

Proof. The proof is a modification of the proof of Theorem 9.27. Pick a σ-order continuous operator $L > 0$ and a compact operator $K > 0$ such that $L \leq S$ and $L \leq K$.

Define the operator $C = QL$ and note that if for some $x > 0$ we have $C^3 x > 0$, then $C^3 x$ is a weak unit. Now notice that $(LQ)^3 = L(QL)(QL)Q$ is a compact strictly positive operator. By Lemma 9.15, the Banach lattice E has the countable sup property, and so the operators Q and L are order continuous. On the other hand, from $C^3 = Q(LQ)(LQ)L$ and the order continuity of L, we see that $C^3 > 0$ and $C^6 > 0$. To complete the proof, invoke the arguments employed in the proof of Theorem 9.27 and use Theorem 9.31 instead of Theorem 9.26. ■

Now let E be a Dedekind complete Banach lattice. Recall that the Dedekind complete Riesz space of all regular operators on E is denoted by $\mathcal{L}_r(E)$ and that the Riesz space of all **abstract integral operators** is the band $(E_n^\sim \otimes E)^{dd}$ generated by the finite-rank operators $E_n^\sim \otimes E$ in $\mathcal{L}_r(E)$. From this definition it follows that for each positive abstract integral operator S there exist a finite-rank operator $K \in \mathcal{L}(E)$ and a positive operator $L \in \mathcal{L}(E)$ satisfying $0 < L \leq S$ and $L \leq K$. In particular, K is a compact operator and so S dominates a compactly dominated order continuous operator. In other words, each positive operator $S \in (E_n^\sim \otimes E)^{dd}$ satisfies condition (2) of Theorem 9.35. We are ready now to formulate and prove the next result.

Theorem 9.36. *Let E be a Dedekind complete Banach lattice. If a σ-order continuous band irreducible positive operator $T\colon E \to E$ commutes with a positive abstract integral operator $S\colon E \to E$, then $r(T) > 0$ and $r(S) > 0$.*

Proof. Fix some $\lambda > r(T)$ and consider the operator $Q = R(\lambda, T)T$. Clearly, Q commutes with S, is σ-order continuous, and, by Corollary 9.14, is expanding. Therefore, the operators Q and S satisfy the hypotheses of Theorem 9.35, and thus $r(QS) > 0$. Finally, using the inequality $r(QS) \leq r\big(R(\lambda, T)\big)r(T)r(S)$, we obtain $r(T) > 0$ and $r(S) > 0$. ■

The abstract formulation of the celebrated Andô–Krieger theorem [**34**, **194**] is a special case of the preceding theorem.

Corollary 9.37 (Andô–Krieger). *If T is a band irreducible abstract integral positive operator on a Dedekind complete Banach lattice, then $r(T) > 0$.*

We close the section with a brief historical comment. The Andô–Krieger theorem above preceded most results in Sections 9.2 and 9.3. It initiated the study of ideal and band irreducible operators on Banach lattices. The first challenging task was to establish an analogue of Corollary 9.37 for σ-order continuous band irreducible positive operators—recall that integral operators are automatically order continuous; see Corollary 5.12. H. H. Schaefer [**292**] was the first to do so assuming that the operator was compact and that the σ-order continuous dual of the space was separating. Later J. J. Grobler [**139**] removed the last assumption, and their final result is Corollary 9.33.

The second challenging problem was whether or not each compact ideal irreducible positive operator has a strictly positive spectral radius. The affirmative answer was provided by the remarkable result of B. de Pagter (Corollary 9.20).

We refer to [**288**] for an example showing that the compactness assumption in the above results is essential and to [**7**] for an example of a nilpotent compact (in fact, rank-one) band irreducible positive operator.

Exercises

1. Show that the conclusion of Lemma 9.30 is false if the "compactness" of T is replaced by "ideal irreducibility" of T. That is, give an example of two commuting positive operators $T, S\colon E \to E$ on a Banach lattice such that:
   (a) T is strictly positive, order continuous, and ideal irreducible.
   (b) S is not order continuous.

2. Let J be an ideal in a Riesz space E. Show that J is a band if and only if the quotient map $x \mapsto [x]$, from E onto E/J, is order continuous. [HINT: Assume that J is a band, that $x_\alpha \downarrow 0$ in E, and that $[x_\alpha] \geq [x]$ in E/J for all α and some $x \in E^+$. Now use the fact that the quotient map is a lattice homomorphism (see Exercise 2 of Section 3.4) to obtain that $0 \leq (x - x_\alpha)^+ \uparrow x$ holds in E.]

3. Prove the following analogue of Corollary 9.28: Let $T\colon E \to E$ be a σ-order continuous positive operator on a Banach lattice. If the commutant of T contains a σ-order continuous band irreducible positive operator and also a σ-order continuous compact positive operator, then $r(T) > 0$.

4. Prove the following analogue of Corollary 9.29: Let $S, T\colon E \to E$ be two commuting positive operators on a Banach lattice. If T is a σ-order continuous band irreducible operator and some power of S dominates a σ-order continuous compactly dominated positive operator, then both operators have positive spectral radii, i.e., $r(S) > 0$ and $r(T) > 0$.
5. Complete the details in the proof of Theorem 9.35.
6. (Schaefer [**292**]; Grobler [**139**]) If $T\colon E \to E$ is a σ-order continuous band irreducible quasinilpotent positive operator and $S\colon E \to E$ is a compact positive operator, then show that $[0, S] \cap [0, T^k] = \{0\}$ for each $k \geq 1$. [HINT: Let $\lambda > r(T)$ and let $Q = R(\lambda, T)$. If $0 < A \leq S$ and $0 < A \leq T^k$, then from Theorem 9.35 it follows that $0 < r(QT^k) \leq r(Q)[r(T)]^k$ and so $r(T) > 0$, a contradiction.]
7. (Caselles [**86**]) A **Harris operator** is a positive operator $T\colon E \to E$ on a Dedekind complete Banach lattice such that some power of T is not disjoint from the band of abstract integral operators.

   Prove that every σ-order continuous band irreducible Harris operator has a positive spectral radius. [HINT: Use Theorem 9.35.]

9.4. Krein Operators on $C(\Omega)$-spaces

We shall discuss here several remarkable properties of Krein operators on (real) $C(\Omega)$-spaces, where Ω denotes a compact Hausdorff space. As usual, the constant function one on Ω will be denoted by $\mathbf{1}$. The vector space $C(\Omega)$ is a Banach lattice under the sup norm. Its complexification is the Banach lattice $C_{\mathbb{C}}(\Omega)$ of all continuous complex-valued functions. Most of the results in this section are due to M. G. Krein. For a general discussion of Krein operators on general partially ordered vector spaces we refer the reader to [**8**].

Recall that a function $u \in C(\Omega)$ is called a **strong unit**, or simply a **unit**, if for each $x \in C(\Omega)$ there exists some $\lambda > 0$ such that $x \leq \lambda u$. Clearly, a function $u \in C(\Omega)$ is a unit if and only if $u(\omega) > 0$ for each $\omega \in \Omega$. The units of $C(\Omega)$ are precisely the interior points of the positive cone $C(\Omega)^+$. Notice that if u is a unit, then so are: αu for each $\alpha > 0$, $u + x$ for each $x \in C(\Omega)^+$, and each $v \in C(\Omega)$ with $v \geq u$. It should also be clear that a non-negative linear functional on $C(\Omega)$ is the zero functional if and only if it vanishes at some unit. The boundary of $C(\Omega)^+$ with respect to the sup norm is denoted $\partial\big[C(\Omega)^+\big]$.

Lemma 9.38. *If $0 \leq x \in C(\Omega)$ and $y \neq 0$, then there exists some scalar α such that $x + \alpha y \in \partial\big[C(\Omega)^+\big]$.*

Proof. If $x \in \partial\big[C(\Omega)^+\big]$, then let $\alpha = 0$. Therefore, we can assume that $x \in \operatorname{Int}(C(\Omega)^+)$, i.e., that x is a unit. Let $A = \big\{\alpha \in \mathbb{R}\colon\ x + \alpha y \in C(\Omega)^+\big\}$, and put

$$\beta = \inf A \quad \text{and} \quad \gamma = \sup A\,.$$

Since $x \in C(\Omega)^+$, we see that $\beta < 0 < \gamma$. We claim that either β or γ is finite. Indeed, if $\beta = -\infty$ and $\gamma = \infty$, then (by the convexity of $C(\Omega)^+$) it follows that $x + \alpha y \in C(\Omega)^+$ for all $\alpha \in \mathbb{R}$. In particular, we have $\pm ny \leq x$ for each n. So $y = 0$, which is a contradiction.

Now note that if $\beta \in \mathbb{R}$, then $x + \beta y \in \partial\big[C(\Omega)^+\big]$, and similarly if $\gamma \in \mathbb{R}$, then $x + \gamma y \in \partial\big[C(\Omega)^+\big]$. ∎

Recall that a positive operator $T\colon C(\Omega) \to C(\Omega)$ is said to be:

- *a **strong Krein operator** whenever T carries non-zero positive elements to units, i.e., if $0 < x \in C(\Omega)$, then Tx is a unit, and*
- *a **Krein operator** whenever for each $x > 0$ there exists some positive integer n (depending on x) such that $T^n x$ is a unit.*

Clearly, every strong Krein operator is a Krein operator. It should also be noticed that every Krein operator is strictly positive.

Example 9.39. Here are some examples of Krein operators.

(1) Let $A = [a_{ij}]$ be a real $n \times n$ matrix such that $a_{ij} > 0$ for all (i, j). Then A is a strong Krein operator from $\mathbb{R}^n$ to $\mathbb{R}^n$.

(2) Assume that $C(\Omega)$ admits a strictly positive linear functional f. (This is equivalent to assuming that there exists a regular Borel measure whose support is Ω.) If u is a unit, then the rank-one operator $T = f \otimes u$ is a strong Krein operator.

More generally, any finite-rank operator $T = \sum_{i=1}^{n} f_i \otimes u_i$, where each f_i is a strictly positive linear functional on $C(\Omega)$, $u_i > 0$ for each i, and $\sum_{i=1}^{n} u_i$ is a unit, is a strong Krein operator.

(3) Let μ be a non-zero regular Borel measure on a compact Hausdorff space Ω, and let $T\colon C(\Omega) \to C(\Omega)$ be a positive integral operator with a continuous strictly positive kernel. That is, there exists a continuous real-valued function K on $\Omega \times \Omega$ satisfying $K(s,t) > 0$ for all $(s,t) \in \Omega \times \Omega$, and

$$Tf(t) = \int_\Omega K(s,t) f(s)\, d\mu(s)$$

for each $f \in C(\Omega)$ and all $t \in \Omega$. Then T is a strong Krein compact operator.

(4) Let $K\colon [a,b] \times [a,b] \to [0, \infty)$ be a continuous function such that $K(s,t) > 0$ for all $|s - t| < \epsilon$, where $\epsilon > 0$ is fixed. Then the

integral operator generated by the function $K(s,t)$, i.e., the operator $T\colon C[a,b] \to C[a,b]$, defined by $Tf(t) = \int_a^b K(s,t)f(s)\,ds$, is a compact Krein operator.

(5) This example is an important special case of the preceding example. Consider a continuous function $K\colon [a,b] \times [a,b] \to [0,\infty)$ satisfying $K(t,t) > 0$ for all $a \le t \le b$. Then the integral operator generated by K on $C[a,b]$ is a compact Krein operator. (See Exercise 4 at the end of the section.)

(6) If $T\colon C(\Omega) \to C(\Omega)$ is a Krein operator and $S\colon C(\Omega) \to C(\Omega)$ is an arbitrary positive operator, then $S+T$ is a Krein operator. ■

Two useful properties of Krein operators are included in the next lemma.

Lemma 9.40. *If $T\colon C(\Omega) \to C(\Omega)$ is a Krein operator, then:*

(1) *T carries units to units.*

(2) *T^n is a Krein operator for each $n \ge 1$ (and so $T^n \ne \lambda I$ for each $n \ge 1$ and all scalars $\lambda \ne 0$).*

Proof. (1) Let u be a unit. Since T is a Krein operator, there exists some positive integer $k \ge 1$ such that $v = T^k u$ is also a unit. Next, choose some $\gamma > 0$ such that $T^{k-1}u \le \gamma u$ and note that $v = T^k u \le \gamma T u$ holds. This implies that Tu is a unit.

(2) Using (1) and an inductive argument, we see that T^k carries units to units for each k. Now fix $n \ge 1$ and let $x > 0$. Choose some k such that $T^k x$ is a unit. But then $(T^n)^k x = T^{(n-1)k}(T^k x)$ is a unit, which proves that T^n is a Krein operator. ■

Notice that from (1) above, it follows that if T is a Krein operator and we let $v = T\mathbf{1}$, then the operator $x \mapsto \frac{1}{v}Tx$ is a Markov operator. Regarding the spectral radius of a Krein operator, we have the following.

Theorem 9.41. *If $T\colon C(\Omega) \to C(\Omega)$ is a Krein operator, then:*

(1) *T has a positive spectral radius, i.e., $r(T) > 0$.*

(2) *$r(T) = \lim_{n\to\infty} \|T^n x\|^{\frac{1}{n}}$ for each $x > 0$.*

Proof. (1) By Lemma 9.40, $T\mathbf{1}$ is a unit. This implies $r(T) > 0$; see Exercise 2 at the end of this section.

(2) Let $x > 0$. Fix some k such that $u = T^k x$ is a unit. This implies that the formula $\|y\|_u = \inf\{\lambda \ge 0\colon\ |y| \le \lambda u\}$ defines an equivalent norm

on $C(\Omega)$. In particular, we have $\|T^n\|_u = \|T^n u\|_u$ for each n and so

$$(\|T^n x\|_u)^{\frac{1}{n}} = (\|T^{n-k}u\|_u)^{\frac{1}{n}} = [(\|T^{n-k}\|_u)^{\frac{1}{n-k}}]^{\frac{n-k}{n}} \xrightarrow[n\to\infty]{} r(T)\,.$$

Since $\|\cdot\|$ and $\|\cdot\|_u$ are equivalent norms, we get $\lim_{n\to\infty}\|T^n x\|^{\frac{1}{n}} = r(T)$ and the proof is finished. ■

We now turn our attention to eigenvalues of Krein operators. In light of the fact that Krein operators are ideal irreducible, the next result parallels Theorem 8.22.

Theorem 9.42 (M. G. Krein). *If a Krein operator $T\colon C(\Omega)\to C(\Omega)$ has a positive eigenvector, say $Tx_0 = \lambda_0 x_0$ for some $x_0 > 0$, then:*

1. *The eigenvalue λ_0 coincides with the spectral radius of the operator T, that is, $\lambda_0 = r(T) > 0$.*
2. *The eigenvector x_0 is a unit and (up to a scalar multiple) is the only positive eigenvector of T.*

Proof. (1) Assume that $Tx_0 = \lambda_0 x_0$ for some $x_0 > 0$. Since T is a Krein operator, there exists some positive integer k such that $T^k x_0$ is a unit. From $T^k x_0 = (\lambda_0)^k x_0$, it follows that $\lambda_0 > 0$ and that x_0 is a unit.

Now from Theorem 9.41, we obtain

$$r(T) = \lim_{n\to\infty}\|T^n x_0\|^{\frac{1}{n}} = \lim_{n\to\infty}\|\lambda_0^n x_0\|^{\frac{1}{n}} = \lambda_0 \lim_{n\to\infty}\|x_0\|^{\frac{1}{n}} = \lambda_0\,.$$

(2) Assume $Tx = \lambda x$ for some $x > 0$ and some scalar λ. By part (1), we get that $\lambda = \lambda_0$. Furthermore, by Lemma 9.38, there exists some $\alpha \in \mathbb{R}$ such that $x_0 + \alpha x \in \partial C(\Omega)^+$. Clearly, $\alpha \neq 0$. If $x_0 + \alpha x > 0$, then from $T(x_0+\alpha x) = \lambda_0(x_0+\alpha x)$ and part (1), we infer that $x_0+\alpha x$ is a unit, that is, $x_0+\alpha x \in \mathrm{Int}(C(\Omega)^+)$, which is impossible. Thus, $x_0+\alpha x = 0$ and hence $x = (-\frac{1}{\alpha})x_0$. ■

Corollary 9.43. *If $T\colon C(\Omega)\to C(\Omega)$ is a compact Krein operator, then the spectral radius $r(T)$ is the only eigenvalue of T having a positive eigenvector (which is necessarily a unit).*

Proof. By Theorem 9.41, we have $r(T) > 0$. So, by the Krein–Rutman Theorem 7.10, $r(T)$ is an eigenvalue of T having a positive eigenvector. Our conclusion now follows from Theorem 9.42. ■

We now state a classical result, due to P. Jentzsch [**162**], regarding integral operators.

Theorem 9.44 (Jentzsch). *Let $K\colon L_2[0,1] \to L_2[0,1]$ be an integral operator defined by*

$$Kx(t) = \int_0^1 K(s,t)x(s)\,ds\,,$$

where $K\colon [0,1]\times[0,1] \to \mathbb{R}$ is a strictly positive continuous function. Then:

(1) *The positive operator K is compact and has a positive spectral radius.*

(2) *The spectral radius $r(K)$ is an eigenvalue of K and has a unique (up to a scalar multiple) strictly positive eigenfunction.*

(3) *The peripheral spectrum of K consists only of the point $r(K)$, that is, if $\lambda \in \sigma(K)$ and $\lambda \neq r(K)$, then $|\lambda| < r(K)$.*

Notice that the compactness of K is a consequence of Theorem 7.7. Also, according to Example 9.39(3), the operator $K\colon C[0,1] \to C[0,1]$ is a strong Krein operator. Since the range of the operator $K\colon L_2[0,1] \to L_2[0,1]$ lies in $C[0,1]$, properties (1) and (2) in Theorem 9.44 follow immediately from Theorem 9.42. However, property (3) is not that easy to establish and requires some extra machinery. For a proof of property (3) we refer the reader to [**343**, pp. 630–633], where the reader will also find several generalizations of Jentzsch's theorem to Banach lattices and Banach function spaces.

Our next objective is to discuss interrelationships between fixed points and eigenvectors of families of commuting positive operators on $C(\Omega)$-spaces. We start by clarifying a crucial connection between fixed points and eigenvectors.

Lemma 9.45. *Let $T\colon C(\Omega) \to C(\Omega)$ be a positive operator, let u be a unit in $C(\Omega)$, and consider the non-empty convex and w^*-compact set*

$$\mathcal{C}_u = \big\{\psi \in C(\Omega)^*_+ \colon\; \psi(u) = 1\big\}\,.$$

Also, consider the continuous mapping $F\colon (\mathcal{C}_u, w^*) \to (\mathcal{C}_u, w^*)$ defined by

$$F(\psi) = \tfrac{\psi + T^*\psi}{[\psi + T^*\psi](u)} = \tfrac{\psi + T^*\psi}{1 + \psi(Tu)}\,.$$

Then:

(1) *Up to a scalar multiple, the positive eigenvectors of the adjoint operator T^* are precisely the fixed points of F.*

(2) *The set of fixed points of F is non-empty and w^*-compact.*

Proof. (1) Let $\phi \in \mathcal{C}_u$. If $F(\phi) = \phi$, then $T^*\phi = [T^*\phi(u)]\phi = \phi(Tu)\phi$. Conversely, if $T^*\phi = \lambda\phi$, then $\lambda \geq 0$ and $F(\phi) = \frac{\phi + T^*\phi}{[\phi + T^*\phi](u)} = \frac{\phi + \lambda\phi}{1+\lambda} = \phi$.

(2) Clearly, $\mathcal{C}_u$ is a non-empty, convex, and w^*-compact subset of $C(\Omega)^*$. A direct verification shows that F indeed maps $\mathcal{C}_u$ into $\mathcal{C}_u$ and that $F\colon (\mathcal{C}_u, w^*) \to (\mathcal{C}_u, w^*)$ is a continuous function. Therefore, by Tychonoff's Fixed Point Theorem (see, for instance [**25**, Corollary 14.51, p. 483]) there exists some $\phi \in \mathcal{C}$ such that $F(\phi) = \phi$. ■

Corollary 9.46 (M. G. Krein [**192**]). *The adjoint of an arbitrary positive operator on a $C(\Omega)$-space has a positive eigenvector corresponding to a non-negative eigenvalue.*

A proof of the above result that does not use a fixed point argument can be found in [**185**]. The next result generalizes the previous corollary to a family of commuting operators.

Theorem 9.47 (M. G. Krein). *Let $\{T_\alpha\}_{\alpha\in A}$ be a family of pairwise commuting positive operators on some $C(\Omega)$-space. Then the family of adjoint operators $\{T_\alpha^*\}_{\alpha\in A}$ has a common positive eigenvector. That is, there exist $0 < \phi \in C(\Omega)^*$ and a family $\{\lambda_\alpha\}_{\alpha\in A}$ of non-negative scalars such that $T_\alpha^*\phi = \lambda_\alpha\phi$ for each α.*

Proof. Consider again the non-empty, convex and w^*-compact set

$$\mathcal{C} = \left\{\psi \in C(\Omega)_+^*\colon\ \psi(\mathbf{1}) = 1\right\}.$$

For each index α let $F_\alpha\colon (\mathcal{C}, w^*) \to (\mathcal{C}, w^*)$ denote the continuous mapping defined by

$$F_\alpha(\psi) = \tfrac{\psi + T_\alpha^*\psi}{[\psi + T_\alpha^*\psi](\mathbf{1})} = \tfrac{\psi + T_\alpha^*\psi}{1 + \psi(T_\alpha\mathbf{1})}\,.$$

By Lemma 9.45(1), every fixed point of F_α is an eigenvector of T_α^* corresponding to a non-negative eigenvalue. Let D_α denote the set of all fixed points of F_α, i.e., let

$$D_\alpha = \left\{\phi \in \mathcal{C}\colon\ F_\alpha(\phi) = \phi\right\}.$$

By Lemma 9.45(2) each D_α is a non-empty and w^*-compact subset of $\mathcal{C}$. To complete the proof, we must show that $\bigcap_{\alpha\in A} D_\alpha \neq \emptyset$, or equivalently that the family of w^*-compact sets $\{D_\alpha\colon\ \alpha \in A\}$ has the finite intersection property.

The proof goes by induction. As mentioned above, each D_α is non-empty. So, for the induction step, assume that the intersection of any n members of the family $\{D_\alpha\colon\ \alpha \in A\}$ has a non-empty intersection and let $\alpha_1, \ldots, \alpha_n, \alpha_{n+1}$ be $n+1$ indices. We must show that $\bigcap_{i=1}^{n+1} D_{\alpha_i} \neq \emptyset$.

To this end, start by choosing some $0 < \phi \in D_{\alpha_1} \cap \cdots \cap D_{\alpha_n}$. Consequently, for $\lambda_i = \phi(T_{\alpha_i}\mathbf{1}) \geq 0$ it follows that $T_{\alpha_i}^*\phi = \lambda_i\phi$ for $i = 1, \ldots, n$. Now for each $1 \leq i \leq n$ define the set

$$\mathcal{C}_i = \left\{\psi \in \mathcal{C}\colon\ T_{\alpha_i}^*\psi = \lambda_i\psi\right\}.$$

Clearly, each $\mathcal{C}_i$ is convex, w^*-compact, and non-empty (since $\phi \in \mathcal{C}_i$). Next, we claim that $F_{\alpha_{n+1}}(\mathcal{C}_i) \subseteq \mathcal{C}_i$ for each $1 \le i \le n$. Indeed, if $\psi \in \mathcal{C}_i$, then

$$\begin{aligned} T^*_{\alpha_i}(F_{\alpha_{n+1}}\psi) &= T^*_{\alpha_i}\Big(\tfrac{\psi+T^*_{\alpha_{n+1}}\psi}{1+T^*_{\alpha_{n+1}}\psi(\mathbf{1})}\Big) = \tfrac{T^*_{\alpha_i}\psi+T^*_{\alpha_i}T^*_{\alpha_{n+1}}\psi}{1+T^*_{\alpha_{n+1}}\psi(\mathbf{1})} \\ &= \tfrac{T^*_{\alpha_i}\psi+T^*_{\alpha_{n+1}}T^*_{\alpha_i}\psi}{1+T^*_{\alpha_{n+1}}\psi(\mathbf{1})} = \tfrac{\lambda_i\psi+\lambda_i T^*_{\alpha_{n+1}}\psi}{1+T^*_{\alpha_{n+1}}\psi(\mathbf{1})} \\ &= \lambda_i\Big[\tfrac{\psi+T^*_{\alpha_{n+1}}\psi}{1+\psi(T_{\alpha_{n+1}}\mathbf{1})}\Big] = \lambda_i F_{\alpha_{n+1}}\psi\,, \end{aligned}$$

which shows that $F_{\alpha_{n+1}}\psi \in \mathcal{C}_i$.

Let $G = \bigcap_{i=1}^n \mathcal{C}_i$. Then G is a convex, w^*-compact, and (in view of $\phi \in G$) non-empty subset of $\mathcal{C}$ such that $F_{\alpha_{n+1}}(G) \subseteq G$. By Tychonoff's Fixed Point Theorem there exists some $0 < \phi_0 \in G$ with $F_{\alpha_{n+1}}(\phi_0) = \phi_0$. Since $\phi_0 \in \mathcal{C}_i$, we have $T^*_{\alpha_i}\phi_0 = \lambda_i\phi_0$, and hence $F_{\alpha_i}(\phi_0) = \phi_0$ for each $1 \le i \le n+1$. Therefore, $\phi_0 \in \bigcap_{i=1}^{n+1} D_{\alpha_i}$. Thus, $\bigcap_{i=1}^{n+1} D_{\alpha_i} \neq \emptyset$, and the proof is finished. ■

As mentioned before, for every unit $u \in C(\Omega)$ the formula

$$\|x\|_u = \inf\{\lambda > 0\colon\ |x| \le \lambda u\}$$

defines a lattice norm on $C(\Omega)$ which is equivalent to the sup norm.

We are now ready to establish that if a family of pairwise commuting positive operators on $C(\Omega)$ fixes a unit, then the family of their adjoints also fixes a positive linear functional.

Theorem 9.48 (M. G. Krein). *Let $\{T_\alpha\}_{\alpha\in A}$ be a family of pairwise commuting positive operators on a $C(\Omega)$-space. If for some unit $u \in C(\Omega)$ we have $T_\alpha u = u$ for all α, then there exists some $0 < \phi \in C(\Omega)^*$ such that $T^*_\alpha\phi = \phi$ for all α.*

Proof. Assume that for some unit u we have $T_\alpha u = u$ for each $\alpha \in A$. Since $-tu \le x \le tu$ implies $-tu = -tT_\alpha(u) \le T_\alpha(x) \le tT_\alpha(u) = tu$, it follows that $\|T_\alpha x\|_u \le \|x\|_u$ for each $x \in C(\Omega)$ and each $\alpha \in A$. This implies that if $k_1, \ldots, k_n$ are natural numbers and $\alpha_1, \ldots, \alpha_n$ are arbitrary indices, then

$$\big\|T^{k_1}_{\alpha_1}T^{k_1}_{\alpha_2}\cdots T^{k_n}_{\alpha_n}x\big\|_u \le \|x\|_u \qquad (\star)$$

for each $x \in C(\Omega)$.

Next, we consider the vector subspace Y of $C(\Omega)$ defined by

$$Y = \Big\{x \in C(\Omega)\colon\ \exists\,\alpha_i \in A,\ x_i \in C(\Omega) \quad\text{with}\quad x = \sum_{i=1}^n (T_{\alpha_i}x_i - x_i)\Big\}.$$

We claim that $Y \cap \mathcal{U} = \emptyset$, where $\mathcal{U} = \operatorname{Int}[C(\Omega)^+]$. To see this, assume by way of contradiction that $Y \cap \mathcal{U} \neq \emptyset$. So, there exist indices $\alpha_1, \ldots, \alpha_n$ and functions $x_1, \ldots, x_n \in C(\Omega)$ such that the vector $e = \sum_{i=1}^n \big(T_{\alpha_i} x_i - x_i\big)$ is a unit. Fix these indices $\alpha_1, \ldots, \alpha_n$ and for each m define the operator

$$T_m = \tfrac{1}{m^n} \sum_{k_1=1}^m \sum_{k_2=1}^m \cdots \sum_{k_n=1}^m T_{\alpha_1}^{k_1} T_{\alpha_2}^{k_2} \cdots T_{\alpha_n}^{k_n},$$

where we write $T_\alpha^{k_j}$ instead of $(T_\alpha)^{k_j}$. Clearly, each T_m is a positive operator and $T_m u = u$. Since the family $\{T_\alpha\}_{\alpha \in A}$ consists of pairwise commuting operators, we see that

$$\begin{aligned} T_m e &= \tfrac{1}{m^n} \Big[\sum_{i=1}^n \sum_{k_1=1}^m \sum_{k_2=1}^m \cdots \sum_{k_n=1}^m T_{\alpha_1}^{k_1} T_{\alpha_2}^{k_2} \cdots T_{\alpha_n}^{k_n} \big(T_{\alpha_i} x_i - x_i\big) \Big] \\ &= \tfrac{1}{m^n} \sum_{i=1}^n \sum_{k_1=1}^m \cdots \sum_{k_{i-1}=1}^m \sum_{k_{i+1}=1}^m \cdots \sum_{k_n=1}^m \prod_{j \neq i} T_{\alpha_j}^{k_j} \sum_{k_i=1}^m \big(T_{\alpha_i}^{k_i+1} - T_{\alpha_i}^{k_i}\big) x_i \\ &= \tfrac{1}{m^n} \sum_{i=1}^n \sum_{k_1=1}^m \cdots \sum_{k_{i-1}=1}^m \sum_{k_{i+1}=1}^m \cdots \sum_{k_n=1}^m \prod_{j \neq i} T_{\alpha_j}^{k_j} \big(T_{\alpha_i}^{m+1} - T_{\alpha_i}\big) x_i \,. \end{aligned}$$

If we let $c = \max\big\{ \|x_i\|_u \colon\ i = 1, \ldots, n \big\}$, then from $(\star)$ it follows that

$$\begin{aligned} &\Big\| \prod_{j \neq i} T_{\alpha_j}^{k_j} \big(T_{\alpha_i}^{m+1} - T_{\alpha_i}\big) x_i \Big\|_u \\ &\qquad = \Big\| T_{\alpha_1}^{k_1} \cdots T_{\alpha_{i-1}}^{k_{i-1}} T_{\alpha_{i+1}}^{k_{i+1}} \cdots T_{\alpha_n}^{k_n} \big(T_{\alpha_i}^{m+1} - T_{\alpha_i}\big) x_i \Big\|_u \le 2c \,. \end{aligned}$$

Consequently, from the triangle inequality, we infer that

$$\big\| T_m e \big\|_u \le \tfrac{n}{m^n} m^{n-1} 2c = \tfrac{2cn}{m} \,.$$

Since e is assumed to be a unit, there exists some $\eta > 0$ such that $u \le \eta e$. This implies

$$1 = \|u\|_u = \|T_m(u)\|_u \le \eta \|T_m(e)\|_u \le \tfrac{2\eta cn}{m}$$

for each m, which is impossible. Therefore, $Y \cap \mathcal{U} = \emptyset$.

Since $\mathcal{U}$ is open, the Separation Theorem guarantees the existence of some non-zero $\phi \in C(\Omega)^*$ satisfying $\phi(y) = 0$ for all $y \in Y$ and $\phi(x) \ge 0$ for all $x \in \mathcal{U}$. Taking into account that $\overline{\mathcal{U}} = C(\Omega)^+$, we conclude that $\phi > 0$. From $T_\alpha x - x \in Y$, it follows that

$$\langle x, T_\alpha^* \phi - \phi \rangle = \langle T_\alpha x - x, \phi \rangle = 0 \,,$$

for each $x \in C(\Omega)$. Thus, $T_\alpha^* \phi = \phi$ for each $\alpha \in A$, and this shows that $\phi > 0$ is a fixed point for the family of operators $\{T_\alpha^*\}_{\alpha \in A}$. ■

Corollary 9.49. *Let $T \colon C(\Omega) \to C(\Omega)$ be a positive operator. If some $\lambda > 0$ is an eigenvalue of T having an eigenvector which is a unit, then the adjoint operator T^* also has λ as an eigenvalue with a positive eigenvector.*

If all operators in the previous result are contractions, then we have the following companion to Theorem 9.48.

Theorem 9.50. *Let $\{T_\alpha\}$ be a family of pairwise commuting positive contractions on $C(\Omega)$. If there exists a non-zero vector $v \in C(\Omega)$ such that $T_\alpha v = v$ for each α, then there exists a non-zero positive linear functional $\phi \in C(\Omega)^*$ satisfying $T_\alpha^* \phi = \phi$ for each α.*

Proof. Start by fixing some non-zero vector $v \in C(\Omega)$ such that $T_\alpha v = v$ for each α. Next note that if $k_1, k_2, \ldots, k_n$ are arbitrary natural numbers and $\alpha_1, \alpha_2, \ldots, \alpha_n$ are arbitrary indices, then

$$\left\|T_{\alpha_1}^{k_1} T_{\alpha_2}^{k_2} \cdots T_{\alpha_n}^{k_n} x\right\| \leq \|x\|,$$

where $\|\cdot\|$ denotes the sup norm.

Now we consider the same vector subspace Y of $C(\Omega)$ as in the proof of Theorem 9.48. Assume that $e \in Y \cap \mathcal{U}$, and then for each m construct the operator T_m. If we let $c = \max\{\|x_i\|: \ i = 1, \ldots, n\}$, then it follows that $\|T_m e\| \leq \frac{2cn}{m}$. Now if $\eta > 0$ satisfies $|v| \leq \eta e$, then

$$0 < \|v\| = \|T_\alpha v\| \leq \eta \|T_\alpha e\| \leq \tfrac{2\eta cn}{m}$$

for each m, which is impossible. The rest of the proof is the same as the proof of Theorem 9.48. ■

Corollary 9.51. *If a positive contraction $T\colon C(\Omega) \to C(\Omega)$ has 1 as an eigenvalue, then there exists some $0 < \phi \in C(\Omega)^*$ such that $T^*\phi = \phi$.*

Our next result is a famous theorem of A. A. Markov [**232**] and it is a surprisingly simple consequence of Theorem 9.48. In order to formulate it, we need some preliminary discussion.

Recall that if Δ is an arbitrary non-empty set, then $\ell_\infty(\Delta)$ denotes the Banach lattice of all bounded real-valued functions on Δ. With the sup norm $\ell_\infty(\Delta)$ is an AM-space with unit $\mathbf{1}$ (the constant function one) and, in particular, it is a $C(\Omega)$-space. Also, recall that the dual $\ell_\infty^*(\Delta)$ can be identified with the space of all finitely additive measures on Δ; see [**25**, Corollary 11.40, p. 394]. Now if $\tau\colon \Delta \to \Delta$ is a mapping, then τ defines a positive operator $C_\tau\colon \ell_\infty(\Delta) \to \ell_\infty(\Delta)$ via the formula

$$C_\tau(x)(\delta) = (x \circ \tau)(\delta) = x\big(\tau(\delta)\big), \ x \in \ell_\infty(\Delta), \ \ \delta \in \Delta.$$

The operator C_τ is, of course, the familiar composition operator determined by τ.

Theorem 9.52 (Markov). *Let $\{\tau_\alpha\}_{\alpha \in A}$ be a family of pairwise commuting mappings on an arbitrary non-empty set Δ into itself. Then there exists some $0 < \phi \in \ell_\infty^*(\Delta)$ such that $\phi(C_{\tau_\alpha} x) = \phi(x)$ for all α and all $x \in \ell_\infty(\Delta)$.*

Proof. As mentioned above $\ell_\infty(\Delta) = C(\Omega)$ for some compact Hausdorff space Ω. Also, the family $\{C_{\tau_\alpha}\}_{\alpha\in A}$ consists of pairwise commuting positive operators on $\ell_\infty(\Delta)$ and $C_{\tau_\alpha}(\mathbf{1}) = \mathbf{1}$ for each α. The conclusion now is a direct consequence of Theorem 9.48. ■

Exercises

1. Let $T\colon C(\Omega) \to C(\Omega)$ be a positive operator. Show that for each unit $u \in C(\Omega)$ we have

$$\|T\|_u = \|Tu\|_u \quad \text{and} \quad r(T) = \lim_{n\to\infty} (\|T^n u\|_u)^{\frac{1}{n}}.$$

2. Consider a positive operator $T\colon C(\Omega) \to C(\Omega)$ and put

$$m = \min_{\omega\in\Omega} T\mathbf{1}(\omega) \quad \text{and} \quad M = \max_{\omega\in\Omega} T\mathbf{1}(\omega).$$

Show that $m \le r(T) \le M$ and conclude from this that if $T\mathbf{1} \gg 0$, then the operator T has a positive spectral radius.

3. Show that the operator defined in Example 9.39(3) is indeed a compact Krein operator.

4. Show that the operator defined in Example 9.39(4) is indeed a compact Krein operator.
[HINT: To see this, note first that T^2 is also an integral operator with kernel given by the formula $K_2(t,s) = \int_a^b K(t,u)K(u,s)\,du$. Now observe that $K_2(t,s) > 0$ for all $|s-t| < 2\epsilon$. To this end, let $s,t \in [a,b]$ satisfy $s < t$ and $t - s = |s-t| < 2\epsilon$. Let $u_0 = \frac{s+t}{2}$, the midpoint of the closed interval $[s,t]$, and put $\delta = \epsilon - \frac{|s-t|}{2}$. Clearly, $0 < \delta < \epsilon$, and if $|u_0 - u| < \delta$, then $|s-u| \le |s-u_0| + |u-u_0| < \frac{|s-t|}{2} + \delta = \epsilon$ and similarly $|t-u| < \epsilon$. So, $K(t,u) > 0$ and $K(u,s) > 0$ for all $u \in (u_0-\delta, u_0+\delta)$. This implies

$$K_2(t,s) = \int_a^b K(t,u)K(u,s)\,du \ge \int_{u_0-\delta}^{u_0+\delta} K(t,u)K(u,s)\,du > 0.$$

Repeating the above process, we see that there exists some positive integer n such that $K_n(t,s) > 0$ holds for all $(t,s) \in [a,b]\times[a,b]$. It follows that T is a compact Krein operator.]

5. By Theorem 9.42 we know that a Krein operator can have only one (up to a scalar multiple) positive eigenvector corresponding to a positive eigenvalue. Give an example of a Krein operator whose eigenvalues are all positive.

6. Show that the product of two commuting Krein operators is also a Krein operator. Give an example of two non-commuting Krein operators whose product is not a Krein operator. [HINT: Notice that the positive matrices

$A = \begin{bmatrix} 0 & 1 \\ 1 & 1 \end{bmatrix}$ and $B = \begin{bmatrix} 1 & 1 \\ 1 & 0 \end{bmatrix}$ both define Krein operators on $\mathbb{R}^2$ while their product $AB = \begin{bmatrix} 1 & 0 \\ 2 & 1 \end{bmatrix}$ is not a Krein operator.]

7. Theorem 9.47 was originally proved by M. G. Krein under the extra hypothesis that $T_\alpha(\mathcal{U}) \subseteq \mathcal{U}$ for each α, where $\mathcal{U} = \text{Int}[C(\Omega)^+]$. Theorem 9.47 implies, of course, that version. Show that in turn the original Krein theorem also implies Theorem 9.47. [HINT: For each $0 < \epsilon < 1$ let $T_{\alpha,\epsilon} = T_\alpha + \epsilon I$, where I is the identity operator on $C(\Omega)$. Clearly, $T_{\alpha,\epsilon}(\mathcal{U}) \subseteq \mathcal{U}$ holds, and so for each $\epsilon > 0$ there exists some $0 < \phi_\epsilon \in \mathcal{C}$ (where $\mathcal{C} = \{\psi \in C(\Omega)^+ \colon \ \psi(\mathbf{1}) = 1\}$) and a family of non-negative scalars $\{\lambda_{\alpha,\epsilon}\}_{\alpha \in A}$ such that

$$T_\alpha^* \phi_\epsilon + \epsilon \phi_\epsilon = \lambda_{\alpha,\epsilon} \phi_\epsilon \,. \tag{$\star$}$$

The net $\{\phi_\epsilon\}_{0<\epsilon<1}$ (where $\epsilon_1 \succeq \epsilon_2$ means $\epsilon_1 \leq \epsilon_2$) of $\mathcal{C}$ has a w^*-convergent subnet. We can assume that $\phi_\epsilon \xrightarrow{w^*} \phi$ holds in $\mathcal{C}$. Now fix $\alpha \in A$ and then pick some $t > 0$ with $T_\alpha \mathbf{1} \leq t\mathbf{1}$. From ($\star$) it follows that

$$0 \leq \lambda_{\alpha,\epsilon} = \lambda_{\alpha,\epsilon}\phi_\epsilon(\mathbf{1}) = T_\alpha^*\phi_\epsilon(\mathbf{1}) + \epsilon\phi_\epsilon(\mathbf{1}) \leq t + 1 \,.$$

This implies that the net $\{\lambda_{\alpha,\epsilon}\}_{0<\epsilon<1}$ has an accumulation point $\lambda_\alpha \geq 0$. Now one more glance at ($\star$) guarantees that $T_\alpha^* \phi = \lambda_\alpha \phi$.]

8. Let $\{T_\alpha\}_{\alpha \in A}$ be a family of pairwise commuting positive operators on a $C(\Omega)$-space. If there exist some $\lambda > 0$ and some unit u such that $T_\alpha u = \lambda u$ for each α, then show that there exists some $0 < \phi \in C(\Omega)^*$ such that $T_\alpha^* \phi = \lambda \phi$ for each α. [HINT: Consider the family of positive operators $\{\frac{1}{\lambda} T_\alpha\}_{\alpha \in A}$ and use Theorem 9.48.]

9. Show that Theorem 9.50 does not follow from Theorem 9.48. [HINT: If $C(\Omega) = \mathbb{R}^2$ and $T \colon C(\Omega) \to C(\Omega)$ is the positive projection represented by the matrix $\begin{bmatrix} 1 & 0 \\ 0 & 0 \end{bmatrix}$, then T has 1 as an eigenvalue with eigenvector $\begin{bmatrix} 1 \\ 0 \end{bmatrix}$. However, T does not have any unit as an eigenvector.]

10. Show that the commutativity assumption regarding the family of operators in Theorems 9.48 and 9.50 cannot be dropped. [HINT: Let $\mathcal{P}$ denote the family of all (Borel) probability measures on $[0,1]$. For each measure $\mu \in \mathcal{P}$ consider the positive operator $T_\mu \colon C[0,1] \to C[0,1]$ defined by $T_\mu f(t) = \int_0^1 f(s)\, d\mu(s)$, i.e., $T_\mu = \mu \otimes \mathbf{1}$. Clearly, $T_\mu T_\nu = T_\nu$ holds for all $\mu, \nu \in \mathcal{P}$. In particular, it follows that $\{T_\mu \colon \ \mu \in \mathcal{P}\}$ is a family of non-commuting positive projections on $C[0,1]$. In addition, notice that $T_\mu \mathbf{1} = \mathbf{1}$ for each $\mu \in \mathcal{P}$. Assume now by way of contradiction that there exists some $0 < \phi \in \mathcal{M}[0,1] = C(\Omega)^*$ such that $T_\mu^* \phi = \phi$ holds for each μ; we can suppose that $\phi([0,1]) = 1$. Hence, $\phi = T_\mu^* \phi = (\mathbf{1} \otimes \mu)\phi = \langle \mathbf{1}, \phi \rangle \mu = \mu$ for each $\mu \in \mathcal{P}$, which is impossible.]

Chapter 10

Invariant Subspaces

One of the most famous and still unsettled problems in functional analysis and modern mathematics is the so-called **invariant subspace problem** that is described by the following simple question:

- *When does a bounded operator on a Banach space have a non-trivial closed invariant subspace?*

To recall some basic definitions related to the above question, let us suppose that $T: X \to X$ is a bounded operator on a Banach space and V is a subspace of X.[1] Then V is **non-trivial** if $V \neq \{0\}$ and $V \neq X$. We say that V is **invariant under** T (or simply T**-invariant**) if $T(V) \subseteq V$. Also, V is said to be **hyperinvariant for** T (or simply T**-hyperinvariant**) whenever V is invariant under every bounded operator on X that commutes with T, i.e., if $S \in \mathcal{L}(X)$ and $ST = TS$ imply that $S(V) \subseteq V$. If X is a Banach lattice and T is a positive operator, then V is called **lattice hyperinvariant for** T (or simply ℓ**-hyperinvariant**) if V is invariant under every positive operator on X that commutes with T.

It is very easy to show (see the beginning of the next section) that for non-separable Banach spaces the invariant subspace problem has an immediate affirmative answer. For finite dimensional complex Banach spaces the invariant subspace problem has an affirmative answer too. To see this, let $T: X \to X$ be a bounded operator on a finite dimensional complex Banach space (of dimension greater than one). There is nothing to prove if T is a multiple of the identity operator, since then each subspace is invariant. So,

[1] Unless otherwise stated, throughout this chapter X will denote an arbitrary real or complex Banach space of dimension greater than one. Similarly, E will denote a real Banach lattice of dimension greater than one.

assume that T is not a multiple of the identity. Now if λ is an eigenvalue of T, then its eigenspace $N_\lambda = \{x \in X\colon\ Tx = \lambda x\}$ is a non-trivial closed hyperinvariant subspace (see Lemma 10.5). In view of the above comments the invariant subspace problem should be formulated as follows.

- *When does a bounded operator on an infinite dimensional separable Banach space have a non-trivial closed invariant subspace?*

For algebras and semigroups of operators the existence of a common non-trivial closed invariant subspace is of interest even in the non-separable case.

P. Enflo [**121**] was the first to present an example of a bounded operator on a real or complex separable non-reflexive Banach space without non-trivial closed invariant subspaces. Subsequently, C. J. Read [**275**] (see also [**276**] and [**277**]) presented an example of a bounded operator on the real ℓ_1 without non-trivial closed invariant subspaces. Thus, these examples have established that the invariant subspace problem in its general form has a negative answer. However, for various important classes of Banach spaces and operators the invariant subspace problem remains open. For instance, it is not known if every bounded operator on a separable Hilbert space (or on a separable reflexive Banach space) has a non-trivial closed invariant subspace.

There exists a vast literature dedicated to this problem but it is impossible to mention here even the most significant publications. From the multitude of results on the existence of invariant subspaces we have chosen only two topics to discuss in detail in this chapter. Our choice does not necessarily depend on the importance of these topics but rather on their proximity to the main theme of this book—positivity, and to some extent of course on the personal tastes of the authors. The topics chosen are:

(1) Lomonosov's theorem on the existence of hyperinvariant subspaces for compact operators and some of its applications.

(2) The existence of invariant subspaces for positive operators, for operators related to them, and for collections of positive operators. As we will see, the invariant subspaces for such operators very often have a special geometrical structure—they are ideals.

For the invariant subspace problem in Hilbert spaces and its history, see the works of H. Bercovici [**55**], S. Brown [**76**], P. R. Halmos [**148**, pp. 100–101], N. K. Nikolskii [**257**], C. Pearcy and A. L. Shields [**262**], H. Radjavi and P. Rosenthal [**272**], and P. Rosenthal [**284**] and the references therein. For a general treatment of the invariant subspace problem see the monograph by B. Beauzamy [**52**]. The invariant subspace problem for positive operators has been studied extensively by Y. A. Abramovich, C. D. Aliprantis, and O. Burkinshaw [**7, 8, 10, 11, 12, 14**].

10.1. A Smorgasbord of Invariant Subspaces

The purpose of this section is to present several basic theorems about invariant subspaces of operators on general Banach spaces. It should be clear that if a vector subspace is invariant or hyperinvariant for a bounded operator, then so is its norm closure. We start with some useful terminology.

If $T\colon X \to X$ is an operator on a Banach space and $x \in X$, then the **orbit** (more precisely the T-**orbit**) of x under T is the set of vectors $\{x, Tx, T^2x, \ldots\}$. The T-**orbit space** of x is the vector subspace generated by the orbit of x and it will be denoted by $\mathcal{O}_T(x)$. Clearly, $\mathcal{O}_T(x) \neq \{0\}$ for each $x \neq 0$. A vector x is said to be **cyclic** (more precisely T-**cyclic**) whenever its orbit space is dense. That is, a vector x is cyclic if $\overline{\mathcal{O}_T(x)} = X$.

It should be obvious that each orbit of T is invariant under T. So, if $\overline{\mathcal{O}_T(x)} \neq X$ for some $x \neq 0$, then $\overline{\mathcal{O}_T(x)}$ is a non-trivial closed T-invariant subspace. Therefore, if V is a non-trivial closed T-invariant subspace of X, then for each non-zero $x \in V$ the closure of the orbit space, $\overline{\mathcal{O}_T(x)}$, is included in V and thus it is a non-trivial closed T-invariant subspace. In other words, we have the following simple result.

- *A bounded operator on a Banach space has a non-trivial closed invariant subspace if and only if it has a non-zero vector that is not cyclic.*

Since each space $\overline{\mathcal{O}_T(x)}$ is clearly separable, it follows that every bounded operator on a non-separable Banach space has a non-trivial closed invariant subspace.

Definition 10.1. *A bounded operator on a Banach space is said to be a* ***scalar operator*** *if it is a multiple of the identity operator. A bounded operator which is not a scalar operator is called a* ***non-scalar operator****.*

Every non-scalar operator is automatically non-zero and different from the identity operator. Also, notice that only Banach spaces of dimension greater than one can admit non-scalar operators.

We now continue with an elementary result describing some special invariant subspaces.

Lemma 10.2. *If $T\colon X \to X$ is a bounded operator on a Banach space, then its kernel and its range are T-hyperinvariant subspaces.*

Proof. Assume that an operator $S \in \mathcal{L}(X)$ commutes with the operator T, i.e., $ST = TS$. If $x \in \operatorname{Ker} T$, then from $T(Sx) = S(Tx) = 0$, we see that $S(\operatorname{Ker} T) \subseteq \operatorname{Ker} T$, i.e., $\operatorname{Ker} T$ is a closed T-hyperinvariant subspace. Next, notice that the equation $S(Tx) = T(Sx)$ valid for each $x \in X$ shows that $S\big(T(X)\big) \subseteq T(X)$, which means that the range of T is T-hyperinvariant. ■

Corollary 10.3. *If a bounded operator T commutes with a non-zero bounded operator which fails either to be one-to-one or to have a dense range, then T has a non-trivial closed invariant subspace.*

In particular, each bounded operator which is not one-to-one or does not have a dense range has a non-trivial closed invariant subspace.

The preceding conclusion shows that as far as the invariant subspace problem is concerned, not only can we suppose that a given operator T is one-to-one and has dense range but also that every non-zero operator commuting with T is one-to-one and has dense range. That is, Corollary 10.3 can be restated as follows.

Corollary 10.4. *If an operator $T \in \mathcal{L}(X)$ does not have a non-trivial closed invariant subspace, then every operator that commutes with T (including T itself) is one-to-one and has a dense range.*

Recall that an **eigenspace** of a bounded operator $T\colon X \to X$ is any subspace of the form $N_\lambda = \{x \in X\colon\ Tx = \lambda x\}$, where λ is an eigenvalue of T. If zero is an eigenvalue of T, then N_0 is precisely the kernel of T, i.e., $\operatorname{Ker} T = N_0$. Clearly, the eigenspaces are non-zero closed subspaces and when T is a non-scalar operator, they are different from X. The eigenspaces are also hyperinvariant.

Lemma 10.5. *Every eigenspace of any operator $T \in \mathcal{L}(X)$ is T-hyperinvariant. Moreover, if T is non-scalar, then every eigenspace of T is non-trivial.*

Proof. Let λ be an eigenvalue of a bounded operator $T\colon X \to X$, and let

$$N_\lambda = \big\{x \in X\colon\ Tx = \lambda x\big\} = \operatorname{Ker}(T - \lambda)\,.$$

By our hypothesis $\operatorname{Ker}(T-\lambda) \neq X$. Since every operator which commutes with T also commutes with $T - \lambda$, it follows from Lemma 10.2 that the closed subspace N_λ is T-hyperinvariant. ∎

The next two results deal with the existence of invariant subspaces for operators on finite dimensional vector spaces. As we shall see, the situation is different between the real and complex cases.

Corollary 10.6. *Every non-scalar operator on a finite dimensional complex Banach space has a non-trivial hyperinvariant subspace.*

Proof. Let $T\colon X \to X$ be a non-scalar operator on a finite dimensional complex Banach space, and consequently T has an eigenvalue $\lambda \in \mathbb{C}$. Using that T is a non-scalar operator, we infer that the eigenspace N_λ is non-trivial. By Lemma 10.5, this eigenspace N_λ is T-hyperinvariant. ∎

Corollary 10.7. *For a non-scalar operator $T\colon X \to X$ on a finite dimensional real Banach space we have the following.*

(1) *If $\dim X$ is an odd number, then T has a non-trivial hyperinvariant subspace.*

(2) *If $\dim X$ is an even number greater than two, then T has non-trivial invariant subspaces that need not be hyperinvariant.*

(3) *If $\dim X = 2$, then T need not have a non-trivial invariant subspace.*

Proof. Let $T\colon X \to X$ be a non-scalar operator on a finite dimensional real Banach space.

(1) Since X is a finite dimensional real Banach space of odd dimension, it follows that T has a real eigenvalue $\lambda \in \mathbb{R}$. But then, as in Corollary 10.6, the eigenspace N_λ is a non-trivial T-hyperinvariant subspace. In particular, $\dim X > 1$.

(2) According to Corollary 10.6 the operator $T_{\mathbf{C}}\colon X_{\mathbf{C}} \to X_{\mathbf{C}}$ has an eigenvalue $\alpha \in \mathbb{C}$. Pick a non-zero vector $z = x + \imath y$ such that $T_{\mathbf{C}} z = \alpha z$. If we consider the vector subspace V of X generated by x and y, then the assumption $\dim X > 2$ guarantees that V is a non-trivial vector subspace of X. It can be easily checked that V is T-invariant.

For an example of an operator on $\mathbb{R}^4$ without non-trivial hyperinvariant subspaces see Exercise 2 at the end of the section.

(3) Let us consider on $\mathbb{R}^2$ a rotation operator through an angle ϕ satisfying $0 < \phi < \pi$. That is, R_ϕ is the operator represented by the orthogonal matrix $\begin{bmatrix} \cos\phi & -\sin\phi \\ \sin\phi & \cos\phi \end{bmatrix}$. It is easy to check (since the non-trivial subspaces of $\mathbb{R}^2$ are the straight lines passing through the origin) that R_ϕ does leave any non-trivial subspace invariant. ■

If the adjoint of an operator has an eigenvalue, then the operator itself has a non-trivial closed invariant subspace.

Theorem 10.8. *Let $T\colon X \to X$ be a non-scalar operator. If T or its adjoint T^* has an eigenvalue, then both T and T^* have non-trivial closed hyperinvariant subspaces.*

Proof. Notice that T is a non-scalar operator if and only if its adjoint T^* is likewise non-scalar. If T has an eigenvalue λ, then by Lemma 10.5 its eigenspace $N_\lambda = \{x \in X\colon\ Tx = \lambda x\}$ is a non-trivial closed T-hyperinvariant subspace.

Now assume that $T^*x^* = \lambda x^*$ for some scalar λ and some non-zero x^* in X^*. Let $V = \overline{(T-\lambda)(X)}$, the closure of the range of the operator $T - \lambda$.

Since $T \neq \lambda I$, we see that $V \neq \{0\}$. Moreover, since $x^* \neq 0$ and since

$$\big\langle (T-\lambda)x, x^*\big\rangle = \big\langle x, (T^*-\lambda)x^*\big\rangle = \langle x, 0\rangle = 0$$

for each $x \in X$, it follows that $V \neq X$. That is, V is a non-trivial closed subspace of X. We claim that the subspace V is T-hyperinvariant. Indeed, if an operator $S \in \mathcal{L}(X)$ satisfies $ST = TS$, then for each $x \in X$ we have $S[(T-\lambda)x] = (T-\lambda)(Sx) \in V$, which implies that $S(V) \subseteq V$.

To complete the proof, notice that if λ is an eigenvalue of T, then λ is also an eigenvalue of T^{**}, and therefore the preceding case implies that T^* has a non-trivial closed hyperinvariant subspace. ■

Recall that a closed subspace V of a Banach space X is said to be **complemented** if there exists another closed subspace W of X such that $V \oplus W = X$. The subspace W is called a **complement** of V. As usual, an operator $P\colon X \to X$ is said to be a **projection** if $P^2 = P$. A projection P is **proper** if $P \neq 0$ and $P \neq I$. The Closed Graph Theorem easily implies that a closed subspace V of a Banach space X is complemented if and only if V is the range of a continuous projection on X.

The following simple lemma describes a necessary and sufficient condition for a complemented subspace to be invariant under a bounded operator.

Lemma 10.9. *The range of a projection $P \in \mathcal{L}(X)$ is invariant under an operator $T \in \mathcal{L}(X)$ if and only if $TP = PTP$.*

Proof. Let $T \in \mathcal{L}(X)$ and $Y = P(X)$, where $P \in \mathcal{L}(X)$ is a projection. Assume first that Y is T-invariant. If $x \in X$, then $TPx \in Y$, and so $TPx = P(TPx)$. That is, $TP = PTP$.

For the converse, suppose that $TP = PTP$ and let $y \in Y$. It follows that $Ty = TPy = P(TPy) \in Y$. This shows that Y is T-invariant. ■

Let $T\colon X \to X$ be a continuous operator on a Banach space. Recall that a non-trivial closed T-invariant subspace V of X is said to be **reducing** the operator T if V has a closed complement W which is also T-invariant. In this case the pair (V, W) is called a pair of **reducing subspaces** for T. The existence of reducing pairs was characterized in Theorem 2.22. We restate this theorem below for convenience.

Theorem 10.10. *A continuous operator $T\colon X \to X$ on a Banach space has a reducing (closed) subspace if and only if there exists a proper continuous projection $P\colon X \to X$ that commutes with T, i.e., $TP = PT$.*

A vector subspace $\mathcal{A}$ of $\mathcal{L}(X)$, where X is a Banach space, is called an **algebra of operators** (or a **subalgebra** of $\mathcal{L}(X)$) whenever $S, T \in \mathcal{A}$ imply $ST \in \mathcal{A}$. An algebra of operators $\mathcal{A}$ is said to be **unital** if the identity operator I belongs to $\mathcal{A}$. The **dual algebra** of an algebra of operators $\mathcal{A}$ in $\mathcal{L}(X)$ is the subalgebra $\mathcal{A}^*$ of $\mathcal{L}(X^*)$ defined by $\mathcal{A}^* = \{T^*\colon\ T \in \mathcal{A}\}$. An algebra of operators $\mathcal{A}$ is said to be **closed** if $\mathcal{A}$ is closed in $\mathcal{L}(X)$.

Every subset of operators $\mathcal{S}$ is contained in a smallest (with respect to inclusion) subalgebra, called the **algebra generated by** $\mathcal{S}$. This algebra coincides with the intersection of all subalgebras containing $\mathcal{S}$ (note that $\mathcal{L}(X)$ is one of them). The algebra generated by a single operator $T \in \mathcal{L}(X)$ will be denoted by $\mathcal{A}_T$. A moment's thought reveals that

$$\mathcal{A}_T = \{p(T)\colon\ p \ \text{ is a polynomial without constant term}\}.$$

The **unital algebra** generated by a set $\mathcal{S}$ is the algebra generated by the set $\mathcal{S} \cup \{I\}$. The unital algebra generated by a single operator $T \in \mathcal{L}(X)$ is precisely the collection of operators

$$\{p(T)\colon\ p \ \text{ is an arbitrary polynomial}\}.$$

If $\mathcal{A}$ is a subalgebra of $\mathcal{L}(X)$, then a subspace V of X is said to be **$\mathcal{A}$-invariant** whenever V is invariant under every operator in $\mathcal{A}$, i.e., if $A(V) \subseteq V$ for each $A \in \mathcal{A}$.

Recall that a collection $\mathcal{F}$ of functions from a topological space Ω to itself is called **transitive** if for each pair $\omega_1, \omega_2 \in \Omega$ and any neighborhood V of ω_2 there exists some $f \in \mathcal{F}$ such that $f(\omega_1) \in V$. Moreover, $\mathcal{F}$ is said to be **strictly transitive** if for any two points $\omega_1, \omega_2 \in \Omega$ there exists some $f \in \mathcal{F}$ such that $f(\omega_1) = \omega_2$. It is easy to see that an algebra of operators $\mathcal{A} \subseteq \mathcal{L}(X)$ does not have non-trivial closed invariant subspaces if for any pair $x, y \in X$ with $x \neq 0$ and every $\epsilon > 0$ there exists some operator $A \in \mathcal{A}$ such that $\|Ax - y\| < \epsilon$. This discussion motivates the following definition.

Definition 10.11. *An algebra $\mathcal{A}$ of operators is said to be:*

(1) ***transitive*** *if it has no non-trivial closed $\mathcal{A}$-invariant subspaces and*

(2) ***non-transitive*** *if it has a non-trivial closed $\mathcal{A}$-invariant subspace.*

It should be clear that a subspace V of X is invariant under an operator $T \in \mathcal{L}(X)$ if and only if V is invariant under the algebra (unital or not) generated by T in $\mathcal{L}(X)$. As usual, for an algebra $\mathcal{A}$ and any $x \in X$ we let

$$\mathcal{A}x = \{Ax\colon\ A \in \mathcal{A}\}.$$

Obviously, $\mathcal{A}x$ is a subspace of X which is also $\mathcal{A}$-invariant. Moreover, it is easy to see that a unital algebra $\mathcal{A}$ is non-transitive if and only if there exists some $x \in X$ such that the subspace $\overline{\mathcal{A}x}$ is non-trivial.

The next result is a folklore characterization of the non-transitive algebras.

Theorem 10.12. *For an algebra $\mathcal{A}$ of bounded operators on a Banach space X the following statements are equivalent.*

(1) *The algebra $\mathcal{A}$ is non-transitive.*

(2) *There exist a non-zero vector $x \in X$ and a non-zero linear functional $x^* \in X^*$ such that for each $A \in \mathcal{A}$ we have*

$$\langle Ax, x^* \rangle = \langle x, A^*x^* \rangle = 0 .$$

Proof. (1)$\Longrightarrow$(2) Assume that W is a non-trivial closed $\mathcal{A}$-invariant subspace. Fix a non-zero vector $x \in W$ and consider the closed vector subspace $V = \overline{\mathcal{A}x}$. Clearly, $V \subseteq W$, and so $V \neq X$. Therefore, there exists some non-zero $x^* \in X^*$ that annihilates V, i.e., $\langle Ax, x^* \rangle = 0$ for all $A \in \mathcal{A}$.

(2)$\Longrightarrow$(1) Suppose that $\langle Ax, x^* \rangle = 0$ for all $A \in \mathcal{A}$ and some non-zero vectors $x \in X$ and $x^* \in X^*$. It follows that the closed subspace $\overline{\mathcal{A}x}$ is not equal to X. Now, if $\mathcal{A}x \neq \{0\}$, then $\overline{\mathcal{A}x}$ is a non-trivial closed $\mathcal{A}$-invariant subspace. On the other hand, if $\mathcal{A}x = \{0\}$ holds, then note that the one-dimensional vector subspace $V = \{\lambda x\colon\ \lambda \text{ scalar}\}$ is non-trivial, closed, and $\mathcal{A}$-invariant. ■

Corollary 10.13. *If an algebra of operators $\mathcal{A}$ on a Banach space X is non-transitive, then its dual algebra $\mathcal{A}^*$ is likewise non-transitive.*

Proof. There exist non-zero $x \in X$ and $x^* \in X^*$ satisfying $\langle Ax, x^* \rangle = 0$ for all $A \in \mathcal{A}$. Viewing x as a vector in X^{**}, we get $\langle x^*, (A^*)^*x \rangle = 0$ for each $A \in \mathcal{A}$. Now Theorem 10.12 guarantees that the subalgebra $\mathcal{A}^*$ of $\mathcal{L}(X^*)$ is non-transitive. ■

The next result describes the hyperinvariant subspaces of an operator in terms of an algebra of operators. Its straightforward proof is omitted.

Lemma 10.14. *If $T \in \mathcal{L}(X)$, then:*

(1) *Its commutant $\{T\}' = \{S \in \mathcal{L}(X)\colon\ ST = TS\}$ is a closed unital subalgebra of $\mathcal{L}(X)$.*

(2) *T has a non-trivial closed hyperinvariant subspace if and only if $\{T\}'$ is a non-transitive algebra.*

The existence of a non-trivial closed invariant subspace for each compact operator on an (infinite dimensional) Hilbert space was established by J. von Neumann in 1935. N. Aronszajn and K. T. Smith [**42**] generalized this to Banach spaces in 1954. It took another nineteen years until V. I. Lomonosov [**211**] improved this by replacing an *invariant subspace* with a *hyperinvariant subspace*. In fact, as we will see in the next section, V. I. Lomonosov [**211**] proved more, but at this point we want to present a rather elementary proof of this result. The proof below is a simplification of Lomonosov's original proof [**211**] and is due to M. Hilden [**241**].

Theorem 10.15 (Aronszajn–Smith–Lomonosov). *Each non-zero compact operator on an infinite dimensional real or complex Banach space has a non-trivial closed hyperinvariant subspace.*

Proof. Let $K\colon X \to X$ be a non-zero compact operator on an infinite dimensional real or complex Banach space. We can assume without loss of generality that $\|K\| = 1$. First, we shall establish the result for the special case when the operator K is quasinilpotent, i.e., we shall assume that $r(K) = \lim\limits_{n\to\infty} \|K^n\|^{\frac{1}{n}} = 0$.

Put $\mathcal{A} = \{K\}'$. We must show that the algebra $\mathcal{A}$ is non-transitive. For this it suffices to show that there exists some $x \neq 0$ such that the subspace $\mathcal{A}x = \{Ax\colon\ A \in \mathcal{A}\}$ is not dense in X. So, assume by way of contradiction that

$$\overline{\mathcal{A}x} = X \quad \text{for each} \quad x \neq 0\,. \tag{$\star$}$$

Since K is non-zero, it easy to see that there exists some $x_0 \in X$ such that $\|x_0\| > 1$ and $\|Kx_0\| > 1$. Let U_0 denote the closed unit ball centered at x_0, i.e., $U_0 = \big\{x \in X\colon\ \|x - x_0\| \leq 1\big\}$. Since $\|K\| = 1$, $\|x_0\| > 1$ and $\|Kx_0\| > 1$, we see that

$$0 \notin U_0 \qquad \text{and} \qquad 0 \notin \overline{K(U_0)}\,. \tag{$\star\star$}$$

In view of ($\star$) we know that for each $x \neq 0$ there exists an operator $A \in \mathcal{A}$ such that $Ax \in U_0$, i.e., $\|Ax - x_0\| < 1$ (and clearly $A \neq 0$). Consequently, when A runs over $\mathcal{A}$, the open sets $\mathcal{O}_A = \big\{y \in X\colon\ \|Ay - x_0\| < 1\big\}$ cover $X\setminus\{0\}$ and, in particular, they cover the compact set $\overline{K(U_0)}$. So, we can find a finite number of operators $A_1, \ldots, A_k \in \mathcal{A}$ such that $\overline{K(U_0)} \subseteq \bigcup_{j=1}^{k} \mathcal{O}_{A_j}$. Let $c = \max_{1\leq j\leq k} \|A_j\|$ and note that $c > 0$.

Since $Kx_0 \in K(U_0)$, there exists some $1 \leq j_1 \leq k$ such that $Kx_0 \in \mathcal{O}_{A_{j_1}}$, and thus $x_1 = A_{j_1}Kx_0 \in U_0$. Then $Kx_1 \in K(U_0)$ and so there exists some $1 \leq j_2 \leq k$ such that $x_2 = A_{j_2}Kx_1 \in U_0$. Proceeding inductively, we can select a sequence of indices $\{j_n\}$ in $\{1, 2, \ldots, k\}$ such that the sequence $\{x_n\}$ of vectors defined via the recursive formula

$$x_{n+1} = A_{j_{n+1}}Kx_n\,, \quad n = 0, 1, 2, \ldots\;,$$

lies in U_0. Recalling now that each operator A_j commutes with K, we can rewrite the previous expression for x_{n+1} as follows:

$$\begin{aligned} x_{n+1} &= A_{j_{n+1}}A_{j_n}\cdots A_{j_1}K^{n+1}x_0 \\ &= (c^{-1}A_{j_{n+1}})(c^{-1}A_{j_n})\cdots(c^{-1}A_{j_1})(cK)^{n+1}x_0\,. \end{aligned}$$

The quasinilpotence of K implies clearly the quasinilpotence of cK, and so $\|(cK)^{n+1}\| \to 0$. Since $\|c^{-1}A_{j_m}\| \le 1$ for each $m = 1, \ldots, n$, we infer from the above identity that the sequence $\{x_n\} \subseteq U_0$ satisfies $\|x_n\| \to 0$, contrary to the first condition in $(\star\star)$. This contradiction proves that $\mathcal{A}x$ cannot be dense for all $x \neq 0$ and so, in this case, K has a non-trivial hyperinvariant subspace.

It remains to consider the case where the compact operator K is not quasinilpotent. We consider the complex case first. In this case, we know that the spectrum of K has some non-zero point $\lambda \in \mathbb{C}$. By Theorem 7.3, we also know that λ is an eigenvalue of K and since K equals λI on the eigenspace N_λ, it follows that N_λ is necessarily finite dimensional. Now Lemma 10.5 guarantees that N_λ is a non-trivial closed K-hyperinvariant subspace of X.

Finally, assume X to be real and consider the operator $K_{\mathbf{C}}\colon X_{\mathbf{C}} \to X_{\mathbf{C}}$. Clearly, $K_{\mathbf{C}}$ is a non-quasinilpotent compact operator. By the preceding case, there is an eigenspace N_λ in $X_{\mathbf{C}}$ that is $K_{\mathbf{C}}$-hyperinvariant. Since N_λ is finite dimensional, there exists a basis $\{z_1, z_2, \ldots, z_n\}$ for N_λ. Write $z_k = x_k + \imath y_k$ for each k and let V be the finite dimensional vector subspace generated in X by the vectors $\{x_1, \ldots, x_n, y_1, \ldots, y_n\}$. Since X is infinite dimensional, V is a non-trivial closed vector subspace of X. To complete the proof, it remains to check that V is also K-hyperinvariant. ■

Corollary 10.16 (Lomonosov). *If a bounded operator T on an infinite dimensional real or complex Banach space commutes with a non-zero compact operator, then T has a non-trivial closed invariant subspace.*

Recall that a bounded operator $T\colon X \to X$ is said to be **polynomially compact** whenever there exists a non-zero polynomial p such that $p(T)$ is a compact operator. A. R. Bernstein and A. Robinson [**60**] proved using non-standard analysis that polynomially compact operators have non-trivial closed invariant subspaces; a "standard" proof of this result was provided later by P. R. Halmos [**146**].

Corollary 10.17 (Bernstein–Robinson [**60**]). *Each polynomially compact operator on an infinite dimensional real or complex Banach space has a non-trivial closed invariant subspace.*

Proof. Let $T\colon X \to X$ be a bounded operator on an infinite dimensional Banach space. If T is a scalar operator, then T has non-trivial closed invariant subspaces. So, suppose that T is a non-scalar polynomially compact operator. Pick a non-zero polynomial $p(t) = t^n + a_{n-1}t^{n-1} + \cdots + a_1 t + a_0$ such that $p(T)$ is a compact operator.

If $p(T) \neq 0$, then (according to the Aronszajn–Smith–Lomonosov Theorem 10.15) the non-zero compact operator $p(T)$ has a non-trivial closed

$p(T)$-hyperinvariant subspace, say V. Since T commutes with $p(T)$, it follows that V is also T-invariant.

Now assume that $p(T) = 0$. In this case, we have $T^n = \sum_{i=0}^{n-1}(-a_i)T^i$. Fix any non-zero vector $x \in X$ and let V denote the non-zero closed subspace generated by the set $\{x, Tx, \ldots, T^{n-1}x\}$. Since X is infinite dimensional, we have $V \neq X$, and a simple verification shows that V is T-invariant. ■

Exercises

1. (Schaefer [**287**]) Show that every linear operator on an infinite dimensional vector space has a non-trivial invariant subspace. [HINT: Let $T\colon X \to X$ be a linear operator on an infinite dimensional vector space. Fix some $x \neq 0$ and consider the T-orbit space $\mathcal{O}_T$ of x. If $\mathcal{O}_T \neq X$, then $\mathcal{O}_T$ is a non-trivial T-invariant subspace. So, assume that $\mathcal{O}_T = X$. In this case, it follows that the set of vectors $\{x, Tx, T^2x, \ldots\}$ is linearly independent. This means that every vector $y \in X$ has a unique representation of the form $y = \sum_{i=0}^{\infty} \lambda_i T^i x$ with the λ_i equal to zero for all but finitely many i. Now define the non-zero linear functional ϕ on X by $\phi(y) = \sum_{i=0}^{\infty} \lambda_i$. It should be easy to verify that $\operatorname{Ker}\phi$ is a non-trivial T-invariant subspace.]

2. Give an example of a non-zero operator on $\mathbb{R}^4$ without non-trivial hyperinvariant subspaces. [HINT: Let θ be a non-zero angle that is an irrational multiple of 2π and consider the operator $T\colon \mathbb{R}^4 \to \mathbb{R}^4$ defined by $T(x_1, x_2, x_3, x_4) = (R_\theta(x_1, x_2), R_\theta(x_3, x_4))$, where $R_\theta\colon \mathbb{R}^2 \to \mathbb{R}^2$ denotes the rotation operator through the angle θ. Show that the invariant subspaces of the operator T are
$$V_\ell = \big\{(x_1, x_2, 0, 0)\colon\ x_1, x_2 \in \mathbb{R}\big\} \text{ and } V_r = \big\{(0, 0, x_1, x_2)\colon\ x_1, x_2 \in \mathbb{R}\big\}$$
and the ones given by
$$V_{\beta,\gamma} = \big\{(x_1, x_2, \gamma R_\beta(x_1, x_2))\colon\ x_1, x_2 \in \mathbb{R}\big\},$$
where $\beta \in [0, 2\pi)$ and $\gamma > 0$.

   If we consider the operator $S\colon \mathbb{R}^4 \to \mathbb{R}^4$ defined by
$$S(x_1, x_2, x_3, x_4) = (x_3, x_4, x_1, x_2),$$
then S commutes with T and leaves neither V_ℓ nor V_r invariant. On the other hand, if α is a non-zero angle such that $\alpha + \beta$ is an irrational multiple of 2π, then the operator $R\colon \mathbb{R}^4 \to \mathbb{R}^4$ defined by
$$R(x_1, x_2, x_3, x_4) = \big(x_1, x_2, R_\alpha(x_3, x_4)\big)$$
commutes with T and $V_{\beta,\gamma}$ is not R-invariant for all $\gamma > 0$.]

3. Let $T\colon X \to X$ be a bounded operator on a Banach space. Denote by $\operatorname{Lat}(T)$ the set of all closed T-invariant subspaces of X. Show that, under the standard set inclusion, $\operatorname{Lat}(T)$ is an order complete lattice with smallest and largest elements.

4. For a bounded operator $T\colon X \to X$ on an infinite dimensional real Banach space establish the following.
(a) The operator T has a non-zero finite dimensional invariant subspace if and only if the operator $T_{\mathbf{c}}\colon X_{\mathbf{c}} \to X_{\mathbf{c}}$ likewise has a non-zero finite dimensional invariant subspace.
(b) The operator T has a non-zero finite dimensional hyperinvariant subspace if and only if the operator $T_{\mathbf{c}}\colon X_{\mathbf{c}} \to X_{\mathbf{c}}$ likewise has a non-zero finite dimensional hyperinvariant subspace.

5. Let $T\colon X \to X$ be a bounded operator on a Banach space. If X is not reflexive, then show that the double adjoint $T^{**}\colon X^{**} \to X^{**}$ has a non-trivial norm closed T^{**}-invariant subspace.

6. Let Ω be a compact Hausdorff space. Fix some $\phi \in C(\Omega)$ and consider the multiplication operator M_ϕ on $C(\Omega)$. If Ω_0 is a non-empty subset of Ω and $V = \{f \in C(\Omega)\colon\ f(\omega) = 0 \text{ for all } \omega \in \Omega_0\}$, then show that V is a closed M_ϕ-invariant subspace of $C(\Omega)$.

7. Let Ω be a compact Hausdorff space. Fix a continuous mapping $\tau\colon \Omega \to \Omega$ and consider the composition operator C_τ on $C(\Omega)$ defined, as usual, by $C_\tau f = f \circ \tau$. If Ω_0 is a nonempty τ-invariant subset of Ω (i.e., $\tau(\Omega_0) \subseteq \Omega_0$), then show that the vector subspace

$$V = \big\{f \in C(\Omega)\colon\ f(\omega) = 0 \text{ for all } \omega \in \Omega_0\big\}$$

is a closed C_τ-invariant subspace of $C(\Omega)$.

8. Prove Lemma 10.14.

9. If X is a Banach space, then show that the only I-hyperinvariant subspaces are $\{0\}$ and X.

10. Let $\mathcal{A}$ be an arbitrary algebra of operators on a vector space. Show that $\mathcal{A}_1 = \{\alpha I + A\colon \alpha \text{ is a scalar and } A \in \mathcal{A}\}$ is the unital algebra generated by $\mathcal{A}$. (This implies that any algebra of operators $\mathcal{A}$ and the unital algebra generated by $\mathcal{A}$ have the same invariant subspaces.)

11. Show that an algebra of operators $\mathcal{A} \subseteq \mathcal{L}(X)$ has no non-trivial closed invariant subspaces if and only if for any pair $x, y \in X$ with $x \neq 0$ and every $\epsilon > 0$ there exists some operator $A \in \mathcal{A}$ such that $\|Ax - y\| < \epsilon$.

12. Let $\mathcal{A}$ be a unital algebra of continuous operators on a Banach space X and, as usual, let $\mathcal{A}x = \{Ax\colon\ A \in \mathcal{A}\}$. Show that $\mathcal{A}$ is non-transitive if and only if $\mathcal{A}$ has a non-trivial closed vector subspace of the form $\overline{\mathcal{A}x}$.

13. Let $T\colon X \to X$ be a bounded operator on a Banach space and consider the set

$$V = \big\{x \in X\colon\ \lim_{n\to\infty} \|T^n x\| = 0\big\}.$$

Establish the following.
(a) V is a T-hyperinvariant subspace.
(b) If T is power bounded, then V is also closed.

14. Let $S, T\colon X \to X$ be a pair of bounded operators on a Banach space. As usual, the **direct sum** of S and T is the operator $S \oplus T\colon X \oplus X \to X \oplus X$ defined by $(S \oplus T)(x \oplus y) = Sx \oplus Ty$. If a bounded operator $R\colon X \to X$ satisfies $RS = TR$, then show that its graph $G_R = \{x \oplus Rx\colon\ x \in X\}$ is a closed $S \oplus T$-invariant subspace of $X \oplus X$.

15. Show that the shift operator on any ℓ_p-space has a non-trivial closed hyperinvariant subspace.

16. Show that the converse of Corollary 10.13 is not true. That is, show that the dual algebra $\mathcal{A}^*$ of a transitive algebra of operators $\mathcal{A}$ can be non-transitive.

17. Let $\phi\colon \Omega \to \mathbb{R}$ be a continuous function separating the points of a compact Hausdorff space Ω and let M_ϕ denote the usual multiplication operator on $C(\Omega)$ defined by $M_\phi(f) = \phi f$. Show that $\{M_\phi\}' = \{M_g\colon\ g \in C(\Omega)\}$. [HINT: Let $T \in \{M_\phi\}'$ and put $h = T(\mathbf{1})$. Since $TM_\phi = M_\phi T$ implies $T(\phi f) = \phi T(f)$ for each $f \in C(\Omega)$, it follows that $T(\phi) = h\phi$ and by induction we see that $T(\phi^n) = h\phi^n$ for each $n \geq 0$. The latter implies $T(p(\phi)) = hp(\phi)$ for each polynomial p. By the Stone–Weierstrass Approximation Theorem the algebra $\{p(\phi)\colon\ p \text{ a polynomial}\}$ is norm dense in $C(\Omega)$ and from this we conclude that $Tf = hf$ for each $f \in C(\Omega)$.]

10.2. The Lomonosov Invariant Subspace Theorem

Many well-known invariant subspace theorems are for operators that are related in one way or another with the compact operators. It is generally acclaimed that the most famous of these results is Lomonosov's Invariant Subspace Theorem [**211**]. It asserts that:

- *If an operator $T\colon X \to X$ on a complex Banach space commutes with a non-scalar operator $S \in \mathcal{L}(X)$ which in turn commutes with a non-zero compact operator, then T has a non-trivial closed invariant subspace.*

The objective of this section is to prove this theorem. To do so, we need the following result which is of importance in its own right.

Theorem 10.18 (Lomonosov [**211**])**.** *If $\mathcal{A}$ is a transitive algebra of operators on a real or complex Banach space, then for each non-zero compact operator K there exists some $A \in \mathcal{A}$ such that the compact operator AK has a non-zero fixed point, i.e., $AKu = u$ for some $u \neq 0$.*

Proof. Let $\mathcal{A} \subseteq \mathcal{L}(X)$ be a transitive algebra and let K be a non-zero compact operator. We can assume $\|K\| = 1$.

Pick some $x_0 \in X$ such that $\|Kx_0\| > 1$ and let U_0 denote the closed unit ball centered at x_0. That is, $U_0 = \{x \in X\colon\ \|x - x_0\| \leq 1\}$. An easy argument shows that

$$0 \notin U_0 \quad \text{and} \quad 0 \notin \overline{K(U_0)}\,. \tag{$\star$}$$

Since $\mathcal{A}$ is an algebra, $\overline{\mathcal{A}x}$ (the norm closure of $\mathcal{A}x$) is $\mathcal{A}$-invariant for each $x \in X$. We claim that $\overline{\mathcal{A}x} \neq \{0\}$ for each $x \neq 0$. Indeed, if $\overline{\mathcal{A}x} = \{0\}$ for some $x \neq 0$, then the non-trivial closed subspace $\{\alpha x\colon \alpha \text{ scalar}\}$ is clearly $\mathcal{A}$-invariant, contrary to the hypothesis. Moreover, the absence of non-trivial closed $\mathcal{A}$-invariant subspaces guarantees that $\overline{\mathcal{A}x} = X$ for each $x \neq 0$.

In particular, for each $x \in \overline{K(U_0)}$ there exists some $A \in \mathcal{A}$ such that $\|Ax - x_0\| < 1$. So, the family of open sets $\{y \in X\colon\ \|Ay - x_0\| < 1\}_{A \in \mathcal{A}}$ is an open cover of $\overline{K(U_0)}$, that is,

$$\overline{K(U_0)} \subseteq \bigcup_{A \in \mathcal{A}} \{y \in X\colon\ \|Ay - x_0\| < 1\}\,.$$

Since (in view of the compactness of K) the set $\overline{K(U_0)}$ is compact, there exist operators $A_1, \ldots, A_m \in \mathcal{A}$ such that

$$\overline{K(U_0)} \subseteq \bigcup_{i=1}^{m} \{y \in X\colon\ \|A_i y - x_0\| < 1\}\,. \tag{$\star\star$}$$

Next, for each i define the continuous function $f_i\colon X \to [0, \infty)$ by

$$f_i(z) = \max\{0, 1 - \|A_i z - x_0\|\}\,.$$

Clearly, $f_i(z) > 0$ if and only if $\|A_i z - x_0\| < 1$. From ($\star\star$), we see that $f(z) = \sum_{i=1}^m f_i(z) > 0$ for each $z \in \overline{K(U_0)}$, and so the formulas $g_i(z) = \frac{f_i(z)}{f(z)}$ define non-negative continuous functions on $\overline{K(U_0)}$ such that $\sum_{i=1}^m g_i(z) = 1$ for each $z \in \overline{K(U_0)}$. In particular, $\sum_{i=1}^m g_i(Kx) = 1$ for each $x \in U_0$.

Now consider the function $\phi\colon U_0 \to X$ defined by

$$\phi(x) = \sum_{i=1}^{m} g_i(Kx) A_i K x\,.$$

Since $x \in U_0$ and $g_i(Kx) > 0$ imply $A_i K x \in U_0$, it follows from the convexity of U_0 that $\phi(U_0) \subseteq U_0$. In addition, because $\phi(x)$ is a convex combination of the vectors $A_1 Kx, \ldots, A_m Kx$ and $A_i Kx \in A_i K(U_0)$, we see that $\phi(x)$ belongs to $C = \overline{\text{co}}\left[\bigcup_{i=1}^m A_i K(U_0)\right]$. But $\bigcup_{i=1}^m A_i K(U_0)$ is a norm totally bounded set (since each $A_i K$ is a compact operator), and so by Mazur's theorem its closed convex hull C is likewise a compact set; see, for example [**30**, Theorem 9.4, p. 131].[2] Hence, $\phi(U_0) \subseteq C \cap U_0$ and $C \cap U_0$ is a non-empty convex compact subset of X.

---

[2] If X is a complex Banach space, then when taking the closed convex hull of C, we consider X as a real vector space.

By Tychonoff's Fixed Point Theorem, the map $\phi\colon C \cap U_0 \to C \cap U_0$ has a fixed point, say u. Since $u \in U_0$, it follows from $(\star)$ that $u \neq 0$. To complete the proof, consider the operator $A = \sum_{i=1}^{m} g_i(Ku)A_i \in \mathcal{A}$, and note that $AKu = \sum_{i=1}^{m} g_i(Ku)A_iKu = \phi(u) = u$. ■

And now we are ready to state and prove the two famous Lomonosov invariant subspace theorems.

Theorem 10.19 (Lomonosov [**211**]). *If a non-scalar operator T on a complex Banach space commutes with a non-zero compact operator, then T has a non-trivial closed hyperinvariant subspace.*

Proof. Assume that $T \in \mathcal{L}(X)$ is a non-scalar operator and that for some non-zero compact operator $K \in \mathcal{L}(X)$ we have $TK = KT$. To establish the existence of a non-trivial closed T-hyperinvariant subspace, according to Lemma 10.14, it suffices to show that the commutant of T is non-transitive.

To verify this, assume by way of contradiction that the commutant $\{T\}'$ is a transitive algebra. So, by Theorem 10.18, there exists some $A \in \{T\}'$ such that the operator AK has a non-zero fixed point. Let

$$F = \{x \in X\colon\ AKx = x\}.$$

Clearly, F is a non-zero closed AK-hyperinvariant subspace. Since AK is a compact operator and it is the identity on F, it follows that F is finite dimensional. Since T commutes with AK and F is AK-hyperinvariant, we see that F is T-invariant. So, from $T(F) \subseteq F$ and the fact that F is a finite dimensional complex Banach space, it follows that the operator $T\colon F \to F$ has an eigenvalue $\lambda \in \mathbb{C}$.[3] Now let

$$N_\lambda = \{x \in X\colon\ Tx = \lambda x\},$$

and note that N_λ is a non-zero closed $\{T\}'$-invariant subspace which is different from X since T is a non-scalar operator. However, this contradicts our assumption and the proof of the theorem is complete. ■

The assumption that the Banach space X is complex is essential. As we know from Corollary 10.7, on a finite dimensional real Banach space of even dimension there exist operators with non-trivial invariant subspaces but without non-trivial hyperinvariant subspaces. It should be noted that for complex Banach spaces Theorem 10.19 implies immediately Theorem 10.15 and Corollary 10.16.

[3] This is the place where we use that X is a complex Banach space. When X is a real Banach space, there is no guarantee that the operator $T\colon F \to F$ has a real eigenvalue.

Theorem 10.20 (The Lomonosov Invariant Subspace Theorem). *Assume that for a continuous operator $T: X \to X$ on a complex Banach space there exists an operator $S \in \mathcal{L}(X)$ satisfying the following three properties:*

(1) *S is a non-scalar operator.*

(2) *S commutes with T.*

(3) *S commutes with a non-zero compact operator.*

Then T has a non-trivial closed invariant subspace.

Proof. Assume that the operators $T, S \in \mathcal{L}(X)$ satisfy the conditions of the theorem. Then S satisfies the hypotheses of Theorem 10.19. Therefore, S has a non-trivial closed hyperinvariant subspace. Since T commutes with S, this subspace is automatically invariant under T. ■

Surprisingly, Theorem 10.20 fails to be true if the "chain" connecting the given operator with a compact operator contains more than one intermediate link. V. G. Troitsky [**321**] exhibited an example of three operators $T, S_1, S_2 \in \mathcal{L}(\ell_1)$ such that:

(a) T does not have a non-trivial closed invariant subspace.

(b) Neither S_2 nor S_3 is a scalar operator.

(c) T commutes with S_1 and S_1 commutes with S_2, i.e., $S_1T = TS_1$ and $S_1S_2 = S_2S_1$.

(d) S_2 commutes with a non-zero compact operator.

We shall say that an operator $T \in \mathcal{L}(X)$ is a **Lomonosov operator** if it satisfies the assumptions of Theorem 10.20. That is, T is a Lomonosov operator if there exists an operator $S \in \mathcal{L}(X)$ such that:

(1) S is a non-scalar operator.

(2) S commutes with T.

(3) S commutes with a non-zero compact operator.

In this terminology Theorem 10.20 can be stated as follows.

- *Every Lomonosov operator on a complex Banach space has a non-trivial closed invariant subspace.*

It should be immediate from Theorem 10.20 that operators without non-trivial closed invariant subspaces fail to be Lomonosov operators. After the appearance of the Lomonosov Invariant Subspace Theorem it was not clear whether the Lomonosov operators were the only operators on Banach spaces with non-trivial closed invariant subspaces. It turns out that not every operator with a non-trivial closed invariant subspace is a Lomonosov operator. Subtle examples of non-Lomonosov operators on complex Banach spaces with non-trivial closed invariant subspaces can be found in [**143**] and [**144**].

Exercises

1. Give an example of a polynomially compact operator which is not compact.
2. If Ω is a compact Hausdorff space without isolated points and a function $\phi \in C(\Omega)$ is one-to-one, then show that the multiplication operator M_ϕ does not commute with any non-zero weakly compact operator.
3. Show that every multiplication operator on a $C(\Omega)$-space is a Lomonosov operator.
4. Assume that (Ω, Σ, μ) is a σ-finite measure space such that there exist two disjoint sets $A, B \in \Sigma$ with $0 < \mu(A) < \infty$ and $0 < \mu(B) < \infty$. Show that every multiplication operator on an $L_p(\mu)$-space with $1 \le p \le \infty$ is a Lomonosov operator.
5. Show that every Markov operator T on an AM-space with unit commutes with a rank-one positive operator—and hence T is a Lomonosov operator.
6. Consider the operator $T\colon C[-1,1] \to C[-1,1]$ defined by $Tx(t) = x(-t)$ for each $x \in C[-1,1]$ and all $t \in [-1,1]$.
   (a) Show that T is a non-scalar surjective lattice isometry which is also a Markov operator—and so, by Exercise 5 above, T commutes with a rank-one positive operator and is a Lomonosov operator.
   (b) Exhibit a rank-one operator on $C[-1,1]$ that commutes with T.
   (c) Show that the compact operator $S\colon C[-1,1] \to C[-1,1]$, defined by
   $$Sx(t) = \int_{-1}^{1} x(s)\sin(st)\,ds\,,$$
   commutes with the operator T.
   (d) Show that 1 and -1 are the only eigenvalues of T.
   (e) Exhibit non-trivial closed invariant subspaces different from the eigenspaces N_1 and N_{-1}.
7. This exercise is a combination of results obtained by W. B. Arveson and J. Feldman [**43**], C. Pearcy and N. Salinas [**261**], and P. Meyer-Nieberg [**239**]. Let X be a complex Banach space and for an operator $T \in \mathcal{L}(X)$ let $\mathcal{R}(T)$ denote the norm closure in $\mathcal{L}(X)$ of the collection of all operators $q(T)$, where $q(\lambda)$ is a rational function (i.e., a quotient of two polynomials with complex coefficients) that is analytic in a neighborhood of the spectrum $\sigma(T)$. Clearly, $\mathcal{R}(T)$ is a commutative unital algebra of operators.

   Show that if an operator $T \in \mathcal{L}(X)$ is non-scalar and $\mathcal{R}(T)$ contains a non-zero compact operator, then $\mathcal{R}(T)$ is non-transitive.

10.3. Invariant Ideals for Positive Operators

The invariant subspace problem when specialized to positive operators on Banach lattices acquires some special and desirable features. These special features will be the subject of our discussion in this section. As we shall see, not only do some broad classes of positive operators have non-trivial closed invariant subspaces but these invariant subspaces also have a very nice geometrical structure—they are ideals. As a matter of fact, a first result of this type has already been proved in Theorem 9.19 asserting the existence of a non-trivial closed ℓ-hyperinvariant ideal for every compact positive operator which is quasinilpotent at a non-zero positive vector.

We begin by formulating from the outset the main (and still open) conjecture regarding the invariant subspace problem for positive operators.

- **Conjecture:** *Every positive operator on a Banach lattice of dimension greater than two has a non-trivial closed invariant subspace.*

The results in this section (as well as the results in the subsequent sections) will provide a considerable amount of evidence to expect the affirmative answer to this conjecture. For a while it was thought that the modulus of some of the quasinilpotent operators on (the real) ℓ_1 without non-trivial closed invariant subspaces constructed by Read [**275, 276, 277**] might provide a counterexample to this conjecture. However, V. G. Troitsky [**319**] has shown (see Corollary 10.33 below) that the modulus of any quasinilpotent Read operator on ℓ_1 has an eigenvalue, and so its modulus necessarily has a non-trivial closed invariant subspace.

Definition 10.21. *If $A, B\colon X \to X$ are two bounded operators on a Banach space, then their* ***commutator*** *is the operator $[A, B]$ defined by*

$$[A, B] = AB - BA\,.$$

Clearly, two operators $A, B \in \mathcal{L}(X)$ commute if and only if $[A, B] = 0$. The commutant of an operator $T \in \mathcal{L}(X)$ can be described in terms of commutators as follows: $\{T\}' = \big\{S \in \mathcal{L}(X)\colon\ [T, S] = 0\big\}$.

If the operators act on a Banach lattice E, then the existence of the order structure allows us to introduce some new and important collections of operators related to the commutator. Namely, we can relax the condition $[A, B] = 0$ by considering operators whose commutators are positive or negative. To do this, we will define the super right- and super left-commutants of a positive operator.

Definition 10.22. *Let $B\colon E \to E$ be a positive operator on a Banach lattice. The* **super right-commutant** $[B\rangle$ *is the collection of all positive operators $A\colon E \to E$ such that $[A, B] \geq 0$. That is,*

$$[B\rangle = \{A \in \mathcal{L}(E)_+ : \ AB - BA \geq 0\}.$$

Similarly, the **super left-commutant** $\langle B]$ *is the collection of all positive operators $A\colon E \to E$ such that $[A, B] \leq 0$. That is,*

$$\langle B] = \{A \in \mathcal{L}(E)_+ : \ AB - BA \leq 0\}.$$

Notice that $A \in [B\rangle$ if and only if $B \in \langle A]$, and also $\langle B] \cap [B\rangle = \{B\}'_+$, the positive part of the commutant of B.

Consider a positive operator $B\colon E \to E$ on a Banach lattice. As usual, the **null ideal** of B is defined by $N_B = \{x \in E\colon \ B|x| = 0\}$. Clearly, N_B is a closed ideal in E. The **range ideal** of B is the ideal generated by the range of B and it will be denoted by R_B. A moment's thought reveals that

$$R_B = \{y \in E\colon \ \exists \ x \in E^+ \ \text{ such that } \ |y| \leq Bx\}.$$

We are now ready to state an analogue of Lemma 10.2 for positive operators. Its proof is straightforward and is omitted. (See also Exercise 4 at the end of the section.)

Lemma 10.23. *If $B\colon E \to E$ is a positive operator on a Banach lattice and $A \in [B\rangle$, then:*

(1) *The null ideal N_B is A-invariant.*

(2) *The range ideal R_A is B-invariant.*

In particular, the ideals N_B and R_B are both ℓ-hyperinvariant.

The next basic result describes a new large class of operators having non-trivial closed ideals.

Theorem 10.24 (Abramovich–Aliprantis–Burkinshaw [**11**]). *Let $B\colon E \to E$ be a positive operator on a Banach lattice and assume that there exists a positive operator $S\colon E \to E$ such that:*

(1) *$S \in \langle B]$, that is, $SB \leq BS$.*

(2) *S is quasinilpotent at some $x_0 > 0$, i.e., $\lim_{n\to\infty} \|S^n x_0\|^{\frac{1}{n}} = 0$.*

(3) *S dominates a non-zero compact operator.*

Then the operator B has a non-trivial closed invariant ideal.

Proof. The proof is similar to that of Theorem 10.15. Let B, S, and x_0 satisfy the properties stated in the theorem, and let K be a non-zero compact operator dominated by S, i.e., $|Kx| \leq S(|x|)$ for each $x \in E$. Obviously, we can assume that $\|B\| < 1$. Therefore, the series $A = \sum_{n=0}^{\infty} B^n$ converges and defines a positive operator on E. Let us show next that the hypothesis $SB \le BS$ implies that the inequalities $SB^k \le B^k S$ and $SA^k \le A^k S$ hold for each k. The first inequality follows by induction. Indeed, if $SB^k \le B^k S$ is true for some k, then $SB^{k+1} = SB^k B \le B^k SB \le B^k BS = B^{k+1} S$. Next note that

$$SA = S\Big(\sum_{n=0}^{\infty} B^n\Big) = \sum_{n=0}^{\infty} SB^n \le \sum_{n=0}^{\infty} B^n S = \Big(\sum_{n=0}^{\infty} B^n\Big) S = AS\,.$$

This implies at once that $SA^k \le A^k S$ for all $k \ge 0$.

For each $x > 0$ we shall denote by J_x the principal ideal generated by Ax. That is,

$$J_x = \big\{ y \in E\colon\ |y| \le \lambda Ax \text{ for some } \lambda > 0 \big\}\,.$$

Since $x \in J_x$, the ideal J_x is non-zero. We claim that J_x is B-invariant. Indeed, if $y \in J_x$, then we have $|y| \le \lambda Ax$ for some $\lambda > 0$, and so

$$|By| \le B|y| \le \lambda B(Ax) = \lambda \sum_{n=1}^{\infty} B^n x \le \lambda Ax\,.$$

This implies $By \in J_x$. So, $\overline{J_x}$ is a non-zero closed B-invariant ideal.

The proof will be finished if we can show that $\overline{J_x} \ne E$ for some $x > 0$. To establish this, assume by way of contradiction that

$$\overline{J_x} = E \quad \text{for each} \quad x > 0\,. \tag{$\star$}$$

Next, we claim that without loss of generality, we can suppose $Kx_0 \ne 0$. To see this, consider the ideal J_{x_0}. If $Ky = 0$ for each $0 < y \in J_{x_0}$, then we have $K = 0$ on J_{x_0}, and consequently by ($\star$), we get $K = 0$, a contradiction. Therefore, $Ky_0 \ne 0$ for some $0 < y_0 \in Jx_0$. Since the operator S is quasinilpotent at x_0 and $y_0 \le \lambda Ax_0$ for some $\lambda > 0$, it follows that S is also quasinilpotent at y_0. Indeed, $S^n y_0 \le \lambda S^n Ax_0 \le \lambda AS^n x_0$ implies $\|S^n y_0\|^{\frac{1}{n}} \le \lambda^{\frac{1}{n}} \|A\|^{\frac{1}{n}} \|S^n x_0\|^{\frac{1}{n}} \to 0$. Hence, replacing x_0 by y_0 if necessary, we can assume that $Kx_0 \ne 0$.

By scaling, we can suppose that $\|K\| = 1$. Also we can suppose (replacing x_0 by ax_0 for an appropriate scalar $a > 1$) that $\|x_0\| > 1$ and $\|Kx_0\| > 1$. Now let $U_0 = \{z \in E\colon\ \|x_0 - z\| \le 1\}$, the closed unit ball centered at $x_0 > 0$. By our choice of x_0, we have

$$0 \notin U_0 \qquad \text{and} \qquad 0 \notin \overline{K(U_0)}\,. \tag{$\star\star$}$$

From ($\star$), we know that $\overline{J_{|x|}} = E$ for each $x \ne 0$. In other words, $A(|x|)$ is a quasi-interior vector in E and consequently for each $y \ge 0$ the sequence $\big\{y \wedge nA(|x|)\big\}$ is norm convergent and $\lim_{n\to\infty} y \wedge nA(|x|) = y$;

see Lemma 4.15. In particular, for each $x \neq 0$ there is some n such that $\|x_0 - x_0 \wedge nA(|x|)\| < 1$.

Since the function $z \mapsto x_0 \wedge nA(|z|)$ is continuous, we see that the set $\{z \in E\colon \|x_0 - x_0 \wedge nA(|z|)\| < 1\}$ is open for each n. In view of $0 \notin \overline{K(U_0)}$, the above arguments guarantee that

$$\overline{K(U_0)} \subseteq \bigcup_{n=1}^{\infty} \{z \in E\colon \|x_0 - x_0 \wedge nA(|z|)\| < 1\}.$$

Since the sets $\{z \in E\colon \|x_0 - x_0 \wedge nA(|z|)\| < 1\}$ are increasing as n increases, the compactness of $\overline{K(U_0)}$ implies that

$$\overline{K(U_0)} \subseteq \{z \in E\colon \|x_0 - x_0 \wedge mA(|z|)\| < 1\}$$

for some m. In other words, there exists some fixed m such that $x \in \overline{K(U_0)}$ implies that $x_0 \wedge mA(|x|) \in U_0$.

In particular, we have $x_1 = x_0 \wedge mA(|Kx_0|) \in U_0$. So, $Kx_1 \in K(U_0)$ and this implies $x_2 = x_0 \wedge mA(|Kx_1|) \in U_0$. Proceeding inductively, we obtain a sequence $\{x_n\}$ of positive vectors in U_0 defined by $x_{n+1} = x_0 \wedge mA(|Kx_n|)$. Now we claim that the inequality

$$0 \leq x_n \leq m^n A^n S^n x_0$$

holds for each n. The proof is by induction. For $n = 1$ the desired inequality $x_1 = x_0 \wedge mA(|Kx_0|) \leq mA(Sx_0)$ is trivially true. For the induction step assume that $0 \leq x_n \leq m^n A^n S^n x_0$ is true for some n. From this and the inequality $SA^n \leq A^n S$ we obtain:

$$\begin{aligned} 0 \leq x_{n+1} &= x_0 \wedge mA(|Kx_n|) \leq mA(|Kx_n|) \\ &\leq mA(Sx_n) \leq m^{n+1} A(SA^n S^n x_0) \\ &\leq m^{n+1} A^{n+1} S^{n+1} x_0\,, \end{aligned}$$

as desired.

Therefore, $\|x_n\| \leq m^n \|A\|^n \|S^n x_0\|$, and so $\|x_n\|^{\frac{1}{n}} \leq m\|A\| \cdot \|S^n x_0\|^{\frac{1}{n}}$ for each n. Since $\lim_{n\to\infty} \|S^n x_0\|^{\frac{1}{n}} = 0$, it follows that $\lim_{n\to\infty} \|x_n\|^{\frac{1}{n}} = 0$, and hence $\lim_{n\to\infty} \|x_n\| = 0$. However, $\{x_n\} \subseteq U_0$ implies that $0 \in \overline{U_0} = U_0$, contrary to $(\star\star)$, and the proof of the theorem is finished. ■

Recall that an operator $T\colon E \to F$ between two Banach lattices is said to be AM-**compact**, provided that T maps order bounded sets to norm totally bounded sets, i.e., if $T[a,b]$ is a norm totally bounded set for each order interval $[a,b]$. It should be clear that each compact operator is necessarily AM-compact, but the converse is not true; see Exercise 7 at the end of the section. Theorem 10.24 can be easily generalized to AM-compact operators.

Moreover, condition (2) can also be relaxed. The proof of the next result is very similar to that of Theorem 10.24 and is omitted.

Theorem 10.25. *Let $B\colon E \to E$ be a positive operator on a Banach lattice and assume that there exists a positive operator $S\colon E \to E$ such that:*

(1) $S \in \langle B]$.

(2) *There exists some $x_0 > 0$ satisfying* $\liminf_{n\to\infty} \|S^n x_0\|^{\frac{1}{n}} = 0$.

(3) *S dominates a non-zero AM-compact operator.*

Then the operator B has a non-trivial closed invariant ideal.

The assumption in Theorems 10.24 and 10.25 that $S \in \langle B]$ is considerably weaker than the assumption of commutativity of operators B and S; see Exercise 3 at the end of the section. It is still an open question if an analogue of either of these two theorems is true for $S \in [B\rangle$, that is, if we replace the inequality $SB \leq BS$ by the reverse inequality $SB \geq BS$. If we assume that S is quasinilpotent (rather than just being locally quasinilpotent) then, as our next theorem demonstrates, such a version is possible. The proof is quite similar to that of Theorem 10.24 and is left as an exercise.

Theorem 10.26. *Let $B\colon E \to E$ be a positive operator on a Banach lattice and assume that there exists a positive operator $S\colon E \to E$ such that:*

(1) *$S \in [B\rangle$, that is, $SB \geq BS$.*

(2) *S is quasinilpotent.*

(3) *S dominates a non-zero AM-compact operator.*

Then the operator B has a non-trivial closed invariant ideal.

The next result is a strong companion of Theorems 10.24–10.26. It shows that we can distribute properties (2) and (3) between the commuting operators.

Theorem 10.27. *If $B, S\colon E \to E$ are two commuting non-zero positive operators on a Banach lattice such that:*

(a) *one of them is quasinilpotent at a non-zero positive vector, and*

(b) *the other dominates a non-zero AM-compact operator,*

then B and S have a common non-trivial closed invariant ideal.

Proof. Assume that S is locally quasinilpotent at some vector $x_0 > 0$ and that B dominates a non-zero AM-compact operator K. If S is not strictly positive, then the null ideal N_S is a desired common non-trivial closed invariant ideal. So, suppose that S is strictly positive. Without loss of generality, we can also assume that $\|B + S\| < 1$. Put $A = \sum_{n=0}^{\infty}(B + S)^n$ and let J denote the ideal generated by Ax_0, i.e.,

$$J = \{y \in E\colon\ \exists\, \lambda > 0 \text{ such that } |y| \le \lambda A x_0\}.$$

Clearly, J is a non-zero $(B+S)$-invariant ideal. Since $0 \le B, S \le S+B$, it follows that this ideal is invariant under both B and S.

If $\overline{J} \ne E$, then $\overline{J}$ is a non-trivial closed ideal invariant under both B and S. So, it remains to consider the case when $\overline{J} = E$. In this case, since $K \ne 0$, there exists some $0 < y_0 \in J$ such that $Ky_0 \ne 0$. A straightforward verification shows that S remains quasinilpotent at y_0.

Since $|Ky_0| \le By_0 \in J$, it follows from Lemma 4.16 that there exists an operator $M\colon E \to E$ such that

$$MKy_0 > 0 \quad \text{and} \quad |Mx| \le |x| \text{ for each } x \in E.$$

Now consider the operator SMK. Clearly, this operator is AM-compact because K is AM-compact. Since S is strictly positive and $MKy_0 > 0$, it follows that $SMKy_0 > 0$ and so SMK is non-zero. Also, we have

$$|SMKx| \le S|MKx| \le S(|Kx|) \le SB|x|.$$

Thus, the positive operator SB dominates the non-zero AM-compact operator SMK, is quasinilpotent at y_0, and commutes with the positive operator $S+B$. By Theorem 10.25, there exists a non-trivial closed $(S+B)$-invariant ideal. Clearly, this ideal is invariant under both B and S. ■

Corollary 10.28. *Assume that a positive operator $B\colon E \to E$ on a Banach lattice satisfies the following two conditions:*

(1) *B is locally quasinilpotent at a non-zero positive vector.*

(2) *Some power of B dominates a non-zero AM-compact operator.*

Then B has a non-trivial closed invariant ideal.

Proof. It suffices to apply Theorem 10.27 to the operators B and $S = B^m$, where B^m dominates a non-zero AM-compact operator. ■

For operators on discrete spaces Theorem 10.24 can be improved.

Theorem 10.29. *Let $B\colon E \to E$ be a positive operator on a discrete Banach lattice and assume that there exists a non-zero operator $S \in \langle B]$ such that S is locally quasinilpotent at a non-zero positive vector. Then B has a non-trivial closed invariant ideal.*

Proof. As E is discrete, there exists a set Γ such that E is an order dense Riesz subspace of $\mathbb{R}^\Gamma$ ($\mathbb{C}^\Gamma$ in the complex case) and the atoms $e_\gamma = \chi_{\{\gamma\}}$ belong to E for each $\gamma \in \Gamma$.

If S is not strictly positive, then consider the null ideal N_S. It is a non-trivial closed S-invariant ideal in E. Since $S \in \langle B]$, we know that $B \in [S\rangle$ and so N_S is also B-invariant; see Lemma 10.23.

Consequently, we can assume that S is strictly positive. Therefore, for any $\gamma \in \Gamma$ the operator $K = SP_\gamma$ is non-zero, where P_γ denotes the standard rank-one projection onto the one-dimensional subspace spanned by e_γ. Clearly, K is compact and $0 < K \leq S$. Now apply Theorem 10.24. ■

Theorem 10.29 will be improved even further in Corollary 10.48.

Corollary 10.30. *Let $B\colon \ell_p \to \ell_p$ $(1 \leq p < \infty)$ be a positive operator. If there exists a non-zero positive operator $S\colon \ell_p \to \ell_p$ such that:*

(1) *S commutes with B, and*

(2) *S is locally quasinilpotent at a non-zero positive vector,*

then B has a non-trivial closed invariant ideal.

Corollary 10.31. *If a positive operator $B\colon \ell_p \to \ell_p$ $(1 \leq p < \infty)$ is quasinilpotent at some positive vector, then B has a non-trivial closed invariant ideal.*

We are now ready to present an elegant result of V. G. Troitsky [**319**] regarding the modulus of regular operators on ℓ_p-spaces.

Theorem 10.32 (Troitsky). *If $B\colon \ell_p \to \ell_p$ $(1 \leq p < \infty)$ is an ideal irreducible regular quasinilpotent operator such that B^- is compact, then:*

(1) *The modulus $|B|$ of B is not quasinilpotent.*

(2) *The spectral radius of $|B|$ is an eigenvalue—and so $|B|$ has a non-trivial closed invariant subspace.*

(3) *The positive part B^+ of B is not quasinilpotent and the spectral radius of B^+ is an eigenvalue—consequently B^+ has a non-trivial closed invariant subspace.*

Proof. We begin by observing that the operator $|B|$ cannot be quasinilpotent. Indeed, if it were, then by Corollary 10.31 the operator $|B|$ would have a non-trivial closed invariant ideal. This ideal would also be B-invariant contrary to our assumptions. Thus, $r(|B|) > 0$. In view of Theorem 7.9 this guarantees that $r(|B|)$ belongs to the spectrum of the operator $|B|$. It is also clear that $|B|$ cannot be a scalar operator, since otherwise the operator B would again have a non-trivial closed invariant ideal.

We will show next that $r_{\text{ess}}(|B|) = 0$, that is, $|B|$ is an essentially quasinilpotent operator. To prove this, notice that $|B| = B^+ + B^- = B + 2B^-$, and so for each n we have

$$|B|^n = (B + 2B^-)^n = B^n + RB^- ,$$

where R is some polynomial in B and B^-. Hence, RB^- is compact, whence

$$\big\||B|^n\big\|_{\text{ess}} = \|B^n + RB^-\|_{\text{ess}} \le \|B^n\| .$$

Consequently,

$$r_{\text{ess}}(|B|) = \lim_{n\to\infty}\big(\big\||B|^n\big\|_{\text{ess}}\big)^{\frac{1}{n}} \le \lim_{n\to\infty}\sqrt[n]{\|B^n\|} = r(B) = 0 .$$

This implies that $|B|$ is an essentially quasinilpotent operator and, as shown above, $0 < r(|B|) \in \sigma(|B|)$.

Now a glance at Corollary 7.49 guarantees that $r(|B|)$ is an eigenvalue of the operator $|B|_{\mathbf{c}}$ in $E_{\mathbf{c}}$. This implies that $r(|B|)$ is also an eigenvalue of $|B|$ in E, and so the corresponding eigenspace in E is a desired non-trivial closed invariant subspace of $|B|$.

Finally, consider the operator B^+. If it is quasinilpotent, then Corollary 10.31 guarantees that B^+ has a non-trivial closed invariant ideal. So, it remains to consider the case when B^+ is not quasinilpotent. Since $B^+ = B + B^-$, the same argument as in the first part of the proof shows that $r_{\text{ess}}(B^+) = 0$. So we have $0 = r_{\text{ess}}(B^+) < r(B^+)$, and this again implies that B^+ has a non-trivial closed invariant subspace. ■

It is worth mentioning that, in fact, there exists a positive eigenvector corresponding to the eigenvalue $r(|B|)$. This follows from the following useful version of the Krein–Rutman Theorem 7.10 established by P. P. Zabreĭko and S. V. Smitskikh [**345**] and independently by R. Nussbaum [**259**]:

- *If S is a positive operator on a Banach lattice and $r_{\text{ess}}(S) < r(S)$, then $r(S)$ is an eigenvalue of S having a positive eigenvector.*

Theorem 10.32 implies immediately that the modulus of each quasinilpotent Read operator on ℓ_1 has a non-trivial closed invariant subspace.

Corollary 10.33 (Troitsky [**319**]). *If $T\colon \ell_1 \to \ell_1$ is a quasinilpotent operator without non-trivial closed invariant subspaces, then:*

(1) *Neither $|T|$ nor T^+ has a non-trivial closed invariant ideal.*

(2) *Each of the operators $|T|$ and T^+ has an eigenvector—and hence each of these operators has a non-trivial closed invariant subspace.*

Proof. According to Theorem 3.9 the operator T is regular. The validity of the first statement is now immediate from the hypothesis that T does not have any non-trivial closed invariant subspace. The second statement follows from Theorem 10.32. ■

We shall close the section with two interesting cases where the existence of an invariant subspace of a positive operator can be obtained without any compactness assumptions.

Theorem 10.34. *If Ω is a compact Hausdorff space, then every non-scalar positive operator on $C(\Omega)$ has a non-trivial closed hyperinvariant subspace. In particular, each Krein operator has a non-trivial closed hyperinvariant subspace.*

Proof. Let $T\colon C(\Omega) \to C(\Omega)$ be a non-scalar positive operator. By Corollary 9.46 the adjoint operator T^* has an eigenvalue, and the first conclusion follows from Theorem 10.8.

The second conclusion follows from the fact that a Krein operator on a $C(\Omega)$-space cannot be a scalar operator. ■

Recall that if Ω is a topological space, then $C_0(\Omega)$ denotes the Banach lattice of all continuous functions $f\colon \Omega \to \mathbb{R}$ such that for each $\epsilon > 0$ there exists a compact subset K of Ω (depending on ϵ) with $|f(\omega)| < \epsilon$ for all $\omega \in \Omega \setminus K$. Clearly, $C_0(\Omega)$ is a closed ideal in $C_b(\Omega)$, the Banach lattice of all bounded continuous real-valued functions on Ω (as usual $C_0(\Omega)$ and $C_b(\Omega)$ are equipped with the sup norm). If Ω is a compact space, then $C_0(\Omega) = C(\Omega)$. If Ω is a locally compact Hausdorff space and Ω_∞ is the one-point compactification of Ω, then $C_0(\Omega) = \big\{x \in C(\Omega_\infty)\colon\ x(\infty) = 0\big\}$.

When Ω is a locally compact Hausdorff space, an extension of the previous theorem to positive quasinilpotent operators on $C_0(\Omega)$ was obtained by H. Schaefer [**288**]. His original proof was rather involved and used the resolvent operator and the Spectral Mapping Theorem. The proof below is more direct and requires only local quasinilpotence.

Theorem 10.35 (Schaefer)**.** *Let $B\colon C_0(\Omega) \to C_0(\Omega)$ be a positive operator, where Ω is a locally compact Hausdorff space. If B is locally quasinilpotent at some non-zero positive vector, then B has a non-trivial closed invariant ideal.*

Proof. Let B be locally quasinilpotent at some $x_0 > 0$. If $Bx_0 = 0$, then there is nothing to prove, since the null ideal $N_B = \{x \in C_0(\Omega)\colon\ B|x| = 0\}$ is a non-trivial closed invariant subspace.

So, assume that $Bx_0 > 0$. Without loss of generality we can assume that the support set $F = \overline{\{t \in \Omega\colon x_0(t) > 0\}}$ of the function x_0 is compact. To see this, fix a compact subset K of Ω on which x_0 does not vanish. Since Ω is a locally compact Hausdorff space, there exists a continuous function $y\colon \Omega \to [0,1]$ with compact support such that $y(\omega) = 1$ for each $\omega \in K$. If we let $y_0 = yx_0$, then $0 < y_0 \le x_0$, the function y_0 has compact support, and the operator B is quasinilpotent at y_0.

We claim next that there exists a point $t_0 \in \Omega$ such that

$$\big[B^n x_0\big](t_0) = 0 \tag{$\star$}$$

for all $n \in \mathbb{N}$. To see this, assume to the contrary that ($\star$) does not hold. Then for each $t \in \Omega$ there exists some $n_t \in \mathbb{N}$ such that $\big[B^{n_t} x_0\big](t) \neq 0$. In particular, there exist some $\epsilon_t > 0$ and an open neighborhood G_t of t such that

$$\big[B^{n_t} x_0\big](s) \geq \epsilon_t > 0 \quad \text{for each } s \in G_t\,.$$

Pick a finite cover $\{G_{t_1}, \ldots, G_{t_k}\}$ of the compact set F, and let $n_1, n_2, \ldots, n_k$ be the corresponding powers of B for these points, i.e., $n_i = n_{t_i}$. Now consider the operator

$$S = B^{n_1} + B^{n_2} + \cdots + B^{n_k}\,, \tag{$\star\star$}$$

and note that for each $t \in F$ we have

$$\big[Sx_0\big](t) = \sum_{i=1}^{k} \big[B^{n_i} x_0\big](t) \geq \min\{\epsilon_{t_1}, \ldots \epsilon_{t_k}\} > 0\,.$$

Therefore, the function Sx_0 is strictly positive on the compact set F. In particular, there exists some $\gamma > 0$ such that $Sx_0 \geq \gamma x_0$. The latter implies easily that S is not quasinilpotent at x_0; see Exercise 1 at the end of the section. On the other hand, ($\star\star$) guarantees that S is quasinilpotent at x_0. Indeed, from ($\star\star$) we see that

$$S = B\big(B^{n_1 - 1} + B^{n_2 - 1} + \cdots + B^{n_k - 1}\big) = BR = RB\,,$$

where $R = B^{n_1-1} + B^{n_2-1} + \cdots + B^{n_k-1}$. Since B and R commute, we have $S^n x_0 = (BR)^n x_0 = R^n B^n x_0$, and so

$$\|S^n x_0\| \leq \|R^n\| \cdot \|B^n x_0\| \leq \|R\|^n \cdot \|B^n x_0\|$$

for each $n \in \mathbb{N}$. This implies $\|S^n x_0\|^{\frac{1}{n}} \leq \|R\| \cdot \|B^n x_0\|^{\frac{1}{n}} \to 0$, which is impossible. Consequently, there exists a point $t_0 \in \Omega$ that satisfies ($\star$).

Finally, consider the following ideal in $C_0(\Omega)$:

$$J = \Big\{x \in C_0(\Omega)\colon\ \exists\ \lambda \in \mathbb{R}^+ \ \text{ and } \ m_1, m_2, \ldots, m_\ell \in \mathbb{N} \\ \text{such that } \ |x| \leq \lambda\big[\textstyle\sum_{i=1}^{\ell} B^{m_i} x_0\big]\Big\}\,.$$

Obviously, $J \neq \{0\}$ as $0 \neq Bx_0 \in J$. It is also obvious that J is B-invariant. It remains to notice that J cannot be dense in $C_0(\Omega)$ since $x(t_0) = 0$ for each $x \in J$. Therefore, the norm closure $\overline{J}$ of J is a non-trivial closed B-invariant ideal, and the proof is finished. ■

Most of the results in this section were obtained by Y. A. Abramovich, C. D. Aliprantis and O. Burkinshaw [**10, 11, 16**]. For quasinilpotent positive operators, Corollary 10.31 was also proven in [**91**].

Exercises

1. Let $S\colon E \to E$ be a positive operator on an arbitrary Banach lattice such that $Sx_0 \geq \gamma x_0$ holds for some vector $x_0 > 0$ and some scalar $\gamma > 0$. Show that S is not locally quasinilpotent at x_0.

2. Let J be an ideal in a Banach lattice E. If J is invariant under some positive operator $B\colon E \to E$, then J is also invariant under every operator dominated by B.

3. Give an example of a pair of non-commuting positive operators A and B such that $A \in [B\rangle$, that is, $AB - BA > 0$.

4. For a positive operator $B\colon E \to E$ on a Banach lattice establish the following.

(a) The **range ideal** R_B of B (i.e., the ideal generated by the range of the operator B) is given by

$$R_B = \big\{ y \in E\colon\ \exists\ x \in E^+ \ \text{ such that } \ |y| \leq Bx \big\}.$$

(b) The range ideal R_B is ℓ-hyperinvariant.

(c) For each $A \in [B\rangle$ the range ideal R_A is B-invariant.

(d) The null ideal N_B is A-invariant for each $A \in [B\rangle$. In particular, N_B is ℓ–hyperinvariant.

(e) Present an example of an operator $A \in \langle B]$ for which the ideal N_B fails to be A-invariant.

(f) If B is non-zero and $[B\rangle$ contains an ideal irreducible operator, then B is strictly positive.

5. Assume that $K\colon E \to E$ is a compact operator on a Banach lattice such that $Kx_0 \neq 0$ for some $x_0 \in E$. Show that there exists a compact operator C such that $Cx_0 > 0$ and $|Cx| \leq |Kx|$ for each $x \in E$. [HINT: Consider the range ideal R_K of K. Using the compactness of K, establish that the norm closure of R_K has a quasi-interior point. Then apply Lemma 4.16.]

6. For a positive operator $T\colon E \to E$ on a Banach lattice let

$$\mathfrak{A}_T = \big\{ A \in \mathcal{L}(E)\colon\ \exists\ B \in \{T\}'_+ \text{ such that } |Ax| \leq B|x| \text{ for all } x \subset E \big\},$$

where $\{T\}'_+$ denotes as usual the set of all positive operators in the commutant of T. Show that:

(a) The center of E is included in $\mathfrak{A}_T$, i.e., $\mathcal{Z}(E) \subseteq \mathfrak{A}_T$.

(b) $\mathfrak{A}_T$ is a unital subalgebra of $\mathcal{L}(E)$.

(c) An ideal J is ℓ-hyperinvariant for T if and only if J is $\mathfrak{A}_T$-invariant.

7. Give an example of an AM-compact operator which is not a compact operator.

8. Prove Theorem 10.25. [HINT: The proof is similar to that of Theorem 10.24. In this case we cannot claim that the set $\overline{K(U_0)}$ is compact. However, it follows that the set $\overline{K(U_0 \cap [0, x_0])}$ is compact. The inclusion

$$\overline{K(U_0 \cap [0, x_0])} \subseteq \bigcup_{n=1}^{\infty} \{ z \in E\colon\ \|x_0 - x_0 \wedge nA(|z|)\| < 1 \}$$

remains true, and so, in view of the compactness of $\overline{K(U_0 \cap [0, x_0])}$, there exists some m such that

$$K(U_0 \cap [0, x_0]) \subseteq \{z \in E\colon \ \|x_0 - x_0 \wedge mA(|z|)\| < 1\}.$$

Now the sequence $\{x_n\}$, defined recursively by $x_1 = x_0 \wedge mA(|Kx_0|)$ and

$$x_{n+1} = x_0 \wedge mA(|Kx_n|), \ n = 1, 2, \dots ,$$

satisfies $x_n \in U_0 \cap [0, x_0]$ for each n. The rest of the proof remains basically the same.]

9. Prove Theorem 10.26.

10. Recall that a bounded operator $T\colon X \to Y$ between Banach spaces is a **Dunford–Pettis operator** if $x_n \xrightarrow{w} 0$ in X implies $\|Tx_n\| \to 0$ in Y.

Show that every positive Dunford–Pettis operator on a Banach lattice which is quasinilpotent at a non-zero positive vector has a non-trivial closed invariant ideal. [HINT: Let $B\colon E \to E$ be a positive Dunford–Pettis operator on a Banach lattice which is quasinilpotent at a non-zero positive vector. We can assume that B is strictly positive—otherwise N_B is a non-trivial closed invariant ideal. Now note that B carries order intervals onto relatively weakly compact sets [**30**, Theorem 19.12, p. 339], and so B^2 is a non-zero AM-compact operator.]

10.4. Invariant Subspaces of Families of Positive Operators

The objective of this section is to present several invariant subspace results for collections of positive operators on Banach lattices. In particular, several theorems obtained in Section 10.3 will be extended to semigroups of operators on Banach lattices. Among other things, it will be shown that each multiplicative semigroup of quasinilpotent compact positive operators has a common non-trivial closed invariant subspace. Though many ideas of the proofs remain the same as in Section 10.3, to apply them to semigroups of operators, some additional techniques must be developed. This has been accomplished by R. Drnovšek [**114**] whose interesting work we adopt here. As noted in [**114**], some of these methods are based on the works by V. S. Shulman [**301**] and Y. V. Turovskii [**324**]. As before, throughout the section X will denote a real or a complex Banach space and E will denote a real Banach lattice.

We begin our discussion with the definition of a semigroup of operators.

Definition 10.36. *A collection $\mathcal{S}$ of bounded operators on a Banach space is said to be a* ***multiplicative*** *(resp. an* ***additive****)* ***semigroup*** *if for each pair $S, T \in \mathcal{S}$ the operator ST (resp. $S + T$) also belongs to $\mathcal{S}$.*

To avoid trivialities, we tacitly assume in this section that all collections or families of operators under consideration in this section are non-empty.

We continue by introducing some notation and terminology regarding a collection $\mathcal{C}$ of operators in $\mathcal{L}(X)$. For a subset D of a Banach space, we let $\|D\| = \sup_{x\in D}\|x\|$. Accordingly, $\|\mathcal{C}\| = \sup_{C\in\mathcal{C}}\|C\|$. As usual, for each $x \in X$ we let $\mathcal{C}x = \{Cx\colon\ C \in \mathcal{C}\}$, and therefore $\|\mathcal{C}x\| = \sup\{\|Cx\|\colon\ C \in \mathcal{C}\}$.

For any $A \in \mathcal{L}(X)$ we define

$$A\mathcal{C} = \{AC\colon\ C \in \mathcal{C}\} \quad \text{and} \quad \mathcal{C}A = \{CA\colon\ C \in \mathcal{C}\}.$$

For each $n \in \mathbb{N}$, we shall also use the notation

$$\mathcal{C}^n = \{C_1C_2\cdots C_n\colon\ C_1,\ldots,C_n \in \mathcal{C}\}$$

and hope that whenever this notation is used it will not cause any confusion with the standard Cartesian product notation. The **commutant** of $\mathcal{C}$ is the unital algebra of operators defined by

$$\mathcal{C}' = \big\{A \in \mathcal{L}(X)\colon\ AC = CA \text{ for all } C \in \mathcal{C}\big\}.$$

A subspace V of X is said to be **$\mathcal{C}$-invariant** if V is C-invariant for each $C \in \mathcal{C}$. A collection $\mathcal{C}$ of operators is said to be **non-transitive** if there exists a non-trivial closed $\mathcal{C}$-invariant subspace. Otherwise, the family $\mathcal{C}$ is called **transitive**.

We now extend the concept of local quasinilpotence to collections of operators.

Definition 10.37. *A family $\mathcal{C}$ of operators in $\mathcal{L}(X)$ is said to be:*

(1) ***(locally) quasinilpotent at a point*** $x \in X$ *if* $\lim_{n\to\infty}\sqrt[n]{\|\mathcal{C}^n x\|} = 0$ *and*

(2) ***finitely quasinilpotent at a point*** $x \in X$ *if every finite subcollection of $\mathcal{C}$ is (locally) quasinilpotent at x.*

Finitely quasinilpotent algebras of operators were considered by V. S. Shulman [**301**]. Recall that if $T\colon X \to X$ is a bounded operator on a Banach space, then $\mathcal{Q}_T$ denotes the subset of X consisting of all vectors at which T is locally quasinilpotent, i.e.,

$$\mathcal{Q}_T = \Big\{x \in X\colon\ \lim_{n\to\infty}\|T^n x\|^{\frac{1}{n}} = 0\Big\}.$$

It was shown in Exercise 5 of Section 9.2 that $\mathcal{Q}_T$ is a T-hyperinvariant subspace of X. To generalize this to a collection of operators $\mathcal{C}$, we let

$$\mathcal{Q}^f_{\mathcal{C}} = \{x \in X\colon\ \mathcal{C} \text{ is finitely quasinilpotent at } x\}.$$

As in the aforementioned exercise, we can prove the following result.

Lemma 10.38. *The set $\mathcal{Q}^f_{\mathcal{C}}$ is a vector subspace of X and this subspace is both $\mathcal{C}$-invariant and $\mathcal{C}'$-invariant.*

For the rest of the discussion in this section we shall assume that:

- *$\mathcal{C}$ denotes a non-empty collection of positive operators on a Banach lattice E.*

The presence of the order structure on E leads naturally to a modification of the set $\mathcal{Q}^f_{\mathcal{C}}$. Namely, we let

$$\widehat{\mathcal{Q}}^f_{\mathcal{C}} = \{x \in E\colon\ |x| \in \mathcal{Q}^f_{\mathcal{C}}\}.$$

It follows immediately from Lemma 10.38 that $\widehat{\mathcal{Q}}^f_{\mathcal{C}}$ is an ideal in E.

Recall that for a positive operator $C\colon E \to E$ on a Banach lattice, we denote by $[C\rangle$ the collection of all positive operators $A\colon E \to E$ such that $[A, C] \geq 0$. That is,

$$[C\rangle = \{A \in \mathcal{L}(E)_+\colon\ AC - CA \geq 0\}.$$

In accordance with this notation we also let

$$[\mathcal{C}\rangle = \{A \in \mathcal{L}(E)_+\colon\ AC - CA \geq 0 \ \text{ for each } \ C \in \mathcal{C}\}.$$

That is, $[\mathcal{C}\rangle = \bigcap_{C\in\mathcal{C}}[C\rangle$. Similarly, we let

$$\langle\mathcal{C}] = \bigcap_{C\in\mathcal{C}}\langle C] = \{A \in \mathcal{L}(E)_+\colon\ AC - CA \leq 0 \ \text{ for each } \ C \in \mathcal{C}\}.$$

Lemma 10.39. *For any family $\mathcal{C}$ of positive operators in $\mathcal{L}(E)$ the set $[\mathcal{C}\rangle$ is a norm closed additive and multiplicative semigroup in $\mathcal{L}(E)$ and contains the zero and the identity operators.*

Proof. The fact that $[\mathcal{C}\rangle$ is norm closed is obvious. Also, it should be clear that the operators 0 and I belong to $[\mathcal{C}\rangle$. Now take two arbitrary operators S, T in $[\mathcal{C}\rangle$. Then for each operator $C \in \mathcal{C}$ we have $SC \geq CS$ and $TC \geq CT$. Adding up the two inequalities, we get $(S+T)C \geq C(S+T)$, that is, $S + T \in [\mathcal{C}\rangle$. Finally, note that

$$STC = S(TC) \geq SCT = (SC)T \geq CST.$$

Therefore, $ST \in [\mathcal{C}\rangle$ and the proof is finished. ■

Lemma 10.40. *The ideal $\widehat{\mathcal{Q}}^f_{\mathcal{C}}$ is both $\mathcal{C}$-invariant and $[\mathcal{C}\rangle$-invariant.*

Proof. Fix $x \in \widehat{\mathcal{Q}}^f_{\mathcal{C}}$. This means that $\big\|\mathcal{G}^n|x|\big\|^{\frac{1}{n}} \to 0$ for each finite subset $\mathcal{G}$ of $\mathcal{C}$. We must prove that Cx and Tx belong to $\widehat{\mathcal{Q}}^f_{\mathcal{C}}$ for each $C \in \mathcal{C}$ and each $T \in [\mathcal{C}\rangle$. To this end, fix $C \in \mathcal{C}$, $T \in [\mathcal{C}\rangle$ and let $\mathcal{F} = \{C_1, \ldots, C_k\}$ be a finite subset of $\mathcal{C}$.

If we let $\mathcal{G} = \{C_1, \ldots, C_k, C\}$, then $\mathcal{F}^n C = \{C_1, \ldots, C_k\}^n C \subseteq \mathcal{G}^{n+1}$ for each $n \in \mathbb{N}$. Therefore, from

$$\big\|\mathcal{F}^n|Cx|\big\|^{\frac{1}{n}} \leq \big\|\mathcal{F}^n C|x|\big\|^{\frac{1}{n}} \leq \big\|\mathcal{G}^{n+1}|x|\big\|^{\frac{1}{n}} = \Big[\big\|\mathcal{G}^{n+1}|x|\big\|^{\frac{1}{n+1}}\Big]^{\frac{n+1}{n}} \to 0,$$

we get $\big\|\mathcal{F}^n|Cx|\big\|^{\frac{1}{n}} \to 0$. This shows that $Cx \in \widehat{\mathcal{Q}}^f_{\mathcal{C}}$, and hence the ideal $\widehat{\mathcal{Q}}^f_{\mathcal{C}}$ is $\mathcal{C}$-invariant.

Now note that $C_iT \le TC_i$ for each $1 \le i \le k$. This implies at once that for each operator $A \in \mathcal{F}^n$ we have $AT \le TA$, and therefore

$$\big\|\mathcal{F}^n|Tx|\big\|^{\frac{1}{n}} \le \big\|\mathcal{F}^nT|x|\big\|^{\frac{1}{n}} \le \big\|T\mathcal{F}^n|x|\big\|^{\frac{1}{n}} \le \|T\|^{\frac{1}{n}} \cdot \big\|\mathcal{F}^n|x|\big\|^{\frac{1}{n}} \to 0\,.$$

Consequently, $\big\|\mathcal{F}^n|Tx|\big\|^{\frac{1}{n}} \to 0$, and so $Tx \in \widehat{\mathcal{Q}}^f_{\mathcal{C}}$. This proves that the ideal $\widehat{\mathcal{Q}}^f_{\mathcal{C}}$ is also $[\mathcal{C}\rangle$-invariant. ∎

We need to introduce two additional collections associated with an arbitrary collection $\mathcal{C}$ of positive operators on E. The first of these collections is the **multiplicative semigroup generated by** $\mathcal{C}$ in $\mathcal{L}(E)$. It is the smallest (with respect to inclusion) semigroup of operators that contains $\mathcal{C}$ and it will be denoted by $\mathcal{S}_{\mathcal{C}}$. It should be clear that $\mathcal{S}_{\mathcal{C}}$ consists of all finite products of operators in $\mathcal{C}$. In other words,

$$\mathcal{S}_{\mathcal{C}} = \bigcup_{n=1}^{\infty} \mathcal{C}^n\,.$$

The second collection, denoted by $\mathcal{D}_{\mathcal{C}}$, is also a large collection of positive operators that is defined as follows:

$$\mathcal{D}_{\mathcal{C}} = \Big\{D \in \mathcal{L}(E)_+\colon\ \exists\ \{T_1,\dots,T_k\} \subseteq [\mathcal{C}\rangle \text{ and } \{S_1,\dots,S_k\} \subseteq \mathcal{S}_{\mathcal{C}} \text{ such that } D \le \textstyle\sum_{i=1}^k T_iS_i\Big\}.$$

Some simple properties of $\mathcal{D}_{\mathcal{C}}$ and $\mathcal{S}_{\mathcal{C}}$ are presented in the next lemmas.

Lemma 10.41. *If $\mathcal{C}$ is a family of positive operators, then the collection $\mathcal{D}_{\mathcal{C}}$ is an additive and multiplicative semigroup in $\mathcal{L}(E)$.*

Proof. Pick any two operators D_1 and D_2 in $\mathcal{D}_{\mathcal{C}}$. Hence, $D_j \le \sum_{i=1}^{n_j} T_{j,i}S_{j,i}$ for some $T_{j,i} \in [\mathcal{C}\rangle$ and $S_{j,i} \in \mathcal{S}_{\mathcal{C}}$, where $j = 1,2$. The fact that $D_1 + D_2$ belongs to $\mathcal{D}_{\mathcal{C}}$ follows immediately from the definition of $\mathcal{D}_{\mathcal{C}}$. Let us verify that $D_1D_2 \in \mathcal{D}_{\mathcal{C}}$. Indeed,

$$D_1D_2 \le \Big[\sum_{k=1}^{n_1} T_{1,k}S_{1,k}\Big]\Big[\sum_{i=1}^{n_2} T_{2,i}S_{2,i}\Big] = \sum_{k=1}^{n_1}\sum_{i=1}^{n_2} T_{1,k}S_{1,k}T_{2,i}S_{2,i}\,.$$

Since $T_{j,i} \in [\mathcal{C}\rangle$, it follows that $T_{j,i} \in [\mathcal{S}_{\mathcal{C}}\rangle$ (see Exercise 5 at the end of the section) and hence $S_{1,k}T_{2,i} \le T_{2,i}S_{1,k}$. Therefore,

$$D_1D_2 \le \sum_{i=1}^{n_1}\sum_{k=1}^{n_2} T_{1,k}T_{2,i}S_{1,k}S_{2,i}\,.$$

Since $[\mathcal{C}\rangle$ and $\mathcal{S}_c$ are semigroups, we have that $T_{1,k}T_{2,i} \in [\mathcal{C}\rangle$, $S_{1,k}S_{2,i} \in \mathcal{S}_c$ and this proves that the operator D_1D_2 belongs to $\mathcal{D}_c$. ■

Recall that for $x \in E$ the symbol $\mathcal{D}_c x$ denotes the orbit of x under the action of $\mathcal{D}_c$, that is, $\mathcal{D}_c x = \{Dx\colon\ D \in \mathcal{D}_c\}$. To continue our discussion, we shall introduce one more notation.

- *For each $x \in E$ we will denote by $[\mathcal{D}_c x]$ the ideal generated by $\mathcal{D}_c x$.*

As we shall see, the ideals $[\mathcal{D}_c x]$ will provide a source of invariant subspaces.

Lemma 10.42. *Each ideal $[\mathcal{D}_c x]$ is both $\mathcal{C}$-invariant and $[\mathcal{C}\rangle$-invariant.*

Proof. Take any $y \in [\mathcal{D}_c x]$. Since $\mathcal{D}_c$ is an additive semigroup, it follows that $|y| \le \lambda Dx$ for some scalar λ and some $D \in \mathcal{D}_c$. By the definition of $\mathcal{D}_c$ there exist operators $T_i \in [\mathcal{C}\rangle$ and $S_i \in \mathcal{S}_c$ $(i = 1, 2, \ldots, n)$ such that $D \le \sum_{i=1}^n T_iS_i$, and so $|y| \le \lambda \sum_{i=1}^n T_iS_ix$.

Next, we shall show that $[\mathcal{D}_c x]$ is $\mathcal{C}$-invariant. To this end, fix $C \in \mathcal{C}$ and consider the vector Cy. From $CT_i \le T_iC$ for each i, we see that

$$|Cy| \le C|y| \le \lambda \sum_{i=1}^n CT_iS_ix \le \lambda \sum_{i=1}^n T_iCS_ix\,.$$

Since $CS_i \in \mathcal{S}_c$ for each i, we see that $A = \sum_{i=1}^n T_i(CS_i) \in \mathcal{D}_c$. Therefore, $|Cy| \le \lambda \sum_{i=1}^n T_i(CS_i)x = \lambda Ax$ and thus $Cy \in [\mathcal{D}_c x]$.

To get that $[\mathcal{D}_c x]$ is $[\mathcal{C}\rangle$-invariant, let $T \in [\mathcal{C}\rangle$. Since $[\mathcal{C}\rangle$ is a multiplicative semigroup, $TT_i \in [\mathcal{C}\rangle$ for each i, and hence the operator $B = \sum_{i=1}^n (TT_i)S_i$ belongs to $\mathcal{D}_c$. Now notice that $|Ty| \le T|y| \le \lambda \sum_{i=1}^n TT_iS_ix = \lambda Bx$. Consequently, $Ty \in [\mathcal{D}_c x]$ and the proof is finished. ■

Lemma 10.43. *If $\mathcal{C}$ is finitely quasinilpotent at some $x_0 \in E^+$, then both $\mathcal{S}_c$ and $\mathcal{D}_c$ are likewise finitely quasinilpotent at x_0.*

Proof. We begin by showing that the semigroup $\mathcal{S}_c$ is finitely quasinilpotent at x_0. To this end, fix $0 < \epsilon < 1$ and a finite subset $\mathcal{F}$ of $\mathcal{S}_c$. By the definition of $\mathcal{S}_c$, there exists a finite subset $\mathcal{G} \subseteq \mathcal{C}$ and an $m \in \mathbb{N}$ such that $\mathcal{F} \subseteq \bigcup_{k=1}^m \mathcal{G}^k$. Since $\mathcal{C}$ is finitely quasinilpotent at x_0 and the set $\mathcal{G}$ is finite, we have $\|\mathcal{G}^n x_0\| \le \epsilon^n$ for each $n \ge n_0$. Now for $n \ge n_0$ we have

$$\begin{aligned} \|\mathcal{F}^n x_0\| &\le \Big\| \Big[\bigcup_{k=1}^m \mathcal{G}^k \Big]^n x_0 \Big\| = \Big\| \Big[\bigcup_{k=n}^{mn} \mathcal{G}^k \Big] x_0 \Big\| \\ &= \max_{n \le k \le mn} \big\| \mathcal{G}^k x_0 \big\| \le \max_{n \le k \le mn} \epsilon^k = \epsilon^n\,. \end{aligned}$$

Consequently, $\big\|\mathcal{F}^n x_0\big\|^{\frac{1}{n}} \le \epsilon$ for all $n \ge n_0$. This shows that $\mathcal{S}_c$ is finitely quasinilpotent at x_0.

Next, we shall verify that each operator $D \in \mathcal{D}_{\mathcal{C}}$ is quasinilpotent at x_0. By the definition of $\mathcal{D}_{\mathcal{C}}$ we know that $0 \leq D \leq \sum_{i=1}^{m} T_i S_i$, where $T_i \in [\mathcal{C}\rangle$ and $S_i \in \mathcal{S}_{\mathcal{C}}$ for each $i = 1, \ldots, m$. Put $c = \max_{1 \leq i \leq m} \|T_i\|$ and $\mathcal{F} = \{S_1, \ldots, S_m\}$, and then a direct verification shows that

$$\|D^n x_0\| \leq \Big\| \Big[\sum_{i=1}^{m} T_i S_i\Big]^n x_0 \Big\| \leq c^n m^n \|\mathcal{F}^n x_0\| .$$

This implies $\|D^n x_0\|^{\frac{1}{n}} \leq cm \|\mathcal{F}^n x_0\|^{\frac{1}{n}} \to 0$, since $\mathcal{S}_{\mathcal{C}}$ is finitely quasinilpotent at x_0. Hence, $\|D^n x_0\|^{\frac{1}{n}} \to 0$ and so D is quasinilpotent at x_0.

The fact that $\mathcal{D}_{\mathcal{C}}$ is finitely quasinilpotent at x_0 can be demonstrated in a similar manner. ■

Theorem 10.44 (Drnovšek). *If a collection $\mathcal{C}$ of positive operators*

(1) *is finitely quasinilpotent at some positive non-zero vector, and*

(2) *some operator in $\mathcal{C}$ dominates a non-zero AM-compact operator,*

then $\mathcal{C}$ and $[\mathcal{C}\rangle$ have a common non-trivial closed invariant ideal.

Proof. Let $C_0 \in \mathcal{C}$ be an operator that dominates a non-zero AM-compact operator K. The basic steps of the proof follow those of Theorem 10.24. By the hypotheses the ideal $\widehat{\mathcal{Q}}^f_{\mathcal{C}}$ is non-zero, and by Lemma 10.40 we know that $\widehat{\mathcal{Q}}^f_{\mathcal{C}}$ is invariant under $\mathcal{C}$ and $[\mathcal{C}\rangle$. If $\widehat{\mathcal{Q}}^f_{\mathcal{C}}$ is not dense, then the norm closure of $\widehat{\mathcal{Q}}^f_{\mathcal{C}}$ is a common non-trivial closed invariant ideal. So, we can assume that $\widehat{\mathcal{Q}}^f_{\mathcal{C}}$ is a norm dense ideal.

From $K \neq 0$, it follows that there exists some $0 < x_0 \in \widehat{\mathcal{Q}}^f_{\mathcal{C}}$ such that $Kx_0 \neq 0$. Without loss of generality we can assume that

$$\|K\| = 1 , \quad \|x_0\| > 1 , \quad \text{and} \quad \|Kx_0\| > 1 .$$

Let U_0 be the closed unit ball centered at x_0 and let $F = U_0 \cap [0, x_0]$. The set F is order bounded and hence $\overline{K(F)}$ is compact since the operator K is AM-compact. It should be clear that $0 \notin \overline{K(F)}$.

Recall that for each $x \geq 0$ we denote by $[\mathcal{D}_{\mathcal{C}} x]$ the ideal generated by the orbit $\mathcal{D}_{\mathcal{C}} x = \{Ax \colon A \in \mathcal{D}_{\mathcal{C}}\}$ of the semigroup $\mathcal{D}_{\mathcal{C}}$. Since the identity operator belongs to $\mathcal{D}_{\mathcal{C}}$, it follows that $x \in [\mathcal{D}_{\mathcal{C}} x]$ and so $[\mathcal{D}_{\mathcal{C}} x]$ is non-zero provided $x > 0$. As shown in Lemma 10.42, this ideal $[\mathcal{D}_{\mathcal{C}} x]$ is invariant under both $\mathcal{C}$ and $[\mathcal{C}\rangle$, and so we are done if there exists some $[\mathcal{D}_{\mathcal{C}} x]$ which is not norm dense in E.

Therefore, we can assume without loss of generality that $[\mathcal{D}_{\mathcal{C}} x]$ is dense in E for each $x > 0$. This implies that for each $y \neq 0$ there exists an operator

$A \in \mathcal{D}_{\mathcal{C}}$ such that $\|x_0 - x_0 \wedge A(|y|)\| < 1$. Because $0 \notin \overline{K(F)}$, it follows that

$$\overline{K(F)} \subseteq \bigcup_{A \in \mathcal{D}_{\mathcal{C}}} \{y \in E\colon\ \|x_0 - x_0 \wedge A(|y|)\| < 1\}.$$

All sets in this union are obviously open, and so we can find a finite subcover. That is, there exist operators $A_1, \ldots, A_m \in \mathcal{D}_{\mathcal{C}}$ such that

$$\overline{K(F)} \subseteq \bigcup_{i=1}^{m} \{y \in E\colon\ \|x_0 - x_0 \wedge A_i(|y|)\| < 1\}.$$

Let $A = A_1 + \cdots + A_m + C_0$. Since $I \in [\mathcal{C}\rangle$, $C_0 \in \mathcal{C}$, and $C_0 = IC_0$, we see that $A \in \mathcal{D}_{\mathcal{C}}$ and thus, in fact, $\overline{K(F)}$ is covered by a single set. That is,

$$\overline{K(F)} \subseteq \{y \in E\colon\ \|x_0 - x_0 \wedge A(|y|)\| < 1\}.$$

Now consider $y = Kx_0$. Then $y \in K(F)$ so that $\|x_0 - x_0 \wedge A(|Kx_0|)\| < 1$, and also $0 \leq x_0 \wedge A(|Kx_0|) \leq x_0$. Hence the vector $x_1 = x_0 \wedge A(|Kx_0|)$ belongs to F, and therefore $Kx_1 \in K(F)$. Repeating this argument with $y = Kx_1$, we will get $x_2 = x_0 \wedge A(|Kx_1|) \in F$. Proceeding by induction, at the $n+1$ step we define $x_{n+1} = x_0 \wedge A(|Kx_n|)$ and obtain $x_{n+1} \in F \cap U_0$. Moreover, we have

$$x_{n+1} \leq A(|Kx_n|) \leq AC_0x_n \leq A^2x_n \leq \cdots \leq A^{2n}x_0.$$

Since $x_0 \in \widehat{\mathcal{Q}}^f_{\mathcal{C}}$, Lemma 10.43 guarantees that the operator $A \in \mathcal{D}_{\mathcal{C}}$ is quasinilpotent at x_0, and so $\|x_{n+1}\|^{\frac{1}{n+1}} \leq \|A^{2n}x_0\|^{\frac{1}{n+1}} \to 0$. In particular, $\|x_n\| \to 0$. However, the latter conclusion contradicts that $\{x_n\} \subseteq U_0$ and $0 \notin \overline{U}_0 = U_0$, and the proof of the theorem is complete. ■

Let $B\colon E \to E$ be a positive operator satisfying the conditions of Theorem 10.24. That is, assume that some positive operator $S \in \langle B]$ is quasinilpotent at a positive non-zero vector and dominates a non-zero compact operator. Consider the collection $\mathcal{C} = \{S\}$ consisting of a single operator S. Then clearly $\mathcal{C}$ satisfies the hypotheses of Theorem 10.44 and hence there exists a non-trivial closed ideal invariant under $\mathcal{C}$ and $[\mathcal{C}\rangle$. The assumption $S \in \langle B]$ implies that $B \in [\mathcal{C}\rangle$, and so this ideal is certainly B-invariant. In other words, Theorem 10.44 implies Theorem 10.24. However, we have decided to keep both proofs in order to separate the details essential for an individual operator from the details essential for collections of operators.

The semigroups of positive operators having a common non-trivial closed invariant ideal are often referred to as **ideal reducible semigroups**. Accordingly, the semigroups without a common non-trivial closed invariant ideal are referred to as **ideal irreducible semigroups**. This is in agreement with the terminology of Chapter 9. Using Theorem 10.44 we can easily prove that various semigroups of positive operators are ideal reducible. We begin with additive semigroups.

Corollary 10.45. *If for an additive semigroup $\mathcal{S}$ of positive operators*

(1) *each operator in $\mathcal{S}$ is quasinilpotent at some vector $x_0 > 0$ and*

(2) *some operator in $\mathcal{S}$ dominates a non-zero AM-compact operator,*

then $\mathcal{S}$ and $[\mathcal{S}\rangle$ have a common non-trivial closed invariant ideal.

Proof. We will apply Theorem 10.44 to the collection $\mathcal{S}$ of positive operators. The only condition we need to verify is that $\mathcal{S}$ is finitely quasinilpotent at $x_0 > 0$. To this end, let $\mathcal{F} = \{S_1, \dots, S_k\}$ be a finite subset of $\mathcal{S}$ and consider the operator $S = S_1 + \cdots + S_k$. Since $\mathcal{S}$ is an additive semigroup, $S \in \mathcal{S}$ and so S is quasinilpotent at x_0. That is, $\|S^n x_0\|^{\frac{1}{n}} \to 0$.

Since each operator $A \in \mathcal{F}^n$ is the product of n operators from $\mathcal{F}$, it follows that $A \le S^n$ and hence $\|Ax_0\| \le \|S^n x_0\|$. Therefore, we have

$$\big\|\mathcal{F}^n x_0\big\|^{\frac{1}{n}} = \sup_{A \in \mathcal{F}^n} \|Ax_0\|^{\frac{1}{n}} \le \big\|S^n x_0\big\|^{\frac{1}{n}} \to 0\,,$$

and so $\big\|\mathcal{F}^n x_0\big\|^{\frac{1}{n}} \to 0$. This implies that the additive semigroup $\mathcal{S}$ is finitely quasinilpotent at x_0. By Theorem 10.44 the semigroups $\mathcal{S}$ and $[\mathcal{S}\rangle$ have a common non-trivial closed invariant ideal. ■

To get an analogue of Corollary 10.45 for multiplicative semigroups, we need the following deep result due to Y. V. Turovskii in [**324**, Corollary 7].

Theorem 10.46 (Turovskii)**.** *Each multiplicative semigroup of compact quasinilpotent operators on a Banach space is finitely quasinilpotent.*

Corollary 10.47. *If $\mathcal{S}$ is a non-zero multiplicative semigroup of quasinilpotent compact positive operators on a Banach lattice, then $\mathcal{S}$ and $[\mathcal{S}\rangle$ have a common non-trivial closed invariant ideal.*

Proof. According to Theorem 10.46 the multiplicative semigroup $\mathcal{S}$ is finitely quasinilpotent. Now apply Theorem 10.44. ■

Corollary 10.47 can be viewed as an extension of Theorem 10.27 to semigroups of compact positive operators. As we shall see later (Example 10.53), the compactness assumption in Corollary 10.47 is essential.

Corollary 10.48. *Let $B\colon E \to E$ be a non-zero positive operator on a discrete Banach lattice. If B is locally quasinilpotent at some non-zero positive vector, then B has a non-trivial closed ℓ-hyperinvariant ideal.*

Proof. In the proof of Theorem 10.29 we indicated that the operator B dominates a non-zero compact operator. Therefore, the singleton collection $\mathcal{C} = \{B\}$ satisfies the hypotheses of Theorem 10.44. Consequently, there exists a non-trivial closed ideal J that is invariant under $\mathcal{C}$ and $[\mathcal{C}\rangle$. But each positive operator that commutes with B belongs to $[\mathcal{C}\rangle$. Thus, J is ℓ-hyperinvariant for B. ■

Some special cases of Corolaries 10.45, 10.47, and 10.48 have been obtained by M. Jahandideh [**158**]. Our next goal is to extend Theorem 10.27 to collections of positive operators. This requires some preliminary work. Recall that a subset $\mathcal{S}_0$ of a multiplicative semigroup $\mathcal{S}$ is said to be a **(two-sided algebraic) semigroup ideal** in $\mathcal{S}$ if for each $T \in \mathcal{S}_0$ and for each $S \in \mathcal{S}$ the operators TS and ST belong to $\mathcal{S}_0$. Clearly, any algebraic semigroup ideal is a multiplicative semigroup in its own right. Since the intersection of any family of algebraic semigroup ideals is again a semigroup ideal, it follows that for any collection of operators $\mathcal{C}$ there exists a smallest algebraic semigroup ideal containing this subset—this semigroup is called the **semigroup ideal generated by** $\mathcal{C}$.

The following simple result provides a well-known method of proving the existence of invariant subspaces for semigroups of operators; see [**271**, Lemma 1] and [**114**, Lemma 4.6].

Lemma 10.49. *Let $\mathcal{S}$ be a multiplicative group of continuous operators on a Banach space and let $\mathcal{S}_0$ be a non-zero semigroup ideal in $\mathcal{S}$. If $\mathcal{S}_0$ has a (common) non-trivial closed invariant subspace, then $\mathcal{S}$ also has a non-trivial closed invariant subspace.*

Moreover, if $\mathcal{S}$ consists of positive operators on a Banach lattice and the non-zero semigroup ideal $\mathcal{S}_0$ has a non-trivial closed invariant ideal, then $\mathcal{S}$ also has a non-trivial closed invariant ideal.

Proof. Let V be a non-trivial closed subspace invariant under a non-zero semigroup ideal $\mathcal{S}_0$ of the semigroup $\mathcal{S}$. Consider the closed vector subspace W of V spanned by the set $G = \{Tx\colon\ x \in V \text{ and } T \in \mathcal{S}_0\}$.

First, we claim that W is invariant under $\mathcal{S}$. To establish this, choose $S \in \mathcal{S}$, $T \in \mathcal{S}_0$, and $x \in V$. We must show that the vector $S(Tx)$ remains in W. But the operator ST belongs to $\mathcal{S}_0$ since $\mathcal{S}_0$ is a semigroup ideal in $\mathcal{S}$, and hence $S(Tx) = STx \in G \subseteq W$. This shows that if $W \neq \{0\}$, then W is a non-trivial closed $\mathcal{S}$-invariant subspace.

So, let us assume that $W = \{0\}$. This means that each operator in $\mathcal{S}_0$ is zero on V. Consider $N = \bigcap_{T \in \mathcal{S}_0} \operatorname{Ker} T$. Clearly, N is a closed subspace, $V \subseteq N$ and also since $\mathcal{S}_0$ is non-zero, N is not equal to the whole space. That is, N is a non-trivial closed subspace. It remains to notice that this subspace is invariant under $\mathcal{S}$. Indeed, take any $x \in N$ and any $S \in \mathcal{S}$. We want to show that $Sx \in N$, that is, $TSx = 0$ for each $T \in \mathcal{S}_0$. But this is obvious, since the operator TS belongs to the semigroup ideal $\mathcal{S}_0$ and thus indeed $TSx = 0$.

The second statement can be proven in a similar manner. ∎

Theorem 10.50 (Drnovšek [**114**]). *Assume that a non-zero collection $\mathcal{C}$ of positive operators on a Banach lattice:*

(1) *is finitely quasinilpotent at a non-zero positive vector and*

(2) *its commutant $\mathcal{C}'$ contains an operator that dominates a non-zero compact operator.*

Then $\mathcal{C}$ and $[\mathcal{C}\rangle$ have a common non-trivial closed invariant ideal.

Proof. It follows that the ideal $\widehat{\mathcal{Q}}^f_{\mathcal{C}}$ is non-zero and, by Lemma 10.40, we know that $\widehat{\mathcal{Q}}^f_{\mathcal{C}}$ is invariant under $\mathcal{C}$ and $[\mathcal{C}\rangle$. Therefore, we can assume that $\widehat{\mathcal{Q}}^f_{\mathcal{C}}$ is norm dense in E (otherwise its closure is a non-trivial closed ideal invariant under $\mathcal{C}$ and $[\mathcal{C}\rangle$).

Now consider the intersection $N_{\mathcal{C}}$ of all null ideals generated by the operators in $\mathcal{C}$, that is, $N_{\mathcal{C}} = \bigcap_{C \in \mathcal{C}} N_C$. Clearly, this is a closed ideal and a straightforward verification shows that it is invariant under both $\mathcal{C}$ and $[\mathcal{C}\rangle$; see Exercise 4 at the end of the section. Therefore, if $N_{\mathcal{C}}$ is non-zero, then it is a common non-trivial closed ideal invariant under $\mathcal{C}$ and $[\mathcal{C}\rangle$. Consequently, we can also assume that $N_{\mathcal{C}} = \{0\}$.

Let K be a non-zero compact operator that is dominated by some positive operator $T_0 \in \mathcal{C}'$. Since $\widehat{\mathcal{Q}}^f_{\mathcal{C}}$ is norm dense in E, we can find a vector $0 < x_0 \in \mathcal{Q}^f_{\mathcal{C}}$ such that $Kx_0 \neq 0$.

Next, we consider the semigroups $\mathcal{S}_{\mathcal{C}}$ and $\mathcal{D}_{\mathcal{C}}$ associated with our collection $\mathcal{C}$. Recall that $\mathcal{S}_{\mathcal{C}}$ is the multiplicative semigroup generated by $\mathcal{C}$ and $\mathcal{D}_{\mathcal{C}}$ is the multiplicative and additive semigroup defined as follows:

$$\mathcal{D}_{\mathcal{C}} = \Big\{ D \in \mathcal{L}(E)_+ \colon \ \exists\, \{T_1, \dots, T_k\} \subseteq [\mathcal{C}\rangle \text{ and } \{S_1, \dots, S_k\} \subseteq \mathcal{S}_{\mathcal{C}} \text{ such that } D \leq \textstyle\sum_{i=1}^k T_i S_i \Big\}.$$

We will also need the multiplicative semigroup $\mathcal{S}_{\mathcal{C}\cup[\mathcal{C}\rangle}$ generated by the collection $\mathcal{C} \cup [\mathcal{C}\rangle$. We claim that $\mathcal{S}_{\mathcal{C}\cup[\mathcal{C}\rangle} \subseteq \mathcal{D}_{\mathcal{C}}$. Indeed, take an arbitrary operator $A \in \mathcal{S}_{\mathcal{C}\cup[\mathcal{C}\rangle}$ and recall that the elements of the semigroup $\mathcal{S}_{\mathcal{C}\cup[\mathcal{C}\rangle}$ are just the finite products of the operators in $\mathcal{C} \cup [\mathcal{C}\rangle$. From the definition of $[\mathcal{C}\rangle$ we also know that for any operators $C \in \mathcal{C}$ and $T \in [\mathcal{C}\rangle$ we have $CT \leq TC$. This immediately implies that the operator A is dominated by an operator of the form $T_1 \cdots T_m C_1 \cdots C_k$ for some $T_1, \dots, T_m \in [\mathcal{C}\rangle$ and $C_1, \dots, C_k \in \mathcal{C}$. But the operator $T = T_1 \cdots T_m$ belongs to $[\mathcal{C}\rangle$ since $[\mathcal{C}\rangle$ is a semigroup, and similarly the operator $S = C_1 \cdots C_k$ belongs to the semigroup $\mathcal{S}_{\mathcal{C}}$. In other words, we have shown that each operator A in $\mathcal{S}_{\mathcal{C}\cup[\mathcal{C}\rangle}$ is dominated by an operator of the form TS with $T \in [\mathcal{C}\rangle$ and $S \in \mathcal{S}_{\mathcal{C}}$, and consequently A belongs to $\mathcal{D}_{\mathcal{C}}$.

Notice next that for each $C \in \mathcal{C}$ the operator T_0C belongs to $\mathcal{S}_{\mathcal{C}\cup[\mathcal{C}\rangle}$, and therefore we can consider the (two-sided) semigroup ideal $\mathcal{S}_0$ in $\mathcal{S}_{\mathcal{C}\cup[\mathcal{C}\rangle}$ generated by $T_0\mathcal{C}$. That is, $\mathcal{S}_0$ is the semigroup ideal generated by the set $\{T_0C\colon\ C \in \mathcal{C}\}$.

From $\mathcal{S}_{\mathcal{C}\cup[\mathcal{C}\rangle} \subseteq \mathcal{D}_{\mathcal{C}}$, we obtain $\mathcal{S}_0 \subseteq \mathcal{D}_{\mathcal{C}}$. Since $\mathcal{C}$ is finitely quasinilpotent at x_0, we know from Lemma 10.43 that $\mathcal{D}_{\mathcal{C}}$ is also finitely quasinilpotent at x_0, and therefore $\mathcal{S}_0$ is likewise finitely quasinilpotent at x_0.

Let us verify that $\mathcal{S}_0$ contains some operator that dominates a non-zero compact operator. As we know (see Exercise 5 in Section 10.3), there exists a compact operator K_1 such that $|K_1x| \leq |Kx|$ for each $x \in E$ and $K_1x_0 > 0$. Since $N_{\mathcal{C}} = \{0\}$, we can find some $C \in \mathcal{C}$ satisfying $CK_1x_0 > 0$. Hence, CK_1 is a non-zero compact operator on E. Moreover, for each $x \in E$ we have $|CK_1x| \leq C|K_1x| \leq C|Kx| \leq CT_0|x|$, establishing our claim since $CT_0 \in \mathcal{S}_0$.

We have shown by now that the collection of positive operators $\mathcal{S}_0$ is finitely quasinilpotent at $x_0 > 0$ and contains some operator dominating a non-zero compact operator. That is, $\mathcal{S}_0$ satisfies all the hypotheses of Theorem 10.44. Therefore, according to this theorem, $\mathcal{S}_0$ has a non-trivial closed invariant ideal. But $\mathcal{S}_0$ is a semigroup ideal in $\mathcal{S}_{\mathcal{C}\cup[\mathcal{C}\rangle}$, and so from Lemma 10.49 we see that $\mathcal{S}_{\mathcal{C}\cup[\mathcal{C}\rangle}$ also has a non-trivial closed invariant ideal. Recalling that $\mathcal{C} \cup [\mathcal{C}\rangle \subseteq \mathcal{S}_{\mathcal{C}\cup[\mathcal{C}\rangle}$, we infer that $\mathcal{C}$ and $[\mathcal{C}\rangle$ have a common non-trivial closed invariant ideal. This completes the proof. ■

Recall that in Theorem 10.35 we have proved that if a positive operator $B\colon C_0(\Omega) \to C_0(\Omega)$ is locally quasinilpotent at some vector $x_0 > 0$, where Ω is a locally compact Hausdorff space, then B has a non-trivial closed invariant ideal. The next theorem is a semigroup version of this result. The proof is very similar to that of Theorem 10.35 and we will just outline it.

Theorem 10.51. *Let Ω be a locally compact Hausdorff space and let $\mathcal{S}$ be a non-zero multiplicative semigroup of positive operators on $C_0(\Omega)$ which are quasinilpotent at some $x_0 > 0$. If $\mathcal{S}$ is also an additive semigroup, then $\mathcal{S}$ has a common non-trivial closed invariant ideal.*

Proof. If $Sx_0 = 0$ for each $S \in \mathcal{S}$, then there is nothing to prove, since the intersection of the null ideals $J = \bigcap_{S\in\mathcal{S}}\{x \in C_0(\Omega)\colon\ S|x| = 0\}$ is a non-trivial closed $\mathcal{S}$-invariant ideal.

So, we assume that $S_0x_0 > 0$ for some $S_0 \in \mathcal{S}$. Without loss of generality we may assume that the support set $F = \overline{\{t \in \Omega\colon\ x_0(t) > 0\}}$ of the function x_0 is compact. Next, as in the proof of Theorem 10.35, it is easy to show that there exists a point $t_0 \in \Omega$ such that

$$\big[Sx_0\big](t_0) = 0$$

for each $S \in \mathcal{S}$. Finally, consider the ideal

$$J = \Big\{x \in C_0(\Omega)\colon\ |x| \le \lambda\Big[\sum_{i=1}^{m} S_i x_0\Big],\ \text{ where } S_i \in \mathcal{S} \text{ and } \lambda \in \mathbb{R}^+\Big\}.$$

Clearly, $J \neq \{0\}$ as $0 \neq S_0x_0 \in J$. It should also be obvious that the ideal J is $\mathcal{S}$-invariant. Furthermore, J cannot be dense in $C_0(\Omega)$ since $\big[Sx_0\big](t_0) = 0$ for each operator $S \in \mathcal{S}$. Therefore, the norm closure $\overline{J}$ of J is a non-trivial closed $\mathcal{S}$-invariant ideal. ■

It comes as a surprise that the assumption of additivity in Theorem 10.51 is essential. Moreover, without this assumption even the quasinilpotence of each operator in the semigroup is not enough to guarantee the existence of a common non-trivial closed invariant ideal. To present such an example, we need some preliminary discussion. Below, Γ will denote the unit circle, and $z = e^{\imath\phi} \in \Gamma$ will be a fixed complex number whose argument ϕ is an irrational multiple of 2π. It is well known that for each $t \in \Gamma$ the orbit $\{tz^n\colon\ n \in \mathbb{N}\}$ is dense in Γ; see Exercise 2 in Section 7.3. With each function $w \in C(\Gamma)$ we associate the weighted composition operator $T_w\colon C(\Gamma) \to C(\Gamma)$ defined for each $x \in C(\Gamma)$ by

$$\big[T_w x\big](t) = w(t)x(tz)\,, \quad t \in \Gamma\,. \tag{$\star$}$$

We will make use of the following remarkable formula, due to A. K. Kitover [**186**] and independently A. B. Antonevich and A. V. Lebedev [**38**], for calculating the spectral radius of the operator T_w.

Theorem 10.52 (Antonevich–Lebedev–Kitover)**.** *If T_w is the operator defined by ($\star$) above, then its spectral radius is given by*

$$r(T_w) = e^{\frac{1}{2\pi}\int_0^{2\pi} \ln|w(e^{\imath\theta})|\,d\theta}\,.$$

Proof. For simplicity let $T = T_w$. Assume first that there exists some $\epsilon > 0$ such that $|w(t)| \ge \epsilon$ for each $t \in \Gamma$. So, the function $t \mapsto \ln|w(t)|$ is continuous. Clearly, for $n \in \mathbb{N}$, $x \in C(\Gamma)$, and $t \in \Gamma$ we have

$$T^n x(t) = w(t)w\big(tz\big)\cdots w\big(tz^{n-1}\big)x\big(tz^n\big)\,.$$

This implies that

$$\begin{aligned}\|T^n\| &= \sup_{\|x\|_\infty \le 1} \|T^n x\|_\infty = \max_{t\in\Gamma} \big|w(t)w(tz)\cdots w(tz^{n-1})\big| \\ &= \big|w(t_n)w(t_n z)\cdots w(t_n z^{n-1})\big|\end{aligned}$$

for some $t_n \in \Gamma$. If we consider the probability measure μ_n on Γ defined by $\mu_n = \frac{1}{n}\sum_{i=0}^{n-1}\delta_{t_n z^i}$, then it is easy to see that

$$\|T^n\|^{\frac{1}{n}} = e^{\int_\Gamma \ln|w|\,d\mu_n}. \tag{$\star$}$$

Now notice that the sequence $\{\mu_n\}$ lies in the closed unit ball of $C(\Gamma)^*$, which is w^*-compact. So, there exists a subnet $\{\mu_{n_\alpha}\}$ of $\{\mu_n\}$ such that $\mu_{n_\alpha} \xrightarrow{w^*} \mu$ for some measure μ. This and ($\star$) imply at once that

$$r(T) = \lim_{n\to\infty}\|T^n\|^{\frac{1}{n}} = \lim_\alpha \|T^{n_\alpha}\|^{\frac{1}{n_\alpha}} = e^{\int_\Gamma \ln|w|\,d\mu}.$$

We will establish next that the measure μ is the normalized Lebesgue measure on Γ. To this end consider the operator $S\colon C(\Gamma)\to C(\Gamma)$ defined by $Sx(t) = x(tz)$. From $\langle x, S^*\mu_n - \mu_n\rangle = \langle Sx - x, \mu_n\rangle = \frac{1}{n}\big[x(t_n z^n) - x(t_n)\big]$, it follows that $\|S^*\mu_n - \mu_n\| = \frac{2}{n}$. This implies $\|S^*\mu_n - \mu_n\| \to 0$, and in particular we get $S^*\mu_n - \mu_n \xrightarrow{w^*} 0$. From $\mu_{n_\alpha} \xrightarrow{w^*} \mu$, we get $S^*\mu_{n_\alpha} \xrightarrow{w^*} S^*\mu$, and so $S^*\mu = \mu$. The latter is equivalent to

$$\int_\Gamma Sx\,d\mu = \int_\Gamma x\,d\mu \tag{$\star\star$}$$

for each $x \in C(\Gamma)$.

Using ($\star\star$) we can easily show that the measure μ is rotation invariant. To this end fix an arbitrary $s\in\Gamma$. Since the argument ϕ of $z = e^{\imath\phi}$ is an irrational multiple of 2π, there exists an increasing sequence $\{k_n\}$ of natural numbers such that $z^{k_n}\to s$. It follows that $S^{k_n}x(t) = x(tz^{k_n}) \to x(st)$, and so from ($\star\star$) and from the Lebesgue Dominated Convergence Theorem we get

$$\int_\Gamma x(st)\,d\mu(t) = \int_\Gamma x(t)\,d\mu(t) \tag{$\dagger$}$$

for all $x \in C(\Gamma)$. Clearly ($\dagger$) is also true for each $x \in L_1(\Gamma,\mu)$. In particular, it follows that

$$\mu(sA) = \mu(A)$$

for each Borel subset A of Γ. But then, by the uniqueness of the Haar measure, we conclude that μ must be the normalized Lebesgue measure on Γ. Thus, in the special case when $|w|$ is separated away from zero, we have established that

$$r(T_w) = e^{\int_\Gamma \ln|w(t)|\,d\mu(t)} = e^{\frac{1}{2\pi}\int_0^{2\pi}\ln|w(e^{\imath\theta})|\,d\theta}.$$

For the general case, consider the sequence of continuous functions $\{w_k\}$ given by $w_k = |w| + \frac{1}{k}$ and let $T_k = T_{w_k}$. From $\|T^n\| \le \|T_k^n\|$ for all n we get $r(T) \le r(T_k) = e^{\int_\Gamma \ln|w_k|\,d\mu}$ for each k. Letting $k \to \infty$ yields

$$r(T) \le e^{\int_\Gamma \ln|w|\,d\mu}. \qquad (\dagger\dagger)$$

In particular, if the measurable function $t \mapsto \ln|w(t)|$ is not Lebesgue integrable, i.e., if $\int_\Gamma \ln|w|\,d\mu = -\infty$, then $r(T) = e^{\int_\Gamma \ln|w|\,d\mu} = 0$.

It remains to consider the case when the function $t \mapsto \ln|w(t)|$ is integrable. In this case, using ($\dagger$) we get

$$\begin{aligned}
\ln\big[\|T^n\|^{\frac{1}{n}}\big] &= \tfrac{1}{n}\ln\big[\max_{t\in\Gamma}\big|w(t)w(tz)\cdots w\big(tz^{n-1}\big)\big|\big] \\
&\ge \tfrac{1}{n}\int_\Gamma \ln\big[\big|w(t)w(tz)\cdots w\big(tz^{n-1}\big)\big|\big]\,d\mu(t) \\
&= \tfrac{1}{n}\sum_{i=0}^{n-1}\int_\Gamma \ln|w(tz^i)|\,d\mu(t) \\
&= \int_\Gamma \ln|w(t)|\,d\mu(t).
\end{aligned}$$

Consequently, $\|T^n\|^{\frac{1}{n}} = e^{\ln[\|T^n\|^{\frac{1}{n}}]} \ge e^{\int_\Gamma \ln|w(t)|\,d\mu(t)}$ for each n and hence

$$r(T) \ge e^{\int_\Gamma \ln|w(t)|\,d\mu(t)}.$$

Combining this with ($\dagger\dagger$), we see that the formula $r(T) = e^{\int_\Gamma \ln|w|\,d\mu}$ is true in this case too, and the proof is finished. ■

It is worth pointing out that Theorem 10.52 is a special case of the following result obtained by the same authors.

- *Let $\phi\colon \Omega \to \Omega$ be a homeomorphism of a compact Hausdorff space, let $w \in C(\Omega)$, and consider the operator $T_w\colon C(\Omega) \to C(\Omega)$ defined by $T_w x(\omega) = w(\omega)x\big(\phi(\omega)\big)$. Then the spectral radius of the operator T_w is given by*

  $$r(T_w) = \sup_\mu e^{\int_\Omega \ln|w|\,d\mu},$$

  where the supremum is taken over all ϕ-invariant Borel probability measures μ on Ω.

From Theorem 10.52 it follows that T_w is quasinilpotent if and only if the Lebesgue measurable function $t \mapsto \ln|w(t)|$ is not integrable over Γ.

For our final result let us fix some function $w_0 \in C(\Gamma)$ with the following properties:

(a) $0 \leq w_0(t) \leq 1$ for each $t \in \Gamma$.

(b) $t = 1$ is the only point at which w_0 vanishes.

(c) $w_0(t) = 1$ for all t outside of a small neighborhood of $t = 1$.

(d) The function w_0 drops rapidly enough at $t = 1$, so that the function $t \mapsto \ln w_0(t)$ is not integrable.[4]

These conditions imply that the operator T_{w_0} is positive and quasinilpotent.

Example 10.53. *There exists an ideal irreducible multiplicative semigroup $\mathcal{S}$ of quasinilpotent positive operators on $C(\Gamma)$.*

Using the function w_0 satisfying the properties (a)–(d) preceding the statement of the theorem, we introduce the following collection of weights:

$$W = \big\{ w \in C(\Gamma) \colon \ 0 \leq w(t) \leq w_0(ts) \text{ for each } t \in \Gamma \text{ and some } s \in \Gamma \big\}.$$

Next, for each $w \in W$ and $n \in \mathbb{N}$ define the operator $T_{w,n} \colon C(\Gamma) \to C(\Gamma)$ by

$$\big[T_{w,n} f\big](t) = w(t) f(tz^n), \quad f \in C(\Gamma), \ \ t \in \Gamma,$$

where $z = e^{\imath\phi} \in \Gamma$ is a fixed complex number whose argument ϕ is an irrational multiple of 2π. Let $\mathcal{S}$ denote the collection of all these operators. That is,

$$\mathcal{S} = \big\{ T_{w,n} \colon \ w \in W \ \text{ and } \ n \in \mathbb{N} \big\}.$$

A glance at $(\star)$ shows that each operator $T_{w,n} \in \mathcal{S}$ is quasinilpotent. To verify that $\mathcal{S}$ is a multiplicative semigroup, let $T_{w_1,n_1}, T_{w_2,n_2} \in \mathcal{S}$, and so for each $f \in C(\Gamma)$ and each $t \in \Gamma$ we have

$$\big[T_{w_i,n_i} f\big](t) = w_i(t) f(tz^{n_i}),$$

where $i = 1, 2$. A direct computation shows that

$$\big[T_{w_2,n_2} T_{w_1,n_1} f\big](t) = w_2(t) w_1(tz^{n_2}) f(tz^{n_1+n_2}) = w(t) f(tz^{n_1+n_2}),$$

where $w(t) = w_2(t) w_1(tz^{n_2})$. Clearly $w \in W$ since $0 \leq w \leq w_2$, and therefore $T_{w_2,n_2} T_{w_1,n_1} \in \mathcal{S}$.

We claim that this semigroup $\mathcal{S}$ is ideal irreducible. If not, then there exists a non-trivial closed ideal J in $C(\Gamma)$ which is T-invariant for each $T \in \mathcal{S}$. Consequently, there exists a closed set $F \subseteq \Gamma$ such that J consists of all functions that vanish on this set F. Clearly, F cannot be dense in Γ, otherwise J would equal $\{0\}$. Also, F is non-empty, otherwise $J = E$. So, there exists a point $t_0 \in \Gamma$ such that some neighborhood of t_0 does not meet F. Pick a non-negative function $f \in C(\Gamma)$ such that f vanishes on F (and therefore f belongs to J) and f is greater than zero on some open set Δ containing t_0.

---

[4] For instance, the function $w_0(t) = w_0(e^{\imath\theta}) = e^{-\frac{1}{\theta^2}}$ about $t = 1$ (or $\theta = 0$) satisfies this non-integrability condition.

Now fix some $t_1 \in F$. Since the set $\{t_1 z^n \colon n \in \mathbb{N}\}$ is dense in Γ, there exists some $k \in \mathbb{N}$ such that the number $t_2 = t_1 z^k$ belongs to Δ. Next choose $s \in \Gamma$ such that $w_0(t_1 s) = 1$, and let $w(t) = w_0(ts)$ for $t \in \Gamma$. Then $w \in C(\Gamma)$, $0 \leq w(t) \leq w_0(ts)$, and $w(t_1) = 1$. Finally, consider the operator $T_{w,k} \in \mathcal{S}$. Since J if T-invariant, $T_{w,k} f \in J$, and so $[T_{w,k} f](t_1) = 0$. On the other hand

$$[T_{w,k} f](t_1) = w(t_1) f(t_1 z^k) = f(t_2) > 0\,,$$

a contradiction. This establishes that the multiplicative semigroup $\mathcal{S}$ is ideal irreducible. ∎

An example of an irreducible multiplicative semigroup of nilpotent positive operators on $L_2[0,1]$ was constructed by Y. Zhong [**349**]. Moreover, as shown in [**115**], it is possible to obtain in addition that the square of each operator in the semigroup is zero and to construct such a semigroup on any $L_p[0,1]$ with $1 \leq p < \infty$. Both Example 10.53 and the above mentioned examples show also that the assumption of compactness in Corollary 10.47 is essential.

Exercises

1. Let $\mathcal{S}$ be a multiplicative semigroup of operators on a Banach space and let $\mathcal{S}_1 = \mathcal{S} \cup \{I\}$, where I is the identity operator. Show that $\mathcal{S}_1$ is again a multiplicative semigroup. (Clearly, $\mathcal{S}$ and $\mathcal{S}_1$ have the same invariant subspaces and therefore, when dealing with invariant subspaces of a multiplicative semigroup, we can always assume that the semigroup is unital.)
2. Let $\mathcal{S}$ be an additive semigroup of operators on a Banach space and let $\mathcal{S}_0 = \mathcal{S} \cup \{0\}$, where 0 is the zero operator. Show that $\mathcal{S}_0$ is again an additive semigroup. (Clearly, $\mathcal{S}$ and $\mathcal{S}_0$ have the same invariant subspaces and therefore, when dealing with invariant subspaces of an additive semigroup, we can always assume that the semigroup contains the zero operator.)
3. Prove Lemma 10.38.
4. Let $\mathcal{C}$ be a collection of positive operators on a Banach lattice E and let $N_{\mathcal{C}} = \bigcap_{C \in \mathcal{C}} N_C$. Show that $N_{\mathcal{C}}$ is invariant under $\mathcal{C}$ and $[\mathcal{C}\rangle$.
5. Let $\mathcal{C}$ be a collection of positive operators on a Banach lattice and let $\mathcal{S}_{\mathcal{C}}$ be the multiplicative semigroup generated by $\mathcal{C}$. Establish the following identities: $[\mathcal{C}\rangle = [\mathcal{S}_{\mathcal{C}}\rangle$, $\langle\mathcal{C}] = \langle\mathcal{S}_{\mathcal{C}}]$, and $\mathcal{C}' = \mathcal{S}_{\mathcal{C}}'$.
6. Let $\mathcal{C}$ be a commutative collection of positive operators on a Banach lattice. If $\mathcal{C}$ is quasinilpotent at some vector $x_0 > 0$, then show that $\mathcal{C}$ is finitely quasinilpotent at x_0.
7. Finish the details of the proof of Theorem 10.51.

Consequently, to proceed with the proof of our theorem, we can assume from now on that

$$\overline{J_x} = E \tag{$\star$}$$

for each $x > 0$, that is, Ax is a quasi-interior point in E for each $x > 0$. (In the terminology of Chapter 9, A is a strongly expanding operator.)

Fix three non-zero operators $R, K, C \colon E \to E$ with K, R positive, K compact, and satisfying

$$RB = BR, \quad |Cx| \le R(|x|) \quad \text{and} \quad |Cx| \le K(|x|) \ \text{ for each } \ x \in E.$$

Since $C \neq 0$, there exists some $x_1 > 0$ such that $Cx_1 \neq 0$. Then $A|Cx_1|$ is a quasi-interior point satisfying $A|Cx_1| \ge |Cx_1|$, and it follows from Lemma 4.16(1), that there exists an operator $M_1 \colon E \to E$ dominated by the identity operator such that $x_2 = M_1Cx_1 > 0$. Put $\Pi_1 = M_1C$, and note that the operator Π_1 is dominated by the compact positive operator K and by the operator R.

By ($\star$) we have $\overline{J_{x_2}} = E$ and consequently, since $C \neq 0$, there exists $0 < y \le Ax_2$ such that $Cy \neq 0$. Hence (again by ($\star$)) the element $A|Cy|$ is a quasi-interior point. Since $|Cy| \le A|Cy|$, there exists (by Lemma 4.16(1) again) an operator $M_2 \colon E \to E$ dominated by the identity operator such that $x_3 = M_2Cy > 0$. Since $|y| \le Ax_2$ and Ax_2 is a quasi-interior point, it follows from Lemma 4.16(2) that there exists an operator $M \colon E \to E$ dominated by the identity operator such that $MAx_2 = y$. So, $x_3 = M_2Cy = M_2CMAx_2$. Let $\Pi_2 = M_2CMA$ and note that Π_2 is dominated by the compact positive operator KA and by the operator RA.

If we repeat the preceding arguments with the vector x_2 replaced by x_3, then we obtain one more operator $\Pi_3 \colon E \to E$ that satisfies $\Pi_3x_3 > 0$ and that is dominated by the compact positive operator KA and by RA.

Consider the operator $\Pi_3\Pi_2\Pi_1$. From $\Pi_3\Pi_2\Pi_1x_1 = \Pi_3x_3 > 0$, we see that $\Pi_3\Pi_2\Pi_1$ is a non-zero operator. Moreover, as shown above, each operator Π_i ($i = 1, 2, 3$) is dominated by a compact operator and, therefore, Theorem 2.34 guarantees that the operator $\Pi_3\Pi_2\Pi_1$ is necessarily compact. A straightforward verification also shows that

$$\big|\Pi_3\Pi_2\Pi_1x\big| \le RARAR|x|$$

for each $x \in E$.

Let $S = RARAR$. We will verify that the operator $S \in [B\rangle$. To see this recall that $A \in [B\rangle$ and that the operator R also belongs to $[B\rangle$ because it commutes with B. Since $[B\rangle$ is a multiplicative semigroup, the operator $S = RARAR$ necessarily belongs to $[B\rangle$.

It remains to notice that the collection $\mathcal{C} = \{B\}$ consisting of the single operator B satisfies the conditions of Theorem 10.50. Indeed, $\mathcal{C}$ is quasinilpotent at $x_0 > 0$ and $[\mathcal{C}\rangle = [B\rangle$ contains the operator S that dominates a non-zero compact operator. By this theorem, there exists a non-trivial closed ideal that is invariant under $[B\rangle$. This completes the proof. ■

It might be worth mentioning that some additional assumption on a compact-friendly operator is necessary if one wants to establish the existence of an ℓ-hyperinvariant ideal. Indeed, the identity operator is compact-friendly but does not have any non-trivial closed ℓ-hyperinvariant ideal.

Corollary 10.56. *Let $S\colon E \to E$ be a positive operator on a Banach lattice and suppose that $\langle S]$ contains a non-zero compact-friendly operator that is quasinilpotent at some positive vector. Then S has a non-trivial closed invariant ideal.*

Proof. Let $B \in \langle S]$ be a non-zero compact-friendly operator that is quasinilpotent at some positive vector. As we know, the condition $B \in \langle S]$ is equivalent to the condition $S \in [B\rangle$ and consequently the desired conclusion follows from Theorem 10.55. ■

For Dedekind complete Banach lattices we can improve Theorem 10.55 by proving that there always exists a non-trivial closed $[B\rangle$-invariant ideal. The proof is a modification of that of Theorem 10.55.

Theorem 10.57. *If a non-zero compact-friendly operator $B\colon E \to E$ on a Dedekind complete Banach lattice is quasinilpotent at some $x_0 > 0$, then there exists a non-trivial closed ideal that is invariant under $[B\rangle$.*

Proof. For each $x > 0$ we denote by J_x the ideal generated by the orbit $[B\rangle x$, that is,

$$J_x = \big\{y \in E\colon\ |y| \le Ax \text{ for some } A \in [B\rangle\big\}.$$

It is obvious that J_x is a non-zero ideal. Observe next that the ideal J_x is $[B\rangle$-invariant. Indeed, if $y \in J_x$, then $|y| \le Ax$ for some $A \in [B\rangle$ and so for any $A_1 \in [B\rangle$ we have

$$|A_1 y| \le A_1|y| \le A_1 A x\,.$$

Since $[B\rangle$ is a multiplicative semigroup, the operator A_1A belongs to $[B\rangle$ and we see that $A_1 y \in J_x$, as claimed. Therefore, if for some vector $x > 0$ the ideal J_x is not norm dense in E, then its norm closure $\overline{J_x}$ is a desired non-trivial closed $[B\rangle$-invariant ideal.

Consequently, to proceed with the proof of our theorem, we can assume from now on that

$$\overline{J}_x = E \tag{$\star$}$$

for each $x > 0$. Our next step is to show that the assumption ($\star$) implies that $[B\rangle$ contains an operator that dominates a non-zero compact operator.

Fix three non-zero operators $R, K, C\colon E \to E$ with K, R positive, K compact, and satisfying

$$RB = BR, \quad |Cx| \le R(|x|) \quad \text{and} \quad |Cx| \le K(|x|) \ \text{ for each } \ x \in E.$$

Since $C \neq 0$, there exists some $x_1 > 0$ such that $Cx_1 \neq 0$. This means that at least one of the vectors $(Cx_1)_+$ or $(Cx_1)_-$ is non-zero, and therefore, since E is Dedekind complete, there exists an operator M_1 dominated by the identity operator such that $x_2 = M_1Cx_1 > 0$. Put $\Pi_1 = M_1C$ and note that the operator Π_1 is dominated by the compact positive operator K and by the operator R.

Since J_{x_2} is norm dense in E and C is non-zero, there exists some $y \in J_{x_2}$ and an operator $A_1 \in [B\rangle$ such that $0 < y \le A_1x_2$ and $Cy \neq 0$. Using again the hypothesis that E is Dedekind complete, we can find operators M and M_2 dominated by the identity operator and such that $y = MA_1x_2$ and $x_3 = M_2Cy = M_2CMA_1x_2 > 0$. Let $\Pi_2 = M_2CMA_1$ and note that the operator Π_2 is dominated by the compact operator KA_1 and by the operator RA_1.

If we repeat the preceding arguments with the vector x_2 replaced by x_3, then we obtain an operator $A_2 \in [B\rangle$ and an operator $\Pi_3\colon E \to E$ which satisfies $\Pi_3x_3 > 0$ and which is dominated by the compact operator KA_2 and by the operator RA_2.

Consider the operator $\Pi_3\Pi_2\Pi_1$. It is non-zero as $\Pi_3\Pi_2\Pi_1x_1 = \Pi_3x_3 > 0$. Moreover, as shown above, each operator Π_i $(i = 1, 2, 3)$ is dominated by a compact operator and therefore Theorem 2.34 guarantees that the operator $\Pi_3\Pi_2\Pi_1$ is necessarily compact. A straightforward verification also shows that

$$\big|\Pi_3\Pi_2\Pi_1x\big| \le RA_2RA_1R(|x|)$$

for each $x \in E$.

Finally, consider the operator $S = RA_2RA_1R$. We claim that $S \in [B\rangle$. To see this, recall that $A_1, A_2 \in [B\rangle$ and also $R \in [B\rangle$ because R commutes with B. Since $[B\rangle$ is a multiplicative semigroup, the operator $S = RA_2RA_1R$ necessarily belongs to $[B\rangle$.

It remains to notice that the collection $\mathcal{C} = \{B\}$ consisting of a single operator B satisfies the conditions of Theorem 10.50. Indeed, $\mathcal{C}$ is quasinilpotent at $x_0 > 0$ and $[\mathcal{C}\rangle = [B\rangle$ contains the operator S that dominates a non-zero compact operator. By this theorem there exists a non-trivial closed ideal that is invariant under $[B\rangle$. This completes the proof. ∎

In order to proceed further, let us recall briefly some terminology and notation introduced in Sections 4.1 and 5.2. A (continuous) **rank-one operator** on a Banach space X is any operator of the form $x^* \otimes u \in \mathcal{L}(X)$, where $x^* \in X^*$, $u \in X$ (and $x^* \otimes u(x) = x^*(x)u$ for $x \in X$). Any operator of the form $\sum_{i=1}^n x_i^* \otimes u_i$ is called a **finite-rank operator**. As usual, $X^* \otimes X$ denotes the vector space of all continuous finite-rank operators on X.

Suppose now that E is a Dedekind complete Banach lattice. Then, by the Riesz–Kantorovich Theorem 1.16, the space of all regular operators $\mathcal{L}_r(E)$ is a Dedekind complete vector lattice. For an arbitrary vector subspace Φ of E^* the operators in the band $(\Phi \otimes E)^{\mathrm{dd}}$ generated by $\Phi \otimes E$ in $\mathcal{L}_r(E)$ are called **Φ-integral operators**. When $\Phi = E_{\mathrm{n}}^*$, the E_{n}^*-integral operators are referred to as **abstract** (or **almost**) **integral operators**. The terms are justified by Lozanovsky's Theorem 5.28 asserting that on a Banach function space an operator is almost integral if and only if it is a regular integral operator.

The next characterization of E^*-integral operators follows immediately from their definition:

- *A positive operator $T\colon E \to E$ on a Dedekind complete Banach lattice is an E^*-integral operator if and only if there exists a net $\{T_\alpha\}$ of positive operators on E such that each T_α is dominated by a finite-rank operator and $0 \le T_\alpha x \uparrow Tx$ for each $x \ge 0$.*

Letting $R = T$ and $C = T_\alpha$ in the above characterization and taking for K any finite-rank operator dominating T_α, we obtain the following result.

Lemma 10.58. *Every positive E^*-integral operator is compact-friendly.*

It is useful to know that the third power of a strictly positive E^*-integral operator dominates a non-zero compact positive operator.

Lemma 10.59. *If $S\colon E \to E$ is a strictly positive E^*-integral operator on a Dedekind complete Banach lattice, then for each $x_0 > 0$ there exists a compact positive operator $K\colon E \to E$ such that $0 \le K \le S^3$ and $Kx_0 > 0$.*

Proof. Fix any $x_0 > 0$. Since $S\colon E \to E$ is a positive E^*-integral operator, there exists a net $\{S_\alpha\}$ of positive operators such that $0 \le S_\alpha \uparrow S$ and each S_α is dominated by a positive finite-rank operator. By Corollary 2.35, each S_α^3 is a compact operator.

Because S is strictly positive, we have $Sx_0 > 0$. From this and from $S_\alpha x_0 \uparrow Sx_0$, we see that there is some index α_1 so that $S_\alpha x_0 > 0$ for all $\alpha \ge \alpha_1$. Likewise from $S(S_{\alpha_1}x_0) > 0$ and $S_\alpha(S_{\alpha_1}x_0) \uparrow_\alpha S(S_{\alpha_1}x_0)$ we get $S_\alpha(S_{\alpha_1}x_0) > 0$ for all $\alpha \ge \alpha_2 \ge \alpha_1$. Finally, from $S\big(S_{\alpha_2}S_{\alpha_1}x_0\big) > 0$ and $S_\alpha\big(S_{\alpha_2}S_{\alpha_1}x_0\big) \uparrow_\alpha S\big(S_{\alpha_2}S_{\alpha_1}x_0\big)$, it follows that $S_\alpha\big(S_{\alpha_2}S_{\alpha_1}x_0\big) > 0$ for all $\alpha \ge \alpha_3 \ge \alpha_2 \ge \alpha_1$.

Now let $K = S^3_{\alpha_3}$ and note that $K\colon E \to E$ is a compact positive operator satisfying $0 \le K \le S^3$ and $Kx_0 > 0$. ∎

We are now ready to present an invariant subspace theorem for E^*-integral operators.

Theorem 10.60. *Let $S\colon E \to E$ be a non-zero positive E^*-integral operator on a Dedekind complete Banach lattice and let $B\colon E \to E$ be any non-zero positive operator commuting with S. If either S or B is quasinilpotent at some non-zero positive vector, then the operators S and B have a common non-trivial closed invariant ideal. Moreover, this ideal can be chosen to be invariant either under $[S\rangle$ or under $[B\rangle$.*

Proof. If S is not strictly positive, then we know that the non-trivial null ideal N_S is invariant under $[S\rangle$; see Exercise 4(d) in Section 10.3. So, we can suppose that S is strictly positive. In this case, it follows from Lemma 10.59 that S^3 dominates a non-zero compact positive operator.

Assume first that S is quasinilpotent at some $x_0 > 0$. Consider the collection $\mathcal{C} = \{S\}$ consisting of the single operator S. Then $\mathcal{C}$ satisfies the conditions of Theorem 10.50. Indeed, $\mathcal{C}$ is quasinilpotent at $x_0 > 0$ and $[S\rangle$ contains the operator S^3 that dominates a non-zero compact operator. By this theorem there exists a non-trivial closed ideal that is invariant under $[S\rangle$. As S and B commute, the operator B belongs to $[S\rangle$.

Assume now that B is quasinilpotent at x_0. Then consider the collection $\mathcal{C} = \{B\}$ consisting of the single operator B. Again $\mathcal{C}$ satisfies the conditions of Theorem 10.50 since $\mathcal{C}$ is quasinilpotent at x_0 and $[B\rangle$ also contains the operator S^3. A second application of Theorem 10.50 guarantees the existence of a non-trivial closed ideal that is invariant under $[B\rangle$. ∎

Corollary 10.61. *Every positive E^*-integral operator that is quasinilpotent at a non-zero positive vector has a non-trivial closed ℓ-hyperinvariant ideal.*

T. Andô [**34**] and H. J. Krieger [**194**] proved that each positive irreducible integral operator on an L_p-space has a positive spectral radius. In the "invariant subspace" terminology this means that every positive quasinilpotent integral operator on an L_p-space has a non-trivial closed invariant subspace; see Corollary 9.37. This was the first result on the invariant subspace problem in the framework of Banach lattices. It should be clear that Corollary 10.61 is a substantial improvement of the Andô–Krieger theorem. For Banach function spaces Corollary10.61 specializes to the following result.

Corollary 10.62. *Let E be a Banach function space associated with a σ-finite measure space (Ω, Σ, μ) and let $S\colon E \to E$ be a positive integral operator with kernel $S(\cdot,\cdot)$. If S is quasinilpotent at some positive vector, then* *the family of integral operators $T\colon E \to E$, defined by*

$$Tx(t) = \int_\Omega w(s,t)S(s,t)x(s)\,d\mu(s)\,, \quad x \in E,$$

where $w(\cdot,\cdot)$ is an arbitrary bounded $\mu\times\mu$-measurable function, has a common non-trivial closed invariant ideal.

Proof. Observe that each integral operator T defined by a kernel $wS(\cdot,\cdot)$ is dominated by a multiple of the operator S, and the conclusion follows from Corollary 10.61 and Exercise 2 in Section 10.3. ■

We shall conclude the section with a characterization obtained in [**17**] of compact-friendly multiplication operators on spaces of continuous functions. To simplify the exposition, we need to introduce the following terminology taken from [**15**].

Definition 10.63. *A continuous function $\phi\colon \Omega \to \mathbb{R}$, where Ω is a topological space, has a* ***flat*** *if there exists a non-empty open set G such that ϕ is constant on G.*

Similarly, a measurable function $\phi\colon \Omega \to \mathbb{R}$, where (Ω,Σ,μ) is now a measure space, has a ***flat*** *if ϕ is constant on some $A \in \Sigma$ with $\mu(A) > 0$.*

We continue with some useful properties of operators dominated by compact positive operators.

Lemma 10.64. *For a norm bounded sequence $\{e_n\}$ in a Banach lattice E and a bounded operator $A\colon E \to E$ that is dominated by a compact operator we have following.*

1. *(1) The sequence $\{Ae_n\}$ has an order bounded subsequence.*
2. *(2) If E has order continuous norm and the sequence $\{Ae_n\}$ is disjoint, then $\|Ae_n\| \to 0$.*

Proof. (1) Let $K\colon E \to E$ be a compact positive operator dominating A, i.e., $|Ax| \le K|x|$ for each $x \in E$. Since K is compact and $\{e_n\}$ is norm bounded, we can extract from $\{K(|e_n|)\}$ a convergent subsequence. Without loss of generality we can assume that the sequence $\{K(|e_n|)\}$ itself converges, that is, there exists some $y \in E$ such that $K|e_n| \to y$. By passing to another subsequence if necessary, we can also assume that $\big\|K|e_n| - y\big\| < 2^{-n}$ for each n. Therefore, the series $e = \sum_{n=1}^{\infty}\big|K|e_n| - y\big|$ is norm convergent and defines a positive vector. Clearly, $\big|K|e_n| - y\big| \le e$, whence $K|e_n| \le e + |y|$ for each n. This implies $|Ae_n| \le K|e_n| \le e + |y|$.

(2) Assume that $\{Ae_n\}$ is a disjoint sequence and let $\{f_n\}$ be a subsequence of $\{e_n\}$. By part (1) there exists a subsequence $\{g_n\}$ of $\{f_n\}$ such

that the disjoint sequence $\{Ag_n\}$ is order bounded. Since E has order continuous norm, Theorem 2.26 guarantees that $\|Ag_n\| \to 0$. Thus, we have shown that every subsequence of $\{Ae_n\}$ has a subsequence that is norm convergent to zero. This implies that $\|Ae_n\| \to 0$. ■

Theorem 10.65. *A positive multiplication operator M_ϕ on a $C(\Omega)$-space, where Ω is a compact Hausdorff space, is compact-friendly if and only if the multiplier ϕ has a flat.*

Proof. Assume first that a non-negative multiplier $\phi \in C(\Omega)$ is constant on a non-empty open subset of Ω. Then, as shown in Theorem 4.20, the operator M_ϕ commutes with a non-zero positive rank-one operator. This implies immediately that the operator M_ϕ is compact-friendly.

For the converse, assume that M_ϕ is compact-friendly. Therefore, there exist three non-zero operators $R, K, A\colon C(\Omega) \to C(\Omega)$ with R, K positive, K compact, $M_\phi R = RM_\phi$ and such that for each $x \in C(\Omega)$ we have

$$|Ax| \leq R|x| \quad \text{and} \quad |Ax| \leq K|x| \, .$$

Taking adjoints, we obtain that $M_\phi^* R^* = R^* M_\phi^*$ and that

$$|A^*x^*| \leq R^*|x^*| \quad \text{and} \quad |A^*x^*| \leq K^*|x^*|$$

for each $x^* \in C(\Omega)^*$; see Exercise 2 in Section 2.4. Now the following three properties can be verified in a rather straightforward way.

(1) For each $\omega \in \Omega$ the support of the measure $R^*\delta_\omega$ is contained in the set $W_\omega = \phi^{-1}(\phi(\omega))$, where δ_ω denotes the δ-measure concentrated at the point ω. This follows immediately from the identity

$$M_\phi^* R^* \delta_\omega = R^* M_\phi^* \delta_\omega = \phi(\omega) R^* \delta_\omega \, .$$

(2) Since A^* is dominated by R^*, it follows immediately from (1) that for each $\omega \in \Omega$ the signed measure $A^*\delta_\omega$ is also supported by W_ω.

(3) Fix any function $h \in C(\Omega)$ such that $\|h\| = 1$ and $Ah \neq 0$. Next, choose a non-empty open set U on which $|Ah(\omega)| \geq \epsilon$ for some $\epsilon > 0$. Then for each $\omega \in U$ we have $\|A^*\delta_\omega\| \geq \epsilon$. To see this, note that

$$\|A^*\delta_\omega\| \geq \big|\langle A^*\delta_\omega, h\rangle\big| = \big|\langle \delta_\omega, Ah\rangle\big| = \big|Ah(\omega)\big| \geq \epsilon \, . \tag{$\star$}$$

From the definition of the sets W_ω we see that $\bigcup_{\omega\in\Omega} W_\omega = \Omega$. To complete the proof, assume by way of contradiction that for each $\omega \in \Omega$ the set W_ω has an empty interior. Then the non-empty open set U, chosen in (3), must meet infinitely many sets W_ω. Pick a sequence $\{\omega_n\}$ in U satisfying $\phi(\omega_m) \neq \phi(\omega_n)$ for $m \neq n$, and let $e_n = |A^*\delta_{\omega_n}|$ for each n. Then $\|e_n\| \geq \epsilon$. Furthermore, since each e_n is supported by the set W_{ω_n} and the sequence $\{W_{\omega_n}\}$ is pairwise disjoint, the sequence $\{e_n\}$ in $C(\Omega)^*$ is also disjoint. Therefore, since the norm in $C(\Omega)^*$ is order continuous, it follows from part (2) of Lemma 10.64 that $\|e_n\| \to 0$. However, the last conclusion contradicts ($\star$), and the proof of the theorem is complete. ■

A similar characterization of compact-friendly multiplication operators on L_p-spaces was obtained in [**17**]. G. Sirotkin [**307**] has managed to extend Theorem 10.65 to arbitrary Banach function spaces.

Exercises

1. Prove that the following operators are all compact-friendly:
   (a) compact positive operators,
   (b) positive operators commuting with non-zero positive compact operators,
   (c) positive operators that dominate non-zero compact positive operators,
   (d) positive E^*-integral operators (in particular, abstract integral operators), where E is a Dedekind complete Banach lattice,
   (e) positive integral operators.

2. Let $B\colon E \to E$ be a positive operator on a Banach lattice. If there exists a non-trivial B-invariant projection band Δ in E such that $Bx = 0$ for each $x \in \Delta^{\mathrm{d}}$, then show that B is compact-friendly.

3. (Sirotkin) For a non-zero positive operator $T\colon E \to E$ on a Banach lattice establish the following.
   (a) If $K \in \mathcal{L}(E)$ is a non-zero compact positive operator, then $T + K$ is compact-friendly.
   (b) There exists a compact-friendly operator $B \in \mathcal{L}(E)$ such that $T \le B$.
   (c) If E is Dedekind complete, then there exists a compact-friendly operator S such that $0 < S \le T$.

4. Let ϕ be a positive one-to-one function in $C(\Omega)$, where Ω is a compact Hausdorff space without isolated points. Show directly, i.e., without using Theorem 10.65, that the multiplication operator M_ϕ is not compact-friendly.

5. A positive operator $B\colon E \to E$ on a Dedekind complete Banach lattice is called a **generalized Harris operator** if some power of B is not disjoint from the band $(E^* \otimes E)^{\mathrm{dd}}$ of E^*-integral operators on E, i.e., if some power of B dominates a non-zero positive E^*-integral operator. Recall that if some power of B is not disjoint from the band $(E^*_{\mathrm{n}} \otimes E)^{\mathrm{dd}}$ of abstract integral operators, then B is called a Harris operator. Of course, each Harris operator is a generalized Harris operator.

For a generalized Harris operator $B\colon E \to E$ establish the following.
   (a) B is compact-friendly.
   (b) If B is quasinilpotent at a non-zero positive vector, then B has a non-trivial closed ℓ-hyperinvariant ideal.

6. Show that Theorems 10.55 and 10.60 remain true if we replace the assumption $\lim_{n\to\infty} \|T^n x_0\|^{\frac{1}{n}} = 0$ with $\liminf_{n\to\infty} \|T^n x_0\|^{\frac{1}{n}} = 0$.

7. Let $T\colon \ell_p \to \ell_p$ $(1 \le p < \infty)$ be a continuous operator with modulus. Assume that there exists a non-zero positive operator $S\colon \ell_p \to \ell_p$ such that:
(a) S commutes with the modulus of T.
(b) S is quasinilpotent at a non-zero positive vector.
Show directly, i.e., without using Theorem 10.24, that T has a non-trivial closed invariant ideal.

8. (Sirotkin) Let $B\colon E \to E$ be a compact-friendly operator on a Banach lattice such that no positive operator commuting with B dominates a non-zero compact operator. Show that for each at most countable family of positive operators commuting with B there exists a non-trivial closed ideal that is B-invariant and $\mathcal{F}$-invariant.

9. (Troitsky [**319**]) Show that every essentially quasinilpotent compact-friendly operator on a Banach lattice has a non-trivial closed invariant subspace.

10. (Troitsky [**319**]) Let $S\colon E \to E$ be a quasinilpotent integral operator on a Banach function space such that S^- is compact. Then each of the operators $|S|$ and S^+ has a non-trivial closed invariant subspace.

11. Show that the set of all compact-friendly operators on a Banach lattice E need not be closed or open in $\mathcal{L}(E)$.

12. This exercise presents an example of a Hilbert–Schmidt operator on $L_2[0,1]$ which is one-to-one, positive, locally quasinilpotent at a positive vector, but fails to be quasinilpotent. The construction of such an operator will be done in three steps as follows.
(a) Let $0 \le a < b \le 1$ and $0 \le c < d \le 1$ [see Figure 1(a)] and consider the integral operator $T\colon L_2[a,b] \to L_2[c,d]$ defined by

$$Tx(t) = \int_a^b \sin(st)x(s)\,ds\,, \quad c \le t \le d\,.$$

Then the operator T is one-to-one and compact.
(b) Assume that $0 \le a < b \le 1$. Then the positive integral operator $T\colon L_2[a,b] \to L_2[a,b]$, defined by

$$Tx(t) = \int_a^b \sin(st)x(s)\,ds, \quad a \le t \le b\,,$$

has a positive spectral radius—and so T is a one-to-one compact positive integral operator which is not quasinilpotent.
(c) Consider the function $K\colon [0,1] \times [0,1] \to \mathbb{R}$ defined as follows: the value $K(s,t)$ is equal to $\sin(st)$ if (s,t) lies in each one of the shaded rectangles shown in Figure 1(b) and is zero everywhere else. The

points $a_0, a_1, a_2, \ldots$ shown in Figure 1(b) are defined by

$$a_0 = 1 \quad \text{and} \quad a_n = 1 - \tfrac{1}{2e} \sum_{k=0}^{n-1} \tfrac{1}{k!} \quad \text{for} \quad n \geq 1\,.$$

The length of the interval (a_{n+1}, a_n) is equal to $a_n - a_{n+1} = \frac{1}{n!2e}$. As usual, we shall denote by K the operator on $L_2[0,1]$ defined be the kernel $K(s,t)$. That is, the operator $K\colon L_2[0,1] \to L_2[0,1]$ is defined by

$$Kx(t) = \int_0^1 K(s,t)x(s)\,ds, \;\; 0 \leq t \leq 1\,.$$

Then the integral operator $K\colon L_2[0,1] \to L_2[0,1]$ is a one-to-one compact positive operator which is quasinilpotent at some positive vectors but fails to be quasinilpotent.

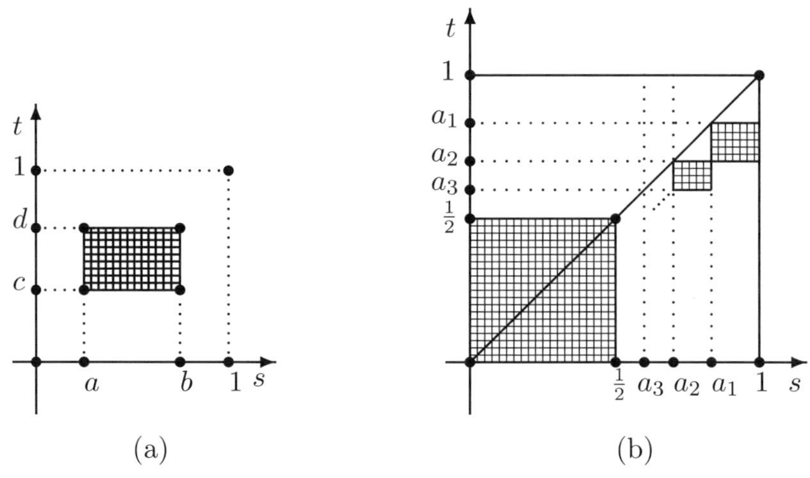

Figure 1. The kernel of the operator K

10.6. Positive Operators on Banach Spaces with Bases

Some of the results on the existence of invariant subspaces of positive operators on Banach lattices can be extended to Banach spaces ordered by the cone generated by a basis. This study was initiated in [**12**] and in this section we present the main results from that paper.

Recall that a sequence $\{x_n\}$ in a Banach space X is called a **Schauder basis** (or simply a **basis**) of X if for every $x \in X$ there exists a unique sequence of scalars $\{\alpha_n\}$ such that $x = \sum_{n=1}^{\infty} \alpha_n x_n$, where the series is assumed to converge in the norm. Every basis $\{x_n\}$ gives rise to a natural closed cone C defined by

$$C = \Big\{ x = \sum_{n=1}^{\infty} \alpha_n x_n \colon \ \alpha_n \geq 0 \ \text{ for each } \ n = 1, 2, \ldots \Big\}.$$

The cone C will be referred to as the (**closed**) **cone generated by the basis** $\{x_n\}$. For an extensive discussion concerning the cone generated by a basis see [**305**].

As discussed in Section 1.3, with every basis we associate the standard sequence of coefficient functionals. Let $\{x_n\}$ be a basis of a Banach space X. Then the linear functional c_n^*, defined for each $x = \sum_{i=1}^{\infty} \alpha_i x_i$ by

$$c_n^*(x) = \alpha_n \,,$$

is a continuous linear functional on X. Observe that each c_n^* is automatically positive with respect to the cone generated by the basis $\{x_n\}$. The sequence of continuous linear functionals $\{c_n^*\}$ satisfies $c_n^*(x_m) = \delta_{nm}$.

An operator $T \colon X \to X$ on a Banach space with a basis $\{x_n\}$ is said to be **positive** (with respect to this basis) if $T(C) \subseteq C$, where C is the closed cone generated by $\{x_n\}$.

If we fix a basis $\{x_n\}$ for a Banach space X, then every continuous operator $T \colon X \to X$ can be identified in the usual manner with an infinite matrix $[t_{ij}]$. In this context, we can also say that an infinite matrix $[t_{ij}]$ defines an operator on X. Note that an operator $T \colon X \to X$ with matrix $[t_{ij}]$ is a positive operator if and only if $t_{ij} \geq 0$ for each pair (i, j).

We are ready to present a version of Theorem 10.24 that holds true for positive operators on a Banach space with a basis.

Theorem 10.66. *Let X be a Banach space with a basis and let $T \colon X \to X$ be a continuous positive operator. If T commutes with a non-zero positive operator that is quasinilpotent at a non-zero positive vector, then T has a non-trivial closed invariant subspace.*

Proof. Let $\{x_n\}$ be a basis of the Banach space X and let $\{c_n^*\}$ be the sequence of coefficient functionals associated with the basis $\{x_n\}$.

Assume that a non-zero positive operator $A \colon X \to X$ satisfies $TA = AT$ and is quasinilpotent at a non-zero positive vector y_0, i.e., $\lim_{n\to\infty} \big\| A^n y_0 \big\|^{\frac{1}{n}} = 0$. If $Ay_0 = 0$, then the kernel of A is a non-trivial closed vector subspace that is invariant under T. Thus, we can suppose that Ay_0 is non-zero. Hence, $Ax_k \neq 0$ for some k. By an appropriate scaling of y_0, we can assume that $0 \leq x_k \leq y_0$.

Now let $P \colon X \to X$ denote the continuous projection onto the vector subspace generated by x_k defined by $Px = c_k^*(x) x_k$. Clearly, $0 \leq Px \leq x$

for each $0 \leq x \in X$. We claim that

$$PT^m Ax_k = 0 \tag{$\star$}$$

for each $m \geq 0$. To see this, fix $m \geq 0$ and let $PT^m Ax_k = \alpha x_k$ for some non-negative scalar $\alpha \geq 0$. Since P is a positive operator and the composition of positive operators is a positive operator, it follows that

$$0 \leq \alpha^n x_k = \big(PT^m A\big)^n x_k \leq \big(T^m A\big)^n x_k = T^{mn} A^n x_k \leq T^{mn} A^n y_0 .$$

Using the positivity of the linear functional c_k^*, we obtain

$$0 \leq \alpha^n = c_k^*\big(\alpha^n x_k\big) \leq c_k^*\big(T^{mn} A^n y_0\big) .$$

Consequently, $0 \leq \alpha^n \leq \|c_k^*\| \, \big\|T\big\|^{mn} \cdot \big\|A^n y_0\big\|$, and so

$$0 \leq \alpha \leq \|c_k^*\|^{\frac{1}{n}} \big\|T\big\|^m \cdot \big\|A^n y_0\big\|^{\frac{1}{n}} .$$

From $\lim_{n\to\infty} \big\|A^n y_0\big\|^{\frac{1}{n}} = 0$, we conclude that $\alpha = 0$, and thus condition $(\star)$ must hold.

Finally, consider the subspace Y generated by $\big\{T^m Ax_k \colon \ m = 0, 1, \ldots\big\}$. Clearly, Y is invariant under T and since $0 \neq Ax_k \in Y$, we see that $Y \neq \{0\}$. Also, it follows from $(\star)$ that for each $y \in Y$ we have

$$c_k^*(y) = c_k^*(Py) = 0 ,$$

and consequently $c_k^*(y) = 0$ for all $y \in \overline{Y}$. The latter shows that $\overline{Y}$ is a non-trivial closed subspace of X invariant under the operator T. ■

Corollary 10.67. *If $T\colon X \to X$ is a quasinilpotent positive operator on a Banach space with a basis, then T has a non-trivial closed invariant subspace.*

For the next result we need to recall the notions of the strong operator topology on $\mathcal{L}(X,Y)$. The **strong operator topology** on $\mathcal{L}(X,Y)$ is the Hausdorff locally convex topology τ_s generated by the family of seminorms $\{p_x\}_{x\in X}$ defined by $p_x(T) = \|Tx\|$ for each $T \in \mathcal{L}(X,Y)$. Notice that a net $\{T_\lambda\} \subseteq \mathcal{L}(X,Y)$ satisfies $T_\lambda \xrightarrow{\tau_s} T$ in $\mathcal{L}(X,Y)$ if and only if $\|T_\lambda x - Tx\| \to 0$ for each $x \in X$.

One can add arbitrary weights to the matrix representing a quasinilpotent positive operator and still be guaranteed that a non-trivial closed invariant subspace exists.

Theorem 10.68. *Let X be a Banach space with a basis. Assume that a positive matrix $A = [a_{ij}]$ defines a continuous operator on X that is quasinilpotent at a non-zero positive vector. If for a double sequence $\{w_{ij}\}$ of scalars, the weighted matrix $\big[w_{ij}a_{ij}\big]$ defines a continuous operator B on X, then the operator B has a non-trivial closed invariant subspace.*

Proof. Let $\{x_n\}$ be a basis of the Banach space X and let $\{c_n^*\}$ be the sequence of coefficient functionals associated with the basis $\{x_n\}$. Assume that the positive operator $A = [a_{ij}]$ satisfies $\lim_{n\to\infty} \|A^n y_0\|^{\frac{1}{n}} = 0$ for some non-zero positive vector y_0. An appropriate scaling of y_0 shows that there exists some k satisfying $0 < x_k \le y_0$. If $Ax_k = 0$, then an easy argument shows that $Bx_k = 0$, and thus the kernel of B is a non-trivial closed invariant subspace (here we assume, of course, that $B \neq 0$). Thus, we can suppose that Ax_k is non-zero.

Now let P be the positive projection on X defined by $Px = c_k^*(x)x_k$. Arguing as in the proof of Theorem 10.66, we can establish that $PA^m x_k = 0$ for each $m \ge 1$. In particular, we have $c_k^*(A^m x_k) = 0$. So, for $m \ge 1$ and each positive operator $S\colon X \to X$ satisfying $0 \le S \le A^m$ we have

$$0 \le c_k^*(Sx_k) \le c_k^*(A^m x_k) = 0\,. \tag{$\star\star$}$$

Next, consider the vector subspace Y of X generated by the set

$$\big\{Sx_k\colon\ S \in \mathcal{L}(X) \text{ and } 0 \le S \le A^m \text{ for some } m \ge 1\big\}\,.$$

Clearly, Y is invariant under each operator R that satisfies $0 \le R \le A$. Also, it follows from ($\star\star$) that

$$c_k^*(y) = 0$$

for all $y \in \overline{Y}$. The latter shows that $\overline{Y}$ is a non-trivial closed vector subspace of X that is invariant under each operator $R\colon X \to X$ satisfying $0 \le R \le A$.

For each pair (i, j) consider the operator A_{ij} defined by $A_{ij}(x_j) = a_{ij}x_j$ and $A_{ij}(x_m) = 0$ for $m \neq j$. Since this operator satisfies $0 \le A_{ij} \le A$, it follows that $\overline{Y}$ is invariant under each operator A_{ij}. Therefore, for each $n \in \mathbb{N}$ the vector subspace $\overline{Y}$ is invariant under the operator

$$B_n = \sum_{i=1}^{n}\sum_{j=1}^{n} w_{ij}A_{ij}\,.$$

It remains to notice that the sequence of operators $\{B_n\}$ converges in the strong operator topology to B. Therefore, $B(\overline{Y}) \subseteq \overline{Y}$, and thus the operator B has a non-trivial closed invariant subspace.[5] ■

Corollary 10.69. *Let X be a Banach space with a basis. Assume that a positive matrix $A = [a_{ij}]$ defines a continuous operator on X which is quasi-nilpotent at a non-zero positive vector. If a continuous operator $T\colon X \to X$ is defined by a matrix $T = [t_{ij}]$ satisfying $t_{ij} = 0$ whenever $a_{ij} = 0$, then the operator T has a non-trivial closed invariant subspace.*

[5] As a matter of fact, $\overline{Y}$ is a common invariant subspace for all operators generated by the weighted matrices $[w_{ij}a_{ij}]$ satisfying the condition of the theorem. Compare this with Exercise 9 in Section 10.4.

A further generalization to Banach spaces with a Markushevich basis can be found in [**122**]. The paper [**49**] contains some results on invariant subspaces of positive operators on general ordered Banach spaces.

Exercises

1. Let $\{x_n\}$ be a basis in a Banach space X and let

$$C = \Big\{x = \sum_{n=1}^{\infty} \alpha_n x_n \colon\ \alpha_n \geq 0 \text{ for each } n = 1, 2, \ldots\Big\}.$$

Show that C is a closed cone in X.

2. If a Banach space X has an unconditional basis $\{u_n\}$, then show that:
(a) The cone C generated by the basis $\{u_n\}$ is a lattice cone.
(b) The Riesz space (X, C) has weak units.
(c) After an equivalent renorming the ordered Banach space (X, C) becomes a Banach lattice with order continuous norm.

3. Let X be a Banach space with an unconditional basis $\{u_n\}$. If a linear operator $T \colon X \to X$ is positive with respect to the cone generated by the basis $\{u_n\}$, then show that T is a bounded operator.

4. (Spalsbury [**310**]) Consider the Volterra operator defined in Example 7.8. Show that there is no basis in $L_2[0,1]$ with respect to which the Volterra operator is positive. [HINT: According to [**168**] every non-trivial closed subspace of $L_2[0,1]$ which is invariant under the Volterra operator is of the form $M_\alpha = \{f \in L_2[0,1] \colon\ f = 0 \text{ a.e. on } [0,\alpha]\}$ for some $0 < \alpha < 1$. Use this result in conjunction with Theorem 10.66.]

10.7. A Characterization of Non-transitive Algebras

Recall that an algebra $\mathcal{A}$ of operators on a Banach space is said to be **non-transitive** if it has a non-trivial closed $\mathcal{A}$-invariant subspace.

The objective of this section is to present some characterizations of non-transitive algebras of operators obtained by Y. A. Abramovich, C. D. Aliprantis, and O. Burkinshaw [**14**]. These characterizations are based on the works of V. I. Lomonosov [**212**] and L. de Branges [**75**].

Throughout this section X will denote an arbitrary real or complex Banach space. As usual, the closed unit ball of X^* will be denoted by U^*. That is,

$$U^* = \big\{x^* \in X^* \colon\ \|x^*\| \leq 1\big\}.$$

We assume that U^* is equipped with its weak* topology. Recall that U^* is a weak* compact subset of X^*.

Definition 10.70. *The vector space of all continuous functions from U^* into X^*, when both U^* and X^* are equipped with the weak* topology, will be denoted by $C(U^*, X^*)$. Occasionally, $C(U^*, X^*)$ will also be denoted by Y, i.e., $Y = C(U^*, X^*)$.*

Clearly, the vector space $C(U^*, X^*)$ equipped with the norm

$$\|f\| = \sup_{x^* \in U^*} \|f(x^*)\|, \quad f \in C(U^*, X^*),$$

is a Banach space. The Banach space $C(U^*, X^*)$ was studied first in [**212**] and [**75**]. Observe that for each operator $T \in \mathcal{L}(X)$ the restriction of the adjoint operator T^* to U^* is an element of $C(U^*, X^*)$.

V. I. Lomonosov [**212**], inspired by L. de Branges' proof of the classical Stone–Weierstrass theorem [**73**], characterized the extreme points of the closed unit ball of the norm dual of $C(U^*, X^*)$. Subsequently, L. de Branges [**75**] presented an analysis of these extreme points and obtained an abstract version of the Stone–Weierstrass Theorem. The Lomonosov–de Branges analysis will be employed later on in our characterization of the invariant subspace problem.

As usual, $C(U^*)$ denotes the Banach space of all continuous real-valued (or complex-valued if X is a complex Banach space) functions defined on U^*. Each function $\alpha \in C(U^*)$ defines a multiplication operator on $C(U^*, X^*)$ via the formula

$$(\alpha f)(x^*) = \alpha(x^*) f(x^*), \quad f \in C(U^*, X^*), \ x^* \in U^*.$$

Clearly,

$$\|\alpha f\| \le \|\alpha\|_\infty \|f\|.$$

Every pair $(x^{**}, x^*) \in X^{**} \times U^*$ gives rise to a linear functional $x^{**} \otimes x^*$ on $C(U^*, X^*)$ defined for each $f \in C(U^*, X^*)$ by

$$\big\langle f, x^{**} \otimes x^* \big\rangle = (x^{**} \otimes x^*)(f) = \big\langle f(x^*), x^{**} \big\rangle = x^{**}\big(f(x^*)\big).$$

Clearly, $x^{**} \otimes x^*$ is a norm continuous linear functional on $C(U^*, X^*)$. The vector space generated by the family $\{x^{**} \otimes x^* \colon \ x^{**} \in X^{**} \text{ and } x^* \in U^*\}$ in the norm dual of $Y = C(U^*, X^*)$ will be denoted by Y', i.e.,

$$Y' = \Big\{ \sum_{i=1}^{n} x_i^{**} \otimes x_i^* \colon \ x_i^{**} \in X^{**} \text{ and } x_i^* \in U^* \text{ for each } i = 1, \ldots, n \Big\}.$$

Clearly, Y' separates the points of Y, and so $\langle Y, Y' \rangle$ is a dual system. Besides the norm topology on Y, we shall also consider two other locally convex topologies. They are the topologies τ_w and τ_s that are defined as follows.

(1) The topology τ_w on Y is the locally convex topology generated by the family of seminorms $\{\rho_{x^{**},x^*}\colon\ x^{**} \in X^{**} \text{ and } x^* \in U^*\}$, where

$$\rho_{x^{**},x^*}(f) = \big|(x^{**} \otimes x^*)(f)\big| = \big|\langle x^{**}, f(x^*)\rangle\big|$$

for $f \in Y = C(U^*, X^*)$, i.e., τ_w is simply the weak topology $\sigma(Y, Y')$.

(2) The topology τ_s on Y is the locally convex topology generated by the family of seminorms $\{\rho_{x^*}\colon\ x^* \in U^*\}$, where

$$\rho_{x^*}(f) = \big\|f(x^*)\big\|$$

for $f \in Y = C(U^*, X^*)$.

The alert reader will recognize immediately that these topologies are similar to the usual weak and strong operator topologies on $\mathcal{L}(X)$ and this justifies our choice of the subscripts w and s. Moreover, in analogy with the classical weak and strong operator topologies [**118**, Theorem 4, p. 477], the topologies τ_w and τ_s have the same continuous linear functionals. That is, they are both consistent with the dual system $\langle Y, Y'\rangle$.

Theorem 10.71. *The locally convex topology τ_s is consistent with the dual system $\langle Y, Y'\rangle$. That is, we have the inclusions*

$$\tau_w = \sigma(Y, Y') \subseteq \tau_s \subseteq \tau(Y, Y'),$$

where, as usual, $\tau(Y, Y')$ denotes the Mackey topology on Y.

Proof. Clearly, $\sigma(Y, Y') \subseteq \tau_s$. So, it suffices to establish $\tau_s \subseteq \tau(Y, Y')$. To this end, fix $x^* \in U^*$. We must show that the set

$$\{f \in Y\colon\ \rho_{x^*}(f) = \|f(x^*)\| \le 1\}$$

is a $\tau(Y, Y')$-neighborhood of zero. Consider the operator

$$R\colon \big(X^{**}, \sigma(X^{**}, X^*)\big) \longrightarrow \big(Y', \sigma(Y', Y)\big)\,,$$

defined by $Rx^{**} = x^{**} \otimes x^*$. We claim that R is continuous. Indeed, if $x^{**}_\alpha \xrightarrow{\sigma(X^{**},X^*)} 0$ in X^{**}, then $\langle f, Rx^{**}_\alpha\rangle = \langle f, x^{**}_\alpha \otimes x^*\rangle = \langle f(x^*), x^{**}_\alpha\rangle \to 0$ for each $f \in Y$, and so $Rx^{**}_\alpha \xrightarrow{\sigma(Y,Y')} 0$. Since U^{**} is a $\sigma(X^{**}, X^*)$-compact set, it follows that the convex circled set $D = R(U^{**}) = \{x^{**} \otimes x^*\colon\ x^{**} \in U^{**}\}$ is $\sigma(Y', Y)$-compact. So, its polar

$$\begin{aligned} D^{\circ} &= \{f \in Y\colon\ |\langle x^{**} \otimes x^*, f\rangle| \le 1 \text{ for all } x^{**} \in U^{**}\} \\ &= \{f \in Y\colon\ |x^{**}(f(x^*))| \le 1 \text{ for all } x^{**} \in U^{**}\} \\ &= \{f \in Y\colon\ \|f(x^*)\| \le 1\} \end{aligned}$$

is a $\tau(Y, Y')$-neighborhood of zero, and the proof is finished. ■

The standard duality theory result yields the following.

Corollary 10.72. *The topologies τ_w and τ_s on $C(U^*, X^*)$ have the same closed convex sets.*

We are now ready to establish that if a subspace $\mathcal{M}$ of $C(U^*, X^*)$ is $C(U^*)$-invariant, then the elements outside of the closure of $\mathcal{M}$ can be separated by linear functionals of the form $x^{**} \otimes x^*$.

Lemma 10.73. *Let $\mathcal{M}$ be a vector subspace of $C(U^*, X^*)$ which is invariant under multiplication by elements of $C(U^*)$. Then a function $f_0 \in C(U^*, X^*)$ does not belong to the τ_s-closure of $\mathcal{M}$ if and only if there exist $x^{**} \in X^{**}$ and $x^* \in U^*$ such that*

$$\langle x^{**} \otimes x^*, f_0 \rangle = 1 \quad \textit{and} \quad \langle x^{**} \otimes x^*, f \rangle = 0$$

for all $f \in \mathcal{M}$.

Proof. The "only if" part needs verification. Therefore, suppose that f_0 does not belong to the τ_s-closure of $\mathcal{M}$. Then there exists some τ_s-continuous linear functional ϕ on $C(U^*, X^*)$ such that $\phi(f_0) \neq 0$ and $\phi(f) = 0$ for each $f \in \mathcal{M}$. By Theorem 10.71 we know that the linear functional ϕ is of the form $\phi = \sum_{i=1}^n x_i^{**} \otimes x_i^*$, where $x_i^* \neq x_j^*$ for $i \neq j$. From $\phi(f_0) = \sum_{i=1}^n \langle x_i^{**}, f_0(x_i^*) \rangle \neq 0$, it follows that there exists some k satisfying $\langle x_k^{**}, f_0(x_k^*) \rangle \neq 0$. We can assume $\langle x_k^{**} \otimes x_k^*, f_0 \rangle = \langle x_k^{**}, f_0(x_k^*) \rangle = 1$.

Now, by Urysohn's lemma, there is some $\alpha \in C(U^*)$ such that $\alpha(x_k^*) = 1$ and $\alpha(x_i^*) = 0$ for $i \neq k$. Since $\alpha f \in \mathcal{M}$ for each $f \in \mathcal{M}$, we have

$$\langle x_k^{**} \otimes x_k^*, f \rangle = \langle x_k^{**}, f(x_k^*) \rangle = \sum_{i=1}^n \langle x_i^{**}, \alpha(x_i^*) f(x_i^*) \rangle = \phi(\alpha f) = 0$$

for all $f \in \mathcal{M}$, and the proof is finished. ■

Now let $\mathcal{A}$ be a subalgebra of $\mathcal{L}(X)$. The next result gives a necessary and sufficient condition for the existence of a common non-trivial $\mathcal{A}^*$-invariant closed subspace.

Theorem 10.74. *For an arbitrary subalgebra $\mathcal{A}$ of $\mathcal{L}(X)$, the following two statements are equivalent.*

(1) *There exists a non-trivial $\mathcal{A}^*$-invariant closed subspace.*

(2) *There exists an operator $B \in \mathcal{L}(X)$ such that its adjoint operator B^* does not belong to the τ_s-closure in $C(U^*, X^*)$ of the vector space generated by the set $\{\alpha T^* \colon \alpha \in C(U^*) \text{ and } T \in \mathcal{A}\}$.*

Proof. Let $\mathcal{M}$ denote the vector subspace of $C(U^*, X^*)$ generated by the collection of functions $\{\alpha T^* \colon \alpha \in C(U^*) \text{ and } T \in \mathcal{A}\}$, and let $\overline{\mathcal{M}}$ denote the closure of $\mathcal{M}$ in the topology τ_s.

(1) $\Longrightarrow$ (2) By Theorem 10.12 there exist non-zero $x^{**} \in X^{**}$ and $x^* \in X^*$ satisfying $\langle T^*x^*, x^{**}\rangle = 0$ for each $T \in \mathcal{A}$. Without loss of generality we can suppose that $x^* \in U^*$.

Pick $b^* \in X^*$ and $b \in X$ such that $\langle b^*, x^{**}\rangle = 1$ and $\langle b, x^*\rangle = 1$. We claim that the rank-one operator $B = b^* \otimes b \in \mathcal{L}(X)$ satisfies $B^* \notin \overline{\mathcal{M}}$. To see this, note that

$$\bigl\langle x^{**} \otimes x^*, \alpha T^*\bigr\rangle = \alpha(x^*)\bigl\langle T^*x^*, x^{**}\bigr\rangle = 0$$

for each $T \in \mathcal{A}$ and each $\alpha \in C(U^*)$. That is, the τ_s-continuous linear functional $x^{**} \otimes x^*$ vanishes on $\mathcal{M}$. On the other hand, we have

$$\bigl\langle x^{**} \otimes x^*, B^*\bigr\rangle = \bigl\langle x^{**} \otimes x^*, b \otimes b^*\bigr\rangle = \langle b^*, x^{**}\rangle\langle b, x^*\rangle = 1,$$

and so Lemma 10.73 implies that $B^* \notin \overline{\mathcal{M}}$.

(2) $\Longrightarrow$ (1) Pick some $B \in \mathcal{L}(X)$ such that $B^* \notin \overline{\mathcal{M}}$. Since $\mathcal{M}$ is invariant under multiplication by elements of $C(U^*)$, it follows from Lemma 10.73 that there exist $x^{**} \in X^{**}$ and $x^* \in U^*$ such that

$$\bigl\langle x^{**} \otimes x^*, B^*\bigr\rangle = \langle B^*x^*, x^{**}\rangle = 1 \quad \text{and} \quad \langle T^*x^*, x^{**}\rangle = 0$$

for all $T \in \mathcal{A}$. Since $\langle B^*x^*, x^{**}\rangle = 1$, it follows that $x^{**} \neq 0$ and $x^* \neq 0$, and by Theorem 10.12 the algebra $\mathcal{A}^*$ is non-transitive. ■

Our next goal is to obtain a similar characterization for the invariant subspace problem in terms of the norm topology. To accomplish this, we need to introduce a new class of functions.

Definition 10.75. *A function $f \in Y = C(U^*, X^*)$ is said to be* ***completely continuous*** *if it is continuous for the weak* topology on U^* and the norm topology on X^*.*

The vector subspace of all completely continuous functions of the Banach space $C(U^*, X^*)$ will be denoted by $\mathcal{K}(U^*, X^*)$.

In other words, a function $f \in C(U^*, X^*)$ is completely continuous if and only if $x^*_\alpha \xrightarrow{w^*} x^*$ in U^* implies $\|f(x^*_\alpha) - f(x^*)\| \to 0$. It is easy to see that if $T\colon X \to X$ is a compact operator, then $T^*\colon U^* \to X^*$ (the restriction of T^* to U^*) is completely continuous, i.e., $T^* \in \mathcal{K}(U^*, X^*)$.

Clearly, $\mathcal{K}(U^*, X^*)$ is a norm closed subspace of $C(U^*, X^*)$. As a vector subspace of $C(U^*, X^*)$, the space $\mathcal{K}(U^*, X^*)$ inherits the three topologies considered on $C(U^*, X^*)$; the norm topology, the topology τ_w, and the topology τ_s. There is no reason to believe that τ_w or τ_s is consistent with the norm topology on $\mathcal{K}(U^*, X^*)$. Nevertheless, for $C(U^*)$-invariant subspaces of $\mathcal{K}(U^*, X^*)$ the situation is different. Using the Lomonosov–de Branges technique, we can show that for any $C(U^*)$-invariant subspace of $\mathcal{K}(U^*, X^*)$ its closures in these topologies coincide.

Theorem 10.76. *For each $C(U^*)$-invariant subspace $\mathcal{M}$ of $\mathcal{K}(U^*, X^*)$ the following statements are equivalent.*

(1) *$\mathcal{M}$ is τ_w-closed in $\mathcal{K}(U^*, X^*)$.*

(2) *$\mathcal{M}$ is τ_s-closed in $\mathcal{K}(U^*, X^*)$.*

(3) *$\mathcal{M}$ is norm closed.*

Proof. Clearly, $(1) \Longrightarrow (2) \Longrightarrow (3)$. It remains to establish $(3) \Longrightarrow (1)$.

To this end, let $f_0 \in \mathcal{K}(U^*, X^*)$ belong to the τ_w-closure of $\mathcal{M}$. To show that $f_0 \in \mathcal{M}$, it suffices to prove that f_0 belongs to the norm closure of $\mathcal{M}$. By the Hahn–Banach theorem, it suffices to verify that for any norm continuous linear functional ϕ on $C(U^*, X^*)$ the equality $\langle \phi, f \rangle = 0$ for all $f \in \mathcal{M}$ implies $\langle \phi, f_0 \rangle = 0$. So, let $\langle \phi, f \rangle = 0$ for all $f \in \mathcal{M}$. We can assume that ϕ has norm one.

For each $\alpha \in C(U^*)$, we define on $Y = C(U^*, X^*)$ the continuous linear functional ϕ_α by $\phi_\alpha(f) = \phi(\alpha f)$, where $f \in Y$. Since $\mathcal{M}$ is invariant under multiplication by elements of $C(U^*)$, it follows that $\langle \phi_\alpha, f \rangle = 0$ for all $f \in \mathcal{M}$. This certainly guarantees $\langle \psi, f \rangle = 0$ for all $\psi \in \mathcal{V}$ and $f \in \mathcal{M}$, where $\mathcal{V}$ is the weak* closed subspace of Y^* generated by the functionals ϕ_α. We denote by $\mathcal{U}$ the intersection of $\mathcal{V}$ with the closed unit ball of Y^*. Obviously $\phi \in \mathcal{U}$. Now let ψ be any extreme point of $\mathcal{U}$. By the characterization of extreme points obtained in [**75**, Theorem 1], it follows that there exist an element $x^* \in U^*$ and an element $x^{**} \in X^{**}$ such that

$$\psi(f) = \big\langle x^{**} \otimes x^*, f \big\rangle = \big\langle f(x^*), x^{**} \big\rangle$$

for every $f \in \mathcal{K}(U^*, X^*)$. In particular, $\psi(f) = \big\langle f(x^*), x^{**} \big\rangle = 0$ for all $f \in \mathcal{M}$ and $\psi(f_0) = \big\langle f_0(x^*), x^{**} \big\rangle$. Since f_0 is in the τ_w-closure of $\mathcal{M}$, it follows that $\psi(f_0) = \langle f_0(x^*), x^{**} \rangle = 0$.

That is, we have proved that $\psi(f_0) = 0$ for each extreme point ψ of $\mathcal{U}$. Since $\phi \in \mathcal{U}$ and since, by the Krein–Milman theorem, $\mathcal{U}$ is the weak* closed convex hull of its extreme points, we can conclude that $\phi(f_0) = 0$. This completes the proof. ■

We are now ready to improve Theorem 10.74.

Theorem 10.77. *For a subalgebra $\mathcal{A}$ of $\mathcal{L}(X)$ the following statements are equivalent.*

(1) *The dual algebra $\mathcal{A}^*$ is non-transitive.*

(2) *There exist operators $B, K \in \mathcal{L}(X)$ with K compact such that the operator B^*K^* does not belong to the norm closure in $C(U^*, X^*)$ of the vector subspace generated by $\big\{\alpha T^* K^* \colon \alpha \in C(U^*) \text{ and } T \in \mathcal{A}\big\}$.*

Proof. (1) $\Longrightarrow$ (2) By Theorem 10.12, there exist non-zero $x^{**} \in X^{**}$ and $x^* \in X^*$ satisfying $\langle T^*x^*, x^{**}\rangle = 0$ for each $T \in \mathcal{A}$. We can suppose that $x^* \in U^*$. Now, pick elements $b \in X$ and $b^* \in X^*$ such that $\langle x^*, b\rangle = 1$ and $\langle b^*, x^{**}\rangle = 1$.

Next, consider the rank-one operators $K = x^* \otimes b$ and $B = b^* \otimes b$, and note that $K^*x^* = \langle x^*, b\rangle x^* = x^*$ and $B^*x^* = \langle x^*, b\rangle b^* = b^*$. Therefore, $B^*K^*x^* = b^*$. We claim that B^*K^* does not belong to the norm closure $\overline{\mathcal{M}}$ of the space $\mathcal{M}$ generated by the set $\{\alpha T^*K^*\colon\ \alpha \in C(U^*) \text{ and } T \in \mathcal{A}\}$ in the space Y.

To see this, observe first that the norm continuous linear functional $\phi = x^{**} \otimes x^*$ on $C(U^*, X^*)$ satisfies

$$\langle \phi, \alpha T^*K^*\rangle = \alpha(x^*)\langle x^{**}, T^*K^*x^*\rangle = \alpha(x^*)\langle T^*x^*, x^{**}\rangle = 0$$

for each $T \in \mathcal{A}$ and each $\alpha \in C(U^*)$. That is, ϕ vanishes on $\mathcal{M}$. On the other hand, $\phi(B^*K^*) = \langle B^*K^*x^*, x^{**}\rangle = \langle b^*, x^{**}\rangle = 1$ and this guarantees that $B^*K^* \notin \overline{\mathcal{M}}$.

(2) $\Longrightarrow$ (1) Assume that $B, K \in \mathcal{L}(X)$ satisfy the stated properties. Let $\overline{\mathcal{M}}$ denote the norm closure in $C(U^*, X^*)$ of the vector space $\mathcal{M}$ generated by the set $\{\alpha T^*K^*\colon\ \alpha \in C(U^*) \text{ and } T \in \mathcal{A}\}$. Clearly, $\mathcal{M}$ is $C(U^*)$-invariant, B^*K^* belongs to $\mathcal{K}(U^*, X^*)$ and $\mathcal{M} \subseteq \mathcal{K}(U^*, X^*)$.

Since B^*K^* does not belong to the norm closure of $\mathcal{M}$, it follows from Theorem 10.76 that B^*K^* is not in the τ_w-closure of $\mathcal{M}$ in $\mathcal{K}(U^*, X^*)$. Therefore, by Lemma 10.73, there exist $x^{**} \in X^{**}$ and $x^* \in U^*$ satisfying

$$\langle B^*K^*x^*, x^{**}\rangle = 1 \quad \text{and} \quad \langle T^*K^*x^*, x^{**}\rangle = 0$$

for all $T \in \mathcal{A}$. The former condition implies that $K^*x^* \neq 0$, and hence the latter condition shows that Theorem 10.12 is applicable to $\mathcal{A}^*$. Thus, the algebra $\mathcal{A}^*$ is non-transitive. ■

The preceding arguments also allow us to obtain a necessary and sufficient condition for the existence of a non-trivial closed $\mathcal{A}$-invariant subspace. The only difference is that we must apply the compact operator on the left, as opposed to the multiplication on the right in the preceding theorem.

Theorem 10.78. *For a subalgebra $\mathcal{A}$ of $\mathcal{L}(X)$ the following statements are equivalent.*

(1) *$\mathcal{A}$ is non-transitive.*

(2) *There exist operators $B, K \in \mathcal{L}(X)$ with K compact such that the operator K^*B^* does not belong to the norm closure in $C(U^*, X^*)$ of the subspace generated by $\{\alpha K^*T^*\colon\ \alpha \in C(U^*) \text{ and } T \in \mathcal{A}\}$.*

Proof. (1) $\Longrightarrow$ (2) By Theorem 10.12 there exist non-zero $x \in X$ and x^* in X^* satisfying $\langle x^*, Tx\rangle = 0$ for each $T \in \mathcal{A}$. We can suppose that $x^* \in U^*$.

Take any $b^* \in X^*$ such that $\langle x, b^*\rangle = 1$, and then consider the rank-one operator $K = b^* \otimes x \in \mathcal{L}(X)$. Observe that $Kx = \langle x, b^*\rangle x = x$. Next choose any element b in X such that $\langle x^*, b\rangle = 1$ and then define the operator $B = b^* \otimes b \in \mathcal{L}(X)$. Clearly, the adjoint operator B^* satisfies $B^*x^* = \langle b, x^*\rangle b^* = b^*$.

The vector subspace generated by $\{\alpha K^*T^* \colon\ \alpha \in C(U^*) \text{ and } T \in \mathcal{A}\}$ will be denoted by $\mathcal{N}$. We claim that the operator K^*B^* is not in the norm closure of $\mathcal{N}$. To see this, note that the linear functional $\phi = x \otimes x^*$ is norm continuous on $C(U^*, X^*)$ and satisfies

$$\begin{aligned}\big\langle \phi, \alpha K^*T^*\big\rangle &= \alpha(x^*)\big\langle x, K^*T^*x^*\big\rangle = \alpha(x^*)\big\langle TKx, x^*\big\rangle \\ &= \alpha(x^*)\big\langle Tx, x^*\big\rangle = 0\end{aligned}$$

for each $T \in \mathcal{A}$ and each $\alpha \in C(U^*)$. That is, $\phi = x \otimes x^*$ vanishes on $\mathcal{N}$. On the other hand, the equality $B^*x^* = b^*$ yields

$$\phi(K^*B^*) = \big\langle x, K^*B^*x^*\big\rangle = \big\langle x, K^*b^*\big\rangle = \big\langle Kx, b^*\big\rangle = \big\langle x, b^*\big\rangle = 1,$$

and this shows that K^*B^* does not belong to the norm closure of $\mathcal{N}$.

(2) $\Longrightarrow$ (1) Assume that the operators B and K satisfy the stated properties. Again, let $\mathcal{N}$ denote the vector space generated in $C(U^*, X^*)$ by the collection of functions $\{\alpha K^*T^* \colon\ \alpha \in C(U^*) \text{ and } T \in \mathcal{A}\}$. Clearly, $\mathcal{N}$ is $C(U^*)$-invariant, K^*B^* belongs to $\mathcal{K}(U^*, X^*)$ and $\mathcal{N} \subseteq \mathcal{K}(U^*, X^*)$.

Since the compact operator K^*B^* does not belong to the norm closure of $\mathcal{N}$, it follows from Theorem 10.76 that K^*B^* is not in the τ_s-closure of $\mathcal{N}$ in $\mathcal{K}(U^*, X^*)$. Consequently, by Lemma 10.73, there exist $x^{**} \in X^{**}$ and $x^* \in U^*$ such that

$$\big\langle K^*B^*, x^{**} \otimes x^*\big\rangle = \big\langle K^*B^*x^*, x^{**}\big\rangle = 1 \quad \text{and} \quad \big\langle K^*T^*x^*, x^{**}\big\rangle = 0$$

for all $T \in \mathcal{A}$. Since $\langle K^*B^*x^*, x^{**}\rangle = 1$, we see that $x_0 = K^{**}x^{**} \neq 0$. Moreover, the compactness of K implies that $x_0 = K^{**}x^{**} \in X$. So, for each $T \in \mathcal{A}$, we have

$$\big\langle x^*, Tx_0\big\rangle = \big\langle x^*, TK^{**}x^{**}\big\rangle = \big\langle T^*x^*, K^{**}x^{**}\big\rangle = \big\langle K^*T^*x^*, x^{**}\big\rangle = 0\,,$$

and consequently Theorem 10.12 guarantees that the algebra $\mathcal{A}$ is non-transitive. The proof is complete. ■

We close the section by emphasizing the following remarkable fact that has been present throughout our discussion on the invariant subspace problem. As soon as we ask about the existence of a non-trivial $\mathcal{A}$-invariant closed subspace of X (or about the existence of a non-trivial $\mathcal{A}^*$-invariant closed subspace of X^*), a compact operator seems to emerge from nowhere though the given algebra $\mathcal{A}$ were not connected with any compactness properties in any way from the outset!

Exercises

1. If Ω is a compact topological space and X is a Banach space, then show that the space $C(\Omega, X)$ of all X-valued continuous functions on Ω equipped with the norm

$$\|f\| = \max_{\omega \in \Omega} \|f(\omega)\|$$

is a Banach space—which is a Banach lattice if X is a Banach lattice.

2. Let X be a Banach space, and let the closed unit ball U^* of X^* be equipped with the w^*-topology. Establish the following.
   (a) If $T\colon X \to X$ is a compact operator, then $T^*\colon U^* \to X^*$ (the restriction of T^* to U^*) is a completely continuous function.
   (b) The collection of all completely continuous functions $\mathcal{K}(U^*, X^*)$ is a norm closed subspace of $C(U^*, X^*)$, and so it is a Banach space in its own right.

3. Assume that for a subalgebra $\mathcal{A}$ of $\mathcal{L}(X)$ there exists a non-zero operator $T_0 \in \mathcal{L}(X)$ such that:
   (a) T_0 commutes with each operator in $\mathcal{A}$.
   (b) For some $B \in \mathcal{L}(X)$ its adjoint operator B^* is not in the τ_s-closure of the vector subspace spanned by $\{\alpha T^* T_0^*\colon\ \alpha \in C(U^*) \text{ and } T \in \mathcal{A}\}$ in $C(U^*, X^*)$.

   Show that $\mathcal{A}^*$ is non-transitive. [HINT: By Lemma 10.73, there exist two vectors $x^{**} \in X^{**}$ and $x^* \in U^*$ satisfying

$$\langle x^{**} \otimes x^*, B^* \rangle = x^{**}(B^* x^*) = 1 \text{ and } \langle x^{**} \otimes x^*, T^* T_0^* \rangle = \langle x^{**}, T^* T_0^* x^* \rangle = 0$$

   for all $T \in \mathcal{A}$. If $y^* = T_0^* x^* \neq 0$, then the relation $\langle T^* y^*, x^{**} \rangle = 0$ for each $T \in \mathcal{A}$ coupled with Theorem 10.12 shows that $\mathcal{A}^*$ admits a non-trivial closed $\mathcal{A}^*$-invariant subspace. Now assume that $T_0^* x^* = 0$, and let N denote the null space of T_0^*. Since T_0 commutes with every member of $\mathcal{A}$, it follows that N is an $\mathcal{A}^*$-invariant closed subspace. From $x^{**}(B^* x^*) = 1$, we infer that $x^* \neq 0$. Since $x^* \in N$, we see that N is non-zero. On the other hand, if $N = X^*$, then $T_0^* = 0$, and so $T_0 = 0$, a contradiction. Therefore, N is a non-trivial $\mathcal{A}^*$-invariant closed subspace.]

4. Recall that a Banach space X is said to have the **Schur property** if $x_n \xrightarrow{w} 0$ implies $\|x_n\| \to 0$.

   If X is a Banach space with the Schur property and $\mathcal{A}$ is a commutative subalgebra of $\mathcal{L}(X)$, then show that the dual algebra $\mathcal{A}^*$ of $\mathcal{L}(X^*)$ is non-transitive. [HINT: If each element in $\mathcal{A}$ is a scalar operator, then the conclusion is trivial. So, we can assume that there exists some non-scalar operator $T \in \mathcal{A}$. Fix some λ_0 in the approximate point spectrum of T and consider the operator $T_0 = T - \lambda_0 I$. Also, select a sequence $\{x_n\}$ of unit vectors in X satisfying

$$\|T_0 x_n\| = \|Tx_n - \lambda_0 x_n\| \to 0\,. \tag{$\star$}$$

Clearly, T_0 commutes with every member of $\mathcal{A}$. Next, consider the subspace $\mathcal{M}$ of $C(U^*, X^*)$ generated by $\{\alpha T^* T_0^* \colon\ \alpha \in C(U^*) \text{ and } T \in \mathcal{A}\}$. We claim that the identity operator $I^* \colon X^* \to X^*$ (restricted to U^*) does not belong to the τ_s-closure of $\mathcal{M}$. To see this, assume by way of contradiction that I^* belongs to the τ_s-closure of $\mathcal{M}$. Therefore, if $\epsilon > 0$ and $x^* \in U^*$ are fixed, then there exist $\alpha_i \in C(U^*)$ and $T_i \in \mathcal{A}$ $(i = 1, \ldots, m)$ satisfying $\big\|\big[\,I^* - \sum_{i=1}^m \alpha_i T_i^* T_0^*\,\big](x^*)\big\| < \epsilon$. This implies

$$\Big|\langle x, x^*\rangle - \sum_{i=1}^{m} \alpha_i(x^*)\big\langle x, T_i^* T_0^*(x^*)\big\rangle\Big| < \epsilon \tag{$\star\star$}$$

for all $x \in X$ with $\|x\| \le 1$. Taking into account that $\mathcal{A}$ is a commutative algebra, it follows from $(\star\star)$ that $\big|\langle x, x^*\rangle - \sum_{i=1}^m \alpha_i(x^*)\langle T_0 x, T_i^* x^*\rangle\big| < \epsilon$. In particular, for each n we have $\big|\langle x_n, x^*\rangle - \sum_{i=1}^m \alpha_i(x^*)\langle T_0 x_n, T_i^* x^*\rangle\big| < \epsilon$, which, in view of $(\star)$, yields $|\langle x_n, x^*\rangle| < \epsilon$ for all sufficiently large n. In other words, the sequence $\{x_n\}$ converges weakly to zero in X. Since X has the Schur property, it follows that $\|x_n\| \to 0$, contrary to $\|x_n\| = 1$ for each n. Therefore, the identity operator I^* does not belong to the τ_s-closure of $\mathcal{M}$ and the conclusion follows from the previous exercise.]

5. (Honor [**156**]) If $\mathcal{A}$ is a commutative algebra of bounded operators on ℓ_1, then show that the dual algebra $\mathcal{A}^*$ of operators on ℓ_∞ is non-transitive.

10.8. Comments on the Invariant Subspace Problem

The existence of non-trivial closed invariant subspaces for a given operator presents the opportunity to analyze the structure of the operator by studying its restrictions to the invariant subspaces. In particular, if an operator on a Banach space has a reducing pair, then the operator can be written as a direct sum of two operators and, as expected, the two summands might provide some crucial information about the action of the operator. This rationale is behind the importance of the invariant subspace problem that has been studied so extensively over the last eighty years.

In this chapter, we have presented a number of invariant subspace results which clearly reflect the authors' interest in the invariant subspace problem. There are, however, many other directions regarding the invariant subspace problem that lie outside the scope of this monograph. In this section, we shall highlight several remarkable invariant subspace results that were considered as major breakthroughs at the time of their discovery.

We start by mentioning again that the first example of a bounded operator on a separable non-reflexive complex Banach space without non-trivial closed invariant subspaces was constructed by P. Enflo [**121**]. Enflo's construction was simplified considerably by B. Beauzamy [**53**]. C. J. Read [**275, 276, 277**] presented examples of bounded operators (some of which are quasinilpotent) on the real ℓ_1 without non-trivial closed invariant subspaces. An example of a continuous operator on a nuclear Fréchet space without non-trivial closed invariant subspaces was constructed by A. Atzmon [**47**]. It is still an open question whether a bounded operator on a separable reflexive Banach space (in particular, on a separable Hilbert space) admits a non-trivial closed invariant subspace.

The first result that we would like to mention here deals with linear isometries. Let $T\colon X \to X$ be a linear isometry on a Banach space. The range $R(T)$ of T is obviously closed and, therefore, if T is not surjective, then $R(T)$ is a non-trivial closed T-hyperinvariant subspace. For surjective isometries the conclusion remains true according to the following theorem, that is basically due to E. R. Lorch [**214**]. Recall that an invertible operator $T\colon X \to X$ is said to be **doubly power bounded** (or a **Cesaro operator**) if $\sup_{n\in Z} \|T^n\| < \infty$. It is obvious that each surjective isometry is doubly power bounded.

Theorem 10.79 (Lorch). *Every non-scalar doubly power bounded operator on a complex Banach space has a non-trivial closed hyperinvariant subspace.*

J. Wermer [**331**] produced an exceptionally strong generalization of Lorch's theorem for the class of non-quasianalytic operators. An invertible operator $T\colon X \to X$ on a Banach space is called **quasianalytic** if $\sum_{n=-\infty}^{\infty} \frac{|\log\|T^n\||}{1+n^2} = \infty$. Accordingly, the invertible operator T is **non-quasianalytic** if

$$\sum_{n=-\infty}^{\infty} \frac{|\log\|T^n\||}{1+n^2} < \infty\,. \tag{$\star$}$$

Clearly, each doubly power bounded operator is non-quasianalytic.

Theorem 10.80 (Wermer). *If T is non-quasianalytic and its spectrum $\sigma(T)$ is not a singleton, then T has a non-trivial closed hyperinvariant subspace.*

This theorem generated an avalanche of subsequent extensions and improvements. In this direction, we mention the papers by Ju. I. Ljubič and V. I. Macaev [**210**], E. M. Dynkin [**119**] and A. Atzmon [**48**]. It is still unknown if Theorem 10.80 is true for non-quasianalytic operators with singleton spectra. We have some reasons to believe that Wermer's theorem is true for non-quasianalytic positive operators.

For operators with a norm estimate on the norms of their powers (in particular, for doubly power bounded operators) Wermer's theorem is true.

Theorem 10.81. *If the spectrum of an invertible operator $T \in \mathcal{L}(X)$ is a singleton and there is a $k \in \mathbb{N}$ such that $\|T^n\| \leq n^k$ for each $n \in \mathbb{Z}$, then T has eigenvalues.*

A much sharper result than Theorem 10.81 was obtained by A. Atzmon [**46**]. The second spectacular result that we shall discuss next is also related to Wermer's theorem. Let Γ be the unit circle in the complex plane and let μ be the normalized Lebesgue measure on Γ. It was shown by A. M. Davie [**100**] that for each $1 \leq p \leq \infty$ the weighted composition operator (called the **Bishop operator**) $T\colon L_p(\Gamma,\mu) \to L_p(\Gamma,\mu)$ defined by the formula

$$Tx(\gamma) = w(\gamma)x(\gamma+\alpha)\,, \quad 0 \leq \gamma \leq 2\pi\,,$$

is non-quasianalytic provided that the weight $w \in C(\Gamma)$ is a Lipschitz function on Γ and $\frac{\alpha}{\pi}$ is an irrational number which cannot be approximated "too fast" by rational numbers—in the sense that $\frac{\alpha}{\pi}$ is a non-Liouville number.[6] Using some additional ingenious arguments, A. M. Davie [**100**] also proved the following.

Theorem 10.82 (Davie)**.** *If $\frac{\alpha}{\pi}$ is not a Liouville number, then every Bishop operator $T\colon L_p(\Gamma,\mu) \to L_p(\Gamma,\mu)$, defined by*

$$Tx(\gamma) = \gamma x(\gamma+\alpha)\,, \quad 0 \leq \gamma \leq 2\pi\,,$$

has a non-trivial closed hyperinvariant subspace.

This result of Davie was refined further by A. L. Gromov [**140**] and by G. W. MacDonald [**227**].

There are a few major recent generalizations of Lomonosov's Invariant Subspace Theorem to algebras and semigroups of compact operators on Banach spaces. A very important breakthrough was obtained in 1984 by V. S. Shulman [**301**] who proved the following result for Volterra algebras.[7]

Theorem 10.83 (Shulman)**.** *Each non-zero Volterra algebra has a non-trivial closed hyperinvariant subspace.*

It took more than a decade to generalize this result to Volterra semigroups of operators. This significant contribution has been accomplished by Y. V. Turovskii [**324**] who proved the following.

---

[6] Recall that a real number u is said to be a **Liouville number** if there exists a sequence $\{\frac{m_n}{k_n}\}$ of pairwise distinct rational numbers such that $(m_n,k_n)=1$ and $\left|u-\frac{m_n}{k_n}\right| < \frac{1}{n(k_n)^n}$ holds for each n. For instance, the real number $u=\sum_{n=1}^{\infty} 10^{-n!} = 0.110001\cdots$ is a Liouville number. Every Liouville number is transcendental.

[7] A collection of operators on a Banach space is said to be a **Volterra collection** if each operator in the collection is compact and quasinilpotent.

Theorem 10.84 (Turovskii). *Each non-zero multiplicative Volterra semigroup has a non-trivial closed hyperinvariant subspace.*

The major step needed to prove Theorem 10.84 was the verification that the algebra generated by a multiplicative Volterra semigroup remained likewise a multiplicative Volterra semigroup. This verification is very technical and uses several subtle ideas. After that, the existence of the hyperinvariant subspace is a simple consequence of Shulman's Theorem 10.83. For a complete exposition of Theorem 10.84 we refer the reader to the excellent monograph by H. Radjavi and P. Rosenthal [**273**].

Recall that if X is a finite dimensional Banach space, then the classical Burnside theorem states the following.

Theorem 10.85 (Burnside). *If X is a finite dimensional complex Banach space, then each proper subalgebra $\mathcal{A}$ of $\mathcal{L}(X)$ is non-transitive.*

For a proof of this theorem see [**273**, Section 1.2]. It is worth pointing out that Burnside's theorem is not true for finite dimensional real Banach spaces. To see this, let $X = \mathbb{R}^{2n}$, $n \geq 2$, and take any non-scalar operator T on X without a non-trivial hyperinvariant subspace—we refer to Exercise 2 in Section 10.1 for the construction of such an operator. Consider the algebra $\mathcal{A} = \{T\}'$. It is transitive, since T does not have a non-trivial hyperinvariant subspace. At the same time a straightforward verification shows that this algebra $\mathcal{A}$ is necessarily proper, since T is a non-scalar operator.

We will mention two elegant ways of extending the Burnside theorem to infinite dimensional Banach spaces. The first result is due to G. L. Litvinov and V. I. Lomonosov [**209**] and it is true for locally convex spaces as well. Notice that while discussing the transitivity of a subalgebra $\mathcal{A}$ of $\mathcal{L}(X)$ we can always assume that $\mathcal{A}$ is weakly closed, that is, closed in the weak operator topology (otherwise we can replace $\mathcal{A}$ by its weak closure which remains non-transitive if $\mathcal{A}$ were non-transitive).

Theorem 10.86 (Litvinov–Lomonosov). *Let X be a complex Banach space and let $\mathcal{A}$ be a weakly closed subalgebra of $\mathcal{L}(X)$. If $\mathcal{A}$ is transitive and contains a non-zero compact operator, then $\mathcal{A} = \mathcal{L}(X)$.*

If X is a finite dimensional Banach space and $\mathcal{A}$ is a proper subalgebra of $\mathcal{L}(X)$, then $\mathcal{A}$ is certainly closed and each operator in $\mathcal{A}$ is compact. Therefore, $\mathcal{A}$ cannot be transitive, that is, Theorem 10.86 does imply the Burnside theorem.

According to Theorem 10.12, Theorem 10.85 is equivalent to the existence of a non-zero $x \in X$ and a non-zero $x^* \in X^*$ such that $\langle Ax, x^* \rangle = 0$ for each $A \in \mathcal{A}$. It was this form of the Burnside theorem that was used by

V. I. Lomonosov [**212**] to produce another generalization to infinite dimensional Banach spaces.

Theorem 10.87 (Lomonosov). *Let X be a complex Banach space and let $\mathcal{A}$ be a proper weakly closed subalgebra of $\mathcal{L}(X)$. Then there exist a non-zero $x^{**} \in X^{**}$ and a non-zero $x^* \in X^*$ such that*

$$\left|\langle A^* x^*, x^{**}\rangle\right| \leq \|A\|_{\text{ess}}$$

for each $A \in \mathcal{A}$, where $\|A\|_{\text{ess}}$ denotes the essential norm of the operator A.

If X is finite dimensional, then $\|A\|_{\text{ess}} = 0$ and so Theorem 10.87 does imply Burnside's theorem. The reader is referred to [**212**] and to [**90**] for proofs and some interesting additional developments.

V. I. Lomonosov also proposed in [**212**] the following conjecture regarding the invariant subspace problem.

- *The adjoint of a bounded operator on a Banach space has a non-trivial closed invariant subspace.*

A Read operator $T \in \mathcal{L}(\ell_1)$ without non-trivial closed invariant subspaces seemed again to be a candidate for disproving this conjecture. However, V. G. Troitsky [**320**] has shown that this operator is not $\sigma(\ell_1, c_0)$-continuous, and therefore T is not the adjoint to any bounded operator on c_0. Moreover, T. Schlumprecht and V. G. Troitsky [**295**] have recently shown that T is not the adjoint of any bounded operator on any predual of ℓ_1.

We close the section with a few results dealing with operators on Hilbert spaces. An excellent introductory book on Hilbert space operators is [**94**]. Let H be a complex Hilbert space. Recall that the **Hilbert space adjoint** of an operator $T \in \mathcal{L}(H)$ is the unique operator $T^* \in \mathcal{L}(H)$ that satisfies the identity

$$\big(Tx, y\big) = \big(x, T^* y\big)$$

for all $x, y \in H$.[8] An operator $T \in \mathcal{L}(H)$ is said to be **normal** if it commutes with T^*, i.e., $TT^* = T^*T$. From the well-known classical theorem of J. von Neumann [**255**] on the spectral decomposition of normal operators, we have the following invariant subspace result. (For details see [**280**, p. 286] and [**317**, p. 430].)

Theorem 10.88. *Every non-scalar normal operator has a reducing hyperinvariant subspace.*

An operator A on a Hilbert space H is called **subnormal** if there exists a Hilbert space $\mathcal{H}$ such that H can be identified with a closed subspace of

[8] The notation T^* for the Hilbert space adjoint of the operator T is standard in Hilbert space theory and although there is a conflict with the notation of the Banach space adjoint of T, we hope that it will cause no problem here.

$\mathcal{H}$ and there exists a normal operator B on $\mathcal{H}$ that leaves the subspace H invariant and $A = B|_H$, that is, the restriction of B to H conicides with the initial operator A.

Regarding subnormal operators, S. Brown [**76**] proved the following remarkable result—that was viewed as a major achievement following a long period of partial successes by many mathematicians working on the invariant subspace problem for Hilbert space operators.

Theorem 10.89 (S. Brown). *Every subnormal operator has a non-trivial closed invariant subspace.*

The following interesting invariant subspace result was proved by B. Sz.-Nagy and C. Foias [**249**] and independently by L. de Branges [**74**].

Theorem 10.90 (Sz.-Nagy–Foias–de Branges). *If a non-scalar operator T in $\mathcal{L}(H)$ is power bounded such that neither T^n nor $(T^*)^n$ converges strongly to zero, then T has a non-trivial closed hyperinvariant subspace.*

We conclude the section by mentioning the important class of quasitriangular operators. An operator $T \in \mathcal{L}(H)$ is said to be **quasitriangular** if there exists a sequence $\{P_n\}$ of orthogonal projections on H that converges strongly to the identity operator and

$$\lim_{n\to\infty} \|P_nTP_n - TP_n\| = 0\,.$$

These operators were introduced by P. R. Halmos [**147**], and it seemed for a while that for quasitriangular operators the invariant subspace problem might be more manageable. However, a remarkable result by C. Apostol, C. Foias, and D. Voiculescu [**39**] has established just the opposite.

Theorem 10.91 (Apostol–Foias–Voiculescu). *Every non-quasitriangular operator $T \in \mathcal{L}(H)$ has a non-trivial closed invariant subspace.*

From Theorem 10.91 it follows that if every quasitriangular operator on a Hilbert space H has a non-trivial closed invariant subspace, then the invariant subspace problem has an affirmative solution for Hilbert spaces, that is, each bounded operator on H has a non-trivial closed invariant subspace.

Chapter 11

The Daugavet Equation

In 1963, I. K. Daugavet [**98**] proved that each compact operator T on the Banach space $C[0,1]$ satisfies the equation

$$\|I+T\| = 1 + \|T\|.$$

It turns out that various classes of operators on many other Banach spaces satisfy this equation, which is known today as the **Daugavet equation**. In 1965, C. Foias and I. Singer [**126**] extended Daugavet's result to arbitrary atomless $C(\Omega)$-spaces and to several classes of operators (including the weakly compact ones), and in 1966 G. Ya. Lozanovsky [**219**] established that compact operators on $L_1[0,1]$ satisfy the Daugavet equation.

During the next fifteen years or so these results were left without much attention approximately until the beginning of the 1980's. At that time a new wave of interest in this topic surfaced and the Daugavet equation has been studied by many authors in various contexts. The major results of these studies will be reported in this chapter. The introduced techniques vary from purely order-theoretical to analytical and we intend to familiarize the reader with the majority of these methods. However, this has its price. Our presentation is not as efficient as it could be, since we prove several basic results repeatedly to demonstrate different methods. For the most part these results deal with various classes of operators on $C(\Omega)$- and $L_1(\mu)$-spaces. These spaces are Banach lattices and we have an option to utilize their order structures.

We hope that the resulting comprehensive treatment of the currently available methods on the subject is an acceptable trade-off for the occasional redundancy of a few proofs. The last section of the chapter demonstrates how the Daugavet equation, a purely isometric property, can be used to obtain some topological conclusions regarding Banach spaces. For instance, a Banach space cannot have an unconditional basis provided each finite-rank operator on this space satisfies the Daugavet equation.

11.1. The Daugavet Equation and Uniform Convexity

The purpose of this section is to study the basic properties of operators that have the Daugavet property. The material in this section is taken from [**6**].

Definition 11.1. *A continuous operator $T\colon X \to X$ on a Banach space is said to satisfy the* **Daugavet equation** *provided*

$$\|I + T\| = 1 + \|T\|\,.$$

It should be clear that a bounded operator $T \in \mathcal{L}(X)$ satisfies the Daugavet equation if and only if $\|I+T\| \geq 1+\|T\|$. The following result should also be obvious.

Lemma 11.2. *A bounded operator satisfies the Daugavet equation if and only if its norm adjoint likewise satisfies the Daugavet equation.*

Lemma 11.3. *If the norm of a bounded operator $T\colon X \to X$ on a Banach space belongs to the spectrum of T, i.e., $\|T\| \in \sigma(T)$, then T satisfies the Daugavet equation.*

Proof. Assume $\|T\| \in \sigma(T)$. Since the spectrum of T lies inside the disk with center at zero and radius $\|T\|$, it follows that $\|T\|$ is a boundary point of the spectrum. In particular, $\|T\| \in \sigma_a(T)$, the approximate point spectrum of T. Pick a sequence of vectors $\{x_n\}$ in X such that $\|x_n\| = 1$ for each n and $\lim_{n\to\infty}\big\| Tx_n - \|T\|x_n \big\| = 0$. Now note that

$$\begin{aligned} \big\|I+T\big\| &\geq \big\|(I+T)x_n\big\| \\ &\geq \big\|x_n + \|T\|x_n\big\| - \big\|\,\|T\|x_n - Tx_n\big\| \\ &= 1+\|T\| - \big\|\,\|T\|x_n - Tx_n\big\|\,. \end{aligned}$$

Taking the limit, we see that $\|I+T\| \geq 1+\|T\|$, and so T satisfies the Daugavet equation. ■

We continue with a simple geometric property of normed spaces which guarantees that if an operator satisfies the Daugavet equation, then so does each positive multiple of the operator.

Lemma 11.4. *If two vectors u and v in a normed space satisfy the equality*

$$\|u+v\| = \|u\| + \|v\|\,,$$

then $\|\alpha u + \beta v\| = \alpha\|u\| + \beta\|v\|$ for all $\alpha, \beta \geq 0$.

Proof. Assume that the vectors u and v satisfy $\|u+v\| = \|u\| + \|v\|$ and let $\alpha,\ \beta \geq 0$. By the symmetry of the situation, we can assume that $\alpha \geq \beta \geq 0$. Then we have

$$\begin{aligned} \alpha\|u\| + \beta\|v\| &\geq \|\alpha u + \beta v\| = \|\alpha(u+v) - (\alpha-\beta)v\| \\ &\geq \alpha\|u+v\| - (\alpha-\beta)\|v\| \\ &= \alpha\big(\|u\| + \|v\|\big) - (\alpha-\beta)\|v\| \\ &= \alpha\|u\| + \beta\|v\|, \end{aligned}$$

and the desired equality follows. ■

Corollary 11.5. *If a bounded operator T on a Banach space satisfies the Daugavet equation, then for each $\alpha \geq 0$ the operator αT likewise satisfies the Daugavet equation.*

The next lemma describes two equivalent norm conditions the first of which is usually taken for the definition of uniform convexity. As usual, we denote by U_X the closed unit ball of a Banach space X.

Lemma 11.6. *For a Banach space X the following are equivalent.*

(1) *For each $0 < \epsilon \leq 2$ there exists some $0 < \delta < 1$ such that if $x, y \in U_X$ satisfy $\|x - y\| \geq \epsilon$, then $\left\|\frac{x+y}{2}\right\| < 1 - \delta$.*

(2) *If two sequences $\{x_n\}$ and $\{y_n\}$ in U_X satisfy $\lim_{n\to\infty}\|x_n + y_n\| = 2$, then $\lim_{n\to\infty}\|x_n - y_n\| = 0$.*

Proof. (1) $\Longrightarrow$ (2) Let two sequences $\{x_n\}$ and $\{y_n\}$ belong to U_X and satisfy $\lim_{n\to\infty}\|x_n + y_n\| = 2$. Now fix $0 < \epsilon \leq 2$ and then pick some $0 < \delta < 1$ that satisfies (1). Next, choose some n_0 such that $\left\|\frac{x_n+y_n}{2}\right\| > 1 - \delta$ for all $n \geq n_0$. But then it follows from (1) that $\|x_n - y_n\| < \epsilon$ for each $n \geq n_0$, i.e., $\lim_{n\to\infty}\|x_n - y_n\| = 0$.

(2) $\Longrightarrow$ (1) If (1) were not true, then there would exist some $\epsilon > 0$ and $x_n, y_n \in U_X$ with $\|x_n - y_n\| > \epsilon$ and $\left\|\frac{x_n+y_n}{2}\right\| > 1 - \frac{1}{n}$. This implies $\lim_{n\to\infty}\|x_n + y_n\| = 2$ and hence, by (2), we have $\lim_{n\to\infty}\|x_n - y_n\| = 0$, contrary to $\|x_n - y_n\| > \epsilon$ for each n. ■

We are now ready to introduce the notions of uniformly convex and uniformly smooth Banach spaces.

Definition 11.7. *A Banach space that satisfies either one of the equivalent statements of Lemma* 11.6 *is said to be* ***uniformly convex*** (*or* ***uniformly rotund***).

Definition 11.8. *A Banach space is called* ***uniformly smooth*** *if for each $\epsilon > 0$ there is some $\delta > 0$ such that $\|x\| \geq 1$, $\|y\| \geq 1$, and $\|x - y\| < \delta$ imply $\|x + y\| \geq \|x\| + \|y\| - \epsilon\|x - y\|$.*

Uniform convexity and uniform smoothness are dual notions.

Lemma 11.9 (Šmulian [**308**]). *A Banach space is uniformly convex if and only if its norm dual is uniformly smooth.*

Similarly, a Banach space is uniformly smooth if and only if its norm dual is uniformly convex.

For a proof of the preceding result see Exercises 13 and 14 at the end of the section. For more information about uniformly convex and uniformly smooth Banach spaces see [**102, 105, 118**] and [**189**, pp. 353–369].

For uniformly convex Banach spaces the converse of Lemma 11.3 is true.

Theorem 11.10. *If X is uniformly convex and $T \in \mathcal{L}(X)$ satisfies the Daugavet equation, then $\|T\| \in \sigma(T)$.*

Proof. Assume that X is uniformly convex and that $T \in \mathcal{L}(X)$ is non-zero and satisfies the Daugavet equation. By Corollary 11.5, the operator $S = \frac{T}{\|T\|}$ also satisfies the Daugavet equation, i.e.,

$$\|I + S\| = \sup_{\|x\|=1} \|x + Sx\| = 1 + \|S\| = 2\,.$$

Thus, there exists a sequence of unit vectors $\{x_n\}$ satisfying

$$\lim_{n\to\infty} \|x_n + Sx_n\| = 2\,.$$

Since X is uniformly convex, we infer that $\lim_{n\to\infty} \|Sx_n - x_n\| = 0$. Consequently, $\lim_{n\to\infty} \big\|\frac{T}{\|T\|}x_n - x_n\big\| = 0$ or $\lim_{n\to\infty} \big\|Tx_n - \|T\|x_n\big\| = 0$. This shows that $\|T\|$ is in the approximate point spectrum of T. ■

Since an operator satisfies the Daugavet equation if and only if its adjoint satisfies the Daugavet equation and since the spectrum of an operator coincides with that of its adjoint (Theorem 6.14), we have the following consequence of Theorem 11.10.

Corollary 11.11. *A continuous operator T on a uniformly convex or on a uniformly smooth Banach space satisfies the Daugavet equation if and only if its norm $\|T\|$ lies in the spectrum $\sigma(T)$.*

From Theorem 7.11 we know that the non-zero points of the spectrum of a strictly singular operator on a Banach space are eigenvalues. Therefore, the condition that the norm of an operator is an eigenvalue characterizes the strictly singular operators that satisfy the Daugavet equation.

Corollary 11.12. *A strictly singular (in particular, compact) operator T on a uniformly convex or a uniformly smooth Banach space satisfies the Daugavet equation if and only if its norm $\|T\|$ is an eigenvalue of T.*

It is well known that every L_p-space with $1 < p < \infty$ is uniformly convex; see Exercise 16 at the end of the section. Therefore, Corollary 11.12 applies and it characterizes the strictly singular operators that satisfy the Daugavet equation on L_p-spaces for $1 < p < \infty$.

Corollary 11.13. *A strictly singular operator $T\colon L_p(\mu) \to L_p(\mu)$, where $1 < p < \infty$, satisfies the Daugavet equation if and only if its norm $\|T\|$ is an eigenvalue of T.*

For the boundary values $p = 1$ and $p = \infty$ the situation is drastically different and it will be discussed in Section 11.2. As we shall see, on atomless L_1- and L_∞-spaces each compact operator satisfies the Daugavet equation.

Theorem 11.14. *Assume that a continuous operator $T\colon X \to X$ on a uniformly convex or a uniformly smooth Banach space satisfies the Daugavet equation and let $f(\lambda) = \sum_{n=0}^{\infty} a_n\lambda^n$ be a power series with $a_n \geq 0$ for each n. If $f(\|T\|) < \infty$, then the continuous operator $f(T)$ satisfies the Daugavet equation and $\|f(T)\| = f(\|T\|)$, i.e., we have*

$$\|I + f(T)\| = 1 + \|f(T)\| = 1 + f(\|T\|)\,.$$

In particular, we have the following consequences:

1. *For each $n = 0, 1, 2, \ldots$ the operator T^n satisfies the Daugavet equation and $\|T^n\| = \|T\|^n$, i.e.,*

$$\|I + T^n\| = 1 + \|T^n\| = 1 + \|T\|^n\,.$$

2. *For any polynomial $p(\lambda) = a_0 + a_1\lambda + \cdots + a_n\lambda^n$ with non-negative coefficients, the operator $p(T)$ satisfies the Daugavet equation and $\|p(T)\| = p(\|T\|)$.*

Proof. The conclusions of the theorem follow easily from the Spectral Mapping Theorem. However, we prefer to present an independent and direct proof. Assume that the Banach space X, the continuous operator $T\colon X \to X$, and the function $f(\lambda)$ satisfy the stated properties. By Corollary 11.11, we know that there exists a sequence $\{x_n\}$ of unit vectors satisfying $\lim_{n\to\infty}\big\|Tx_n - \|T\|x_n\big\| = 0$. Using the identity

$$T^{k+1}x_n - \|T\|^{k+1}x_n = T\big(T^k x_n - \|T\|^k x_n\big) + \|T\|^k\big(Tx_n - \|T\|x_n\big)$$

and an easy inductive argument, we see that

$$\lim_{n\to\infty}\big\|T^k x_n - \|T\|^k x_n\big\| = 0$$

for each $k = 0, 1, 2, \ldots$. Next, we claim that

$$\lim_{n\to\infty}\big\|f(T)x_n - f(\|T\|)x_n\big\| = 0\,.$$

To see this, let $\epsilon > 0$. Fix some m such that $\sum_{i=m+1}^{\infty} a_i \|T\|^i < \epsilon$ and then pick some n_0 such that $\sum_{i=0}^{m} a_i \big\| T^i x_n - \|T\|^i x_n \big\| < \epsilon$ holds for all $n \geq n_0$. Therefore, for $n \geq n_0$ we have

$$\begin{aligned}
\big\| f(T)x_n - f(\|T\|)x_n \big\| &= \Big\| \sum_{i=0}^{\infty} a_i \big(T^i x_n - \|T\|^i x_n \big) \Big\| \\
&\leq \sum_{i=0}^{m} a_i \big\| T^i x_n - \|T\|^i x_n \big\| + \sum_{i=m+1}^{\infty} a_i \big\| T^i x_n - \|T\|^i x_n \big\| \\
&\leq \epsilon + 2 \sum_{i=m+1}^{\infty} a_i \|T\|^i < 3\epsilon ,
\end{aligned}$$

which shows that $\lim_{n\to\infty} \big\| f(T)x_n - f(\|T\|)x_n \big\| = 0$. Consequently, the real number $f(\|T\|)$ lies in the approximate point spectrum of $f(T)$. Since $\|f(T)\| \leq f(\|T\|)$ and the spectrum of $f(T)$ lies inside the closed disk with center at zero and radius $\|f(T)\|$, we infer that the equality $\|f(T)\| = f(\|T\|)$ must hold. Now Corollary 11.11 guarantees that the operator $f(T)$ satisfies the Daugavet equation. ∎

We now turn our attention to operators acting on locally uniformly convex Banach spaces. To discuss this notion, we need the next result whose proof is similar to that of Lemma 11.6 and is left as an exercise.

Lemma 11.15. *For a unit vector x in a Banach space X the following statements are equivalent.*

1. *For each $0 < \epsilon \leq 2$ there exists some $0 < \delta < 1$ such that $\|y\| \leq 1$ and $\|x - y\| \geq \epsilon$ imply $\big\| \frac{x+y}{2} \big\| < 1 - \delta$.*
2. *If $\{x_n\} \subseteq U_X$ and $\lim\limits_{n\to\infty} \big\| \frac{x+x_n}{2} \big\| = 1$, then $\lim\limits_{n\to\infty} \|x - x_n\| = 0$.*

We are ready to define the locally uniformly convex Banach spaces.

Definition 11.16. *A Banach space is said to be* ***locally uniformly convex at a unit vector*** *x if it satisfies either one of the equivalent statements of Lemma 11.15.*

A ***locally uniformly convex*** *Banach space is a Banach space that is locally uniformly convex at every unit vector.*

It is well known that there exist locally uniformly convex Banach spaces that are not uniformly convex. Therefore, it is natural to ask whether the characterization of operators satisfying the Daugavet equation on uniformly convex Banach spaces given in Corollary 11.11 remains valid for locally uniformly convex spaces. Surprisingly, as the next example shows, the answer is negative.

Example 11.17. Fix a sequence $\{p_n\} \subseteq \mathbb{R}$ such that $\lim_{n\to\infty} p_n = 1$ and $p_n > 1$ for each n; for instance let $p_n = 1 + \frac{1}{n}$. For each n let $X_n = \ell^2_{p_n}$. That is, X_n is $\mathbb{R}^2$ equipped with the ℓ_{p_n}-norm $\|(x,y)\|_{p_n} = \big(|x|^{p_n} + |y|^{p_n}\big)^{\frac{1}{p_n}}$. Clearly, X_n is a reflexive uniformly convex Banach lattice. If $u_n = (1,0)$ and $v_n = (0,1)$, then

$$\|u_n\|_{p_n} = \|v_n\|_{p_n} = 1 \qquad \text{and} \qquad \|u_n + v_n\|_{p_n} = 2^{\frac{1}{p_n}}\,.$$

For each n define the positive operator $T_n\colon X_n \to X_n$ by

$$T_n(x,y) = (y,0)\,.$$

Observe that $T_n(u_n) = 0$, $T_n(v_n) = u_n$ and $\|T_n\| = 1$ for each n.

Next, consider the Banach lattice $X = (X_1 \oplus X_2 \oplus \cdots)_2$, that is, X is the ℓ_2-sum of the Banach lattices X_n, and define the positive operator $T\colon X \to X$ by

$$T(x_1, x_2, \ldots) = (T_1x_1, T_2x_2, \ldots)\,.$$

The conclusion that the operator T and the space X have the desired properties will be obtained from the following three statements.

(1) *The Banach lattice X is reflexive, fails to be uniformly convex but is locally uniformly convex.*

(2) *The operator T satisfies the Daugavet equation.*

(3) *The norm $\|T\| = 1$ does not belong to the spectrum of T.*

The reflexivity of X follows from the identity

$$X^{**} = (X_1^{**} \oplus X_2^{**} \oplus \cdots)_2 = (X_1 \oplus X_2 \oplus \cdots)_2\,,$$

established in Exercise 3 of Section 7.2. To see that the Banach space X is not uniformly convex, let

$$\mathbf{x}_n = (0,\ldots,0,u_n,0,\ldots) \quad \text{and} \quad \mathbf{y}_n = (0,\ldots,0,v_n,0,\ldots)\,, \tag{$\star$}$$

where the vectors u_n and v_n occupy the nth coordinates. Then we have $\|\mathbf{x}_n\|_2 = \|\mathbf{y}_n\|_2 = 1$. Moreover,

$$\begin{aligned} \big\|\tfrac{\mathbf{x}_n+\mathbf{y}_n}{2}\big\|_2 &= \tfrac{1}{2}\big\|u_n + v_n\big\|_{p_n} = 2^{\frac{1}{p_n}-1} \longrightarrow 1\,, \text{ and} \\ \big\|\mathbf{x}_n - \mathbf{y}_n\big\|_2 &= \big\|u_n - v_n\big\|_{p_n} = 2^{\frac{1}{p_n}} \longrightarrow 2\,. \end{aligned}$$

Therefore, X is not uniformly convex.

The local uniform convexity of X follows from the well-known fact that the ℓ_2-sum of a sequence of locally uniformly convex Banach spaces is locally uniformly convex; see, for instance [**218**]. Now we shall verify statement (2). If $\mathbf{x} = (x_1, x_2, \ldots) \in X$ satisfies $\|\mathbf{x}\|_2 \le 1$, then note that

$$\|T\mathbf{x}\|_2^2 = \sum_{n=1}^{\infty} \|T_n x_n\|_{p_n}^2 \le \sum_{n=1}^{\infty} \|T_n\|^2 \cdot \|x_n\|_{p_n}^2 = \sum_{n=1}^{\infty} \|x_n\|_{p_n}^2 \le 1\,,$$

and so $\|T\| \le 1$. On the other hand, if $\mathbf{y}_n$ is the unit vector defined in $(\star)$, then

$$\|T\| \ge \|T\mathbf{y}_n\|_2 = \|T_n v_n\|_{p_n} = \|u_n\|_{p_n} = 1\,,$$

and so $\|T\| \ge 1$. Hence, $\|T\| = 1$.

Similarly, if $I_n \colon X_n \to X_n$ denotes the identity operator on X_n, then

$$\begin{aligned} \|I+T\| &\ge \|(I+T)\mathbf{y}_n\|_2 = \|(I_n + T_n)v_n\|_{p_n} \\ &= \|v_n + T_n v_n\|_{p_n} = \|v_n + u_n\|_{p_n} = 2^{\frac{1}{p_n}}\,. \end{aligned}$$

From $\lim_{n\to\infty} 2^{\frac{1}{p_n}} = 2$, it follows that $\|I+T\| = 2$, and so T satisfies the Daugavet equation.

Finally, note that $T^2 = 0$, i.e., the operator T is nilpotent, and so $\sigma(T) = \{0\}$. Hence, $\|T\| = 1$ does not belong to the approximate point spectrum of T. ∎

The situation for locally uniformly convex Banach spaces is not as bad as one might expect from the previous example. For compact operators on locally uniformly spaces the conclusion of Corollary 11.12 remains true.

Theorem 11.18. *A compact operator $T\colon X \to X$ on a locally uniformly convex Banach space satisfies the Daugavet equation if and only if its norm $\|T\|$ is an eigenvalue of T.*

Proof. The "if" part is true for each Banach space and is independent of the compactness of the operator. Indeed, if $\|T\|$ is an eigenvalue of T, then by Theorem 11.10 we get $\|I+T\| = 1 + \|T\|$.

For the converse, assume that T is non-zero and satisfies the Daugavet equation. By Corollary 11.5, we know that the compact operator $S = \frac{T}{\|T\|}$ also satisfies the Daugavet equation, and so $\sup_{\|x\|=1} \|x+Sx\| = 1+\|S\| = 2$. Pick a sequence of unit vectors $\{x_n\}$ such that

$$\lim_{n\to\infty} \|x_n + Sx_n\| = 2\,. \qquad (\star\star)$$

Using the compactness of S (and passing to a subsequence if necessary), we can assume that $\lim_{n\to\infty} \|Sx_n - x\| = 0$ holds for some $x \in X$. From

$$\|x_n + Sx_n\| \le \|x_n\| + \|Sx_n\| = 1 + \|Sx_n\| \le 2$$

and $(\star\star)$, we infer that $\|x\| = \lim_{n\to\infty}\|Sx_n\| = 1$. Furthermore, from the inequality

$$\|x_n + Sx_n\| - \|Sx_n - x\| \le \|x_n + x\| \le 2\,,$$

we see that $\lim_{n\to\infty}\|x_n + x\| = 2$, or $\lim_{n\to\infty}\left\|\frac{x+x_n}{2}\right\| = 1$. By the local uniform convexity of X, we conclude that $\lim_{n\to\infty}\|x - x_n\| = 0$. So, $\lim_{n\to\infty}\|Sx - Sx_n\| = 0$, and hence $Sx = x$, or equivalently $\frac{T}{\|T\|}x = x$. Thus, $Tx = \|T\|x$, which shows that $\|T\|$ is an eigenvalue of T. ■

Exercises

1. Show that if the vectors $x_1, \dots, x_n$ in a normed space satisfy

$$\Big\|\sum_{k=1}^{n} x_k\Big\| = \sum_{k=1}^{n}\|x_k\|\,,$$

then $\big\|\sum_{k=1}^{n}\alpha_k x_k\big\| = \sum_{k=1}^{n}\alpha_k\|x_k\|$ holds for each choice of non-negative scalars $\alpha_1, \dots, \alpha_n$. [Hint: The proof is by induction. For $n = 1$, the claim is trivially true. So, assume the claim true for some n and let the vectors $x_1, \dots, x_n, x_{n+1}$ satisfy $\|\sum_{k=1}^{n+1}x_k\| = \sum_{k=1}^{n+1}\|x_k\|$. Also, let $\alpha_1, \dots, \alpha_n, \alpha_{n+1}$ be non-negative real numbers; we can suppose $\alpha_1 \ge \alpha_k$ for each k. From

$$\sum_{k=1}^{n+1}\|x_k\| = \Big\|\sum_{k=1}^{n+1}x_k\Big\| \le \|x_1\| + \Big\|\sum_{k=2}^{n+1}x_k\Big\| \le \sum_{k=1}^{n+1}\|x_k\|\,,$$

it follows that $\|\sum_{k=2}^{n+1}x_k\| = \sum_{k=2}^{n+1}\|x_k\|$. Now notice that the identity

$$\sum_{k=1}^{n+1}\alpha_k x_k = \alpha_1\Big[\sum_{k=1}^{n+1}x_k\Big] - \sum_{k=2}^{n+1}(\alpha_1 - \alpha_k)x_k\,,$$

in conjunction with the induction hypothesis, yields

$$\begin{aligned}\Big\|\sum_{k=1}^{n+1}\alpha_k x_k\Big\| &\ge \alpha_1\Big\|\sum_{k=1}^{n+1}x_k\Big\| - \Big\|\sum_{k=2}^{n+1}(\alpha_1-\alpha_k)x_k\Big\|\\ &= \alpha_1\Big[\sum_{k=1}^{n+1}\|x_k\|\Big] - \sum_{k=2}^{n+1}(\alpha_1-\alpha_k)\|x_k\| = \sum_{k=1}^{n+1}\alpha_k\|x_k\|.]\end{aligned}$$

2. This exercise presents an isomorphic version of Lemma 11.4. Let X be a normed space and assume that a vector $u \in X$ and some unit vector v satisfy $\|u + v\| \ge 2 - \epsilon$ for some $\epsilon > 0$. Show that for all $\alpha, \beta \in \mathbb{R}_+$ we have

$$\|\alpha u + \beta v\| \ge (1-\epsilon)(\alpha + \beta)\,.$$

3. Let V be a vector subspace of a normed space X, and assume that there exist a vector $u \in X$ and some $\epsilon > 0$ such that $\|u + v\| \ge 2 - \epsilon$ holds for each $v \in V$ with $\|v\| = 1$. Show that for each $v \in V$ and all scalars λ we also have $\|\lambda u + v\| \ge (1-\epsilon)\big[|\lambda| + \|v\|\big]$.

4. (C.-S. Lin [**202**]) Assume that two sequences $\{u_n\}$ and $\{v_n\}$ in a normed space satisfy $\lim_{n\to\infty}[\|u_n+v_n\| - (\|u_n\|+\|v_n\|)] = 0$. Show that

$$\lim_{n\to\infty}[\|\alpha u_n+\beta v_n\| - (\alpha\|u_n\|+\beta\|v_n\|)] = 0$$

for all $\alpha \geq 0$ and $\beta \geq 0$. [HINT: If $\alpha \geq \beta \geq 0$, then note that

$$\begin{aligned} 0 &\geq \|\alpha u_n+\beta v_n\| - (\alpha\|u_n\|+\beta\|v_n\|) \\ &= \|\alpha u_n+\alpha v_n-(\alpha-\beta)v_n\| - (\alpha\|u_n\|+\beta\|v_n\|) \\ &\geq \alpha\|u_n+v_n\| - (\alpha-\beta)\|v_n\| - (\alpha\|u_n\|+\beta\|v_n\|) \\ &= \alpha\|u_n+v_n\| - \alpha(\|u_n\|+\|v_n\|) \to 0\,.] \end{aligned}$$

5. (C.-S. Lin [**202**]) Generalize the preceding exercise as follows. Assume that m sequences $\{u_n^k\}$ $(k=1,2,\ldots,m)$ in a normed space satisfy

$$\lim_{n\to\infty}\Big[\Big\|\sum_{k=1}^{m} u_n^k\Big\| - \Big(\sum_{k=1}^{m}\|u_n^k\|\Big)\Big] = 0\,.$$

Then show that $\lim_{n\to\infty}[\|\sum_{k=1}^m \alpha_k u_n^k\| - (\sum_{k=1}^m \alpha_k\|u_n^k\|)] = 0$ for any choice of non-negative scalars $\alpha_1,\ldots,\alpha_m$.

6. (C.-S. Lin [**202**]) Generalize Theorem 11.10 by showing that for a uniformly convex Banach space X and two operators $S,T \in \mathcal{L}(X)$ the following statements are equivalent.

(a) $\|I+S+T\| = 1+\|S\|+\|T\|$.

(b) There exists a sequence $\{x_n\}$ of unit vectors such that

$$\lim_{n\to\infty}\big\|(I+S+T)x_n\big\| = 1+\|S\|+\|T\|\,.$$

7. (C.-S. Lin [**202**]) If $T\colon X\to X$ is an invertible operator on a uniformly convex or uniformly smooth Banach space, then show that $\|I-2T\|=1$ implies $\|I-T\|<1$. [HINT: If $\|I-2T\|\in\sigma(I-2T)$, then the operator $I-(I-2T)=2T$ is not invertible, contradicting our assumption that T is invertible. So, $I-2T$ does not satisfy the Daugavet equation, and hence

$$2\|I-T\| = \|2I-2T\| = \|I+(I-2T)\| < 1+\|I-2T\| = 2\,.$$

This implies $\|I-T\|<1$.]

8. Let $T\colon X\to X$ be a non-zero bounded operator on a uniformly convex or uniformly smooth Banach space. Show that $\|T\|I-T$ is non-invertible if and only if T satisfies the Daugavet equation. [HINT: Note that $\|T\|I-T$ is non-invertible if and only if $\|T\|\in\sigma(T)$.]

9. If a continuous operator $T\colon X\to X$ on a uniformly convex or uniformly smooth Banach space satisfies the Daugavet equation, then show that for each $\alpha\geq 0$ the continuous operator $e^{\alpha T}$ satisfies the Daugavet equation and that $\|e^{\alpha T}\| = e^{\alpha\|T\|}$ holds.

10. Let (Ω, Σ, μ) be an arbitrary σ-finite measure space. A bounded operator $T\colon L_2(\mu) \to L_2(\mu)$ is called **symmetric** if

$$\langle Tf, g \rangle = \langle f, Tg \rangle$$

for all $f, g \in L_2(\mu)$, where $\langle \cdot, \cdot \rangle$ denotes the usual inner product in $L_2(\mu)$ defined by $\langle f, g \rangle = \int_\Omega f(\omega)g(\omega)\, d\mu(\omega)$. Establish the following.

(a) Symmetric operators are bounded.

(b) An integral operator T on $L_2(\mu)$ with a kernel $K \in L_2(\mu \times \mu)$ is symmetric if and only if $K(s,t) = K(t,s)$ holds for $\mu \times \mu$-almost all $(s,t) \in \Omega \times \Omega$.

(c) The integral operator $T\colon L_2[0,1] \to L_2[0,1]$, defined by

$$Tx(t) = \int_0^1 \sin(st)x(s)\, ds\,,$$

is symmetric.

(d) Every symmetric operator $T\colon L_2(\mu) \to L_2(\mu)$ satisfies $r(T) = \|T\|$.

(e) Every symmetric positive operator $T\colon L_2(\mu) \to L_2(\mu)$ satisfies the Daugavet equation, i.e., $\|I + T\| = 1 + \|T\|$.

[HINT: For (c) note that $\|Tf\|^2 = \langle Tf, Tf \rangle = \langle T^2 f, f \rangle \le \|T^2\| \cdot \|f\|^2$ implies $\|T\|^2 \le \|T^2\| \le \|T\|^2$ or $\|T^2\| = \|T\|^2$. Since T^k is also a symmetric operator for each k, an inductive argument guarantees that $\|T^{2^n}\| = \|T\|^{2^n}$ for each n. Hence $r(T) = \lim_{n\to\infty} \|T^{2^n}\|^{\frac{1}{2^n}} = \|T\|$. For (d) invoke Theorems 7.9 and 11.10.]

11. Let $\{x_n\}$ and $\{y_n\}$ be two sequences of unit vectors in a uniformly convex Banach space X, and let $\{\phi_n\}$ be a sequence of unit vectors in X^*. If $\phi_n(x_n) \to 1$ and $\phi_n(y_n) \to 1$, then show that $\|x_n - y_n\| \to 0$.

12. (Milman [**242**]) Show that uniformly convex Banach spaces are reflexive.

13. Show that a Banach space X is uniformly smooth if and only if for each $\epsilon > 0$ there exists some $\gamma > 0$ such that for each unit vector u and for each $v \in X$ with $\|v\| \le \gamma$ we have

$$\|u + v\| + \|u - v\| \le 2 + \epsilon\|v\|\,.$$

14. (Šmulian [**308**]) For a Banach space X establish the following.

(a) X is uniformly convex if and only if its dual X^* is uniformly smooth.

(b) X is uniformly smooth if and only if its dual X^* is uniformly convex.

15. Show that a locally uniformly convex Banach space X always satisfies the **Kadets–Klee property**. That is, show that in a locally uniformly convex Banach space a sequence $\{x_n\}$ is norm convergent to some vector x if and only if $x_n \xrightarrow{w} x$ and $\|x_n\| \to \|x\|$.

16. Show that each L_p-space with $1 < p < \infty$ is uniformly convex and uniformly smooth. Also show that the infinite dimensional L_1- and L_∞-spaces are neither uniformly convex nor uniformly smooth.

17. This problem discusses various differentiability properties of the norm. Let x be an arbitrary non-zero vector in a Banach space X. Establish the following.

(a) If $0 < t_1 < t_2$, then for each $v \in X$ we have

$$-\|v\| \le \frac{\|x + t_1 v\| - \|x\|}{t_1} \le \frac{\|x + t_2 v\| - \|x\|}{t_2} \le \|v\| .$$

This shows that the limit

$$\phi_x(v) = \lim_{t \to 0+} \frac{\|x + tv\| - \|x\|}{t}$$

exists and $|\phi_x(v)| \le \|v\|$. Also, $\phi_x(x) = \|x\|$ and $\phi_x(-x) = -\|x\|$.

(b) The function $v \mapsto \phi_x(v)$ from X to $\mathbb{R}$ is sublinear.

(c) A linear functional $x^* \in X^*$ is said to be a **supporting functional** at x if $\|x^*\| = 1$ and $x^*(x) = \|x\|$. We denote by J_x the non-empty collection of all supporting functionals at x. Show that J_x consists of all linear functionals on X that are dominated by $\phi_x(\cdot)$, i.e.,

$$J_x = \big\{x^* \in X^* \colon\ x^*(v) \le \phi_x(v) \ \text{ for all } \ v \in X\big\} . \tag{$\star$}$$

(d) For the function $v \mapsto \phi_x(v)$ the following statements are equivalent.

(i) $\phi_x(\cdot)$ is linear.

(ii) $\phi_x(-v) = -\phi_x(v)$ for each $v \in X$.

(iii) The norm of the Banach space X is **Gateaux differentiable** at x, i.e., $\lim_{t\to 0} \frac{\|x+tv\|-\|x\|}{t}$ exists in $\mathbb{R}$ for each $v \in X$.

(iv) There exists a unique supporting functional at x, that is, the set J_x is a singleton.

(v) $\phi_x(\cdot)$ is linear and is the only supporting functional at x.

(e) The norm of a Banach space X is said to be **Fréchet differentiable** at x if the limit $\lim_{t\to 0} \frac{\|x+tv\|-\|x\|}{t}$ exists uniformly in v in the unit sphere $S_X = \{v \in X : \|v\| = 1\}$. Establish the following.

(α) If the norm is Fréchet differentiable at x, then it is also Gateaux differentiable at x.

(β) The norm is Fréchet differentiable at x if and only if there exists a (necessarily unique) linear functional $x^* \in X^*$ such that: for every $\epsilon > 0$ there is some $\delta > 0$ so that for each $u \in X$ with $\|u\| < \delta$ we have

$$\big|\, \|x + u\| - \|x\| - x^*(u) \,\big| \le \epsilon \|u\| . \tag{$\star\star$}$$

18. Show that a Banach space X is uniformly smooth if and only if its norm is uniformly Fréchet differentiable over the unit sphere of X, that is, the limit

$$\phi_x(y) = \lim_{t \to 0} \frac{\|x + ty\| - \|x\|}{t}$$

exists in $\mathbb{R}$ uniformly for all unit vectors $x, y \in X$.

11.2. The Daugavet Property in AL- and AM-spaces

We begin with a definition that generalizes the Daugavet property for families of operators. It was introduced first in [**166**] and [**167**].

Definition 11.19. *Let X be a (not necessarily closed) vector subspace of a Banach space Y and let $J\colon X \to Y$ denote the natural inclusion operator. We say that the pair (X, Y) has (or satisfies) the* ***Daugavet property*** *for a collection $\mathcal{S}$ of bounded operators from X to Y provided*

$$\|J + T\| = 1 + \|T\| \tag{$\star$}$$

for each $T \in \mathcal{S}$.

If $X = Y$, then J coincides with the identity operator I and ($\star$) simply means that each T in $\mathcal{S}$ satisfies the usual Daugavet equation. The most important families of operators that we will consider are the vector space $\mathcal{K}(X)$ of all compact operators on X, the vector space $\mathcal{W}(X)$ of all weakly compact operators on X, the vector space $\mathcal{F}(X)$ of all finite-rank operators on X, and (when X is a Banach lattice) the vector space I^{d} of all regular operators disjoint from the identity operator or, equivalently, disjoint from all central operators.

In this section we shall study the Daugavet property when X is either an AL- or an AM-space and $Y = X$. Our approach will be based on the lattice structure of the spaces.

It follows from Corollary 11.13 that L_p-spaces with $1 < p < \infty$ do not satisfy the Daugavet property for compact operators—and hence do not for weakly compact operators either. However, as we shall see next, for $p = 1$ and $p = \infty$ the situation is quite different.

Theorem 11.20. *The atomless Dedekind complete AM-spaces with units and the atomless AL-spaces satisfy the Daugavet property for weakly compact operators.*[1]

The proof of this theorem will be a consequence of the next two lemmas. First of all, we wish to point out that neither AM- nor AL-spaces with atoms can satisfy the Daugavet property; see Exercise 2 at the end of this

[1] A few remarks are in order regarding the long history of this theorem. As mentioned in the introduction to the chapter, for compact operators on $C[0,1]$ the theorem was discovered by I. K. Daugavet [**98**]. It was extended to weakly compact operators on $C(\Omega)$-spaces, where Ω is an arbitrary compact topological space without isolated points, by C. Foias and I. Singer [**126**]. For compact operators on $L_1[0,1]$ the theorem was established by G. Ya. Lozanovsky [**219**] in 1966 and reproved in 1981 by V. F. Babenko and S. A. Pichugov [**50**]. Further generalizations and different proofs were obtained by Y. A. Abramovich, C. D. Aliprantis, and O. Burkinshaw [**6**], P. Chauveheid [**88**], U. U. Diallo and P. P. Zabreĭko [**104**], J. R. Holub [**153, 154**], H. Kamowitz [**171**], M. A. Krasnoselsky [**190**], K. D. Schmidt [**297**], and J. Synnatzschke [**312, 313, 314**]. The proof presented here is taken from [**6**].

section. It is also important to keep in mind that if a Banach lattice E is either an AL-space or a Dedekind complete AM-space with unit, then every continuous operator $T\colon E \to E$ is order bounded—and hence, its modulus exists; see Theorem 3.9.

Two classes of operators that satisfy the Daugavet equation are presented in the next lemma.

Lemma 11.21. *Let E be either an AL-space or a Dedekind complete AM-space with unit and let $T\colon E \to E$ be a continuous operator. If T is either*

(1) *positive or*

(2) *disjoint from the identity operator,*

then T satisfies the Daugavet equation.

Proof. Assume first that $E = C(\Omega)$ is a Dedekind complete AM-space with unit. If T is positive, then from Lemma 3.2 it follows that

$$\begin{aligned}\|I+T\| &= \|\mathbf{1}+T\mathbf{1}\| = \sup_{\omega\in\Omega}\big[1+T\mathbf{1}(\omega)\big] = 1+\sup_{\omega\in\Omega} T\mathbf{1}(\omega)\\ &= 1+\|T\mathbf{1}\| = 1+\|T\|\,.\end{aligned}$$

If T is disjoint from the identity operator, then using Theorem 3.9, Corollary 3.10 and the preceding conclusion, we see that

$$\|I+T\| = \|\,|I+T|\,\| = \|\,I+|T|\,\| = 1+\|\,|T|\,\| = 1+\|T\|\,.$$

If E is an AL-space, then E^* is a Dedekind complete AM-space with unit. In addition, the adjoint operator $T^*\colon E^* \to E^*$ is positive if and only if T is positive. Furthermore, if T is disjoint from the identity operator on E, then Theorem 2.28 implies that T^* is also disjoint from the identity operator on E^*. So, the preceding case yields

$$\|I+T\| = \|I^*+T^*\| = 1+\|T^*\| = 1+\|T\|\,,$$

and the proof is finished. ■

Lemma 11.21 is, of course, false for general non-atomic L_p-spaces; see Exercise 6 at the end of the section. The next lemma is what is missing to complete the proof of Theorem 11.20.

Lemma 11.22. *If E is either an atomless AL-space or an atomless Dedekind complete AM-space with unit and $T\colon E \to E$ is a weakly compact operator, then T is disjoint from the identity operator—and hence it satisfies the Daugavet equation.*

Proof. Let $T\colon E \to E$ be a weakly compact operator. Assume first that E is an atomless Dedekind complete AM-space with unit. Then, by Theorem 3.12, we know that the modulus $|T|$ of T is a weakly compact operator.

Now from the inequalities $0 \leq I \wedge |T| \leq I$, $0 \leq I \wedge |T| \leq |T|$ and Theorem 2.40, we see that $I \wedge |T|$ is a weakly compact central operator. But then Lemma 4.18 guarantees that $I \wedge |T| = 0$.

Next, assume that E is an atomless AL-space. Since E has order continuous norm, it follows from Lemma 2.31 that the norm dual E^* is an atomless Dedekind complete AM-space with unit. Moreover, $T^*\colon E^* \to E^*$ is a weakly compact operator and so, by the preceding case, we have $I^* \wedge |T^*| = 0$. Finally, using that the mapping $S \mapsto S^*$ is a lattice isometry (Theorem 2.28), we see that

$$\big(I \wedge |T|\big)^* = I^* \wedge |T|^* = I^* \wedge |T^*| = 0\,,$$

and so $I \wedge |T| = 0$ holds in this case too. ■

We conclude this section with one more interesting property related to the Daugavet equation for operators on AL- and AM-spaces.

Theorem 11.23. *If $T\colon E \to E$ is a continuous operator on an AL- or an AM-space, then either T or $-T$ satisfies the Daugavet equation.*[2]

Proof. Let $T\colon E \to E$ be a bounded operator on an AL- or an AM-space. Since T satisfies the Daugavet equation if and only if its adjoint operator $T^*\colon E^* \to E^*$ satisfies the Daugavet equation and since the norm dual of an AL-space and the second norm dual of an AM-space are both Dedekind complete AM-spaces with units, we can assume without loss of generality that E is a Dedekind complete AM-space with unit. That is, we can assume that $E = C(\Omega)$, where Ω is an extremally disconnected compact Hausdorff space.

From Theorem 3.9, we know that

$$\|T\| = \big\|\,|T|\,\big\| = \|\,|T|(\mathbf{1})\,\| = \sup\big\{(|T|\mathbf{1})(\omega)\colon\ \omega \in \Omega\big\}\,.$$

This implies that for any given $\epsilon > 0$ there exists a non-empty clopen subset V of Ω such that

$$(|T|\mathbf{1})(\omega) > \|T\| - \epsilon \qquad (\star)$$

for all $\omega \in V$. Furthermore, by Theorem 1.17, we have

$$|T|\mathbf{1} = \sup\big\{T(2e - \mathbf{1})\colon\ e \in E \ \text{ and } \ e \wedge (\mathbf{1} - e) = 0\big\}\,.$$

So, there exist $e = \chi_W$, where W is a clopen set, and some $\omega_0 \in V$ such that

$$T(2e - \mathbf{1})(\omega_0) > \|T\| - \epsilon\,. \qquad (\star\star)$$

[2] This theorem has a long history. For $E = C[0,1]$, it was proven first by J. Duncan, C. M. McGregor, J. D. Price, and A. J. White [**116**] and was rediscovered later by J. R. Holub [**153**]. It was generalized and proven in its present form by Y. A. Abramovich [**4**] and K. D. Schmidt [**297**]. The proof presented here is taken from [**4**].

If we put $x = 2e - \mathbf{1} = \chi_W - \chi_{W^c}$, then $x \in C(\Omega)$ and ($\star\star$) can be rewritten as

$$(Tx)(\omega_0) > \|T\| - \epsilon.$$

Clearly, $|x| = \mathbf{1}$ and thus $x(\omega_0) = 1$ or $x(\omega_0) = -1$. If $x(\omega_0) = 1$, then

$$\begin{aligned} \|I+T\| &\geq \|(I+T)(x)\| = \|x+Tx\| \\ &\geq x(\omega_0) + (Tx)(\omega_0) = 1 + (Tx)(\omega_0) \\ &\geq 1 + \|T\| - \epsilon, \end{aligned}$$

and if $x(\omega_0) = -1$, then

$$\|I - T\| = \| - I + T\| \geq -x(\omega_0) + (Tx)(\omega_0) \geq 1 + \|T\| - \epsilon.$$

Thus, we have shown that for each $\epsilon > 0$ we have either

$$\|I+T\| \geq 1 + \|T\| - \epsilon \quad \text{or} \quad \|I-T\| \geq 1 + \|T\| - \epsilon,$$

and our conclusion follows. ■

Exercises

1. Show that a compact Hausdorff space Ω has an isolated point if and only if $C(\Omega)$ has an atom.

2. This exercise shows that a compact operator on an AM- or an AL-space with atoms need not satisfy the Daugavet equation. Let $E = C(\Omega)$, where Ω is a compact Hausdorff space having an isolated point, say ω_0. Consider the multiplication operator $T\colon E \to E$, defined by $Tf = -\chi_{\{\omega_0\}} f$. Establish that:
   (a) T is compact.
   (b) $\|T\| = 1$.
   (c) $\|I+T\| \leq 1 < 1 + \|T\| = 2$.

3. If $J\colon E \to E$ is an arbitrary isometry on an AL-space, then show that $\|J+T\| = 1 + \|T\|$ for each operator $T \in \mathcal{L}(E)$ which is disjoint from J. [HINT: If $x \in E$ satisfies $\|x\| = 1$, then note that

$$\begin{aligned} \|J+T\| &= \|\,|J+T|\,\| = \|\,|J| + |T|\,\| \geq \|\,|J||x| + |T||x|\,\| \\ &= \|\,|J||x|\,\| + \|\,|T||x|\,\| \geq \|Jx\| + \|Tx\| = 1 + \|Tx\|\,.] \end{aligned}$$

4. Take two arbitrary non-zero bounded real measurable functions f and g on $[0,1]$ and consider the rank-one operator $T = f \otimes g$ acting on each $L_p[0,1]$-space. That is, the operator $T\colon L_p[0,1] \to L_p[0,1]$ is defined via the formula

$$Th = (f \otimes g)h = \langle h, f\rangle g = \Big[\int_0^1 h(s)f(s)\,ds\Big] g$$

for each $h \in L_p[0,1]$.

(a) Verify directly (i.e., without using Theorem 11.20) that the operator $T\colon L_\infty[0,1] \to L_\infty[0,1]$ satisfies the Daugavet equation.
(b) Show that $T\colon L_1[0,1] \to L_1[0,1]$ satisfies the Daugavet equation.
(c) If $\langle g, f\rangle = 0$, then show that $T\colon L_p[0,1] \to L_p[0,1]$ does not satisfy the Daugavet equation for any $1 < p < \infty$.

5. If μ is a σ-finite non-atomic measure, then show that every finite-rank operator on $L_p(\mu)$, where $1 \le p < \infty$, is disjoint from the identity operator. (And so, in this case, by Lemma 11.21 every finite-rank operator on $L_1(\mu)$ satisfies the Daugavet equation.) [HINT: Assume that (Ω, Σ, μ) is a non-atomic finite measure space, $f \in L_q^+(\mu)$ and $u \in L_p^+(\mu)$, where $1 \le p < \infty$ and $1 < q \le \infty$ satisfy $\frac{1}{p} + \frac{1}{q} = 1$. To show $I \wedge (f \otimes u) = 0$, it suffices to establish that $I \wedge (f \otimes u)(\mathbf{1}) = 0$.

To verify this, assume that some $g \in L_p^+(\mu)$ satisfies $g \le I \wedge (f \otimes u)(\mathbf{1})$. Fix $\epsilon > 0$. The absolute continuity of the integral guarantees the existence of some $\delta > 0$ such that $\int_A f(t)\,d\mu(t) < \epsilon$ holds for each measurable set A with $\mu(A) < \delta$. Next, choose some n with $\frac{\mu(\Omega)}{n} < \epsilon$ and then select pairwise disjoint measurable sets $A_1, \ldots, A_n$ with $\mu(A_i) = \frac{\mu(\Omega)}{n}$ for each i. Now note that $g \le \chi_{A_i^c} + [\int_{A_i} f(t)\,d\mu(t)]u \le \chi_{A_i^c} + \epsilon u$ for each i. This implies $g \le \epsilon u$. Since $\epsilon > 0$ is arbitrary, the latter shows that $g = 0$ and from this it follows that $I \wedge (f \otimes u) = 0$.]

6. Show that Lemma 11.21 is false for L_p-spaces with $1 < p < \infty$. [HINT: Consider, for instance, the rank-one operator $T\colon L_2[0,1] \to L_2[0,1]$ defined by $Tf = (\int_0^1 \sqrt{2s} f(s)\,ds)\mathbf{1}$, where $\mathbf{1}$ is the constant function one. Then T is positive, compact, disjoint from the identity and (since $\|T\| = 1$ is not an eigenvalue) it fails to satisfy—by Corollary 11.13—the Daugavet equation.]

11.3. The Daugavet Property in Banach Spaces

It is well known that the classical spaces $L_1(\mu)$ and $L_\infty(\mu)$ often play an exceptional role in many problems of functional analysis. In particular, as we have seen in the previous section, they play a special role regarding the Daugavet property. It even seemed quite natural to expect that the Daugavet property for compact or weakly compact operators would characterize the atomless *AM*- and *AL*-spaces. However, as it turns out, this is not the case. The purpose of this section is to present some new classes of Banach spaces different from the atomless *AL*- and *AM*-spaces that also satisfy the Daugavet property for large classes of operators.

If $\{X_\lambda\}_{\lambda\in\Lambda}$ is a family of Banach spaces, then as usual we denote by $(\sum_\lambda \oplus X_\lambda)_1$ (resp. $(\sum_\lambda \oplus X_\lambda)_\infty$) the Banach space consisting of all vectors $x = (x_\lambda)_{\lambda\in\Lambda}$, with $x_\lambda \in X_\lambda$ for each λ and $\|x\|_1 = \sum_\lambda \|x_\lambda\| < \infty$ (resp. $\|x\|_\infty = \sup_\lambda \|x_\lambda\| < \infty$). In the case of a finite set $\Lambda = \{1, 2, \ldots, n\}$, we shall also employ the notation

$$(X_1 \oplus X_2 \oplus \cdots \oplus X_n)_1 \quad \text{and} \quad (X_1 \oplus X_2 \oplus \cdots \oplus X_n)_\infty .$$

The Banach spaces $(\sum_\lambda \oplus X_\lambda)_1$ and $(\sum_\lambda \oplus X_\lambda)_\infty$ are usually referred to as the ℓ_1-**sum** and ℓ_∞-**sum**, respectively, of the family of Banach spaces $\{X_\lambda\}_{\lambda\in\Lambda}$. The direct sums of sequences of Banach spaces were already introduced and discussed in Section 7.2.

For each index $\lambda \in \Lambda$ we denote by $P_\lambda \colon X \to X_\lambda$ the natural projection $x \mapsto P_\lambda(x) = x_\lambda$ from either one of the Banach spaces $X = (\sum_\lambda \oplus X_\lambda)_1$ or $X = (\sum_\lambda \oplus X_\lambda)_\infty$ onto X_λ.

As we shall see later, the introduced Banach spaces satisfy the Daugavet property for a given family of operators if and only if each summand X_λ satisfies the Daugavet property for this family. We start our discussion with several auxiliary lemmas.

Lemma 11.24. *If u and v are arbitrary vectors in a normed space and $\alpha, \beta \in [0,1]$, then $\|\alpha u + \beta v\| \le \max\{\|u+v\|, \|u-v\|\}$.*

Proof. The desired inequality follows immediately from the identity

$$\alpha u + \beta v = \tfrac{\alpha+\beta}{2}(u+v) + \tfrac{\alpha-\beta}{2}(u-v)$$

and the triangle inequality. Notice also that if $0 \le \beta \le \alpha \le 1$, then the above identity implies $\|\alpha u + \beta v\| \le \alpha \max\{\|u+v\|, \|u-v\|\}$. ∎

Corollary 11.25. *If $T\colon X \to Z$ and $S\colon Y \to Z$ are bounded operators between normed spaces, then*

$$\sup\{\|Tx+Sy\|\colon\ \|x\|\le 1\,,\ \|y\|\le 1\} = \sup\{\|Tx+Sy\|\colon\ \|x\|=\|y\|=1\}\,.$$

We continue by introducing the concept of an operator ideal.

Definition 11.26. *Let $\mathcal{B}$ denote the class of all Banach spaces and let $\mathcal{O}$ denote the class of all bounded operators between Banach spaces. An* ***operator ideal*** *$\mathcal{I}$ is a "mapping" $\mathcal{I}\colon \mathcal{B}\times\mathcal{B} \to 2^{\mathcal{O}}$ such that:*

1. *For each pair of Banach spaces X and Y the collection of operators $\mathcal{I}(X,Y)$ is a (not necessarily closed) linear subspace of $\mathcal{L}(X,Y)$ containing all finite-rank operators from X into Y.*
2. *If in a scheme of bounded operators between Banach spaces*

$$V \xrightarrow{S_1} X \xrightarrow{T} Y \xrightarrow{S_2} W$$

we have $T \in \mathcal{I}(X,Y)$, then $S_2TS_1 \in \mathcal{I}(V,W)$.

As usual, we write $\mathcal{I}(X)$ instead of $\mathcal{I}(X,X)$. In the operator ideal terminology we can talk about the operator ideals of compact operators, weakly compact operators, Dunford–Pettis operators, and strictly singular operators. For details regarding operator ideals we refer the reader to [**266**].

Definition 11.27. *A Banach space X is said to satisfy the* ***Daugavet property with respect to an operator ideal*** $\mathcal{I}$ *if each operator in* $\mathcal{I}(X)$ *satisfies the Daugavet equation.*

In the sequel we shall deal with operators acting on either $(X \oplus Y)_\infty$ or $(X \oplus Y)_1$. Recall that the natural projections P_1 and P_2 on either one of these two spaces are defined by $P_1(x, y) = x$ and $P_2(x, y) = y$. The purpose of the next two lemmas is to establish some inequalities that will be used to approximate $\|I_{X \oplus Y} + T\|$ via the norm $\|I_X + S\|$ for an appropriate operator $S\colon X \to X$.

Lemma 11.28. *Let $T\colon (X \oplus Y)_\infty \to (X \oplus Y)_\infty$ be a continuous operator, where X and Y are Banach spaces. Consider the operator*

$$A = P_1 T\colon (X \oplus Y)_\infty \to X\,,$$

and assume that $\|A\| \geq \|T\| - \epsilon$ for some $\epsilon > 0$. Then there exists an operator $S\colon X \to X$ such that:

(1) $\|A\| - \epsilon \leq \|S\| \leq \|A\|$.

(2) $\|I_X + S\| - \epsilon \leq \|I_{X \oplus Y} + T\|$.

(3) *If for some operator ideal $\mathcal{I}$ the operator T belongs to $\mathcal{I}\big((X \oplus Y)_\infty\big)$, then S belongs to $\mathcal{I}(X)$.*

Proof. If we consider the operators $S_1\colon X \to X$ and $S_2\colon Y \to X$ defined by $S_1 x = A(x, 0)$ and $S_2(y) = A(0, y)$, then $A(x, y) = S_1 x + S_2 y$. But then from

$$\|A\| = \sup\big\{\|S_1 x + S_2 y\|\colon\ \|x\| \leq 1 \text{ and } \|y\| \leq 1\big\}$$

and Corollary 11.25, it follows that there exist two unit vectors $x_0 \in X$ and $y_0 \in Y$ such that $\|S_1 x_0 + S_2 y_0\| = \|A(x_0, y_0)\| \geq \|A\| - \epsilon$.

Next, choose a linear functional $\phi \in X^*$ satisfying $\phi(x_0) = \|\phi\| = 1$, and then define the bounded operator $S\colon X \to X$ by $Sx = A(x, \phi(x) y_0)$. This implies $\|S\| \geq \|S x_0\| = \|A(x_0, y_0)\| \geq \|A\| - \epsilon$. On the other hand, we have

$$\begin{aligned} \|S\| &= \sup\big\{\|A(x, \phi(x) y_0)\|\colon\ \|x\| \leq 1\big\} \\ &\leq \sup\big\{\|A(x, y)\|\colon\ \|x\| \leq 1 \text{ and } \|y\| \leq 1\big\} \\ &= \|A\|\,, \end{aligned}$$

and the validity of (1) has been established.

To prove (2), we begin by choosing a unit vector $u \in X$ such that $\|u + Su\| > \|I_X + S\| - \epsilon$. Then we have

$$\begin{aligned} \|I_{X \oplus Y} + T\| &\geq \big\|(I_{X \oplus Y} + T)(u, \phi(u) y_0)\big\| \geq \big\|P_1[(I_{X \oplus Y} + T)(u, \phi(u) y_0)]\,\big\| \\ &\geq \|u + A(u, \phi(u) y_0)\| = \|u + Su\| \\ &> \|I_X + S\| - \epsilon\,. \end{aligned}$$

For (3) consider the bounded operator $R\colon X \to (X \oplus Y)_\infty$, defined via the formula $Rx = \big(x, \phi(x)y_0\big)$, and note that in the scheme of bounded operators

$$X \xrightarrow{R} (X \oplus Y)_\infty \xrightarrow{T} (X \oplus Y)_\infty \xrightarrow{P_1} X\,,$$

we have $S = P_1TR$. So, if $T \in \mathcal{I}\big((X \oplus Y)_\infty\big)$ for some operator ideal $\mathcal{I}$, then it follows from Definition 11.26 that $S \in \mathcal{I}(X)$. ∎

An ℓ_1-version of Lemma 11.28 is as follows.

Lemma 11.29. *Let $T\colon (X \oplus Y)_1 \to (X \oplus Y)_1$ be an operator, where X and Y are Banach spaces, and consider the operator $A\colon X \to X \oplus_1 Y$ defined by $Ax = T(x, 0)$. Assume that $\|A\| \geq \|T\| - \epsilon$ for some $\epsilon > 0$. Then there exists an operator $S\colon X \to X$ such that:*

(1) $\|A\| - \epsilon < \|S\| \leq \|A\|$.

(2) $\|I_X + S\| - \epsilon \leq \|I_{X \oplus Y} + T\|$.

(3) *If T belongs to $\mathcal{I}\big((X \oplus Y)_\infty\big)$ for some operator ideal $\mathcal{I}$, then S belongs to $\mathcal{I}(X)$.*

Proof. Define the operators $A_1 = P_1A\colon X \to X$ and $A_2 = P_2A\colon X \to Y$. Clearly, $\|Ax\| = \|A_1x\| + \|A_2x\|$ for each $x \in X$. Now fix $x_0 \in X$ with $\|x_0\| = 1$ and

$$\|Ax_0\| = \|A_1x_0\| + \|A_2x_0\| \geq \|A\| - \epsilon\,,$$

and then choose some $\psi \in Y^*$ with $\|\psi\| = 1$ and $\psi(A_2x_0) = \|A_2x_0\|$.

Assume first that $A_1x_0 = 0$. In this case, we claim that the rank-one operator $S\colon X \to X$, defined by $Sx = \psi(A_2x)x_0$, satisfies the desired properties. To see this, note that

$$\|A\| - \epsilon < \|Ax_0\| = \|A_2x_0\| = |\psi(A_2x_0)| = \|Sx_0\| \leq \|S\| \leq \|A\|\,.$$

For (2), we have

$$\begin{aligned}
\|I_{X \oplus Y} + T\| &\geq \|(x_0, 0) + T(x_0, 0)\| = \|(x_0 + A_1x_0, A_2x_0)\| = \|(x_0, A_2x_0)\| \\
&= \|x_0\| + \|A_2x_0\| = 1 + \|A_2x_0\| = 1 + \|Ax_0\| \\
&\geq 1 + \|A\| - \epsilon \geq 1 + \|S\| - \epsilon \\
&\geq \|I_X + S\| - \epsilon\,.
\end{aligned}$$

For (3), consider the scheme of bounded operators

$$X \xrightarrow{R_1} (X \oplus Y)_1 \xrightarrow{T} (X \oplus Y)_1 \xrightarrow{R_2} X\,,$$

where $R_1x = (x, 0)$, $R_2(x, y) = \psi^*(P_2(x, y))x_0$, and note that $S = R_2TR_1$.

Now assume $A_1x_0 \neq 0$. In this case, we define the bounded operator $S\colon X \to X$ by $Sx = A_1x + \psi(A_2x)v_0$, where $v_0 = \frac{A_1x_0}{\|A_1x_0\|}$. We claim that

S satisfies the desired properties. Let us verify first (1). For an arbitrary $x \in X$, we have

$$\|Sx\| \leq \|A_1x\| + |\psi(A_2x)| \leq \|A_1x\| + \|A_2x\| = \|Ax\| \leq \|A\|\cdot\|x\|,$$

whence $\|S\| \leq \|A\|$. For the other inequality note that

$$\begin{aligned} \|S\| \geq \|Sx_0\| &= \big\|A_1x_0 + \psi(A_2x_0)v_0\big\| = \big\|A_1x_0 + \|A_2x_0\|v_0\,\big\| \\ &= \|A_1x_0\| + \|A_2x_0\| = \|Ax_0\| \\ &\geq \|A\| - \epsilon. \end{aligned}$$

For (2) consider the operator $I_X + S\colon X \to X$ and choose some $u \in X$ with $\|u\| = 1$ and $\|u + Su\| > \|I_X + S\| - \epsilon$. Now from the definition of S, we see that $u + Su = u + A_1u + \psi(A_2u)v_0$ and so

$$\begin{aligned} \|I_{X\oplus Y} + T\| &\geq \|(I_{X\oplus Y} + T)(u,0)\| = \|u + Au\| = \|u + A_1u\| + \|A_2u\| \\ &\geq \|u + A_1u\| + |\psi(A_2u)| = \|u + A_1u\| + \|\psi(A_2u)v_0\| \\ &\geq \|u + A_1u + \psi(A_2u)v_0\| = \|u + Su\| \\ &> \|I_X + S\| - \epsilon. \end{aligned}$$

Finally, to see (3) in this case, consider the scheme of bounded operators

$$X \xrightarrow{R_1} (X \oplus Y)_1 \xrightarrow{T} (X \oplus Y)_1 \xrightarrow{R_2} X\,,$$

where $R_1x = (x,0)$ and $R_2(x,y) = P_1(x,y) + \psi(P_2T(x,y))v_0$, and note that $S = R_2TR_1$. ■

We are now ready to establish that ℓ_1-sums and ℓ_∞-sums of Banach spaces that satisfy the Daugavet property also satisfy the Daugavet property.

Theorem 11.30 (Wojtaszczyk [**339**]). *Let $\{X_\lambda\}_{\lambda\in\Lambda}$ be a family of Banach spaces such that each X_λ satisfies the Daugavet property with respect to some operator ideal $\mathcal{I}$. Then the Banach spaces $(\sum_\lambda \oplus X_\lambda)_\infty$ and $(\sum_\lambda \oplus X_\lambda)_1$ also satisfy the Daugavet property with respect to this operator ideal $\mathcal{I}$.*

Proof. Assume that $\{X_\lambda\}_{\lambda\in\Lambda}$ is a family of Banach spaces such that each X_λ satisfies the Daugavet property with respect to a given operator ideal $\mathcal{I}$. Put $X = (\sum_\lambda \oplus X_\lambda)_\infty$ and let T belong to $\mathcal{I}\big((\sum_\lambda \oplus X_\lambda)_\infty\big)$. Consider the operators $A_\lambda = P_\lambda T\colon (\sum_\lambda \oplus X_\lambda)_\infty \to X_\lambda$ and note that $\|T\| = \sup_\lambda \|A_\lambda\|$.

Fix $\epsilon > 0$ and an index $\mu \in \Lambda$ for which $\|A_\mu\| > \|T\| - \epsilon$ and let $Y = \big(\sum_{\lambda\in\Lambda\setminus\{\mu\}} X_\lambda\big)_\infty$. Then $T\colon X_\mu \oplus_\infty Y \to X_\mu \oplus_\infty Y$ and, moreover, the operator $A_\mu = P_\mu T\colon X_\mu \oplus_\infty Y \to X_\mu$ satisfies the hypotheses of Lemma 11.28. Therefore, there exists an operator $S \in \mathcal{I}(X_\mu)$ such that $\|S\| \geq \|A_\mu\| - \epsilon$ and $\|I_X + T\| \geq \|I_{X_\mu} + S\| - \epsilon$. Since X_μ satisfies the Daugavet property with respect to the ideal $\mathcal{I}$, it follows that $\|I_{X_\mu} + S\| = 1 + \|S\|$.

Consequently,

$$\|I_X + T\| \geq 1 + \|S\| - \epsilon \geq 1 + \|A_\mu\| - 2\epsilon \geq 1 + \|T\| - 3\epsilon \,.$$

As $\epsilon > 0$ is arbitrary, $\|I_X + T\| \geq 1 + \|T\|$. This shows that T satisfies the Daugavet property.

For the ℓ_1-sum consider an operator $T \in \mathcal{I}\big((\sum_\lambda \oplus X_\lambda)_1\big)$ and define the operators $A_\lambda = T \circ \pi_\lambda \colon X_\lambda \to (\sum_\lambda \oplus X_\lambda)_1$, where π_λ is the natural embedding of X_λ into $(\sum_\lambda \oplus X_\lambda)_1$. Clearly, $\|T\| = \sup_\lambda \|A_\lambda\|$. The remaining arguments for $(\sum_\lambda \oplus X_\lambda)_1$ are similar to those for $(\sum_\lambda \oplus X_\lambda)_\infty$ with the exception that instead of Lemma 11.28 we must invoke Lemma 11.29. ■

Consider now the space $X = (\sum_n \oplus X_n)_\infty$, where $X_n = L_1[0,1]$ for each $n \in \mathbb{N}$. By Theorem 11.20 each X_n satisfies the Daugavet property for weakly compact operators and so, by Theorem 11.30, X also satisfies the Daugavet property for weakly compact operators. At the same time X is not lattice isomorphic to an AL- or to an AM-space; see Exercise 3 at the end of the section. This demonstrates that, indeed, there exist "non-classical" Banach lattices satisfying the Daugavet property. The first examples of such Banach lattices were obtained in [**5**], where finite sums of AL- and AM-spaces were considered.

Exercises

1. Prove Corollary 11.25.
2. (Abramovich [**2**]) Let E be a σ-Dedekind complete Banach lattice and let U_E be its closed unit ball. Consider the set D of all vectors x for which there exists some collection $\{x_1, \ldots, x_n\}$ of pairwise disjoint positive vectors in U_E such that $|x| = x_1 \vee x_2 \vee \cdots \vee x_n$, and let

$$M(E) = \sup\big\{\|x\| \colon \ x \in D\big\} \,.$$

Show that E is lattice isomorphic to an AM-space if and only if $M(E) < \infty$.

3. Let $E_n = L_1[0,1]$ and $F_n = L_\infty[0,1]$ for each n. For the Banach lattices $E = (\sum_{n=1}^\infty \oplus E_n)_\infty$ and $F = (\sum_{n=1}^\infty \oplus F_n)_1$ establish the following.
   (a) Neither E nor F is lattice isomorphic to an AL- or an AM-space.
   (b) Neither $(E \oplus F)_1$ nor $(E \oplus F)_\infty$ is lattice isomorphic to an AL- or an AM-space.
   (c) The Banach spaces $(E \oplus F)_1$ and $(E \oplus F)_\infty$ satisfy the Daugavet property for weakly compact operators.

11.4. The Daugavet Property in $C(\Omega)$-spaces

In this section Ω will denote a compact Hausdorff space. We will prove here several results on the Daugavet equation in $C(\Omega)$-spaces. These results either provide new classes of operators satisfying the Daugavet equation or they illustrate various techniques of establishing the validity of the Daugavet equation.

We start by describing a new class of operators satisfying the Daugavet equation that was found by J. R. Holub [**155**]. In this work, he proved that every continuous operator on $C[0,1]$ that factors[3] through c_0 satisfies the Daugavet equation. Later S. I. Ansari [**37**] extended this result to arbitrary non-atomic $C(\Omega)$-spaces.

One of the major order related problems in dealing with operators on a $C(\Omega)$-space for an arbitrary compact space Ω (as opposed to a Stonean space Ω) is caused by the fact that in general the ordered Banach space $\mathcal{L}(C(\Omega))$ is not a Dedekind complete Riesz space. Even worse, it fails to be a Riesz space, and so the modulus of an operator does not necessarily exist. Therefore, one cannot expect to obtain the validity of the Daugavet equation for a given class of operators on a $C(\Omega)$-space by simply proving that any operator from this class is disjoint from the identity operator. A way to bypass this obstacle is of interest in itself and is described next.

Definition 11.31. *Let $S, T\colon C(\Omega) \to C(\Omega)$ be two continuous operators. We shall say that S and T are* ***essentially disjoint*** *if there exists a Stonean space Q and a lattice isomorphism $J\colon C(\Omega) \to C(Q)$ such that:*

(1) *$J(\mathbf{1}_\Omega) = \mathbf{1}_Q$, where $\mathbf{1}_\Omega$ and $\mathbf{1}_Q$ denote the constant one functions in $C(\Omega)$ and $C(Q)$, respectively.*

(2) *The operators JT and JS are disjoint in $\mathcal{L}_r(C(\Omega), C(Q))$.*

(3) *$\|JT\| = \|T\|$ and $\|JS\| = \|S\|$.*

If the mapping J above is merely a lattice homomorphism (rather than a lattice isomorphism), then we shall call T and S ***almost essentially disjoint*** *operators.*

Some remarks are in order. First, notice that—since $C(Q)$ is a Dedekind complete AM-space with unit—each bounded operator $R\colon C(\Omega) \to C(Q)$ is regular and that $\|R\| = \|R\|_r$. Furthermore, note that J is a Markov operator, that condition (1) implies $\|J\| = 1$, and that in view of (2) we have $|JT + JS| = |JT| + |JS|$.

[3] Recall that a continuous operator $T\colon X \to Y$ between Banach spaces **factors through** a Banach space Z if there exist two continuous operators $R\colon X \to Z$ and $S\colon Z \to Y$ such that $T = SR$.

Lemma 11.32. *If Ω is Stonean, then for bounded operators on $C(\Omega)$:*

(a) *The notions of essential disjointness and disjointness coincide.*

(b) *Almost essential disjointness does not imply essential disjointness.*

Proof. (a) Assume that two operators $S, T \in \mathcal{L}\big(C(\Omega)\big)$ satisfy $S \wedge T = 0$. If we take $Q = \Omega$ and let $J\colon C(\Omega) \to C(Q)$ be the identity operator, then clearly J is a lattice isomorphism such that $J(\mathbf{1}_\Omega) = \mathbf{1}_Q$, $\|JT\| = \|T\|$, $\|JS\| = \|S\|$, and JT is disjoint from JS in $\mathcal{L}_r\big(C(\Omega), C(Q)\big)$.

For the converse, assume that S and T are essentially disjoint. That is, there exist a Stonean space Q and a lattice isomorphism $J\colon C(\Omega) \to C(Q)$ satisfying the properties of the definition of essential disjointness. Since Ω is Stonean, the operator $S \wedge T$ exists in $\mathcal{L}(C(\Omega))$. And now from the inequalities $0 \le J(S \wedge T) \le (JS) \wedge (JT) = 0$, it follows that $J(S \wedge T) = 0$. This implies $S \wedge T = 0$ in $\mathcal{L}\big(C(\Omega)\big)$.

(b) To show that the essential disjointness may be stronger than the almost essential disjointness let us fix three arbitrary disjoint non-empty clopen subsets A, B_1, B_2 of Ω such that $\Omega = A \cup B_1 \cup B_2$. (We assume here that Ω contains more than two points and so such A, B_1, B_2 do exist.) Let $E = C(B_1 \cup B_2)$ and let $J\colon C(\Omega) \to E$ be the natural restriction operator, i.e., for $f \in C(\Omega)$ we let $Jf = f|_{B_1 \cup B_2}$. Finally, define $S, T\colon C(\Omega) \to C(\Omega)$ by $Tf = f\chi_{A \cup B_1}$ and $Sf = f\chi_{A \cup B_2}$. It is easy to see that T and S are not disjoint in $\mathcal{L}_r(C(\Omega))$. Indeed, using Theorem 1.17, we see that

$$\begin{aligned}
(T \wedge S)\mathbf{1}_\Omega &= \inf\big\{T\chi_V + S\chi_{V^c}\colon\ V \text{ is a clopen subset of } \Omega\big\}\\
&= \inf\big\{\chi_{V \cap (A \cup B_1)} + \chi_{V^c \cap (A \cup B_2)}\colon\ V \text{ clopen}\big\}\\
&\ge \chi_A > 0\,.
\end{aligned}$$

Thus, we have established that $T \wedge S > 0$. Finally, a direct verification shows that the operators JT and JS are disjoint in $\mathcal{L}_r(C(\Omega), E)$ and also that $\|JT\| = \|JS\| = \|T\| = \|S\| = 1$. ■

As shown above, the almost essential disjointness is a rather weak property, but nevertheless, it is sufficient to imply an analogue of Lemma 11.21.

Lemma 11.33. *If a bounded operator T on $C(\Omega)$ is almost essentially disjoint from the identity operator, then T satisfies the Daugavet equation.*

Proof. Let Q be a Stonean space and $J\colon C(\Omega) \to C(Q)$ be a lattice isomorphism satisfying the properties of Definition 11.31 for the operators I and T. Clearly, $\|J(I+T)\| \le \|I+T\|$ as $\|J\| \le 1$. On the other hand, we have

$$\begin{aligned}
\|JI + JT\| &= \|JI + JT\|_r = \big\||JI + JT|(\mathbf{1}_\Omega)\big\| = \big\|\mathbf{1}_Q + |JT|(\mathbf{1}_\Omega)\big\|\\
&= 1 + \big\||JT|(\mathbf{1}_\Omega)\big\| = 1 + \big\||JT|\big\| = 1 + \|T\|\,.
\end{aligned}$$

Hence, $\|I+T\| \geq 1+\|T\|$, and we are done. ■

If μ is a regular Borel measure on Ω, then by J_μ we will denote the natural embedding of $C(\Omega)$ into $L_\infty(\mu)$. That is, for each $x \in C(\Omega)$ its image in $L_\infty(\mu)$ is denoted by $J_\mu x$. Obviously, J_μ is a lattice homomorphism and $\|J_\mu\| = 1$ (provided $\mu \neq 0$).

We will need below the well-known fact (due to W. Rudin [**286**] and A. Pelczynski and Z. Semadeni [**263**]) that a perfect compact topological space Ω admits a non-atomic measure. For other proofs see [**300**, Theorem 19.7.6], [**24**, Theorem 5.3] or [**25**, Theorem 10.20]. As usual, a measure μ is called **non-atomic** if $\mu(\{\omega\}) = 0$ for each $\omega \in \Omega$.

Theorem 11.34. *If Ω has no isolated points, then it admits a non-atomic regular Borel probability measure.*

From now on, until the end of this section, if μ is a measure on Ω, then we shall denote the μ-essential supremum of a function $f \in L_\infty(\mu)$ by $\|f\|_{\mu,\infty}$.

Lemma 11.35. *If Ω has no isolated points and $T\colon C(\Omega) \to C(\Omega)$ is a bounded operator, then there exists a non-atomic regular Borel probability measure μ on Ω such that $\|J_\mu T\| = \|T\|$.*

Proof. For each k choose a function $x_k \in C(\Omega)$ such that $\|x_k\|_\infty = 1$ and $\|Tx_k\|_\infty > \|T\| - \frac{1}{k}$. Put $U_k = \big\{\omega \in \Omega\colon\ |Tx_k(\omega)| > \|Tx_k\|_\infty - \frac{1}{k}\big\}$, and note that each U_k is a non-empty open set.

Since Ω does not have isolated points, it follows that each $\overline{U}_k$ does not have isolated points either; see Exercise 4 at the end of this section. Therefore, by Theorem 11.34 there exists a non-atomic regular Borel probability measure μ_k on $\overline{U}_k$. We extend μ_k to a non-atomic measure on all of Ω by letting $\mu_k(\Omega \setminus \overline{U}_k) = 0$. Clearly, $\|Tx_k\|_{\mu_k,\infty} > \|T\| - \frac{1}{k}$.

Now let $\mu = \sum_{k=1}^\infty 2^{-k}\mu_k$. Then μ is a non-atomic regular Borel probability measure on Ω and $\|Tx_k\|_{\mu,\infty} \geq \|Tx_k\|_{\mu_k,\infty} > \|T\| - \frac{1}{k}$ holds for each k. This implies $\|J_\mu T\| \geq \|T\|$, and consequently $\|J_\mu T\| = \|T\|$. ■

If Ω is a metrizable compact space without isolated points, then there exists a non-atomic regular Borel probability measure on Ω such that the embedding $J_\mu : C(\Omega) \to L_\infty(\mu)$ is a lattice isometry; see Exercise 5 at the end of this section.

Corollary 11.36. *If Ω has no isolated points, then for every sequence $\{T_n\}$ of bounded operators on $C(\Omega)$ there exists a non-atomic regular Borel probability measure μ on Ω such that $\|J_\mu T_n\| = \|T_n\|$ for each n.*

Proof. By Lemma 11.35, for each n there exists a non-atomic regular Borel probability measure μ_n on Ω such that $\|J_{\mu_n} T_n\| = \|T_n\|$. Then the measure

$\mu = \sum_{n=1}^{\infty} 2^{-n}\mu_n$ is a non-atomic regular Borel probability measure on Ω with the desired property. ■

Lemma 11.37. *If μ is a regular Borel measure on Ω, then for each linear operator $S\colon C(\Omega) \to L_\infty(\mu)$ satisfying $0 \le S \le J_\mu$ there exists a non-negative function $h \in L_\infty(\mu)$ such that $Sx = hx$ for each $x \in C(\Omega)$.*

Proof. The image $G = J_\mu(C(\Omega))$ is a Riesz subspace of the Dedekind complete Banach lattice $L_\infty(\mu)$. We define an operator $R\colon G \to L_\infty(\mu)$ by letting $R(g) = Sx$, where $x \in C(\Omega)$ and $J_\mu x = g$. It is easy to see that R is well defined and satisfies $0 \le Rg \le g$ for each $g \in G^+$. Therefore, by [**30**, Theorem 8.15] the operator R can be extended to a positive central operator on $L_\infty(\mu)$, which we shall denote by R again. However, as we already know from Theorem 3.36, each central operator on $L_\infty(\mu)$ is a multiplication operator. That is, there exists a function $h \in L_\infty(\mu)$ such that $Rg = hg$ for each $g \in L_\infty(\mu)$. Since R is positive, the function h is non-negative. Finally, for each $x \in C(\Omega)$ we have that $Sx = RJ_\mu x = Rx = hx$. ■

Recall that the series $\sum_{n=1}^{\infty} x_n$ of elements of a (real or complex) Banach space X is said to be **weakly unconditionally Cauchy** if for each $x^* \in X^*$ the series of real numbers $\sum_{n=1}^{\infty} |x^*(x_n)|$ converges in $\mathbb{R}$. Clearly, every subseries of a weakly unconditionally Cauchy series is also weakly unconditionally Cauchy.

If $\sum_{n=1}^{\infty} h_n$ is a weakly unconditionally Cauchy series in $C(\Omega)$, then by using the functional $x^* = \delta_\omega$, we obtain that $\sum_{n=1}^{\infty} |h_n(\omega)| < \infty$ for each $\omega \in \Omega$. The next result informs us that not only each weakly unconditionally Cauchy series in $C(\Omega)$ converges absolutely at each point of Ω, but also that this convergence can be made uniform on subsets of positive measure.

Lemma 11.38. *Let $\sum_{n=1}^{\infty} h_n$ be a weakly unconditionally Cauchy series in $C(\Omega)$. If μ is a regular Borel measure on Ω, then for each $\epsilon > 0$ there exists a closed subset B of Ω such that $\mu(\Omega \setminus B) < \epsilon$ and the series $\sum_{n=1}^{\infty} h_n$ converges uniformly on B.*

Proof. Since $\sum_{n=1}^{\infty} |h_n(\omega)| < \infty$ holds for each $\omega \in \Omega$, the conclusion follows from Egorov's theorem and the regularity of the measure μ. ■

We continue with a factorization property of operators on $C(\Omega)$-spaces.

Lemma 11.39 (Holub [**155**]). *An operator $T \in \mathcal{L}(C(\Omega))$ factors through c_0 if and only if T admits a representation of the form*

$$T = \sum_{n=1}^{\infty} \mu_n \otimes h_n \,, \tag{$\star$}$$

where:

(a) *The sequence $\{\mu_n\} \subseteq C(\Omega)^*$ is weak* convergent to zero.*

(b) *The series $\sum_{n=1}^{\infty} h_n$ is weakly unconditionally Cauchy in $C(\Omega)$.*

(c) *The series in $(\star)$ converges in the strong operator topology, that is, $\sum_{n=1}^{\infty} \mu_n(x) h_n$ is norm convergent in $C(\Omega)$ for each $x \in C(\Omega)$.*

Proof. Let $\{e_n\}$ denote the standard basis in c_0 and $\{e_n^*\}$ denote its biorthogonal sequence. We begin by establishing that each operator T in $\mathcal{L}(C(\Omega))$ factoring through c_0 admits a $(\star)$-representation. Therefore, assume that there exist two bounded operators $C(\Omega) \xrightarrow{R} c_0 \xrightarrow{S} C(\Omega)$ such that $T = SR$. Let $h_n = Se_n$ and $\mu_n = R^* e_n^*$. As $e_n^* \to 0$ in the topology $\sigma(\ell_1, c_0)$, it follows that $\mu_n = R^* e_n^* \to 0$ in the topology $\sigma(C(\Omega)^*, C(\Omega))$.

To show that $\sum_{n=1}^{\infty} h_n$ is a weakly unconditionally Cauchy series, pick $x^* \in C(\Omega)^*$ and consider the vector $S^* x^* = (\lambda_1, \lambda_2, \ldots) \in c_0^* = \ell_1$. Then

$$\sum_{n=1}^{\infty} |x^*(h_n)| = \sum_{n=1}^{\infty} |x^*(Se_n)| = \sum_{n=1}^{\infty} |(S^* x^*)(e_n)| = \sum_{n=1}^{\infty} |\lambda_n| < \infty \,.$$

To show that the series in $(\star)$ converges in the strong operator topology, fix any $x \in C(\Omega)$, and let $Rx = (\alpha_1, \alpha_2, \ldots) = \sum_{n=1}^{\infty} \alpha_n e_n = \sum_{n=1}^{\infty} \langle Rx, e_n^* \rangle e_n$. It follows that

$$Tx = S(Rx) = S\Big(\sum_{n=1}^{\infty} \langle Rx, e_n^* \rangle e_n\Big) = \sum_{n=1}^{\infty} \langle x, R^* e_n^* \rangle Se_n = \sum_{n=1}^{\infty} \mu_n(x) h_n \,.$$

For the converse, assume that an operator T can be represented in the form $(\star)$. Let $Rx = \sum_{n=1}^{\infty} \mu_n(x) e_n$. Since $\mu_n \to 0$ in the topology $\sigma(C(\Omega)^*, C(\Omega))$, the mapping R is a continuous operator from $C(\Omega)$ to c_0. Next define $S\colon c_0 \to C(\Omega)$ by $S\lambda = \sum_{n=1}^{\infty} \lambda_n h_n$. This series converges in $C(\Omega)$, since $\sum_{n=1}^{\infty} h_n$ is weakly unconditionally Cauchy; see Exercise 3 at the end of the section. It remains to note that the convergence condition (c) implies that $T = SR$. ■

Theorem 11.40 (Ansari [**37**]). *If Ω has no isolated points, then every continuous operator on $C(\Omega)$ factoring through c_0 is almost essentially disjoint from the identity operator.*

Proof. Let $T\colon C(\Omega) \to C(\Omega)$ be a bounded operator that factors through c_0. By Lemma 11.35 there exists a non-atomic Borel measure μ on Ω such that $\|J_\mu T\| = \|T\|$ and $\|J_\mu\| = 1$, where $J_\mu\colon C(\Omega) \to L_\infty(\mu)$ is the natural embedding. Lemma 11.39 guarantees that

$$T = \sum_{n=1}^{\infty} \mu_n \otimes h_n$$

for some weak* null sequence $\{\mu_n\}$ in $C(\Omega)^*$ and a weakly unconditionally Cauchy series $\sum_{n=1}^{\infty} h_n$ in $C(\Omega)$. Let $\gamma = \sup_{n\in\mathbb{N}} \max\{\|\mu_n\|, \|h_n\|\}$. The validity of our theorem will be established if we show that the operator $S = |J_\mu T| \wedge J_\mu \in \mathcal{L}_r\big(C(\Omega), L_\infty(\mu)\big)$ is trivial, that is, $S = 0$.

To this end, assume contrary to our claim that $S \neq 0$. Since $0 \le S \le J_\mu$, by Lemma 11.37 there exists a non-negative function $h \in L_\infty(\mu)$ such that $Sx = hx$ for each $x \in C(\Omega)$. As $S \neq 0$, the multiplier h is non-zero, and hence for some $\epsilon > 0$ we have $\mu(A) > 0$, where $A = \{\omega \in \Omega : h(\omega) > \epsilon\}$. By Lemma 11.38, we can find an integer k and a closed subset B of A with $\mu(B) > 0$ such that

$$\sup_{\omega\in B} \sum_{n=k+1}^{\infty} |h_n(\omega)| < \tfrac{\epsilon}{3\gamma}\,.$$

As μ is non-atomic and each μ_n is regular, we can find an open set U such that $\mu(B \cap U) > 0$ and $|\mu_n|(U) < \frac{\epsilon}{3k\gamma}$ for each $1 \le n \le k$. Using the regularity of μ, we can assume without loss of generality that $B \subseteq U$ (otherwise, we can replace B by a compact subset of $B \cap U$ of positive measure). Also pick a function $x_0 \in C(\Omega)$ such that $x_0|_B = 1$, $x_0|_{\Omega\setminus U} = 0$, and $0 \le x_0 \le \mathbf{1}$.

Notice that $Sx_0 = hx_0$, and so $Sx_0(\omega) > \epsilon$ for each $\omega \in B$. On the other hand, we have

$$\begin{aligned} Sx_0 &= |J_\mu T| \wedge J_\mu(x_0) \le |J_\mu T|(x_0) \\ &= \sup\big\{J_\mu T x\colon\ x \in C(\Omega) \text{ and } |x| \le x_0\big\}, \end{aligned}$$

where the supremum is taken in $L_\infty(\mu)$. Let us denote this supremum by y_0. We claim that $y_0 = G$, where the function $G\colon \Omega \to [0,\infty]$ is defined by $G(\omega) = \sup\{Tx(\omega)\colon\ x \in C(\Omega) \text{ and } |x| \le x_0\}$. Indeed, each Tx above is a continuous function and $\|Tx\| \le \|T\|$. Therefore, G is a bounded lower semicontinuous function, and hence $G \in L_\infty(\mu)$. This implies $G = y_0$, and so $Sx_0 \leq G$. Now using the representation for T, for each $\omega \in B$ we get

$$\begin{aligned}
\epsilon < Sx_0(\omega) &\leq G(\omega) = \sup_{|x| \leq x_0} \Big[\sum_{n=1}^{\infty} \mu_n(x) h_n(\omega) \Big] \\
&= \sup_{|x| \leq x_0} \Big[\sum_{n=1}^{k} \mu_n(x) h_n(\omega) + \sum_{n=k+1}^{\infty} \mu_n(x) h_n(\omega) \Big] \\
&\leq \sum_{n=1}^{k} |\mu_n|(x_0) |h_n(\omega)| + \tfrac{\epsilon}{3} \\
&\leq k\gamma \tfrac{\epsilon}{3k\gamma} + \tfrac{\epsilon}{3} = \tfrac{2}{3}\epsilon,
\end{aligned}$$

which is a contradiction. This completes the proof of the theorem. ■

Corollary 11.41 (Ansari–Holub). *If Ω has no isolated points, then every operator on $C(\Omega)$ that factors through c_0 satisfies the Daugavet equation.*

Proof. If an operator $T \in \mathcal{L}(C(\Omega))$ factors through c_0, then T is almost essentially disjoint from the identity operator by Theorem 11.40. So, by Lemma 11.33, the operator T satisfies the Daugavet equation. ■

Note that there exist non-weakly compact operators on $C(\Omega)$-spaces that factor through c_0; see Exercise 1 at the end of the section. Thus, Corollary 11.41 describes a new class of operators satisfying the Daugavet equation.

We proceed with a few comments regarding the role of the embedding $J_\mu \colon C(\Omega) \to L_\infty(\mu)$ for the method discussed above. Recall that $C(\Omega)^{**}$ is a Dedekind complete Banach lattice and the operator of the canonical embedding $j \colon C(\Omega) \to C(\Omega)^{**}$ is a lattice isometry. Consequently, for each bounded operator T on $C(\Omega)$ the operator jT has a modulus in $\mathcal{L}(C(\Omega), C(\Omega)^{**})$. Moreover, repeating verbatim the proof of Lemma 11.33, one can see that if jT is disjoint from jI, where I is the identity operator on $C(\Omega)$, then T satisfies the Daugavet equation. Therefore, it is natural to ask if it is possible to use the operator j instead of an appropriate isomorphism J_μ. Exercise 9 at the end of this section shows that the answer to this question is negative.

The rest of this section will be devoted to a description of one more approach to the Daugavet equation for operators on $C(\Omega)$-spaces. This will allow us to describe two more classes of operators on $C(\Omega)$-spaces that satisfy the Daugavet equation. These classes of operators have been introduced by L. Weis and D. Werner in **[330]** who used them to generalize Corollary 11.41 by showing that c_0 can be replaced by any Banach space with a separable dual. As we shall see, the proof is surprisingly simple and independent of that of Corollary 11.41.

Recall that each bounded operator $T\colon C(\Omega) \to C(\Omega)$ generates the so-called **representing kernel**, i.e., a family of signed regular Borel measures $\{\mu_\omega\}_{\omega\in\Omega}$ on Ω defined by $\mu_\omega = T^*\delta_\omega$ for each $\omega \in \Omega$. From the estimate $\|\mu_\omega\| = \|T^*\delta_\omega\| \le \|T^*\| \cdot \|\delta_\omega\| = \|T\|$, it follows that the family $\{\mu_\omega\}_{\omega\in\Omega}$ is a norm bounded subset of the norm dual $C(\Omega)^*$.

The following characterization of the operators satisfying the Daugavet equation via the representing kernel is due to D. Werner [**332**]. The proof below is similar in spirit to the original proof of I. K. Daugavet in [**98**]. A more formal algebraic proof is given in [**332**].

Lemma 11.42 (Werner). *If the representing kernel $\{\mu_\omega\}_{\omega\in\Omega}$ of a continuous operator $T : C(\Omega) \to C(\Omega)$ satisfies*

$$\inf_{\omega\in G} |\mu_\omega(\{\omega\})| = 0 \qquad (\dagger)$$

for each non-empty open subset G of Ω, then T satisfies the Daugavet equation.

Proof. Let $T\colon C(\Omega) \to C(\Omega)$ be a bounded operator. Fix $\epsilon > 0$ and choose some $x \in C(\Omega)$ such that $\|x\|_\infty = 1$ and $\|Tx\|_\infty > \|T\| - \epsilon$. The set $G = \big\{\omega \in \Omega\colon\ |Tx(\omega)| > \|T\| - \epsilon\big\}$ is non-empty and open and hence, by ($\dagger$), we can find a point $t_0 \in G$ such that $\big|\mu_{t_0}(\{t_0\})\big| = |\mu_{t_0}|(\{t_0\}) < \epsilon$. By the regularity of the measure $|\mu_{t_0}|$ there exists an open neighborhood V of the point t_0 such that $|\mu_{t_0}|(V) < \epsilon$. We can assume that $V \subseteq G$. Now for any function $y \in C(\Omega)$ whose support lies in V we have

$$\begin{aligned} \big|Ty(t_0)\big| &= \big|\langle Ty, \delta_{t_0}\rangle\big| = \big|\langle y, T^*\delta_{t_0}\rangle\big| = \big|\langle y, \mu_{t_0}\rangle\big| \\ &\le \langle |y|, |\mu_{t_0}|\rangle \le \|y\|_\infty \cdot |\mu_{t_0}|(V) < \epsilon\|y\|_\infty . \end{aligned}$$

Consequently, for each $y \in C(\Omega)$ with $\operatorname{Supp} y \subseteq V$ we have

$$\begin{aligned} \big\|(x+y) + T(x+y)\big\|_\infty &\ge (x+y)(t_0) + Tx(t_0) + Ty(t_0) \\ &\ge (x+y)(t_0) + \|T\| - \epsilon - \epsilon\|y\|_\infty . \end{aligned}$$

Next pick a function $z \in C(\Omega)$ such that $0 \le z \le \mathbf{1}$, $z(t_0) = 1$ and $z = 0$ on V^{c}. Consider the function $y = (\mathbf{1} - x)z \in C(\Omega)^+$ and note that y has support in V and that $\|y\|_\infty \le 2$. From $\|x\|_\infty = 1$, it follows that $-\mathbf{1} \le x \le x + y = x + (\mathbf{1} - x)z \le \mathbf{1}$, and so $\|x + y\|_\infty \le 1$. Hence, from

$$1 \ge \|x+y\|_\infty \ge x(t_0) + y(t_0) = x(t_0) + 1 - x(t_0) = 1 ,$$

it follows that $x(t_0) + y(t_0) = 1$. Finally, note that

$$\begin{aligned} \|I + T\| &\ge \|(I+T)(x+y)\| \ge (x+y)(t_0) + \|T\| - \epsilon - \epsilon\|y\|_\infty \\ &\ge 1 + \|T\| - 3\epsilon . \end{aligned}$$

Since $\epsilon > 0$ is arbitrary, this implies $\|I + T\| \ge 1 + \|T\|$. That is, T satisfies the Daugavet equation. ∎

Theorem 11.43 (Weis–Werner [**330**]). *Assume that Ω is a compact Hausdorff space without isolated points and let $T \in \mathcal{L}(C(\Omega))$ satisfy one of the following two conditions:*

(1) *The adjoint of T has a separable range.*

(2) *T factors through a Banach space with a separable dual.*

Then T satisfies the Daugavet property.

Proof. (1) Let $\{\mu_\omega\}_{\omega\in\Omega}$ be the representing kernel of T. Since the range of T^* is separable, there exists an at most countable subset Ω_0 of Ω such that the collection $\{\mu_\omega\colon\ \omega \in \Omega_0\}$ is norm dense in $\{\mu_\omega\colon\ \omega \in \Omega\}$. Therefore, it is easy to see that the set $A = \big\{\omega \in \Omega\colon\ \mu_\omega(\{\omega\}) \neq 0\big\}$ coincides with the set $\big\{\omega \in \Omega_0\colon \mu_\omega(\{\omega\}) \neq 0\big\}$. Consequently, A is at most countable.

Now for each $a \in A$ let $\Omega_a = \{\omega \in \Omega\colon\ \omega \neq a\} = \{a\}^{\mathrm{c}}$. Since Ω has no isolated points, it follows that each Ω_a is an open dense subset of Ω. From the Baire Category Theorem (see, for instance [**247**, p. 294]), we conclude that the set

$$\bigcap_{a\in A}\Omega_a = A^{\mathrm{c}} = \big\{\omega \in \Omega\colon\ \mu_\omega\big(\{\omega\}\big) = 0\big\}$$

is also dense in Ω. This implies that condition (†) of Lemma 11.42 is satisfied, and so T satisfies the Daugavet equation.

(2) Let $T \in \mathcal{L}\big(C(\Omega)\big)$ and assume that there exist a Banach space Z with a separable dual and bounded operators $C(\Omega) \xrightarrow{R} Z \xrightarrow{S} C(\Omega)$ satisfying $T = SR$. From $T^* = R^*S^*$ and the fact that Z^* is separable, it follows that the range of T^* is necessarily separable. Therefore, by the previous case, T satisfies the Daugavet equation. ■

Exercises 7 and 8 at the end of this section indicate how one can use Theorem 11.43 to obtain an alternative proof of the fact that each (weakly) compact operator on an atomless $C(\Omega)$-space satisfies the Daugavet equation (Theorem 11.20).

We conclude this section by one more elegant application of Theorem 11.43. Recall that a bounded operator $T\colon X \to Y$ between two Banach spaces **fixes a copy** of a Banach space Z if there exists a subspace X_0 of X such that X_0 is isomorphic to Z and $T\colon X_0 \to T(X_0)$ is an isomorphism.

It should be clear that if Ω is uncountable and a bounded operator $T\colon C(\Omega) \to C(\Omega)$ fixes a copy of $C(\Omega)$, then the range of T^* cannot be separable.[4] Moreover, according to a deep theorem of H. P. Rosenthal [**281**], for an uncountable metrizable Ω the converse statement is also true, i.e., if

[4] Notice that if $\omega_1 \neq \omega_2$, then $\|\delta_{\omega_1} - \delta_{\omega_2}\| = 1$. Therefore, if Ω is uncountable, then the norm dual $C(\Omega)^*$ of $C(\Omega)$ is not separable.

for some bounded operator $T\colon C(\Omega) \to C(\Omega)$ the adjoint operator T^* has a non-separable range, then T fixes a copy of $C(\Omega)$. This and Theorem 11.43 yield the following nice result which is also contained in [**330**].

Theorem 11.44 (Weis–Werner). *Assume that Ω is a metrizable compact space without isolated points. If an operator $T \in \mathcal{L}(C(\Omega))$ does not fix a copy of $C(\Omega)$, then T satisfies the Daugavet equation.*

The natural generalizations of the $C(\Omega)$-spaces are the uniform algebras, and it does not come as a surprise that some of them satisfy the Daugavet property for the same collections of operators as $C(\Omega)$-spaces. This was established by Wojtaszczyk in [**339**]. Since the theory of Banach algebras is beyond the scope of this book, we refer the reader to this paper for details.

Exercises

1. Give an example of a continuous operator on $C[0,1]$ that factors through c_0 but that is not weakly compact. [HINT: By a well-known theorem of A. Sobczyk [**309**], we know that every closed subspace X of $C[0,1]$ that is isomorphic to c_0 is complemented in $C[0,1]$. If $P\colon C[0,1] \to C[0,1]$ is a continuous projection with range X, then P is a non-weakly compact operator that factors through c_0.]

2. For a series $\sum_{n=1}^\infty x_n$ in a (real or complex) Banach space the following statements are equivalent.
(a) The series $\sum_{n=1}^\infty x_n$ is weakly unconditionally Cauchy.
(b) The supremum $\sup_{|\epsilon_n|\le 1} \sup_{m\in\mathbb{N}} \left\|\sum_{n=1}^m \epsilon_n x_n\right\|$ is finite.
(c) The supremum $\sup_{|\epsilon_n|=1} \sup_{m\in\mathbb{N}} \left\|\sum_{n=1}^m \epsilon_n x_n\right\|$ is finite.

3. Prove that a series $\sum_{n=1}^\infty x_n$ in a Banach space X is weakly unconditionally Cauchy if and only if for each $\lambda = (\lambda_1, \lambda_2, \ldots) \in c_0$ the series $\sum_{n=1}^\infty \lambda_n x_n$ is norm unconditionally convergent. Furthermore, show that the mapping $T\colon c_0 \to X$, defined by

$$T(\lambda_1, \lambda_2, \ldots) = \sum_{n=1}^\infty \lambda_n x_n\,,$$

is a bounded operator that need not be weakly compact.

4. Let S be a topological space without isolated points and let G be an arbitrary open subset of S. Show that neither G nor its closure $\overline{G}$ has isolated points—where, of course, G and $\overline{G}$ are equipped with the induced topology.

5. For a compact Hausdorff space Ω establish the following.
(a) If μ is a regular Borel measure on Ω with full support (that is, $\operatorname{Supp}\mu = \Omega$), then the natural embedding $J_\mu \colon C(\Omega) \to L_\infty(\mu)$ is a lattice isometry.
(b) If Ω is also metrizable and has no isolated points, then there exists a non-atomic regular Borel probability measure on Ω with full support.
[HINT: For (b) mimic the proofs of Lemma 11.35 and Corollary 11.36.]

6. Establish the following separability properties.
(a) If the adjoint of a bounded operator $T \colon X \to Y$ between Banach spaces has a separable range, then T likewise has a separable range.
(b) If a Banach space X has a separable dual, then X itself is separable.
Also, give an example of a separable Banach space whose norm dual is not separable.

7. (Weis–Werner [**330**]) If Ω is a metrizable compact space without isolated points, then show that every weakly compact operator on $C(\Omega)$ satisfies the Daugavet equation. [HINT: Recall that an operator $T \in \mathcal{L}\big(C(\Omega)\big)$ is weakly compact if and only if $T^{**}\big(C(\Omega)^{**}\big) \subseteq C(\Omega)$.]

8. (Weis–Werner [**330**]) The Bartle–Dunford–Schwartz theorem [**118**, p. 306] states that: *A non-empty subset A of $C(\Omega)^* = \mathrm{ca}(\Omega)$ is weakly compact if and only if it is norm bounded and there exists a measure $\mu \in \mathrm{ca}(\Omega)$ such that $\nu(E) \to 0$ uniformly in $\nu \in A$ when $\mu(E) \to 0$.*

Modify, using the Bartle–Dunford–Schwartz theorem, the arguments of the proof of Theorem 11.43 to show that the assumption of metrizability made in Exercise 7 is not necessary. That is, show that if Ω is a compact Hausdorff space without isolated points, then every weakly compact operator on $C(\Omega)$ satisfies the Daugavet equation.

9. (Ansari [**37**]) Fix $q \in [0,1]$ and consider the rank-one positive operator $T = \delta_q \otimes \mathbf{1}$ on $C[0,1]$. That is, $Tx = x(q)\mathbf{1}$ for each $x \in C[0,1]$. Establish the following properties of T.
(a) T factors through c_0 (and hence T satisfies the Daugavet equation).
(b) The operator $jT \colon C[0,1] \to C[0,1]^{**}$ is not disjoint from jI, where the operator $j \colon C[0,1] \to C[0,1]^{**}$ is the canonical embedding and I is the identity operator on $C[0,1]$.
(c) The adjoint operator $T^* \colon C[0,1]^* \to C[0,1]^*$ is not disjoint from I^*, the identity operator on $C[0,1]^*$.

11.5. Slices and the Daugavet Property

The purpose of this section is to establish the following interesting result due to V. M. Kadets, R. V. Shvidkoy, G. G. Sirotkin, and D. Werner [**166, 167**]: *If the rank-one operators on a Banach space satisfy the Daugavet equation, then every weakly compact operator also satisfies the Daugavet equation.* To prove this, we shall follow the approach of the above authors which is based on the notion of a slice of a set.

Definition 11.45. *Let $\langle X, X'\rangle$ be a dual pair of real vector spaces, and let C be a non-empty weakly bounded subset of X. Also, let $x' \in X'$ and $\epsilon > 0$ be given, and put $s = \sup_{c\in C}\langle c, x'\rangle$. Then the* ***slice*** *of C determined by x' and ϵ is the set*

$$\mathit{Slice}(C, x', \epsilon) = \big\{x \in C\colon\ \langle x, x'\rangle \geq s - \epsilon\big\}\,.$$

The geometric illustration of a slice is shown in Figure 1. Clearly, each slice is non-empty, and the smaller the ϵ the smaller the slice.

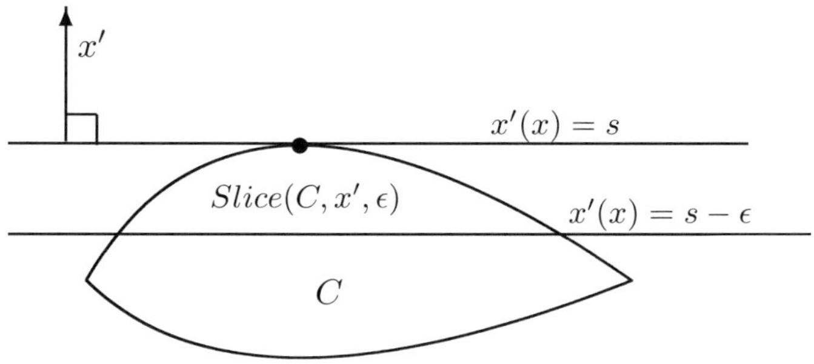

Figure 1

The next lemma presents a convenient geometric characterization in terms of slices of Banach spaces satisfying the Daugavet property with respect to rank-one operators.

In our discussion, the dual pair $\langle X, X'\rangle$ will be, as a rule, $\langle X, X^*\rangle$, where X is a Banach space and X^* is its norm dual. The unit ball of X or of X^* will serve as the set C. Under these conditions we will also assume that the vector $x^* \in X^*$ (resp. $x \in X$) is of norm one. In this case, we have

$$\sup\big\{\langle x, x^*\rangle\colon\ x \in U_X\big\} = \|x^*\| = 1 \ \text{ and } \ \sup\big\{\langle x, x^*\rangle\colon\ x^* \in U_{X^*}\big\} = \|x\| = 1\,.$$

Lemma 11.46. *Let X be a linear subspace of a Banach space Y and let $J\colon X \to Y$ denote the inclusion operator. Then the following statements are equivalent.*

(1) *Every rank-one operator $T\colon X \to Y$ satisfies the Daugavet equation, i.e., $\|J + T\| = 1 + \|T\|$.*

(2) *If $y_0 \in Y$ and $x_0^* \in X^*$ are unit vectors, then for each $0 < \epsilon < 1$ there exist some linear functional $x^* \in X^*$ of norm one and some $0 < \delta < 1$ such that:*
   (a) *$\mathit{Slice}(U_X, x^*, \delta) \subseteq \mathit{Slice}(U_X, x_0^*, \epsilon)$.*

(b) *For each* $x \in Slice(U_X, x^*, \delta)$ *we have* $\|x + y_0\| \geq 2 - \epsilon$.

(3) *If* $y_0 \in Y$ *and* $x_0^* \in X^*$ *are unit vectors, then for each* $0 < \epsilon < 1$ *there exist a unit vector* $y \in Y$ *and some* $0 < \delta < 1$ *such that:*
(a) $Slice(U_{Y^*}, y, \delta) \subseteq Slice(U_{Y^*}, y_0, \epsilon)$.
(b) *For each* $y^* \in Slice(U_{Y^*}, y, \delta)$ *we have* $\|x_0^* + J^*y^*\| \geq 2 - \epsilon$.

Proof. We establish only the equivalence (1) $\iff$ (2). The proof of the equivalence (1) $\iff$ (3) is similar and is left to the reader.

(1) $\Longrightarrow$ (2) Assume that $y_0 \in Y$ and $x_0^* \in X^*$ are unit vectors, and let $0 < \epsilon < 1$. Start by observing that the rank-one operator $T = x_0^* \otimes y_0$ in $\mathcal{L}(X, Y)$ satisfies $\|T\| = 1$. So, by our hypothesis, $\|J + T\| = 1 + \|T\| = 2$, and consequently $\|J^* + T^*\| = 2$. Fix some linear functional $y^* \in Y^*$ of norm one satisfying $\|J^*y^* + T^*y^*\| > 2 - \epsilon > 0$ and $y^*(y_0) \geq 0$, and let

$$x^* = \frac{J^*y^* + T^*y^*}{\|J^*y^* + T^*y^*\|} \quad \text{and} \quad \delta = 1 - \frac{2 - \epsilon}{\|J^*y^* + T^*y^*\|}.$$

Clearly, $\|x^*\| = 1$ and $0 < \delta < 1$.

Fix any $x \in Slice(U_X, x^*, \delta) = \{x \in U_X \colon\ x^*(x) \geq 1 - \delta\}$. We claim that

$$y^*(x) + y^*(y_0)x_0^*(x) \geq 2 - \epsilon. \tag{$\star$}$$

To see this, note that

$$\begin{aligned} y^*(x) + y^*(y_0)x_0^*(x) &= \langle x, (J^* + T^*)y^* \rangle = \langle x, \|J^*y^* + T^*y^*\| x^* \rangle \\ &\geq (1 - \delta)\|J^*y^* + T^*y^*\| = 2 - \epsilon. \end{aligned}$$

Since $y^*(x) \leq 1$ and $0 \leq y^*(y_0) \leq 1$, it follows from ($\star$) that

$$x_0^*(x)y^*(y_0) \geq 2 - \epsilon - y^*(x) \geq 1 - \epsilon > 0.$$

But then $y^*(y_0) \geq 0$ yields $y^*(y_0) > 0$ and $x_0^*(x) > 0$. Therefore,

$$x_0^*(x) \geq x_0^*(x)y^*(y_0) \geq 1 - \epsilon > 0.$$

That is, $x \in Slice(U_X, x_0^*, \epsilon)$, and so $Slice(U_X, x^*, \delta) \subseteq Slice(U_X, x_0^*, \epsilon)$.

Now taking into account $0 < x_0^*(x) \leq 1$ and using ($\star$) once more, we get

$$2 - \epsilon \leq y^*(x) + y^*(y_0)x_0^*(x) \leq y^*(x) + y^*(y_0) = \langle x + y_0, y^* \rangle \leq \|x + y_0\|.$$

(2) $\Longrightarrow$ (1) Let $T = x_0^* \otimes y_0 \in \mathcal{L}(X, Y)$ be a rank-one operator. We must verify that $\|J + T\| \geq 1 + \|T\|$. Without loss of generality we can assume that $\|x_0^*\| = \|y_0\| = 1$ (and so $\|T\| = 1$).

Fix $0 < \epsilon < 1$. By our assumption there exist a linear functional $x^* \in X^*$ of norm one and some $0 < \delta < 1$ such that if $x \in Slice(U_X, x^*, \delta)$, then

$x \in \mathit{Slice}(U_X, x_0^*, \epsilon)$ and $\|x + y_0\| \geq 2 - \epsilon$. In particular, if we take any $x \in \mathit{Slice}(U_X, x^*, \delta)$, then

$$\begin{aligned} \|J + T\| &\geq \|x + Tx\| = \big\|x + x_0^*(x)y_0\big\| \\ &= \big\|[x_0^*(x)x + x_0^*(x)y_0] + [1 - x_0^*(x)]x\big\| \\ &\geq \big\|x_0^*(x)(x + y_0)\big\| - \big\|[1 - x_0^*(x)]x\big\| \\ &\geq (1-\epsilon)(2-\epsilon) - \epsilon \end{aligned}$$

for each $0 < \epsilon < 1$. Letting $\epsilon \downarrow 0$ yields $\|J + T\| \geq 2$, as desired. ■

Corollary 11.47. *Let X be a linear subspace of a Banach space Y. If every rank-one operator from X to Y satisfies the Daugavet equation, then any slice of the unit balls U_X and U_{Y^*} has diameter equal to 2.*

Proof. We shall show the result only for the slices of U_X. So, let $x_0^* \in X^*$ be a unit vector, let $0 < \eta < 1$, and consider the slice $\mathit{Slice}(U_X, x_0^*, \eta)$. Also, fix some $0 < \epsilon < \eta$ and note that $\mathit{Slice}(U_X, x_0^*, \epsilon) \subseteq \mathit{Slice}(U_X, x_0^*, \eta)$.

Pick an arbitrary unit vector $u \in \mathit{Slice}(U_X, x_0^*, \epsilon)$ and let $y_0 = -u$. Now apply the second statement of Lemma 11.46 to y_0, x_0^* and ϵ to get a unit vector $x^* \in X^*$ and some $\delta > 0$ such that $\mathit{Slice}(U_X, x^*, \delta) \subseteq \mathit{Slice}(U_X, x_0^*, \epsilon)$ and $\|x + y_0\| \geq 2 - \epsilon$ for each $x \in \mathit{Slice}(U_X, x^*, \delta)$. That is, $\|x - u\| \geq 2 - \epsilon$ for each $x \in \mathit{Slice}(U_X, x^*, \delta)$. Since both x and u belong to $\mathit{Slice}(U_X, x_0^*, \eta)$, it follows that the diameter of $\mathit{Slice}(U_X, x_0^*, \eta)$ is greater than or equal to $2 - \epsilon$ for all $0 < \epsilon < \eta$. Hence, the diameter of $\mathit{Slice}(U_X, x_0^*, \epsilon)$ is 2, as claimed. ■

In order to proceed further, we need to recall a few facts regarding extreme points of convex sets in Banach spaces. First, let us recall their definition.

Definition 11.48. *Let C be a non-empty convex subset of a real Banach space Z. A point $c \in C$ is said to be:*

1. *an **extreme point** if $c = \alpha u + (1-\alpha)v$ for some $u, v \in C$ and some $\alpha \in (0,1)$ imply that $u = v = c$,*
2. *an **exposed point** if there exists some $z^* \in Z^*$ (called an **exposing linear functional**) satisfying $z^*(c) > z^*(z)$ for all $z \in C \setminus \{c\}$, and*
3. *a **strongly exposed point** if it is an exposed point and there exists an exposing linear functional $z^* \in Z^*$ such that a sequence $\{z_n\}$ in C satisfies $\|z_n - c\| \to 0$ if and only if $z^*(z_n) \to z^*(c)$.*[5]

It should be clear that an exposed point is an extreme point. However, an extreme point need not be an exposed point; see Figure 2. Also, an

[5] In this case, we also say that the linear functional $z^* \in Z^*$ **strongly exposes** the point c.

exposed point need not be strongly exposed. For instance, if $Z = \ell_2$ and $C = \ell_2^+$, then 0 is an exposed point of C but it fails to be strictly exposed.

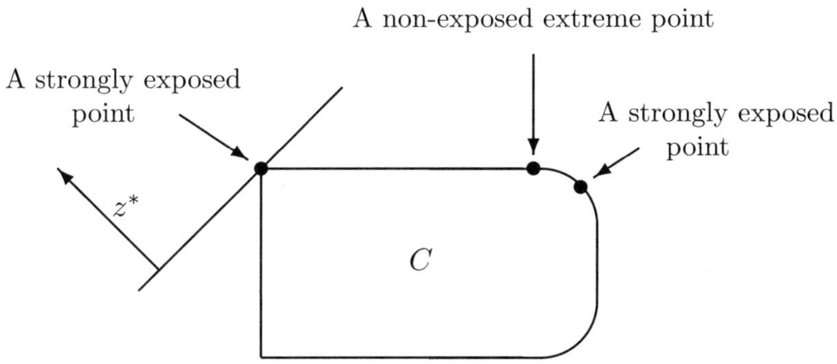

Figure 2

Regarding strongly exposed points we have the following result of J. Lindenstrauss [**203**] and S. L. Troyanski [**323**]. (For a geometric proof see J. Bourgain [**69**].)

Theorem 11.49 (Lindenstrauss–Troyanski). *Each convex weakly compact non-empty subset of a Banach space is the closed convex hull of its strongly exposed points.*

Now we are ready to prove the promised characterization of the Daugavet property for weakly compact operators by means of rank-one operators.

Theorem 11.50 (Kadets–Shvidkoy–Sirotkin–Werner). *Let X be a linear subspace of a Banach space Y and let $J\colon X \to Y$ denote the inclusion operator. If every rank-one operator from X to Y satisfies the Daugavet equation, then every weakly compact operator in $\mathcal{L}(X, Y)$ also satisfies the Daugavet equation.*

Proof. Assume that every rank-one operator from X to Y satisfies the Daugavet equation, and let $T\colon X \to Y$ be a weakly compact operator with $\|T\| = 1$. Fix $0 < \epsilon < 1$. We shall show that $\|J + T\| \geq 2 - 3\epsilon$.

Start by observing that the set $K = \overline{T(U_X)}$, the norm closure of $T(U_X)$, is weakly compact. Therefore, according to Theorem 11.49, K coincides with the closed convex hull of its strongly exposed points. This implies that there exists a strongly exposed point y_0 in K such that

$$\|y_0\| > \sup_{y \in K} \|y\| - \epsilon = \|T\| - \epsilon = 1 - \epsilon\,.$$

From $\|T\| = 1$, it follows that $\|y_0\| \leq 1$. Next pick a linear functional $y^* \in Y^*$ that strongly exposes y_0 and satisfies $y^*(y_0) = \max_{y\in K} y^*(y) = 1$. We claim that there exists some $0 < \delta < \epsilon$ such that the slice of K

$$B = \mathit{Slice}(K, y^*, \delta) = \big\{y \in K\colon\ y^*(y) \geq 1 - \delta\big\}$$

is included in the open ball centered at y_0 with radius ϵ, i.e., $B \subseteq B(y_0, \epsilon)$. To establish the existence of such a δ, assume by way of contradiction that this is not the case. This means that there exist a null sequence $\{\delta_n\}$ and a sequence $\{y_n\} \subseteq K$ satisfying $0 < \delta_n < \epsilon$, $y^*(y_n) \geq 1 - \delta_n$, and $\|y_n - y_0\| \geq \epsilon$ for each n. This implies $y^*(y_n) \to 1 = y^*(y_0)$, and so (since y^* strongly exposes y_0) we have $\|y_n - y_0\| \to 0$, which is a contradiction. This contradiction establishes the existence of a desired δ.

Next, let $x^* = T^*y^*$ and note that

$$\|x^*\| = \sup\big\{y^*(Tx)\colon\ x \in U_X\big\} = \max\big\{y^*(y)\colon\ y \in K\big\} = 1\,.$$

Moreover, the slice $A = \mathit{Slice}(U_X, x^*, \delta)$ of U_X satisfies $T(A) \subseteq B$. Indeed,

$$\begin{aligned} T(A) &= \big\{Tx\colon\ x \in U_X \text{ and } x^*(x) \geq 1 - \delta\big\} \\ &= \big\{Tx\colon\ x \in U_X \text{ and } y^*(Tx) \geq 1 - \delta\big\} \subseteq B\,. \end{aligned}$$

In particular, for each $x \in A$ we have $\|y_0 - Tx\| < \epsilon$.

Now statement (2) of Lemma 11.46, applied to the unit vector $\frac{y_0}{\|y_0\|}$ and the slice A, guarantees the existence of some vector $x_0 \in A$ satisfying $\big\|\, x_0 + \frac{y_0}{\|y_0\|} \,\big\| \geq 2 - \delta > 2 - \epsilon$. This implies

$$\begin{aligned} \|x_0 + y_0\| &= \big\|\big(x_0 + \tfrac{y_0}{\|y_0\|}\big) + \big(1 - \tfrac{1}{\|y_0\|}\big)y_0\big\| \\ &\geq \big\|x_0 + \tfrac{y_0}{\|y_0\|}\big\| - \big(\tfrac{1}{\|y_0\|} - 1\big)\|y_0\| \\ &\geq 2 - \epsilon - 1 + \|y_0\| \\ &\geq 2 - \epsilon - 1 + 1 - \epsilon > 2 - 2\epsilon\,. \end{aligned}$$

Finally, note that

$$\begin{aligned} \|J + T\| &\geq \|x_0 + Tx_0\| = \big\|(x_0 + y_0) - (y_0 - Tx_0)\big\| \\ &\geq \|x_0 + y_0\| - \|y_0 - Tx_0\| \geq 2 - 2\epsilon - \epsilon = 2 - 3\epsilon\,, \end{aligned}$$

and the proof is finished. ■

Exercises

1. Prove the equivalence of statements (1) and (3) in Lemma 11.46.
2. ([**167**]) Let X be a subspace of a Banach space Y and let $J\colon X \to Y$ denote the inclusion operator. Show that the following statements are equivalent.
   (a) Every rank-one operator $T\colon X \to Y$ satisfies the Daugavet equation, i.e., $\|J+T\| = 1 + \|T\|$.
   (b) For any pair of unit vectors $y \in Y$ and $x^* \in X^*$ and each $0 < \epsilon < 1$ there is a unit vector $x \in X$ satisfying $x^*(x) \geq -\epsilon$ and $\|x+y\| \geq 2-\epsilon$.
   (c) For every pair of unit vectors $y \in Y$ and $x^* \in X^*$ and each $0 < \epsilon < 1$ there is some unit linear functional $y^* \in Y^*$ satisfying $y^*(y) \geq 1-\epsilon$ and $\|x^* + J^*y^*\| \geq 2-\epsilon$.
3. If X is a reflexive Banach space, then X does not satisfy the Daugavet property for rank-one operators.
4. Let $C = \ell_2^+$, the positive cone of ℓ_2. Show that 0 is an exposed point of C that fails to be strongly exposed.
5. Use Theorem 11.50 to present an alternative proof of the second part of Theorem 11.20: *If E is an atomless AL-space, then E satisfies the Daugavet property for weakly compact operators.* [HINT: According to Theorem 11.50 we must show that each rank-one operator satisfies the Daugavet equation. Without loss of generality, we can assume that there is an atomless measure μ such that $E = L_1(\mu)$. By Lemma 11.21 it suffices to show that if $0 \leq \phi \in L_1^*(\mu)$ and $0 \leq u \in L_1(\mu)$, then $I \wedge (\phi \otimes u) = 0$. Since the functions in $L_1(\mu)$ have σ-finite support, we can assume that μ is also σ-finite.]
6. ([**167, 181**]) Let (Ω, Σ, μ) be a non-atomic measure space and let Z be a Banach space. Show that the Banach space $X = L_1(\mu, Z)$ of Z-valued Bochner integrable functions satisfies the Daugavet property for weakly compact operators.
7. (Kadets [**165**]; Khalil [**181**]) Let K be a compact Hausdorff space without isolated points and let Z be an arbitrary Banach space. Show that the Banach space $X = C(K, Z)$ of Z-valued continuous functions satisfies the Daugavet property for weakly compact operators.

11.6. Narrow Operators

In this section we shall discuss the relatively new class of narrow operators and we shall study its relationship with the Daugavet property. Though it is not clear to us who introduced the class of narrow operators, it should be mentioned that the works of J. Bourgain, N. Ghoussoub, and H. P. Rosenthal [**71, 136, 282, 283**] on sign-embedding operators are closely related to narrow operators. In our presentation we follow [**267**], where the reader can also find many other interesting results.

Throughout this section, we shall assume that (Ω, Σ, μ) is an atomless probability measure space and that E is a real Banach function space on (Ω, Σ, μ) with order continuous norm and that Y is an arbitrary Banach space. We will also assume that $L_\infty(\mu) \subseteq E \subseteq L_1(\mu)$. The last assumption is quite mild since by the Twist Theorem 5.34 of G. Lozanovsky for any Banach function space E there exists a measurable function h such that $L_\infty(\mu) \subseteq hE = \{hx\colon\ x \in E\} \subseteq L_1(\mu)$. Moreover, if E is a rearrangement invariant Banach function space,[6] then the inclusions $L_\infty(\mu) \subseteq E \subseteq L_1(\mu)$ hold automatically.

We start with a simple result that will be needed later.

Lemma 11.51. *If E has order continuous norm, then for each $x \in E$ and for each $\epsilon > 0$ there exists some $\delta > 0$ such that for each measurable set B with $\mu(B) < \delta$ we have $\|x\chi_B\| < \epsilon$.*

Proof. If this is not the case, then there exist some $\epsilon > 0$ and a sequence $\{B_n\} \subseteq \Sigma$ satisfying $\mu(B_n) < 2^{-n}$ and $\|x\chi_{B_n}\| \geq \epsilon$ for each n. Letting $A_n = \bigcup_{k=n}^\infty B_k$, we see that $|x|\chi_{A_n} \downarrow 0$ and $\|x\chi_{A_n}\| \geq \|x\chi_{B_n}\| \geq \epsilon$ for each n, which is impossible. ■

We are now ready to introduce the notion of a narrow operator.

Definition 11.52. *A bounded operator $T\colon E \to Y$ is said to be* ***narrow*** *if for each $A \in \Sigma$ and each $\epsilon > 0$ there exist some $B \in \Sigma$ with $B \subseteq A$ and $\mu(B) \geq \frac{1}{2}\mu(A)$ and a function $x \in E$ satisfying $|x| = \chi_B$ and $\|Tx\| < \epsilon$.*

Note that since we deal with real Banach spaces, any measurable function x that satisfies $|x| = \chi_B$ is necessarily of the form $x = \chi_C - \chi_{B\setminus C}$ for some measurable subset C of B. Clearly, scalar multiples of narrow operators are again narrow operators. However, the sum of two narrow operators need not be a narrow operator; see Exercise 7 at the end of this section.

A useful characterization of narrow operators is as follows.

Theorem 11.53. *A bounded operator $T\colon E \to Y$ is narrow if and only if for each $A \in \Sigma$ and each $\epsilon > 0$ there exists a function $x \in E$ such that*

$$|x| = \chi_A, \quad \int_\Omega x\, d\mu = 0, \quad \textit{and} \quad \|Tx\| < \epsilon\,.$$

Proof. The "only if" part is obvious. We shall prove the "if" part. So, assume that T is a non-zero narrow operator. We shall show first that for

[6] A Banach function space E is called **rearrangement invariant** if for each $y \in L_0(\mu)$ that is equimeasurable to some $x \in E$ we have $y \in E$ and $\|y\| = \|x\|$. As usual, two functions $x, y \in L_0(\mu)$ are said to be **equimeasurable** if they generate the same distribution function, i.e., $\mu(\{\omega \in \Omega\colon\ x(\omega) \leq t\}) = \mu(\{\omega \in \Omega\colon\ y(\omega) \leq t\})$ for each $t \in \mathbb{R}$.

any $D \in \Sigma$ and any $\eta > 0$ there exists some $y \in E$ satisfying $|y| = \chi_D$ and $\|Ty\| < \eta$.

To see this, fix $D \in \Sigma$ and $\eta > 0$. We claim that there exist a sequence $\{y_1, y_2, \ldots\}$ of pairwise disjoint measurable functions and a sequence $\{C_0, C_1, C_2, \ldots\}$ of pairwise disjoint measurable subsets of D such that for each $n \geq 1$ we have:

(a) $C_n \subseteq D_{n-1} := D \setminus \bigcup_{i=0}^{n-1} C_i$.

(b) $\mu(C_n) \geq \frac{1}{2}\mu(D_{n-1})$.

(c) $\mu(D_n) \leq \frac{1}{2}\mu(D_{n-1})$.

(d) $|y_n| = \chi_{C_n}$.

(e) $\|Ty_n\| < \frac{\eta}{2^n}$.

If this is done, then for each n let $z_n = \sum_{i=n}^{\infty} |y_i|$ and note that $z_n \in E$ and $z_n \downarrow 0$. From $\left|\sum_{i=n}^{n+p} y_i\right| \leq \sum_{i=n}^{n+p} |y_i| \leq z_n$ and the order continuity of the norm, it follows that the series $y = \sum_{n=1}^{\infty} y_n$ is norm convergent in E. Clearly, $\|Ty\| \leq \sum_{n=1}^{\infty} \|Ty_n\| < \sum_{n=1}^{\infty} \frac{1}{2^n}\eta = \eta$. Moreover, from

$$\begin{aligned} 0 \leq \mu(D) - \mu\Big(\bigcup_{i=1}^{n} C_i\Big) &= \mu\Big(D \setminus \bigcup_{i=0}^{n} C_i\Big) = \mu(D_{n-1} \setminus C_n) \\ &= \mu(D_{n-1}) - \mu(C_n) \leq \mu(D_{n-1}) - \tfrac{1}{2}\mu(D_{n-1}) \\ &= \tfrac{1}{2}\mu(D_{n-1}) \leq \tfrac{1}{2^n}\mu(D)\,, \end{aligned}$$

it follows that $\mu(D) = \sum_{n=1}^{\infty} \mu(C_n)$, that is, $D = \bigcup_{n=1}^{\infty} C_n$. Consequently, $|y| = \sum_{n=1}^{\infty} |y_n| = \sum_{n=1}^{\infty} \chi_{C_n} = \chi_D$.

The existence of $\{y_1, y_2, \ldots\}$ and $\{C_0, C_1, \ldots\}$ will be established by induction. Start by letting $D_0 = D$ and $C_0 = \emptyset$. Since T is a narrow operator, there exist a measurable subset C_1 of D with $\mu(C_1) \geq \frac{1}{2}\mu(D)$ and a measurable function $y_1 \in E$ such that $|y_1| = \chi_{C_1}$ and $\|Ty_1\| < \frac{\eta}{2}$. Clearly,

$$\mu(D_1) = \mu(D \setminus C_1) = \mu(D) - \mu(C_1) \leq \mu(D) - \tfrac{1}{2}\mu(D) = \tfrac{1}{2}\mu(D) = \tfrac{1}{2}\mu(D_0)\,.$$

Now, for the induction step, suppose that the measurable functions $y_1, \ldots, y_k$ and the measurable subsets $C_1, \ldots, C_k$ of D have been chosen to satisfy (a), (b), (c), (d), and (e) for each $n = 0, 1, \ldots, k$. Since T is a narrow operator, there exists a measurable subset C_{k+1} of $D_k = D \setminus \bigcup_{i=1}^{k} C_i$ with $\mu(C_{k+1}) \geq \frac{1}{2}\mu(D_k)$ and some measurable function $y_{k+1} \in E$ satisfying $|y_{k+1}| = \chi_{C_{k+1}}$ and $\|Ty_{k+1}\| < \frac{\eta}{2^{k+1}}$. Moreover, we have

$$\begin{aligned} \mu(D_{k+1}) &= \mu(D_k \setminus C_k) = \mu(D_k) - \mu(C_k) \\ &\leq \mu(D_k) - \tfrac{1}{2}\mu(D_k) = \tfrac{1}{2}\mu(D_k)\,. \end{aligned}$$

This completes the induction.

We are now ready to finish the proof of the theorem. To this end, fix $A \in \Sigma$ and $\epsilon > 0$. Since $\mathbf{1} \in E$, there exists (by Lemma 11.51) some $\delta > 0$ such that $\|\chi_B\| = \|\mathbf{1}\chi_B\| < \frac{\epsilon}{4\|T\|}$ whenever $B \in \Sigma$ satisfies $\mu(B) < \delta$. Fix some n with $\frac{\mu(A)}{n} < \delta$, and then partition A into n subsets $B_1, \ldots, B_n$ each of measure $\frac{\mu(A)}{n}$; this is possible since μ is non-atomic. By the first part, for each $k = 1, \ldots, n$ we can find some function $u_k \in E$ such that $|u_k| = \chi_{B_k}$ and $\|Tu_k\| < \frac{\epsilon}{2n}$. Let $a_k = \int_\Omega u_k \, d\mu$. Clearly $|a_k| \leq \frac{\mu(A)}{n}$, and therefore we can find signs $\theta_k = \pm 1$ such that $\left|\sum_{k=1}^n \theta_k a_k\right| \leq \frac{\mu(A)}{n}$.

Let $u = \sum_{k=1}^n \theta_k u_k$. Clearly, $|u| = \chi_A$, $\|Tu\| < \frac{\epsilon}{2}$ and $\left|\int_\Omega u \, d\mu\right| \leq \frac{\mu(A)}{n}$. So, if $\Omega_1 = \{\omega \in A\colon\ u(\omega) = 1\}$ and $\Omega_2 = \{\omega \in A\colon\ u(\omega) = -1\}$, then the last inequality implies $\left|\mu(\Omega_1) - \mu(\Omega_2)\right| \leq \frac{\mu(A)}{n}$. We can assume that $\mu(\Omega_1) \geq \mu(\Omega_2)$. Now select any measurable subset Ω_3 of Ω_1 such that $\mu(\Omega_3) = \frac{\mu(\Omega_1) - \mu(\Omega_2)}{2}$, and consider the function $x \in E$ defined by $x(\omega) = -1$ if $\omega \in \Omega_3$ and $x(\omega) = u(\omega)$ if $\omega \in \Omega \setminus \Omega_3$. It follows that $|x| = \chi_A$, $\int_\Omega x \, d\mu = 0$, and

$$\|Tx\| \leq \|T(x-u)\| + \|Tu\| \leq \|T\| \cdot \|2\chi_{\Omega_3}\| + \tfrac{\epsilon}{2} < \tfrac{\epsilon}{2} + \tfrac{\epsilon}{2} = \epsilon\,.$$

Thus, x is a function with the desired properties. ∎

Corollary 11.54. *If $T \in \mathcal{L}(E, Y)$ is a narrow operator, then for each A in Σ and each n there is a measurable subset B of A such that $\mu(B) = 2^{-n}\mu(A)$ and the function $h = (2^n - 1)\chi_B - \chi_{A\setminus B} \in E$ satisfies $\|Th\| < \epsilon$.*

Proof. The proof is by induction. For $n = 1$ the conclusion follows immediately from Theorem 11.53. For the induction step, suppose that the statement is true for some n. We must prove that it is also true for $n + 1$.

By the induction hypothesis, there exists a measurable subset B_n of A such that $\mu(B_n) = 2^{-n}\mu(A)$ and the function $h_n = (2^n - 1)\chi_{B_n} - \chi_{A\setminus B_n} \in E$ satisfies $\|Th_n\| < \frac{\epsilon}{2}$. By Theorem 11.53 there exists some $x \in E$ such that $|x| = \chi_{B_n}$, $\int_\Omega x \, d\mu = 0$ and $\|Tx\| < \frac{\epsilon}{2^{n+1}}$.

Now consider the measurable set $B_{n+1} = \{\omega \in B_n\colon\ x(\omega) = 1\}$. Clearly, $\mu(B_{n+1}) = 2^{-(n+1)}\mu(A)$ and the function $h_{n+1} = (2^{n+1} - 1)\chi_{B_{n+1}} - \chi_{A\setminus B_{n+1}}$ in E satisfies $h_{n+1} - h_n = 2^n x$. This implies

$$\|Th_{n+1}\| \leq \|T(h_{n+1} - h_n)\| + \|Th_n\| \leq \|2^n Tx\| + \tfrac{\epsilon}{2} < \tfrac{\epsilon}{2} + \tfrac{\epsilon}{2} = \epsilon\,,$$

and the induction is complete. ∎

To continue our discussion, we need the concept of a generalized sequence of Rademacher functions supported by an arbitrary measurable subset of Ω. To define such a sequence, let A be a measurable subset of Ω. We let $r_1 = r_1(A) = \chi_A$. Since μ is assumed to be non-atomic, we can split A into two disjoint measurable subsets of equal measure. That is, there are disjoint measurable subsets $A_1, A_2 \in \Sigma$ such that $A = A_1 \cup A_2$ and $\mu(A_1) = \mu(A_2) = \frac{1}{2}\mu(A)$. We let $r_2 = r_2(A) = \chi_{A_1} - \chi_{A_2}$.

The sequence $\{r_n\}$ will be constructed by induction following the preceding process as follows. Assume that the nth Rademacher function r_n has been constructed so that

$$r_n = r_n(A) = \sum_{k=1}^{2^n} (-1)^{k+1} \chi_{B_k},$$

where $B_i \cap B_j = \emptyset$ if $i \neq j$, $A = \bigcup_{k=1}^{2^n} B_k$, and $\mu(B_k) = \frac{1}{2^n}\mu(A)$ for each $1 \leq k \leq 2^n$. For each $1 \leq k \leq n$ choose two disjoint measurable subsets $B_{k,1}$ and $B_{k,2}$ of B_k satisfying $B_{k,1} \cup B_{k,2} = B_k$ and $\mu(B_{k,i}) = \frac{1}{2}\mu(B_k) = \frac{1}{2^{n+1}}\mu(A)$ for each $i = 1, 2$. Now the $(\mathrm{n}+1)^{\mathrm{th}}$ Rademacher function is defined by

$$r_{n+1} = r_{n+1}(A) = \sum_{k=1}^{2^n} \left(\chi_{B_{k,1}} - \chi_{B_{k,2}}\right).$$

Clearly, each function $r_n(A)$ satisfies $|r_n(A)| = \chi_A$ and so, since $\mathbf{1} \in E$, each $r_n(A)$ necessarily belongs to E.

Definition 11.55. *Any sequence of functions $\{r_n\}$ constructed as above is called a (**generalized**) **sequence of Rademacher functions** supported by A.*

Generalized sequences of Rademacher functions converge weakly to zero.

Lemma 11.56. *Every generalized sequence of Rademacher functions $\{r_n(A)\}$ is weakly convergent to zero in E, i.e., $r_n(A) \xrightarrow{\sigma(E,E^*)} 0$.*

Proof. This proof is due to Y. A. Abramovich [**3**, Theorem 3.1]. Fix some measurable subset A of Ω with $\mu(A) > 0$, and let $\{r_n\}$ be a generalized sequence of Rademacher functions supported by A. We must show that $f(r_n) \to 0$ for each $f \in E^*$.

Fix some $f \in E^*$. Since $L_\infty(\mu)$ is an ideal in E and E has order continuous norm, it follows that f is an order continuous linear functional on $L_\infty(\mu)$. Taking into account that the order continuous dual of $L_\infty(\mu)$ coincides with $L_1(\mu)$, we see that f restricted to $L_\infty(\mu)$ belongs to $L_1(\mu)$. Now observe that $\{r_n\} \subseteq L_\infty(\mu)$. So, in order to establish $r_n \xrightarrow{w} 0$, it suffices to prove that $\int_\Omega f r_n \, d\mu \to 0$ for all $f \in L_1(\mu)$. In view of the $\|\cdot\|_1$-denseness of $L_\infty(\mu)$ in $L_1(\mu)$, to establish the latter we must show that $\int_\Omega f r_n \, d\mu \to 0$ for all $f \in L_\infty(\mu)$. So, assume that $f \in L_\infty(\mu)$.

Next, consider the sequence $\{r_n\}$ as a sequence in the Hilbert space $L_2(\mu)$. The construction of $\{r_n\}$ guarantees that

$$\langle r_n, r_m \rangle = \int_\Omega r_n r_m \, d\mu = 0$$

holds for all $n \neq m$ and $\|r_n\|_2 = \sqrt{\mu(A)}$ for each n. Therefore, the sequence $\{\frac{r_n}{\|r_n\|_2}\}$ is an orthonormal sequence in $L_2(\mu)$. Since $f \in L_2(\mu)$, Parseval's inequality yields $\sum_{n=1}^{\infty}|\langle f, r_n\rangle|^2 \leq \mu(A)\|f\|_2^2 < \infty$. This implies

$$\int_{\Omega} f r_n \, d\mu = \langle f, r_n\rangle \to 0\,,$$

and the proof is finished. (For a generalization of this result see Exercise 10 at the end of the section.) ■

Recall that a bounded operator $T\colon X \to Y$ between two Banach spaces is a **Dunford–Pettis operator** if $x_n \xrightarrow{w} 0$ in X implies $\|Tx_n\| \to 0$. We are now ready to establish that Dunford–Pettis and compact operators are narrow operators.

Theorem 11.57. *Each Dunford–Pettis operator—and hence each compact operator—from E to Y is narrow.*

Proof. Assume that $T \in \mathcal{L}(E, Y)$ is a Dunford–Pettis operator. Fix any $A \in \Sigma$ and let $\epsilon > 0$. Consider a Rademacher sequence $\{r_n\}$ supported by A. By Lemma 11.56, we know that $r_n \xrightarrow{w} 0$. Since T is Dunford–Pettis, it follows that $\|Tr_n\| \to 0$. In particular, there exists some k such that $\|Tr_k\| < \epsilon$. But then the function $x = r_k$ satisfies the required condition of Definition 11.52. This shows that T is a narrow operator. ■

It should be noted that a narrow operator need not be compact. For an example see Exercise 7 at the end of this section.

And now we can state and prove the main result of this section.

Theorem 11.58. *If (Ω, Σ, μ) is a non-atomic probability measure space, then every narrow operator on $L_1(\mu)$ satisfies the Daugavet equation.*

Proof. Let $T\colon L_1(\Omega, \Sigma, \mu) \to L_1(\Omega, \Sigma, \mu)$ be a non-zero narrow operator, and let $\epsilon > 0$. Pick a step function $x = \sum_{k=1}^{m} a_k \chi_{A_k}$ such that $\|x\|_1 = 1$ and $\|Tx\|_1 > \|T\| - \epsilon$. We can assume that $A_i \cap A_j = \emptyset$ for $i \neq j$. Since Tx is absolutely continuous, there exists some $\delta > 0$ such that

$$\|\chi_B Tx\|_1 < \epsilon \qquad (\star)$$

for each $B \in \Sigma$ with $\mu(B) < \delta$. Fix $n \in \mathbb{N}$ such that $2^{-n} < \delta$.

Since T is narrow, for each $1 \leq k \leq m$ there exists (by Corollary 11.54) some measurable subset B_k of A_k such that $\mu(B_k) = 2^{-n}\mu(A_k)$ and the function $h_k = (2^n - 1)\chi_{B_k} - \chi_{A_k \setminus B_k}$ satisfies $\|Th_k\|_1 < \frac{\epsilon}{(|a_k|+1)m}$. Now consider the function $h = \sum_{k=1}^{m} a_k h_k$. Then $\|Th\|_1 < \epsilon$ and it is easy to check that $x + h = \sum_{k=1}^{m} 2^n a_k \chi_{B_k}$. This implies

$$\|x+h\|_1 = \sum_{k=1}^{m} |a_k| 2^n \mu(B_k) = \sum_{k=1}^{m} |a_k| \mu(A_k) = \|x\|_1 = 1\,.$$

Next, let $B = \Omega \setminus \bigcup_{k=1}^{m} B_k$ and $y = \chi_B Tx$. Then $Tx - y = \chi_{\Omega\setminus B} Tx$. Furthermore, since

$$\mu(\Omega \setminus B) = \mu\Big(\bigcup_{k=1}^{m} B_k\Big) = \sum_{k=1}^{m} \mu(B_k) = 2^{-n} \sum_{k=1}^{m} \mu(A_k) \le 2^{-n}\mu(\Omega) < \delta\,,$$

it follows from $(\star)$ that $\|Tx - y\|_1 < \epsilon$. Consequently,

$$\|x+h+Tx\|_1 \ge \|x+h+y\|_1 - \|Tx-y\|_1 \ge \|x+h+y\|_1 - \epsilon\,.$$

Observe now that the functions $x + h$ and y are disjoint, and hence

$$\|x+h+y\|_1 = \|x+h\|_1 + \|y\|_1 = 1 + \|y\|_1 \ge 1 + \|Tx\|_1 - \epsilon\,.$$

This implies $\|x + h + Tx\|_1 \ge 1 + \|Tx\|_1 - 2\epsilon$. Finally,

$$\begin{aligned} \|I+T\| &\ge \big\|(I+T)(x+h)\big\|_1 \ge \|x+h+Tx\|_1 - \|Th\|_1 \\ &\ge \|x+h+Tx\|_1 - \epsilon \ge 1 + \|Tx\|_1 - 3\epsilon \\ &\ge 1 + \|T\| - 4\epsilon\,. \end{aligned}$$

Since $\epsilon > 0$ is arbitrary, it follows that $\|I+T\| \ge 1 + \|T\|$, and so T satisfies the Daugavet equation. ■

We conclude the section by stating without a proof an analogue of Theorem 11.44 which is due to J. Bourgain and H. P. Rosenthal [**71**]. Its proof is quite involved and can be found in [**71**].

Theorem 11.59 (Bourgain–Rosenthal). *If (Ω, Σ, μ) is a non-atomic probability measure space, then every operator on $L_1(\mu)$ that does not fix a copy of $L_1(\mu)$ is narrow—and hence it satisfies the Daugavet equation.*

Exercises

1. Give an example of a weakly compact operator that is not narrow.
2. Show that each weakly compact operator on $L_1(\mu)$ is a narrow operator.
3. If in a scheme of bounded operators $E \xrightarrow{S} Y \xrightarrow{T} Z$ the operator S is narrow, then show that the operator $TS\colon E \to Z$ is narrow.
4. Give an example of a narrow operator that is not regular.
5. Let $T\colon E \to Y$ be a narrow operator. Show that for each $A \in \Sigma$ there exists a sequence of Rademacher functions $\{r_n\}$ supported by A such that $\|Tr_n\| \to 0$.

6. (Plichko–Popov [**267**]) Show that the sum of a narrow and a compact operator is narrow.

7. (Plichko–Popov [**267**]) Let $E = L_p[0,1]$, where $1 < p < \infty$. Show that the identity operator I is the sum of two narrow projections—and conclude from this that the sum of two narrow operators need not be a narrow operator. [HINT: Use the Haar system (see Exercise 3 in Section 10.6 for a precise definition of the Haar functions) to split E into a direct sum of two complemented subspaces E_1 and E_2 such that each of these subspaces has infinite many "packs" of the consecutive Haar functions with pairwise disjoint supports covering $[0,1]$.]

8. (Plichko–Popov [**267**]) Show that each bounded operator on $L_p[0,1]$, where $1 < p < \infty$, is the sum of two narrow operators. [HINT: Use Exercises 3 and 7 above.]

9. Let (Ω, Σ, μ) be a non-atomic probability space. Show that every generalized sequence of Rademacher functions $\{r_n(A)\}$ satisfies the conclusion of the Riemann–Lebesgue lemma, i.e., $\int_\Omega x r_n \, d\mu \to 0$ for each $x \in L_1(\mu)$.

10. Let (Ω, Σ, μ) be a non-atomic probability space, and let E be a Banach function space with order continuous norm such that $L_\infty(\mu) \subseteq E \subseteq L_1(\mu)$. Also, let $\{r_n\}$ be a sequence of Rademacher functions supported by a measurable subset A of Ω. Establish the following.
   (a) For each $x \in E$ we have $x r_n \xrightarrow{w} 0$ in E.
   (b) For each $x \in L_1(\mu)$ we have $\|x + x r_n\|_1 \to \|x\|_1$.

11.7. Some Applications of the Daugavet Equation

Recall that by its very definition the Daugavet property is isometric in nature. Therefore, it comes as a surprise that this isometric property has purely isomorphic consequences. Some of them will be discussed in this section.

The first result of this type is due to V. Kadets [**165**] and it informs us that a Banach space with the Daugavet property cannot have an unconditional basis.

Theorem 11.60 (Kadets). *If a Banach space satisfies the Daugavet property with respect to rank-one operators, then it does not have an unconditional basis.*

Proof. Let X be a Banach space satisfying the Daugavet property with respect to rank-one operators. Assume by way of contradiction that X has an unconditional basis $\{e_n\}$.

For each finite subset A of natural numbers let P_A and Q_A denote the natural projections on the closed subspaces generated by $\{e_k\colon\ k \in A\}$ and $\{e_k\colon\ k \notin A\}$, respectively. Clearly, $P_A + Q_A = I$ and P_A is a finite-rank operator. By Theorem 11.50, the operator $-P_A$ satisfies the Daugavet equation and therefore

$$\|Q_A\| = \|I - P_A\| = 1 + \|P_A\|\,. \tag{$\star$}$$

Next, consider the subsets of real numbers

$$\begin{aligned} V &= \big\{\|P_A\|\colon\ A \text{ is a finite subset of } \mathbb{N}\big\} \text{ and} \\ W &= \big\{\|Q_A\|\colon\ A \text{ is a finite subset of } \mathbb{N}\big\}\,. \end{aligned}$$

If A is an arbitrary finite subset of $\mathbb{N}$ and $A_n = \{1, \ldots, n\} \cap A^c$, then clearly $\|P_{A_n}x - Q_A x\| \to 0$ for each $x \in X$. This implies that $\|Q_A\| \le \sup V$. Therefore, $\sup W \le \sup V$. On the other hand, it follows from ($\star$) that $\sup W \ge 1 + \sup V$. Consequently, $\sup V = \sup W = \infty$. However, according to Lemma 1.53, $\sup V$ must be finite, a contradiction. This contradiction establishes that X cannot have an unconditional basis. ■

Since we know that the Banach lattices $C[0,1]$ and $L_1[0,1]$ satisfy the Daugavet property for rank-one operators, Theorem 11.60 provides immediately a new and simple proof of the following well-known fact.

Corollary 11.61. *Neither $C[0,1]$ nor $L_1[0,1]$ has an unconditional basis.*

Another surprising isomorphic consequence forced by the Daugavet property on a Banach space X is due to V. Kadets, R. Shvidkoy, G. Sirotkin and D. Werner [**167**] and is presented next.

Theorem 11.62 (Kadets–Shvidkoy–Sirotkin–Werner)**.** *If a Banach space X satisfies the Daugavet property for rank-one operators, then X contains a copy of ℓ_1 (and hence X cannot be reflexive).*

Proof. Let V be a finite dimensional subspace of X and let U be the closed unit ball of X. If $0 < \epsilon < 1$ is given, then we claim that there exists a $\mathit{Slice}(U, x^*, \delta)$ such that

$$\|x + v\| \ge (1 - \epsilon)(\|x\| + \|v\|)$$

for each $x \in \mathit{Slice}(U, x^*, \delta)$ and for each $v \in V$.

To establish this claim, choose a finite set $\{v_1, \ldots, v_n\}$ of unit vectors that are $\frac{\epsilon}{2}$-dense in the unit sphere of V. An inductive argument in conjunction with Lemma 11.46 guarantees the existence of n slices satisfying

$$\mathit{Slice}(U, x_1^*, \delta_1) \supseteq \mathit{Slice}(U, x_2^*, \delta_2) \supseteq \cdots \supseteq \mathit{Slice}(U, x_n^*, \delta_n)$$

such that

$$\|x + v_k\| \ge 2 - \tfrac{\epsilon}{2} \tag{$\star\star$}$$

for each k and for all $x \in \mathit{Slice}(U, x_k^*, \delta_k)$.

In particular, for the last (smallest) slice $Slice(U, x_n^*, \delta_n)$ the inequality $(\star\star)$ holds for each $k = 1, \dots, n$. Moreover, if we recall that the set $\{v_1, \dots, v_n\}$ is $\frac{\epsilon}{2}$-dense in the unit sphere of V, then $(\star\star)$ shows that

$$\|x + v\| \geq 2 - \epsilon$$

for each $x \in Slice(U, x_n^*, \delta_n)$ and for each $v \in V$ with $\|v\| = 1$. Therefore, by Exercise 3 in Section 11.1 for any $x \in Slice(U, x_n^*, \delta_n)$ we have

$$\|\lambda x + v\| \geq (1 - \epsilon)(\|\lambda x\| + \|v\|)$$

for each $v \in V$ and all scalars λ.

Now it is easy to construct a sequence in X whose closed linear span is isomorphic to ℓ_1. Indeed, using inductively the previous part of the proof, we can construct a sequence of unit vectors $e_1, e_2, \dots$ and a decreasing sequence of slices $\{Slice(U, x_n^*, \epsilon_n)\}$ such that $e_n \in Slice(U, x_n^*, \epsilon_n)$ and

$$\|\lambda e_{n+1} + v\| \geq (1 - \epsilon_{n+1})(|\lambda| + \|v\|)$$

for each λ and for each v in the linear span of $\{e_1, \dots, e_n\}$. Without loss of generality, we can assume that $0 < \epsilon_n \leq \frac{1}{2^n}$ for each n. (This condition guarantees that $\gamma = \prod_{k=1}^{\infty}(1 - \epsilon_k) \in (0, 1)$.)

Next, we claim that the inequality

$$\Big\|\sum_{k=1}^{n} \lambda_k e_k\Big\| \geq \Big[\prod_{k=1}^{n}(1 - \epsilon_k)\Big] \cdot \sum_{k=1}^{n} |\lambda_k| \tag{$\dagger$}$$

holds for all n and any choice of scalars $\lambda_1, \dots, \lambda_n$. The proof is by induction. For $n = 1$ notice that $\|\lambda_1 e_1\| = |\lambda_1| \geq (1 - \epsilon_1)|\lambda_1|$. Now assume that $(\dagger)$ is true for some n. Then for any choice of scalars $\lambda_1, \dots, \lambda_n, \lambda_{n+1}$ we have

$$\begin{aligned}
\Big\|\sum_{k=1}^{n+1} \lambda_k e_k\Big\| &\geq (1 - \epsilon_{n+1})\Big[|\lambda_{n+1}| + \Big\|\sum_{k=1}^{n} \lambda_k e_k\Big\|\Big] \\
&\geq (1 - \epsilon_{n+1})\Big(|\lambda_{n+1}| + \Big[\prod_{k=1}^{n}(1 - \epsilon_k)\Big] \cdot \sum_{k=1}^{n} |\lambda_k|\Big) \\
&\geq \Big[\prod_{k=1}^{n+1}(1 - \epsilon_k)\Big] \cdot \sum_{k=1}^{n+1} |\lambda_k| .
\end{aligned}$$

This completes the induction and establishes the validity of $(\dagger)$.

Since $\prod_{k=1}^{n}(1 - \epsilon_k) \geq \gamma$ for each n, it follows from $(\dagger)$ that for all n and any choice of scalars $\lambda_1, \dots, \lambda_n$ we have

$$\Big\|\sum_{k=1}^{n} \lambda_k e_k\Big\| \geq \gamma \sum_{k=1}^{n} |\lambda_k| .$$

This shows that the sequence $\{e_n\}$ is equivalent to the standard unit basis in ℓ_1, and so X contains a copy of ℓ_1. ■

The relationships between the Daugavet property and the geometric properties of the unit ball of a Banach space are still far from being understood completely. Only a few results in this direction are available. The next theorem and its consequences are due to P. Wojtaszczyk [**339**].

Theorem 11.63 (Wojtaszczyk). *If a Banach space X satisfies the Daugavet property with respect to the rank-one operators, then the closed unit ball of X does not have strongly exposed points.*

Proof. Assume by way of contradiction that the closed unit ball of a Banach space X has a strongly exposed point u_0. Pick a linear functional $f \in X^*$ that strongly exposes u_0 such that $f(u_0) = \|f\| = 1$. Also, fix any non-zero vector $v \in X$ with $f(v) = 0$.

By our assumption, the rank-one operator $T = f \otimes v$ satisfies the Daugavet property, i.e., $\|I + T\| = 1 + \|T\| = 1 + \|f\| \cdot \|v\| = 1 + \|v\|$. Now take a sequence of unit vectors $\{x_n\}$ such that

$$\|x_n + Tx_n\| = \|x_n + f(x_n)v\| \longrightarrow 1 + \|v\| .$$

From $\|x_n + Tx_n\| \le 1 + |f(x_n)| \cdot \|v\| \le 1 + \|v\|$, it follows that $|f(x_n)| \to 1$. Replacing $\{x_n\}$ by $\{-x_n\}$ and passing to a subsequence if necessary, we can assume that $f(x_n) \to 1$. Since f strongly exposes u_0, we see that $x_n \to u_0$. Therefore,

$$\|u_0 + v\| = \lim_{n\to\infty} \|x_n + f(x_n)v\| = \lim_{n\to\infty} \|x_n + Tx_n\| = 1 + \|v\| ,$$

and this is true for all $v \in X$ with $f(v) = 0$.

The above show that X is isometric to the ℓ_1-sum $(\operatorname{Ker} f \oplus \mathbb{R}u_0)_1$, where $\mathbb{R}u_0$ denotes the one-dimensional vector subspace spanned by u_0. Finally, consider the rank-one operator $S = f \otimes u_0$ and note that for each vector $x = y + \lambda u_0 \in (\operatorname{Ker} f \oplus \mathbb{R}u_0)_1$, we have

$$\begin{aligned} \big\|(I - S)x\big\| &= \big\|y + \lambda u_0 - f(y)u_0 - \lambda f(u_0)u_0\big\| \\ &= \|y\| \le \|y\| + \|\lambda u_0\| = \|y + \lambda u_0\| = \|x\| . \end{aligned}$$

This shows that $\|I - S\| \le 1 < 1 + \|S\|$, and consequently the rank-one operator S does not satisfy the Daugavet equation, which is impossible. This contradiction establishes that the closed unit ball of X has no strongly exposed points. ■

Recall that a Banach space Z has the **Radon–Nikodym property** if every continuous operator $T\colon L_1[0,1] \to Z$ can be represented as a Bochner integral, i.e., there exists an essentially bounded strongly measurable function $\phi\colon [0,1] \to Z$ such that

$$Tx = \int x\phi \, d\mu$$

for all $x \in L_1[0,1]$. We refer to [**107**] for a comprehensive treatment of the Radon–Nikodym property that plays an important role in many problems in analysis.

Corollary 11.64. *If a Banach space X has the Radon–Nikodym property, then X does not satisfy the Daugavet property for rank-one operators.*

Proof. The conclusion follows from Theorem 11.63 by invoking the following well-known theorem of R. Phelps: *The closed unit ball of a Banach space with the Radon–Nikodym property has strongly exposed points.* (See [**107**, Chapter VII, Theorem 3].) ∎

Corollary 11.65. *Every Banach space can be renormed so that in the new norm it fails the Daugavet property for rank-one operators.*

Proof. It suffices to establish that a given Banach space X can be renormed in such a way that its closed unit ball has strongly exposed points. To see that this is possible, take any x_0 that does not belong to the original closed unit ball U and then renorm X by taking for the new closed unit ball the closure of the circled convex hull of $U \cup \{x_0\}$. It remains to be noticed that x_0 is a strongly exposed point of the new closed unit ball. ∎

Exercises

1. Let X be a Banach space and assume that there exists an increasing continuous function $\psi\colon \mathbb{R}_+ \to \mathbb{R}$ and a scalar $\gamma > 0$ such that for every finite-rank projection P we have

$$\psi(\|I - P\|) \geq \gamma + \psi(\|P\|)\,.$$

Show that X does not have an unconditional basis.

2. Complete the details of the proof of Corollary 11.65.

Bibliography

1. Y. A. Abramovich, Injective envelopes of normed lattices, *Soviet Math. Dokl.* **12** (1971), 511–514.
2. Y. A. Abramovich, Some theorems on normed lattices, *Vestnik Leningr. Univ. Mat. Meh. Astronom.*, no. 13 (1971), 5–11. (English translation: *Vestnik Leningr. Univ. Math.* **4** (1977), 153–159.)
3. Y. A. Abramovich, Weakly compact sets in Dedekind complete topological vector lattices, *Teor. Funkcii Funkcional. Anal. i Prilozen.* (*Kharkov*), no. 15 (1972), 27–35.
4. Y. A. Abramovich, A generalization of a theorem of J. Holub, *Proc. Amer. Math. Soc.* **108** (1990), 937–939.
5. Y. A. Abramovich, New classes of Banach spaces on which compact operators satisfy the Daugavet equation, *J. Operator Theory* **25** (1991), 331–345.
6. Y. A. Abramovich, C. D. Aliprantis, and O. Burkinshaw, The Daugavet equation in uniformly convex Banach spaces, *J. Funct. Anal.* **97** (1991), 215–230.
7. Y. A. Abramovich, C. D. Aliprantis, and O. Burkinshaw, On the spectral radius of positive operators, *Math. Z.* **211** (1992), 593–607. Corrigendum: *Math. Z.* **215** (1994), 167–168.
8. Y. A. Abramovich, C. D. Aliprantis, and O. Burkinshaw, Positive operators on Krein spaces, *Acta Appl. Math.* **27** (1992), 1–22.
9. Y. A. Abramovich, C. D. Aliprantis, and O. Burkinshaw, An elementary proof of Douglas' theorem on contractive projections on L_1-spaces, *J. Math. Anal. Appl.* **177** (1993), 641–644.
10. Y. A. Abramovich, C. D. Aliprantis, and O. Burkinshaw, Invariant subspaces of operators on ℓ_p-spaces, *J. Funct. Anal.* **115** (1993), 418–424.
11. Y. A. Abramovich, C. D. Aliprantis, and O. Burkinshaw, Invariant subspaces for positive operators, *J. Funct. Anal.* **124** (1994), 95–111.
12. Y. A. Abramovich, C. D. Aliprantis, and O. Burkinshaw, Invariant subspaces for positive operators acting on a Banach space with basis, *Proc. Amer. Math. Soc.* **123** (1995), 1773–1777.
13. Y. A. Abramovich, C. D. Aliprantis, and O. Burkinshaw, Local quasinilpotence, cycles and invariant subspaces, in: N. Kalton, E. Saab, and S. Montgomery-Smith, eds.,

Interaction Between Functional Analysis, Harmonic Analysis, and Probability (Lecture Notes in Pure and Applied Mathematics, **175**, Marcel Dekker, New York, 1996), 1–12.

14. Y. A. Abramovich, C. D. Aliprantis, and O. Burkinshaw, Another characterization of the invariant subspace problem, in: C. B. Huijsmans, M. A. Kaashoek, W. A. J. Luxemburg, and B. de Pagter, eds., *Operator Theory in Function Spaces and Banach Lattices* (Operator Theory Advances and Applications, **75**, Birkhäuser Verlag, Berlin, 1995), 15–31.
15. Y. A. Abramovich, C. D. Aliprantis, and O. Burkinshaw, Multiplication and compact-friendly operators, *Positivity* **1** (1997), 171–180.
16. Y. A. Abramovich, C. D. Aliprantis, and O. Burkinshaw, The invariant subspace problem: Some recent advances, *Rend. Istit. Mat. Univ. Trieste* **29** (1998), 1–76.
17. Y. A. Abramovich, C. D. Aliprantis, O. Burkinshaw, and A. W. Wickstead, A characterization of compact-friendly multiplication operators, *Indag. Math. N. S.* **10** (1999), 161–171.
18. Y. A. Abramovich, C. D. Aliprantis, and I. A. Polyrakis, Positive projections and lattice-subspaces, *Proc. Royal Irish Acad. Sect. A* **94** (1994), 237–253.
19. Y. A. Abramovich, C. D. Aliprantis, and I. A. Polyrakis, Some remarks on surjective and bounded below operators, *Atti Sem. Mat. Fis. Univ. Modena* **44** (1996), 455–464.
20. Y. A. Abramovich, C. D. Aliprantis, and W. R. Zame, A representation theorem for Riesz spaces and its applications to economics, *Econom. Theory* **5** (1995), 527–535.
21. Y. A. Abramovich, E. L. Arenson, and A. K. Kitover, *Banach $C(K)$-modules and Operators Preserving Disjointness*, Pitman Research Notes in Mathematical Series, # 277, Longman Scientific & Technical, New York, 1992.
22. Y. A. Abramovich and A. K. Kitover, Inverses of disjointness preserving operators, *Mem. Amer. Math. Soc.*, **679**, Providence, RI, 2000.
23. Y. A. Abramovich, A. I. Veksler, and A. V. Koldunov, On operators preserving disjointness, *Soviet Math. Dokl.* **20** (1979), 1089–1093.
24. Y. A. Abramovich and A. W. Wickstead, Remarkable classes of unital AM-spaces, *J. Math. Anal. Appl.* **180** (1993), 398–411.
25. C. D. Aliprantis and K. C. Border, *Infinite Dimensional Analysis: A Hitchhikers Guide*, 2nd Edition, Springer–Verlag, Heidelberg and New York, 1999.
26. C. D. Aliprantis, D. J. Brown, and O. Burkinshaw, *Existence and Optimality of Competitive Equilibria*, Springer–Verlag, Heidelberg and New York, 1990.
27. C. D. Aliprantis and O. Burkinshaw, Positive compact operators on Banach lattices, *Math. Z.* **174** (1980), 289–298.
28. C. D. Aliprantis and O. Burkinshaw, On weakly compact operators on Banach lattices, *Proc. Amer. Math. Soc.* **183** (1981), 573–578.
29. C. D. Aliprantis and O. Burkinshaw, *Locally Solid Riesz Spaces*, Academic Press, New York and London, 1978.
30. C. D. Aliprantis and O. Burkinshaw, *Positive Operators*, Academic Press, New York and London, 1985.
31. C. D. Aliprantis and O. Burkinshaw, *Principles of Real Analysis*, 3rd Edition, Academic Press, New York and London, 1998.
32. C. D. Aliprantis and O. Burkinshaw, *Problems in Real Analysis*, 2nd Edition, Academic Press, New York and London, 1998.

33. C. D. Aliprantis, O. Burkinshaw, and P. Kranz, On lattice properties of the composition operator, *Manuscripta Math.* **36** (1981), 19–31.
34. T. Andô, Positive operators in semi-ordered linear spaces, *J. Fac. Sci. Hokkaido Univ., Ser.* 1, **13** (1957), 214–228.
35. T. Andô, Contractive projections in L_p-spaces, *Pacific J. Math.* **17** (1966), 391–405.
36. T. Andô, Banachverbände und positiver Projektionen, *Math. Z.* **109** (1969), 121–130.
37. S. I. Ansari, Essential disjointness and the Daugavet equation, *Houston J. Math.* **19** (1993), 587–601.
38. A. B. Antonevich and A. V. Lebedev, Spectral properties of operators with shift, *Izv. Akad. Nauk USSR Ser. Mat.* **47** (1983), 915–941.
39. C. Apostol, C. Foias, and D. Voiculescu, Some results on non-quasitriangular operators IV, *Rev. Romaine Math. Pures Appl.* **18** (1973), 487–514.
40. W. Arendt, Über das spektrum regularer operatoren, Ph.D. Dissertation, University of Tübingen, 1979.
41. W. Arendt, A. Grabosch, G. Greiner, U. Groh, H. P. Lotz, U. Moustakas, R. Nagel, F. Neubrander, and U. Schlotterbeck, *One-parameter Semigroups of Positive Operators*, Lecture Notes in Mathematics, **1184**, Springer–Verlag, Berlin and New York, 1986.
42. N. Aronszajn and K. T. Smith, Invariant subspaces of completely continuous operators, *Ann. of Math.* **60** (1954), 345–350.
43. W. B. Arveson and J. Feldman, A note on invariant subspaces, *Michigan Math. J.* **15** (1968), 61–64.
44. F. V. Atkinson, The normal solubility of linear equations in normed spaces, *Mat. Sb. (N.S.)* **28** (1951), 3–14.
45. A. Atzmon, Power regular operators, *Trans. Amer. Math. Soc.* **347** (1995), 3101–3109.
46. A. Atzmon, On the existence of hyperinvariant subspaces, *J. Operator Theory* **11** (1984), 3–40.
47. A. Atzmon, An operator without invariant subspaces on a nuclear Fréchet space, *Ann. of Math.* (2) **117** (1983), 669–694.
48. A. Atzmon, Operators which are annihilated by analytic functions and invariant subspaces, *Acta Math.* **144** (1980), 27–63.
49. A. Atzmon and G. Godefroy, An application of the smooth variational principle to the existence of nontrivial subspaces, *C. R. Acad. Sci. Paris Sér. I Math.* **332** (2001), 151–156.
50. V. F. Babenko and S. A. Pichugov, A property of compact operators in Banach spaces, *Ukrainian Math. J.* **33** (1981), 374–376.
51. S. Banach, *Théorie des Opérations Linéaires*, Monografje Matematyczne, Warsaw, 1932. (Reprinted by Chelsea Publishing Co., New York, 1955.)
52. B. Beauzamy, *Introduction to Operator Theory and Invariant Subspaces*, North-Holland Mathematical Library, **42**, North-Holland, Amsterdam, 1988.
53. B. Beauzamy, Un opérateur sans sous-espace invariant: simplification de l'exemple de P. Enflo (French) [An operator with no invariant subspace: simplification of the example of P. Enflo], *Integral Equations Operator Theory* **8** (1985), 314–384.
54. A. Benedek and R. Panzone, The spaces with mixed norm, *Duke Math. J.* **28** (1961), 301–324.
55. H. Bercovici, Notes on invariant subspaces, *Bull. Amer. Math. Soc. (N.S.)* **23** (1990), 1–36.

56. J. Bergh and J. Löfstrem, *Interpolation Spaces. An Introduction*, Springer–Verlag, Heidelberg and New York, 1976.

57. A. Berman, M. Neumann, and R. J. Stern, *Nonnegative Matrices in Dynamical Systems*, John Wiley and Sons, New York, 1989.

58. A. Berman and R. J. Plemmons, *Nonnegative Matrices in the Mathematical Sciences*, Academic Press, New York, 1979.

59. S. J. Bernau and H. E. Lacey, The range of a contractive projection on an L_p-space, *Pacific J. Math.* **53** (1974), 21–41.

60. A. R. Bernstein and A. Robinson, Solution of an invariant subspace problem of K. T. Smith and P. R. Halmos, *Pacific J. Math.* **16** (1966), 421–431.

61. C. Bessaga and A. Pelczynski, On bases and unconditional convergence of series in Banach spaces, *Studia Math.* **17** (1958), 151–164.

62. A. Bigard and K. Keimel, Sur les endomorphismes conversant les polaires d'un group réticulé Archimédien, *Bull. Soc. Math. France* **97** (1969), 381–398.

63. G. Birkhoff, Moyennes de fonctions, algebre et theorie des nombrees, *Coloques Internationaux du Centre National de la Recherche Scientifique*, no. **24** (1950), 143–153.

64. G. Birkhoff, Reactor criticality in transport theory, *Proc. Nat. Acad. Sci. U.S.A.* **45** (1959), 567–569.

65. G. Birkhoff, Positivity and criticality, Proc. Sympos. Appl. Math. **10**, 116–126. Amer. Math. Soc., Providence, RI, 1961.

66. E. Bishop and R. R. Phelps, The support functionals of a convex set, in: V. Klee ed., *Convexity* (Proceedings of Symposia in Pure Mathematics, **7**, Amer. Math. Soc., Providence, RI, 1963), 27–35.

67. R. P. Boas, *Invitation to Complex Analysis*, The Random House–Birkhaüser Math. Series, New York, 1987.

68. F. F. Bonsall, Linear operators in complete positive cones, *Proc. London. Math. Soc.* **8** (1958), 53–75.

69. J. Bourgain, Real isomorphic complex Banach spaces need not be complex isomorphic, *Proc. Amer. Math. Soc.* **96** (1986), 221–226.

70. J. Bourgain, Strongly exposed points in weakly compact sets in Banach spaces, *Proc. Amer. Math. Soc.* **58** (1976), 197–200.

71. J. Bourgain and H. P. Rosenthal, Applications of the theory of semi-embeddings to Banach space theory, *J. Funct. Anal.* **52** (1983), 149–188.

72. M. Boyle and D. Handelman, The spectra of nonnegative matrices via symbolic dynamics, *Ann. of Math.* **133** (1991), 249–316.

73. L. de Branges, The Stone–Weierstrass theorem, *Proc. Amer. Math. Soc.* **10** (1959), 822–824.

74. L. de Branges, Factorization and invariant subspaces, *J. Math. Anal. Appl.* **129** (1970), 163–200.

75. L. de Branges, A construction of invariant subspaces, *Math. Nachr.* **163** (1993), 163–175.

76. S. Brown, Invariant subspaces for subnormal operators, *Integral Equations Operator Theory* **1** (1978), 310–333.

77. A. V. Bukhvalov, Integral representation of linear operators, *Zap. Nauchn. Sem. Leningrad Otdel. Mat. Inst. Steklov.* (*LOMI*) **47** (1974), 5–14. Also, in: *J. Soviet Mathematics* **8** (1978), 129–137.

78. A. V. Bukhvalov, V. B. Korotkov, A. G. Kusraev, S. S. Kutateladze, and B. M. Makarov, *Vector Lattices and Integral Operators*, Nauka, Novosibirsk, 1992. (English translation edited by S. S. Kutateladze was published by Kluwer Academic Publishers, Dordrecht and Boston, 1996.)

79. J. J. Buoni, R. Harte, and A. W. Wickstead, Upper and lower Fredholm spectra, I, *Proc. Amer. Math. Soc.* **66** (1977), 309–314.

80. R. Bürger, Mutation-selection and continuum-of-Alleles models, *Math. Biosci.* **91** (1988), 67–83.

81. G. Buskes, Extension of Riesz homomorphisms. I, *J. Austral. Math. Soc. Ser.* **A 39** (1985), 107–120.

82. G. Buskes, On extension of Riesz homomorphisms. II, *J. Miss. Acad. Sci.* **33** (1988), 29–47.

83. G. Buskes, Extension of Riesz homomorphisms. III, *J. Austral. Math. Soc. Ser.* **A 43** (1987), 35–46.

84. J. W. Calkin, Two-sided ideals and congruences in the ring of bounded operators in Hilbert spaces, *Ann. of Math.* **42** (1941), 839–873.

85. S. R. Caradus, W. E. Pfaffenberger, and B. Yood, *Calkin Algebras and Algebras of Operators on Banach Spaces*, Lecture Notes in Pure and Applied Mathematics, **9**, Marcel Dekker, New York, 1974.

86. V. Caselles, On irreducible operators on Banach lattices, *Indag. Math.* **48** (1986), 11–16.

87. J. J. M. Chadwick and A. W. Wickstead, A quotient of ultrapowers of Banach spaces and semi-Fredholm operators, *Bull. London Math. Soc.* **9** (1977), 321–325.

88. P. Chauveheid, On a property of compact operators in Banach spaces, *Bull. Soc. Roy. Sci. Liège* **51** (1982), 371–378.

89. Z. L. Chen and A. W. Wickstead, Some applications of Rademacher sequences in Banach lattices, *Positivity* **2** (1998), 171–191.

90. B. Chevreau, W. S. Li, and C. Pearcy, A new Lomonosov lemma, *Integral Equations Operator Theory* **40** (1998), 409–417.

91. M.-D. Choi, E. A. Nordgren, H. Radjavi, P. Rosenthal, and Y. Zhong, Triangularizing semigroups of quasinilpotent operators with non-negative entries, *Indiana Univ. Math. J.* **42** (1993), 15–25.

92. J. A. Clarkson, Uniformly convex spaces, *Trans. Amer. Math. Soc.* **40** (1936), 396–414.

93. P. F. Conrad and J. E. Diem, The ring of polar preserving endomorphisms of an Abelian lattice-ordered group, *Illinois J. Math.* **15** (1971), 222–240.

94. J. B. Conway, *A Course in Operator Theory*, Graduate Studies in Mathematics, **21**, Amer. Math. Soc., Providence, RI, 2000.

95. H. G. Dales, Automatic continuity: a survey, *Bull. London Math. Soc.* **10** (1978), 129–183.

96. H. G. Dales and J. Esterle, Discontinuous homomorphisms from $C(X)$, *Bull. Amer. Math. Soc.* **83** (1977), 257–259.

97. H. G. Dales, R. J. Loy, and G. A. Willis, Homomorphisms and derivatives from $\mathcal{B}(E)$, *J. Funct. Anal.* **120** (1994), 201–219.

98. I. K. Daugavet, On a property of compact operators in the space C, *Uspekhi Mat. Nauk* **18** (1963), 157–158. (Russian)

99. K. R. Davidson and H. A. Domingo, Decomposition of Banach space operators, *Indiana Univ. Math. J.* **35** (1986), 333–343.

100. A. M. Davie, Invariant subspaces for Bishop's operators, *Bull. London Math. Soc.* **6** (1974), 343–348.

101. M. M. Day, The spaces L^p with $0 < p < 1$, *Bull. Amer. Math. Soc.* **46** (1940), 816–823.

102. M. M. Day, *Normed Linear Spaces*, 2nd edition, Springer–Verlag, Heidelberg and New York, 1973.

103. D. Deckard, R. G. Douglas, and C. Pearcy, On invariant subspaces of quasitriangular operators, *Amer. Math. J.* **91** (1969), 637–647.

104. U. U. Diallo and P. P. Zabreĭko, On the Daugavet–Krasnoselsky theorem, *Vestnik Acad. Nauk BSSR Ser. Fiz. Mat. Nauk*, no. 3 (1987), 26–31. (Russian)

105. J. Diestel, *Sequences and Series in Banach Spaces*, Graduate Texts in Mathematics, **92**, Springer–Verlag, Heidelberg and New York, 1984.

106. J. Diestel, H. Jarchow, and A. Tong, *Absolutely Summing Operators*, Cambridge Studies in Advanced Mathematics, **43**, Cambridge University Press, 1995.

107. J. Diestel and J. J. Uhl, Jr., *Vector Measures*, Math. Surveys, **15**, Amer. Math. Soc., Providence, Rhode Island, 1977.

108. J. Dieudonné, *Foundations of Modern Analysis*, Academic Press, Pure and Applied Math Series, **X**, New York and London, 1960.

109. P. G. Dodds and D. H. Fremlin, Compact operators in Banach lattices, *Israel J. Math.* **34** (1979), 287–320.

110. P. G. Dodds, C. B. Huijsmans, and B. de Pagter, Characterizations of conditional expectation-type operators, *Pacific J. Math.* **141** (1990), 55–77.

111. J. L. Doob, *Measure Theory*, Graduate Texts in Mathematics, **143**, Springer–Verlag, Heidelberg and New York, 1994.

112. R. G. Douglas, Contractive projections on an L_1-space, *Pacific J. Math.* **15** (1965), 443–462.

113. H. R. Dowson, *Spectral Theory of Linear Operators*, Academic Press, New York and London, 1978.

114. R. Drnovšek, Common invariant subspaces for collections of operators, *Integral Equations Operator Theory* **39** (2001), 253–266.

115. R. Drnovšek, D. Kokol-Bukovšek, L. Livshits, G. MacDonald, M. Omladiič, and H. Radjavi, An irreducible semigroup of non-negative square-zero operators, *Integral Equations Operator Theory*, forthcoming.

116. J. Duncan, C. M. McGregor, J. D. Price, and A. J. White, The numerical index of a normed space, *J. London Math. Soc.* (2) **2** (1970), 481–488.

117. N. Dunford, Integration and linear operations, *Trans. Amer. Math. Soc.* **40** (1936), 474–494.

118. N. Dunford and J. T. Schwartz, *Linear Operators I*, Wiley (Interscience), New York, 1958.

119. E. M. Dynkin, An operator calculus based on the Cauchy–Green formula, *Zap. Nauchn. Sem. Leningrad Otdel. Mat. Inst. Steklov. (LOMI)* **30** (1972), 33–39.

120. P. Enflo, A counterexample to the approximation property in Banach spaces, *Acta Math.* **130** (1973), 309–317.

121. P. Enflo, On the invariant subspace problem for Banach spaces, *Seminaire Maurey–Schwarz* (1975–1976); *Acta Math.* **158** (1987), 213–313.

122. Z. Ercan and S. Onal, Invariant subspaces for positive operators acting on a Banach space with Markushevich basis, *Positivity*, forthcoming.

123. J. Flores and F. L. Hernández, Domination by positive strictly singular operators, *J. London Math. Soc.*, forthcoming.

124. J. Flores and F. L. Hernández, Domination by positive disjointly strictly singular operators, *Proc. Amer. Math. Soc.* **129** (2001), 1979–1986.

125. J. Flores and F. L. Hernández, On positive strictly singular operators and domination, preprint.

126. C. Foias and I. Singer, Points of diffusion of linear operators and almost diffuse operators in spaces of continuous functions, *Math. Z.* **87** (1965), 434–450.

127. I. Fredholm, Sur une classe d'équations fonctionnelles, *Acta Math.* **27** (1903), 365–390.

128. D. H. Fremlin, *Topological Riesz Spaces and Measure Theory*, Cambridge University Press, London and New York, 1974.

129. H. Freudenthal, Teilweise geordnete moduln, *Nederl. Akad. Wetensch. Proc. Ser. A* **39** (1936), 641–651.

130. G. Frobenius, Über die Darstellung der endlichen Gruppen durch lineare Substitutionen, *S.-B. K. Preuss. Akad. Wiss. Berlin* (1897), 994–1015.

131. G. Frobenius, Über Matrizen aus nicht-negativen Elementen, *Sitz. Berichte Kgl. Preuß. Akad. Wiss. Berlin*, 456–477, 1912.

132. V. Gantmacher, Über schwache totalstetige operatoren, *Mat. Sb. (N.S.)* **7** (49) (1940), 301–308.

133. I. M. Gelfand, Normierte Ringe, *Mat. Sb. (N.S.)*, **9**(51) (1941), 3–24.

134. S. A. Gershgorin, Über die Albrenzung Eigenwerte einer Matrix, *Izv. Akad. Nauk SSSR Ser. Fiz.-Mat.* **6** (1931), 749–754.

135. N. Ghoussoub, Positive embeddings of $C(\Delta)$, L_1, $\ell(\Gamma)$, and $\left(\sum_n \oplus\ell_\infty^n\right)_{\ell_1}$, *Math. Ann.* **262** (1983), 461–472.

136. N. Ghoussoub and H. P. Rosenthal, G_δ-embeddings and quotients of L_1, *Math. Ann.* **264** (1983), 321–332.

137. W. T. Gowers and B. Maurey, The unconditional sequence problem, *J. Amer. Math. Soc.* **6** (1993), 851–874.

138. W. T. Gowers and B. Maurey, Banach spaces with small spaces of operators, *Math. Ann.* **307** (1997), 541–568.

139. J. J. Grobler, Band irreducible operators, *Indag. Math.* **48** (1986), 405–409.

140. A. L. Gromov, Invariant subspaces of weighted permutation operators, *Funktsional. Anal. i Prilozhen* **22** (1988), 75–76. (Russian)

141. A. Grothendieck, Produits tensoriels toplogiques et espaces nuclaires, *Mem. Amer. Math. Soc.*, **16**, Providence, RI, 1965.

142. A. Goullet de Rugy, La structure idéale des M-espaces, *J. Math. Pures et Appl.* **51** (1972), 331–373.

143. D. W. Hadwin, An operator still not satisfying Lomonosov's hypothesis, *Proc. Amer. Math. Soc.* **123** (1995), 3039–3041.

144. D. W. Hadwin, E. A. Nordgren, H. Radjavi, and P. Rosenthal, An operator not satisfying Lomonosov's hypothesis, *J. Funct. Anal.* **38** (1980), 410–415.

145. P. R. Halmos, *Finite-Dimensional Vector Spaces*, 2nd Edition, Van Nostrand Co., New York and London, 1958. Also, reprinted in the series: Undergraduate Texts in Mathematics, Springer-Verlag, New York and Heidelberg, 1974.

146. P. R. Halmos, Invariant subspaces of polynomially compact operators, *Pacific J. Math.* **16** (1966), 433–437.

147. P. R. Halmos, Quasitriangular operators, *Acta Sci. Math. (Szeged)* **29** (1968), 283–293.

148. P. R. Halmos, *A Hilbert Space Problem Book*, 2nd Edition, Graduate Texts in Mathematics, **19**, Springer–Verlag, New York and Heidelberg, 1982.

149. P. R. Halmos and V. S. Sunder, *Bounded Integral Operators on L^2 Spaces*, Springer–Verlag, Berlin and New York, 1978.

150. F. L. Hernández and B. Rodriguez-Salinas, On ℓ^p-complimented copies in Orlicz spaces II, *Israel J. Math.* **68** (1989), 27–55.

151. W. Hildenbrand and H. Sonnenschein, Eds., *Handbook of Mathematical Economics*, Vol. IV, North-Holland, Amsterdam and New York, 1991.

152. E. Hille and R. S. Phillips, *Functional Analysis and Semigroups*, Amer. Math. Soc. Colloq. Publ., **36**, Providence, RI, 1974.

153. J. R. Holub, A property of weakly compact operators on C[0,1], *Proc. Amer. Math. Soc.* **97** (1986), 396–398.

154. J. R. Holub, Daugavet's equation and operators on $L^1(\mu)$, *Proc. Amer. Math. Soc.* **100** (1987), 295–300.

155. J. R. Holub, Daugavet's equation and ideals of operators on $C[0,1]$, *Math. Nachr.* **141** (1989), 177–181.

156. R. B. Honor, Density and transitivity results on ℓ^∞ and ℓ^1, *J. London Math. Soc.* **32** (1985), 521–527.

157. R. A. Horn and C. R. Johnson, *Matrix Analysis*, Cambridge University Press, London and New York, 1985.

158. M. T. Jahandideh, *Decomposability and Triangularizability of Positive Operators on Banach Lattices*, Ph.D. Dissertation, Dalahousie University, Halifax, Canada,1997.

159. R. C. James, Bases and reflexivity in Banach spaces, *Ann. of Math.* **52** (1950), 518–527.

160. R. C. James, Bases in Banach spaces, *Amer. Math. Monthly* **89** (1982), 625–640.

161. G. Jameson, *Ordered Linear Spaces*, Lecture Notes in Mathematics, **141**, Springer–Verlag, Berlin and New York, 1970.

162. P. Jentzsch, Über Integralgleichungen mit positivem Kern, *J. Reine Angew. Math.* **141** (1912), 235–244.

163. C. L. Jiang, Stampfli's theorem for operators on some Banach spaces, *Northeast. Math. J.* **11** (1995), 56–60.

164. W. B. Johnson and J. Lindenstrauss, Eds., *Handbook of the Geometry of Banach Spaces*, Vol. **1**, North-Holland, Amsterdam and New York, 2001.

165. V. M. Kadets, Some remarks concerning the Daugavet equation, *Quaestiones Math.* **19** (1996), 225–235.

166. V. M. Kadets, R. V. Shvidkoy, G. G. Sirotkin, and D. Werner, Espaces de Banach ayant la propriété de Daugavet, *C. R. Acad. Sci. Paris Sér. I Math.* **325** (1997), 1291–1294.

167. V. M. Kadets, R. V. Shvidkoy, G. G. Sirotkin, and D. Werner, Banach spaces with the Daugavet property, *Trans. Amer. Math. Soc.* **352** (2000), 855–873.

168. G. K. Kalisch, On similarity, reducing manifolds, and unitary equivalences of certain Volterra operators, *Ann. of Math.* **66** (1957), 481–494.

169. N. J. Kalton, An elementary example of a Banach space not isomorphic to its complex conjugate, *Canadian Math. Bull.* **38** (1995), 218–222.

170. N. J. Kalton, N. T. Peck, and W. Roberts, *An F-Sampler*, London Math. Soc. Lecture Notes Series, **89**, Cambridge Univ. Press, London and New York, 1984.

171. H. Kamowitz, A property of compact operators, *Proc. Amer. Math. Soc.* **91** (1984), 231–236.

172. L. V. Kantorovich, On partially ordered linear spaces and their applications in the theory of linear operators, *Dokl. Akad. Nauk SSSR* **4** (1935), 13–16. (Russian)

173. L. V. Kantorovich, Sur les propriétés des espaces semi-ordonnés linéaires, *C. R. Acad. Sci. Paris Ser. A-B* **202** (1936), 813–816.

174. L. V. Kantorovich and G. P. Akilov, *Functional Analysis*, Pergamon Press, Oxford and New York, 1982.

175. L. V. Kantorovich, B. Z. Vulikh, and A. G. Pinsker, *Functional Analysis in Partially Ordered Spaces*, Gostekhizdat, Moscow and Leningrad, 1950.

176. S. Kaplan, *The Bidual of $C(X)$, I*, North-Holland, Amsterdam and New York, 1985.

177. S. Karlin, Positive operators, *J. Math. Mech.* **8** (1959), 907–937.

178. T. Kato, Perturbation theory for nullity deficiency and other quantities of linear operators, *J. Analyse Math.* **6** (1958), 273–322.

179. Y. Katznelson and L. Tzafriri, On power-bounded operators, *J. Funct. Anal.* **68** (1986), 313–328.

180. J. L. Kelley, Averaging operators on $C_\infty(X)$, *Illinois J. Math.* **2** (1958), 214–223.

181. R. Khalil, The Daugavet equation in vector-valued function spaces, *Panamer. Math. J.* **6** (1996), 51–53.

182. M. A. Khan and N. C. Yannelis (Editors), *Equilibrium Theory in Infinite Dimensional Spaces*, Studies in Economic Theory, **1**, Springer–Verlag, Heidelberg and New York, 1991.

183. J. Kim, The characterization of a lattice homomorphism, *Canadian J. Math.* **27** (1975), 172–175.

184. K. H. Kim, N. S. Ormes, and F. W. Roush, The spectra of nonnegative integer matrices via formal power series, *J. Amer. Math. Soc.* **13** (2000), 773–806.

185. A. K. Kitover, The spectral properties of weighted homomorphisms in algebras of continuous functions and their applications, *Zap. Nauchn. Sem. Leningrad Otdel. Mat. Inst. Steklov. (LOMI)* **107** (1982), 89–103.

186. A. K. Kitover, On the spectrum of weighted automorphisms and on a Kamowittz–Shineberg theorem, *Funct. Anal. Appl.* **13** (1979), 70–71.

187. A. N. Kolmogorov, *Grundbegriffe der Wahrscheinlichkeitrechnung*, Ergebnisse der Mathematik, Vol. 2, no. 3, 1933. [English Translation: *Foundations of the Theory of Probability*, Chelsea Publ. Co., New York, 1950.]

188. V. B. Korotkov, *Integral Operators*, Nauka, Novosibirsk, 1983.

189. G. Köthe, *Topological Vector Spaces I*, Springer–Verlag, New York and Heidelberg, 1969.

190. M. A. Krasnoselsky, A class of linear operators in the space of abstract continuous functions, *Math. Notes* **2** (1967), 856–858.

191. M. A. Krasnoselsky, P. P. Zabreĭko, E. I. Pustilnik, and P. E. Sobolevsky, *Integral Operators in Spaces of Summable Functions*, Nauka, Moscow, 1966. (Translated from the Russian by T. Andô: Monographs and Textbooks on Mechanics of Solids and Fluids, Mechanics: Analysis, Noordhoff International Publishing, Leiden, 1976.)

192. M. G. Krein and M. A. Rutman, Linear operators leaving invariant a cone in a Banach space, *Uspekhi Mat. Nauk* **3** (1948), 3–95 (Russian). Also, *Amer. Math. Soc. Transl. Ser.*, 2, **26**, Providence, RI, 1950.

193. U. Krengel, Über den Absolutbertrag stetiger linearer Operatoren und seine Anwendung auf ergodische Zerlegungen, *Math. Scand.* **13** (1963), 151–187.

194. H. J. Krieger, Beiträge zur Theorie positiver Operatoren, Schriftenreihe der Institute für Math. Reihe A, Heft 6; Akad.-Verlag, Berlin, 1969.

195. A. G. Kusraev, *Dominated Operators*, Mathematics and its Applications, **519**, Kluwer Academic Publishers, Dordrecht & London, 2000.

196. S. Kwapien, On the form of a linear operator in the space of all measurable functions, *Bull. Acad. Polon. Sci.* **21** (1973), 951–954.

197. H. E. Lacey, *The Isometric Theory of Classical Banach Spaces*, Springer–Verlag, Berlin and New York, 1974.

198. P. Lancaster and M. Tismenetsky, *The Theory of Matrices with Applications*, 2nd Edition, Academic Press, Orlando and New York, 1985.

199. K. B. Laursen and M. M. Neumann, *An Introduction to Local Spectral Theory*, Clarendon Press, Oxford, 2000.

200. N. J. Laustsen, Maximal ideals in the algebra of certain Banach spaces, *University of Leeds Preprint Series*, no. 1, 2000.

201. L. Le Cam, *Asymptotic Methods in Statistical Decision Theory*, Springer–Verlag, New York and Berlin, 1986.

202. C.-S. Lin, Generalized Daugavet equations and invertible operators on uniformly convex Banach spaces, *J. Math. Anal. Appl.* **197** (1996), 518–528.

203. J. Lindenstrauss, On operators which attain their norm, *Israel J. Math.* **5** (1963), 139–148.

204. J. Lindenstrauss and L. Tzafriri, On the complemented subspaces problem, *Israel J. Math.* **9** (1971), 263–269.

205. J. Lindenstrauss and L. Tzafriri, *Classical Banach Spaces I*, Springer–Verlag, Berlin and New York, 1977.

206. J. Lindenstrauss and L. Tzafriri, *Classical Banach Spaces II*, Springer–Verlag, Berlin and New York, 1979.

207. Z. Lipecki, Extension of vector-lattice homomorphisms, *Proc. Amer. Math. Soc.* **79** (1980), 247–248.

208. Z. Lipecki, D. Plachky, and W. Thomsen, Extension of positive operators and extreme points I, *Colloq. Math.* **42** (1979), 279–284.

209. G. L. Litvinov and V. I. Lomonosov, Density theorems in locally convex spaces and applications, *Selecta Math. Soviet* **8** (1989), 323–339.

210. Ju. I. Ljubič and V. I. Macaev, Operators with separable spectrum, *Mat. Sb. (N.S.)* **56** (1962), 433–468.

211. V. I. Lomonosov, Invariant subspaces of the family of operators that commute with a completely continuous operator, *Funktsional. Anal. i Prilozhen* **7** (1973), *No.* 3, 55–56. (Russian)

212. V. I. Lomonosov, An extension of Burnside's theorem to infinite-dimensional spaces, *Israel J. Math.* **75** (1991), 329–339.

213. V. I. Lomonosov, A counterexample to the Bishop–Phelps theorem in complex spaces, *Israel J. Math.* **115** (2000), 25–28.

214. E. R. Lorch, The integral representation of weakly almost-periodic transformations in reflexive vector spaces, *Trans. Amer. Math. Soc.* **49** (1941), 18–40.

215. H. P. Lotz, Über das Spektrum positiver Operatoren, *Math. Z.* **108** (1968), 15–32.

216. H. P. Lotz, Minimal and reflexive Banach lattices, *Trans. Amer. Math. Soc.* **211** (1975), 85–100.

217. H. P. Lotz and H. H. Schaefer, Über einen Satz von F. Niiro und I. Sawashima, *Math. Z.* **108** (1968), 33–36.

218. A. R. Lovaglia, Locally uniformly convex Banach spaces, *Trans. Amer. Math. Soc.* **78** (1955), 225–238.

219. G. Ya. Lozanovsky, On almost integral operators in KB-spaces, *Vestnik Leningrad Univ. Mat. Mekh. Astronom.*, no. 7 (1966), 35–44. (Russian)

220. G. Ya. Lozanovsky, On some Banach lattices, *Siberian Math. J.* **10** (1969), 584–599.

221. G. Ya. Lozanovsky, On a theorem of N. Dunford, *Izv. Vyssh. Uchebn. Zaved. Mat.* **8** (1974), 58–59. (Russian)

222. W. A. J. Luxemburg and B. de Pagter, Maharam extensions of positive operators and f-modules, *Positivity* **6** (2002), 147–190.

223. W. A. J. Luxemburg and A. R. Schep, An extension theorem for Riesz homomorphisms, *Indag. Math.* **41** (1979), 145–154.

224. W. A. J. Luxemburg and A. C. Zaanen, Notes on Banach function spaces, IX, *Indag. Math.* **26** (1964), 360–376.

225. W. A. J. Luxemburg and A. C. Zaanen, *Riesz Spaces I*, North-Holland, Amsterdam, 1971.

226. W. A. J. Luxemburg and A. C. Zaanen, The linear modulus of an order bounded linear transformation, *Indag. Math.* **33** (1973), 422–447.

227. G. W. MacDonald, Invariant subspaces for Bishop-type operators, *J. Funct. Anal.* **91** (1990), 287–311.

228. D. Maharam, The representation of abstract integrals, *Trans. Amer. Math. Soc.* **75** (1953), 154–184.

229. D. Maharam, On kernel representation of linear operators, *Trans. Amer. Math. Soc.* **79** (1955), 229–255.

230. D. Maharam, On positive operators. Conference in modern analysis and probability (New Haven, Conn., 1982), 263–277, *Contemp. Math.*, **26**, Amer. Math. Soc., Providence, RI, 1984.

231. P. Mankiewicz, A superreflexive Banach space X with $L(X)$ admitting a homomorphism onto the Banach algebra $C(\beta\mathbb{N})$, *Israel J. Math.* **65** (1989), 1–16.

232. A. A. Markov, Some theorems on Abelian collections, *Dokl. Akad. Nauk USSR* **10** (1936), 311–313.

233. M. Mbekhta, Généralisation de la décomposition de Kato aux opérateurs paranormeaux et spectraux, *Glasgow Math. J.* **29** (1987), 159–175.

234. M. Mbekhta, Sur la théorie spectrale locale et limite des nilpotents, *Proc. Amer. Math. Soc.* **110** (1990), 621–631.

235. R. E. Megginson, *An Introduction to Banach Space Theory*, Graduate Texts in Mathematics, **183**, Springer–Verlag, New York and Heidelberg, 1998.

236. A. A. Mekler, On a representation of stochastic projections, *Systems of Processing and Transferring of Information, Trudy* ($LIAP$), no. 87 (1974), 115–118.

237. A. A. Mekler, *Operators Averaging Over σ-Subalgebras on Ideals in $L_1(\mu)$*, Dissertation, Leningrad Institute of Airapparatus Construction (LIAP), Leningrad, 1997.

238. A. A. Mekler, Intermediate spaces and bistochastic projections, *Math. Issledovaniya* **10** (1975), 270–275. (Kishinev).

239. P. Meyer-Nieberg, Quasitriangulierbare Operatoren und invariante Untervekrorräume stetiger linearer Operatoren, *Arch. Math.* (*Basel*) **22** (1971), 186–199.

240. P. Meyer-Nieberg, *Banach Lattices*, Springer–Verlag, Berlin and New York, 1991.

241. A. J. Michaels, Hilden's simple proof of Lomonosov's invariant subspace theorem, *Adv. in Math.* **25** (1977), 56–58.

242. D. P. Milman, On some criteria for the regularity of spaces of the type (B), *Dokl. Akad. Nauk SSSR, N. S.,* **20** (1938), 243–246. (Russian)

243. H. Minc, *Nonnegative Matrices*, John Wiley and Sons, Berlin and New York, 1988.

244. B. S. Mityagin and I. C. Edelstein, Homotopy of linear groups for two classes of Banach spaces, *Funkt. Analiz Priloz.,* (3), **4** (1970), 61–72. (Russian)

245. S. Miyajima, Structure of Banach quasi-sublattices, *Hokkaido Math. J.* **12** (1983), 83–91.

246. S-T. C. Moy, Characterization of conditional expectation as a transformation on function spaces, *Pacific J. Math.* **4** (1954), 47–63.

247. J. R. Munkres, *Topology*, Prentice-Hall, Englewood Cliffs, NJ, 1975.

248. R. J. Nagel, Order in pure and applied functional analysis, *Trends in Functional Analysis and Approximation Theory* (Acquafredda di Maratea, 1989), 87–101, Univ. Modena, Modena, 1991.

249. B. Sz.-Nagy and C. Foias, Quasi-similitude des opérateurs et sous-espaces invariants *C. R. Acad. Sci. Paris Sér. I Math.* **261** (1965), 3938–3940.

250. B. Sz.-Nagy and C. Foias, *Harmonic Analysis of Operators on Hilbert Space*, North-Holland, Amsterdam and New York, 1970.

251. H. Nakano, *Modern Spectral Theory*, Maruzen Co., Tokyo, 1950.

252. H. Nakano, *Modulared Semi-Ordered Linear Spaces*, Maruzen Co., Tokyo, 1950.

253. H. Nakano, Product spaces of semi-ordered linear spaces, *J. Math. Fac. Sci. Hokkaido* **12** (1953), 163–210.

254. C. Neumann, Untersuchungen über das logarithmische und Newtonsche Potential, Teubner, Leipzig, 1877.

255. J. von Neumann, Zur Algebra der Funktionaloperationen und Theorie der normalen Operatoren, *Math. Ann.* **102** (1929), 370–427.

256. J. Neveu, *Mathematical Foundations of the Calculus of Probability*, Holden & Day, San Fransisco & London, 1965. (English translation from the 1963 French edition.)

257. N. K. Nikolskii, Invariant subspaces in operator theory and function theory, *Mathematical Analysis* **12** (1974), 199–412, Akad. Nauk SSSR Vsesojuz. Inst. Naučn. i Tehn. Informacii, Moscow, 1974. (Russian)

258. R. D. Nussbaum, The radius of the essential spectrum, *Duke Math. J.* **37** (1970), 473–478.

259. R. D. Nussbaum, Eigenvectors of nonlinear positive operators and the linear Krein–Rutman theorem, in: E. Fadell and G. Fourier eds., *Fixed Point Theory* (Sherbrooke, Que., 1980), **886**, Springer–Verlag, Berlin and New York, 1981, 309–330.

260. B. de Pagter, Irreducible compact operators, *Math. Z.* **192** (1986), 149–153.

261. C. Pearcy and N. Salinas, An invariant-subspace theorem, *Michigan Math. J.* **20** (1973), 21–31.

262. C. Pearcy and A. L. Shields, A survey of the Lomonosov technique in the theory of invariant subspaces, in *Topics in Operator Theory*, 219–229, *Math. Surveys*, **13**, Amer. Math. Soc., Providence, RI, 1974.

263. A. Pelczynski and Z. Semadeni, Spaces of continuous functions. III. Spaces $C(\Omega)$ for Ω without perfect subsets, *Studia Math.* **18** (1959), 211–222.

264. A. L. Peressini, *Ordered Topological Vector Spaces*, Harper and Row, New York and London, 1967.

265. O. Perron, Zur Theorie der Matrizen, *Math. Ann.* **64** (1907), 248–263.

266. A. Pietsch, *Operator Ideals*, VEB, Deutscher Verlag der Wissenschaften, Berlin, 1978.

267. A. M. Plichko and M. M. Popov, *Symmetric function spaces on atomless probability spaces*, Dissertationes Mathematicae, **306**, 1990.

268. I. A. Polyrakis, Schauder bases in locally solid lattice Banach spaces, *Math. Proc. Cambridge Philos. Soc.* **101** (1987), 91–105.

269. I. A. Polyrakis, Lattice subspaces of $C[0,1]$ and positive bases, *J. Math. Anal. Appl.* **184** (1994), 1–18.

270. I. A. Polyrakis, Finite-dimensional lattice-subspaces of $C(\Omega)$ and curves of $\mathbb{R}^n$, *Trans. Amer. Math. Soc.* **348** (1996), 2793–2810.

271. H. Radjavi, On reducibility of semigroups of compact operators, *Indiana Univ. Math. J.* **39** (1990), 499–515.

272. H. Radjavi and P. Rosenthal, *Invariant Subspaces*, Springer–Verlag, Berlin and New York, 1973.

273. H. Radjavi and P. Rosenthal, *Simultaneous Triangularization*, Springer–Verlag, Berlin and New York, 2000.

274. B. Randrianantoanina, Norm-one projections in Banach spaces, *Taiwanese J. Math.* **5** (2001), 35–95.

275. C. J. Read, A solution to the invariant subspace problem on the space ℓ_1, *Bull. London Math. Soc.* **17** (1985), 305–317.

276. C. J. Read, A short proof concerning the invariant subspace problem, *J. London Math. Soc.* (2) **34** (1986), 335–348.

277. C. J. Read, Quasinilpotent operators and the invariant subspace problem, *J. London Math. Soc.* (2) **56** (1997), 595–606.

278. F. Riesz, Sur la décomposition des opérations linéaires, *Atti. Congr. Internaz. Mat. Bologna* **3** (1930), 143–148.

279. F. Riesz, Sur quelques notions fondamentals dans la theorie générale des opérations linéaires, *Ann. of Math.* **41** (1940), 174–206.

280. F. Riesz and B. Sz.-Nagy, *Functional Analysis*, Frederick Ungar Publishing Co., New York, 1965.

281. H. P. Rosenthal, On factors of $C([0,1])$ with non-separable dual, *Israel J. Math.* **13** (1972), 361–378; Correction. Ibid. **21** (1975), 93–94.

282. H. P. Rosenthal, Sign-embeddings of L^1, *Banach spaces, Harmonic Analysis, and Probability Theory* (Storrs, Conn., 1980/1981), 155–165, Lecture Notes in Mathematics, **995**, Springer–Verlag, Berlin and New York, 1983.

283. H. P. Rosenthal, Some remarks concerning sign-embedding, *Seminar on the geometry of Banach spaces* (Paris, 1982), 129–132; *Publ. Math. Univ. Paris VII* **16**, Univ. Paris VII, Paris, 1983.

284. P. Rosenthal, Equivalents of the invariant subspace problem, in *Paul Halmos: Celebrating 50 Years of Mathematics*, Springer–Verlag, Berlin and New York, 1991, 179–188.

285. G. C. Rota, On the representing of averaging operators, *Rend. Padova* **30** (1960), 52–64.

286. W. Rudin, Continuous functions on compact spaces without perfect subsets, *Proc. Amer. Math. Soc.* **8** (1957), 39–42.

287. H. H. Schaefer, Spekraleigenschaften positwer linearer Operatoren, *Math. Z.* **82** (1963), 303–313.

288. H. H. Schaefer, Topologische Nilpotenz irreduzibler Operatoren, *Math. Z.* **117** (1970), 135–140.

289. H. H. Schaefer, *Topological Vector Spaces*, Springer–Verlag, Berlin and New York, 1974.

290. H. H. Schaefer, *Banach Lattices and Positive Operators*, Springer–Verlag, Berlin and New York, 1974.

291. H. H. Schaefer, On the o-spectrum of order bounded operators, *Math. Z.* **154** (1977), 79–84.

292. H. H. Schaefer, On theorems of de Pagter and Andô–Krieger, *Math. Z.* **192** (1986), 155–157.

293. E. Scheffold, Das Spektrum von Verbandsoperatoren in Banachverbänden, *Math. Z.* **123** (1971), 177–190.

294. A. R. Schep, Kernel operators, *Indag. Math.* **41** (1979), 39–53.

295. T. Schlumprecht and V. G. Troitsky, On quasi-affine transforms of Read's operator, *Proc. Amer. Math. Soc.*, forthcoming.

296. K. D. Schmidt, On the modulus of weakly compact operators and strongly additive vector measures, *Proc. Amer. Math. Soc.* **102** (1988), 862–866.

297. K. D. Schmidt, Daugavet's equation and orthomorphisms, *Proc. Amer. Math. Soc.* **108** (1990), 905–911.

298. H.-U. Schwarz, *Banach Lattices and Operators*, Teübner Texte, **71**, Leipzig, 1984.

299. G. L. Seever, Nonnegative projections on $C_0(X)$, *Pacific J. Math.* **17** (1966), 159–166.

300. Z. Semadeni, *Banach Spaces of Continuous Functions*, Polish Scientific Publishers, Warsaw, 1971.

301. V. S. Shulman, On invariant subspaces of Volterra operators, *Funktsional Anal. i Prilozhen* **18** (1984), 84-85.

302. A. Simonič, An extension of Lomonosov's techniques to non-compact operators, *Trans. Amer. Math. Soc.* **348** (1995), 975–995.

303. A. Simonič, A construction of Lomonosov functions and applications to the invariant subspace problem, *Pacific J. Math.* **175** (1996), 257–270.

304. A. M. Sinclair, *Automatic Continuity of Linear Operators*, Cambridge University Press, Cambridge, 1976.

305. I. Singer, *Bases in Banach Spaces*, Vol. 1, Springer–Verlag, Berlin and New York, 1970.

306. I. Singer, *Bases in Banach Spaces*, Vol. 2, Springer–Verlag, Berlin and New York, 1981.

307. G. G. Sirotkin, Compact-friendly multiplication operators on Banach function spaces, *J. Funct. Analysis*, forthcoming.

308. V. L. Šmulian, Sur la structure de la sphere unitaire dans l'espace de Banach, *Mat. Sb. (N.S.)* **9** (1941), 545–561.

309. A. Sobczyk, Projection of the space m on its subspace c_0, *Bull. Amer. Math. Soc.* **47** (1941), 938–947.

310. A. Spalsbury, Operators not positive with respect to any basis, *Quaestiones Math.* **23** (2000), 489–494.

311. J. G. Stampfli, Compact perturbations, normal eigenvalues and a problem of Salinas, *J. London Math. Soc.* (2) **9** (1974/75), 165–175.

312. J. Synnatzschke, The almost integral operators in K-spaces, *Vestnik Leningrad Univ. Math. Mekh. Astronom.* **13** (1971), 81–89.

313. J. Synnatzschke, On the adjoint of a regular operator and some of its applications to the question of complete continuity and weak continuity of regular operators, *Vestnik Leningrad Univ. Math. Mekh. Astronom.* **5** (1978), 71–81.

314. J. Synnatzschke, Über eine additive Normgleichheit für Operatoren in Banachverbänden, *Math. Nachr.* **117** (1984), 175–180.

315. A. Szankowski, A Banach lattice without the approximation property, *Israel J. Math.* **24** (1976), 329–337.

316. B.-S. Tam, A cone-theoretic approach to the spectral theory of positive linear operators: the finite dimensional case, *Taiwanese J. Math.* **5** (2001), 207–277.

317. A. Taylor and D. C. Lay, *Introduction to Functional Analysis*, R. E. Krieger, Malabar, Florida, 1986.

318. T. Terzioglu, A characterization of compact linear mappings, *Arch. Math. (Basel)* **22** (1971), 76–78.

319. V. G. Troitsky, On the modulus of C. J. Read's operator, *Positivity* **3** (1998), 257–264.

320. V. G. Troitsky, *Invariant Subspace Problem and Spectral Properties of Bounded Linear Operators on Banach Spaces, Banach Lattices, and Topological Vector Spaces*, Ph.D. Dissertation, University of Illinois, Urbana-Champaign, Illinois, 1999.

321. V. G. Troitsky, Lomonosov's theorem cannot be extended to chains of four operators, *Proc. Amer. Math. Soc.* **128** (2000), 527–540.

322. V. G. Troitsky, Measures of non-compactness of operators on Banach lattices, preprint.

323. S. L. Troyanski, On locally uniformly convex and differentiable norms in certain non-separable Banach spaces, *Studia Math.* **37** (1970/71), 173–180.

324. Y. V. Turovskii, Volterra semigroups have invarinat subspaces, *J. Funct. Anal.* **162** (1999), 313–322.

325. L. Tzafriri, Remarks on contractive projections in L_p-spaces, *Israel J. Math.* **7** (1969), 9–15.

326. L. Tzafriri, An isomorphic characterization of L_p- and c_0-spaces, II, *Michigan Math. J.* **18** (1971), 21–31.

327. J. Voight, The projection onto the center of operators in a Banach lattice, *Math. Z.* **199** (1988), 115–117.

328. B. Z. Vulikh, *Introduction to the Theory of Partially Ordered Spaces*, Wolter–Hoordhoff, Groningen, Netherlands, 1967. (English translation from the Russian.)

329. B. Z. Vulikh and G. Ya. Lozanovsky, Representation of order continuous and regular functionals in partially ordered spaces, *Math. USSR-Sb.* **13** (1971), 323–343.

330. L. Weis and D. Werner, The Daugavet equation for operators not fixing a copy of $C(S)$, *J. Operator Theory* **39** (1998), 89–98.

331. J. Wermer, The existence of invariant subspaces, *Duke Math. J.* **19** (1952), 615–622.

332. D. Werner, An elementary approach to the Daugavet equation, in: N. Kalton, E. Saab, and S. Montgomery-Smith, eds., *Interaction Between Functional Analysis, Harmonic Analysis, and Probability* (Lecture Notes in Pure and Applied Mathematics, **175**, Marcel Dekker, New York, 1996), 449–454.

333. T. T. West, The decomposition of Riesz operators, *Proc. London Math. Soc.* (3) **16** (1966), 737–752.

334. A. W. Wickstead, Representation and duality of multiplication operators on Archimedean Riesz spaces, *Compositio Math.* **35** (1977), 225–238.

335. A. W. Wickstead, Extremal structure of cones of operators, *Quart. J. Math. Oxford Ser.* (2) **32** (1981), 239–253.

336. A. W. Wickstead, Banach lattices with trivial center, *Proc. Roy. Irish Acad. Sect. A* **88** (1988), 71–83.

337. A. Wilansky, Subalgebras of $B(X)$, *Proc. Amer. Math. Soc.* **29** (1971), 355–360.

338. W. Wnuk, *Banach Lattices with Order Continuous Norms*, Polish Scientific Publishers, Warsaw, 1999.

339. P. Wojtaszczyk, Some remarks on the Daugavet equation, *Proc. Amer. Math. Soc.* **115** (1992), 1047–1052.

340. M. P. H. Wolff, An introduction to nonstandard functional analysis. In *Nonstandard Analysis* (*Edinburgh, 1966*), pp. 121–151. Kluwer Acad. Publ., Dordrecht, 1977.

341. D. E. Wulbert, Averaging projections, *Illinois J. Math.* **13** (1969), 689–693.

342. A. C. Zaanen, Examples of orthomorphisms, *J. Approx. Theory* **13** (1975), 192–204.

343. A. C. Zaanen, *Riesz Spaces II*, North-Holland, Amsterdam, 1983.

344. A. C. Zaanen, *Introduction to Operator Theory in Riesz Spaces*, Springer–Verlag, New York and Heidelberg, 1997.

345. P. P. Zabreĭko and S. V. Smitskikh, A theorem of M. G. Kreĭn and M. A. Rutman, *Funktsional. Anal. i Prilozhen.* **13** (1979), 81–82.

346. H. J. Zhong, Riesz operators on the spaces L_p have West decompositions, *Northeast. Math. J.* **4** (1988), 282–288. (In Chinese with an English summary.) MR 90c:47031

347. H. J. Zhong, B-convex spaces and West decomposition of Riesz operators on them, *Acta Math. Sinica* **37** (1994), 563–569.

348. H. J. Zhong, Tsirelson's space and West decomposition of Riesz operators on it, *Sci. China Ser. A* **39** (1996), 491–500. (In Chinese with an English summary.) MR 97j:46003

349. Y. Zhong, Irreducible semigroups of functionally positive nilpotent operators, *Trans. Amer. Math. Soc.* **347** (1995), 3093–3100.

Index

AL-space, 94, 101
AM-compact operator, 401
AM-space, 94, 101
  complex, 106
  with unit, 94
$\alpha(T)$, ascent of operator T, 80
$\mathcal{A}$-invariant subspace, 387
$\mathcal{A}_T$, algebra generated by T, 387
$\mathcal{A}x$, the orbit of x under $\mathcal{A}$, 387
a.e. equal functions, 52
absolute integral operator, 187
absolute value of vector, 15
abstract integral operator, 193, 368, 430
additive semigroup, 409
adjoint operator, 4, 244
  Hilbert space, 453
algebra, 3, 386
  $\mathcal{A}x$, 387
  Calkin, 299
  closed, 387
  dual, 387
  generated by set of operators, 387
  non-transitive, 387, 440
  transitive, 387
  unital, 3, 386, 387
algebra of operators, 386
  generated by a set, 387
  non-transitive, 387, 440
  transitive, 387
algebraic complement of subspace, 169
algebraic homomorphism, 144
almost essentially disjoint operators, 477
almost integral operator, 193, 430
analytic function, 45
annihilator, 8
  of a set, 328
  of a vector subspace, 8, 328
annulus, 49
antisymmetry, 13
approximate eigenvalue, 249
approximate point spectrum, 249
approximation property, 125
Archimedean ordered vector space, 15
ascent of operator, 80, 270
atom, 86
atomic measure space, 193
atomless Riesz space, 86
automatic continuity, 147, 154
averaging operator, 213
averaging property, 213

B_x, band generated by x, 20
$[B\rangle$, super right-commutant of B, 399
$\langle B]$, super left-commutant of B, 399
backward shift operator, 251
Banach function space, 194
  associated with $L_0(\pi)$ or π, 194
  rearrangement invariant, 223, 494
Banach lattice, 22
  AL, 94
  AM, 94
  complex, 105
  with order continuous norm, 84
Banach space, 2
  direct sum, 5, 10
  function, 194
  locally uniformly convex, 460
  quotient, 8
  reflexive, 465
  uniformly convex, 12, 457, 465
  uniformly rotund, 457

uniformly smooth, 457
Banach–Steinhaus theorem, 4
band, 20
generated by a set, 20
of abstract integral operators, 193, 430
of integral operators, 188
principal, 20, 118
projection, 21
band irreducible operator, 349
band preserving operator, 112
band projection, 21
basic sequence, 30
unconditional, 35
basis, 28, 436
in $L_p[0, 1]$, 36
monotone, 29
normalized, 29
unconditional, 32, 500
basis constant, 29
basis problem, 28
biorthogonal functional, 31
Bishop operator, 451
bistochastic projection, 221
bound, 15
greatest lower, 15
least upper, 15
lower, 15
upper, 15
bounded function, 64
from above, 64
bounded operator, 2
from below, 69

$C^\infty(Q)$, extended continuous functions, 115
$C_0(\Omega)$, functions vanishing at ∞, 406
$\mathbb{C}$, complex numbers, 2
$\mathfrak{C}(X)$, the Calkin algebra, 299
$\mathcal{C}$-invariant collection, 410
$x_n \xrightarrow{\mathcal{U}} x$, convergence along an ultrafilter, 38
Calkin algebra, 299
canonical projection, 299
Carathéodory extension of measure, 61
Carathéodory's theorem, 61
Carleman operator, 192
carrier of ideal, 27
in L_0, 60
Cayley–Hamilton theorem, 318
center of Banach function space, 209
center of Banach lattice, 112
trivial, 116
central operator, 112
Cesaro operator, 247, 450
chain of open sets, 51
characteristic polynomial, 317, 327
of operator, 321
characterization of narrow operator, 494
closed algebra, 387
Closed Graph Theorem, 5
closest point to a set, 12
codimension of vector subspace, 155
cofinal set, 66
collection
strictly transitive, 387
transitive, 387
commutant, 138, 357, 388, 398, 410
of family of operators, 410
super left, $\langle B]$, 399
super right, $[B\rangle$, 399
commutator of two operators, 398
compact operator, 88, 272
compact-friendly operator, 425
complement
algebraic of subspace, 169
of a closed subspace, 386
complemented subspace, 134, 386, 486
complete measure, 66
completely continuous function, 444
completion
Dedekind, 26
Maeda–Ogasawara–Vulikh, 116
universal, 116
complex AM-space, 106
complex Banach lattice, 105
complex lattice isometry, 105
complexification, 5
of Banach lattice, 105, 110
component of a vector, 16
composition operator, 142, 378
conditional expectation of function, 212
conditional expectation operator, 212
cone, 14, 440
generated by a basis, 437, 440
generating, 14
conjugate of matrix, 317
continuous spectrum, 249
contraction, 78
control function, 315
convergence
$\star$, 67
weak* , 3
in measure, 53
notions in L_0, 209
of Laurent series, 49
order, 16
pointwise, 53
uniform, 53
weak, 3
convolution of kernels, 210
corank of operator, 155
countable sup property, 22, 354
cyclic point of a set, 288
cyclic set, 288
cyclic vector, 328, 383

$\delta(T)$, descent of operator T, 80
$\|D\|$, norm of set D, 410
$d(T)$, defect of operator T, 155
Daugavet equation, 456, 470
Daugavet property, 467, 473
decreasing net, 16
Dedekind complete Riesz space, 15
Dedekind completion, 26
defect of operator, 155
derivative of a function, 45
descent of operator, 80, 270
diagonal matrix, 118
diagonal part of operator, 117
diagonal projection, 117
diagonal property, 56
diagonalizable matrix, 319
diagonalizing matrix, 319
differentiable function, 45
direct sum Banach space, 10
direct sum normed space, 5
direct sum of Banach spaces, 247
direct sum of operators, 393
direct sum operator, 247
direct sum property, 208
Dirichlet's theorem, 236
discrete vector, 86
disjoint complement, 21
disjoint operators, 16
  almost essentially, 477
  essentially, 477
disjointly strictly singular operator, 178
disk, 44
domain of analyticity of function, 258
domination of operators, 89, 109
double adjoint, 4
double dual, 3
doubly power bounded operator, 247, 450
dual, 3
  double, 3
  Köthe, 198
  norm, 3
  order, 21
  order continuous, 21, 193
  second, 3
dual algebra, 387
Dunford–Pettis operator, 409, 498
Dunford–Pettis property, 360

E_x, ideal generated by x, 20
$E^{\sim}$, order dual of E, 21
eigenspace, 384
eigenvalue, 248
  approximate, 249
eigenvector, 248
entire function, 47
equation
  Daugavet, 456
equimeasurable functions, 223, 494
essential property, 299
essential singularity, 264, 270
essential spectral radius, 302
essential spectrum of an operator, 299
essential sup, 64
essential sup norm, 64
essential upper bound, 64
essentially bounded function, 64
  from above, 64
essentially disjoint operators, 477
essentially invertible operator, 299
essentially quasinilpotent operator, 302
Euclidean norm, 316, 321
  of matrix, 318
Euclidean topology, 321
expanding operator, 350
exposed point, 490
exposing linear functional, 490
extremally disconnected space, 96
extreme point, 490

$\mathcal{F}red(X,Y)$, Fredholm operators, 163
Φ-integral operators, 193, 430
$\Phi\otimes Y$, Φ-finite-rank operators, 125
$\hat{\mathcal{F}}red(X,Y)$, semi-Fredholm operators, 166
$\mathcal{F}(X,Y)$, finite-rank operators from X to Y, 125
$f(T)$, 258
factorization of an operator, 477
family of operators
  finitely quasinilpotent at a point, 410
  non-transitive, 410
  quasinilpotent at a point, 410
  transitive, 410
Fatou property, 65
  σ-, 65
filter, 37
  Fréchet, 38
Finite Dimensional Separation Theorem, 13
finite quasinilpotence, 410
finite-rank operator, 91, 124, 430, 471
  rank, 124
  representation, 124
fixed ultrafilter, 38
flat of a function, 432
Fréchet differentiability of the norm, 466
Fréchet filter, 38
Fredholm alternative, 169
Fredholm operator, 156
Fredholm spectrum, 299
free ultrafilter, 38
Fubini's theorem, 63
function
  σ-order continuous, 18
  analytic, 45
  bounded, 64

completely continuous, 444
control, 315
defining an integral operator, 181
differentiable at a point, 45
entire, 47
essentially bounded above, 64
harmonic, 236
holomorphic, 45
index, 163
jointly measurable, 63
measurable, 52
onto, 69
order continuous, 18
resolvent, 238, 245
retraction, 13
state, 315
surjective, 69
weakly analytic, 47
function space, 194
normal, 194
functional
biorthogonal, 31
multiplicative, 148
supporting, 466
Functional Calculus, 258

Gantmacher's theorem, 89
Gateaux differentiability of the norm, 466
generalized Harris operator, 434
generating cone, 14
graph of function, 4

Haar system, 36
harmonic function, 236
Harris operator, 370
hereditarily indecomposable B-space, 177
Hermitian matrix, 317
Hilbert space adjoint, 453
Hilbert–Schmidt operator, 193
holomorphic function, 45
homogeneity of the norm, 2
homomorphism
algebraic, 144
lattice, 88
hyperinvariant subspace, 348, 381

$\mathcal{I}(X, Y)$, 472
ideal, 20, 110
generated by a set, 20
null, 119, 348, 399, 408
null of an ℓ-seminorm, 122
null of functional, 27
operator, 472
principal, 20, 95, 119
range of operator, 399, 408
semigroup, 417
ideal irreducible operator, 111, 349
ideal irreducible semigroup, 415
ideal reducible semigroup, 415
implication scheme, 351
indecomposable matrix, 333
index function, 163
index of nilpotence, 82, 323
index of operator, 155
Index Theorem, 158
inequality
Jensen, 227
triangle, 2
infimum, 15
inner product preserving matrix, 317
input-output matrix, 315
integral
iterated, 62
integral operator, 102, 180
absolute, 187
internal point, 101
interpolation space, 224
interval preserving operator, 24, 154
invariant measure, 355
invariant subspace, 10, 348, 381
invariant subspace problem, 381
invertible operator, 10, 298
irreducible matrix, 333
isolated point of the spectrum, 265
isometric normed spaces, 2
isometry, 2, 11
lattice, 23, 105
linear, 2, 11
isomorphic Banach spaces, 71
isomorphism between Banach spaces, 71
iterated integral, 62

Jensen's inequality, 227
joint spectral radius, 425
jointly measurable function, 63
Jordan contour, 257
surrounding a set, 257
Jordan form of a matrix, 326

KB-space, 85
$\mathcal{K}(X, Y)$, compact operators from X to Y, 89
Köthe dual, 198
Kadets–Klee property, 465
Kantorovich–Banach space, 85
kernel of integral operator, 181
kernel operator, 181
Krein operator, 350, 371
Krein–Rutman theorem, 277
Kronecker's theorem, 291

L-space, 122
$L_0(\pi)$, 52
$L_{p,1}$, 210

$L_{p,\infty}$, 210
ℓ_1-sum of Banach spaces, 472
ℓ_∞-sum of Banach spaces, 472
ℓ_p-sum of Banach spaces, 281
$\mathcal{L}(X)$, bounded operators on X, 2
$\mathcal{L}(X,Y)$, bounded operators from X to Y, 2
$\mathcal{L}_\kappa(E,F)$, regular integral operators from E to F, 188
$\mathcal{L}_b(X,Y)$, order bounded operators from X to Y, 15
$\mathcal{L}_r(X,Y)$, regular operators from X to Y, 14
Laplace equation, 236
lattice homomorphism, 23, 88, 154
lattice hyperinvariant subspace, 348, 381
lattice isometric Banach lattices, 23
lattice isometric complex B-lattices, 105
lattice isometry, 23, 105
  complex, 105
lattice isomorphism, 23
lattice operations, 15
lattice seminorm, 22
lattice-subspace, 229
Laurent series, 49
  expansion of function, 49
Laurent series expansion, 266
limit
  $\mathcal{U}$, 38
  functional, 155
  order, 17
linear functional
  exposing a point, 490
  strongly exposing a point, 490
linear isometry, 2, 11
linear operator, 2
Liouville number, 451
Liouville's theorem, 47
local quasinilpotence, 357, 410
locally uniformly convex B-space, 460
Lomonosov operator, 396

M-space, 122
$M_n(\mathbb{R})$, space of $n \times n$ square matrices, 318
Maeda–Ogasawara–Vulikh completion, 116
majorizing vector space, 96
mapping
  nearest point, 12
  quotient, 8
  sublinear, 149
Markov matrix, 103, 332
Markov operator, 100
Markov projection, 213
matrix
  diagonal, 118
  diagonalizable, 319
  diagonalizing, 319
  Hermitian, 317
  indecomposable, 333
  inner product preserving, 317
  input-output, 315
  irreducible, 333
  Markov, 103, 332
  non-negative, 329
  norm preserving, 317
  orthogonal, 317
  permutation, 333
  positive, 329
  positive semidefinite, 319
  primitive, 343
  production, 315
  reducible, 333
  representing, 321
  stochastic, 103
  strictly positive, 329
  strongly positive, 329
  symmetric, 317
  unitary, 317
  upper triangularizable, 326
  upper triangularizing, 326
Mazur's lemma, 30
measurable function, 52
measure, 61
  atomic, 193
  complete, 66
  invariant, 355
  non-atomic, 479
  on a semiring, 61
  product, 62
measure convergence, 53
measure of non-compactness
  of operator, 309
  of set, 309
minimal polynomial, 327
mixed norm, 210
Miyajima projection, 235
modulus, 16, 104, 107
  of integral operator, 102, 186, 187
  of operator, 16, 96, 107, 113
  of vector, 104
monotone basis, 29
Morera's theorem, 50
multiplication operation, 3
multiplication operator, 136
multiplicative functional, 148
multiplicative operator, 144

N_λ, eigenspace with eigenvalue λ, 384
$\mathbb{N}$, the natural numbers, $\{1, 2, \ldots\}$, 2
$n(T)$, nullity of operator T, 155
narrow operator, 494
nearest point mapping, 12
nearest point to a set, 12
negative part of vector, 15
neighborhood of set, 257

net, 16
  decreasing, 16
  order convergent, 16
Neumann series, 239
nilpotent operator, 82, 244, 323
non-atomic measure, 479
non-atomic Riesz space, 86
non-negative matrix, 329
non-quasianalytic, 450
non-scalar operator, 383
non-transitive algebra, 387, 440
non-transitive family of operators, 410
non-trivial vector subspace, 348, 381
norm, 2
  essential sup, 64
  Euclidean, 316, 321
  Fréchet differentiable, 466
  Gateaux differentiable, 466
  mixed, 210
  of a set, 410
  of integral operator, 102
  of matrix, 318
  order continuous, 84
  quotient, 8
  strictly monotone, 216
  sup, 64
  uniformly convex, 12
norm dual, 3
norm preserving matrix, 317
normal function space, 194
normal integral, 198
normal operator, 453
normal Riesz space, 21, 193
normalized basis, 29
normed function space, 194
normed Riesz space, 22
normed space, 2
  function, 194
norming subspace, 51
null ideal, 27, 348, 399, 408
  of ℓ-seminorm, 122
  of functional, 119
null space of operator, 79, 155
nullity of operator, 155

onto function, 69
open disk, 44
Open Mapping Theorem, 4
operation
  multiplication, 3
operator, 2
  AM-compact, 401
  $f(T)$, 258
  abstract integral, 193, 368, 430
  adjoint, 4, 244, 453
  almost integral, 193, 430
  averaging, 213
  backward shift, 251
  band irreducible, 349
  band preserving, 112
  Bishop, 451
  bounded, 2
  bounded below, 69
  Carleman, 192
  central, 112
  Cesaro, 247, 450
  compact, 88, 272
  compact-friendly, 425
  composition, 142, 378
  conditional expectation, 212
  disjointly strictly singular, 178
  double adjoint, 4
  doubly power bounded, 247, 450
  Dunford–Pettis, 409, 498
  essentially invertible, 299
  essentially quasinilpotent, 302
  expanding, 350
  finite-rank, 91, 124, 430, 471
  fixing a copy of a B-space, 485
  Fredholm, 156
  generalized Harris, 434
  Harris, 370
  Hilbert–Schmidt, 193
  ideal irreducible, 111, 349
  integral, 102, 180
  interval preserving, 24, 154
  invertible, 10, 298
  isometry, 11
  kernel, 181
  Krein, 350, 371
  lattice homomorphism, 23
  lattice isometry, 23
  lattice isomorphism, 23
  linear, 2
  Lomonosov, 396
  Markov, 100
  multiplication, 136
  multiplicative, 144
  narrow, 494
  nilpotent, 82, 244, 323
  non-quasianalytic, 450
  non-scalar, 383
  normal, 453
  nuclear, 134
  of rank k, 124
  order adjoint, 200
  order bounded, 15, 296
  polynomially compact, 390
  positive, 14, 437
  power compact, 272
  projection, 21, 386
  quasianalytic, 450
  quasinilpotent, 244
  quasitriangular, 454

rank-one, 123, 430
real on a complex B-space, 6
regular, 14, 107
residue, 266
Riesz, 302
Riesz homomorphism, 23
Riesz isomorphism, 23
rotation, 385
scalar, 383
semi-Fredholm, 156
shift, 251
stochastic, 100
strictly positive, 14, 408
strictly singular, 170
strong Krein, 350, 371
strongly expanding, 350
subnormal, 453
symmetric, 465
unbounded, 2
Volterra, 275
weakly compact, 89
weakly expanding, 350
operator domination, 89, 109, 425
operator ideal, 472
operator norm, 2
operator norm of matrix, 318
operator of rank k, 124
orbit, 383
orbit space, 383
order adjoint operator, 200
order bounded operator, 15, 296
order bounded set, 14
order closed set, 18
order complete Riesz space, 15
order continuous dual, 21, 193
order continuous function, 18
order continuous norm, 84
order continuous topology, 84
order convergence, 16
order convergence of sequences, 18
order convergent net, 16
order dense Riesz subspace, 58, 66
order dual, 21
order interval, 84
order limit, 17
order relation, 13
order spectrum, 292
order unit, 20, 94
ordered vector space, 14
Archimedean, 15
orthogonal matrix, 317
orthomorphism, 112
outer measure generated by a measure, 61

pair of reducing subspaces, 386
partially ordered set, 14
partially ordered vector space, 14
Archimedean, 15
partition of set, 227
peripheral spectrum, 245
permutation matrix, 333
permutation of a set, 31
Perron–Frobenius theorem, 341
perturbation, 163
of operator, 163
point
closest to a set, 12
cyclic, 288
exposed, 490
extreme, 490
internal, 101
nearest to a set, 12
quasi-interior, 135, 349
strongly exposed, 490, 503
support, 13
point spectrum, 248
pointwise bounded set, 4
pointwise convergence, 53
pole, 264
of the resolvent, 264, 268
simple, 264
polynomial
characteristic, 317
minimal, 327
polynomially compact operator, 390
positive matrix, 329
positive operator, 14, 437
on a B-space, 436
positive part of vector, 15
positive semidefinite matrix, 319
positively complemented subspace, 230
positivity of the norm, 2
power compact operator, 272
power series expansion of function, 46
predual of ideal generated by a functional, 119
primitive matrix, 343
principal band, 20, 118
principal ideal, 20, 95, 119
product σ-algebra, 62
product measure, 62
product rule for derivatives, 50
production matrix, 315
projection, 21, 386
band, 21
bistochastic, 221
canonical, 299
diagonal, 117
Markov, 213
Miyajima, 235
proper, 386
spectral, 262
projection band, 21
proper projection, 386

property
  antisymmetry, 13
  approximation, 125
  countable sup, 22, 354
  Daugavet, 467, 473
  direct sum, 208
  Dunford–Pettis, 360
  essential, 299
  Kadets–Klee, 465
  Radon–Nikodym, 503
  reflexivity, 13
  Schur, 448
  transitivity, 14
pseudoinverse of operator, 167

$\mathbb{Q}$, the rational numbers, 2
quasi-interior point, 135, 349
quasianalytic operator, 450
quasinilpotent operator, 244
quasitriangular operator, 454
quotient Banach space, 8
quotient map, 8
quotient norm, 8
quotient vector space, 8

$\mathbb{R}$, the real numbers, 2
$\mathcal{R}(A)$, vector sublattice generated by A, 233
r-norm of operator, 22
$r_{\text{ess}}(T)$, essential spectral radius of T, 302
Rademacher functions, 497
  generalized, 497
radius of convergence, 44
Radon–Nikodym property, 503
range, 69
  of function, 69
  of operator, 79, 155
range ideal of operator, 399, 408
rank
  of finite-rank operator, 124
  of matrix, 328
  of operator, 155
rank-one operator, 123, 430
real operator on a complex B-space, 6
real part of complex functional, 7
rearrangement invariant B-function space, 223, 494
rectangle, 62
reducible matrix, 333
reducing pair of subspaces, 81, 262, 322
reducing subspace, 81, 386
reflexive Banach space, 465
reflexivity property, 13
regular operator, 14, 107
regulator of convergence, 53
representation of finite-rank operator, 124
representing kernel, 484
representing matrix, 321
residual spectrum, 248
residue operator, 266
resolvent, 238, 245
resolvent function, 238, 245, 265
resolvent identity, 240
  second, 246
resolvent set, 238
retract of a space, 12
retraction, 13
Riesz homomorphism, 23
Riesz isomorphism, 23
Riesz operator, 302
Riesz space, 15
  σ-Dedekind complete, 16
  σ-order complete, 16
  atomless, 86
  Dedekind complete, 15
  non-atomic, 86
  normal, 21, 193
  normed, 22
  order complete, 15
Riesz subspace, 84
  order dense, 58, 66
Riesz–Kantorovich formula, 16, 107
Riesz–Thorin theorem, 224
rotation invariant set, 340
rotation invariant spectrum, 290
rotation operator, 385

σ-Dedekind complete Riesz space, 16
σ-order closed set, 18
σ-order complete Riesz space, 16
σ-order continuous function, 18
$\sigma(T)$, spectrum of operator T, 238
$\sigma_a(T)$, approximate spectrum of T, 249
$\sigma_c(T)$, continuous spectrum of T, 249
$\sigma_p(T)$, point spectrum of T, 248
$\sigma_r(T)$, residual spectrum of T, 248
$\sigma_w(T)$, Weyl spectrum of T, 312
$\sigma_{\text{ess}}(T)$, essential spectrum of T, 299
$\star$-convergence, 67
scalar operator, 383
Schauder basis, 28, 436
Schauder system, 36
Schur property, 448
second dual, 3
second resolvent identity, 246
semi-Fredholm operator, 156
semigroup
  ideal irreducible, 415
  ideal reducible, 415
semigroup ideal, 417
  generated by a set, 417
semigroup of operators, 409
  additive, 409
  generated by a family, 412
  multiplicative, 409

Volterra, 451
seminorm, 22, 126
lattice, 22
semiring of sets, 60
separation of points, 209
separation of sets, 13
Separation Theorem
Finite Dimensional, 13
series
convergent unconditionally, 31
Laurent, 49
Neumann, 239
weakly unconditionally Cauchy, 480, 486
set
σ-order closed, 18
cofinal, 66
cyclic, 288
order bounded, 14
order closed, 18
partially ordered, 14
pointwise bounded, 4
resolvent, 238
rotation invariant, 340
solid, 20
spectral, 261
shift operator, 251
similar matrices, 318
similar operators, 253
simple pole, 264
slice of a set, 488
solid set, 20
space
AL, 94
AM, 94
KB, 85
L, 122
M, 122
Banach, 2
Kantorovich–Banach, 85
normed, 2
orbit, 383
Riesz, 15
Stonean, 115
structure of B-lattice, 136
Spectral Mapping Theorem, 260
spectral projection, 262
spectral radius, 243, 245
essential, 302
joint, 425
Rota–Strang, 425
spectral set, 261
spectrum, 245
spectrum of operator, 238
approximate point, 249
continuous, 249
essential, 299
Fredholm, 299
order, 292
peripheral, 245
point, 248
residual, 248
rotation invariant, 290
Weyl, 312
spectrum of vector in a B-algebra, 245
state function, 315
stochastic matrix, 103
stochastic operator, 100
Stone space of a Riesz space, 116
Stonean topological space, 96, 115
strictly monotone norm, 216
strictly positive matrix, 329
strictly positive operator, 14, 408
strictly positive vector, 135, 316
strictly singular operator, 170
strictly transitive collection, 387
strong Krein operator, 350, 371
strong operator topology, 438
strong unit, 20, 94, 370
strongly expanding operator, 350
strongly exposed point, 490, 503
strongly exposing linear functional, 490
strongly positive matrix, 329
structure space, 136
subalgebra of operators, 386
subgroup of the unit circle, 291
dense, 291
sublinear mapping, 149
subnormal operator, 453
subspace
ℓ-hyperinvariant, 348, 381
complemented, 134, 386
hyperinvariant, 348, 381
invariant, 10, 348, 381
lattice, 229
non-trivial, 348, 381
norming, 51
positively complemented, 230
reducing an operator, 81, 386
subspace invariant under an operator, 10
sum of Banach spaces, 247, 281, 472
ℓ_1, 472
ℓ_∞, 472
sum of vector spaces closed, 121
sup norm, 64
super left-commutant, 399
super right-commutant, 399
support
of function, 58
of ideal in L_0, 60
point of a set, 13
supporting functional, 466
supremum, 15
surjective function, 69
symmetric difference of two sets, 66

symmetric matrix, 317
symmetric operator, 465
system
  Haar, 36
  Schauder, 36

T^*, the adjoint of operator T, 4
T^{**}, the double adjoint of operator T, 4
tensor-product notation, 124
Tonelli's theorem, 63
topological space
  extremally disconnected, 96
  Stonean, 96, 115
topology
  Euclidean, 321
  of uniform convergence on compact sets, 126
  order continuous, 84
  strong operator, 438
trace, 133, 318
  of finite-rank operator, 133
  of matrix, 318
transitive algebra, 387
transitive collection, 387
transitive family of operators, 410
transitivity property, 14
transpose of matrix, 317
triangle inequality, 2
trivial center, 116
twist of a set, 207

U, U^*, U^{**}, closed unit balls, 3
$\mathcal{U}$-limit of sequence, 38
ultrafilter, 37
  fixed, 38
  free, 38
ultrapower of B-space, 40
unbounded component of $\rho(T)$, 256
unbounded operator, 2
unconditional basic sequence, 35
unconditional basis, 32, 500
unconditional constant, 33
unconditionally convergent series, 31
uniform boundedness principle, 4
uniform convergence, 53
uniformly convex B-space, 465
uniformly convex Banach space, 12, 457
uniformly convex norm, 12
uniformly rotund Banach space, 457
uniformly smooth Banach space, 457
unit, 3, 20, 94, 370
  order, 20, 94
  weak, 21, 135, 350
unital algebra, 3, 386, 387
unitary matrix, 317
universal completion, 116
upper triangularizable matrix, 326
upper triangularizing, 326

vector
  cyclic, 328, 383
  discrete, 86
  strictly positive, 135, 316
vector lattice, 15
vector space
  majorizing, 96
  ordered, 14
  partially ordered, 14
  quotient, 8
vector space order, 14
vector sublattice, 84
vector sublattice generated by a set, 233
Volterra collection, 451
Volterra operator, 275

$\mathcal{W}(X, Y)$, weakly compact operators from X to Y, 89
weak* convergence, 3
weak convergence, 3
weak unit, 21, 135, 350
weakly analytic function, 47
weakly compact operator, 89
weakly expanding operator, 350
weakly unconditionally Cauchy series, 480, 486
West decomposition, 312
Weyl spectrum of operator, 312

$X_{\mathcal{U}}$, ultrapower of B-space X, 40
X^*, norm dual of X, 3
X^{**}, double norm dual of X, 3
$x_n \xrightarrow{\mathcal{U}} x$, convergence along an ultrafilter, 38

$\mathbb{Z}$, the integers, 2